SPACE ENGINEERING

ASTROPHYSICS AND
SPACE SCIENCE LIBRARY

A SERIES OF BOOKS ON THE RECENT DEVELOPMENTS

OF SPACE SCIENCE AND OF GENERAL GEOPHYSICS AND ASTROPHYSICS

PUBLISHED IN CONNECTION WITH THE JOURNAL

SPACE SCIENCE REVIEWS

VOLUME 15

SPACE ENGINEERING

Proceedings of the Second International Conference
on Space Engineering

HELD AT THE FONDAZIONE GIORGIO CINI,
ISOLA DI SAN GIORGIO, VENICE, ITALY, MAY 7–10, 1969

ORGANIZED BY
THE CENTRO STUDI TRASPORTI MISSILISTICI, STM, ROMA, ITALY
AND THE
ASSOCIATION POUR L'ÉTUDE ET LA RECHERCHE
ASTRONAUTIQUE ET COSMIQUE, AERA, PARIS, FRANCE

Edited by

G. A. PARTEL
Centro Studi Trasporti Missilistici, Rome, Italy

SPRINGER-VERLAG NEW YORK INC. / NEW YORK

D. REIDEL PUBLISHING COMPANY / DORDRECHT-HOLLAND

SOLE-DISTRIBUTOR FOR NORTH AND SOUTH AMERICA
SPRINGER-VERLAG NEW YORK INC. / NEW YORK

Library of Congress Catalog Card Number 79-105673
Title No.: 882005

ISBN-13: 978-94-011-7553-1   e-ISBN-13: 978-94-011-7551-7
DOI: 10.1007/978-94-011-7551-7

Softcover reprint of the hardcover 1st edition 1970

PREFACE

The 2nd International Conference on Space Engineering took place May 7–10, 1969, at Venice, Italy, under the organization of the Centro Studi Trasporti Missilistici and the Association pour l'Étude et la Recherche Astronautique et Cosmique.

Its purpose was to bring together those interested in the technological development of space components, to exchange information by the presentation of papers and to discuss present problems and future trends, and to this end forty-eight papers were presented by distinguished experts from all over the world.

The papers were selected from as wide a background as possible, approximately an equal number coming from the academic and research establishments as from industry. The principal criterion for their selection was that they should contribute to the knowledge of Space Engineering, and have application either to the improvement of current technologies or to the design of more advanced systems for the future.

Six pertinent sessions were planned which covered the major areas of interest: (1) Structures and Materials, where three important papers were presented; (2) Guidance and Control Systems, in which six valuable papers were presented, including problems of controlling space ships, details of the inertial guidance system of the ELDO launch vehicle, the attitude control system of the "Europa 2"; (3) Propellants and Combustion, where eleven papers described recent work on solid and liquid rocket engines, advanced fuels and oxidizers, effects of additives, propellant injection, propellant expulsion techniques; (4) Propulsion, in which session ten papers analyzed low-cost launch- and re-entry vehicles, reduction in cost of the first stage, a booster-ramrocket combination, hybrid motors in sounding rockets, reaction control engines, solid propellant grain design, small bipropellant rocket engines for communication satellites, the "Apollo" service module rocket engine, problems associated with tanks partially filled with liquids; (5) Auxiliary Power Systems, where four papers were presented, describing ZrH reactor development status, batteries for satellites and balloons, low power nuclear-energy conversion; (6) Testing, Support and Accessories, where fourteen papers were presented on selected aspects of ground testing apparatus, simulation of Mars landing, the deep space network for the "Mariner" vehicles, antenna spacing, space programs in India and Canada, and also papers on satellite drag measurements, gravitational stabilization systems for satellites, and a new transducer. These papers represent a most important contribution to the literature in this field.

G. A. PARTEL

TABLE OF CONTENTS

PART X / SPEECHES DELIVERED AT THE BANQUET

OPENING CEREMONY

ADDRESS

Au moment où la deuxième Conférence internationale du Génie spatial, réunie auprès de la méritante fondation G. Cini, commence ses travaux, le Souverain Pontife est heureux d'adresser ses salutations aux représentants hautement qualifiés de toutes les nations qui y participent. Il désire leur exprimer ses vœux fervents, et les assurer du vif intérêt avec lequel il suit l'activité des Organismes engagés dans le développement de la science, au bénéfice de l'humanité.

La préoccupation vigilante de l'Eglise catholique est bien connue, en effet, à l'égard des implications morales et sociales liées aux énormes progrès qui ont été atteints dans tous les domaines des conquêtes scientifiques. Ses appels répétés sont bien connus aussi, pour contribuer efficacement au vrai bien des hommes et écarter le plus possible tout péril, même le plus minime, que les vaisseaux spatiaux puissent servir à des buts nuisibles pour les peuples. L'Eglise a toujours souligné l'importance des principes d'ordre moral et spirituel, ainsi que l'objectif urgent de la paix, suprême préoccupation de tous les esprits soucieux du destin du monde.

Aussi, en adressant ses vœux cordiaux à votre Conférence, le Souverain Pontife se plaît-il à penser que ces appels seront présents à l'esprit des illustres hommes de science et des éminentes personnalités qui sont réunies pour cette importante rencontre. Ainsi pourront se vérifier heureusement les préalables d'une entente fraternelle pour l'emploi et la mise en valeur bénéfiques des merveilleuses découvertes de la science.

Avec ces vœux, le Souverain Pontife invoque de grand cœur sur les organisateurs comme sur tous les congressistes les plus abondantes grâces du Tout-Puissant.

A. G. CARDINAL CICOGNANI

Du Vatican, le 1ᵉʳ mai 1969

PART I

STRUCTURES AND MATERIALS

Chairman: B. H. Goethert, U.S.A.

VIBRATIONS OF THIN ELASTIC PLATES SUBJECT TO RANDOM DRIVING FORCES

WILLIAM A. NASH and F. CHOU

University of Massachusetts, Amherst, Mass., U.S.A.

Summary. A stationary Gaussian random load is simulated digitally and employed as the forcing function in the equations of motion of a damped, elastic square plate whose resistance to deformation is due to bending and stretching. The power-residue method for generating pseudo-random numbers is employed in a technique presented for constructing the random forcing function. The nonlinear equation of motion and the compatibility equation are solved in finite-difference form for the case of a forcing function representing a time-random concentrated load applied at the center of a plate. From the numerical solution, statistical measures of the response at the center of the plate are obtained. The analysis holds for finite amplitude oscillations of the plate.

1. Background

The problem of thin elastic structural elements subject to various types of random driving forces has received considerable attention during the past decade. The recent book by Lin offers an extensive list of references pertaining to small amplitude lateral vibrations of strings, beams, and plates [1]. In addition, that reference lists several works devoted to large amplitude vibrations. Most of these, such as that due to Herbert [2, 3] are restricted to Gaussian white-noise excitation which is spatially homogeneous and uncorrelated. Random vibrations of thin elastic shells have been investigated by Dimentberg [4] and Jullien [5] for the case of small amplitudes of oscillation. It is the objective of the present paper to study the response of a clamped edge square elastic plate to a central concentrated random driving force. Finite deflections involving stretching of the middle surface of the plate are considered. The stationary, Gaussian, random load is simulated digitally and employed as the forcing function of the plate.

2. Simulation of Random Forcing Function

When random numbers are required in large quantities for use in a digital computer, it is convenient to have a program generate, deterministically, a set of numbers which are *operationally random*, i.e. which are indistinguishable from true random numbers under statistical tests of randomness. Such numbers are termed 'pseudo-random numbers'.

The power residue method for generating pseudo-random numbers has been employed here. In brief, that method consists of letting S_0 be any odd integer and

$$S_i = KS_{i-1}.$$

If b is the word size of a binary computer, K is chosen such that $K \approx 2^{b/2}$. Calculation

G. A. Partel (ed.), Proceedings of the Second International Conference on Space Engineering. All rights reserved.

of KS_0 produces a product $2b$ bits long. The multiplication, performed with double precision, permits the b least significant digits to be retained as the next random integer, while the b higher bits are discarded. The sequence of random integers generated by continuation of the procedure is converted to a set, S, of random numbers distributed uniformly over the unit interval bounded below by zero by floating the random integers and establishing a decimal point to the immediate left of each integer.

From the elements, S_k of set S, a set of random numbers, B, having a nearly Gaussian probability density function may be computed:

$$B_j = (1/n) \sum_{k=j}^{j+n} S_k \quad j = 1, n+2, 2n+3, \dots . \tag{1}$$

By the central limit theorem of statistics, the B_j have a mean, μ_B, approximately equal to the mean, μ_S of set S ($\mu_B = \mu_S = \frac{1}{4}$) and a variance $\sigma_B^2 = (1/n) \sigma_S^2$ or $\sigma_B = \frac{1}{12}n$. From the set of number, B, a continuous, piecewise linear function $Q(t)$, was constructed; pairs of Gaussian random numbers,

$$\begin{aligned} Q_i &= B_j & i &= 1, 2, \dots q \\ T_i &= \mu_T + \sigma_T B_{j-1} & j &= 1, 3, \dots r \end{aligned} \tag{2}$$

were defined in terms of the B_j. Here μ_T and σ_T are pre-assigned values of mean and standard deviation of the T_i, which are interpreted as random increments of a continuous variable, t, i.e.

$$t_m = \sum_{i=1}^{m} T_i \tag{3}$$

where t_m is a particular value of t; corresponding to t_m is a value of $Q(t)$ denoted by Q_m and determined from Equation (2). By linearly interpolating between successive pairs $Q_1, t_1, \dots, Q_q, t_q$ a continuous, piecewise linear function, $Q(t)$, $0 \leqslant t \leqslant T$, was obtained. A random function $F(t)$, having a predetermined mean, μ_F and variance σ_F^2 was obtained from $Q(t)$ by letting

$$F_g = \beta Q_g + \alpha; \quad g = 1, \dots q . \tag{4}$$

The mean and variance may thus be expressed as

$$\begin{aligned} \mu_F &= \beta \mu_Q + \alpha \\ \sigma_F^2 &= \beta^2 \sigma_Q^2 . \end{aligned} \tag{5}$$

By pre-assigning $\mu_F = 0$, $F(t)$ can be written as

$$F(t) = (\sigma_F/\sigma_Q) [Q(t) - \mu_Q] .$$

The percentage of statistical confidence with which it may be said that F has a Gaussian probability density function may be calculated by use of the chi-squared test. For a system corresponding to $n = 25$ this percentage was found to exceed 98%.

3. Analysis

Let us consider a thin, elastic square plate of side length a, thickness h, Young's modulus E, and Poisson's ratio v. An x–y coordinate system is introduced along two adjacent edges of the plate and lying in the middle surface of the plate and a z-coordinate is taken normal to the x–y plane. Let $\bar{W}$ denote the z-component of displacement of a point in the middle surface of the plate and let an Airy force function F be defined by the relations

$$\bar{N}_x = hF_{,yy}$$
$$\bar{N}_y = hF_{,xx}$$
$$\bar{N}_{xy} = -hF_{,xy} \tag{6}$$

where $\bar{N}_x$, $\bar{N}_y$, and $\bar{N}_{xy}$ are the normal and shearing forces per unit length of the middle surface of the plate. Here, commas denote partial differentiation with respect to the variables following the comma. Further, ϱ denotes the density of the plate material, $\bar{p}$ the intensity of lateral load, $\bar{c}$ is a velocity-type damping factor, $\bar{t}$ is time, and $D = Eh^3/12(1 - v^2)$. The dimensionless form of the nonlinear equation governing the finite-amplitude motions of such a plate is

$$W_{,tt} + cW_{,t} = -\nabla^4 W + 12(1 - v^2)(\bar{N}_x W_{,xx} + N_y W_{,yy} + 2N_{xy} W_{,xy})$$
$$+ 12(1 - v^2)p \tag{7}$$

and the dimensionless compatibility equation is

$$\nabla^4 F = W_{,xy}^2 - W_{,xx}\, W_{,yy} \tag{8}$$

where ∇^4 is the biharmonic operator and the scaling factors between the dimensionless quantities (i.e. those lacking bars) and the corresponding dimensional quantities are defined as follows:

$$
\begin{aligned}
W &= \bar{W}/h & F &= \bar{F}/Eh^2 \\
p &= \bar{p}a^4/Eh^4 & N_x &= \bar{N}_x a^2/Eh^3 \\
x &= \bar{x}/a & t &= \bar{t}(D/\varrho a^4)^{1/2} \\
y &= \bar{y}/a & c &= \bar{c}/(D/\varrho a^4)^{1/2}.
\end{aligned}
\tag{9}
$$

The time scale employed is based upon the natural frequency of a plate with zero membrane tension, i.e.

$$\omega^2 = D/\varrho h a^4.$$

The dimensionless x and y components of displacement of a point in the middle surface of the plate are denoted by u and v respectively.

For a clamped edge plate, the boundary conditions pertaining to displacement are $u = v = W = 0$ along all edges and

$$W_{,x}]_{x=0,\,1} = 0; \qquad W_{,y}]_{x=0,\,1} = 0. \tag{10}$$

Let us examine the boundary conditions along the edges $x=0$, 1. Since the non-linear strain-displacement relation is

$$\varepsilon_y = v_{,y} + (1/2) W_{,y}^2 \tag{11}$$

and along these edges both v and W vanish, then ε_y vanishes everywhere along these edges. Now, since

$$\varepsilon_y = (F_{,xx} - vF_{,yy})/E \tag{12}$$

then along $x=0$, $x=1$ we have

$$(F_{,xx} - vF_{,yy})_{x=0,\,1} = 0. \tag{13}$$

It is to be noted that Equation (13) is not an independent boundary condition but essentially a re-formulation of (10). For the case of a first-mode type response, it is necessary to consider the deflections of only one-eighth of the plate and hence it is not necessary to examine boundary conditions along the other edges $y=0$, $y=1$ because they are images of those already considered.

Next, let us examine the stretching of the middle surface along the line $y=$ constant from $x=0$ to $x=1$. The extension can be determined from the relation

$$\int_0^1 \varepsilon_x \, dx \big]_{y=\text{constant}} = \int_0^1 \frac{1}{E}(F_{,yy} - vF_{,xx})]_{y=\text{constant}} \, dx \tag{14}$$

$$= \int_0^1 [u_{,x} + \tfrac{1}{2}W_{,x}^2]_{y=\text{constant}} \, dx .$$

But since $\int_0^1 u_{,x} \, dx$ vanishes by virtue of (10), we have

$$\int_0^1 \frac{1}{E}[F_{,yy} - vF_{,xx}]_{y=\text{constant}} \, dx = \int_0^1 [\tfrac{1}{2}W_{,x}^2]_{y=\text{constant}} \, dx \tag{15}$$

as an additional relation between W and F.

Next, we shall separate the equation of motion (7) into two second order differential equations by the following procedure. First, we define two new variables U and V as follows:

$$U = (W_{,xx} + W_{,yy}) + aW_{,x} + bW_{,y} - kW \tag{16}$$

$$V = W_{,t} \tag{17}$$

where a, b, and k are three arbitrary constants. With these new variables, the Equation (7) is equivalent to the two equations

$$U_{,t} = (V_{,xx} + V_{,yy}) + aV_{,x} + bV_{,y} - kV \tag{18}$$

$$V_{,t} + cV = -(U_{,xx} + U_{,yy}) + aU_{,x} + bU_{,y} + kU + k^2U + 12(1 - v^2)p \tag{19}$$

provided we take

$$a^2 = \tfrac{1}{2}[N_x - N_y + \sqrt{(N_x - N_y)^2 + 4N_{xy}^2}]\,12(1 - v^2)$$
$$b^2 = \tfrac{1}{2}[-N_x + N_y + \sqrt{(N_x - N_y)^2 + 4N_{xy}^2}]\,12(1 - v^2) \qquad (20)$$
$$k = \tfrac{1}{4}[N_x + N_y - \sqrt{(N_x - N_y) + 4N_{xy}^2}]\,12(1 - v^2).$$

The two Equations (18) and (19) may conveniently be written in matrix form as

$$\begin{pmatrix} 0 & 0 \\ 0 & C \end{pmatrix} U + U_{,t} = \begin{pmatrix} 0 & 1 \\ -1 & 0 \end{pmatrix} \left(U_{,xx} + U_{,yy} \right) + \begin{pmatrix} 0 & 1 \\ 1 & 0 \end{pmatrix} \left(aU_{,x} + bU_{,y} \right)$$
$$+ \begin{pmatrix} 0 & -1 \\ 1 & 0 \end{pmatrix} kU + (12(1 - v^{2^0})p + kU) \qquad (21)$$

where (U/V) is defined as a vector $\mathbf{U}$. The advantage of this procedure is that Equation (21) is well-suited to being put in finite-difference form. In the interest of brevity that form will not be presented here but is denoted as Equation (22). Analogously, the compatibility Equation (8) may be written in the following finite-difference form:

$$(F_{10} + F_{11} + F_{12} + F_{13}) + 2(F_6 + F_7 + F_8 + F_9) - 8(F_1 + F_2 + F_3 + F_4)$$
$$+ 20F_5/\Delta x^2 \Delta y^2$$
$$= (W_6 - W_7 + W_8 - W_9/4\Delta x \Delta y)^2 - (W_1 + W_3 - 2W_5/\Delta x^2)$$
$$(W_2 + W_4 - 2W_5/\Delta y^2) \qquad (23)$$

where the subscripts refer to the network points shown in Figure 1.

The boundary conditions may be expressed in finite difference form as follows:

$$W_5 = 0$$
$$W_1 = W_3$$
$$(F_1 + F_3 - 2F_5/\Delta x^2) - v(F_2 + F_4 - 2F_5/\Delta y^2) = 0$$
$$\int_0^1 [(F_2 + F_4 - 2F_5/\Delta y^2) - v(F_1 + F_3 - 2F_5/\Delta x^2)]_{y=c}\, dx$$
$$= \int_0^1 [\tfrac{1}{2}(W_1 - W_3/2)^2]_{y=c}\, dx$$

Equations (22) and (23) may be solved by digital computer techniques using the following procedure:

1. Specify the initial conditions of the plate. This determines the initial membrane stresses.

2. Integrate the equation of motion along the time axis from t to $(t + \Delta t)$.

3. Substitute the results of step 2 into the right-hand-side of the compatibility Equation (23). Solve this equation by the matrix inverse method.

4. Substitute the membrane stresses obtained in step 3 into the right-hand side of

the motion Equation (22) and solve for U, then find W at each grid point. Repeat the computation starting from step 2.

The grid system employed to study a first-mode type of response of the plate is shown in Figure 2. The fictitious points 16 through 20 lie outside the boundary of the plate

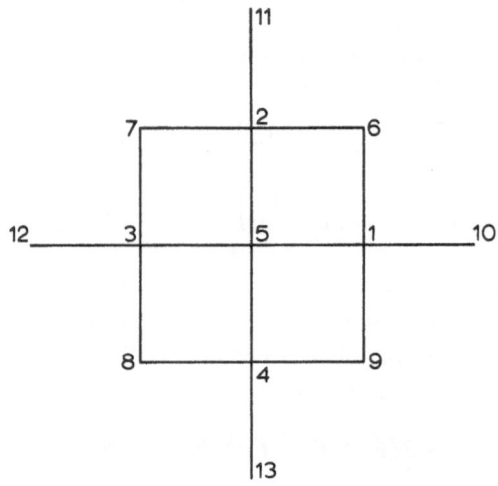

Fig. 1.   Basic network notation.

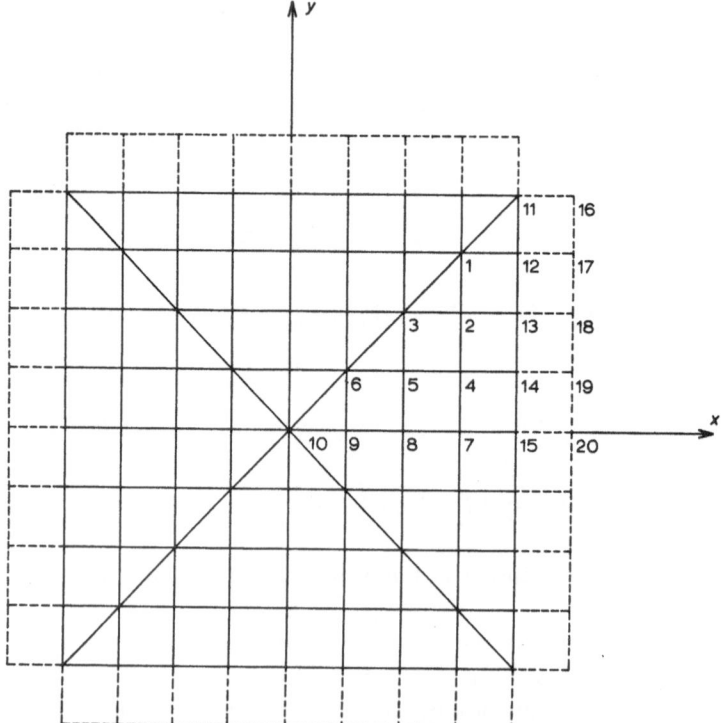

Fig. 2.   Network notation for one-eighth of plate.

and are introduced to make possible the consideration of the true boundary points 11 through 15 in the difference process. The finite difference Equations (22) and (23) are written for each point shown in Figure 2 and these equations solved by a matrix inverse technique. Numerically it was found necessary to employ a grid with $x = 0.125$ and a dimensionless time increment $t = 0.002$ to bring about numerical convergence.

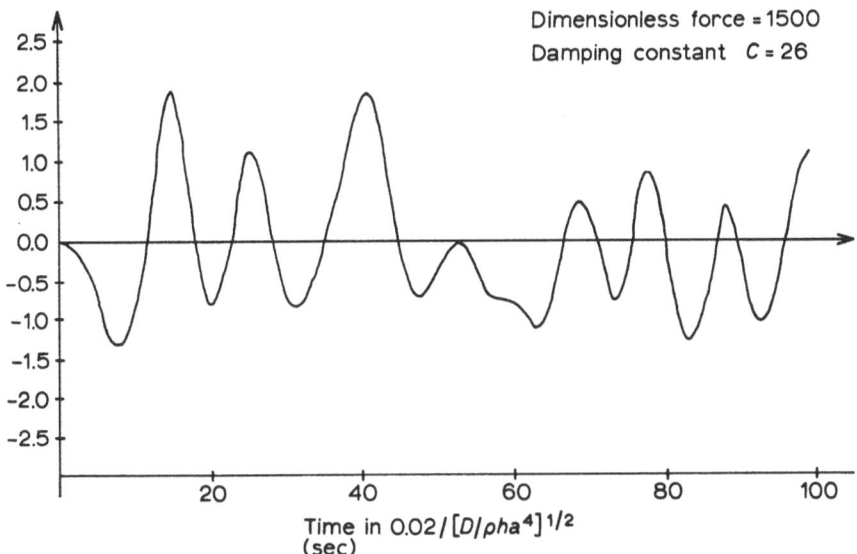

Fig. 3.   Deflection as a function of time.

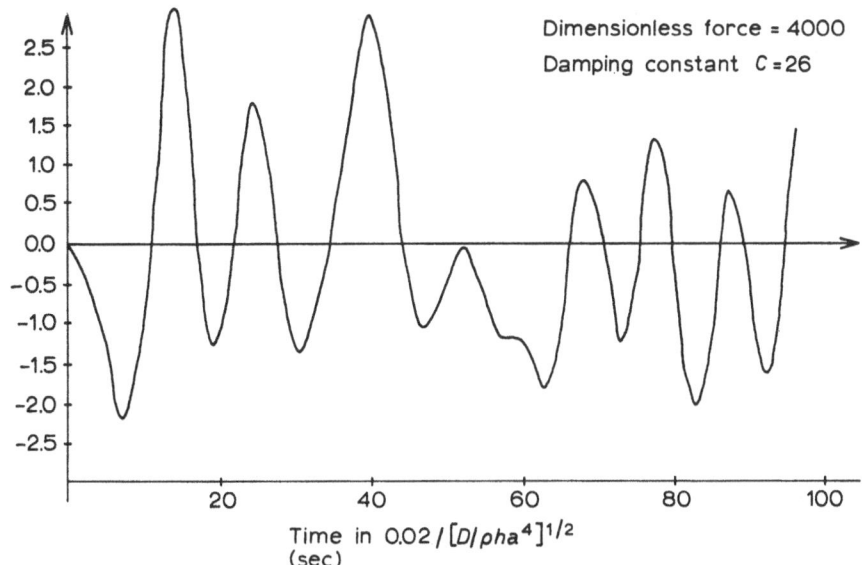

Fig. 4.   Deflection as a function of time.

4. Results

Numerical solutions were obtained on the CDC 3600 digital computer for the Equations (22) and (23) subject to the boundary conditions (24). A dimensionless damping constant $c = 26$ was selected for all cases and three cases of centrally applied concentrated random forces were investigated. One corresponds to a peak dimensionless value of 1500, another to a peak value of 3500, and the third to a peak of 4000. Digital solutions of the governing equations in finite difference form yielded the dimensionless central deflection as a function of time as indicated in Figures 3, 4, and 5 respectively. The standard deviation of dimensionless response amplitude for the case of a peak value of 4000 was found to be 1.31. The mean value of central response was found to approach zero, which is reasonable in view of the zero mean value of the random driving force. For this same case the probability density function of the central response was plotted and found to be nearly the probability density function of a Gaussian process.

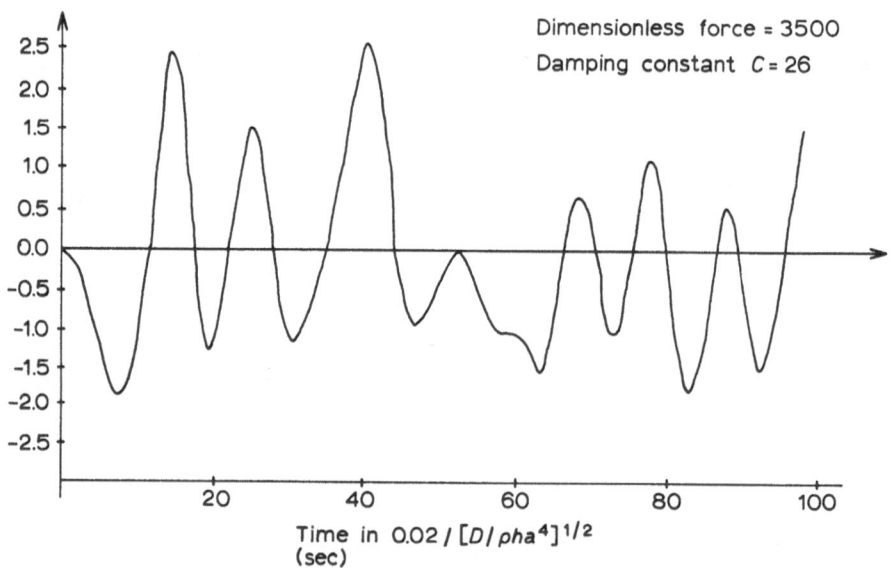

Fig. 5.   Deflection as a function of time.

Acknowledgement

This research was sponsored by the Air Force Office of Scientific Research under Grant AFOSR 68-1527.

References

[1] Lin, Y. K., *Probability Theory of Structural Dynamics*, McGraw-Hill, 1967.
[2] Herbert, R. E., 'Random Vibrations of a Nonlinear Elastic Beam', *J. Acoust. Soc. Am.* **36** (1964) 11.
[3] Herbert, R. E., 'Random Vibrations of Plates with Large Amplitudes', *J. Appl. Mech.* **32** (1965) 3.
[4] Dimentberg, M. F., 'On the Nonlinear Vibrations of Elastic Shells with Random Loads' (in Russian), in *Teoriya obolochek i plastin*, Armenian Academy of Science Publishing House, 1964.
[5] Jullien, Y., 'Vibrations aléatoires de calottes sphériques élastiques', *J. Mecan.* **5** (1966) 4.

DESIGNING COST-EFFECTIVE PRESSURE VESSELS
BASED ON FRACTURE MECHANICS *

L. RAYMOND

Aerospace Corporation, Los Angeles, Calif., U.S.A.

Abstract. Fracture mechanics provides the framework to couple materials properties with the design requirements for pressure vessel applications. Materials can be rated with regard to fracture-safe performance and their ability to fulfill the service life requirements. The approach results in a material-selection process, which, when coupled with fabrication costs, can be analyzed from a cost-effective viewpoint. The approach is primarily related to the design of large-diameter pressure vessels.

1. Introduction

The fracture mechanics concept of the "stress intensity parameter" can be used to couple design and material parameters with cost parameters in order to provide a total systems approach to design that does not sacrifice reliability. The material selection process, specification of material quality, specification of weld effectiveness, specification of non-destructive testing (NDT) requirements, and the fulfilment of service-life requirements can all be coupled with cost factors by defining a critical flaw size. The results can then be integrated into a minimum cost design for the total system.

This presentation emphasizes the design and material aspects of fabricating large diameter pressure vessels wherein the cost per pound of additional weight is relatively low. The variables are selected such that the fabricated vessel is cost effective in making the total system costs a minimum. Although this application is being used in the aerospace industry, similar analytical techniques can be used in large diameter pressure vessels for the power, chemical, and transportation industries.

2. Cost Effectiveness

The term cost-effective design is used in preference to minimum cost design in order to differentiate between the total cost of a system, which should be minimized, and the cost of a subsystem, which need not be a minimum. Each subsystem is designed to a cost per unit weight value that is effective in making the total cost of the system a minimum.

In commercial tankage for storage of fluids or gases, standard ASME 'boiler plate' techniques are employed. The result is a heavy structure which is designed to minimum cost and also minimum cost/weight since the tank represents the entire system. The cost/weight value decreases as the weight of the structure increases (Figure 1). In this case, increasing the weight with a given design criterion implies that the volume con-

---

* With permission from the editor and publisher of the Second Tewksbury Symposium – Fracture.

tained by the tank increases. Minimum weight designs for pressure vessels used in spacecraft also produce cost/weight values which decrease with weight at about the same slope as commercial tankage, but the absolute cost/weight value is much greater. The cost/weight curve for commercial aircraft is slightly less than spacecraft and the difference reflects on the cost and availability of propulsion. If no limit is placed on availability, a new region of technology requirements evolves. This technology gap is indicated by the cross-hatched region of Figure 1 and it reflects technology requirements based on cost-weight trade-offs as compared to current minimum-weight pressure-vessel design. In comparison to storage tanks, which represent the total system, the pressure vessel in spacecraft or aircraft is only a part of the total system or subsystem.

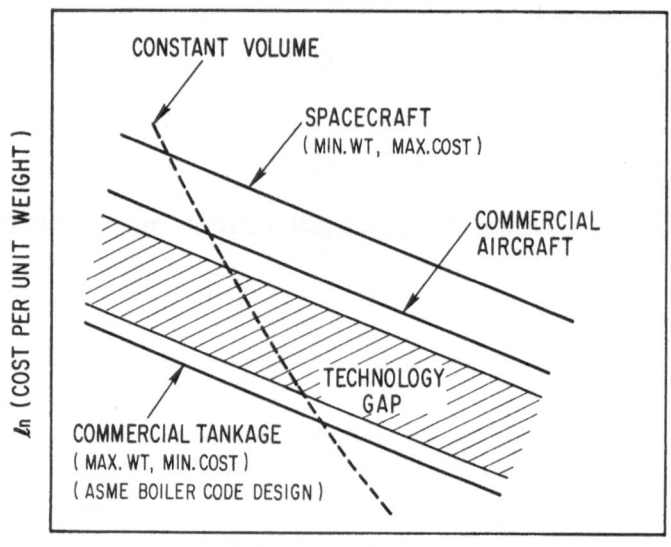

Fig. 1.   Effect of design criteria on cost (after [1]).

Lift or propulsion is another subsystem. In Figure 2, the situation is simplified for the sake of illustration. The cost for each subsystem is plotted separately as a function of pressure-vessel weight, keeping the volume constant (note the constant-volume line in Figure 1). The points along the vessel curve represent the total cost of a fabricated pressure vessel employing different design criteria from minimum weight to minimum cost. Propulsion costs behave in the opposite manner. They increase as the weight of the vessel increases. In minimizing the toal system cost, the cost-effective (dollar/pound) values of a vessel and propellant are determined (Figure 2). Since the cost figures used are based on current materials, fabrication, and design technology, these cost-effective values represent a specific baseline design in which material, design, and manufacturing practices have been selected.

   To evaluate any design or material alternatives, corresponding weight changes must be established, and the cost in dollars of an additional pound of weight must be calculated by perturbation of the analysis for the entire system. In this way, a cost-weight trade-off relationship is determined. Therefore, any design changes should be evaluated in terms of the resulting weight change and the resulting cost when compared to the baseline design. An example would be the selection of design alternatives such as the use of web and land construction instead of a constant thickness boiler-plate construction. If one were to use land construction at the welded joint to account for the difference between base plate and weld properties, the thickness of the web could then be reduced to an appropriate value less than the land thickness, but not without the additional expense of removing metal either by chemical milling or machining. One must then trade-off the additional cost and reduced weight against the baseline design construction.

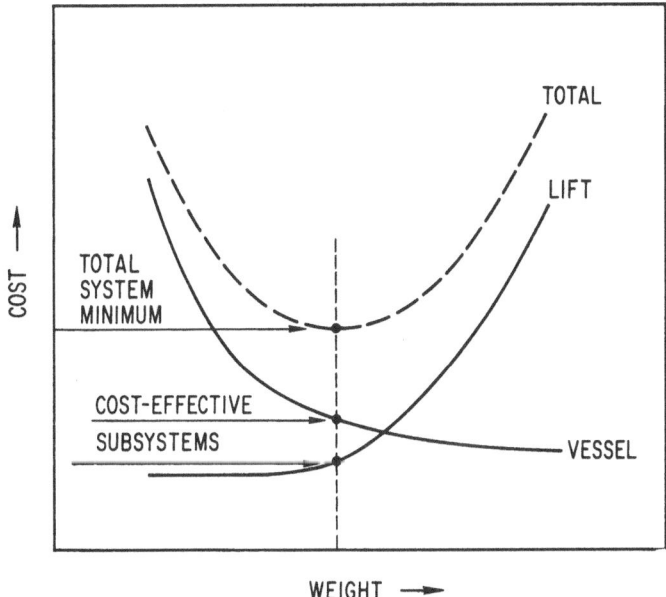

Fig. 2.   Cost-effective value of subsystem as related to minimum cost of total system.

Tolerances are another example of a basis for conducting weight-cost trade-off studies. As the thickness tolerance for a plate of steel purchased from a mill is minimized, cost per pound of material increases. Reduction of mismatch tolerances in welding also increases cost at the expense of decreasing weight. All of these items have a breakoff point. For example, the mismatch tolerance according to boiler code standards might be 0.030 in. for a given design. A mismatch of 0.010 in. is easily attainable with little or no increase in cost, but a request for 0.005 in. becomes extremely expensive because of the extensive rigging required to maintain such a small tolerance.

Excessively restrictive specifications for materials can increase the cost of steel by as much as $ 0.50/pound; vacuum melting, instead of air melting, can increase costs

18   L. RAYMOND

by as much as $ 1.00/pound. In large pressure vessels, these incremental cost factors
are significant when one talks of a material with a base price below $ 0.50/pound.
Other considerations are welding methods, processing methods (e.g., shear spinning
versus roll and weld construction), quality control, and logistics, i.e. transportation
costs of the structural segments.

If the baseline design is perturbated by alteration of material selection, joining
fabrication or any of the other items discussed, the resulting cost-weight trade-off
must be evaluated relative to the entire system. Figure 3 is an example of the ultimate
goals of such cost-weight trade-off for materials as related to a given baseline design.
The cost-effective value for the baseline design is indicated as $ 11/pound for a vessel
constructed from a 150 ksi yield strength material. Other materials would then become

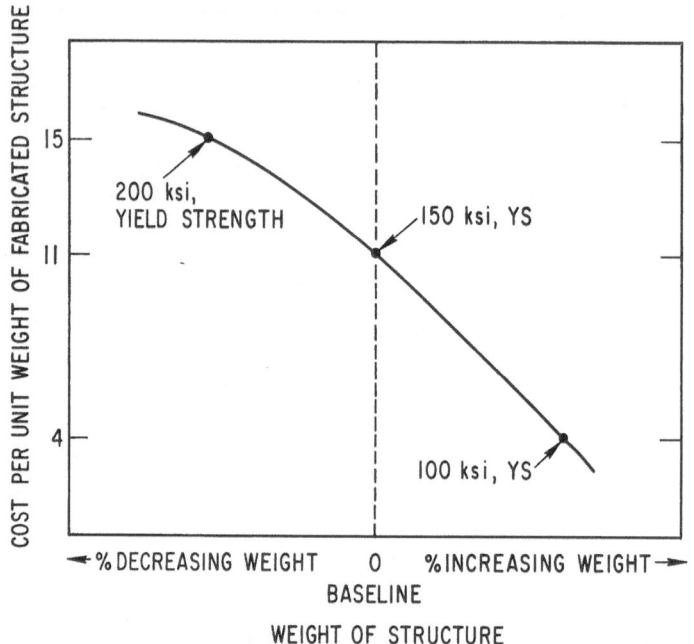

Fig. 3.   Cost-weight break-even relationships.

cost competitive, if for example, a 100 ksi yield strength steel, which would make the
vessel heavier, could be fabricated into a vessel for less than $ 4/pound or a 200 ksi
yield strength steel, which would make the vessel lighter, could be fabricated into a
vessel for less than $ 15/pound. This approach establishes materials fabrication and
design goals. If attainable, the baseline value would be decreased, thereby allowing
iteration of the analysis about the new baseline value.

Such an analytical approach on large diameter pressure vessels has been proposed
by Schnitt [2]. The complete analysis takes into account recurring and non-recurring
costs, transportation costs for structural segments, on-site fabrication, and intangibles
such as learning curves and cost trends. The complete analysis is beyond the scope of

this presentation, but it does indicate the coupling required in the fields of design, materials, joining, fabrication and economics, and it can be used to point to those fields of engineering design and materials research that would be economically fruitful.

Some significant conclusions of this analysis which are retained throughout because they inherently minimize cost are:

(1) The loss of vessels during proof test cannot be tolerated because of the large expense involved, especially in large size vessels. This condition necessitates fracture-safe design.

(2) Maximum utilization must be made of a material's strength. This condition means that a designer should work with a minimum factor of safety. In fact, it will be shown that from the point of view of safety and performance, it is better to design with a low factor of safety in low-strength, high-toughness materials than with a large factor of safety in high-strength, low toughness materials.

(3) Expenses for non-destructive inspection must be minimized with no decrease in the level of structural reliability. Too often, the inspection requirements are unnecessarily rigorous; yet oftentimes the most rigorous inspection is inadequate because it cannot be achieved with 100 per cent reliability.

3. Fracture Mechanics

The requirement of fracture-safe design can be fulfilled by using the approach of Tiffany and Masters [3]. Since the most critical flaw in a pressure vessel would be a crack that does not extend through the thickness of the material, the test specimen configuration that best reproduces this condition is the part-through-crack (PTC) specimen shown in Figure 4, where W=width of specimen, t_w=thickness of wall, a=depth of crack, and $2c$=length of crack. For this configuration, the geometric constant is calculated from linear elastic theory, resulting in the fracture criterion given by

$$K_{Ic} = 1 \cdot 1 M_K \sigma_{op} (\pi a_{cr}/Q)^{1/2}, \tag{1}$$

where K_{Ic} is the stress intensity at the tip of a sharp crack under plane strain conditions, M_K is the stress magnification factor for deep flaw behaviour $(a > t_w/2)$, σ_{op} is the operating stress, Q is the shape factor which is a function of $a/2c$, and a_{cr} designates the critical crack depth. K_{Ic} is also considered to be a material property, the fracture toughness. Comprehensive analytical solutions for M_K are not available; therefore, for the purpose of this presentation, it is assumed that $M_K = 1$ (valid for $a < t_w/2$) and $2c \gg a$ or Q is slightly less than unity, therefore,

$$K_{Ic} \cong 2 \sigma_{op} a_{cr}^{\frac{1}{2}}. \tag{2}$$

Material properties can now be coupled with the design requirements for pressure vessels to provide fracture-safe design conditions. Fracture-safe design conditions are imposed by reducing t_w until it is less than a_{cr}, thereby ensuring that the critical crack depth cannot be obtained without first causing the pressure vessel to leak. Equation

(2) is combined with the stress equation for pressure vessels,

$$\sigma = PR/t_w \tag{3}$$

where P is the pressure, and R is the radius of the vessel, to give

$$\frac{PR}{\sigma_{op}} = t_w = a_{cr} = (K_{Ic}/2\sigma_{op})^2. \tag{4}$$

Rearranging,

$$PR = K_{Ic}^2/4\sigma_{op} \equiv (PR)_{FS}. \tag{5}$$

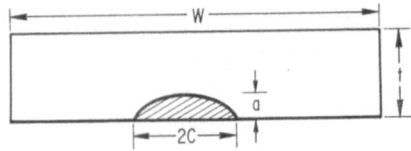

SPECIMEN CROSS SECTION

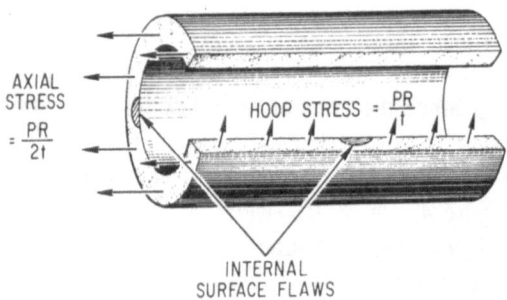

Fig. 4.   Test specimen configuration used to simulate internal surface flaws in pressure vessels.

Since the designer selects the variables P and R from the volumetric requirements and internal pressure for a given application, Equation (5) expresses the capability of the material to meet these requirements under fracture-safe conditions. Since $PR = \sigma t$, the shaded area of Figure 5 represents $(PR)_{FS}$. The figure is bounded by the yield strength (YS) as an upper limit on stress and t_w as a boundary on crack depth. A plot of Equation (2) confines the area consistent with fracture-safe design. As the wall thickness is reduced, the point of intersection moves up the curve $\sigma = K_{Ic}/2\sqrt{a}$. $(PR)_{FS}$ decreases to a minimum value when $t_w = t_{YS}$, the point of intersection with YS. This minimum value is designated as $(PR)^*$. Using this approach, fracture mechanics provides a method of quantitatively:

(1) rating material performance in fracture-safe pressure-vessel design;

(2) specifying NDT requirements in order to proof test with a minimum probability of loss during proof testing;

(3) specifying the amount of toughness required for fracture-safe pressure-vessel design;

(4) predicting the fulfilment of service-life requirements.

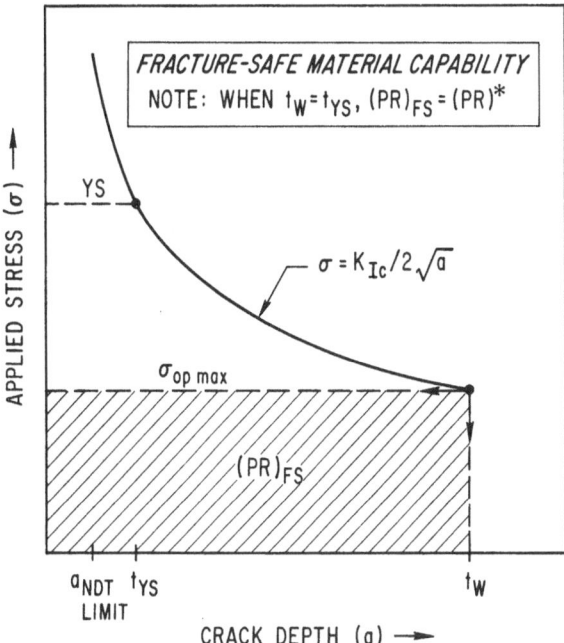

Fig. 5. Graphical representation of the capability of a material to meet pressure-vessel requirements under fracture-safe design conditions.

4. Material Rating System

The proposed method of rating materials for cost-effective pressure-vessel design assumes that it is desirable to use the material near its yield strength in order to obtain maximum utilization of the material's strength. Also, it is common design practice to proof-test pressure vessels near the yield strength of the material. Before the proof test is attempted, rigorous and expensive non-destructive testing is employed to eliminate critical flaws, in order to minimize the possibility of bursting the vessel. Therefore, to be consistent with fracture-safe design, it would be more efficient and economical for the material to be fracture-safe up to its yield strength.

The material rating system uses Equation 5 to calculate $(PR)^*$ by substituting YS for σ_{op}. The results are tabulated in Table I for five high-toughness, low-carbon steels. A titanium alloy and an aluminium alloy are also included to demonstrate the inability of these alloys to compete with steels in large-diameter pressure-vessel applications. Table I also shows that low-strength, high-toughness steels are better for large pressure-vessel applications than the high-strength low-toughness steels.

As an example, an application requires a vessel with a 60-in. radius to operate at a

TABLE I

Material capability of candidate steels for fracture-safe pressure-vessel design to YS
(After [4])

Material	Composition	YS (ksi)	K_{Ic} (ksi–$\sqrt{}$in.)	(PR)* (ksi–in.)
	Low-carbon steels			
HY 140	5Ni–Cr–Mo–V	140	280	140
T-1	Ni–Cr–Mo–V	100	180	80
HY 180	12Ni–5Cr–3Mo	180	180	45
Marage 200	18Ni–8Co–4Mo–Ti–Al	190	110	16
Marage 250	18Ni–8Co–4.5Mo–Ti–Al	240	90	8
Ti-6A1-4V	–	145	45	4
2014-T6	Al–4.5Cu–Mn–Si–Mg	68	35	5

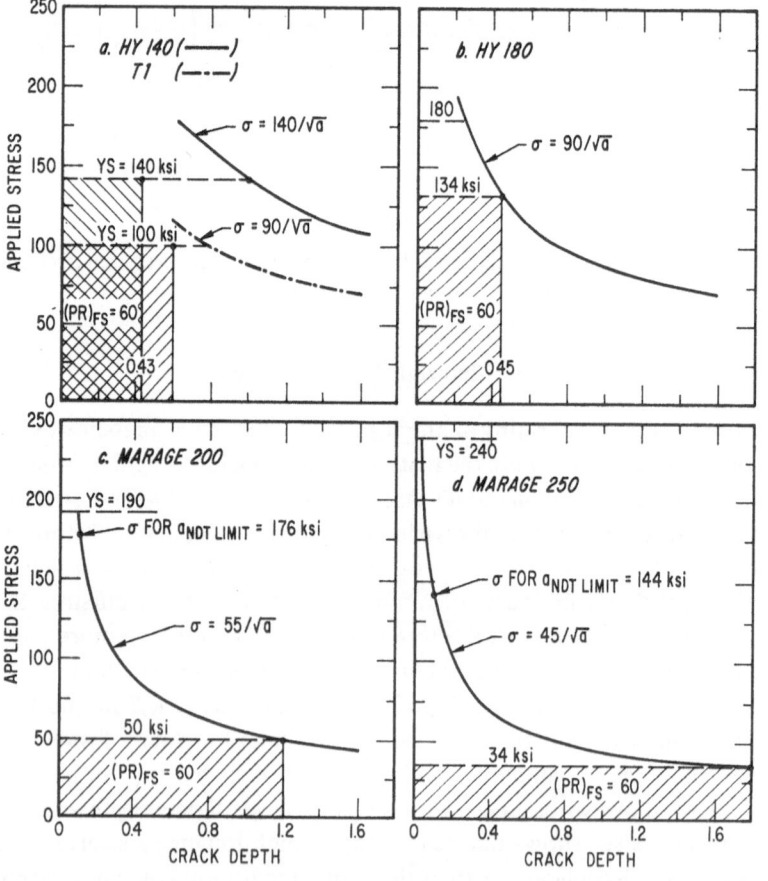

Fig. 6.   Performance of five high-toughness low-C steels under fracture-safe design conditions
(PR in units of ksi-in.).

pressure of 1000 psi, then the design requirements are a PR of 60 ksi-in. Referring to Table I, only two steels, T-1 and HY 140, could fulfil this requirement because $(PR)^* > PR$. Materials selected in this manner would result in a structure that requires a minimum of non-destructive testing, while maximum reliability is guaranteed because of the assurance that the structure will not burst at stresses below the yield strength of the material.

To be more precise, the rating system should incorporate the lowest values of YS and K_{Ic} existing in the fabricated structure. For example, if the structure is to be of welded construction, the lowest properties should be used and, in all probability, these values would be associated with the heat-affected zone or the weld metal and not with the base metal. The properties used in Table I are typical base metal properties.

To use the other candidate steels at 60 ksi-in. under fracture-safe design conditions, a limit must be placed on the operating stress ($\sigma_{op\ max}$). This maximum can be calculated by rearranging Equation (5),

$$\sigma_{op\ max} = K_{Ic}^2/4(PR)_{FS}.$$
(6)

As a result of the calculation, $\sigma_{op\ max} = 134$ ksi, 50 ksi, and 34 ksi for HY 180, Marage 200, and Marage 250 and thicknesses of 0.45, 1.2 and 1.8 in., respectively (Fig. 6).

Since $\sigma_{op\ max}$ is so low compared to YS for the latter steels, a design approach that makes better use of the available strength must be employed. This approach is a function of the limits of non-destructive testing as described in the next section.

5. Requirements in Non-Destructive Testing

From the previous section, it becomes apparent that the high-strength steels are toughness-limited. To obtain better use of the strength of these steels, non-destructive testing must be employed. Equation (2) may be analysed from another point of view, that of quantitatively specifying the resolution of flaw detection limits required by non-destructive testing in order to proof test to the YS of a material with a minimum probability of loss during proof testing. Rearranging Equation (2)

$$a_{NDT} = (K_{Ic}/2YS)^2.$$
(7)

Table II shows the result for candidate steels. It is again apparent that high-strength steels cannot be proof tested to their YS without entertaining the possibility of burst, because current non-destructive testing methods do not have adequate reliability to eliminate all flaws less than 0.100 in. If a limit of 0.100 in. is assumed for a_{NDT}, then $\sigma_{op\ max} = K_{Ic}/2\sqrt{a_{NDT}} = 1.6\ K_{Ic}$; i.e., HY 180 could be proof tested to its YS, but Marage 200 could only be proof tested to 176 ksi and Marage 250 could only be proof tested to 144 ksi with minimum probability of burst. These values are indicated on Figure 6 (c and d).

The corresponding PR values would be 18 ksi-in., 17.6 ksi-in. and 14.4 ksi-in., which are not adequate to meet the requirements set forth in the example, i.e., $PR = 60$ ksi-in. Therefore, these high strength alloy steels are non-destructive testing limited

TABLE II

Method of quantitatively specifying non-destructive testing requirements for proof to YS with minimum probability of loss during proof test

$a_{NDT} = (K_{Ic}/2YS)^2$

Material	YS	K_{Ic}	a_{NDT} (inches)
T-1	100	180	0.77
HY 140	140	280	1.00
HY 180	180	180	0.25
Marage 200	190	110	0.084
Marage 250	240	90	0.035

and could only be used to meet the requirement of $PR=60$ ksi-in. at the risk of entertaining the possibility of burst during service. To guarantee service life, no flaws of depth less than 0.100 inch can exist. This condition is only achieved by monitoring flaw sizes during service through rigorous non-destructive testing inspection methods.

Again, the cost factor must be introduced, for more rigorous methods such as radiography are an additional expense, and there always remains the doubt of achieving 100 % reliability. This is another reason why the lower-strength, higher-toughness steels are more attractive for cost-effective design.

6. Fracture-Safe Material Requirements

Similar to the previous section, Equation (2) can be employed to specify the amount of toughness required to proof test a vessel of given thickness to its yield strength without

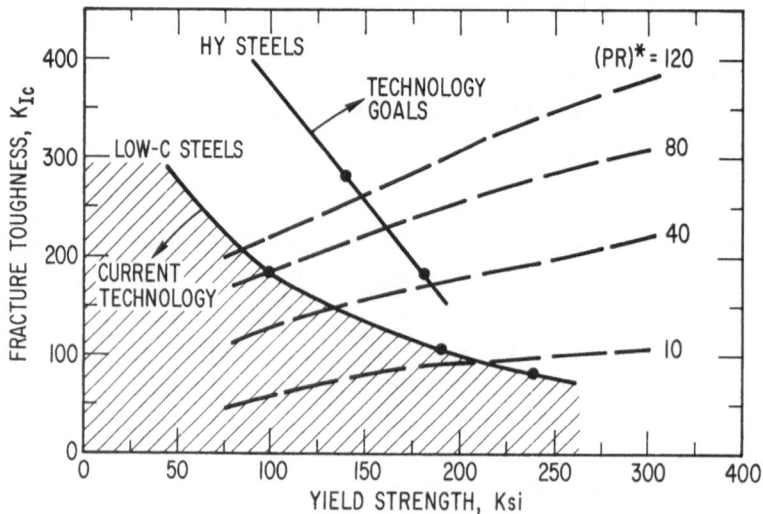

Fig. 7.   Material requirements to fulfil needs of fracture-safe design in pressure vessels (*PR* in units of ksi-in.).

entertaining the possibility of bursting the vessel. This is achieved by substituting the wall thickness (t_w) for a_{cr} and YS for σ_{op} in Equation (2). Figure 7 is a result of these calculations. Combined with the data from Table I, the ability of commercial low-C steels to meet fracture-safe pressure-vessel requirements is indicated. The HY steels, which are still in the developmental stages, demonstrate significant improvement in meeting the demands of large diameter pressure vessels. For example, to meet a $(PR)^*$ requirement of 40 ksi-in. in a 250 ksi steel, the fracture toughness (K_{Ic}) should be 200 ksi-$\sqrt{}$in. The data may also be interpreted to specify the weld efficiencies necessary for maintaining fracture-safe design. For example, if HY 140 were used in the application of $PR = 60$ ksi-in. and 100% efficiency were obtained for the yield strength of the weld, degradation of the fracture toughness of the weld much below 200 ksi-$\sqrt{}$in. could not be tolerated because it would introduce the possibility of bursting the tank.

7. Service-Life Requirements

Fracture toughness defines the critical flaw size that will cause the pressure vessel to burst at operating stresses below the yield strength of the material. Elimination of the critical flaws by non-destructive testing or proof testing does not ensure that subcritical flaws $(a < a_{cr})$ will not grow to critical size during the service life of the vessel. The two mechanisms of subcritical flaw growth during the service life of a pressure vessel are fatigue and/or stress-corrosion.

A common way to characterize the crack growth behaviour of a material subjected to stress-corrosion is the recording of time-to-failure for precracked specimens loaded in an aggressive environment to fractions of the K_{Ic} measured in an innocuous environment. For fatigue testing, the loading could be such that the applied stress intensity ranged from zero to K_{Ii}, $(K_{Ii} = 2\sigma_{applied}\sqrt{a})$ and for stress-corrosion, a static K_{Ii} would be used [4, 5]. There is a fundamental difference in the dependence of time-to-failure between fatigue and stress-corrosion (Figure 8). For fatigue, there is no obvious lower limit to the damaging K_{Ii}, whereas for stress-corrosion there is usually a very definite lower limit on damaging initial stress intensity. The stress intensity factor

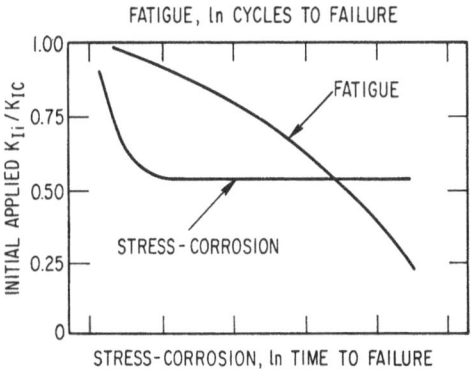

Fig. 8.   Service-life effects which induce subcritical flaw growth.

K_{ISER} is introduced at this point to describe the ability of the material to survive the required design life. It represents the combined effects of stress corrosion and/or fatigue, and it is used to define the subcritical flaw size $[a_{\mathrm{sc}} = (K_{\mathrm{ISER}}/2\sigma_{\mathrm{operating}})^2]$.

In stress-corrosion testing, the specimens that fail usually do so fairly rapidly (Figure 8). Because of this and the sensitivity of failure time to geometric variables (namely, the rate of increase of stress intensity with crack length at a given gross stress), it is best to let K_{ISER} be the threshold value of K_{Ii} for a given environment. Also, the threshold value of K_{Ii} has been shown to be independent of specimen geometry.

For fatigue testing, one must first determine the number of expected cycles and the range of applied stress intensity for each in-service cycle and the expected crack-growth rate for the part-through crack specimen. Fortunately, crack propagation laws are available that may be used to describe crack growth rate in terms of stress intensity. The so called 4th-power law usually associated with Paris [5] is given by

$$\frac{\mathrm{d}a}{\mathrm{d}N} = C(\Delta K)^4. \tag{8}$$

The value of C has to be evaluated for each material. Using Equation (8), the reduction in flaw size necessary to ensure reliability for a given operating stress can be determined.

Figure 9 indicates three different regions that define the performance of pressure vessels in service. The figure shows t_w to be greater than t_{YS}. If t_w were reduced to

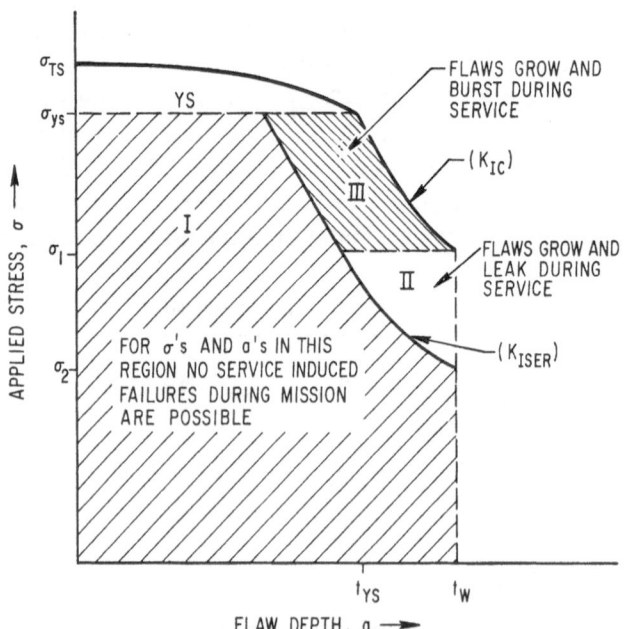

Fig. 9.   Regions of applied stress and flaw size which define the performance of pressure vessels in service.

t_{YS}, then region III would be eliminated (F-S design). As mentioned previously, fracture-safe conditions can also be maintained for $t_w < t_{YS}$ if the operating stress is limited to σ_1 as a maximum. σ_2 is then a maximum operating stress for which service life is guaranteed.

For the conditions of Figure 9, proof testing to YS would experimentally verify that no flaws $> t_{YS}$ exist. This would permit operating at a stress slightly below σ_1 in order to ensure service life. If $t_w = t_{YS}$, proof testing to YS would not ensure that the vessel would leak during service unless the operating stress was slightly below YS. The exact stress value in the latter case would depend on the degree of degradation. If we assume $K_{ISER} = 0.8\ K_{Ic}$, then $\sigma_{op\ max} = 0.8$ YS. To operate at higher stresses, one again has to depend on non-destructive testing.

8. Material Selection Process

As a result of the previous discussions, a material selection process for cost-effective pressure-vessel design can now be outlined:

(1) Define the design PR requirements.

(2) Identify the materials that will fulfil these requirements under fracture-safe design conditions up to the yield strength of the material; i.e., $(PR)^* > (PR)$.

(3) Evaluate K_{ISER} for these materials for the specific environment and fatigue life anticipated in service and eliminate those that display significant degradation.

(4) Establish the maximum operating stress, which will guarantee service life for these materials, given by $\sigma_{op\ max} = K^2_{ISER}/4(PR)$. For the example selected ($PR = 60$ ksi-in.), assume $K_{ISER} = 0.8 K_{Ic}$ for both the T-1 and HY. 140 steels that qualify in step 2 and find $\sigma_{op\ max} = $ YS for HY.140, and 86 ksi for T-1; i.e., HY.140 would be 0.43-in. thick and T-1 would be 0.70-in. thick, or T-1 must be almost twice as thick as NY.140 to perform in the same manner as HY.140. Proof testing to the yield strength in both cases would be adequate inspection to guarantee service life. Both steels would be fracture-safe up to their respective yield strengths.

The final step in selecting a material would depend on the total cost of the fabricated vessel which would reflect on the vendor's existing capabilities and experiences in forming and joining the remaining competitive materials. These cost figures would then be introduced into a cost-effective analysis and evaluated as illustrated in Figure 3.

9. Summary

In a system that consists of the propulsion of large diameter pressure vessels, cost-effective analysis uses a design criterion between the minimum-weight, maximum-cost spacecraft design concepts and the minimum-cost, maximum-weight 'boiler plate' design concepts.

Fracture mechanics provides the framework for quantitatively:

(1) rating material performance in fracture-safe design;

(2) specifying non-destructive testing requirements by defining a critical flaw dimension;

(3) specifying the amount of toughness required for fracture-safe pressure vessel design up to the yield strength of a material;

(4) predicting the fulfilment of service-life requirements;

(5) generating a process of material selection for cost-effective pressure-vessel design, which not only takes into account material costs but the total cost of the fabricated vessel, thereby providing the framework for a total systems approach to design.

Material selection, specification of non-destructive testing, quality control, and specific joining methods can be related to cost by definition of toughness requirements or critical flaw dimensions.

From a safety and performance viewpoint, the value of using a low factor of safety in high-toughness, low-strength materials instead of a large factor of safety in high-strength, low-toughness materials is emphasized.

Acknowledgements

The author extends his gratitude to A. Schnitt, of the Aerospace Corporation, who has been pioneering the minimum-cost systems design approach for many years and to his colleague, Dr. R. J. Usell, whose activity in the laboratory has experimentally validated the fracture-safe design criterion, resulting in discussions which have stimulated this presentation.

References

[1] Allesina, B., and Styer, E. F., 'Studies of Cost Effective Structures Design for Space Future Systems', NASA CR-1068, June 1968.
[2] Schnitt, A., and Kniss, F. W., 'Proposed Minimum Cost Space Launch Vehicle System', presented AIAA 4th Propulsion Joint Special Conf., Cleveland (10–14 June 1968).
[3] Tiffany, C. F., and Masters, J. N., 'Applied Fracture Mechanics', in *ASTM STP* 381, American Society for Testing and Materials, 1965, pp. 245–278.
[4] Novak, S. R., and Rolfe, S. T., Discussion in *ASTM STP* 410, American Society for Testing and Materials, 1966, pp. 126–129.
[5] Paris, P. C., 'The Fracture Mechanics Approach to Fatigue', in *Proceedings, 10th Sagamore Conference*, 1963 (Syracuse U.P.) p. 107.

INVESTIGATION OF LOW α_s/ε-COATINGS IN A
SIMULATED SPACE ENVIRONMENT*

L. PREUSS and W. SCHÄFER

Messerschmidt-Bölkow, Germany

and

H. HÖRSTER and H. KÖSTLIN

Philips Zentrallaboratorium, Germany

Abstract. For the joint U.S.-German solar probe project HELIOS, different experimental studies were conducted. One of these studies deals with environmental testing of the so-called 'Optical Solar Reflectors' (OSR = fused silica with vacuum-deposited metal), as developed and fabricated by the Philips Zentrallaboratorium Aachen. The environmental tests, especially tests with a simulated space environment, are described. UV irradiation of OSR were conducted in small vacuum chambers by different probe temperatures, the α_s/ε-ratio of the probes were measured *in situ*. The UV range mainly investigated was the near UV (0.4–0.2 μ).

Preliminary results from irradiation in the vacuum UV are also given. Test results and α_s/ε-measurements of OSR are very promising, the Ag-OSR showing a high environmental stability and the extremely low α_s/ε-ratio of 0.068.

The simulation and test facilities are described, the measured values are discussed.

Within a joint program of the Federal Republic of Germany and the United States it is planned to develop a solar probe called 'Helios'. This probe will be sent into a solar orbit with a perihelion of 0.3 AU* by an Atlas Centaur and a solid propellant third stage.

Figure 1 shows the solar probe design of the Messerschmitt-Bölkow group. The probe shall be stabilized by spinning at 60 rpm. Around the central body including the experimental and electronic equipment a cylindrical shell is mounted, which is used as a solar cell carrier and as a protective screen against the solar radiation. Around the solar cells there are certain coatings arranged to provide cooling against the intense solar radiation. This paper deals with investigations on these thermal control coatings.

To solve the thermal problem we have to consider the spectral distribution of the heating solar radiation and the cooling thermal radiation of the probe surface. In Figure 2 the intensity is computed for solar radiation at 1 AU, say 1 earth–sun distance, and the intensity of a black body surface at 300 K and 400 K. We see there are two separate spectral regions. Absorption of the coating in the region from 0.2 μm to 3 μm is responsible for the integral solar absorption α_s. Absorption in the region from 3 μm to 50 μm is equivalent with the thermal emissivity ε of the coating. Equilibrium is accomplished at the outer shell of the probe, if the areas below both types of curves multiplied by these optical constants α_s respectively ε are equal. As can be seen in Figure 2, it is possible to get normal temperatures for solar cell arrays at 1 AU, if the surface has equal optical constants in both regions, that means $\alpha_s/\varepsilon = 1$. But at 0.3 AU

* 1 AU = the distance earth–sun.

G. A. Partel (ed.), Proceedings of the Second Int rnational Conference on Space Engineering. All rights reserved.

the solar radiation increases about one order of magnitude and therefore extreme low α_s/ε values are needed. This means low absorption in the sunlight region and large emissivity in the infrared.

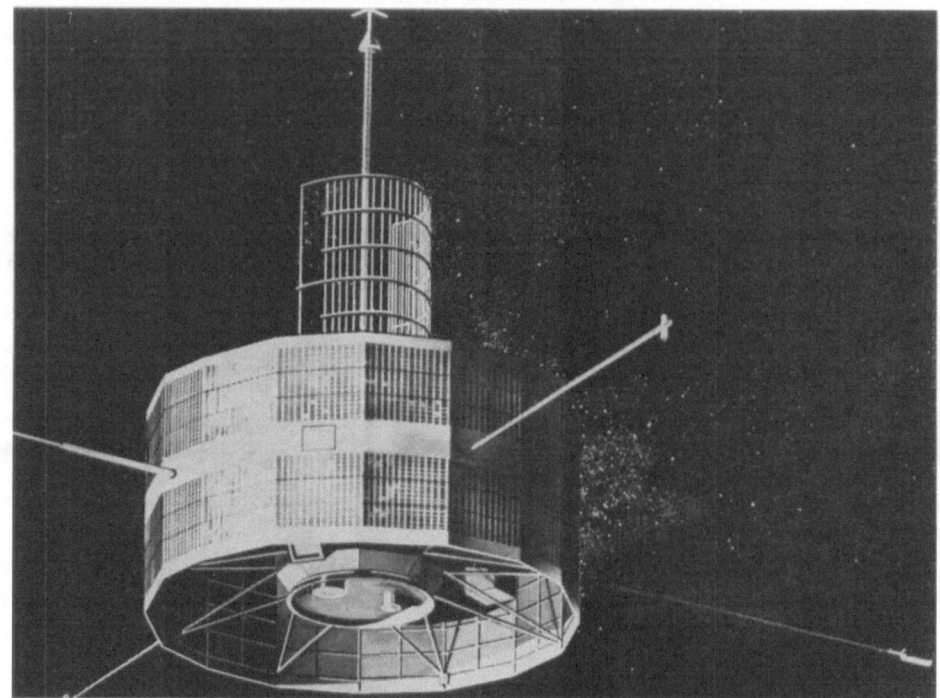

Fig. 1.   Solar probe Helios.

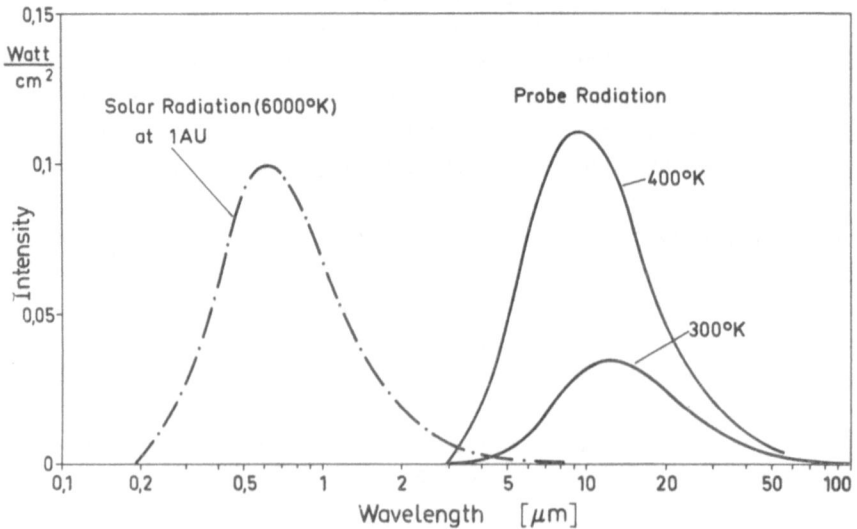

Fig. 2.   Radiation balance.

Figure 3 shows the dependence of the surface temperature as a function of the distance probe-sun. Parameter is α_s/ε. For the calculation of the equilibrium temperature the effect of spinning was taken into account. At a distance of 0.3 AU the probe should have an overall α_s/ε-ratio below 0.5. The maximum temperature limit corresponds to the state of art of high temperature Si-cells. Because solar cells already have an α_s/ε-ratio of about 1, the overall value of the surface must be reduced by special cooling coatings having very small α_s/ε-values.

The investigated systems are shown in Figure 4. Low α_s is obtained by back scattering from white paints or by reflection from special mirrors called optical solar reflectors (OSR). In both cases high ε, that means ε near 1, is provided by the infrared absorption bands of molecular vibrations.

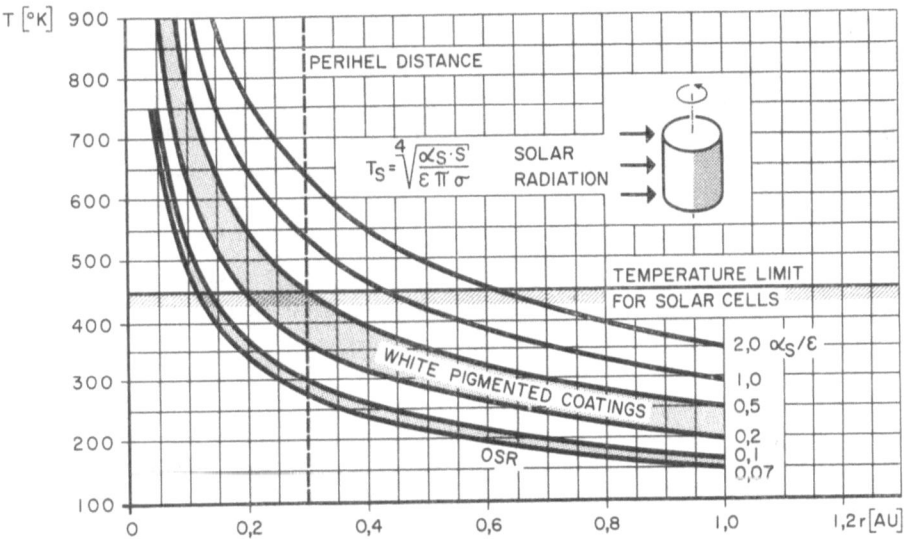

Fig. 3. Surface equilibrium temperature of a spinning cylindrical probe (vertical incidence, back-side insulated).

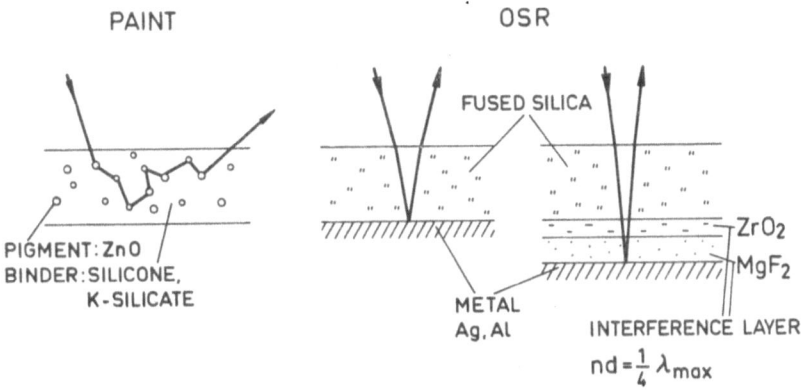

Fig. 4. Thermal control coatings with low α_s/ε.

The white paints were prepared from a suspension of ZnO in silicone or K-silicate, having an α_s/ε ratio of about 0.2.

The OSR were formed of a fused silica sheet, a material having a good transparency for sunlight and a high ε value with an evaporated highly reflecting Ag or Al-film on the rear. Fused silica is used and α_s/ε ratios of 0.07 have been achieved. A modified OSR was equipped with 2 interference layers to improve the reflection of the metal [1]. In the case of Al the α_s could be diminished about 25%.

Selecting the most promising system one has to consider the mechanical stability and the behavior of the optical properties during the mission. Stress by temperature and mechanical vibrations can be controlled without difficulty. But a chemical active atmosphere during storage and the complex solar radiation are of great influence on the optical properties of the coatings. For instance high energy radiation, plasma, γ- and electron radiation cause material defects, which in dielectrics often are associated with characteristic absorption bands. Furthermore the intense UV part of the sunlight causes photochemical processes, which create similar absorption bands. All these processes are very sensitive to impurities. We see that solar radiation is able to produce defects in dielectric materials, which cause an increase of α_s. Because of their characteristic bands these defects are called colour centres. Much work has been done on this subject, but the formation and bleaching of centres is always very complicated.

Under this point of view one expects for the white pigment coating an increase of absorption during solar radiation, because of the complex chemical and physical

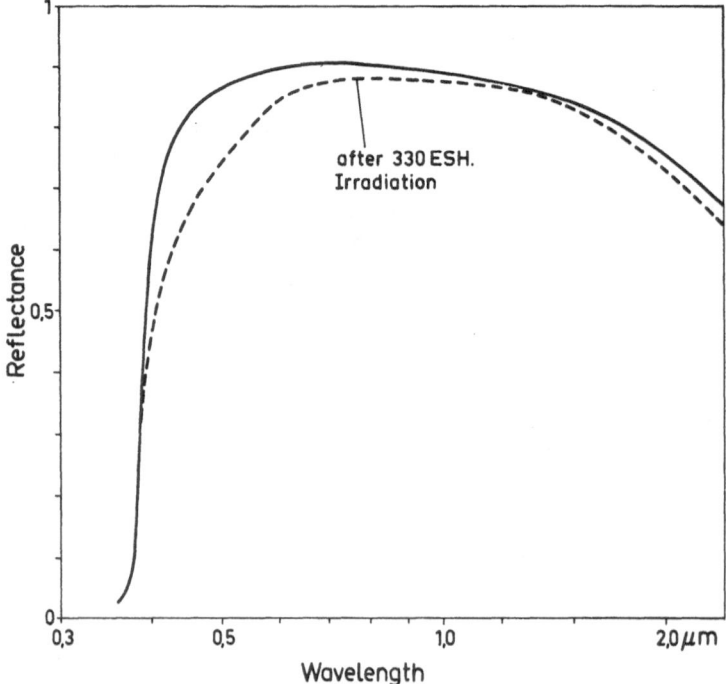

Fig. 5.   Degradation of ZnO/K-silicate under simulated sun irradiation.

structure, the long way of the scattered light and the rough surface against vacuum. Figure 5 shows the spectral reflectance of such a coating. It has a broad maximum in the visible region and a steep decrease in the near UV due to the fundamental absorption of ZnO. Although, according to our experiments, ZnO pigment coatings are the most stable ones [2], the lower curve shows a decrease of the reflectance even after a moderate Xe-lamp irradiation [3] of 330 equivalent sun hours. This effect is equivalent to a 10% enhancement of α_s/ε.

But the OSR behaves quite different during solar irradiation. No change of the reflection properties of the metal film can be distinguished, and also no formation of colour centres in the fused silica sheet is observed. The latter is due to the high purity of the SUPRASIL fused silica used. Only the modified OSR shows a small amount of degradation, probably due to the ZrO_2-interference layer.

Figure 6 shows the reflection spectra of three types of OSR (Ag, Al and Al+interference layers). Ag is for visible and infrared light the best reflector [4], only in the UV Al is the better one [5]. The interference layers raise the Al values, but they can reduce them at harmonic wavelengths too. Nevertheless, the unreflected part of the sunlight, respectively the integral absorption is $\alpha_s = 0.07$ for Ag. This is only half the value of the other coatings. The temperature dependence of the spectra is very small. Thus the integral α_s only rises about 0.01% per deg.

To obtain these curves, optimal conditions are required [6]: very clean surfaces of the substrates, short pre-evaporation with protected substrates, vacuum better than 10^{-5} mm Hg during the whole evaporation process, evaporation rates of about 30 Å/sec for Ag and at least 250 Å/sec for Al. Finally the Ag-films were covered with a protective film of MgF_2 against corrosion.

The spectral contribution to ε can be deduced from the reflection spectra of the

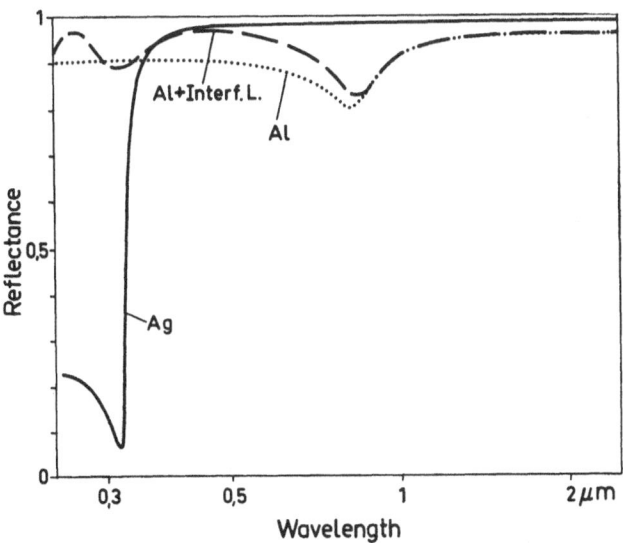

Fig. 6.   Spectral reflectance of three types of OSR in the sunlight region.

OSR in the region of the thermal radiation [4]. Figure 7 shows this for an Ag-OSR. The full IR absorption is reached at 5 μm. The peaks at 10 and 20 μm are Reststrahlen bands. Related to the body radiation at room temperature, one calculates an ε of 0.8. With increasing temperature the emissivity ε decreases about 0.03% per deg. This is caused by the shift of the thermal radiation to shorter wavelengths.

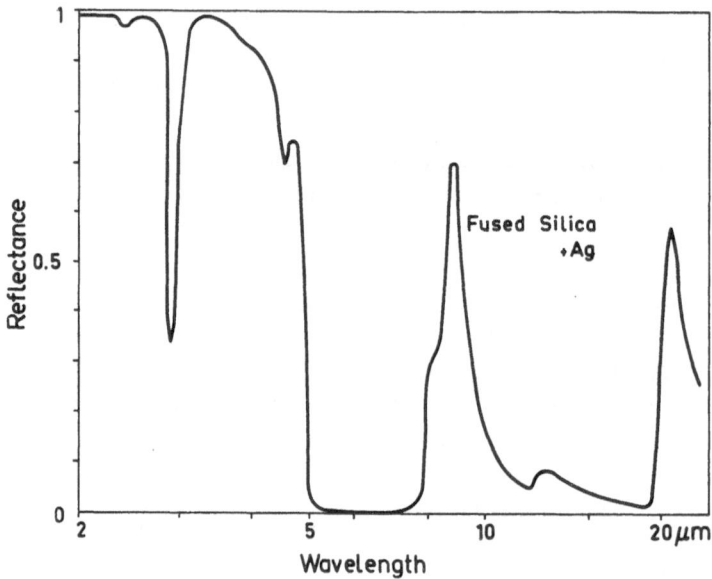

Fig. 7.   Spectral reflectance of an OSR in the region of the thermal probe radiation.

The reflection spectra were measured with conventional spectrometers relative to standards. Absolute values were obtained for some wavelengths with interference filters and a photometer sphere. These values could be used as standards: Ag in the infrared and visible, and Al in the UV-region. The angular dependence of ε was obtained by measuring the thermal radiation of the probe at 200°C. From 0° to 50° it remained constant and decreased about 20% at the upper limit of measurements at about 70°.

Besides these computations [7] based on optical measurements also calorimetric methods were used to get α_s and α_s/ε in a direct way. Sunlight was simulated by a Xenon lamp and an adapting filter.

α_s was measured in a dynamic way by observing the rate of temperature change, when the light was switched on or off. The temperature was detected with a Pt-resistor deposited by evaporation onto the rear of the probe. The 100% point was tested with a layer of a black standard. With this method also the angular dependence of α_s was checked. Up to 70° the OSR systems showed no effect within 5%.

α_s/ε was measured in a stationary way by observing the equilibrium temperature of the probe under simulated sunlight. Figure 8 shows the principle design of the space simulation chamber. The probe is surrounded by a black LN$_2$ cooled cylinder, which absorbs most of the emitted probe radiation and also some of the scattered and reflected sunlight. The probe manipulator is used for in situ measurements in degradation tests, in order to place the probe near the window, where it can be irradiated with a high UV dose from a Hg-lamp. The sample holder consists of two parts, the probe support and the heat shield. The heat shield can be heated separately to the equilibrium temperature of the sample and compensates the unwanted heat transfer.

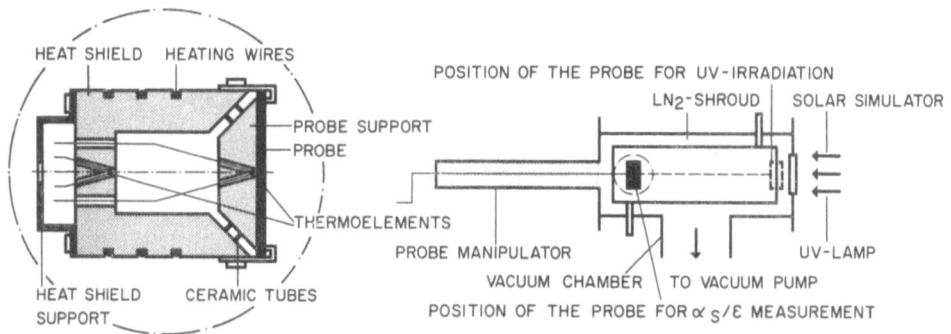

Fig. 8.   Principle of the static α_s/ε measurement.

All measured data of the different systems are collected in Table I. As mentioned before, the Ag-OSR shows the best results. For a mixed module (50% solar cells and 50% Ag-OSR) we measured an overall α_s/ε ratio of 0.37 at 470 K. Thus it is possible, to achieve a temperature below 200 °C under radiation of an intensity of 10 solar constants.

TABLE I

Optical data of thermal control coatings

Coating	α_s	ε	α_s/ε
OSR–Ag	0.05	0.077	0.07
OSR–Al + interference layers	0.09	0.78	0.11
OSR–Al	0.11	0.77	0.15
Paint ZnO/K-silicate	0.17	0.95	0.18

References

[1] Anders, H., *Dünne Schichten für die Optik*, 1965.
[2] Neel, Carr B., *Progr. Astron. Aeron.* **20**, 411.
[3] Miller, W. D. and Bevans, J. T., *International Symposium on Solar Radiation Simulation, Proceedings,* 1965, p. 115.
[4] Marshall, K. N. and Breuch, R. A., *J. Spacecraft* **5**, Nr. 9 (1968) 1051.
[5] Hass, G., *Appl. Optics and Optical Eng.*, Vol. III (ed. by R. Kingslake), Academic Press, New York, 1965, p. 309.
[6] Holland, L., *Vacuum Deposition of Thin Films*, 1961.
[7] Johnson, F. S., *J. Meteor.* **11** (1954) 431.

PART II

GUIDANCE AND CONTROL SYSTEMS

Chairman: B. H. Goethert, U.S.A.

CONTRÔLE AUTOMATIQUE DES CHAÎNES DE TÉLÉMESURES

P. GILLES

S.E.R.E.B., France

Résumé. La complexité des engins balistiques et spatiaux conduit à contrôler avant le lancement un grand nombre d'équipements ou de fonctions. Les opérations de préparation peuvent durer plusieurs jours, mais il faut vérifier rapidement dans les dernières secondes, que tous les paramètres sont corrects. Seul un contrôle automatisé permet de le faire avec la rapidité suffisante, ce qui conduit à sortir tous les paramètres par une prise ombilicale et à les vérifier successivement par un contrôleur commandé dans les solutions les plus évoluées par un calculateur.

Mais ces engins en plus de leurs fonctions opérationnelles ont une installation de télémesure pour retransmettre pendant le vol les données de surveillance ou d'expérimentation.

On se propose d'utiliser l'installation de contrôle automatique pour effectuer une vérification quantitative de chaque mesure en chaîne complète après la mise en route des émetteurs quelques instants avant le lancement. Il ne s'agit donc plus de sortir les mesures par l'intermédiaire de la prise ombilicale mais d'effectuer le contrôle en sortie d'une station de réception de télémesure.

Après un bref rappel du fonctionnement d'une station de contrôle automatique, nous décrivons sommairement les chaînes de mesure à contrôler pour montrer les problèmes particuliers à résoudre pour atteindre le but recherché.

Nous décrivons ensuite l'installation utilisée pour le lanceur de satellite 'DIAMANT', puis une installation plus récente développée pour des programmes d'engins tirés du Centre d'Essais des Landes.

Abstract. The complexity of ballistic missiles and space vehicles demands the checking of many equipments or functions before launching. Preparatory operations may last for several days, but quickly, in the last few seconds one has to verify that all the parameters are correct. Only an automated check-out enables this to be done with the necessary rapidity, and leads to the consecutive extraction of all parameters through an umbilical plug and verifies them with a controller commanded, in the most advanced solutions, by a computer.

In particular, in addition to their operational functions, these missiles have a telemetry system to retransmit the data of monitoring and experimentation.

As the telemetry transmitters are only started a short time before ignition (1 2 min), the final manual check-out is only qualitative. Automation permits a quantitative control of all the measurements in the allowable time.

The telemetry systems used are of the type PAM/FM/PM AJAX derived from the IRIG system.
The constituent elements of the control system are:
(1) a telemetry receiver,
(2) a switchable sub-carrier discriminator,
(3) a decommutator for the PAM channels,
(4) an automatic tester controlled by a programmer and a control console ensuring the selection of the parameters, their code, their comparison with the limiting values.

After the analysis of the functions made by all those diverse elements, some existing systems are described in particular those that have been used for the measurements 'DIAMANT' satellite launchers.

In conclusion, the problem of P.C.M. telemetry equipments is discussed.

1. Principe d'un équipement de contrôle automatique

Le bloc diagramme d'un tel équipement est donné figure 1.

Les fonctions assurées par l'équipement seront plus ou moins nombreuses suivant la définition du programmeur. Celui-ci peut en effet aller d'une simple mémoire sur bande magnétique avec des dispositifs logiques de commande jusqu'au calculateur le plus puissant.

G. A. Partel (ed.), Proceedings of the Second International Conference on Space Engineering. All rights reserved.

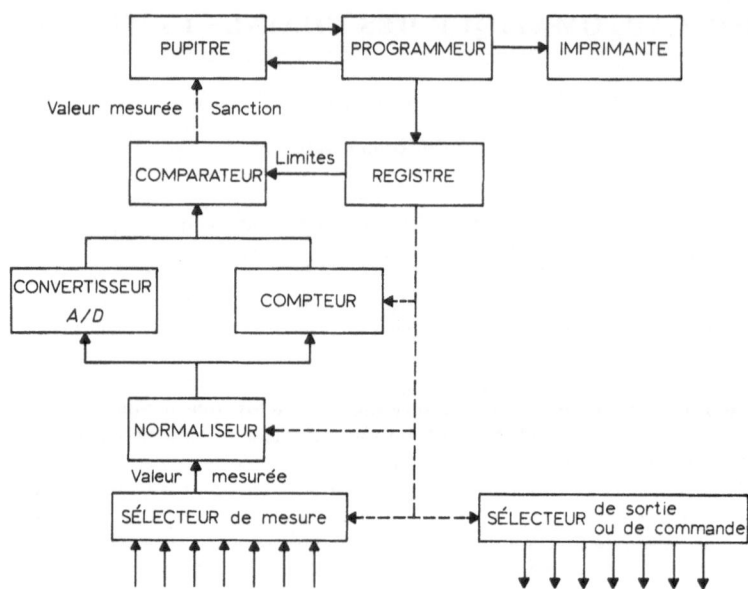

Fig. 1.   Bloc diagramme équipement contrôle automatique.

La programmation d'une séquence doit comporter les informations suivantes:
- numéro de la séquence,
- adresse du sélecteur de mesure,
- adresse du sélecteur de sortie ou de commande,
- type et gamme de la mesure,
- temporisation de la mesure,
- valeur maximale,
- valeur minimale.

Les paramètres à vérifier sont envoyés sur le sélecteur de mesures qui en choisit un suivant les ordres reçus et l'envoie sur un convertisseur analogique digital dont la sortie est comparée aux tolérances mises en mémoire par un comparateur. Le résultat est envoyé au pupitre de commande et de visualisation.

Le sélecteur de sortie ou de commande permet de donner les ordres convenables à l'engin ou aux matériels périphériques.

2. Description des chaînes de mesure et des contrôles a effectuer

Les chaînes de télémesure utilisées sont du type PAM/FM/PM au standard IRIG.

2.1. Chaîne d'émission embarquée

Le schéma de principe est le suivant (figure 2).

Les capteurs transforment les variations du phénomène à mesurer en tension électrique normalisée proportionnelle.

Les traducteurs convertissent les variations de cette tension en variations de fréquence proportionnelle à une fréquence centrale. L'ampli mélangeur sert à multi-

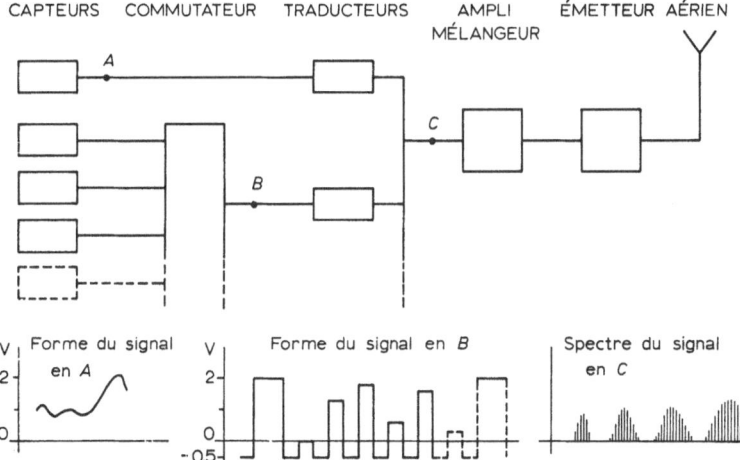

Fig. 2. Schéma de principe d'une chaîne d'émission de télémesure.

plexer plusieurs sous-porteuses ayant des fréquences centrales différentes en un signal vidéo unique qui module en phase l'émetteur. Le spectre de ce signal vidéo est représenté figure 2.

Chaque émetteur comporte ainsi une douzaine de sous-porteuses au standard IRIG. Deux à quatre émetteurs sont utilisés par engin.

Les commutateurs servent à répartir dans le temps une même sous-porteuse entre plusieurs mesures.

La séquence de commutation est du type IRIG PAM 2 indiqué sur la figure 2. Chaque mesure est séparée de la suivante par un interplot dont le niveau est à 25% en dehors de l'étendue de mesure.

La synchronisation de cycle est assurée par la réunion des deux premiers plots et de l'interplot correspondant, la valeur de la tension de ce plot double étant égale au maximum de l'étendue de mesure. Le premier plot suivant est maintenu à une valeur de référence égale au minimum de l'étendue de mesure.

2.2. STATION DE RÉCEPTION

Le schéma d'une station est donné figure 3.

Les récepteurs démodulent les porteuses et restituent les multiplex des sous-porteuses qui sont séparées par filtrage et démodulées par un discriminateur de fréquence restituant une tension proportionnelle à la tension d'entrée du traducteur.

Sur la figure 3 nous avons donc reporté l'allure des signaux de sortie d'un discriminateur pour une voie continue et pour une voie commutée.

2.3. PROBLÈMES PARTICULIERS POSÉS PAR LE CONTRÔLE AUTOMATIQUE DES CHAÎNES DE MESURE

Les opérations que nous voulons effectuer sont la comparaison à une valeur théorique affectée d'une tolérance de chacune des mesures pendant les instants précédant le tir.

Nous devons donc vérifier toutes les tensions en sortie du discriminateur. Pour cela, le sélecteur de commande du contrôleur devra:
  – commander la sélection d'un des quatre récepteurs,
  – commander le sélecteur d'un des filtres de sous-porteuses.
Jusqu'ici nous ne rencontrons aucune difficulté.

S'il s'agit d'une voie continue, la sortie du discriminateur peut être envoyée sur le sélecteur de mesure. Par contre, pour une voie commutée nous devons assurer la séparation des différentes mesures.

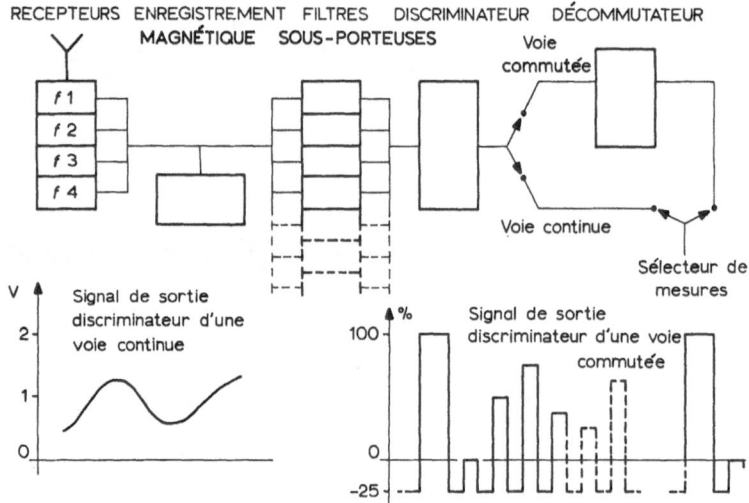

Fig. 3.   Schéma d'une station de réception au sol.

Plusieurs solutions sont possibles.

(a) Utilisation d'un décommutateur, qui par détection du plot double et comptage des plots sélectionne le plot à mesurer et délivre une tension proportionnelle à sa valeur en pourcentage de l'étendue de mesure. Le décommutateur est commandé par le sélecteur de sortie. C'est la solution qui est retenue sur la figure 3.

(b) *Echantillonnage rapide*. Cette méthode nécessite que le programmeur soit un calculateur. On analyse le signal par échantillonnage rapide. Les valeurs recueillies sur un cycle complet sont stockées dans les mémoires du calculateur, qui recherchera le plot long et numérotera les voies.

(c) *Retard prédéterminé de la conversion analogique numérique*. On utilise la partie du discriminateur qui détermine le plot long. Le début de la conversion est retardé d'un temps connu à l'aide d'un monostable présélectionné correspondant à la position du plot à contrôler. Pour contrôler la voie suivante, il faut se synchroniser à nouveau sur le plot long, donc attendre un nouveau cycle.

Nous avons retenu l'utilisation du décommutateur.

3. Fonctionnement du décommutateur

Le schéma synoptique est donné sur la figure 4.

Dans le cas d'une sous-porteuse continue, le décommutateur transmet directement la tension au contrôleur, sans aucun traitement.

Si la sous-porteuse choisie par le contrôleur est commutée, le décommutateur délivre en sortie une tension continue qui représente l'amplitude de l'un des N paramètres du cycle.

Le décommutateur comprend sept parties:

– un dispositif de décommutation qui permet de connaître à chaque instant le numéro du paramètre présent à l'entrée du décommutateur,

– un dispositif de synchronisation qui fournit au décommutateur les signaux de référence à partir du signal analysé,

– un dispositif de calibration qui traduit l'amplitude du paramètre analysé en pourcentage de l'étendue de mesure,

– les commandes,

– la commande rapide qui permet de réduire le temps nécessaire aux commandes en excitant simultanément plusieurs relais,

– les alimentations,

– un simulateur de télémesure commutée.

3.1. Le dispositif de synchronisation élabore à partir du signal de télémesure des signaux H de synchronisation de voie et des signaux R de synchronisation de cycle. En cas d'absence d'un plot ou d'un interplot, un dispositif substitue un signal au signal d'horloge manquant.

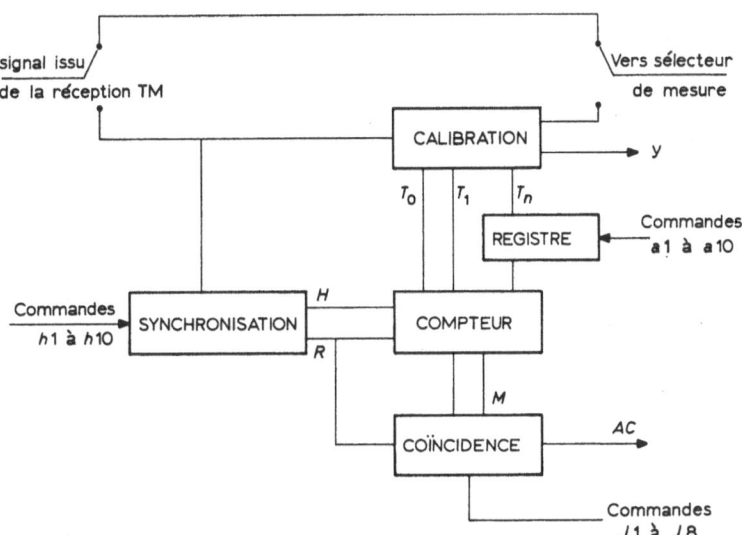

Fig. 4. Bloc diagramme du décommutateur.

3.2. Le dispositif de décommutation délivre les impulsions de prélèvement correspondant aux voies de référence (T_0 et T_1) et à la voie choisie T_n.

Le compteur est attaqué par les impulsions de voie H. Il est remis à zéro à chaque cycle par les signaux de synchronisation R.

La logique de coïncidence assure la sélection du nombre de voies par cycle indiqué par les commandes $J1$ à $J8$ en permettant de repérer le dernier paramètre de chaque cycle par coïncidence entre les commandes et le compteur. Elle permet de contrôler que la décommutation est correcte, la disparition du signal L fourni par cette coïncidence devant être suivie d'un signal de synchronisation de cycle R.

Cette disparition déclenche un monostable délivrant un signal M pendant un intervalle de temps convenable. Si le signal R arrive pendant cet intervalle de temps, un signal AC est émis vers le contrôleur pour valider les signaux.

Une logique de coïncidence attaquée par le compteur permet de détecter le passage de la référence 100% (T_0) et de la référence 0% (T_1) pour les diriger vers la calibration, ces signaux ne pouvant sortir que s'ils sont validés par le signal AC.

La sélection du paramètre à mesurer est assurée par la coïncidence entre le compteur et un registre commandé, soit par le contrôleur, soit manuellement. Le top T_n correspondant est dirigé vers le dispositif de calibration. Le signal n'est disponible que s'il est validé par le signal AC.

3.3. Le dispositif de calibration fournit une tension proportionnelle à l'amplitude du paramètre évaluée en pourcentage de l'étendue de mesure, la voie 0 représentant 100% et la voie 1.0%.

Le principe est le suivant:

– on opère une conversion analogique numérique à T_0, T_1 et T_n,

– une soustraction numérique opère un changement de niveau ramenant à 0 la valeur de référence minimale.

– on opère une conversion numérique analogique par l'intermédiaire d'un réseau de résistances parcourues par un courant constant et on obtient une valeur analogique proportionnelle à l'étendue de mesure.

Un signal Y est envoyé au contrôleur pour le prévenir qu'un paramètre calibré est présent.

3.4. Les commandes reçues du contrôleur par le décommutateur sont les suivantes:

– sélection du type de voie continue ou commutée,

– nombre de voies par cycle (8 valeurs possibles: 16, 18, 24, 30, 32, 45, 48, 64),

– cadences (10 valeurs possibles : 75, 90, 110, 150, 180, 240, 300, 360, 450, 900) avec pour chacune une tolérance de – 30 à +25%.

– numéro du paramètre à contrôler.

Ces diverses commandes nécessitent 40 relais dont l'excitation successive à raison de 16 ms par relais entraîne un temps d'application des commandes important. Un dispositif de commande rapide a été mis au point pour procéder à l'excitation simultanée des relais.

Ce mode de fonctionnement est possible si le contrôleur dispose d'un registre d'au moins 22 digits dans lequel sont stockées les informations nécessaires à la mesure de chaque paramètre. Une mémoire correspondante dans le décommutateur est chargée en parallèle et décodée par des matrices connectées aux commandes.

4. Installation de contrôle automatique des lanceurs de satellite 'diamant'

L'installation est représentée figure 5. Elle comprend:
- un contrôleur-programmeur S.E.T.I.,
- une station de réception télémesure avec un récepteur et un discriminateur télécommandables,
- un adaptateur-décommutateur identique à celui que nous venons de décrire,
- deux répéteurs pour assurer la transmission correcte des signaux sur une certaine distance.

Ces éléments sont reliés avec l'engin par l'intermédiaire de la boîte pied de rampe et de la prise ombilicale et avec le pupitre de lancement qui reçoit les comptes rendus et donne les ordres de mise en route. La distance entre la station de réception et la tour de lancement est de quelques kilomètres, celle entre le poste de tir et la tour de lancement de 300 m.

Cette installation est l'exemple d'un programmeur sommaire constitué essentiellement d'une mémoire sur bande magnétique et des circuits associés.

4.1. DESCRIPTION DU CONTRÔLEUR–PROGRAMMEUR

L'equipement comprend trois parties: pupitre, programmeur et contrôleur.

(a) Le pupitre permet de composer le programme à enregistrer et de commander les différents modes de fonctionnement. Il assure la visualisation par néons du programme, des résultats de la mesure et du numéro de séquence ainsi que l'impression des résultats.

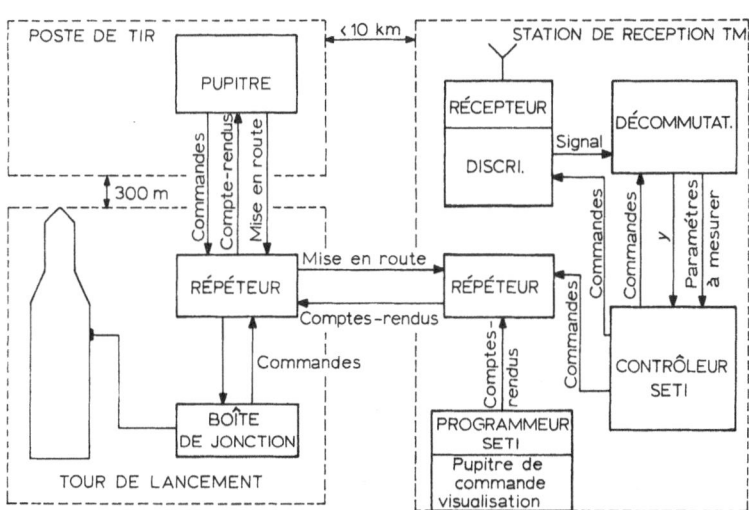

Fig. 5.   Installation de contrôle automatique des mesures du lanceur de satellites 'DIAMANT'.

En position d'enregistrement, on peut effacer l'enregistrement initial et positionner la bande, puis enregistrer les informations affichées sur des digiswitches à sortie sur quatre contacts dans le code 1, 2, 4, 2, ou sur des interrupteurs à deux positions.

En position de lecture, on peut procéder à une lecture continue ou pas à pas, à la recherche d'un bloc dont le numéro est affiché.

En position exploitation on peut:

– lire et exécuter le programme d'un bloc, puis s'arrêter après chaque bloc,

– rechercher un bloc préaffiché et exécuter son programme puis s'arrêter,

– exécuter de façon continue le programme.

L'édition peut être obligatoire, interdite ou effectuée uniquement pour les mesures hors limite. Elle se fait sur une machine à écrire IBM de vitesse 10 caractères par seconde. Une mesure hors limite est imprimée en rouge.

(b) Le programmeur assure les fonctions suivantes:

– enregistrement du programme composé sur le pupitre,

– lecture du programme et aiguillage des informations dans les différents registres,

– transmission des informations avec validation de la lecture par contrôle de parité et du nombre de caractères,

– synchronisation par horloge des opérations de la séquence programme et de la commande du mouvement de la bande programme.

La bande magnétique qui sert de mémoire centrale est une unité C.D.C. PNS 1. Capacité: 50 m – Vitesse: 15 in./sec – départ et arrêt en 5 ms.

Les circuits d'entrée – sortie de la bande comprennent:

– un registre tampon d'entrée,

– l'élaboration de l'information parité p,

– un registre tampon de sortie,

– le contrôle de la parité,

– les signaux de rythme,

– l'adaptation au câble des sorties des amplis de lecture.

Des circuits d'enregistrement permettent d'obtenir:

– l'enregistrement d'une piste rapide,

– le balayage des 25 caractères affichés sur les digiswitches et les interrupteurs,

– l'envoi en série des informations sur le registre tampon d'entrée.

Des circuits de saut et appel de bloc permettant d'effectuer:

– la recherche d'un bloc dont le numéro est affiché,

– le saut automatique à une séquence donnée en cas de mesure hors limite,

– l'élaboration des commandes vers le sous-ensemble synchronisation,

– l'arrêt en fin de bandes.

Des circuits de lecture permettent d'élaborer les ordres à envoyer au sous-ensemble de synchronisation, au décommutateur et au contrôleur ainsi que la validation de la lecture.

(c) Le contrôleur assure les fonctions suivantes:

– conversion analogique numérique des tensions,

– comptage,

– comparaison aux limites,
– sélecteur de sortie,
– sélecteur de mesure,
– commutation des organes de normalisation.

Le bloc diagramme du contrôleur est donné figure 6.

Les sélecteurs de mesure et de sortie sont constitués par des matrices de relais dont le temps de commutation est inférieur à 100 ms, la résistance d'isolement supérieure à 1 M et la résistance de contact inférieure à 0.01.

Le sélecteur de mesure est destiné à ne connecter qu'une voie à la fois. Il comporte 200 voies de mesures avec possibilité d'extension à 400 voies.

Dans notre utilisation une seule voie de sélecteur de mesure est utilisée, sur laquelle est connectée la sortie du décommutateur.

Le normaliseur permet de ramener toutes les tensions dans la gamme 1V–10 V avant l'entrée dans le convertisseur analogique digital. Ce dernier est un codeur de la série M 2 300 S.E.T.I. travaillant dans le code 1, 2, 4, 2.

La durée du codage est de 250 sec, la précision supérieure à 10^{-3}. Le sélecteur de sortie comporte 100 voies de sortie avec possibilité d'extension à 200 voies. Il comporte également 50 entrées de générateurs avec un panneau de câblage permettant de brancher une source sur plusieurs sorties. Ces entrées permettent d'envoyer des tensions aux éléments extérieurs au contrôleur.

La vitesse du système est de 20 séquences par seconde sans édition et d'une séquence par seconde avec édition.

4.2. DESCRIPTION DE LA STATION DE RÉCEPTION DES TÉLÉMESURES

Le récepteur est un récepteur SUD-AVIATION type X 4 901 pouvant recevoir 4 émissions IRIG sur des fréquences différentes avec deux bandes passantes. La com-

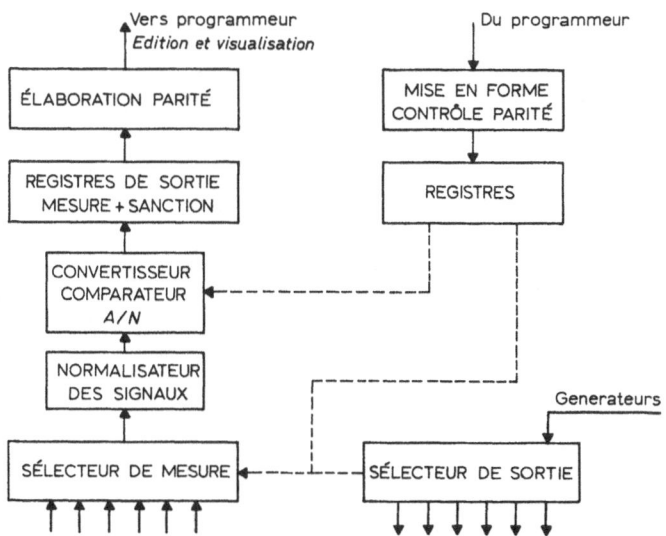

Fig. 6. Bloc diagramme du contrôleur.

mutation des bandes passantes et des frequences peut être manuelle ou commandée par relais à partir du contrôleur. Les commandes provenant du contrôleur sont au nombre de 6:2 de bande passante et 4 de fréquence.

Le discriminateur commutable comprend:
– un discriminateur sans filtres SUD-AVIATION TDM 101,
– une platine avec 21 filtres BF commutables à distance et manuellement,
– une platine avec 21 filtres passe-bande et 21 constantes de temps commutables à distance et manuellement.

La tension de sortie est $+1$ V pour 0% de modulation et $+9$ V pour 100% de modulation.

Le décommutateur est le décommutateur C.S.E.E. décrit précédemment. Il reçoit du contrôleur:
– un ordre sous-porteuse commutée ou continue,
– six ordres   numéro du plot à contrôler,
– huit ordres   nombre de voies du commutateur,
– dix ordres   fréquences de commutation,
– douze ordres   retard programmé,
– un ordre   signal d'autorisation,
– un ordre   mise en route.

5. Installation de contrôle automatique de la base de lancement du C.E.L.

Le bloc diagramme de l'installation est le suivant (figure 7). La différence essentielle avec l'installation précédente est que le programmeur comporte un calculateur.

Les équipements sont répartis dans deux remorques. La première contient le programmeur, la seconde comporte:

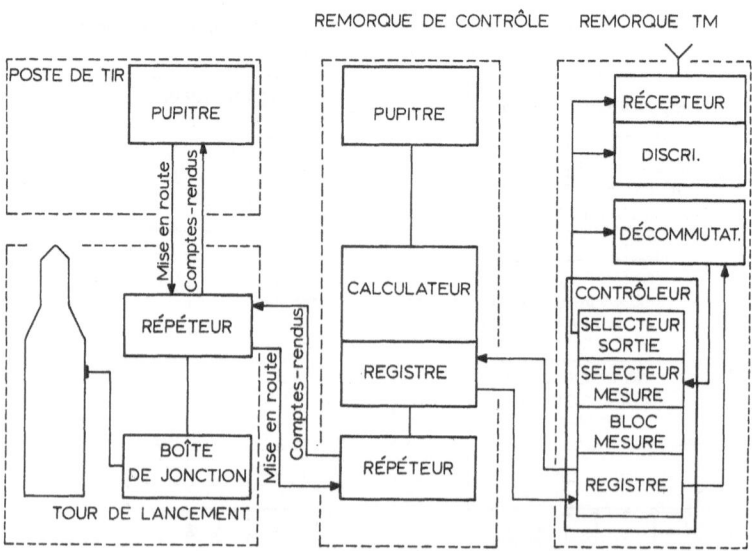

Fig. 7. Installation de contrôle du C.E.L.

– la station de réception de télémesure (récepteur et discriminateur télécommandables identiques à ceux de la station précédente),

– le décommutateur qui sera simplifié par rapport à celui de l'installation précédente,

– le contrôleur.

Des récepteurs assurent une transmission correcte des signaux d'échange (ordre de mise en route et comptes rendus) avec le pupitre de lancement sitée à proximité de la tour de lancement.

5.1. DESCRIPTION DE PROGRAMMEUR

Le programmeur comporte un calculateur C.A.E. 130 avec une unité d'échange 141 sur laquelle sont branchés un lecteur et un perforateur de bandes ainsi qu'une machine à écrire. Les bandes perforées servent de mémoire extérieure pour stocker le programme de contrôle. La machine à écrire sert à l'édition des résultats.

C'est un calculateur digital binaire parallèle basé sur la technique de la logique enregistrée. Les programmes sont combinés en groupes de commandes élémentaires définies par des mots de 15 bits.

Le calculateur comprend une mémoire à noyaux de ferrite comprenant 8:92 mots de 16 bits (15 bits significatifs + 1 bit de parité).

Le temps d'accès est de 3 secondes.

Le calculateur possède un canal d'entrée et un canal de sortie pour les échanges avec le contrôleur.

Dans le cas d'une entrée, le contrôleur place l'information sur les lignes d'entrée, puis avertit le calculateur par une ligne demande d'entrée. Le calculateur vient la prélever quand il est disponible et renvoie un signal d'exécution.

Dans le cas d'une sortie, le contrôleur indique au calculateur qu'il est prêt à recevoir des informations par une ligne demande de sortie. Le calculateur détecte cette demande dès qu'il est libre et met le message en place, ainsi qu'une autorisation de prélèvement.

Le pupitre de commande groupe les organes de commande et de contrôle ainsi que les indicateurs de visualisation.

Les commandes sont les suivantes:

(a) mise en route et arrêt,

(b) mode d'exploitation:

– enregistrement

– exploitation

– lecture

– recherche séquence

– calibration

– déroulement continu

– déroulement pas à pas

– répétition mesure

– arrêt sur séquence affichée

– arrêt sur NO GO,

(c) mode d'édition
- systématique
- hors limite
- programmée
- différée.

Les visualisations donnent le numéro de la séquence et la mesure avec son signe.

5.2. Description du contrôleur

Les fonctions assurées par le contrôleur sont sensiblement les mêmes que celle du contrôleur de l'installation précédente.

Les circuits d'entrée en provenance du programmeur sont évidemment différents pour s'adapter aux caractéristiques propres à ce dernier.

Pour les circuits de sortie il importe de signaler la possibilité d'envoyer du contrôleur vers le décommutateur une commande numérique à 22 bits en parallèle qui permet d'assurer la commande simultanée de tous les relais nécessaires à l'obtention d'une mesure décommutée, possibilité que nous avions signalée dans la description du décommutateur.

Le codeur traduit une tension analogique comprise dans les limites ± 10 V en une expression numérique de 11 chiffres binaires plus le signe.

La précision de la conversion est $0.05\% \pm \frac{1}{2}$ chiffre binaire de plus faible poids.

5.3. Contrôle d'une voie commutée

Si nous nous reportons au schéma du décommutateur (figure 4) nous voyons que le fait de disposer d'un calculateur nous permet d'éliminer toute la partie calibration dont le fonctionnement est assez complexe et dont la présence avait été rendue nécessaire pour corriger les valeurs des paramètres des évolutions des tensions de référence.

Nous utilisons le signal T_n pour commander le convertisseur analogique numérique du contrôleur au moment où le plot n est présent sur la ligne signal.

Une séquence de contrôle du paramètre n commandera 3 prélèvements successifs pour $n=0$, $n=1$ et $n=n$.

Les trois résultats seront mis en mémoire dans le calculateur qui pourra ensuite effectuer le calcul de correction et sortir la valeur vraie du paramètre n.

- *la photo n° 9* donne une vue de la remorque télémesure avec l'antenne à droite
- *les photos n° 10 et11* montrent les armoires du contrôleur
- *la photo n° 12* montre le décommutateur
- *la photo n° 13* montre le récepteur de télémesure
- *la photo n° 14* montre la baie de décommutation au fond et la baie de réception-discrimination au centre
- *la photo n° 15* montre le décommutateur.

6. Conclusions

La conclusion générale que nous pouvons retirer de cet exposé est que l'application du contrôle automatique aux chaînes de télémesure a nécessité au départ des adaptations assez lourdes. L'évolution des contrôleurs automatiques liés à des calculateurs permet une simplification notable de ces adaptateurs. Le passage à des télémesures P.C.M. et les possibilités de plus en plus grandes d'échanges avec les calculateurs devront permettre d'arriver à des solutions encore plus intégrées.

SOME PROBLEMS OF CONTROLLING SPACESHIPS

B. N. PETROV, I. S. UKOLOV, and E. I. MITROSHIN

Academy of Sciences, U.S.S.R.

Abstract. The problem of assuring the safe return of a spaceship is one of the most complicated affairs in the whole complex of problems connected with the realization of a manned flight into space.

Very stringent requirements are demanded of the automatic control systems at reentry into the atmosphere with the escape or superescape velocities (for instance, when returning from a flight to the moon or the planets). The well-known control systems for the descent of spaceships may be divided into two classes: the control systems constructed with the use of the nominal trajectory, and the control systems based on the prediction of the point of landing (terminal systems). For developing systems as the former, problems of the optimal trajectories arise, as well as constructional difficulties of adaptive control systems checking the motion relative to the nominal trajectory, whereby the parametric invariance results from such disturbances as atmosphere density fluctuations.

The terminal systems are most promising from the point of view of flexibility and accuracy of control. They are based on the use of an onboard digital computer, which requires the development of new algorithms and control systems (for instance, stochastic optimal systems).

At the present time the controlling of the trajectories of spaceships in the atmosphere can only be done with the help of changing the angle of roll. However, the control of flight of a spaceship with the help of the simultaneous change of the angle of attack and roll may increase the accuracy of controlling the space motion, may expand the possible corridor of reentry, and may increase the restriction of reentry velocities.

One of the methods enabling realization of control of the angle of attack is the method of changing the centre of gravity with the help of displacing masses within the spaceship during the flight.

1. General Questions

In the flights of manned spaceships, one of the most important problems is to secure the safe return of the spaceship into the dense layers of the atmosphere and its landing in the predetermined region of the Earth's surface. The difficulties of solving this problem greatly increase when the spaceship reenters the atmosphere with escape (on returning from the Moon) or superescape (on returning from the planets) speeds [1].

The return of the piloted spaceship from orbits around the Earth with the orbital velocity attained there, allowed free ballistic descent [2]; however, in the case of ballistic descent with escape velocity, the amount of uncertainty can be dangerously large, due to possible variations in the density of the atmosphere, the mass of the spaceship, aerodynamic coefficients, the duration of physical processes while descending and the initial conditions of reentry.

As a whole, this problem of safe landing of the spaceship must be solved in two stages. In the first stage, before entering the atmosphere, the trajectory is corrected by the application of drag impulses so that it hits a given corridor of reentry. The angle of reentry must be sufficient to ensure that the spaceship is captured by the atmosphere and at the same time the angle of reentry must not exceed a value above which extremely rapid deceleration arises. In the second stage, by means of aerodynamic forces the energy of the spaceship must be dissipated and its landing in the given region ensured.

Because of the difficulties of radio-connection with the spaceship during its flight in the atmosphere when it is surrounded by clouds of the ionized gases, the given system must be self-contained. In this case the main source of information is the acceleration measured in the related or inertial axes of the spaceship.

Due to the rapid recent progress of space technique it is possible to use the manoeuvre of braking in the atmosphere of a planet to reduce the overall expenditure of energy for putting the spaceship into the orbit of the planet's satellite or making an indirect landing on its surface. Therefore it is very important to choose the method of control and the analysis of the influence of atmospheric parameters of a planet and variations of these parameters on the geometry of the possible trajectories of reentry and on the required accuracy of navigation on approaching the planet.

2. Methods of Control

The analysis of all the known works connected with the development and the investigation of the systems controlling the descent makes it possible to divide the methods of control into two classes: the methods of control using the nominal trajectory of descent, and the methods of control based on the prediction of the place of landing [3]. Taking into account the modern tendency toward the development of onboard computers, the methods of terminal control based on the prediction are more promising. The block diagram of this system is given in Figure 1.

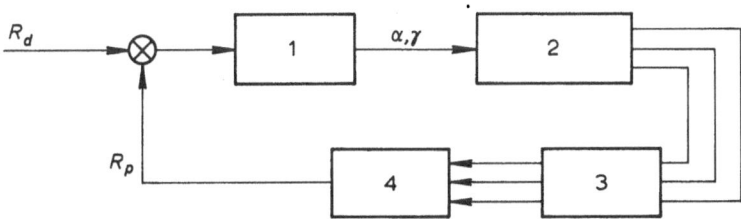

Fig. 1.   1 – Stabilization and control system. 2 – Spaceship. 3 – Inertial measurement unit. 3 – Computer. R_d – Desired range. R_p – Predicted range.

When controlling with a prediction based on measuring the trajectory parameters of descent and estimating the aerodynamic characteristics of the spaceship, the onboard computer integrates a complete system of differential equations of motion of the spaceship and determines the position of the place of landing hundreds of times quicker than the flight itself is proceeding. The computer simultaneously checks and controls the permitted restriction of the maximum acceleration and the thermal conditions of the spaceship, etc. If the spaceship is not on a suitable course to land in the predetermined place or one of the limitations is being exceeded, new values of controlling parameters are determined, which will correct the descent. To find the necessary values of a controlling parameter an iterating process is used.

The use of the predetermined technique is directed to create a universal control system capable of the effective control of the spaceship over a wide range of behaviour during the flight, when the available control resources are known.

Besides implying quick action, on-board computers will make it possible to create a frameless inertial system, which essentially will increase the reliability and accuracy of solving the navigation problems. The main drawbacks of such a method for controlling the trajectory of ascent is that there are great requirements as to the weight, size, the volume of storing, reliability and especially to the quick action of the onboard computer (On. B. C.).

The choice of the equations used for the prediction of trajectory parameters of descent must be done by taking into account the following requirements:

(1) The complexity of the accepted system of equations must correspond to the limited possibilities of On. B. C.

(2) The accuracy of the computations must correspond to the accuracy of the inserted data into the computer and satisfy the requirements necessary to the system of navigation.

(3) Computations must be carried out quickly so that the given results could be used effectively.

The repetitive prediction must be accomplished in very short time intervals (the order of a few seconds) to obtain normal control along the trajectories of landing with fast changing flight conditions. In particular, the limitation of the quick action of On Board Computer would not enable this method of control to be used directly on the spaceship 'Apollo' [4].

The requirement for the quick action of On. B. C. can be reduced at the expense of improving the methods of mathematical programming of On. B. C. For the purpose of prediction one can use simplified equations of motions instead of complete ones and at the same time exactly describing the dynamics equations (for example Chapman's and Jaroshevsky's equations) and using the approximate solutions for determining the whole or the part of possible trajectories. When using this method, the range of the effective control of the trajectories of descent is reduced since the trajectories of ascent are permitted here for which the analytic solutions have been obtained.

Recently much attention has been paid to more simple practical methods of guidance by existing means with using the nominal trajectories. In this case the parameters of motion of the spaceship along the nominal trajectories are computed beforehand and the necessary information about these parameters is stored in the on-board storing device. The information of the magnitude of the deviation of the real values of parameters (for example, accelerations) from nominal ones is used for controlling the motion of the centre of mass regarding the nominal trajectory (Figure 2.)

Among the variations of this method, the method of tracking the nominal trajectory is more easily realized. A more effective but more complicated method of control is the method in which the deviation of the values of the parameters of trajectories from the nominal is used for forming the control and is directed not on eliminating the current deviation of these parameters but on eliminating the terminal deviation.

Using such techniques and keeping the advantages of the method of tracking the nominal trajectory one must secure high accuracy of landing under the considerable variations of moving conditions.

The method of directing to the given point based on using the nominal trajectory and the methods of the theory of sensitivity can prove to be effective. But on using such systems there arise definite difficulties connected with the necessity to have an on-board complicated computer possessing a very quick action while computing the coefficients of influence during the flight or great storing necessary for keeping those coefficients which are precomputed. Moreover the increased demands to the on-board storing device can be due to the necessity of storing the families of the nominal trajectories.

However, by using one of the coordinates of state of a spaceship as an independent variable, it is possible to use one nominal trajectory for the whole range of the initial conditions in question. In this case the problem of producing a regulator providing the high quality of motion along the nominal trajectory is of great importance.

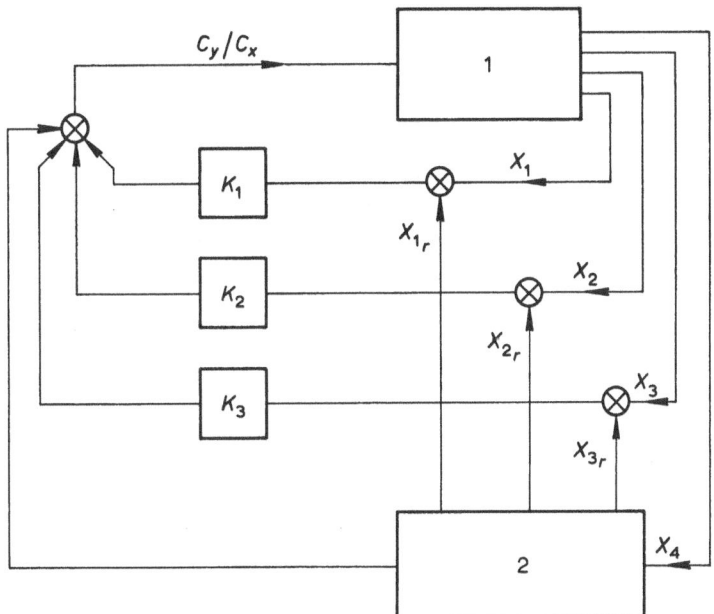

Fig. 2. 1 – Spaceship. 2 – Reference trajectory.

A. ADAPTIVE SYSTEMS

Upon changing the speed of the spaceship on the portion of descending from escape (or even superescape) velocity to a velocity near that of sound and lower, the dynamic characteristics of the control device change greatly – hundreds of times. Therefore, the problem of obtaining the given quality of control over the trajectory can be solved by means of adaptive control systems which enable one to automatically secure in flight, by means of returning the gains of a regulator, the parameter characteristic invariation of loop-characteristic "object regulator".

Such invariation can be realized by means of the systems of variable structure [5, 6, 7, 8]. The automatic control systems with the variable structure implies the systems, (those in which the structure and parameters during the transient process depend specially on the state e.g., on the value of the adjusted magnitude or its derivatives), of the structure and parameters are changed.

The distinctive feature of such systems, belonging to the class of non-linear systems which control the braking function, is that for a particular choice of the breaking boundary one may obtain, in the phase space of the coordinate of the system, the motion of depicting a point after its entering into the given hyperplane occurring in the sliding regime along the phase trajectories belonging to these hyperplanes.

Therefore the motion of the system with an accuracy up to an infinitely small magnitude will be described by the homogeneous linear differential equation which practically does not depend on either the characteristics of the object nor on the imposed disturbances and is determined only by the rule coefficients of control.

To establish a stable sliding regime in effecting the disturbance, for example, the disturbance resulted from by the deviations of atmospheric characteristics from the standard characteristics, it would be sound practice to introduce additional commutated feedbacks (for example, for controlling effect) since it provides quasi-invariance of the control system to the broad class of disturbance [9].

In a more general sense the problem of the adaptation of control systems of spaceship should be considered in the stochastic respect. In the problem of controlling the range of the flight from the point of reentry to the point of landing the reasons resulting in differences of the range are due to random variations of the initial conditions of reentry from predetermined ones and current deviations of the aerodynamic characteristics of the spaceship and atmospheric density from their nominal values.

In general case these variations are unstationary random processes [10]. The dynamic process under study is represented by the following system of equations, given on the portion [O.T.].

$$\dot{x} = X(x, u, t) \tag{1}$$

$$\dot{x}_1 = A(x, u, t)x_1 + B(x, u, t)\vartheta + \varepsilon(t) \tag{2}$$

$$y = H(x, u, t)x_1 + \xi(t) \tag{3}$$

where the equation (1) represents the motion along the nominal (undisturbed) trajectory; the equation (2) represents the disturbed (linearized) motion; the equation (3) represents the process of getting information of the current state of the disturbed motion from the results of observations.

Here:

X – a vector-function of one-valued non-linearity upon moving along the nominal trajectory;

U – a control vector on moving the nominal trajectories belonging to the closed region;

t – an independent variable (time, or one of the state coordinates).

X_1 – a general state vector of system (de viation of the disturbed motion),

y – an observed vector;

ε, ξ – disturbances;

ϑ – a control vector under the disturbed motion;

H, B, H – matrixes of corresponding dimensions, depending on the parameters of nominal trajectories.

The boundary conditions on the terminals are represented in the form:

$$x(T) \in \{x : g_k(x) = 0\}$$
$$x_1(0) \in \{x : g_0(x) = 0\} \tag{4}$$

$X_1(0)$ – a vector of random values with known characteristics.

Besides it is necessary to have the current values of parameter motions along the trajectories of descent (for example, on maximum acceleration) and control parameters. It means formally that at every moment of the time the following relation is fulfilled:

$$g(t, x + x_1, u + v) \leqslant 0 \tag{5}$$

where g – a known vector-function.

In general there arises the problem of minimization of functional equation:

$$f = M\tilde{\omega}[x(T) + x_1(T)] = \omega[x(T)] + M\omega_1[x_1(T)] + f[x(T);$$
$$Mx_1(T)] \tag{6}$$

where f – a scalar function

ω, ω_1 – scalar non-negative functions.

The problem of programming the nominal motion and determining the actual realization is usually investigated separately. Therefore the connection between the nominal and disturbed motion is ignored. However, the control system of the disturbed motion is determined by the trajectory of the nominal motion. Therefore, the latter influences the characteristics of the disturbed motion (controllability, stability, etc.).

On the other hand the benefit on the functional under the optimal programming of the nominal motion and also the accuracy of fulfilling the boundary conditions and current restrictions to a large extent are determined by the disturbed motion. Therefore the two problems are worth considering simultaneously.

B. CONTROLLED ANGLE OF ATTACK

At the present time the main method of controlling the trajectors of flight of spaceships is the method of changing the effective lifting force (projection of the lifting force into the vertical plane) by means of changing the angle of bank. In this case the spaceship is moving under the constant balanced angle of attack. Controlling the lateral range is performed by means of changing at definite moments of time the symbol of the angle bank. These turns from one side to the other side introduce definite disturbances into the look (circuit) for controlling the longitudinal range. Due to great variations in controlling it is impossible to secure enough flexibility for controlling the space motion.

In this case controlling the magnitude of the drag of the spaceship practically remains constant and only the effective aerodynamic quality is being changed which leads to the partial use of its manoeuvrable possibilities.

As shown [11, 12] simultaneous changing of the drag and the effective aerodynamic quality of the spaceship allows essentially to spread the admitted passage of reentry into the atmosphere and for the given width of the passage to increase the value of extremely tolerated velocities of reentry. This problem is of great importance when the speed of reentry into the Earth's atmosphere on returning will be 15–20 km/sec.

The possibilities of controlling the angle of attack may prove to be useful and in saving the spaceship with its crew in case the rocket-carrier is wrecked along the portion of putting into orbit. By means of changing the angle of attack it is possible to find the position of the spaceship, corresponding to the flight along the actual maximum lift-drag ratio and thus to reduce accelerations acting on the spaceship [13].

On producing the spaceships with the changing angle of attack the application of jet controlling units for controlling the angle of attack for known designs of spaceships is not expedient due to expected great expenditures of operating body. In the use of aerodynamic rudders for this purpose there arises difficulties connected with the necessity to avoid overheating the rudder units and essentially changing their effectiveness.

Therefore one of the most hopeful methods of controlling the angle of attack is the method, based on changing the centre of the spaceship at the expense of replacing the mass within the spaceship [13, 14], for example, replacing instrument equipment, repumping the fuel.

As an example one may consider the system of controlling the angle of attack of the spaceship, having the form of an unsharpened, axes-symmetrical body (Figure 3) whose aerodynamic characteristics are given on Figure 4. Here C_N and C – dimensionless coefficients of the longitudinal and normal forming an aerodynamic force. For obtaining the necessary values of the balanced angle of attack (due to corresponding position of the centre of gravity within the spaceship) the replacement of the moving element is directed perpendicular to the longitudinal axes of the spaceship. As the moving element, as it has been noted, that part of the on-board equipment or ballast can be used which often is placed in the front part of the ship for securing the necessary reserves of the statistic stability [15].

Under the synthesis of the system controlling the angle of attack first of all there arises the problem of obtaining the mathematical model of the angle of motion of the spaceship by changing the centre of gravity. In connection with the additional dynamic effects due to the replacement of the mass within the spaceship, the system of equations describing the movement of the spaceship will differ greatly from a more common system at the present time permitted for the spaceship with the constant centre. For obtaining the model one can use the theorem of changing the momentum of the quality motion.

Analysing the angular motion of the system (the body of the spaceship – replaced mass) one can assume that the centre of inertia of the body lies on the geometric axes of the symmetry. Let us introduce the coordinate system $X_0 Y_0 Z_0$ (the main central for the body) and $X_1 Y_1 Z_1$ (the main central for the replaced mass). These systems are collinear but the planes $X_0 Y_0$ and $X_1 Y_1$ coincide.

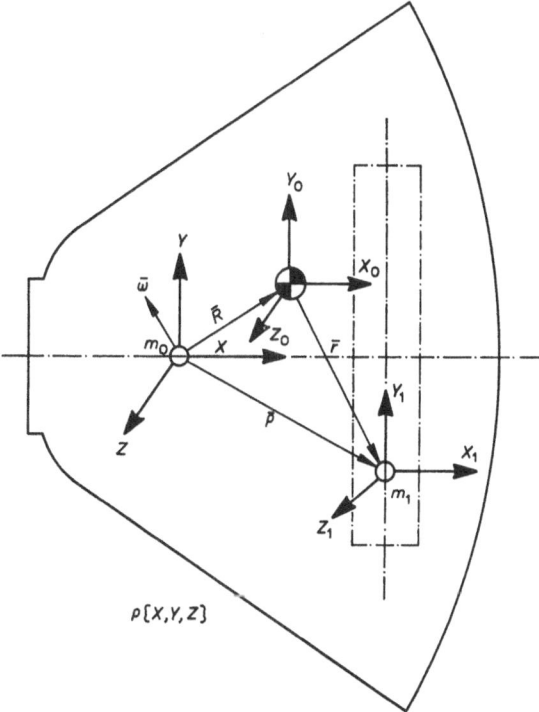

Fig. 3.

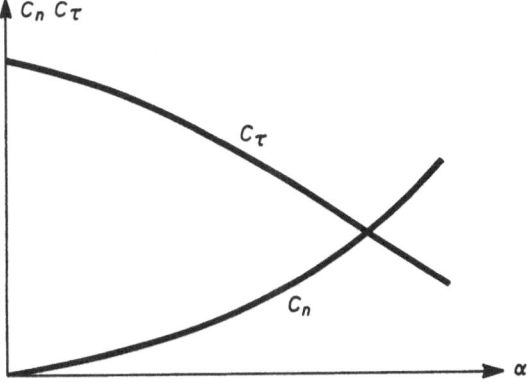

Fig. 4.

Then the angular motions of the spaceship about the system C, Y, Z (the coordinate system whose axes are parallel to the axes of the system $X_0 Y_0 Z_0$ and the origin lies in the centre of the mass of the spaceship) will be presented by the following system of equations:

$$\left(I_{x_1} + I_{x_0} + \frac{m_1 m_0}{m_1 + m_0} y^2\right)\omega_X + \left(I_{z_0} + I_{z_1} - I_{y_0} - I_{y_1} + \frac{m_1 m_0}{m_1 + m_0} y^2\right)\omega_y\omega_z$$

$$+ \frac{m_1 m_0}{m_1 + m_0}[2y\dot{y}\omega_X + xy(\omega_X\omega_z - \dot{\omega}_y)] = M_x$$

$$\left(I_{y_0} + I_{y_1} + \frac{m_1 m_0}{m_1 + m_0} x^2\right)\dot{\omega}_y + \left(I_{x_0} + I_{x_1} - I_{z_0} - I_{z_1} - \frac{m_1 m_0}{m_1 + m_0} x^2\right)\omega_X\omega_z$$

$$+ \frac{m_1 m_0}{m_1 + m_0}[- 2x\dot{y}\omega_X - xy(\omega_z\omega_y + \omega_X)] = M_y$$

$$\left[I_{z_0} + I_{z_1} + \frac{m_1 m_0}{m_1 + m_0}(x^2 + y^2)\right]\dot{\omega}_z + \left[I_{y_0} + I_{y_1} - I_{x_0} - I_{x_1} + \frac{m_1 m_0}{m_1 + m_0}\right.$$

$$\left. \times (x^2 - y^2)\omega_X\omega_y + \frac{m_1 m_0}{m_1 + m_0}\right][2y\dot{y}\omega_z + xy(\omega_y^2 - \omega_x^2 + x\ddot{y}] = M_z$$

where

M_x, M_y, M_z – projections of moments of aerodynamic and jet forces along the axes X, Y, Z.

    ω – angular velocity of the spaceship in the inertial space.

   m_0, m_1 – the mass of the spaceship and the replaced mass accordingly.

As the investigations have shown, a more reasonable system from the point of view of securing the given quality of the transient processes under minimum weights and consumed power is a combined system, consisting of a device for replacing the mass within the spaceship and jet units used for damping the oscillations. Such a regulator can be used both as a hand-operated control of the flight and as the main forced element in the system of the automatic control of descent. The jet units besides damping the angular oscillations and the stability of the angle of roll of the spaceship will be used for obtaining the orientation of the spaceship beyond the atmosphere.

References

[1] Ohocimsky, D. E., Belchansky, G. I., Bukharina, A. P., Golubiev, U. F., Zolotukhina, N. D., and Ivanov, U. N., 'The Optimal Control at the Reentry into the Atmosphere', *Space Invest.* (1968).

[2] Petrov, B. N. and Roushenbah, B. V., 'Summary of Ten Years Work of the Soviet Scientists in the Field of Exploring Space', *The Reports at the Second International Symposium IFAC on the Automatic Control in Space* (1968).

[3] Wingrove, R. C., 'A Survey of Atmosphere Reentry Guidance and Control Methods', *AIAA J.* No. 9 (1963).

[4] Morth, R., 'Reentry Guidance for Apollo', *Proceedings of the Second IFAC Symposium on Automatic Control in Space* (1968).

[5] Petrov, B. N. and Emeljanov, S. V. (ed.), 'The Systems of Variable Structure and its Uses in Problems of Automatization in Flight', *Science* (1968).

[6] Emeljanov, S. V., 'The Systems of Automatic Control of Variable Structure', *Science* (1967).
[7] Ukolov, I. S., Tulin, E. A., and Mitroshin, E. I., 'Controlling of Spaceships along the Portion of Reentry into the Atmosphere by Means of the System of the Variable Structure', *Space Exploration* No. 6 (1967).
[8] Petrov, B. N., Ulanov, G. M., and Emeljanov, S. V., 'Optimization and Invariance in the System of the Automatic Regulation with Rigid and Variable Structure', *The Works of the Second International Congress IFAC on the Automatic Control* (1963).
[9] Petrov, B. N., Emeljanov, S. V., and Utkin, V. I., 'The Principle of Designing Invariant Systems of Automatic Regulation of Variable Structure', *Dokl. Akad. Nauk SSR* **154**, No. 6 (1964).
[10] Vlasov, A. G., Mitroshin, E. I., and Ukolov, I. S., 'The Problems of Optimization of Controlling Reentry of the Spaceship in the Atmosphere', *Space Exploration* No. 2 (1969).
[11] Wingrove, R. C., 'Trajectory Control Problems in the Planetary Entry of Manned Vehicles', AIAA Entry Technology Conference (1964).
[12] Smoljakov, E. R., 'Optimization of Reentry Passage into the Atmosphere', *Space Exploration* No. 1 (1968).
[13] Ageev, G. S., Mitroshin, E. I., and Ukolov, I. S., 'Automatic Optimization of Flight of the Spaceship at the Regime of Maximum Lift Drag Ratio', *Space Exploration* No. 6 (1968).
[14] Shapland, D. I. and Munroo, W. F., 'A Comparative Design Analysis of Tree Configuration Families for Manned Earth Entry at Hyperbolic Speeds', *J. Spacecraft Rockets* No. 6 (1967)
[15] 'NASA Plans to Substitute Depleted Uranium for Lead as Launch Escape System Ballast on Manned Apollo Spacecraft', *Industry Observer, Aviation Week*, **89**, 12/VIII, No. 7 (1968) 23.

NUMERICAL PROCEDURES FOR THE ATTITUDE AND ORBITAL MANEUVERS COMPUTATION OF A SPIN-STABILISED SATELLITE

R. C. MICHELINI and G. ACACCIA

Istituto di Meccanica Applicata alle Macchine, Università di Genova

and

E. GIMELLI and D. DINI

Nuova San Giorgio S.p.A., Div. Servosistemi ed Elettronica

Abstract. One purpose of this paper is to recall the mathematical procedures which can be used for integrating ordinary nonlinear differential equations. Actually, in digital computations matters such as machine memory and running time, have to be weighted against accuracy and stability and different procedures should be employed even for a given system following the class (i.e. the accuracy and running time) of the expected results. Accordingly, the main aspects embedded in integrating methods based on local polynomial approximation are discussed to some extent, and first order stability conditions for the numerical procedures are worked out simply by applying the well known z-transform.

The second purpose of the paper is to present the basic analytical support for the maneuvers of attitude and the orbital corrections for a spin stabilized satellite. To that end the reference dynamic equations are first established and the basic applied control is discussed. This brings an example computation of great importance since solution accuracy and computer running time must both be optimized. An adapted Hamming procedure with variable steps seems the best mate to the problem, being about twice less time-consuming than an equal-accuracy iterative Runge-Kutta algorithm.

List of Symbols

$x, y, z,$	body-fixed axes
$X, Y, Z,$	inertially-referred axes
F_x, F_y, F_z	components of the thrust along the body axes
J_x, J_y, J_z	principal moments of inertia of the satellite referred to the center of mass
$\dot{J}_x, \dot{J}_y, \dot{J}_z$	rate of change of principal moments of inertia
l	distance of the efflux surface from the center of mass
M	mass of the satellite
$\dot{m}$	is the mass flow distribution along the radius r of surface s of the divergent
$\dot{M}$	mass variation of the satellite
M_x, M_y, M_z	moments around the body-fixed axes (principal of inertia)
r	actual radius of the divergent
$\omega_x, \omega_y, \omega_z$	components about the body axes of the angular velocity
φ, ϑ, ψ	yaw, roll and pitch angles
$\Omega_x, \Omega_y, \Omega_z$	angular velocities of the ejected particles referred to the body axes in terms of radius r
$\varepsilon_x, \varepsilon_y, \varepsilon_z$	represents the mass-expulsion effects around the satellite axes.

G. A. Partel (ed.), Proceedings of the Second International Conference on Space Engineering. All rights reserved.

1. Introduction

A great deal of computation for maneuvers during orbital motion has to be done right through the project phase of a telecommunication satellite since the layout is continuously modified and accordingly its inertial characteristics are modified. The problem has been tackled by trying to work out some computation procedures particularly suited to the dynamic equations to be dealt with.

Hereafter the results are discussed in terms of the basic case of a rigid spin-stabilized satellite with time-varying inertia parameters. Obviously more sophisticated models can be established when real satellites provided with a number of nonrigidities are investigated; and this we have done in some recent studies [1, 2]. The dynamics equations in terms of the rigid body picture can be assumed as basic reference and accordingly the reckoning methods here presented are valid provided the real systems have dampers that quench away the modes due to non-rigidities.

Furthermore, the present note gives the basic analytical support for the maneuvers of attitude and orbital corrections. It is thus possible to evaluate the attitude behaviour during the maneuvers in order to check the general design of the attitude control system.

2. Basic Dynamic Equations

The reference coordinate systems are given in Figure 1: x, y and z are body-fixed axes assumed to be principal of inertia for the satellite; ψ, ϑ, and φ are the modified Euler angles: pitch, roll and yaw, they are more suitable for representing erection maneuvers

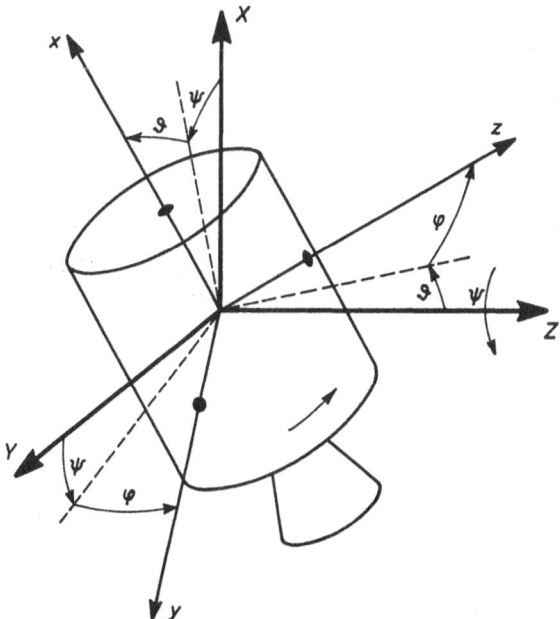

Fig. 1.   Reference coordinate systems.

than conventional Euler angles since they are not affected by gimbal lock during a 90° degrees erection about one of the axes. Euler's equations with reference to the mobile frame are:

$$
\dot{\omega}_z = \frac{M_x}{J_x} - \frac{J_z - J_y}{J_x}\omega_y\omega_z - \frac{\dot{J}_x}{J_x}\omega_x - \frac{\varepsilon_x}{J_x}
$$

$$
\dot{\omega}_y = \frac{M_y}{J_y} - \frac{J_x - J_z}{J_y}\omega_z\omega_x - \frac{\dot{J}_y}{J_y}\omega_y - \frac{\varepsilon_y}{J_y} \tag{1}
$$

$$
\dot{\omega}_z = \frac{M_z}{J_z} - \frac{J_y - J_x}{J_z}\omega_x\omega_y - \frac{\dot{J}_z}{J_z}\omega_z - \frac{\varepsilon_z}{J_z}
$$

and

$$
\left|\begin{matrix}\dot{\varphi} \\ \dot{\vartheta} \\ \dot{\psi}\end{matrix}\right| = B|\omega_x \quad \omega_y \quad \omega_z|'. \tag{2}
$$

The motion of the satellites center of mass, with reference to the fixed frame, is:

$$
\left|\begin{matrix}\ddot{X} \\ \ddot{Y} \\ \ddot{Z}\end{matrix}\right| = A\left|\frac{F_x}{M} \quad \frac{F_y}{M} \quad \frac{F_z}{M}\right|'. \tag{3}
$$

The matrices A and B are given by:

$$
A = \left|\begin{matrix} c\psi c\vartheta & s\varphi s\vartheta c\psi - s\psi c\varphi & c\psi c\varphi s\vartheta + s\varphi s\psi \\ s\psi c\vartheta & s\varphi s\vartheta s\psi + c\psi c\varphi & s\psi c\varphi s\vartheta - s\varphi c\psi \\ -s\vartheta & c\vartheta s\varphi & c\vartheta c\varphi \end{matrix}\right|
$$

and:

$$
B = \frac{1}{c\vartheta}\left|\begin{matrix} c\vartheta & s\vartheta s\varphi & s\vartheta c\varphi \\ 0 & c\vartheta c\varphi & -c\vartheta s\varphi \\ 0 & s\varphi & c\varphi \end{matrix}\right|.
$$

The terms that appear in Equations (1), (2) and (3) are briefly discussed in the following sections.

3. Basic Applied Control

Basically two classes of external excitations can be applied to maneuvers during orbital satellite motion: the first during the apogee motor firing phase, the second during the erection phase and, subsequently, during the attitude and small orbital corrections phase.

3.1. APOGEE MOTOR FIRING PHASE

In this phase the main excitation is the apogee motor thrust whose nominal direction is along the body-fixed x axis.

F_x, F_y and F_z are the components of the thrust along the inertially referred axes X, Y, and Z.

M_x, M_y and M_z are the moments around the body-fixed (principal of inertial) axes: in this phase they represent spurious moments due to the thrust misalignments; their amplitude is a time function of the apogee thrust law.

$J_x, J_y, J_z, \dot{J}_x, \dot{J}_y$ and $\dot{J}_z$ are the principal moments of inertia and their rate of change, they are functions of time according to the law of combustion of the apogee motor.

$(\dot{J}_i \omega_i + \varepsilon_i)$ represent the mass-expulsion effects around the satellite axes. They produce small variations in spin velocity and a little damping-out of the nutation. The ε_i's are function of many parameters and can be expressed by:

$$\varepsilon_x = \int_S \dot{m} r^2 \Omega x \, dS$$

$$\varepsilon_y = \int_S \dot{m} (l^2 + z^2) \Omega_v \, dS \qquad (4)$$

$$\varepsilon_z = \int_S \dot{m} (l^2 + y^2) \Omega_z \, dS$$

where (see Figure 2):

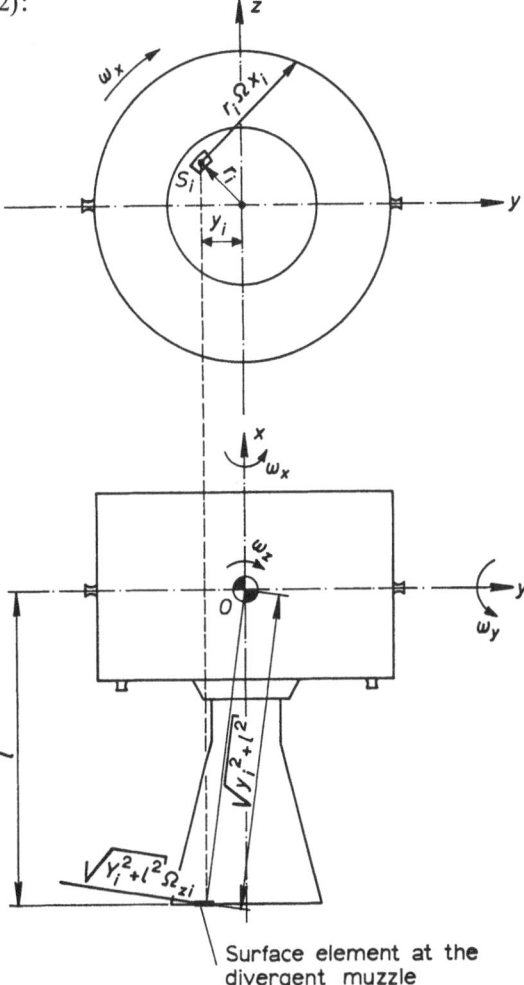

Fig. 2.   Apogee motor firing phase set-up.

– $\dot{m}$ is the mass flow distribution along radius r of surface S of the divergent:
– Ω_x, Ω_y, and Ω_z are the angular velocities of the ejected particles referred to the body axes, in terms of radius r;
– l is the distance of the efflux surface from the center of mass.

The above quantities depend on the motor inner surface shape, the gases viscosity and pressure, the propellant burning rate, and the satellite angular velocity ω_x. To get free from a particular apogee motor, we will presently suppose that the interactions around the spin axis of the gas particles leaving the nozzles is negligeable (i.e. moto inner surface perfectly smooth, and zero gas viscosity are assumed): further we will suppose that the gas particles at the divergent muzzle bear the same angular velocity about the transverse axes on the divergent wall. Thus:

$$\varepsilon_x = - J_x \omega_x$$
$$\varepsilon_y = - \omega_y l^2 \dot{M}$$
$$\varepsilon_z = - \omega_z l^2 \dot{M}. \tag{5}$$

Terms $l^2 + y^2$ and $l^2 + z^2$ have been contracted to l^2 during integration. Within this approximation no hypothesis has to be advanced on the flow distribution at the divergent surface.

3.2. ATTITUDE MANEUVERS PHASES

Both the satellite erection and the fine attitude maneuvers can be obtained by similar control means. The Figure 3 shows the jets arrangement for the attitude control.

The erection maneuver is supposed to take place nominally about the inertial axis Z, with a variation of angle ψ from 90° to 0°.

The excitation is in the form of a train of rectangular pulsed torques, fired twice in a spin revolution with a pulse duration matched to a 90° of rotation around the spin axis. The logic for the train of pulsed is thus derived from the spin angular position.

An alternative program with only one pulse per revolution is considered for fine corrections.

The mathematical model is established with some generality, thus the satellite is supposed to be non symmetrical with respect to the excitation plane, owing to structural layout or to subsequent different propellant consumption from the various tanks.

With reference to the mobile frame (principal of inertia for the satellite) any jet pulse will create a torque with both components M_y and M_z different from zero. Accordingly as the inertia principal axes do not coincide with the jet axes, in order to produce a precession towards the proper direction the angular pulse repetition must enchain with a suitable angle lag φ.

During attitude maneuvers the mass and the moments of inertia of the satellite are slowly time varying: thus constant parameters can be assumed, or, for protracted maneuvers, steplike variations can be introduced.

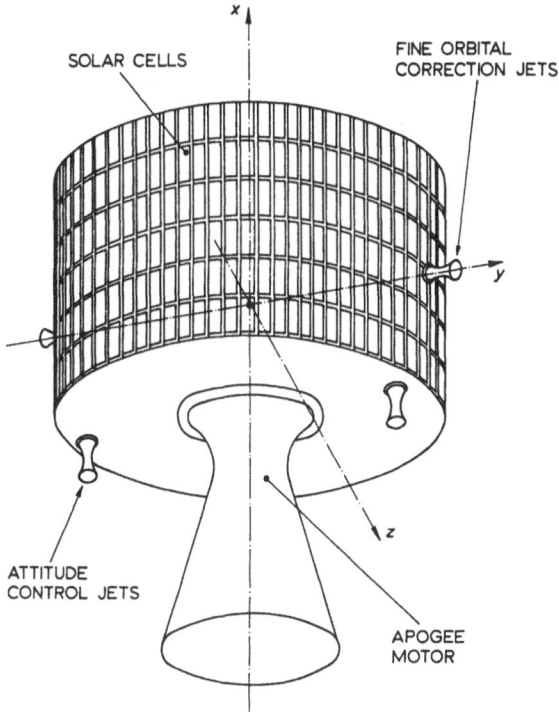

Fig. 3.   Attitude maneuvers phases set-up.

4. Stability of Numerical Solutions

Two drawbacks appear during the numerical computations of dynamic systems trajectories for comparatively long time intervals. On one side single step algorithms (as in the Runge-Kutta methods) easily yield cumulative errors; further they do not offer any estimate of the divergence produced. On the other hand, predictor-corrector algorithms may generate spurious solutions related to the constructed difference-equations and not to the original differential system.

To get rid of both digital misfits, the appropriate way is to apply iteratively the corrector formula up to a smooth convergence to the local solution [5]. This is not the case with normal predictor-corrector algorithms when the corrector formula is applied only once [6] and for these the range of absolute numerical stability is related to the amplitude of the local time increment.

We will now discuss some points concerning the simultaneous integration of systems (1) and (2). It is apparent that the two systems could be treated separately if fixed multisteps algorithms or if variable one-step algorithms are employed. If combined methods are used the simultaneous solution of the six equations has the advantage of offering the evaluation of the variables and the derivatives necessary to bring one step ahead the computation in a straightforward manner even when modified steps are introduced.

The system to be integrated is then:

$$\dot{y}_\alpha = F_\alpha(t, y_\beta) = f_\alpha(y_\beta) - u_\alpha \quad (\alpha = 1, 2, \dots v) \tag{6}$$

being autonomous and subject to an additive control. We can locally construct the associate linear system:

$$\dot{y}_\alpha = Q_{\alpha\beta}\, y_\beta - u_\alpha \quad (\alpha, \beta = 1, 2, \dots v) \tag{7}$$

where coefficients $Q_{\alpha\beta}$ can be thought of as terms of a Jacobian matrix. Following the Lyapunov first method, the local stability of the nonlinear system can be evaluated by inspecting the roots of the characteristic equation:

$$|Q_{\alpha\beta} - s\delta_{\alpha\beta}| = 0 \quad (\alpha, \beta = 1, 2, \dots v) \tag{8}$$

that is the eigenvalues: $s_1, s_2, \dots s_v$, of the complementary system.

A numerical procedure necessarily breaks down the continuous process into a multistep one. In order to evaluate the stability, the numerical integration of system (6), or in vector notation:

$$\dot{y} = f(y) + u \tag{9}$$

let us consider a general predictor-corrector method defined by the formulas:

$$\text{predictor:} \quad y^*_{n+1} = a^*_j\, y_{n+1-j} + hb^*_j\, \dot{y}_{n+1-j}$$
$$\text{corrector:} \quad y_{n-1} = a_j\, y_{n+1-j} + hb_j\, y_{n+1-j}$$

thus:

$$y_{n+1} = \varrho_j\, y_{n+1-j} + h\sigma_j\, \dot{y}_{n+1-j} \quad (j = 0, \dots k) \tag{10}$$

or, defining: $m = n + 1 - k$,

$$y_{m+k} = \varrho_j\, y_{m+j} + h\sigma_j\, \dot{y}_{m+j} \quad (j = 0, \dots k) \tag{11}$$

(m, n) being the step location employed during a cycle computation. The numerical procedure can be regarded in terms of an impulse-modulating process, with sampling time equal to h. During each cycle computation only k sampling instants are involved: k is the stepnumber of the integration procedure, and h the step size.

Taking the z-transform of Equations (11), we have:

$$(z^k - \varrho_j z^j)\, y_m - h\sigma_j z^j \dot{y}_m = 0 \quad (j = 0, 1, \dots k) \tag{12}$$

now letting:

$$S(k) = \frac{z^k - \varrho_j z^j}{h\sigma_j z^j} \quad (j = 0, 1, \dots k) \tag{13}$$

the linear multistep algorithms (11) is represented by the sampling operator S which has an inner subcharacteristic build-up of order k.

With reference to the mth sampling instant the S-operator will behave like the Laplace s-operator, that is:

$$S[y_\alpha]_m = [\dot{y}_\alpha]_m = [f_\alpha(y_\beta)]_m - [u_\alpha]_m \quad (\alpha, \beta = 1, 2, \dots v) \tag{14}$$

or, within the linearising assumption (7),

$$([Q_{\alpha\beta}]_m - S\delta_{\alpha\beta})[y_\beta]_m = [u_\alpha]_m \quad (\alpha, \beta = 1, 2, \dots v). \tag{15}$$

Therefore the S-operator has an outer spectrum defined by the same characteristic Equation (3):

$$|[Q_{\alpha\beta}]_m - S\delta_{\alpha\beta}| = 0 \quad (\alpha, \beta = 1, 2, \dots v). \tag{16}$$

thus: $S_\beta = s_\beta \ (\beta = 1, 2, \dots v)$. The result cannot, in general, be extended to the original nonlinear system (6). However the local stability is governed by Equations (15), and these have a $(v+k)$ root-structure, since substituting each exact solution:

$$y_\alpha = Y_{\alpha\beta} e^{s_\beta t} \tag{17}$$

of the complementary linearized system into relation (12) we have the v equations:

$$z^k - (\varrho_j - [S_\beta]_m h\sigma_j) z^j = 0 \tag{18}$$

each of which has a k-order spectrum $z_{i\beta} \ (i=1, 2, \dots k)$: thus:

$$[y_\alpha]_m = Y_{\alpha\beta}(M_i z_{i\beta}^m) = [Y'_{\alpha\beta}]_i z_{i\beta}^m \quad \begin{matrix}(\beta = 1, 2, \dots v) \\ (i = 1, 2, \dots k).\end{matrix} \tag{19}$$

Among the terms of the (v, k)-array $z_{i\beta}$, we will isolate the principal roots: $z_{s\beta}$ that yield:

$$z_{s\beta}^m = (e^{s_\beta h})^m \cong e^{s_\beta t} \tag{20}$$

from the $(v, k-1)$-array of the spurious roots. Then:

$$[y_\alpha]_m = [Y'_{\alpha\beta}]_i e^{s_\beta h m} + [Y'_{\alpha\beta}]_i z_{i\beta}^m \quad \begin{matrix}(\beta = 1, 2, \dots v) \\ (i = 2, 3, \dots k).\end{matrix} \tag{21}$$

It is then apparent from the second member of relation (16) that for a numerical solution to be called locally stable the first term must converge on the relation (17) and the second terms must be negligible compared to the first.

The consistency of the predictor-corrector method (11) complies with the first condition with regard to system (7) and, locally, to system (6); furthermore if the real parts of all the eigenvalues of the complementary system (7) are less than (or equal to, but not multiples of) zero, and in addition all the characteristic roots $z_{i\beta}$ of relations

(18) individually satisfy:

$$|z_{i\beta}| \leqslant 1 \tag{22}$$

then the numerical integration is stable. These also follow from a well known Dalquist theorem. Conditions different from (22) have to be worked out if the original differential system is unstable (i.e. positive eigenvalues of the complementary system); in this case the conditions of relative numerical stability have to be introduced in order to ensure convergency.

Moreover, inequalities (22) are functions of the step size h: the problem is then reduced to developing methods that are (numerically) stable for as large a value of h as possible; or, conversely, to find, for a given integration procedure, the biggest h that will still keep the stability. The problem is somewhat modified according to the accuracy of the solution.

Now all predictor-corrector methods ought to be consistent, that is coefficients ϱ_j and σ_j have to be chosen so that, the set of the principal roots $z_{s\beta}$ of the difference equations yields solutions (20) that satisfy system (6) sufficiently. For linear systems characterized by discrete spectra the accuracy of the solution is only affected by an eigenvalue, normally the highest. For nonlinear systems the role played by the eigenvalues of the linearized complementary systems is not categorically defined.

5. Evaluation of the Maximum Stable Increment

Within the limits of the mathematical model established in the previous paragraphs, the problem of the stability of the numerical solutions is theoretically reduced to quite simple schedules. When more sophisticated mathematical models are established to match real non-rigid satellites, a number of cross-coupled vibration modes will arise, and – more or less – ought to be considered all excited due to the basic applied controls shape. Thus there is no simple a priori rule for checking the real nature of any tendency to oscillate. A good solution, in our new, is to bypass the problem via a hybrid simulation in order to find the modes that primarily affect the dynamic behaviour of the real satellite.

For systems totally stable in the large, the ones of practical interest, it has been found that higher harmonics due to non-rigidities carried by the main frame rapidly fadeway. Therefore the conclusions stated in the present paper still held: the same integration procedures can be employed and into the overall program layout the only difference is the substitution of the subroutines dealing with the system dynamic equations.

Insofar as the evaluation of the time increment leading to stable integration procedures, let us consider the two systems (1) and (2), when attitude maneuvers are performed. In this case then terms involving variations of inertia parameters as well as the quantities depending on the propellant mass ejection are negligeable; thus the overall system can assume the form:

$$\dot{y}_\alpha = G(y_\beta) y_\alpha + u_\alpha \tag{23}$$

, where:

$$
G = \begin{vmatrix}
0 & \dfrac{J_y - J_z}{J_x}\omega_z & \dfrac{J_y - J_z}{J_x}\omega_y & 0 & 0 & 0 \\[2ex]
\dfrac{J_z - J_x}{J_y}\omega_z & 0 & \dfrac{J_z - J_x}{J_y}\omega_x & 0 & 0 & 0 \\[2ex]
\dfrac{J_x - J_y}{J_z}\omega_y & \dfrac{J_x - J_y}{J_z}\omega_x & 0 & 0 & 0 & 0 \\[2ex]
1 & t\vartheta s\varphi & t\vartheta c\varphi & t\vartheta(\omega_y c\varphi - \omega_z s\varphi) & \dfrac{\omega_y s\varphi + \omega_z c\varphi}{c^2\vartheta} & 0 \\[2ex]
0 & c\vartheta c\varphi & -c\vartheta s\varphi & -c\vartheta(\omega_z c\varphi + \omega_y s\varphi) - s\vartheta(\omega_y c\varphi - \omega_z s\varphi) & 0 \\[2ex]
0 & s\varphi & c\varphi & \omega_y c\varphi - \omega_z s\varphi & 0 & 0
\end{vmatrix} .
$$

The above is the Jacobi matrix characteristic of the nonlinear system: in this simple case it is possible to work out a straightforward solution for the original system: in fact the precession angle ψ does not appear explicitly. Further, the five order matrix G, breaks into a three-order and a two-order matrix, when the characteristic roots are then:

$$
\begin{vmatrix}
-S & \dfrac{J_y - J_z}{J_x}\omega_z & \dfrac{J_y - J_z}{J_x}\omega_y \\[2ex]
\dfrac{J_z - J_x}{J_y}\omega_z & -S & \dfrac{J_z - J_x}{J_y}\omega_x \\[2ex]
\dfrac{J_x - J_y}{J_z}\omega_y & \dfrac{J_x - J_y}{J_z}\omega_x & -S
\end{vmatrix} = 0
$$

and:

$$
\begin{vmatrix}
(\omega_y c\varphi - \omega_z s\varphi)t\vartheta - S & (\omega_y s\varphi + \omega_z c\varphi)^2 c\vartheta \\[2ex]
-(\omega_y s\varphi + \omega_z c\varphi)c\vartheta & -(\omega_y c\varphi - \omega_z s\varphi)s\vartheta - S
\end{vmatrix} = 0 .
$$

The characteristic equations are accordingly:

$$
s^3 - \left[\dfrac{(J_x - J_y)(J_y - J_z)}{J_z J_x}\omega_y^2 + \dfrac{(J_z - J_x)(J_y - J_z)}{J_x J_y}\omega_z^2 \right.
$$
$$
\left. + \dfrac{(J_z - J_x)(J_x - J_y)}{J_y J_z}\omega_x^2 \right]s +
$$
$$
- 2\dfrac{(J_x - J_y)(J_y - J_z)(J_z - J_x)}{J_x J_y J_z}\omega_x \omega_y \omega_z = 0 \quad (25)
$$

$$
s^2 - (t\vartheta - s\vartheta)(\omega_y c\varphi - \omega_z s\varphi)s + [(\omega_y s\varphi + \omega_z c\varphi)^2/c^2\vartheta
$$
$$
- (\omega_y c\varphi - \omega_z s\varphi)^2 t^2\vartheta] = 0 . \quad (26)
$$

As a matter of fact, for small nutation angles $t\vartheta \simeq s\vartheta$ and $\omega_y \simeq \omega_z \cong 0$; we thus have the first-order approximation roots:

$$s = \pm j\omega_x \sqrt{\frac{(J_z - J_x)(J_y - J_x)}{J_y J_z}} \tag{27}$$

and:

$$s = \pm j \frac{\omega_y s\varphi + \omega_z c\varphi}{c\vartheta} \tag{28}$$

however general solutions of Equations (25) and (26) can be parametrically computed as functions of the actual values of ω_x, ω_y, ω_z, ϑ, and φ. As a general rule, the upper values of the characteristic roots are the ones deriving from Equation (25) and have been found slightly different from values (27), when small attitude corrections are considered.

A similar relationship exists for real non-rigid satellites.

The maximum values of the velocity components ω_y and ω_z can – in all cases – be obtained with analogical simulations.

Relation (27) may thus allow us to evaluate the maximum stable increment on the independent variable.

6. Numerical Procedures and Conclusions

Actually several numerical integration procedures are available, differing in accuracy. Accuracy depends on truncation and propagation errors and, further, on amplifications of round-off errors due to particular combinations of coefficients in the finite difference formulas.

In our application the errors due to round-off at each step are drastically reduced by keeping the length of the numbers during the reckoning such that only truncation and propagation errors have to be considered. In order to minimize the truncation errors, it is useful to introduce suitable modifiers acting on the coefficients of the predictor-corrector formulas, i.e. matching best the output variables with the stored informations present in a multistep procedure. And this has been done in our computation.

Insofar as propagation errors are concerned, it is worth recalling that the problem is closely connected with numerical instability. In effect, it can be shown that for globally stable systems the disturbances that arise from solutions of the approximate difference equations (which do not correspond to solutions of the original differential equations) may be considered as perturbations in the motion of the system and the smaller the spurious roots the quicker the cumulative errors fade away, since, any errors in each previous step are negative exponentially weighted. When iterative procedures are employed, convergency is thus of decisive importance for the purpose of minimizing both truncation and propagation errors. The use of the Milne method was rejected because of its intrinsic instability due to the fact that the satellite is a dynamic system with practically pure imaginary roots (at least the principal ones) according to the very low damping even for real systems.

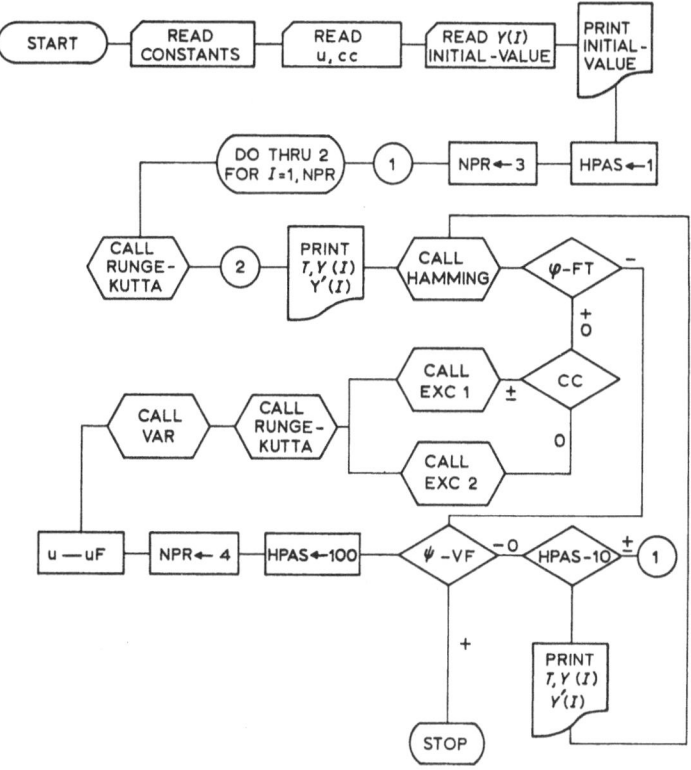

Fig. 4.   Numerical integration procedure general flow diagram.

Conversely, quite good results have been obtained with the Adams [5, 13] Crane [4] and Hamming [3] methods, even if the boundary of the absolutely stable region in the neighborhood of the imaginary axis hardly attains $\frac{1}{4}$ of the one obtained by a fourth-order Runge-Kutta method [4].

After some test computations, it was found that the modified Hamming procedure can bring, in twice less running time, to an accuracy of the same order of a classical fourth order Runge-Kutta method if particular care is employed when choosing the step size h.

Figure 4 gives the example overall flow-diagram of the procedure adopted to the present problem, with reference to the attitude control phase. It should be noticed that the initialising of each subsequent unit region of the on-off control is connected to the spin position angle φ and not to the independent variable t. Errors due to small overlapping of the on-off regions, must be avoided if not a systematic error is introduced.

So a program was constructed for isolating the individual limit angles by getting rid of the fixed step rhythmus of a multistep procedure.

Further at the beginning of each new control region, the memory of the Hamming algorithms has to be rebuilt (and this specially since the modifiers are used) thus, each time, a starting Runge-Kutta procedure is recalled.

References

[1] Rossi, L. C., Michelini, R. C. and Ghigliazza, R., 'Attitude Dynamics and Stability Conditions of a Non-Rigid Spinning Satellite', *Aeron. Quart. Space*, no. 3 (Aug. 1969).

[2] Dini, D. and Michelini, R. C., 'Modèle mathématique de la dynamique d'attitude d'un satellite gyroscopique non-rigide', Evolution d'Attitude et Stabilisation des Satellites – Paris 7–12 (Oct. 1968).

[3] Hamming, R. W., 'Stable Predictor-Corrector Methods for Ordinary Differential Equations', *J.A.C.M.* **6** (Jan. 1959), 37.

[4] Crane, R. L. and Klopfenstein, R. W., 'A Predictor-Corrector-Algorithm with an Increased Range of Absolute Stability', *J.A.C.M.* **12** (Apr. 1965) 227.

[5] Kopal, Z., *Numerical Analysis*, John Wiley, 1955.

[6] Chase, P. E., 'Stability Properties of Predictor-Corrector Methods for Ordinary Differential Equations', *J.A.C.M.* **9** (Oct. 1962) 225.

[7] Abdel Karim, A. I., 'A Theorem for the Stability of General Predictor-Corrector Methods for the Solutions of Systems of Differential Equations', *J.A.C.M.* **15** (Oct. 1968) 706.

[8] Marzulli, P., 'Stabilità massimale nei metodi di integrazione numerica del tipo predittore-correttore', *Calcolo* **III** (Sept. 1966) 339.

[9] Rebolia, L., 'Riduzione del raggio spettrale per la stabilizzazione numerica nella soluzione di sistemi differenziali ordinari', *Rend. Sem. Mat. Padova* **XL** (1968) 311.

[10] Butcher, J. C., 'On the Convergence of Numerical Solutions to Ordinary Differential Equations', *Math. Comp.* **73** (1966) 1.

[11] Gragg, G. B. and Stetter, H. J., 'Generalized Multistep Predictor-Corrector Methods' *J.A.C.M.* **11** (Apr. 1964) 188.

[12] Sconzo, P., 'Formule d'estrapolazione per l'integrazione numerica delle equazioni differenziali ordinarie', *Boll. Un. Mat. Ital.* **9**, 3 (1954) 391.

[13] Romanelli, M. J., 'Runge-Kutta Method for the Solution of Ordinary Differential Equations' in *Mathematical Methods for Digital Computers* (ed. by A. Ralston and H. Wilf), Wiley, 1960, p. 110.

[14] Ceschino, F., 'L'intégration approchée des équations différentielles', *Compt. Rend. Acad. Sci. Paris* **243** (1956) 1478.

[15] Conte, S. D., 'Stable Operators in the Numerical Solution of Second Order Differential Equations', N.N. 112, Space Technology Laboratories, Los Angeles, 1958.

[16] Capra, V., 'Valutazione degli errori nella integrazione numerica dei sistemi di equazioni differenziali ordinarie', *Atti Accad. Sci. Torino, cl. Sci. Fis. Mat. Nat.* **XCI** (1957) 188.

A ROLL AND PITCH ATTITUDE CONTROL SCHEME
FOR A SYNCHRONOUS, FLYWHEEL STABILIZED,
COMMUNICATION SATELLITE

F. LÉORAT, M. GAUVRIT, M. BROSSON

Centre d'Etudes et de Recherches en Automatisme, France

and

R. CAPRARO

Centre National d'Etudes Spatiales, France

Abstract. This paper deals with the problem of synthesising a control scheme well fitted to the long-range mission of the station keeping of a synchronous, flywheel yaw-stabilized communication satellite. This particular type of mission imposes drastic constraints upon on-board equipment and its technology: a minimum number of attitude sensors should be used, while stringent conditions are imposed upon the precision of the attitude by the communication equipment's operation. Furthermore, the fuel consumption should be kept to a minimum, whereas the propulsive equipment should be as light and reliable as possible.

These various considerations have lead the authors to devise a pulse-width-modulated control law operating gas jets: the control scheme, fully non-linear, takes advantage of the interaction properties of the system, and operates in a sequential pattern from the data collected from one angular sensor and one angular accelerometer.

Finally, complete hybrid and digital simulations have been simultaneously conducted to verify that the proposed control scheme indeed allows to meet the specifications in precision of attitude control.

1. Introduction

The growing interest in intercontinental communication channels through synchronous satellites casts a special importance upon the various means of achieving a long-lasting, easily implemented, reliable attitude control scheme well fitted to the specific demands of such spacecrafts.

Under the guidance of the CNES, the authors have focused their interest in synthesising attitude control schemes for the acquisition phase and the long range mission of station keeping in the case of a synchronous, flywheel stabilized satellite. A control scheme suitable to the acquisition phase has been presented in a previous paper, whereas a solution to the station keeping problem is submitted in this communication.

Consider a satellite (cf. Figure 1) exhibiting a symmetry around the Ow-axis, equipped with a flywheel of same axis. The now classical equations describing the dynamics of this satellite under the most general conditions are:

$$
\begin{aligned}
I_{xx}\dot{\omega}_x + (I_{zz} - I_{xx})\omega_y\omega_z + \omega_y C\omega_R &= \Gamma_\psi \\
I_{xx}\dot{\omega}_y - (I_{zz} - I_{xx})\omega_x\omega_z - \omega_x C\omega_R &= \Gamma_\varphi \\
I_{zz}\dot{\omega}_z + C\dot{\omega}_R &= \Gamma_\theta
\end{aligned}
\tag{1}
$$

G. A. Partel (ed.), Proceedings of the Second International Conference on Space Engineering. All rights reserved.

where

I_{xx} = moment of inertia around the Ou and Ov axis
I_{zz} = moment of inertia around the Ow axis
C = moment of inertia of the wheel around the Ow axis
ω_R = angular velocity of the flywheel
$\omega_x, \omega_y, \omega_z$ = components of the instantaneous rotation vector on Ou, Ov, Ow,
$\Gamma_\psi, \Gamma_\varphi, \Gamma_\theta$ = torques applied on Ou, Ov, Ow.

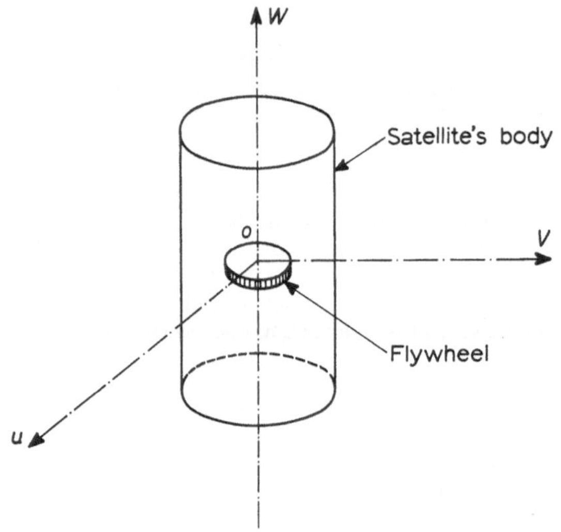

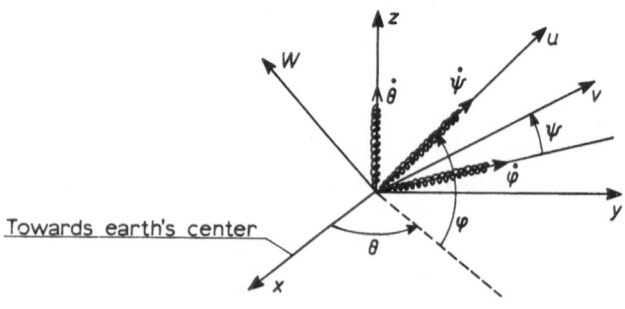

Fig. 1.

The aim of the station keeping system is to maintain in coincidence the $Ouvw$ coordinate system of the satellite with the $Oxyz$ coordinate system defining the ideal attitude of the satellite at every point of its orbit. The $Ouvw$ system is located with respect to the $Oxyz$ system by the angles θ, φ, ψ and their derivatives $\dot\theta$, $\dot\varphi$, $\dot\psi$ according to Figure 2. The angles θ, φ, ψ are respectively the yaw, roll and pitch angles. The equations of the dynamics of the satellite expressed in the $Oxyz$ system are obtained

Fig. 2.

from Equations (1) by noticing that:

$$\omega_x = \dot\psi - \dot\vartheta \sin\varphi$$
$$\omega_y = \dot\varphi \cos\psi + \dot\theta \sin\psi \cos\varphi \qquad (2)$$
$$\omega_z = - \dot\varphi \sin\psi + \dot\theta \cos\psi \cos\varphi .$$

In the station keeping problem, it is dealt only with small variations of θ, φ, ψ around the origin, so that Equations (1) and (2) may be linearized, which yields:

$$\ddot\psi + \omega\dot\varphi + \omega\omega_0\psi = \frac{\Gamma_\psi}{I_{xx}} \qquad (3\text{-}1)$$

$$\ddot\varphi - \omega\dot\psi + \omega\omega_0\varphi = \frac{\Gamma_\varphi}{I_{xx}} \qquad (3\text{-}2) \Bigg\} \quad (3)$$

$$\ddot\theta + \frac{C}{I_{zz}}\dot\omega_R = \frac{\Gamma_\theta}{I_{zz}} \qquad (3\text{-}3)$$

where

$$\omega = \frac{C}{I_{xx}}\omega_R$$

ω_0 = sideral pulsation of the earth's rotation.

Equations (3) show that Equation (3-3) is independent of Equations (3-1) and (3-2) as long as the angular velocity of the flywheel varies very slowly, which is implied by the hypothesis of station keeping.

Therefore the control of the yaw angle may be considered as a problem tackable by classical techniques as long as a measurement of θ is available.

2. Technological Constraints

The question of availability of measurements is crucial in this problem: from Figure 2, it is easily seen that the yaw and roll angles may be readily obtained by means of ordinary infra-red sensors aiming at the earth, whereas the pitch angle can be detected only with the aid of a much more sophisticated star-tracker. For the sake of reducing on-board equipment and improving reliability, it would be highly desirable to eliminate the need for a star-tracker. Furthermore, for the sake of performance, it seems much more attractive to use accelerometers (with a sensitivity of 10^{-7} rd/s^2) rather than gyrometers (whose precision does not go far below 10^{-4} rd/s); again for the sake of reducing on-board equipment, it would be of importance to use only one accelerometer.

As far as the propulsion method is concerned, the technology of constant amplitude, pulse-width modulated gas jets has been retained, for it has proved to be one of the best workable systems every time sensible propulsive torques are required. This choice of technology more or less imposes the actual form of the control law which will be pulse-width modulated. And again for the sake of reducing on-board equipment, we shall try to locate the gas jets on one axis only; various considerations detailed in Appendix III lead to locate this gas propellers so that they create a control torque around the pitch axis.

3. Dynamics of the Autonomous System

The satellite is essentially submitted to the action of solar pressure which induces disturbance torques on the satellite's body and solar panels. In first approximation, these torques may be taken as:

$$\Gamma_\psi = q_0 \sin \omega_0 t$$
$$\Gamma_\varphi = q_0 \cos \omega_0 t.$$

As the calculations detailed in Appendix I show, the proper modes of the satellite are given by:

$$\omega_0{}^* \simeq \omega_0 - \frac{\omega_0^2}{\omega}$$
$$\omega^* \simeq \omega + \omega_0.$$

Since $\omega_0^* \simeq \omega_0$, the sollicitation at pulsation ω_0 by the solar disturbances would induce, if the satellite were not under control, a quasi-resonance of maximum amplitude equal to $2q_0/I_{xx} \omega_0^2$ on either roll and pitch angles, which is not admissible, since the system would be quite out of its domain of linearity and since the station keeping would not be achieved in the first place. As a result, the main goal of the control scheme is to counteract this resonance.

4. Theoretical Approach to the Control Law

Let $U_C(t)$ be the actual control variable. Taking Laplace transform and letting

$$U_1(t) = q_0 \sin \omega_0 t + U_C(t) = \Gamma_\psi + U_C(t)$$
$$U_2(t) = q_0 \cos \omega_0 t \qquad\quad = \Gamma_\varphi$$

we get:

$$\begin{bmatrix} I_{xx}s^2 + C\omega_0\omega_R & C\omega_R s \\ -C\omega_R s & I_{xx}s^2 + \omega_0\omega_R \end{bmatrix} \begin{bmatrix} \psi \\ \varphi \end{bmatrix} = \begin{bmatrix} U_1(s) \\ U_2(s) \end{bmatrix}.$$

Letting $\Delta(s)$ be the characteristic equation of the system and taking inverse Laplace transform yields:

$$\mathcal{L}^{-1}\{\Delta(s)\,\psi(s)\} = [2C\omega_0\omega_R - I_{xx}\omega_0^2]\,q_0 \sin \omega_0 t + I_{xx}\ddot{U}_c(t) + $$
$$+ C\omega_0\omega_R U_c(t)$$
$$\mathcal{L}^{-1}\{\Delta(s)\,\varphi(s)\} = [2C\omega_0\omega_R - I_{xx}\omega_0^2]\,q_0 \cos \omega_0 t + C\omega_R \dot{U}_c(t).$$

In order for the first harmonic of the quasi-resonance at ω_0 to be suppressed, it is necessary that the control variable $U_c(t)$ contain a component in $\sin \omega_0 t$ such that:

$$
\left.
\begin{array}{ll}
I_{xx}\dot{U}_c + C\omega_0\omega_R U_c = q_0\left[I_{xx}\omega_0^2 - 2C\omega_0\omega_R\right]\sin\omega_0 t & (4\text{-}1) \\
C\omega_R\dot{U}_c = q_0\left[I_{xx}\omega_0^2 - 2C\omega_0\omega_R\right]\cos\omega_0 t . & (4\text{-}2)
\end{array}
\right\} \quad (4)
$$

From (4-2), we get:

$$
U_c^0(t) = \left[\frac{I_{xx}}{C}\frac{\omega_0}{\omega_R} - 2\right]q_0\sin\omega_0 t .
$$

With $U_c(t) = U_c^0(t)$, the solution in $\varphi(t)$ for Equation (4-2) is either $\varphi = 0$, or, since $\Delta(s)$ contains imaginary poles, a stationary oscillatory solution depending on the initial conditions.

Now, if $U_c(t) = U_c^0(t)$ is fed into Equation (4-1), it is readily seen that, as long as $(I_{xx}/C)\omega_0/\omega_R$ is negligible with respect to 1, which is absolutely the case, Equation (4-1) is identically verified, which means that the solution in $\psi(t)$ for Equation (4-1) is either $\psi = 0$ or a stationary oscillatory solution depending on the initial conditions. Therefore, a good theoretical control law would be:

$$
U_c(t) = -2q_0\sin\omega_0 t .
$$

In itself, this law has no practical value, since:

(1) it is continuous, whereas the gas jets provide constant amplitude pulses only;

(2) it would require a continuous detection of the actual disturbance torque, which is out of question.

But it shows that any effective pulse width modulated control law will exhibit a first harmonic of the form $-2q_0\sin\omega_0 t$ and that ψ will remain bounded if φ is tightly controlled.

Under this assumption about the control law, it can be shown (see Appendix II-2) that the maximum error, due to the neglected term $(I_{xx}/C)\omega_0/\omega_R$, in station keeping on ψ and φ is reduced to:

$$
\frac{q_0}{I_{xx}\omega\omega_0}
$$

which is quite acceptable, and that for example the behaviour in ψ is that of a modulated sine-wave in opposition of phase with the forcing function Γ_ψ.

This derivation is valid only if $\varphi(t)$ remains unconstrained. But if φ is constrained so that $|\varphi| \leqslant \varphi_s$, one may admit that the expression for φ is:

$$
\varphi(t) = \frac{q_0}{I_{xx}\omega\omega_0}\sin\frac{\omega_0^2 t}{2\omega}\sin\omega_0 t \quad \text{if} \quad |\varphi| < \varphi_s
$$

$$
\varphi(t) = \tfrac{1}{2}\varphi_s \quad \text{if} \quad \left|\frac{q_0}{I_{xx}\omega\omega_0}\sin\frac{\omega_0^2 t}{2\omega}\sin\omega_0 t\right| \geqslant \varphi_s
$$

(see Figure 3).

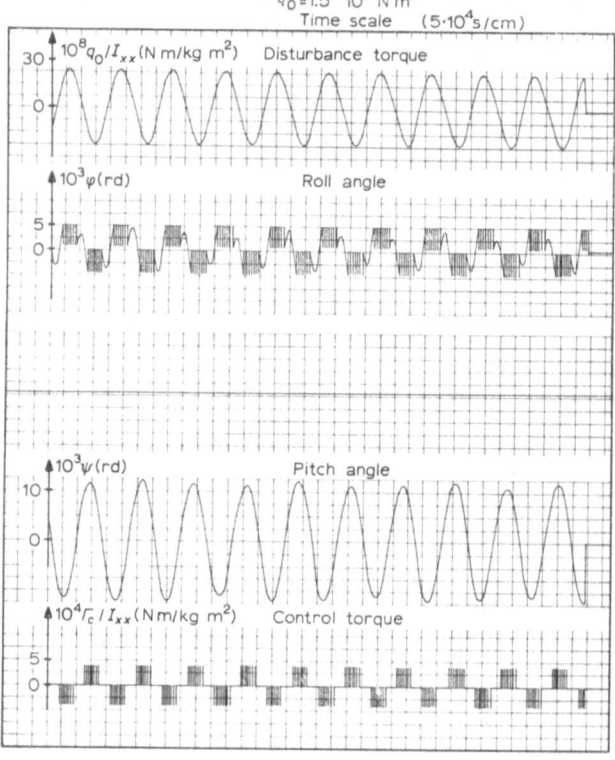

Fig. 3.

As a result, $\varphi(t)$ may be regarded as the output of a non-linear element with input:

$$\frac{q_0}{I_{xx}\omega\omega_0} \sin \frac{\omega_0^2}{2\omega} t \sin \omega_0 t.$$

The first harmonic a_1 of $\varphi(t)$ is readily given by:

$$a_1 = \frac{2}{\pi} a_0 \left[\mathrm{Arcsin} \frac{\varphi_s}{a_0} - \frac{\varphi_s}{a_0} \sqrt{1 - \left(\frac{\varphi_s}{a_0}\right)^2} \right] + \frac{2\varphi_s}{\pi} \left[1 - \left(\frac{\varphi_s}{a_0}\right)^2 \right]^{\frac{1}{2}}$$

where

$$a_0 = \frac{q_0}{I_{xx}\omega\omega_0}.$$

Feeding the approximate expression for $\dot{\varphi}(t)$,

$$\varphi(t) = a_1 \sin \omega_0 t$$

into the differential equation for $\psi(t)$ yields:

$$\ddot{\psi} + \omega\omega_0\psi = -\frac{q_0}{I_{xx}} \sin \omega_0 t - \omega\omega_0 a_1 \cos \omega_0 t$$

whose forced solution is:

$$\psi(t) \simeq -a_0 \sin\omega_0 t - a_1 \cos\omega_0 t$$

so that the maximum amplitude for $\psi(t)$ is:

$$\psi_M = \sqrt{a_0^2 + a_1^2}$$

and the phase angle is:

$$\Phi = \text{Arctg}\frac{a_1}{a_0}.$$

5. Practical Work-Out of a Pulse Width-Modulated Control Law

The roll angle being the only position measurement available, the precision requirements in station keeping will be forced upon φ. For this study, it has been admitted
that a maximum deviation of $5\cdot10^{-3}$ rd from the ideal attitude was admissible.
Therefore a control pulse will be applied every time $|\varphi| > 5\cdot10^{-3}$ rd, and the motion
of the satellite submitted to such a pulse satisfies the equations detailed in Appendix
IV-3. Graphically, an ideal control trajectory takes on the following form:

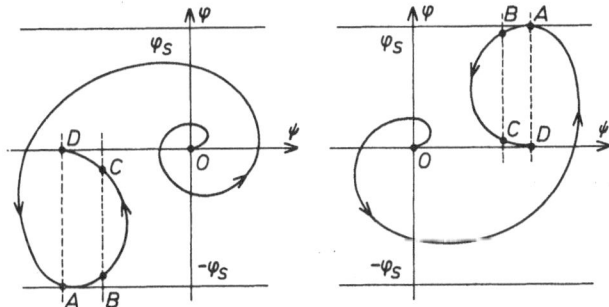

Fig. 4.
Case a: the boundary $\varphi = -\varphi_s$ is attained first.
Case b: the boundary $\varphi = +\varphi_s$ is attained first.

Suppose that the initial conditions are the ideal attitude. The portion OA corresponds to the control-free trajectory of the satellite submitted to the disturbance
torque only. AB corresponds to the first 'acceleration' control pulse communicating
an angular velocity ψ_1 to the pitch axis. BC is a free motion trajectory and CD
corresponds to the second 'deceleration' control pulse taking ψ back to zero. The
velocity ψ_1 is such that point D lie on the ψ axis at the end of the control cycle. It
is clear from the equations in the various Appendices that the slow mode ω_0^* of the
satellite is excited on OA, whereas the fast mode ω^* is predominant along AD.

The above control procedure suggests that it be implemented by means of a
sequential circuit. Two sequential circuits have been devised, one using constant level
signals, and the other impulses only. These circuits and a brief outline of their synthesis are the subject of Appendix IV.

6. Simulation Results and Comments

The simulation of the system's behaviour has been extensively conducted on both high performance analog computer and digital equipment.

As it may be seen on the various charts, the results of the analog simulation show that:

(1) the typical phase trajectory sketched on Figure 4 actually occurs (see photography)

(2) the pitch angle, although not directly controlled, remains bounded at all time

(3) the value of the bound is correctly given by $q_0/I_{xx} \, \omega\omega_0$

(4) the value of the pitch angle is correctly represented by a sine-wave in opposition of phase with Γ_ψ

(5) the pulse width modulated control law definitely exhibits a first harmonic component of the form $-\sin\omega_0 t$.

The main difficulty in the analog simulation residing in the fact that ω_0 and ω_0^* differ only by an extremely tiny amount, it was advisable to test the quantitative validity of the results obtained through the analog simulation by a less descriptive but quantitatively more accurate digital program giving in addition the required control angular velocity $\dot\psi_l$. The Fortran notations, the program and some of its results, which are in complete harmony with the analog simulations, are given in Appendix III.

7. Conclusion

The control scheme presented in this communication seems to be of some interest from both the theoretical and the practical points of view:

– theoretically speaking, the satellite is a multivariable system with strong interaction: the proposed control scheme relies heavily on this interaction, whereas in most applications, interaction is often considered as bothersome. Furthermore, the idea of the first harmonic component of the response together with the fact that the system is completely state controllable allows to realize an ε-controllability quite sufficient for the mission of the satellite.

– practically speaking, the proposed control scheme requires a minimum of equipment:

– an angular sensor to detect φ

– an accelerometer to measure ψ from the relationship

$$\psi = \frac{I_{xx}}{C\omega_R} \ddot\varphi + \omega_0\varphi - \frac{q_0}{C\omega_R} \cos\omega_0 t.$$

Since:

$$\omega_0\varphi < 5\cdot10^{-3} \times 7.268\cdot10^{-5} < 5\cdot10^{-7} \text{ rd/s}$$

$$\frac{q_0}{C\omega_R} \cos\omega_0 t < \frac{10^{-6}}{20} < 10^{-7} \text{ rd/s}.$$

the simple relation $\psi = (I_{xx}/C\omega_R)\ddot\varphi$ is quite sufficient.

Furthermore, as may be seen from the numerical results of the digital simulation, $\dot{\varphi}$ remains inferior to 10^{-4} rd/s when φ crosses the boundary φ_s, so that a measurement of $\dot{\varphi}$ may be omitted without drawbacks.

Finally the propulsive equipment reduces to a set of gas jets able to provide boost in both directions. However, a set of gas jets providing boost in one direction only could even be acceptable, as Figure 5 suggests.

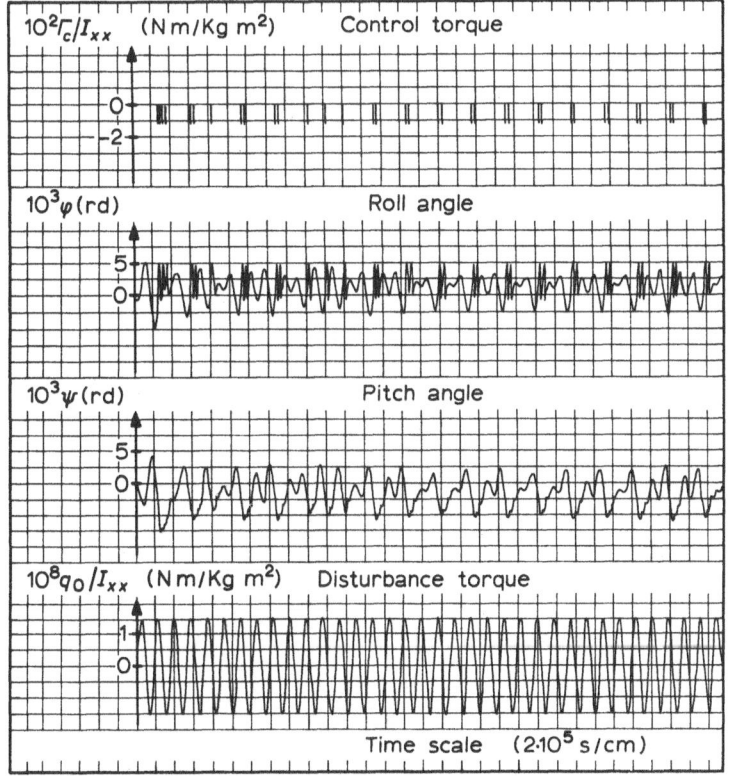

Fig. 5.

However, it may be objected that the representation of the disturbance torques Γ_ψ and Γ_φ by $q_0 \sin\omega_0 t$ and $q_0 \cos\omega_0 t$ respectively is far from realistic, were it only for the irregular form of the satellite and its solar panel: but Γ_ψ and Γ_φ in any case present a first harmonic with pulsation ω_0, for the physical situation of the satellite is periodic at pulsation ω_0. The other harmonics of the disturbances are of very little importance for they do not create a quasi-resonance at pulsation ω_0 as does the first harmonic, which constitutes the heart of the control problem.

Numerical Values

The analog and digital simulations have been conducted with the following numerical values:

$$I_{xx} = 64 \text{ kg} \cdot \text{m}^2$$

$$I_{zz} = 104 \text{ kg} \cdot \text{m}^2$$

$$C = 0.2 \text{ kg} \cdot \text{m}^2$$

$$\omega_R = 100 \text{ rd/s}$$

$$\omega = \frac{C}{I_{xx}} \omega_R = 0.3125 \text{ rd/s}$$

$$\omega_0 = 7{,}268 \, 10^{-5} \text{ rd/s}$$

$$q_0 = 1.5 \cdot 10^{-5} \text{ N} \cdot \text{m}$$

$$\varphi_s = 5 \cdot 10^{-3} \text{ rd}$$

$$\psi_M \simeq \frac{2q_0}{I_{xx}\omega_0^2} = 88.7 \text{ rd}$$

$$\psi_M = \frac{q_0}{I_{xx}\omega\omega_0} = 0.103 \cdot 10^{-1} \text{ rd}$$

$$\frac{I_{xx}}{C} \frac{\omega_0}{\omega_R} = 23.26 \cdot 10^{-5}$$

$$\Gamma c = 2 \cdot 10^{-2} \text{ N} \cdot \text{m}$$

$$\dot{\psi}_l = 0.7030 \text{ rd/s}$$

$$\frac{\omega_0^2}{2\omega} = 8.45 \cdot 10^{-9} \text{ rd/s}.$$

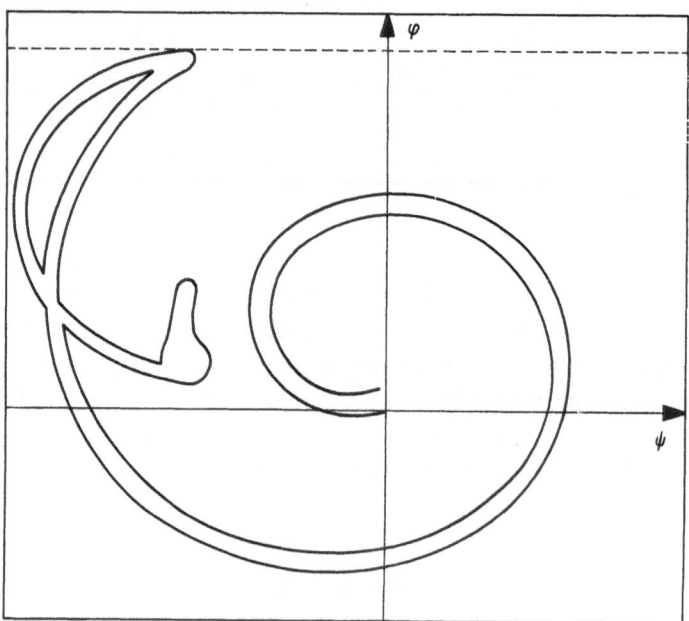

Fig. 6. (Photography)

Appendix I

The signal flow-graph of the system reads as follows:

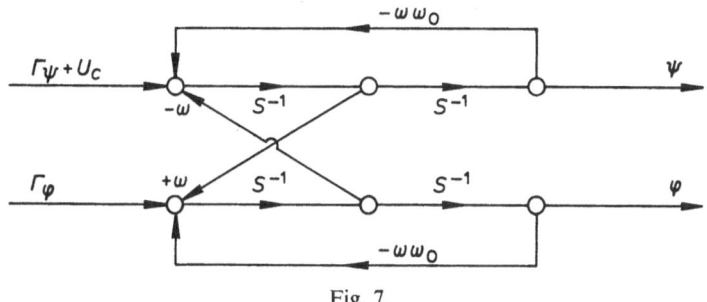

Fig. 7.

The characteristic equation of the system is:

$$\Delta(s) = (s^2 + \omega^{*2})(s^2 + \omega_0^{*2}) = (s^2 + \omega\omega_0)^2 + \omega^2 s^2.$$

Therefore, the modes of the satellite are:

$$\omega^{*2} = \frac{1}{2}\left[\omega^2 + 2\omega\omega_0 + \omega^2\left(1 + \frac{4\omega_0}{\omega}\right)^{\frac{1}{2}}\right]$$

$$\omega_0^{*2} = \frac{1}{2}\left[\omega^2 + 2\omega\omega_0 - \omega^2\left(1 - \frac{4\omega_0}{\omega}\right)^{\frac{1}{2}}\right].$$

In first approximation, we have:

$$\omega^* \simeq \omega + \omega_0$$

$$\omega_0^* \simeq \omega_0 - \frac{\omega_0^2}{\omega} + 2\frac{\omega_0^3}{\omega^2}$$

Appendix II

II-1. Response of the Autonomous System

The response to initial conditions only is given by:

$$\psi_l = A_\psi^0 \cos\omega_0^* t + B_\psi^0 \sin\omega_0^* t + C_\psi^0 \cos\omega^* t + D_\psi^0 \sin\omega^* t$$

$$\varphi_l = A_\varphi^0 \cos\omega_0^* t + B_\varphi^0 \sin\omega_0^* t + C_\varphi^0 \cos\omega^* t + D_\varphi^0 \sin\omega^* t$$

with

$$A_\psi^0 = \frac{1}{D}\left[(\omega^2 + \omega\omega_0 - \omega_0^{*2})\psi_0 - \omega\dot\varphi_0\right]$$

$$B_\psi^0 = \frac{1}{D\omega_0^*}\left[\omega_0\omega^2\varphi_0 + (\omega\omega_0 - \omega_0^{*2})\dot\psi_0\right]$$

$$C_\psi^0 = -\frac{1}{D}[(\omega^2 + \omega\omega_0 - \omega^{*2})\psi_0 - \omega\dot\varphi_0]$$

$$D_\psi^0 = -\frac{1}{D\omega^*}[\omega_0\omega^2\varphi_0 + (\omega\omega_0 - \omega^{*2})\dot\psi_0]$$

$$A_\varphi^0 = \frac{1}{D}[(\omega^2 + \omega\omega_0 - \omega_0^{*2})\varphi_0 + \omega\psi_0]$$

$$B_\varphi^0 = \frac{1}{D\omega_0^*}[-\omega^2\omega_0\psi_0 + (\omega\omega_0 - \omega_0^{*2})\dot\varphi_0]$$

$$C_\varphi^0 = -\frac{1}{D}[(\omega^2 + \omega\omega_0 - \omega^{*2})\varphi_0 + \omega\dot\psi_0]$$

$$D_\varphi^0 = \frac{1}{D\omega^*}[\omega^2\omega_0\psi_0 - (\omega\omega_0 - \omega^{*2})\dot\varphi_0]$$

and

$$D = \omega^{*2} - \omega_0^{*2}.$$

II-2. RESPONSE OF THE FORCED SYSTEM $\Gamma_\psi = q_0 \sin\omega_1 t$, $\Gamma_\varphi = q_0 \cos\omega_1 t$

The response to the disturbance torques Γ_ψ and Γ_φ only is given by:

$$\psi_f = \frac{A_\psi^1}{\omega_0^*}\sin\omega_0^* t + \frac{B_\psi^1}{\omega^*}\sin\omega^* t + \frac{C_\psi^1}{\omega_1}\sin\omega_1 t$$

with:

$$A_\psi^1 = \frac{q_0}{I_{xx}}\frac{\omega\omega_0\omega_1 - \omega_0^{*2}(\omega_1 - \omega)}{(\omega^{*2} - \omega_0^{*2})(\omega_1^2 - \omega_0^{*2})}$$

$$B_\psi^1 = \frac{q_0}{I_{xx}}\frac{\omega\omega_0\omega_1 - \omega_0^{*2}(\omega_1 - \omega)}{(\omega^{*2} - \omega_0^{*2})(\omega_1^2 - \omega^{*2})}$$

$$C_\psi^1 = \frac{q_0}{I_{xx}}\frac{\omega\omega_0\omega_1 - \omega_1^2(\omega_1 - \omega)}{(\omega_0^{*2} - \omega_1^2)(\omega^{*2} - \omega_1^2)}$$

and

$$\varphi_f = A_\varphi^1 \cos\omega_0^* t + B_\varphi^1 \cos\omega^* t + C_\varphi^1 \cos\omega_1 t$$

with:

$$A_\varphi^1 = \frac{q_0}{I_{xx}}\frac{\omega(\omega_0 + \omega_1) - \omega_0^{*2}}{(\omega^{*2} - \omega_0^{*2})(\omega_1^2 - \omega_0^{*2})}$$

$$B_\varphi^1 = \frac{q_0}{I_{xx}}\frac{\omega(\omega_0 + \omega_1) - \omega^{*2}}{(\omega_0^{*2} - \omega^{*2})(\omega_1^2 - \omega^{*2})}$$

$$C_\varphi^1 = \frac{q_0}{I_{xx}}\frac{\omega(\omega + \omega_1) - \omega_1^2}{(\omega_0^{*2} - \omega_1^2)(\omega^{*2} - \omega_1^2)}.$$

By setting $\omega_1 = \omega_0$, one gets the response to the actual disturbance torques, whereas by setting $\omega_1 = -\omega_0$, one gets the response to the ideal control law $U_C = -2q_0 \sin\omega_0 t$.

Taking the approximate values of ω_0^* and ω^* given in Appendix I, a good estimation of the bound of ψ may be computed in both cases $\omega_1 = \omega_0$ and $\omega_1 = -\omega_0$.

For
$$\omega_1 = \omega_0$$
$$\psi_M \simeq \frac{A_\psi + |C_\psi|}{\omega_0} = \frac{2q_0}{I_{xx}\omega_0^2}$$

and for
$$\omega_1 = -\omega_0$$
$$\psi_M \simeq \frac{q_0}{I_{xx}\omega\omega_0}.$$

Furthermore a good approximation for $\psi(t)$ is furnished by:

$$\psi(t) \simeq -\frac{q_0}{I_{xx}\omega\omega_0} \cos \frac{\omega_0^2}{2\omega} t \sin \omega_0 t$$

which shows that $\psi(t)$ is a modulated sine wave in opposition of phase with the disturbance torque on the pitch-axis. Identically $\varphi(t)$ may be represented by:

$$\varphi(t) = \frac{q_0}{I_{xx}\omega\omega_0} \sin \frac{\omega_0^2}{2\omega} t \sin \omega_0 t$$

which shows that $\varphi(t)$ is a modulated sine-wave in phase with the disturbance torque on the pitch-axis.

II-3. RESPONSE OF THE SYSTEM TO A CONTROL TORQUE

In this case, the equations of the system read:

$$(s^2 + \omega\omega_0)\psi + \omega s \varphi = \frac{\Gamma_c}{I_{xx}} \frac{1}{s}$$
$$-\omega s \psi + (s^2 + \omega\omega_0)\varphi = 0$$

where s is the Laplace variable.
    One readily gets:

$$\psi_c(t) = C_1 + A_1 \cos \omega_0^* t + B_1 \cos \omega^* t$$
$$\varphi_c(t) = \frac{B_0}{\omega_0^*} \sin \omega_0^* t - \frac{B_0}{\omega^*} \sin \omega^* t$$

with:

$$A_1 = \frac{\Gamma_c}{I_{xx}} \frac{\omega_0^{*2} - \omega\omega_0}{\omega_0^{*2}(\omega^{*2} - \omega_0^{*2})}$$

$$B_1 = \frac{\Gamma_c}{I_{xx}} \frac{\omega^{*2} - \omega\omega_0}{\omega^{*2}(\omega_0^{*2} - \omega^{*2})}$$

$$C_1 = \frac{\Gamma_c}{I_{xx}} \frac{\omega\omega_0}{\omega^{*2}\omega_0^{*2}}$$

$$B_0 = \frac{\Gamma_c}{I_{xx}} \frac{\omega}{\omega^{*2} - \omega_0^{*2}}.$$

Appendix III

In this Appendix, it will be shown that the best location of the gas jets is that which produces a control torque on the pitch-axis.

Consider point A where the state trajectory of the satellite crosses the boundary $\varphi = \varphi_s$

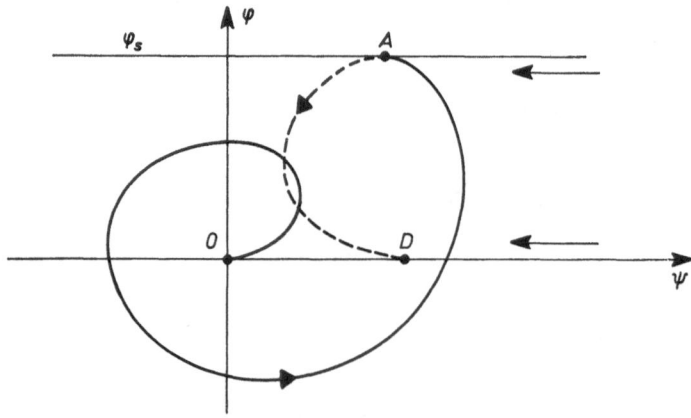

Fig. 8.

Theoretically one has three possible control procedures in A:
  – Either a negative impulse is applied on the ψ axis so that

$$\dot{\psi} = -\dot{\psi}_l = -\frac{\omega\varphi_s}{2}.$$

According to the equations given in A-II, the trajectory is nearly half a circle of radius $\varphi_s/2$. Then in D, the same impulse is applied in order to bring the satellite to a halt.
  – Or a negative impulse is applied on the φ axis so that

$$\varphi = -2\dot{\psi}_l = -\omega\varphi_s.$$

The trajectory is a fourth of a circle of radius φ_s

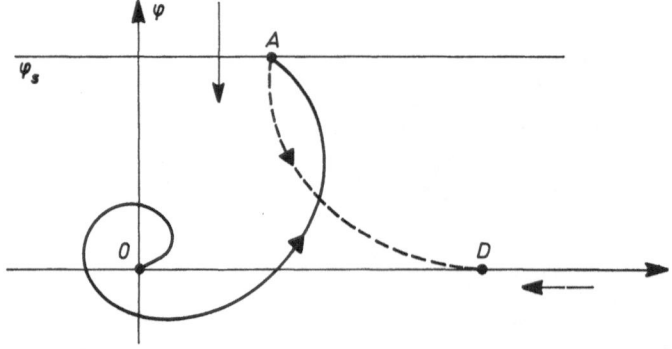

Fig. 9.

Then in D, another negative impulse is applied on the ψ-axis in order to bring the satellite to a halt.

  – Or a negative impulse is applied on the ψ axis so that

$$\dot\psi = -2\dot\psi_1 = -\omega\varphi_s$$

The trajectory is a fourth of a circle of radius φ_s. Then in D, a positive impulse of same magnitude is applied on the φ axis in order to bring the satellite to a halt.

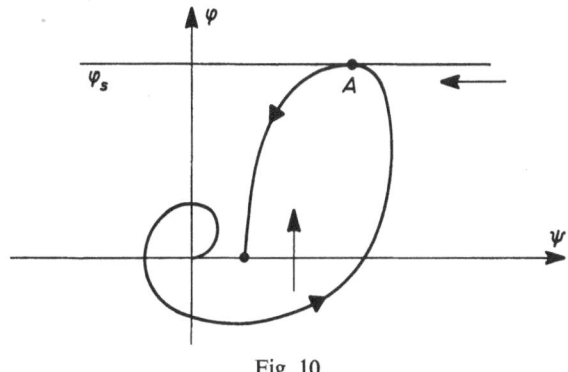

Fig. 10.

The second and third control procedures when compared to the first one, exhibit two main drawbacks:

  – They necessitate a propulsive system on *both* axis;

  – the gas consumption is twice as important, for the required impulses are twice as energetic.

Therefore, the most advantageous solution is to locate the gas jets on the ψ axis. In addition, the accelerometer measuring $\ddot\varphi$ will not be affected by the discontinuities due to the control pulses appearing on the ψ-axis.

Appendix IV

In order to implement the sequential control law, two logical circuits have been devised.

The first one is of the asynchronous type and uses the following input variables, appearing at constant level:

$$
\begin{aligned}
e &= 1 \quad \text{if} \quad \varphi > 0\\
X &= 1 \quad \text{if} \quad |\varphi| > \varphi_s\\
d &= 1 \quad \text{if} \quad |\dot\varphi| > \varepsilon\\
d^* &= 1 \quad \text{if} \quad |\dot\psi| > \varepsilon\\
S &= 1 \quad \text{if} \quad \dot\psi < -\dot\psi_1\\
P &= 1 \quad \text{if} \quad \dot\psi > \dot\psi_1
\end{aligned}
$$

ε denotes the limit of sensibility of the angular velocity sensor.

The output variables are T, order of positive control pulse, and Z, order of negative control pulse.

Heuristically, this sequential circuit operates as follows:

– Nor 1 identifies point A. This information is memorized in Memory 1 and a pulse order appears on T or Z according to the sign of φ.

– When $|\psi| > \psi_l$ (point B) Memory 1 is set back to zero and the pulse disappears.

– When $|\psi| > \psi_l$ (point C) Memory 2 is excited and a pulse order of same sign as the previous one appears.

– When $|\psi| < \varepsilon$ (point D) Memory 2 is set back to zero, the pulse order disappears.

The various signals $|\psi| > \psi_l$ have to be memorized for they appear practically as impulses.

To make sure that under any circumstances the sign of the pulse order is correct, it is necessary to memorize the sign of φ during the control cycle as it is at the beginning of the control sequence. As a result a variable $\varDelta$, 'end of cycle', has to be generated in order to set back e_1, the memorized sign of φ, equal to e at the end of the control sequence.

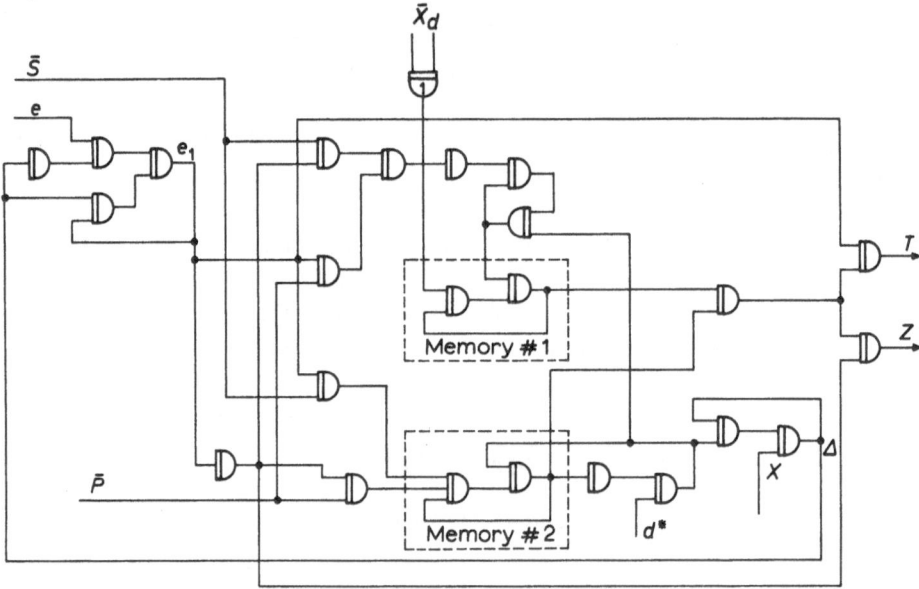

Fig. 11.   Asynchronous logical scheme.

The second circuit that has been devised is of the synchronised asynchronous type and uses the following input variables, appearing as impulses;

$$X_A = 1 \quad \text{if} \quad \varphi > \varphi_s$$
$$X_B = 1 \quad \text{if} \quad \varphi < -\varphi_s$$
$$X_C = 1 \quad \text{if} \quad |\dot{\varphi}| > \varepsilon$$
$$X_D = 1 \quad \text{if} \quad |\psi| > \varepsilon$$
$$X_E = 1 \quad \text{if} \quad |\psi| > \psi_l.$$

The output variables are as before.

This circuit provides the same sequential control law as the previous one, but it is designed so that any random disturbance impulse can have no redhibitory consequence upon the sequel of the control sequence.

The various flow-tables, excitation map, transition map allowing to write the circuit's equations are in Tables I to IV.

TABLE I

Original flow-table

	X_A	X_B	X_C	X_D	X_E
1	2, 0	10, 0	9, 0		
2	2, 0		3, 1		
3	3, 1		3, 1		4, 0
4				5, 0	4, 0
5				5, 0	6, 1
6			6, 1	7, 0	6, 1
7	2, 0	10, 0	8, 0	7, 0	
8	3, 1	11, 2	8, 0	7, 0	
9	3, 1	11, 2	9, 0		
10		10, 0	11, 2		
11		11, 2	11, 2		12, 0
12			1	13, 0	12, 0
13				13, 0	14, 2
14			14, 2	15, 0	14, 2
15	2, 0	10, 0	16, 0	15, 0	
16	11, 2	11, 2	16, 0	15, 0	

TABLE II

Reduced flow-table

		X_A	X_B	X_C	X_D	X_E
$a = (1, 4)$	a	b, 0	g, 0	f, 0	d, 0	a, 0
$b = (2, 12)$	b	b, 0	–, –	c, 1	g, 0	b, 0
$c = (3)$	c	c, 1	–. –	c, 1	–, –	a, 0
$d = (5, 7, 15)$	d	b, 0	g, 0	f, 0	d, 0	e, 1
$e = (6)$	e	–, –	–, –	e, 1	d, 0	e, 1
$f = (8, 9, 16)$	f	c, 1	h, 2	f, 0	d, 0	–, –
$g = (10, 13)$	g	–, –	g, 0	h, 2	g, 0	j, 2
$h = (11)$	h	–, –	h, 2	h, 2	–, –	b, 0
$j = (14)$	j	–, –	–, –	j, 2	d, 0	j, 2

TABLE III
Excitation map: y_1, y_2, y_3, y_4

	$y_1y_2y_3y_4$	X_A	X_B	X_C	X_D	X_E
a	0000	0001	0010	0100	1000	0000
b	0001	0001	–	0101	0010	0001
c	0101	0101	–	0101	–	0000
d	1000	0001	0010	0100	1000	1101
e	1101	–	–	1101	1000	1101
f	0100	0101	0110	0100	1000	–
g	0010	–	0010	0110	0010	1100
h	0110	–	0110	0110	–	0001
j	1100	–	–	1100	1000	1100

TABLE IV
Transition map: $\tau_1, \tau_2, \tau_3, \tau_4$

$y_1y_2y_3y_4$	X_A	X_B	X_C	X_D	X_E
0000	0001	0010	0100	1000	0000
0001	0000	–	0100	0011	0000
0101	0000	–	0000	–	0101
1000	1001	1010	1100	0000	0101
1101	–	–	0000	0101	0000
0100	0001	0010	0000	1100	–
0010	–	0000	0100	0000	1110
0110	–	0000	0000	–	0111
1100	–	–	0000	0100	0000

TABLE V
Output table Z_1, Z_2

$y_1y_2y_3y_4$	$X_AX_BX_CX_DX_E$	X_A	X_B	X_C	X_D	X_E
0000	00	00	00	00	00	00
0001	00	00	–	10	00	00
0101	10	10	–	10	–	00
1000	00	00	00	00	00	10
1101	10	–	–	10	00	10
0100	00	10	01	00	00	–
0010	00	–	00	01	00	01
0110	01	–	01	01	–	01
1100	01	–	–	01	00	00

These tables and maps allow to establish the formulas for the τ_i's and Z_i's:

$$\tau_1 = y_1(X_A + X_B) + y_1\bar{y}_2X_C + \bar{y}_1\bar{y}_3\bar{y}_4X_D + \bar{y}_2y_3X_E$$

$$\tau_2 = \bar{y}_2X_C + y_2X_D + (y_1\bar{y}_2 + \bar{y}_1y_2 + y_3)X_E$$

$$\tau_3 = \bar{y}_3X_B + \bar{y}_1y_4X_D + y_3X_E$$

$$\tau_4 = \bar{y}_4X_A + y_4X_D + (\bar{y}_1y_2 + y_1\bar{y}_2)X_E$$

$$Z_1 = y_2X_A + y_4X_C + y_1X_E(\bar{y}_2 + y_4) + \bar{X}_B\bar{X}_D\bar{X}_Ey_2y_4$$

$$Z_2 = y_2X_B + y_3(X_C + \bar{y}_2X_E) + y_2\bar{y}_4(y_1X_C + y_3X_E) + \bar{X}_D\bar{X}_E(y_2y_3 + y_1y_2\bar{y}_3)$$

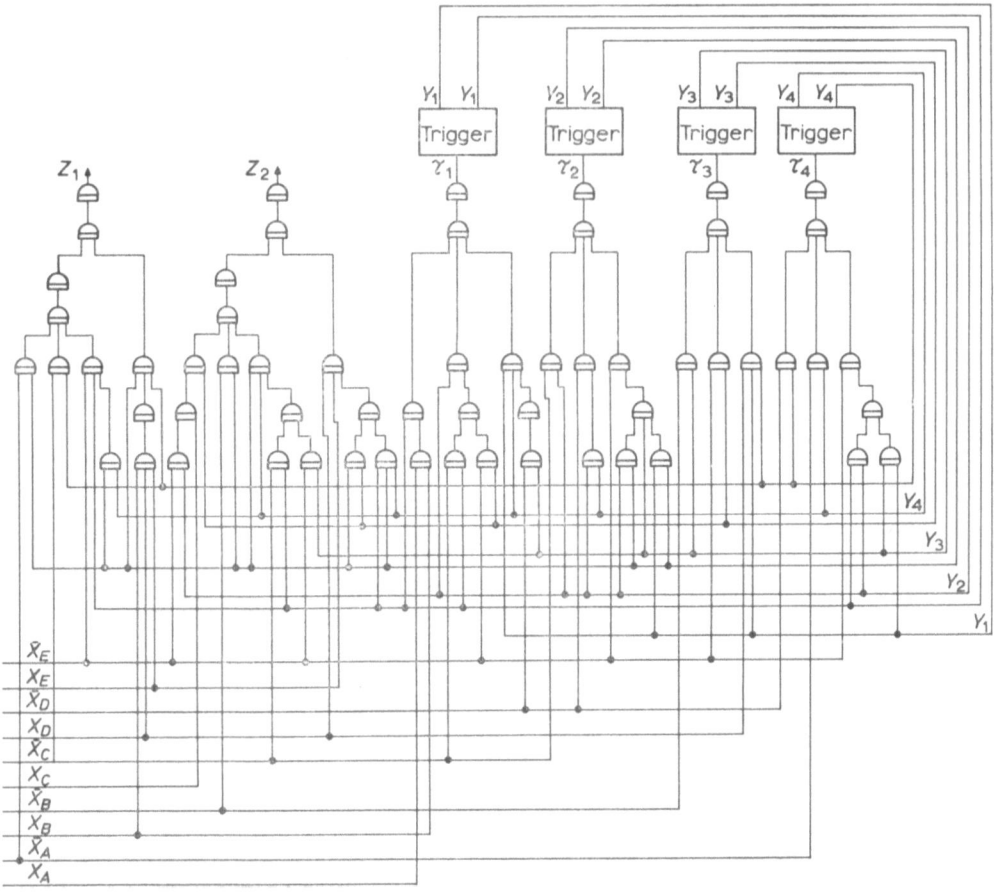

Fig. 12.   Asynchronous synchronized circuit.

Appendix V

Digital Simulation

The FORTRAN program written to analyse the satellite's behaviour during its mission closely follows the logical steps implied by the sequential circuits. The numerical computation involves a combination of the autonomous, forced and controlled solutions to the satellite's equations. In this program, time T3 is used to compute the controlled solution, time T2 denotes the actual time and is therefore fitted to the calculation of the forced solution, whereas time T1, used to compute the autonomous solution, is set back to zero after each control pulse, the new corresponding initial conditions being the state resulting from the control pulse.

A summary of the FORTRAN notations appearing in this program is given below.

Fortran Notations

Variables	FORTRAN Variables	Meaning	Units
φ	PHI	Roll angle	rd
ψ	PSI	Pitch angle	rd
$\dot\varphi$	PHIP	Roll angular velocity	rd/s
$\dot\psi$	PSIP	Pitch angular velocity	rd/s
$\varphi_0\ \psi_0$ $\dot\varphi_0\ \dot\psi_0$	PHIO, PSIO PHIPO, PSIPO	Initial conditions	rd rd/s
$\varphi_l\ \psi_l$ $\dot\varphi_l\ \dot\psi_l$	PHIL, PSIL PHIPL, PSIPL	Solution to the autonomous system	rd rd/s
$\varphi_f\ \psi_f$ $\dot\varphi_f\ \dot\psi_f$	PHIF, PSIF PSIPF, PSIPF	Solution to the perturbed system	rd rd/s
$\varphi_c\ \psi_c$ $\dot\varphi_c\ \dot\psi_c$	PHIC, PSIC PHICP, PSICP	Solution to the controlled system	rd rd/s
φ_s	PHIS	Required roll-attitude precision	rd
$\dot\psi_l$	PSIPLM	Control pitch velocity	rd/s
ω	OM	Gyroscopic pulsation	rd/s
ω^*	OMS	Nutation's pulsation	rd/s
ω_0	RO	Earth's pulsation	rd/s
ω_0^*	ROS	Precession's pulsation	rd/s
$C_0 = \dfrac{\Gamma_c}{I_{xx}}$	CO	Normalised amplitude of control torque	rd/s^2
q_0	QO	Maximal amplitude of solar perturbation	N·m
$C_p = \dfrac{q_0}{I_{xx}}$	CP	Normalised amplitude of solar perturbation	rd/s^2

```
FORTRAN IV    MODEL  44  PS     VERSION 3,  LEVEL 1  DATE 69015

0001                    DOUBLE PRECISION PHI,PSI,PHIP,PSIP,PHIO,
                       1PSIO,PHIPO,PSIPO,D,APS,BPS,CPS,DPS,APH,
                       2BPH,CPH,DPH,PHIL,PSIL,PHIPL,PSIPL,PHIF,
                       3PSIF,PHIFP,PSIFP,APSI,BPSI,CPSI,APHI,
                       4BPHI,CPHI,PHIC,PSIC,PHICP,PSICP,BO,B1,A1,
                       5C1,PHIS,OM,OMS,ROS,RO,RO1,QO,CP,PAS1,
                       6PAS2,TS,X,Y,Z,VALCO,T1,T3,U,S,PSIPLM
0002            100   FORMAT(10X,16,7(3X,E11.4))
0003            102   FORMAT(5X,'RO =',E11.4,1X,'OM =',E11.4,
                       11X,'ROS =',E11.4,1X,'OMS =',E11.4,IX,'PHIS =',
                       2E11.4,1X,'QO =',E11.4,1X,'PSIPLM =',E11.4////)
0004            103   FORMAT(14X,'K',8X,'PSI',11X,'PHI',11X,
                       1'PSIP',10X,'PHIP',11X,'T1',12X,'T2',12X,'T3',///)
0005            200   FORMAT(10X,/)
0006                  PHIS = 0.50D − 2
0007                  QO = 0.1D − 5.
0008                  CP = QO/64.
0009                  RO = 7.268D − 05
0010                  OM = 20./64.
0011                  OMS = DSQRT((OM**2 + 2.*OM*RO + OM**2*
                      DSQRT(1. + 4.*RO/OM))/2.)
0012                  ROS = RO − RO**2/OM + 2.*RO**3/OM**2
0013                  D = OMS**2 − ROS**2
0014                  CO = (2.D − 2)/64.
0015                  A1 = − C0*(RO*OM − ROS**2)/(ROS**2*(OMS**
                      12 − ROS**2))
0016                  B0 = C0*OM/(OMS**2 − ROS**2)
0017                  B1 = − C0*(RO*OM − OMS**2)/(OMS**2*
                      1(ROS**2 − OMS**2))
0018                  C1 = C0*RO*OM/(ROS**2*OMS**2)
0019                  CALL VILIM(PHIS,RO,ROS,OMS.OM,D,A1,
                      1B0,B1,C1,PSIPLM)
0020                  WRITE(6,102)RO,OM,ROS,OMS,PHIS,QO,
                      1PSIPLM
0021                  WRITE(6,103)
0022                  PSIO = 0.
0023                  PHIO = 0.
0024                  PSIPO = 0.
0025                  PHIPO = 0.
0026                  RO1 = RO
```

FORTRAN IV   MODEL   44   PS   VERSION 3,   LEVEL 1   DATE 69015

```
0027            APHI=CP*(OM*(RO+RO1)-ROS**2)/
           1(D*(RO1**2-ROS**2))
0028            BPHI=CP*(OM*(RO+RO1)-OMS**2)/
           1((ROS**2-OMS**2)*(RO1**2-OMS**2))
0029            CPHI=CP*(OM*(RO+RO1)-RO1**2)/
           1((ROS**2-RO1**2)*(OMS**2-RO1**2))
0030            APSI=CP*(OM*RO*RO1-(RO1-OM)*ROS**
           12)/((RO1**2-ROS**2)*D)
0031            BPSI=CP*(OM*RO*RO1-(RO1-OM)*OMS**
           12)/((ROS**2-OMS"*2)*(RO1**2-OMS**12))
0032            CPSI=CP*(OM*RO*RO1-(RO1-OM)*RO1**2)/
           1((ROS**2-RO1**2)*(OMS**2-RO1**12))
0033            PAS1=10.
0034            PAS2=0.1
0035            K=1
0036            T2=0.
0037            T1=0.
0038            T3=0.
0039         4  AFH=((OM**2+RO*OM-ROS**2)*PHIO+
           1OM*PSIPO)/D
0040            BPH=(-RO*OM**2*PSIO+(RO*OM-ROS**2)
           1*PHIPO)/(D*ROS)
0041            CPH=((OM**2+RO*OM-OMS**2)*PHIO+
           1OM*PSIPO)/(-D)
0042            DPH=(RO*OM**2*PSIO-(RO*OM-OMS**2)*
           1*PHIPO)/(D*OMS)
0043            T1=PAS1
0044            T2=T2+PAS1
0045         80 S=ROS*T1
0046            U=OMS*T1
0047            X=RO1*T2
0048            Y=ROS*T2
0049            Z=OMS*T2
0050            PHIL=APH*DCOS(S)+BPH*DSIN(S)+CPH*
           1DCOS(U)+DPH*DSIN(U)
0051            PHIF=APHI*DCOS(Y)+BPHI*DCOS(Z)+
           1CPHI*DCOS(X)
0052            PHI=PHIL+PHIF
0053            VAL=DABS(PHI)
0054            IF(VAL-PHIS)1,2,3
0055         1  T1=T1+PAS1
```

FORTRAN IV   MODEL 44 PS   VERSION 3, LEVEL 1   DATE 69015

```
0056                    T2=T2+PAS1
0057                    GO TO80
0058               3    T1=T1-PAS1
0059                    T2=T2-PAS1
0060                    PAS1=PAS1/100.
0061                    IF(PAS1-0.000009)2,2,1
0062               2    PHIL= -ROS*APH*DSIN(S)+ROS*BPH*DCOS
                        1(S)-OMS*CPH*DSIN(U)+OMS*DPH*DCOS(U)
0063                    PHIFP= -ROS*APHI*DSIN(Y)-OMS*BPHI*
                        1DSIN(Z)-RO1*CPHI*DSIN(X)
0064                    PHIP=PHIPL+PHIFP
0065                    IF(PHI)44,44,45
0066              44    J=0
0067                    PAS1=0.001
0068              46    IF(PHIP+0.1D-03)54,22,22
0069              54    CONTINUE
0070                    T1=T1+PAS1
0071                    T2=T2+PAS1
0072                    S=ROS*T1
0073                    U=OMS*T1
0074                    X=RO1*T2
0075                    Y=ROS*T2
0076                    Z=OMS*T2
0077                    PHIPL= -ROS*APH*DSIN(S)+ROS*BPH*
                        1DCOS(S)-OMS*CPH*DSIN(U)+OMS*DPH
                        2*DCOS(U)
0078                    PHIFP= -ROS*APHI*DSIN(Y)-OMS*BPHI
                        1DSIN(Z)-RO1*CPHI*DSIN(X)
0079                    PHIP=PHIPL+PHIFP.
0080                    GO TO 46
0081              45    J=1
0082                    PAS1=0.001
0083              47    IF(PHIP-0.1D-0.3)22,22,55
0084              55    CONTINUE
0085                    T1=T1+PAS1
0086                    T2=T2+PAS1
0087                    S=ROS*T1
0088                    U=OMS*T1
0089                    X=RO1*T2
0090                    Y=ROS*T2
0091                    Z=OMS*T2
```

FORTRAN IV   MODEL 44 PS   VERSION 3,   LEVEL 1   DATE 69015

```
0092                    PHIPL = − ROS*APH*DSIN(S) + ROS*BPH*
                        1DCOS(S) − OMS*CPH*DSIN(U) + OMS*DPH
                        2*DCOS(U)
0093                    PHIFP = − ROS*APHI*DSIN(Y) − OMS*BPHI
                        1*DSIN(Z) − RO1*CPHI*DSIN(X)
0094                    PHIP = PHIPL + PHIFP
0095                    GO TO 47
0096              22    PHIL = APH*DCOS(S) + BPH*DSIN(S) + CPH
                        1*DCOS(U) + DPH*DSIN(U)
0097                    PHIF = APHI*DCOS(Y) + BPHI*DCOS(Z) +
                        1CPHI*DCOS(X)
0098                    PHI = PHIL + PHIF
0099              24    APS = ((OM**2 + RO*OM − ROS**2)*PSIO − OM
                        1*PHIPO)/D
0100                    BPS = (RO*OM**2*PHIO + (RO*OM − ROS**2)
                        1*PSIPO)/(D*ROS)
0101                    CPS = ((OM**2 + RO*OM − OMS**2)*PSIO −
                        1*PHIPO)/(−D)
0102                    DPS = (RO*OM**2*PHIO + (RO*OM − OMS**2)
                        1*PSIPO)/(−D*OMS)
0103                    PSIL = APS*DCOS(S) + BPS*DSIN(S) + CPS*
                        1DCOS(U) + DPS*DSIN(U)
0104                    PSIF = (APSI/ROS)*DSIN(Y) + (BPSI/OMS)
                        1*DSIN(Z) + (CPSI/RO1)*DSIN(X)
0105                    PSI = PSIL + PSIF
0106                    PSIPL = − ROS*APS*DSIN(S) + ROS*BPS*DCOS
                        1(S) − OMS*CPS*DSIN(U) + OMS*DPS*DCOS(U)
0107                    PSIFP = APSI*DCOS(Y) + BPSI*DCOS(Z) + CPSI
                        1DCOS(X)
0108                    PSIP = PSIPL + PSIFP
0109                    WRITE(6,100)K,PSI,PSIP,PHIP,T1,T2,T3
0110                    T1 = T1 + PAS2
0111                    T2 = T2 + PAS2
0112                    T3 = PAS2
0113                    IF(J)5,5,6
0114               5    CONTINUE
0115                    S = RUS*T1
0116                    U = OMS*T1
0117                    X = RO1*T2
0118                    Y = ROS*T2
0119                    Z = OMS*T2
```

FORTRAN IV   MODEL 44 PS   VERSION 3, LEVEL 1   DATE 69015

0120	PSIPL = − ROS*APS*DSIN(S) + ROS*BPS*DCOS 1(S) − OMS*CPS*DSIN(U) + OMS*DPS*DCOS(U)
0121	PSIFP = APSI*DCOS(Y) + BPSI*DCOS(Z) + 1CPSI*DCOS(X)
0122	PSICP = − ROS*A1*DSIN(ROS*T3) − OMS*B1 1*DSIN(OMS*T3)
0123	PSIP = PSIPL + PSIFP + PSICP
0124	IF(PSIP − PSIPLM)7,8,9
0125	7 T1(T1 + PAS2
0126	T2 = T2(PAS2
0127	T3 = T3 + PAS2
0128	GO TO 5
0129	9 T1 = T1 − PAS2
0130	T2 = T2 − PAS2
0131	T3 = T3 − PAS2
0132	PAS2 = PAS2/10.
0133	IF(PAS2 − 0.0009)8,7,7
0134	6 CONTINUE
0135	S = ROS*T1
0136	U = OMS*T1
0137	X = RO1*T2
0138	Y = ROS*T2
0139	Z = OMS*T2
0140	PSIPL = − ROS*APS*DSIN(S) + ROS*BPS 1*DCOS(S) − OMS*CPS*DSIN(U) + OMS*DPS1 2*DCOS(U)
0141	PSIFP = APSI*DCOS(Y) + BPSI*DCOS(Z) + CPSI 1DCOS(X)
0142	PSICP = − ROS*A1*DSIN(ROS*T3) − OMS*B1 1*DSIN(OMS*T3)
0143	PSIP = PSIPL + PSIFP − PSICP
0144	IF(PSIP + PSIPLM)10,8,12
0145	12 T1 = T1 + PAS2
0146	T2 = T2 + &AS2
0147	T3 = T3 + PAS2
0148	GO TO 6
0149	10 T1 = T1 − PAS2
0150	T2 = T2 − PAS2
0151	T3 = T3 − PAS2
0152	PAS2 = PAS2/10.
0153	IF(PAS2 − 0.0009)8,12,12

FORTRAN IV   MODEL   44   PS   VERSION 3,   LEVEL 1   DATE   69015

```
0154        8   PSIL = APS*DCOS(S) + BPS*DSIN(S) + CPS*DCOS
                1(U) + DPS*DSIN(U)
0155            PSIF = (APSI/ROS)*DSIN(Y) + (BPSI/OMS)
                1*DSIN(Z) + (CPSI/RO1)*DSIN(X)
0156            PSIC = C1 + A1*DCOS(ROS*T3) + B1*DCOS
                1(OMS*T3)
0157            PHIL = APH*DCOS(S) + BPH*DSIN(S) + CPH
                1*DCOS(U) + DPH*DSIN(U)
0158            PHIF = APHI*DCOS(Y) + BPHI*DCOS(Z) +
                1CPHI*DCOS(X)
0159            PHIC = (B0/ROS)*DSIN(ROS*T3) - (BO/OMS)
                1*DSIN(OMS*T3)
0160            PHIPL = - ROS*APH*DSIN(S) + ROS*BPH
                1*DCOS(S) - OMS*CPH*DSIN(U) + OMS*DPH
                1*DCOS(U)
0161            PHIFP = - ROS*APHI*DSIN(Y) - OMS*BPHI
                1*DSIN(Z) - RO1*CPHI*DSIN(X)
0162            PHICP = B0*DCOS(ROS*T3) - B0*DCOS(OMS
                1*T3)
0163            IF(J)38,38,48
0164       38   PSI = PSLI + PSIF + PSIC
0165            PHI = PHIL + PHIF + PHIC
0166            PHIP = PHIPL + PHIPP + PHICP
0167            PSIO = PSIL + PSIC
0168            PHIO = PHIL + PHIC
0169            PSIPO = PSIPL + PSICP
0170            PHIPO = PHIPL + PHICP
0171            GO TO 59
0172       48   PSI = PSIL + PSIF - PSIC
0173            PHI = PHIL + PHIF - PHCI
0174            PHIP = PHIPL + PHIFP - PHICP
0175            PSIO = PSIL - PSIC
0176            PHIO = PHIL - PHIC
0177            PHIPO = PHIPL - PHICP
0178            PSIPO = PSIPL - PSICP
0179       59   CONTINUE
0180       49   WRITE(6,100)K,PSI,PHI,PSIP,PHIP,T1,T2,T3
0181            PAS2 = 0.1
0182            T1 = PAS2
0183            T2 = T2 + PAS2
```

FORTRAN IV   MODEL 44 PS   VERSION 3, LEVEL 1   DATE 69015

```
0184              APS=((OM**2+RO*OM−ROS**2)*PSIO−OM
             1*PHIPO)/D
0185              BPS=(RO*OM**2*PHIO+(RO*OM−ROS**2)
             1*PSIPO)/(D*ROS)
0186              CPS=((OM**2+RO*OM−OMS**2)*PSIO−OM
             1*PHIPO)/(−D)
0187              DPS=(RO*OM**2*PHIO+(RO*OM−OMS**2)
             1*PSIPO)/(−D*OMS)
0188              APH=((OM**2+RO*OM−ROS**2)*PHIO+
             1OM*PSIPO)/D
0189              BPH=(−RO*OM**2*PSIO+(RO*OM−ROS
             1**2)*PHIPO)/(D*ROS)
0190              CPH=((OM**2+RO*OM−OMS**2)*PHIO+OM
             1*PSIPO)/(−D)
0191              DPH=(RO*OM**2*PSIO−(RO*OM−OMS**2)
             1*PHIPO)/(D*OMS)
0192          50  S=ROS*T1
0193              U=OMS*T1
0194              X=RO1*T2
0195              Y=ROS*T2
0196              Z=OMS*T2
0197              PSIPL=−ROS*APS*DSIN(S)+ROS*BPS*DCOS
             1(S)−OMS*CPS*DSIN(U)+OMS*DPS*DCOS(U)
0198              PSIFP=APSI*DCOS(Y)+BPSI*DCOS(Z)+CPSI
             1*DCOS(X)
0199              PSIP=PSIPL+PSIFP
0200              IF(J)14,14,15
0201          14  IF(PSIP+PSIPLM)18,30,16
0202          16  T1=T1+PAS2
0203              T2=T2+PAS2
0204              T3=T3+PAS2
0205              GO TO 50
0206          18  T1=T1−PAS2
0207              T2=T2−PAS2
0208              T3=T3−PAS2
0209              PAS2=PAS2/10.
0210              IF(PAS2−0.0009)30,30,16
0211          15  IF(PSIP−PSIPLM)16,30,18
0212          30  PSIL=APS*DCOS(S)+BPS*DSIN(S)+CPS*DCOS
             1(U)+DPS*DSIN(U)
```

FORTRAN IV   MODEL 44 PS   VERSION 3,   LEVEL 1   DATE 69015

```
0213          PSIF=(APSI/ROS)*DSIN(Y)+(BPSI/OMS)
             1*DSIN(Z)+(CPSI/RO1)*DSIN(X)
0214          PHIL=APH*DCOS(S)+BPH*DSIN(S)+CPH
             1*DCOS(U)+DPH*DSIN(U)
0215          PHIF=APHI*DCOS(Y)+BPHI*DCOS(Z)+
             1CPHI*DCOS(X)
0216          PHIPL=-ROS*APH*DSIN(S)+ROS*BPH
             1*DCOS(S)-OMS*CPH*DSIN(U)+OMS*DPH
             2*DCOS(U)
0217          PHIFP=-ROS*APHI*DSIN(Y)-OMS*BPHI
             1*DSIN(Z)-RO1*CPHI*DSIN(X)
0218          PSI=PSIL+PSIF
0219          PHI=PHIL+PHIF
0220          PHIP=PHIPL+PHIFP
0221          WRITE(6,100)K,PSI,PHI,PSIP,PHIP,T1,T2,T3
0222          PAS2=0.1
0223          IF(J)31,31,41
0224       31 T1=T1+PAS2
0225          T2=T2+PAS2
0226          T3=PAS2
0227       32 S=ROS*T1
0228          U=OMS*T1
0229          X=RO1*T2
0230          Y=ROS*T2
0231          Z=OMS*T2
0232          PSIPL=-ROS*APS*DSIN(S)+ROS*BPS
             1*DCOS(S)-OMS*CPS*DSIN(U)+OMS*DPS
             2*DCOS(U)
0233          PSICP=-ROS*A1*DSIN(ROS*T3)-OMS*B1
             1*DSIN(OMS*T3)
0234          PSIFP=APSI*DCOS(Y)+BPSI*DCOS(Z)+CPSI
             1*DCOS(X)
0235          PSIP=PSIPL+PSIFP+PSICP
0236          IF(PSIP+0.0001)26,60,28
0237       26 T1=T1+PAS2
0238          T2=T2+PAS2
0239          T3=T3+PAS2
0240          GO TO 32
0241       28 T1=T1-PAS2
0242          T2=T2-PAS2
0243          T3=T3-PAS2
```

FORTRAN IV   MODEL 44 PS   VERSION 3, LEVEL 1 DATE 69015

```
0244                      PAS2 = PAS2/10.
0245                      IF(PAS2 − 0.0009)60,60,26
0246              41   T1 = T1 + PAS2
0247                   T2 = T2 + PAS2
0248                   T3 = PAS2
0249              42   S = ROS*T1
0250                   U = OMS*T1
0251                   X = RO1*T2
0252                   Y = ROS*T2
0253                   Z = OMS*T2
0254                   PSIPL = − ROS*APS*DSIN(S) + ROS*BPS
                      1*DCOS(S) − OMS*CPS*DSIN(U) + OMS*DPS
                      2*DCOS(U)
0255                   PSIFP = APSI*DCOS(Y) + BPSI*DCOS(Z) + CPSI
                      1*DCOS(X)
0256                   PSICP = − ROS*A1*DSIN(ROS*T3) − OMS*B1
                      1*DSIN(OMS*T3)
0257                   PSIP = PSIPL + PSIFP − PSICP
0258                   IF(PSIP − 0.0001)29,60,27
0259              27   T1 = T1 + PAS2
0260                   T2 = T2 + PAS2
0261                   T3 = T3 + PAS2
0262                   GO TO 42
0263              29   T1 = T1 − PAS2
0264                   T2 = T2 − PAS2
0265                   T3 = T3 − PAS2
0266                   PAS2 = PAS2/10.
0267                   IF(PAS2 − 0.0009)60,60,27
0268              60   PSIL = APS*DCOS(S) + BPS*DSIN(S) + CPS
                      1*DCOS(U) + DPS*DSIN(U)
0269                   PSIF = (APSI/ROS)*DSIN(Y) + (BPSI/OMS)
                      1*DSIN(Z) + (CPSI/RO1)*DSIN(X)
0270                   PSIC = C1 + A1*DCOS(ROS*T3) + B1*DCOS
                      1(OMS*T3)
0271                   PHIL = APH*DCOS(S) + BPH*DSIN(S) + CPH
                      1*DCOS(U) + DPH*DSIN(U)
0272                   PHIF = APHI*DCOS(Y) + BPHI*DCOS(Z) +
                      1CPHI*DCOS(X)
0273                   PHIC = (B0/ROS)*DSIN(ROS*T3) − (B0/OMS)
                      1*DSIN(OMS*T3)
```

FORTRAN IV   MODEL 44 PS   VERSION 3,   LEVEL 1   DATE 69015

```
0274                    PHIPL = − ROS*APH*DSIN(S) + ROS*BPH
                       1*DCOS(S) − OMS*CPH*DSIN(U) + OMS*DPH
                       2*DCOS(U)
0275                    PHIFP = − ROS*APHI*DSIN(Y) − OMS*BPHI
                       1*DSIN(Z) − RO1*CPHI*DSIN(X)
0276                    PHICPOB0*DCOS(ROS*T3) − B0*DCOS(OMS*T3)
0277                    IF(J)68,68,78
0278              68    PSI = PSIL + PSIF + PSIC
0279                    PHI = PHIL + PHIF + PHIC
0280                    PHIP = PHIPL + PHIFP + PHICP
0281                    PHIPO = PHIPL + PHICP
0282                    PSIPO = PSIPL + PSICP
0283                    PHIO = PHIL + PHIC
0284                    PSIO = PSIL + PSIC
0285                    GO TO 89
0286              78    PSI = PSIL + PSIF − PSIC
0287                    PHI = PHIL + PHIF − PHIC
0288                    PHIP = PHIPL + PHIFP − PHICP
0289                    PHIPO = PHIPL − PHICP
0290                    PSIPO = PSIPL − PSICP
0291                    PHIO = PHIL − PHIC
0292                    PSIO = PSIL − PSIC
0293              89    CONTINUE
0294              79    WRITE(6,100)K,PSI,PHI,PSIP,PHIP,T1,T2,T3
0295                    WRITE(6,200)
0296                    K = K + 1
0297                    PAS1 = 10.
0298                    PAS2 = 0.1
0299                    T1 = 0.
0300                    T3 = 0.
0301                    GO TO 4
0302                    END


0001                    SUBROUTINE VILIM (PHIS,RO,ROS,OMS,
                       1OM,D,A1,B0,B1,C1,PSIPLM)
0002                    DOUBLE PRECISION PHI,PSI,PHIP,PSIP,PHIO,
                       1PSIO,PHIPO,PSIPO,D,APS,BPS,CPS,DPS,APH,
                       2BPH,CPH,DPH,PHIL,PSIL,PHIPL,PSIPL,PHIC,
                       3PSIC,PHICP,PSICP,B0,B1,A1,C1,PHIS,OM,OMS,
                       4ROS,RO,C0,T1,T3,U,S,PSIPLM,PAS,PAPSIP
```

FORTRAN IV   MODEL 44 PS   VERSION 3,   LEVEL 1   DATE 69015

```
0003                        PAPSIP = 1.D − 4
0004                        PSIPLM = 0.6D − 3
0005                   15   CONTINUE
0006                        PHIO = PHIS
0007                        PSIO = 0.
0008                        PSIPO = 0.
0009                        PHIPO = 0.
0010                        T1 = 0.
0011                        T3 = 0.
0012                        PAS = 0.1
0013                        APH = ((OM**2 + RO*OM − ROS**2)*PHIO + OM
                            1*PSIPO)/D
0014                        BPH = (−RO*OM**2*PSIO + (RO*OM − ROS**2)
                            1*PHIPO)/(D*ROS)
0015                        CPH = ((OM**2 + RO*OM − OMS**2)*PHIO + OM
                            1*PSIPO)/(−D)
0016                        DPH = (RO*OM**2*PSIO − (RO*OM − OMS**2)
                            1*PHIPO)/(D*OMS)
0017                        APS = ((OM**2 + RO*OM − ROS**2)*PSIO − OM
                            *PHIPO)/D
0018                        BPS = (RO*OM**2*PHIO + (RO*OM − ROS
                            1**2)*PSIPO)/(D*ROS)
0019                        CPS = ((OM**2 + RO*OM − OMS**2)*PSIO
                            1 − OM*PHIPO)/(−D)
0020                        DPS = (RO*OM**2*PHIO + (RO*OM − OMS**2)
                            1*PSIPO)/(−D*OMS)
0021                    4   CONTINUE
0022                        U = OMS*T1
0023                        S = RUS*T1
0024                        PSIPL = −ROS*APS*DSIN(S) + ROS*BPS
                            1*DCOS(S) − OMS*CPS*DSIN(U) + OMS*DPS
                            2*DCOS(U)
0025                        PSICP = −ROS*A1*DSIN(ROS*T3) − OMS
                            1*B1*DSIN(OMS*T3)
0026                        PSIP = PSIPL − PSICP
0027                        IF(PSIP + PSIPLM)3,2,1
0028                    1   T1 = T1 + PAS
0029                        T3 = T3 + PAS
0030                        GO TO 4
0031                    3   T1 = T1 − PAS
0032                        T3 = T3 − PAS
```

FORTRAN IV   MODEL 44 PS   VERSION 3, LEVEL 1   DATE 69015

```
0033                        PAS = PAS/10.
0034                        IF(PAS − 0.00009)2,1,1
0035                     2  CONTINUE
0036                        PHIPL = − ROS*APH*DSIN(S) + ROS*BPH*
                            1DCOS(S) − OMS*CPH*DSIN(U) + OMS*DPH
                            2*DCOS(U)
0037                        PHICP = B0*DCOS(ROS*T3) − B0*DCOS
                            1(OMS*T3)
0038                        PSIL = APS*DCOS(S) + BPS*DSIN(S) + CPS*
                            1DCOS(U) + DPS*DSIN(U)
0039                        PSIC = C1 + A1*DCOS(ROS*T3) + B1*DCOS
                            1(OMS*T3)
0040                        PHIC = (B0/ROS)*DSIN(ROS*T3) − (B0/OMS)
                            1*DSIN(OMS*T3)
0041                        PHIL = APH*DCOS(S) + BPH*DSIN(S) + CPH
                            1*DCOS(U) + DPH*DSIN(U)
0042                        PHI = PHIL − PHIC
0043                        PSI = PSIL − PSIC
0044                        PHIP = PHIPL − PHICP
0045                        PSIO = PSI
0046                        PSIPO = PSIP
0047                        PHIO = PHI
0048                        PHIPO = PHIP
0049                        DPH = (RO*OM**2*PSIO − (RO*OM − OMS**2)
                            1*PHIPO)/(D*OMS)
0050                        DPS = (RO*OM**2*PHIO + (RO*OM − OMS**2)
                            1*PSIPO)/(−D*OMS)
0051                        CPS = ((OM**2 + RO*OM − OMS**2)*PSIO − OM
                            1*PHIPO)/(−D)
0052                        BPS = (RO*OM**2*PHIO + (RO*OM − ROS**2)
                            1*PSIPO)/(D*ROS)
0053                        APS = ((OM**2 + RO*OM − ROS**2)*PSIO − OM
                            1*PHIPO)/D
0054                        APH = ((OM**2 + RO*OM − ROS**2)*PHIO + OM
                            1*PSIPO)/D
0055                        CPH = ((OM**2 + RO*OM − OMS**2)*PHIO + OM
                            1*PSIPO)/(−D)
0056                        BPH = (−RO*OM**2*PSIO + (RO*OM − ROS**2)
                            1*PHIPO)/(D*ROS)
0057                        PAS = 0.1
0058                        T1 = PAS
```

FORTRAN IV   MODEL 44 PS   VERSION 3, LEVEL 1 DATE 69015

```
0059              8  CONTINUE
0060                 S = ROS*T1
0061                 U = OMS*T1
0062                 PSIPL = − ROS*APS*DSIN(S) + ROS*BPS*DCOS
                     1(S) − OMS*CPS*DSIN(U) + OMS*DPS*DCOS(U)
0063                 IF(PSIPL − PSIPLM)5,6,7
0064              5  CONTINUE
0065                 T1 = T1 + PAS
0066                 GO TO 8
0067              7  CONTINUE
0068                 T1 = T1 − PAS
0069                 PAS = PAS/10.
0070                 IF(PAS − 0.00009)6,6,5
0071              6  CONTINUE
0072                 PHIPL = − ROS*APH*DSIN(S) + ROS*BPH
                     1*DCOS(S) − OMS*CPH*DSIN(U) + OMS*DPH
                     2*DCOS(U)
0073                 PSIL = APS*DCOS(S) + BPS*DSIN(S) + CPS
                     1*DCOS(U) + DPS*DSIN(U)
0074                 PHIL = APH*DCOS(S) + BPH*DSIN(S) + CPH
                     1*DCOS(U) + DPH*DSIN(U)
0075                 PHIP = PHIPL
0076                 PSI = PSIL
0077                 PHI = PHIL
0078                 PSIP = PSIPL
0079                 PAS = 0.1
0080                 T3 = PAS
0081                 T1 = T1 + PAS
0082             20  CONTINUE
0083                 S = ROS*T1
0084                 U = OMS*T1
0085                 PSIPL = − ROS*APS*DSIN(S) + ROS*BPS*DCOS
                     1(S) − OMS*CPS*DSIN(U) + OMS*DPS*DCOS(U)
0086                 PSICP = − ROS*A1*DSIN(ROS*T3) − OMS*B1*
                     1DSIN(OMS*T3)
0087                 PSIP = PSIPL − PSICP
0088                 IF(PSIP)9,10,11
0089              9  CONTINUE
0090                 T1 = T1 − PAS
0091                 T3 = T3 − PAS
0092                 PAS = PAS/10.
```

FORTRAN IV   MODEL 44 PS   VERSION 3,   LEVEL 1   DATE 69015

```
0093                    IF(PAS − 0.00009)10,10,11
0094            11      CONTINUE
0095                    T1 = T1 + PAS
0096                    T3 = T3 + PAS
0097                    GO TO 20
0098            10      CONTINUE
0099                    PHIPL = − ROS*APH*DSIN(S) + ROS*BPH
                       1*DCOS(S) − OMS*CPH*DSIN(U) + OMS*DPH
                       2*DCOS(U)
0100                    PHICP = B0*DCOS(ROS*T3) − B0*DCOS(OMS
                       1*T3)
0101                    PSIL = APS*DCOS(S) + BPS*DSIN(S) + CPS
                       1*DCOS(U) + DPS*DSIN(U)
0102                    PSIC = C1 + A1*DCOS(ROS*T3) + B1*DCOS
                       1(OMS*T3)
0103                    PHIC = (B0/ROS)*DSIN(ROS*T3) − (B0/OMS)
                       1*DSIN(OMS*T3)
0104                    PHIL = APH*DCOS(S) + BPH*DSIN(S) + CPH
                       1*DCOS(U) + DPH*DSIN(U)
0105                    PHI = PHIL − PHIC
0106                    PSI = PSIL − PSIC
0107                    PHIP = PHIPL − PHICP
0108                    IF(PHI)12,13,14
0109            12      CONTINUE
0110                    PSIPLM = PSIPLM − PAPSIP
0111                    PAPSIP = PAPSIP/10.
0112                    IF(PAPSIP − 0.9D − 6)13,14,14
0113            14      CONTINUE
0114                    PSIPLM = PSIPLM + PAPSIP
0115                    GO TO 15
0116            13      CONTINUE
0117                    RETURN
0118                    END
```

SHEST SEVERITY NAS 0

−04 OM=0.3125D 00 RUS=0.7266D−04 OMS=0.3126D 00.PHIS=0.5000D

−02 QO=0.10000−05 PSIPLM=0.7030D−03

Numerical Results

	PSI	PHI	PSIP	PHIP	T1	T2	T3
0001	−0.1452D−02	0.5000D−02	0.3939D−06	0.1308D−06	0.1042D 06	0.1042D 06	0.0
0001	−0.2377D−02	0.4755D−02	−0.7031D−03	−0.2889D−03	0.1042D 06	0.1042D 06	0.2496D 01
0001	−0.2378D−02	0.2575D−03	0.7030D−03	−0.2899D−03	0.7550D 01	0.1042D 06	0.9946D 01
0001	−0.1469D−02	0.1079D−04	0.9969D−04	−0.5701D−05	0.9727D 01	0.1042D 06	0.2177D 01
0002	−0.1738D−02	0.5000D−02	−0.3507D−04	0.9356D−04	0.8831D 05	0.1925D 06	0.0
0002	−0.2555D−02	0.5007D−02	−0.7032D−03	−0.1620D−03	0.8831D 05	0.1925D 06	0.2131D 01
0002	−0.2562D−02	0.5096D−03	0.7030D−03	−0.1646D−03	0.8588D 01	0.1925D 06	0.1062D 02
0002	−0.1756D−02	0.4941D−03	0.9989D−04	0.8726D−04	0.1053D 02	0.1925D 06	0.1945D 01
0003	−0.2827D−02	0.5001D−02	−0.8686D−04	0.1000D−03	0.8505D 05	0.2776D 06	0.0
0003	−0.3626D−02	0.5014D−02	−0.5014D−03	−0.1499D−03	0.8506D 05	0.2776D 06	0.1959D 01
0003	−0.3633D−02	0.5178D−03	0.7030D−03	−0.15288D−03	0.8691D 01	0.2776D 06	0.1055D 02
0003	−0.2836D−02	0.5221D−03	0.9968D−04	0.9643D−04	0.1062D 02	0.2776D 06	0.1926D 01
0004	0.1670D−02	−0.5000D−02	0.1250D−03	−0.5900D−04	0.4973D 05	0.3273D 06	0.0
0004	0.2494D−02	−0.4922D−02	0.7032D−03	0.1985D−03	0.4972D 05	0.3273D 06	0.1928D 01
0004	0.2499D−02	−0.4244D−03	−0.7030D−03	0.2005D−03	0.8279D 01	0.3273D 06	0.1011D 02
0004	0.1665D−02	−0.3469D−03	−0.9993D−04	−0.5993D−04	0.1029D 02	0.3273D 06	0.2007D 01
0005	0.1151D−02	−0.5000D−02	0.9483D−04	−0.6711D−04	0.8510D 05	0.4124D 06	0.0
0005	0.1979D−02	−0.4939D−02	0.7031D−03	0.1919D−03	0.8510D 05	0.4124D 06	0.2007D 01
0005	0.1984D−02	−0.4417D−03	−0.7030D−03	0.1937D−03	0.8336D 01	0.4124D 06	0.1024D 02
0005	0.1155D−02	−0.3762D−03	−0.9989D−04	−0.6504D−04	0.1033D 02	0.4124D 06	0.1995D 01
0006	0.1043D−02	−0.5000D−02	0.6831D−04	−0.9743D−04	0.8491D 05	0.4974D 06	0.0
0006	0.1850D−02	−0.5010D−02	0.7032D−03	0.1548D−03	0.8492D 05	0.4974D 06	0.2021D 01
0006	0.1056D−02	−0.5126D−03	0.7030D−03	−0.1513D−03	0.8651D 01	0.4974D 06	0.1057D 02
0006	0.1055D−02	−0.5094D−03	−0.9979D−04	−0.9290D−04	0.1058D 02	0.4974D 06	0.1933D 01
0007	0.2180D−02	−0.5000D−02	0.1261D−03	−0.4928D−04	0.8341D 05	0.5808D 06	0.0
0007	0.3012D−02	−0.4900D−02	0.7031D−03	0.2110D−03	0.8341D 05	0.5808D 06	0.1944D 01

F. LÉORAT ET AL.

	PSI	PHI	PSIP	PHIP	T1	T2	T3
0007	0.3015D — 02	— 0.4832D — 03	—0.7030D — 03	0.21 23D — 03	0.8177D 01	0.5808D 06	0.1002D 02
0007	0.2172D — 02	— 0.3048D — 03	—0.9993D — 04	—0.5112D — 04	0.1020D 02	0.5808D 06	0.2028D 06
0008	0.2826D — 02	— 0.5061D — 02	0.5090D — 04	—0.1000D — 03	0.8660D 05	0.6674D 06	0.0
0008	0.3634D — 02	— 0.5080D — 02	0.7031D — 03	0.1527D — 03	0.8660D 05	0.6674D 06	0.2068D 01
0008	0.3039D — 02	— 0.5839D — 03	—0.7030D — 03	0.1548D — 03	0.8670D 01	0.6674D 06	0.1064D 02
0008	0.2840D — 02	— 0.5849D — 03	—0.9979D — 04	—0.9486D — 04	0.1060D 02	0.6674D 06	0.1929D 01
0009	—0.1499D — 02	0.5000D — 02	—0.1209D — 03	0.6491D — 04	0.4969D 05	0.7171D 06	0.0
0009	—0.2318D — 02	0.4935D — 02	—0.7032D — 03	—0.1913D — 03	0.4969D 05	0.7171D 06	0.1928D 01
0009	—0.2324D — 02	0.4375D — 03	0.7030D — 03	—0.1935D — 03	0.8339D 01	0.7171D 06	0.1017D 02
0009	—0.1496D — 02	0.3724D — 03	0.9976D — 04	0.6522D — 04	0.1033D 02	0.7171D 06	0.1995D 01
0010	—0.7550D — 03	0.5000D — 02	—0.9352D — 04	0.7334D — 04	0.8510D 05	0.8022D 06	0.0
0010	—0.1577D — 02	0.4953D — 02	—0.7032D — 03	—0.1839D — 03	0.8510D 05	0.8022D 06	0.1997D 01
0010	—0.1583D — 02	0.4556D — 03	0.7030D — 03	—0.1860D — 03	0.8403D 06	0.8022D 06	0.1030D 02
0010	—0.7603D — 03	0.4036D — 03	0.9973D — 04	0.7090D — 04	0.1038D 02	0.8022D 06	0.1982D 01
0011	—0.1161D — 02	0.5000D — 02	—0.1128D — 03	0.4643D — 04	0.8331D 05	0.8855D 06	0.0
0011	—0.2004D — 02	0.4891D — 02	—0.7031D — 03	—0.2169D — 03	0.8331D 05	0.8855D 06	0.1997D 01
0011	—0.2006D — 02	0.3939D — 02	0.7031D — 03	—0.2183D — 03	0.8128D 01	0.8855D 06	0.1002D 02
0011	—0.1158D — 02	0.2847D — 02	0.9987D — 04	0.4663D — 04	0.1017D 02	0.8855D 06	0.2039D 01
0012	—0.1783D — 02	0.5000D — 02	—0.9695D — 04	0.5179D — 04	0.8514D 05	0.9707D 06	0.0
0012	—0.2627D — 02	0.4903D — 02	—0.7032D — 03	—0.2121D — 03	0.8514D 05	0.9707D 06	0.2037D 01
0012	—0.2631D — 02	0.4053D — 03	0.7030D — 03	—0.2140D — 03	0.8166D 01	0.9707D 06	0.1010D 02
0012	—0.1787D — 02	0.3038D — 03	0.9993D — 04	0.4983D — 04	0.1020D 02	0.9707D 06	0.2051D 01
0013	—0.2845D — 02	0.5000D — 02	—0.1104D — 03	0.1401D — 04	0.8542D 05	0.1056D 07	0.0
0013	—0.3726D — 02	0.4610D — 02	—0.7031D — 03	—0.2613D — 03	0.8542D 05	0.1056D 07	0.2087D 01
0013	—0.3728D — 02	0.3130D — 03	0.7030D — 03	—0.7030D — 03	0.7766D 01	0.1056D 07	0.9753D 01
0013	—0.2843D — 02	0.1205D — 03	0.9999D — 04	0.1400D — 04	0.9888D 01	0.1056D 07	0.2122D 01
0013	—0.2903D — 03	— 0.5000D — 02	0.1002D — 03	—0.7795D — 05	0.5548D 05	0.1112D 07	0.0
0013	0.6019D — 03	— 0.4791D — 02	0.7031D — 03	0.2713D — 03	0.5548D 05	0.1112D 07	0.2138D 01
0013	0.6047D — 03	— 0.2921D — 03	—0.7030D — 03	0.2727D — 03	0.7688D 01	0.1112D 07	0.9726D 01
0013	—0.2893D — 03	— 0.7969D — 04	—0.9941D — 04	—0.6590D — 05	0.9830D 01	9.1112D 07	0.2142D 01

Appendix VI

Hybrid Simulation

The problem has been programmed on a hybrid high performance computer in order to get a precise qualitative evaluation of the system's performances. The various results are summarized in the following chart and are in quite good concordance with both the predictions of the theoretical study and the numerical results of the digital simulation.

FORTRAN IV   MODEL 44 PS   VERSION 3, LEVEL 1   DATE 69111

```
0001                    DOUBLE PRECISION C0,IXX,OM,OMO,
                        1PSIM,A0,A1,PHIS,ALFA
0002            200   FORMAT (/////10X,'A0 =',E11.4,10X,'A1 =',E11.4,
                        110X,'PSIM =',E11.4)
0003            300   FORMAT (////10X,'PSIM IS THE MAXIMUM
                        1AMPLITUDE OF THE PITCH ANGLE')
0004                    PHIS = 0.5D−2
0005                    IXX = 64.
0006                    OMU0.3125D0
0007                    C0U1.5D−5
0008                    OMO = 0.7268D−4
0009                    A0 = C0/(IXX*OM*OMO)
0010                    ALFA = PHIS/A0
0011                    A1 = (2./3.14159)*(A0*(DARSIN(ALFA)+ALFA
                        1*DSQRT(1. − ALFA**2)) + PHIS*DSQRT(1. −
                        2ALFA**2))
0012                    PSIM = DSQRT(A1**2 + A0**2)
0013                    WRITE (6,200)A0,A1,PSIM
0014                    WRITE (6,300)
0015                    END
```

A0 = 0.1032D−01   A1 = 0.8892D−02   PSIM = 0.1362D−01
PSIM IS THE MAXIMUM AMPLITUDE OF THE PITCH ANGLE

AUTOPILOT COMPATIBILITY OF THE DIGITAL INERTIAL GUIDANCE SYSTEM USED IN THE ELDO 'EUROPA' LAUNCH VEHICLE

A. H. CAIRNS

*Space and Guided Weapons Division, Elliott Space & Weapon Automation Ltd.,
Frimley, Surrey, England*

Abstract.

1. *Introduction.* – The Digital Inertial Guidance System is designed to guide the CECLES-ELDO satellite launch vehicle from the launch site along an optimal trajectory into the required orbit, and, when this has been achieved, to initiate the cut-off of the third stage motors.

This paper is principally concerned with the problems of compatibility between the digital guidance computer and the analogue attitude control loop of the vehicle.

2. *The Inertial Guidance System.* – The attitude of the vehicle is measured by synchros mounted on the gimbals of an Inertial Platform. Incremental velocity components due to external impressed forces are obtained by integration of the outputs of three accelerometers orthogonally mounted on the Platform central cluster.

The incremental velocity components are input to the computer and used to calculate the current velocity and position of the vehicle and to solve the guidance equations to form demanded steering angles. The synchro outputs are converted to a 12-bit digital format and compared, in the computer, with the demanded steering angles. The resulting differences are output in a 7-bit digital format as autopilot commands.

The computer also calculates the time-to-go, and initiates cut-off of the third stage motors at the appropriate instants.

3. *Autopilot Compatibility Tests.* – The Inertial Guidance Computer forms an integral part of the autopilot and attitude control loop of each stage of the vehicle. The introduction of this digital element is inherently accompanied by both time and angular quantisations.

An updating frequency of 80 Hz was selected for the Autopilot outputs from the guidance Computer. This frequency was chosen as substantially above the significant vehicle harmonics while remaining compatible with the speed of computation of the full control equations on the 920 M Computer.

The 12-bit binary format for the angular inputs to the Computer was selected on the basis of noise free resolution available from contemporary synchro designs. This is typically of the order of 360×2^{-11}. The output decoder bit-size was derived from the input bit-size and the total range of $\pm 10°$ that is required.

The effects of quantisation upon the control stability of each of the three stages were examined by a series of Autopilot compatibility tests performed in conjunction with the appropriate stage design authorities. The Guidance Computer and the associated analogue-to-digital and digital-to-analogue converters were introduced into analogue simulations of each stage Autopilot system.

The 80 Hz control iteration rate was found to be satisfactory for all stages, and the effects of angular quantisation upon the second and third stages were shown to be negligible. Under certain conditions, however, the first stage was found to be susceptible to excessive limit cycling due to angular quantisation.

Close examination of the first stage control problem showed that under ideal conditions of quantisation at the specified magnitudes, the limit cycling was maintained below an acceptable level.

In practice, however, it was found that the characteristics of the synchro-to-digital conversion of angular inputs to the Computer varied with time and temperature. In the worst case this causes an effective doubling of the angular quantisation resulting in a region of local instability at the control loop origin.

Three alternative solutions to the problem were examined:

A. *Dither.* The application of an oscillatory signal to the attitude outputs prior to digital conversion results in a digital waveform, the time average of which forms an interpolated measure of angle. Provided the dither frequency is reasonably high (> 30 Hz), the necessary smoothing is inherently provided by the Autopilot. A dither frequency of 112 Hz was finally selected to eliminate the possibility of low-frequency beating with the 80 Hz sampling frequency.

G. A. Partel (ed.), Proceedings of the Second International Conference on Space Engineering. All rights reserved.

The dither is theoretically capable of very high resolution, and was demonstrated by simulation to provide a completely satisfactory solution to the first stage control problem.

The principal disadvantage of the dither solution was the extent of engineering modification required for its adoption.

B. *Input Characteristic Inversion*. Input characteristic inversion is a method of regularising the bit-size of the angular inputs to the Computer. The characteristics of the synchro-to-digital converter are deliberately inverted on successive samples. This results in an effective halving of the least significant bit-size.

The engineering modifications required for this solution are significantly less than for full dither, and the control behaviour of the first stage was shown to be marginally acceptable with this solution.

C. *Deadzone*. The Deadzone method was developed as a purely software solution to the first stage control problem. Modifications are required only in the computer program.

Although the input quantisation characteristic cannot be controlled by the Computer, the overall transfer function of angular commands may be readily modified by program. The introduction of a narrow Deadzone at the origin provides a stable region even under the worst quantisation conditions. The control behaviour of the first stage was shown to be improved in preliminary tests of this method.

The Preferred Solution. – Despite the extent of engineering modification necessary, dither was adopted for use in the Inertial Guidance System.

The additional modifications to minimise the effects of quantisation of Autopilot commands are also described in the paper.

1. Introduction

The development of an Inertial Guidance System for the European Launcher was agreed upon at the ELDO Ministerial Conference in 1966 as part of the Supplementary Programme of work. The system was required to replace the existing Radio Guidance System which was unsuitable for launches from the new equatorial base at Guyana. It was further agreed that contracts for the Supplementary Programme should be placed directly with industry by ELDO instead of the previous arrangement whereby the governments of Member States acted as agents for ELDO in this matter.

In August, 1966, Elliott were awarded a study contract to define the Inertial Guidance System and to produce a detailed development cost plan and technical proposal. As a result of this study, Elliott were appointed Prime Contractor for the development of the Inertial Guidance System on the 1st January 1967.

Under the terms of this contract, work on the Inertial Guidance System was distributed as follows:

Prime Contractor
Elliott (U.K.)   project management
                                             overall system design
                                             digital computer

Sub-Contractors
CGE-FIAR (Italy)   power supply unit
Ferranti (U.K.)   inertial platform and
                                             associated electronics
Van der Heem (Netherlands)   Interface Unit
BGT (Germany)   field test equipment
                                             for inertial platform
NLR (Netherlands)   system test facilities

In Collaboration
RAE (U.K.)   guidance laws
                                                        advisers to ELDO

2. The Inertial Guidance System

The Digital Inertial Guidance System is required to guide the Europa Launch vehicle from the launch site along an optimal trajectory into the required orbit [1].

The vehicle is steered by deflecting the motors of the currently active stage in an appropriate manner to cause the required vehicle angular motion. The actual motor deflection required at any instant is formed in the Autopilot in response to commands from the guidance system for a change from the current vehicle attitude.

To fulfil its task therefore the guidance system must be able to:

(a) measure the current Vehicle Attitude

(b) calculate the required current Vehicle Attitude

(c) output any differences as commands to the Autopilot for a change in attitude.

In the Inertial Guidance System the current vehicle attitude is measured by synchros mounted on the gimbals of the Inertial Platform. The vehicle acceleration is measured by accelerometers mounted on the central cluster of the Platform.

This information is input via the Interface Unit to the digital computer [2] and is used by a Navigation Program to calculate the current position and velocity of the vehicle. The required current Vehicle Attitude is formed in the computer either as an explicit function of time during 1st Stage flight, or using a guidance law expressed as a function of current velocity and position during 2nd and 3rd stage flight.

The attitude errors from the required value are resolved into demands for rotations about the vehicle pitch, yaw and roll axes, which are then output to the Autopilots via the Interface Unit. (Figure 1.)

3. Digital Quantisation

The Inertial Guidance System forms an integral part of the attitude control loops of all three stages. The Autopilot systems were already at an advanced state of development when the decision to introduce the Inertial Guidance System was taken. Significant changes in Autopilot design to accommodate the new digital elements in the control loop were thus impractical and would have introduced undesirable delays to the overall programme.

The Inertial Guidance System was therefore required to present a control loop interface as similar to that of the existing guidance system as was practicable.

Information must clearly be presented to a digital computer in a digital format. The conversion of essentially analogue data to a digital equivalent inherently introduces quantisation, and in a practical system the degree of quantisation is dependent upon the wordlength of the analogue to digital converter. The wordlength (or number of binary bits) must be sufficiently large for the total range of the input data to be

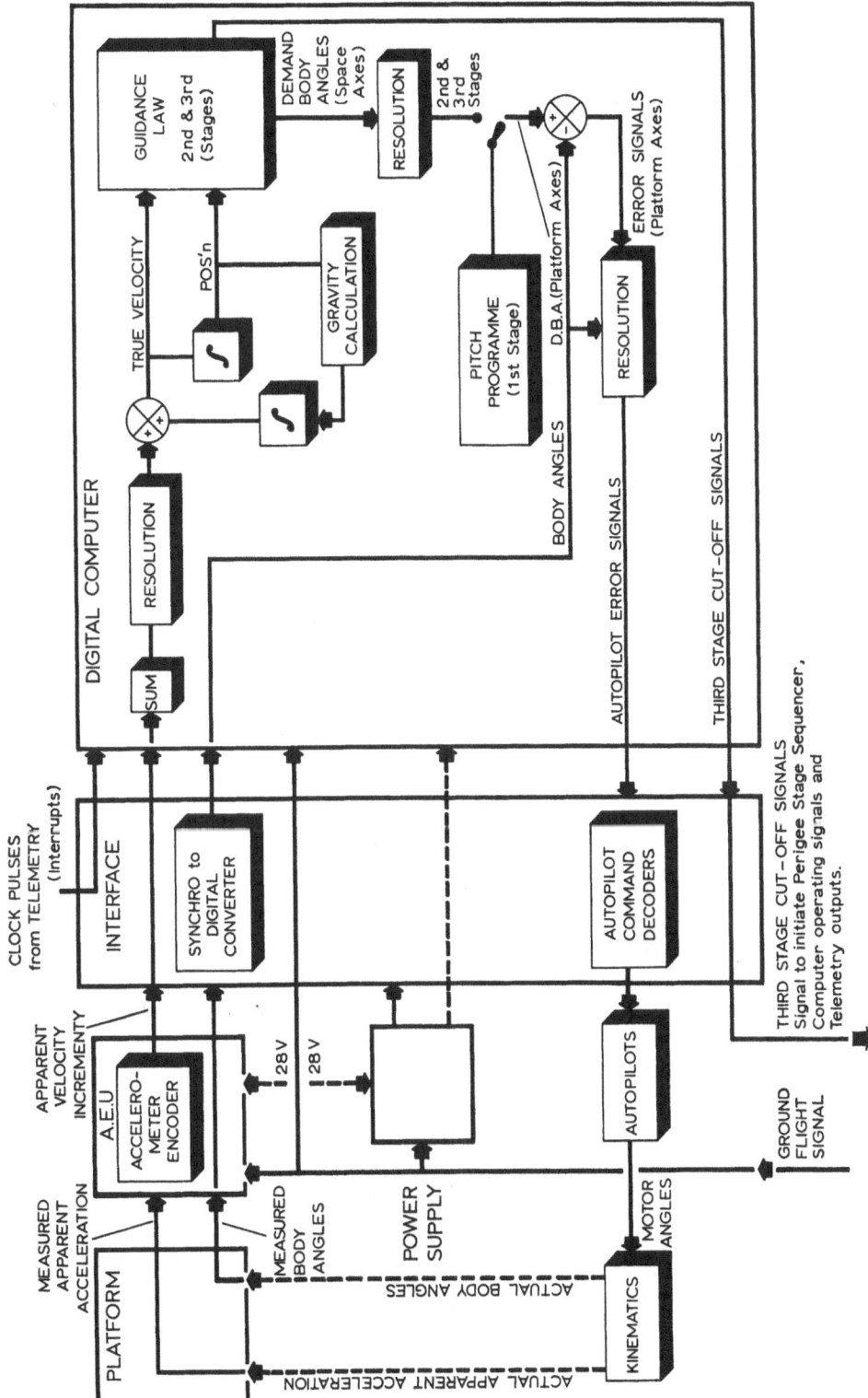

Fig. 1. The inertial guidance system.

represented to sufficient accuracy. The complexity of the converter, however, increases with the wordlength, and a compromise between converter accuracy and complexity must be made.

The conversion of digital data to an analogue equivalent is similarly quantised to a degree dependent on the bitsize of the digital input to the converter.

In the case of the Inertial Guidance System, quantisation in the attitude control loop is introduced at the Gimbal Angle inputs to the computer and at the Autopilot Command outputs from the computer. (Figure 2.) The selection of bit sizes for these

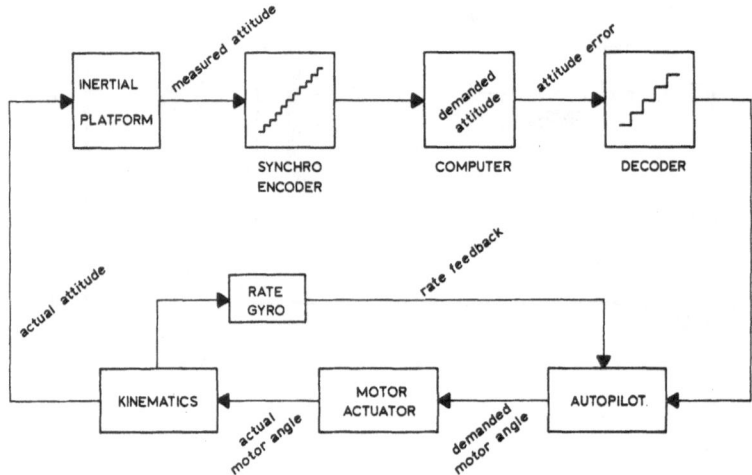

Fig. 2.   Simplified attitude control loop.

two conversions was made taking into consideration the angular resolution available from the platform synchros. Contemporary synchro designs are typically accurate to 5 min. of arc. Random noise may be considered to raise this figure by a factor of 2 giving a resolution of 10 min. of arc. Assuming that full four quadrant coverage (360°) is required, this resolution would necessitate a binary representation of eleven-bits. To include a safety margin, an encoder word-length of 12-bits was therefore selected. This gives a least significant bit representation of 0.09° on the angular input data to the computer.

Calculations are performed in the computer to an accuracy of 18-bits, and may therefore be considered to be of negligible quantisation compared with the data source.

The output decoder bit-size was also selected taking account of the available accuracy of data. In this case the available resolution was determined by the encoder, and no advantage would be gained from having a least significant bit-size less than 0.09°. The decoder was required to cover a range of ±10° and therefore an output wordlength of 7 bits was selected giving a least significant bit representation of 0.16°.

A second, very important source of quantisation is implicitly introduced by the serial processing of a digital computer. Information can only be updated at discreet intervals and therefore the measurement of time is quantised.

The computer calculations must be synchronised to real time, and in the Inertial Guidance System, this is achieved by use of the computer interrupt facilities. Four Priority Levels are available in the 920M and these are allocated as follows:

Level 1	Telemetry Output Program
Level 2	Control Loop Program
Level 3	Navigation and Guidance Program
Level 4	Self Check Program.

Entry to the 3 upper Priority Levels is performed in response to interrupt pulses externally applied to the computer. Upon receipt of an interrupt pulse one cycle of the appropriate operation is performed.

Accurate synchronisation of the various operations may therefore be achieved by deriving interrupt pulses from a master clock. In the Europa Launcher a suitable accurate source of timing pulses is provided by the telemetry clock.

The iteration rates of the control loop program operating on Priority Level 2 and of the Navigation and Guidance program on Priority Level 3 are directly related to the degree of time quantisation introduced into the overall Vehicle Attitude Control Loop. These iteration rates define the updating intervals of the information computed in each program.

The Level 3 program iteration rate was selected from the requirements of the integration routines used for Navigation. A detailed accuracy analysis showed that Navigation errors were negligible for iteration rates greater than 2.5 cps. A rate of 5 cps was selected as consistent with the Navigation requirements and with the computation speed of the 920M Computer.

The Level 2 program directly affects the stability of the control loop, and an iteration rate higher than any significant loop harmonics is necessary. A rate of 80 cps was therefore selected for this program.

The various quantisation levels introduced by the digital Inertial Guidance System may therefore be summarised as follows:

Attitude Inputs	$0.09°$
Autopilot Commands	$0.16°$
Control Rate	12.5 msec
Guidance Rate	200.0 msec.

4. Autopilot Compatibility Tests

The detailed effects of these levels of quantisation upon the attitude control loops of the vehicle are not readily predictable. Theoretical analysis of the probable effects gave inconclusive and often conflicting predictions of performance.

It was therefore necessary to carry out an extensive programme of simulation testing to evaluate the compatibility of the digital system with the remainder of the attitude control loops of each of the three stages.

The compatibility tests were carried out by the respective Autopilot design authorities in close co-operation with Elliott. In each case representative Inertial Guidance units were connected to a detailed analogue simulation of the appropriate stage attitude control loop. This required the construction of synchro simulator units to convert the analogue d.c. outputs representing vehicle attitude to a 3-wire synchro form suitable for input to the synchro encoders.

During the initial testing of the open loop characteristics it was observed that the digital encoders exhibited a tendency to change by 2 quantisation steps as the input was varied. Close examination showed that this effect varied with temperature from one extreme of two-bit 'odd' quantisation through the nominal one-bit quantisation to the other extreme of two-bit 'even' quantisation (Figure 3).

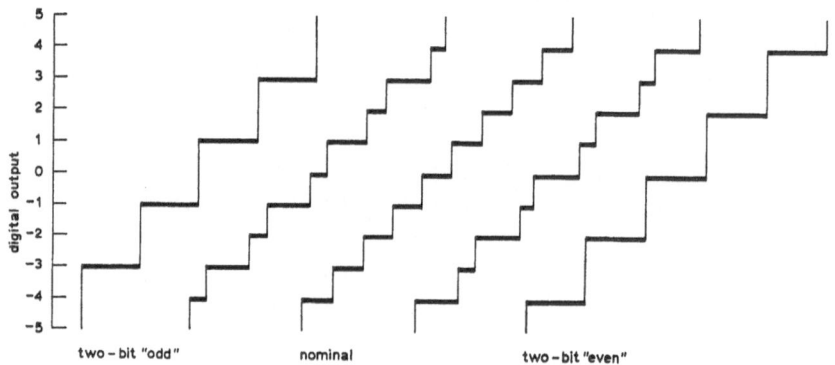

Fig. 3.   The variations of the encoder quantisation characteristics.

This effect is due to the method of synchro-to-digital conversion used in the encoder. Incoming signals are converted to two fixed amplitude waveforms, the phase separation of which is proportional to the synchro angle. These waveforms are used to trigger zero crossing detectors which gate a free running pulse train into a binary counter. A small change in input angle results in a differential phase change of BOTH waveforms relative to the pulse train. The situation can therefore arise where a small change in angle will advance the first zero crossing point sufficiently to admit one extra pulse and at the same time retard the second zero crossing point sufficiently to also admit an extra pulse to the counter. The digital angle, represented by the counter output, will therefore tend to change in steps of two-bits.

From the attitude control loop point of view, this effect not only doubles the degree of angular quantisation but also increases the possibility of local instability at the origin. In the extreme characteristic (2-bit 'odd'), the digital value +0 does not exist and for a nominally zero input, the digital angle will oscillate between −1 and +1. Compatibility tests of the I.G. System with the first stage simulation showed that with the nominal specified angular quantisation characteristics limit cycle amplitudes remained below the maximum acceptable value.

Under the 'two-bit' quantisation conditions however, the control loop behaviour was seriously degraded. The Rigid body and Propellant Sloshing limit cycle amplitudes became unacceptable.

A detailed account of the behaviour of the vehicle control loop under these conditions is given in [3]. From the Inertial Guidance System point of view, however, it was concluded that the quantised input-output characteristics of the system were the source of the problem and, as such, formed the area in which the problem could most readily be solved. It should be noted at this point that the performance of the 2nd and 3rd stage control loops was found to be entirely satisfactory even under the worst case conditions of 2-bit quantisation.

5. Alternative Approaches to the Problem

In assessing the various possible approaches to the quantisation problem the relative timescale implications of each modification to the system required careful consideration.

From this point of view the most attractive solution would require modifications only to the computer program. These could be incorporated with very little delay and could be further modified and improved equally readily. Solutions based on hardware modifications, however, almost inevitably introduce a significant delay to the overall delivery schedule, and therefore require a corresponding development lead time.

The most direct solution would appear to consist of explicitly increasing the wordlength of both the encoders and decoders by hardware modifications. This, however, would have required a radical change of design of the Interface Unit and it was ruled out as completely impracticable in the time available.

The logical extension of the extended wordlength solution to the problem is to obtain the same increase in resolution by implicit methods. This may be achieved by interpolation between the existing quantisation steps on a time averaging basis, and this technique appeared to form the basis for a practical solution of the quantisation problems due to both the encoder and the decoder.

An alternative possible solution was suggested by considering the quantisation effects as causing a more classical instability problem. As such the problem should be amenable to solution by a modification in the digital computer to the overall loop gain over the region of instability. This approach formed the basis for a second possible solution to the quantisation problem. The investigations of the solutions to the problem based on interpolation methods will be considered first.

6. Interpolation of the Decoder Quantisation

To enable interpolation to be carried out, a proportional time distribution of the outputs from the quantiser is required.

In the case of the Autopilot Command Decoders, the quantiser output is directly controlled by the digital computer. The output data is potentially available to the full 18 bit wordlength computer accuracy. In the basic system, the least significant part of the output data was rounded to the nearest decoder least significant bit-size before output.

A relatively straight-forward modification to the computer program was evolved whereby the residual part of the previous sample of output data was added to the current sample, before output to the decoder. The decoder outputs, occurring at 80 Hz, were thus modulated in direct proportion to the full, 18-bit accuracy, output data. The autopilot input filters provided the necessary smoothing to establish a mean interpolated d.c. level related to the workspace ratio of the modulated output.

This modification was shown to be completely effective in eliminating quantisation effects due to the Autopilot Command decoders and it was therefore adopted as a standard feature of the Inertial Guidance Computer program.

7. Interpolation of the Encoder Quantisation

The elimination of the Autopilot Command decoders as an effective source of quantisation reduced the overall problem to one source: the synchro encoder. In the case of the synchro encoder it was apparent from laboratory testing that the digital output to the computer contained a component due to system noise. The relationship of the time averaged output noise distribution with the input to the encoder was examined for synchro position increments of 1 arc min over a test range of 30 arc min. The correlation of averaged digital output with input angle was found to be poor, and it was concluded that the inherent system noise in the laboratory environment was of too low an amplitude to provide an effective basis for interpolation.

8. Encoder Interpolation with Dither

Theoretical considerations of the time averaging method of interpolation, indicated that the principle of using an additional waveform superimposed on the basic input to the encoder was entirely practicable provided that these components were of sufficient amplitude.

The non-linear transfer function of the quantiser transforms the zero time average of the input waveform to a non zero-time average component on the quantiser output; the magnitude of which is directly related to the displacement of the static component of the encoder input. (Figure 4.)

This was confirmed by exploratory tests carried out with the first stage simulation. The d.c. to synchro converter was modified to permit a dither signal to be added to the synchro outputs. Provided that this signal is of a reasonably high frequency, the time-averaging effect is provided by the inherent smoothing of the Autopilot. The tests demonstrated that the introduction of dither was accompanied by a reduction of limit cycle amplitudes to acceptable levels.

An investigation of the relative effects of different dither waveforms was then carried out. It was shown by theoretical analysis, that a triangular waveform is required to give a linear interpolation with an ideal encoder. The use of a sinusoidal waveform,

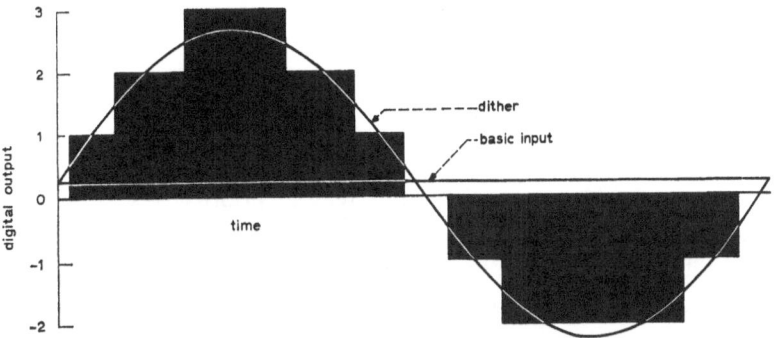

Fig. 4.   Nominal quantisation of sinusoidal dither superimposed on a basic input of ¼ bit.

is accompanied by a small sinusoidal residual component on the interpolated output, while a squarewave is theoretically only capable of halving the quantisation.

Practical tests of these three waveforms, however, indicated that an acceptable first stage performance was achieved with all three. A sinusoidal waveform was therefore selected on engineeiing grounds.

In selecting the optimum frequency of the dither waveform consideration was given to possible harmonic interaction effects with the computer, 80 Hz sampling frequency. The various significant vibration modes of the vehicle during first stage flight cover the frequency range 0–30 Hz. To avoid harmonic interaction occurring in this range a frequency of 112.5 Hz was selected as a convenient binary submultiple of the 2.4 kHz synchio excitation frequency.

The optimum dither amplitude may be readily found by analytic methods for an ideal quantiser by consideration of the variation of interpolation distortion with amplitude.

The deviations of the practical encoder from the ideal characteristic, however, required a more empirical approach and a dither amplitude of 9 arc min. of synchro angle was selected as a result of simulation testing with the complete range of known encodei characteristics.

9. Hardware Modifications for Dither

The development tests of the dither solution were carried out using a standard laboratory oscillator as the dither source. A transformer network was used to mix the dither signal with the synchro signals to the encoder.

Considerations of space and weight indicated that the practical dither source should be an integral part of the Interface Unit. An evaluation of practical methods of generating and mixing dither in the Interface Unit was therefore carried out. As a result of this investigation, a frequency dividing circuit, operating from the 2.4 kHz synchro excitation supply, was developed to provide the 112.5 Hz dither source.

The dither signal must be used to disturb the synchro input angles. In practice the most convenient method of achieving this effect was found to consist of using dither to modify the zero point of the zero crossing detectors in the encoder. This provided a differential phase motion between the two waveforms equivalent to an angular oscillation.

10. Encoder Characteristic Switching

The effectiveness of dither as a total solution to the problems due to digital quantisation was apparent from early tests, and for this reason, priority was given to detailed investigations of this method.

As a result of the hardware design study for the incorporation of dither into the Interface Unit a possible alternative solution to the quantisation problem was also evolved.

As has already been stated, the first-stage attitude control loop performance appeared to be satisfactory under the nominal level of encoder quantisation. It was found that a simple modification to the encoder permitted the characteristics to be deliberately switched on alternate samples. This meant that the time average of several samples tended to conform to the ideal encoder characteristic. The necessary modification was made to the Interface Unit and a set of simulation tests was performed.

The results of these tests showed a significant improvement in attitude control loop performance. Limit Cycle Amplitudes were reduced to acceptable levels in all but a few extreme cases, and it was felt that this method although not as effective as dither could well provide an economic solution to the problems due to encoder quantisation.

11. The Deadzone Solution

The two methods of reducing encoder quantisation effects described so far both require hardware changes in the system. Both methods are furthermore based on the principle of directly improving the quality of the input data used by the computer.

It was recognised from elementary considerations that the characteristics of the synchro encoder could not be improved by numerical processing within the computer. Information below the level of input quantisation was irretrievably lost.

In seeking a purely software solution, the effects of encoder quantisation upon the stability of the attitude loop gain were considered. The contribution of the digital computer to this gain is directly related to the local transfer function as implemented by the computer program.

Examination of the early simulation results, assuming a zero demand showed that the limit cycle amplitudes became unacceptable as the encoder characteristics tended towards the two-bit 'odd' type. In this case the binary input value zero ceases to exist, and over a small angular range the effective loop gain becomes very large.

Although the encoder characteristics cannot be directly modified by software, the overall input-output transfer function of the computer may be readily changed.

A small dead-zone was therefore deliberately introduced in the transfer function (Figure 5) such that, even under unfavourable encoder conditions, the gain near the origin was reduced. The use of the output-modulation technique already described

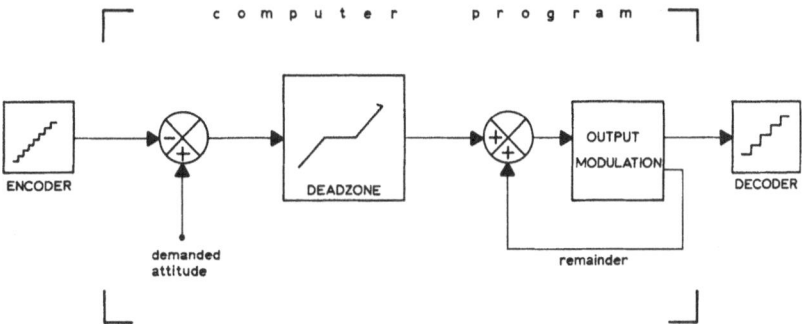

Fig. 5.   Computer program block diagram incorporating a deadzone and output modulation.

to overcome decoder quantisation enabled the dead-zone width to be varied in very small increments.

Simulation tests were carried out with the worst case encoder characteristics for a range of dead-zone widths from approximately ± 1 to ± 9 min of arc.

The results of these tests were encouraging; as the dead-zone width was incieased, the limit cycle amplitudes decreased. A dead-zone width of $\pm \frac{1}{2}$ nominal encoder bit ± 5 min of arc) was found to give marginally acceptable limit cycle amplitudes.

A further development of the dead-zone method was required to accommodate non-zero mean inputs. A long time constant lag was used in the digital computer to establish the mean vehicle attitude and thus the position of the dead-zone in the computer transfer function. The dead-zone method, however had one significant disadvantage. Under the opposite extreme quantisation chaiacteristic, the combined effects of 'two-bit even' quantisation and the dead-zone resulted in a significant region over which the attitude control loop was closed on rate terms only. The vehicle attitude could thus only be controlled to the accuracy of this increased deadzone.

12. The Preferred Solution

In order that the necessary modifications to the Inertial Guidance hardware could be completed in time foi the F9 firing in 1969, a decision on the preferred solution to the quantisation problem had to be made at least one year in advance.

A review of the merits of the three alternative methods showed that full sinusoidal dither provided a completely effective solution under all conditions. Considerably more time and effort had been spent on the full dither investigation than on either of the alternative solutions to the problem, and although dither required the most extensive hardware modifications of the three solutions it was decided to adopt this method as the preferred solution.

The encoder characteristic switching method gave good results except under the most extreme conditions. It was felt, however, that the risk involved in using this solution, although small, was not acceptable.

The dead-zone solution to the problem was potentially the most attractive of the three alternatives as no modifications to the hardware were necessary. It was felt, however that the method was not sufficiently developed at the time to form a practical basis for an effective solution to the problem.

13. Conclusion

The problems of compatibility of the Inertial Guidance System with the First Stage Autopilot of the Europa Launcher provide a good illustration of the complexity of secondary effects of mixing digital and analogue elements in a single control system.

The development of the small general purpose digital computer has not been accompanied by a corresponding development of digital sensors and actuators, and the type of problem described in this paper will require careful consideration in mixed system designs in the immediate future.

The dither method is not the only practical solution, purely software methods appear to be feasible and the use of more sophisticated data handling techniques than those briefly mentioned in this paper may well provide a satisfactory and practical solution to the problems arising from quantisation. The use of dither, however, does provide a relatively straightforward and efficient method of data interpolation.

References

[1] Riley, A., 'The Inertial Guidance System for the Europa Satellite Launcher', Elliott, 1968.
[2] Cairns, A. H., 'The Elliott MCS 920M Computer with Particular Reference to its Application to the Inertial Guidance of the CECLES-ELDO Launcher', Elliott, 1968.
[3] Ewert, D. G., 'Development of the Europa I First Stage Attitude Control System to the Europa II Requirements', Hawker Siddeley Dynamics, 1969.

DEVELOPMENT OF THE 'EUROPA I' FIRST STAGE ATTITUDE CONTROL SYSTEM TO THE 'EUROPA II' REQUIREMENTS

D. G. EWART

Hawker Siddeley Dynamics, England

Abstract. Whilst the 'Europa II' launch vehicle is essentially the 'Europa I' vehicle with additional, solid propellant, perigee and apogee stages, certain system modifications and improvements in design have been adopted, together with the introduction of an inertial guidance system. The effects of these changes upon the design and performance of the first-stage attitude control system have been to raise problems requiring detailed re-examinations of the attitude control system. The investigations that were required are detailed together with the consequent engineering changes that were required.

1. Introduction

According to the original ELDO (European Launcher Vehicle Development Organisation) target plan Blue Streak, the first stage of the three stage Europa vehicle, was to have reached its final operational standard at the F7 launch vehicle – that is after 7 flights of which the first 3 were to be first stage only flights. According to the latest plans the operational first stage standard will now be reached with the F11 vehicle. The delay in reaching the operational standard has arisen not because the first stage is raising obdurate problems but because of the coincidence of a number of separate reasons. The main reasons are:

(1) The original target standard did not exploit the full performance capability of the first stage – subsequent studies showed that this could be achieved at the expense of a few simple modifications which will be introduced in vehicle F9.

(2) Many of the first stage electronics systems were designed in the period 1955–1960 and difficulties arise in obtaining components now becoming obsolescent – the electronics have therefore been re-designed using modern components and increasing reliability and flexibility and the new standards of equipment will be introduced in F9. Minor propulsion system component improvements will also be introduced by the F11 vehicle launch.

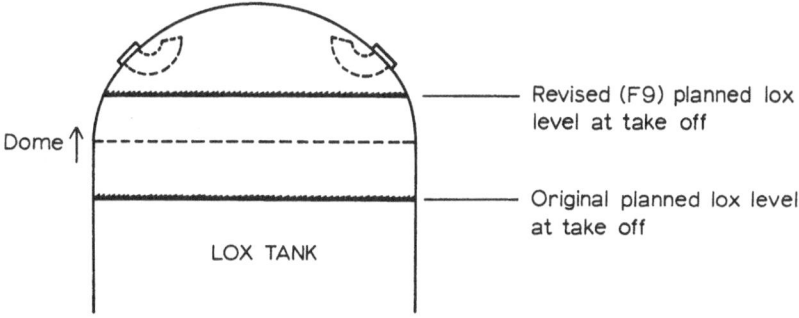

Fig. 1. Lox tank filling levels.

G. A. Partel (ed.), Proceedings of the Second International Conference on Space Engineering. All rights reserved.

(3) The Europa vehicle was originally designed for launch from Woomera under radio guidance. Launching of F11 and subsequent rounds will be from the Guyane launch site under inertial guidance with the guidance computer providing attitude error signals to the first stage attitude control system.

(4) In the course of the flight trials various minor modifications have been found to be desirable to improve the preparation procedures.

The purpose of this paper is to discuss the effects of these changes upon the design and performance of the first stage attitude control system.

2. Effects of Increasing First Stage Propellant Levels at Take-Off

2.1. In the original Europa I overall specification design requirements at lift-off were set as:

Take-off thrust	1 350 kN (300 000 1bf).
Take-off acceleration (apparent)	1.3 g.

These figures implied a vehicle mass, at take-off, of 104 700 kg and performance analyses lead to a required first stage propellant mass (Lox + Kerosene) at take-off amounting to 85 500 kg. The first stage, which in fact was already at an advanced stage of development, was designed to have a take-off propellant mass capacity of 89 600 kg and was thus to be under-filled in its role as the first stage of Europa I.

2.2. The take-off 'G' requirement of 1.3 g was set to allow the launcher release mechanisms already built for Blue Streak to be used for Europa I without modification although performance analyses for Europa I showed that it would be beneficial to increase the first stage propellant mass and reduce the take-off 'g'.

2.3. As a result of testing of the release gear, on simulated launchings, it was found that by minor adjustments the launcher would operate satisfactorily at take-off levels as low as 1.15 g if adverse tolerances of -0.05 g occurred and thus the nominal take-off level could be reduced to 1.20 g allowing an increase in vehicle take-off mass of 8 700 kg.

2.4. Under the original first stage filling proposals the nominal Lox mass at engine start (7 sec approx. before take-off) was 57 200 kg and the Lox surface level was then 14.8 cm below the Lox tank top dome leaving an ullage volume of 2.07×10^7 cm^3. When detailed work began to increase the first stage Lox mass at take-off it was found that, after making certain modifications to Lox tank equipment to prevent Lox leaking through the tank vent valves when the Lox level at engine start was raised, an additional 1.98×10^7 cm^3 (6700 kg) of Lox could be introduced. This had the effect, as shown in Figure 1, of raising the Lox level at take-off by 104 cm so that this level now lay within the lox tank dome instead of in the cylindrical part of the tank. The reduction in the diameter of the liquid surface at take-off had the effect of changing the Lox sloshing characteristics and this required a re-examination of the attitude control system stability in the presence of propellant sloshing at take-off.

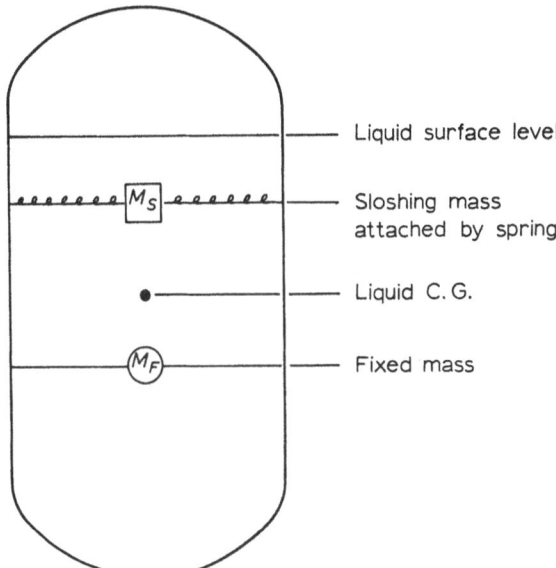

Fig. 2. Propellant tank sloshing simulation.

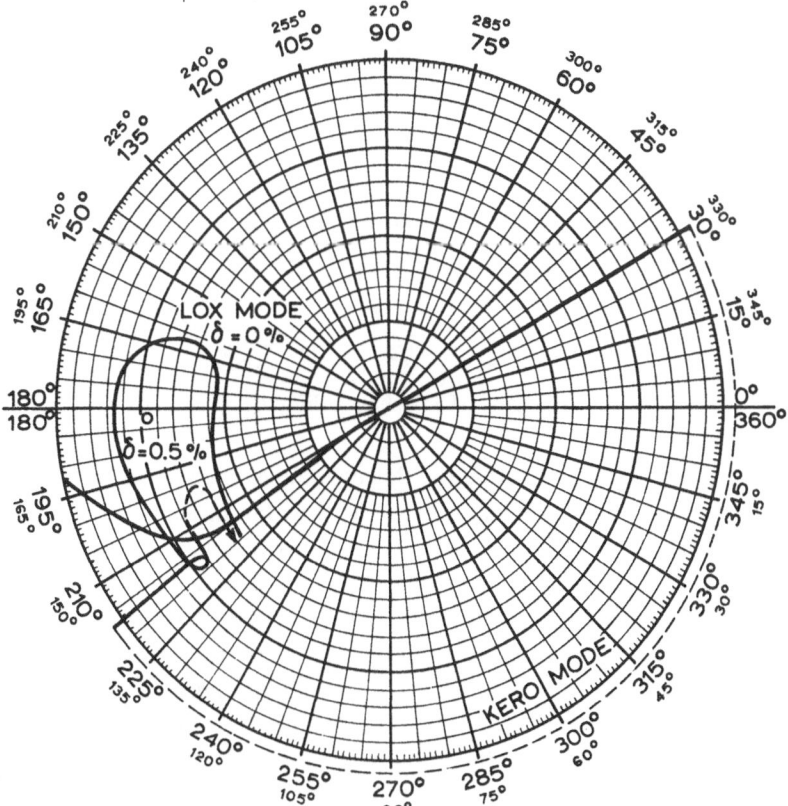

Fig. 3. F9 sloshing instability at lift-off.

2.5.   As described elsewhere each propellant tank may be represented, for sloshing stability analyses, by a mass fixed to the vehicle at a certain point plus a mass attached by a spring at another point (Figure 2). The ratios of the masses, their points of action and the spring characteristics were defined by the tank geometry and the liquid level. For take-off Lox levels lying within the cylindrical portion of the tank

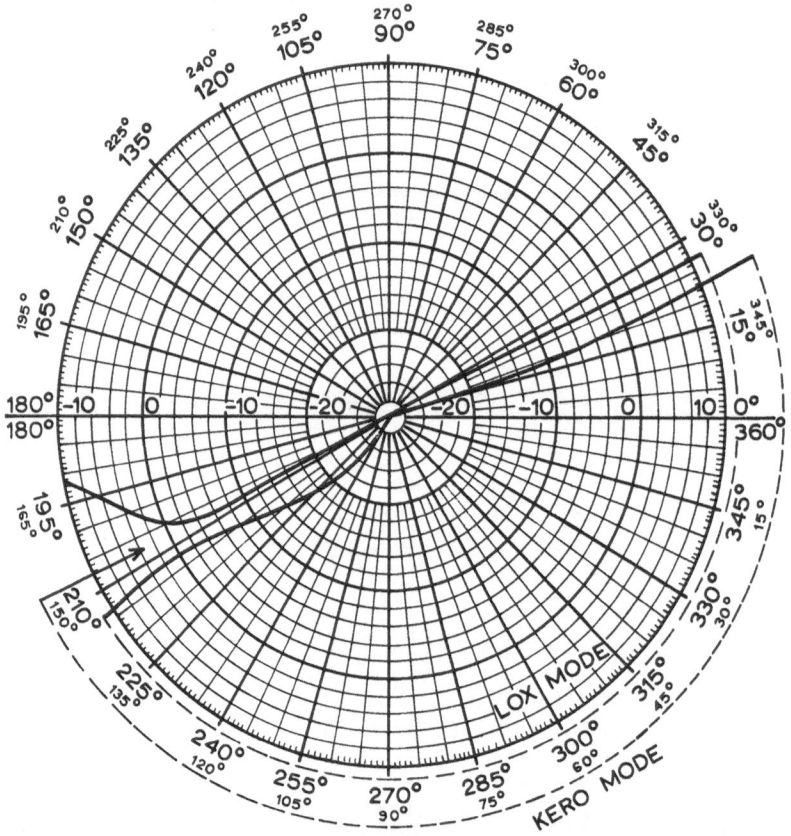

Fig. 4.   F8 sloshing at lift-off.

previous analyses had confirmed that the sloshing modes were phase stable at take-off and thus raised no problems. When the new level within the tank dome was investigated it was found that as a result of the decreased liquid surface diameter the sloshing frequency was higher and that whilst the level lay in the dome the sloshing mode was phase unstable (Figures 3 and 4). In the absence of any slosh damping the pitch and yaw loops were potentially unstable at take-off for the standard Europa I values of forward patch electronic gain. The instability rapidly disappeared as the liquid level fell and was also reduced if the electronic gain was reduced or if any viscous damping arising from liquid motion over the dome surface was present. Since various tests made in the UK and in France had shown that about 0.2% critical damping could be expected from viscous friction and the capability existed in the first stage autopilot electronics to change the electronic gain in the course of flight this problem of sloshing instability at take-off could be cured without any hardware modifications and the full Lox tank fillings could be exploited.

3. Introduction of New Design of Autopilot Electronics Unit

3.1. One of the difficulties experienced with the Europa I vehicle is that the building of the first stage and its systems is often complete a long time before launch and before the build of the other stages is complete. This has meant that building of the first stage control system electronics units (autopilot) was completed before the final overall vehicle characteristics were definitely known. In addition changes to autopilot electronics design could not be made to allow for changes in actual from assumed vehicle characteristics without incurring long delays to launch. Fortunately the variation in vehicle characteristics from the standard assumed when the autopilot was designed have been such that the design criteria for stability margins have always been satisfied but it has not been possible to adjust autopilot parameters to their optimum values for each vehicle.

3.2. An additional factor experienced was that apart from lacking flexibility to change parameter values without causing long delays the build of the autopilot was founded on using components and techniques that were becoming obsolescent and of lower reliability than was desired.

3.3. In view of the above factors the decision was taken to re-design the autopilot electronics units for the operational vehicle and in the redesign to provide the degree of flexibility previously lacking so that selected parameters could be adjusted when the final characteristics of each vehicle were established. Provision was therefore made for each parameter – gain, time constants, etc. – to have a range of discrete values available, for certain of these parameters to be capable of variation at regular intervals during flight, for the circuit boards controlling the time variation programmes to be easily accessible and for the time variation programmes to be easily changed. The decision was also taken in the re-design to allow for the future introduction of new extra inputs to the autopilot should they be required for any successors to the Europa I and II vehicles which retained the existing first stage. As examples of the degree of flexibility now available

(1) The forward path electronic gain in pitch may now take any of 16 values lying between .3 deg/deg and 2.0 deg/deg and may be changed every 8 seconds during flight instead of being restricted to the fixed programme:

$$
\begin{aligned}
0 < t < 90 \quad & 1.43 \\
90 < t < 105 \quad & 1.01 \\
105 < t < 120 \quad & 0.802 \\
t < 120 \quad & 0.506.
\end{aligned}
$$

(2) The rate time constant instead of being fixed throughout flight at 0.5 may now take any of 16 values in the range 0.4 to 0.6 and may be changed every 8 sec during first stage flight.

(3) The 2nd order filter damping factor remains fixed throughout flight but instead of being fixed at 0.5 for all flights may now be set to any of 8 values in the range 0.4 to 0.7.

3.4. The new Operational Standard Autopilot (OSAP) will be introduced in vehicle F9. The immediate advantages are clearly demonstrable – to counter the F9 take-off sloshing problem described above the pitch and yaw gains may be reduced until 8 sec after take-off and then increased – a facility and means of problem solution that could not be considered on any previous vehicle.

4. Attitude Reference System Alignment Procedure

4.1. One of the critical periods of flight for the Europa vehicle is, in fact, the first few metres of vertical motion after release. At the launch site there are several items that lie very close to the vehicle – parts of the launch release mechanism, water deluge pipes, umbilical cable mast, etc. These items place tight limits on the permissible lateral and rotational motions of the vehicle during the initial vertical rise in order to prevent any collisions.

4.2. To satisfy these requirements it is necessary to ensure that the first stage engine gimballing deflections immediately before release do not exceed a given limit (0.5°) in pitch or yaw and a device is fitted to the launcher, the Motor Error Limiter (MEL) to measure the actual deflections and initiate STOP action if the limit is reached. The motor deflections may arise from a number of sources, namely:
(1) Attitude Measurement System Zero Errors and inherent drift.
(2) Vehicle misalignments
(3) Steady Wind deflections of the vehicle
(4) Vortex shedding induced oscillations of the vehicle
(5) Drift and zero errors in the autopilot and motor actuators.
(6) Earth rotation effects on space stabilised attitude measurements system.
(7) Accuracy of the MEL system.

4.3. The limiting motor error angle is set at 0.5° and after allowing for items (3) to (7) it is calculated that the attitude error signals input to the autopilot arising from items (1) and (2) must lie within the following ranges:

Pitch −0.059 to +0.05°
Yaw   −0.020 to 0.129°
Roll −0.29 to +0.29°.

4.4. The attitude error signals for all stages of the vehicle are derived from a 4-axis space stabilised attitude reference platform (ARP) mounted in the third stage. Before the third stage is placed in position the ARP is carefully aligned to the third stage and adjusted to remove zero error outputs and the inherent gyro shift of the ARP is negligible. Thus after erection of the overall vehicle in the tower protecting the vehicle from winds there will only be attitude error signal outputs if the third stage axes are not correctly aligned to the first stage. If the error signals lie outside the permitted ranged quoted above then some corrective action is necessary to reduce the error signals to lie within the permitted bounds.

4.5. On vehicles F6/2 and F7 – the first two vehicles carrying this ARP system in an executive role – the correction procedure adopted was to change the alignment of the platform axis to the third stage axes – by moving the platform cradle relative to its mounted points – until the error signals received by the first stage were acceptable. By these means the danger of triggering the motor error limiter and of collision during initial motion was overcome but there was the consequence that the attitude reference datum was effectively misaligned and the vehicle would deviate from its planned flight path. This situation is in general undesirable and for vehicles F11 onwards where a perigee stage has to be accurately oriented before spin-up and ignition this deliberate platform misalignment is unacceptable.

4.6. To overcome this problem the requirement is to reduce the error signals to acceptable magnitude – by some means other than platform cradle re-alignment – over the first few seconds of flight. Once the vehicle is clear of the launch site the constraint disappears. The requirement can be met by filtering the error signals before input to the first stage autopilot to determine their average values and then applying a bias equal to the average to give an input error signal where average value is zero. The filtering must cease before the vehicle is released so that true attitude error measurements once the vehicle is in flight are not corrupted but the bias must be applied until the vehicle is clear of the launch site.

4.7. Whilst the requirements can and will be satisfied, as indicated above, on vehicle F8 onwards the method of achieving the results will change at vehicle F11. This is because on vehicle F11 a digital inertial guidance computer is to be introduced which will process the attitude reference platform outputs to derive error signals and replace the present system where the error signals are derived by feeding the ARP outputs through differential synchros. On vehicles carrying the Inertial Guidance Computer (IGC) it is a simple matter to provide a software programme operating as follows:

Prior to 15 sec before launch   derive average error signal $\bar{\theta}$ and output $\varepsilon = \theta - \bar{\theta}$
                                  where θ is the measured error signal.
From -15 sec to $+5$ sec   output $\varepsilon = \theta - \bar{\theta}$, where $\bar{\theta}$, is the average error
                                  signal at -15 sec.
From $+5$ secs   output $\varepsilon = \theta$.

4.8. On vehicles before F11 this process must be achieved by hardware. The hardware

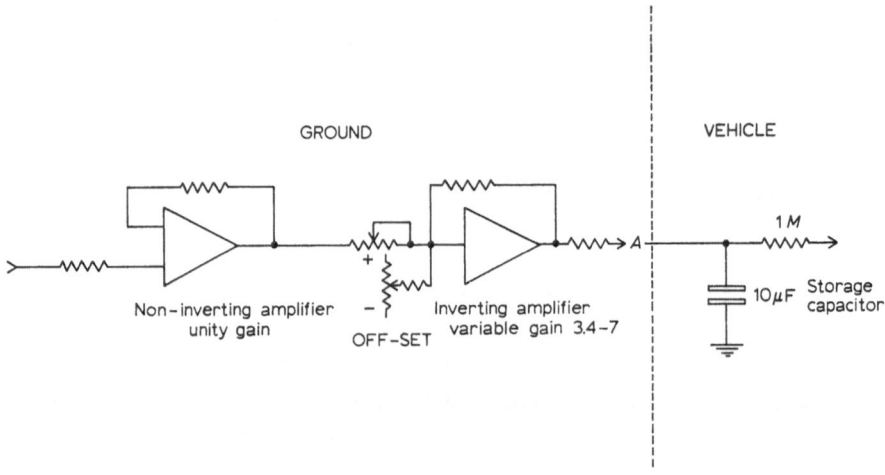

Fig. 5.   Bias circuit schematic.

is simple and consists of two capacitors and a resistance (Figure 5). In the basic first stage autopilot the attitude error signal is normally fed direct to a summing amplifier. In the modified scheme the error signal will be fed in parallel to a ground equipment box which will output the average signal ($\bar{\theta}$). The signal $\bar{\theta}$ is fed back to the summing junction with suitable sign to give an effective signal $\theta - \bar{\theta}$. By connecting a capacitor to the return line at the autopilot input the average signal is stored in the vehicle. At take-off the ground equipment is disconnected but the storage capacitor in the vehicle holds the bias which is thus still applied and as the storage capacitor discharges the bias is gradually removed. The storage capacitor is chosen to give a 10 sec time constant and thus its output varies little from $\bar{\theta}$ over the first, critical, 2 to 4 sec of flight.

5. Introduction of Digital Inertial Guidance Computer

5.1.   From the first stage attitude control systems performance point of view the most significant change between the Europa I and II vehicles is the introduction for the Europa II vehicle of a digital Inertial Guidance Computer (IGC) in place of an analogue system, as the element deriving attitude error signals from the Attitude Reference Platform. Considering the pitch plane, as an example, the system used in vehicles F6/2 to F9 is as follows. The actual vehicle attitude in pitch is measured directly relative to the direction of the local vertical at the time of launch by a gimbal signal from the ARP. The required pitch attitude at any time during first stage flight is generated by an electronic Programme Unit and the two signals – actual and demanded attitude – are combined to give a continuous, analogue, pitch attitude error signal which is then fed directly to the first stage autopilot. An attitude rate signal is also required by the autopilot and this is supplied directly from a suitable rate gyro. The first stage autopilot is designed on the basis of continuous error input signals and the allowable tolerances on zero errors on input and within the unit were also defined

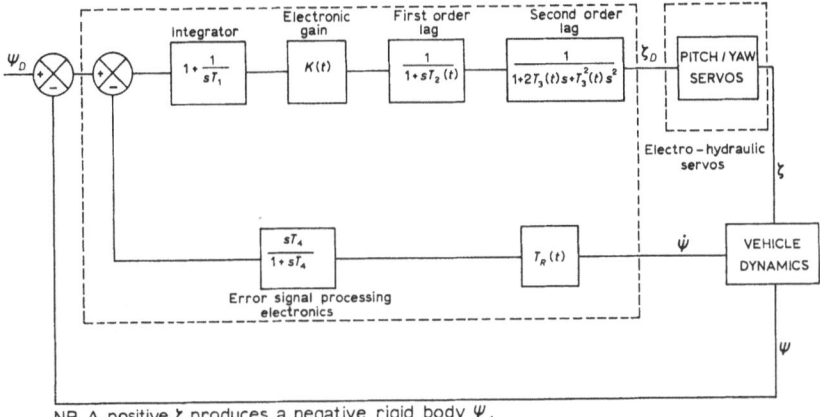

NB A positive ζ produces a negative rigid body Ψ.

Fig. 6.   Lateral (pitch/yaw) loops.

on the assumption of continuous signal inputs. A block diagram of the pitch loop is shown in Figure 6 as illustration.

5.2. In vehicle F11 onwards equipped with the IGC the system will operate in a different manner. The actual attitude is measured, as before, by the ARP but the gimbal output signals are now fed to the IGC by a convertor. This convertor – in fact a synchro-to-digital encoder – samples the attitude signal at 12.5 msec intervals and digitises the signal with a bit size of 0.088 deg (equal to $180/2''$ deg/bit). The demanded attitude is generated by IGC software according to a pre-set programme and the attitude error signal is also formed digitally within the IGC. The error signal is fed to the first stage autopilot through an Autopilot Command Decoder which converts the digital error signal from the IGC to analogue form again at the 12.5 msec intervals. The output bit size is 0.156 deg ($10/2^7$ deg/bit). The attitude error signal to the first stage autopilot is thus no longer continuous but is now sampled and quantised and thus quite different from that assumed in designing the autopilot (Figure 7).

5.3. The decision to introduce the IGC operating in the above mode was taken some time after the redesign of the first stage autopilot electronics had been specified and the majority of the circuit design completed. Any redesign to the first stage autopilot electronics to cater for the new attitude error signal characteristics would thus raise serious problems and major delays to the launching programme. Intensive investigations of the effects of changing the signal characteristics upon the first stage attitude control system performance were thus imperative to determine whether the IGC was compatible with the first stage autopilot.

5.4. When the investigations began representative IGC hardware – computer, encoder and decoder – were not available and had thus to be simulated. Because the output decoder had a much larger bit size than the input encoder the 'perfect' system

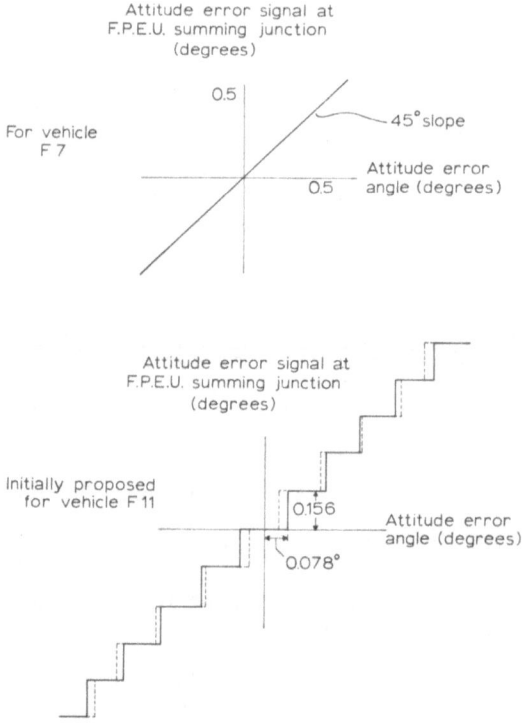

Fig. 7. —— analogue mode; ---- actual characteristic for encoder bit of 0.088°
and decoder bit of 0.156°.

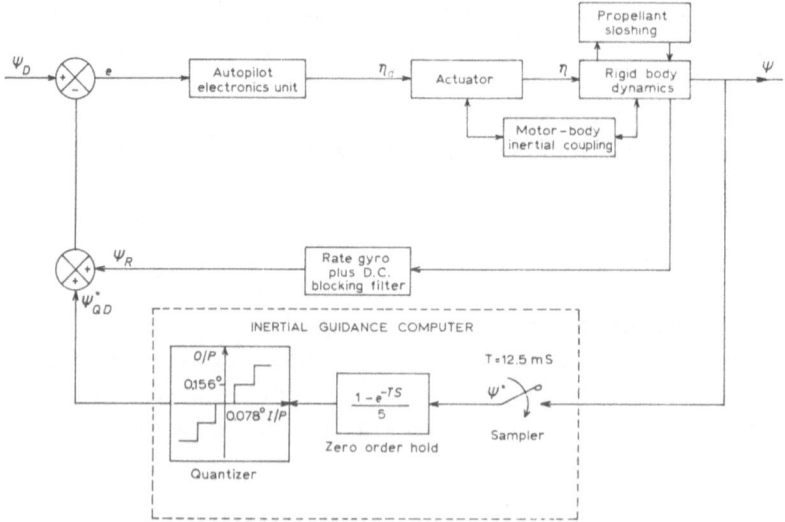

Fig. 8.   Block diagram of pitch plane control loop.

could initially be simulated by a 'sample and hold' circuit with a quantisation level of 0.156 degrees and an input/output characteristic as shown in Figure 7.

5.5. At the beginning analytical methods of study were considered but major difficulties arose. This was because propellant sloshing, body bending and unstable aerodynamics are dominating factors in the first stage attitude control system stability analyses and even when the hydraulic actuator characteristics are linearised a 20th order system had to be considered. It was soon evident that time would not permit analytical study to precede simulation by computer and the decision was taken to change immediately to investigation by analogue simulation.

5.6. For simulating the first stage pitch (and yaw) control system loops two EAL 680 analogue computers are used. The simulation includes:
  (1) autopilot electronics   rate gyro d.c. blocking, integrator, first and second order filter, electronic gain.
  (2) hydraulic actuator   including simulation of the non-linear equations of flow in the jack, the non-linearities due to friction at the gimbals, and the dynamic equations describing motor/body coupling.
  (3) 5 Bending modes.
Because of limitations on equipment capacity inter-coupling of the sloshing and bending modes is omitted – this effect was previously confirmed to be negligible in practice – and time variance of system parameters cannot be represented. Studies have thus to be made at each of several conditions of flight to assess stability over flight.

5.7. To simulate the ICC computer elements were inserted to sample the closed-loop attitude signal and then to quantise and hold the sampled signal. The sampling rate and quantisation level were adjustable to allow examination of variations from their nominal values. To excite the system provision was made to input a step attitude demand either immediately before or after the sample/quantise/hold IGC simulation. Figure 8 shows a logic diagram of the analogue simulation used.

5.8. As had been expected the sampling rate of 80 samples/sec did not cause any degraduations in performance – the critical frequencies for the first stage attitude control system range up to 30 Hz and are thus well separated from the 80 Hz sampling rate. Performance degradation did not arise until the sampling rate was reduced to 10 Hz samples/sec.

5.9. For cases when the demand was injected before the 'IGC' no degradations in performance were found but in cases when the demand was injected after the 'IGC' it was found that major problems could arise in the form of first stage limit cycle operation. The magnitude and frequency of the limit cycle varied with the size of the

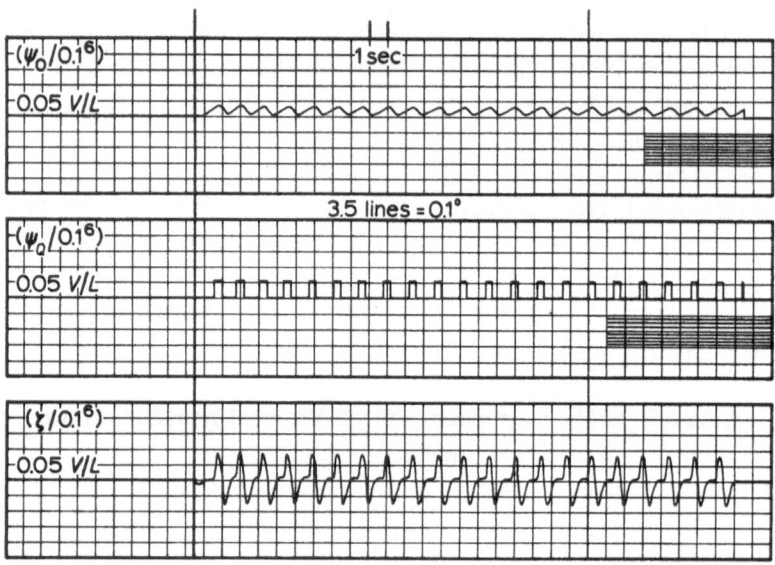

Fig. 9.   Rigid body simulation (no sloshing). Bias $\psi_D = h/4 = 0.039°$.

step demand in a cyclic manner. At some values a rigid body limit cycle occurred and at others first stage propellant sloshing limit cycling was excited. Figures 9 and 10 show examples of system responses and the variation of limit cycle characteristics with demand. The worst cases arose when the demand had a value of $(2n+1)h/2$ degrees when h is the quantisation level. Referring to Figure 7 which shows the input/output characteristic of the simulated IGC it is seen that for demands of $(2n+1)h/2$ degrees the system null lies on switching points of the IGC. The attitude output signal is then either nh or $(n+1)h$ and the error signal fed to the autopilot is thus switching between $h/2$ and $+h/2$. The difference in response to the two kinds of demand lies in that in one case the error signal is formed and then quantised whereas in the second case the error signal is formed after quantising the actual attitude signal.

5.10. The seriousness of this phenomenon lay in the fact that the critical 'demand' value of 0.078 degrees (and odd multiples thereof) when expressed as equivalent voltage (60 millivolts) lay within the range of zero errors permitted in the design of the first stage autopilot electronics and could thus arise in practice.

5.11. It was found that by reducing the quantisation size the seriousness of this phenomenon could be reduced – illustrative examples are shown in Figures 11 to 12 – but once the IGC output bit size was reduced one could no longer neglect the encoder bit size as being much smaller than the decoder bit size.
    It was therefore necessary to extend the simulation to include encoder and decoder before any further studies were made. The problems had arisen because an attitude error signal forming device with a linear input/output characteristic had been replaced by a device with a significantly non-linear characteristic. The surest cure looked to be

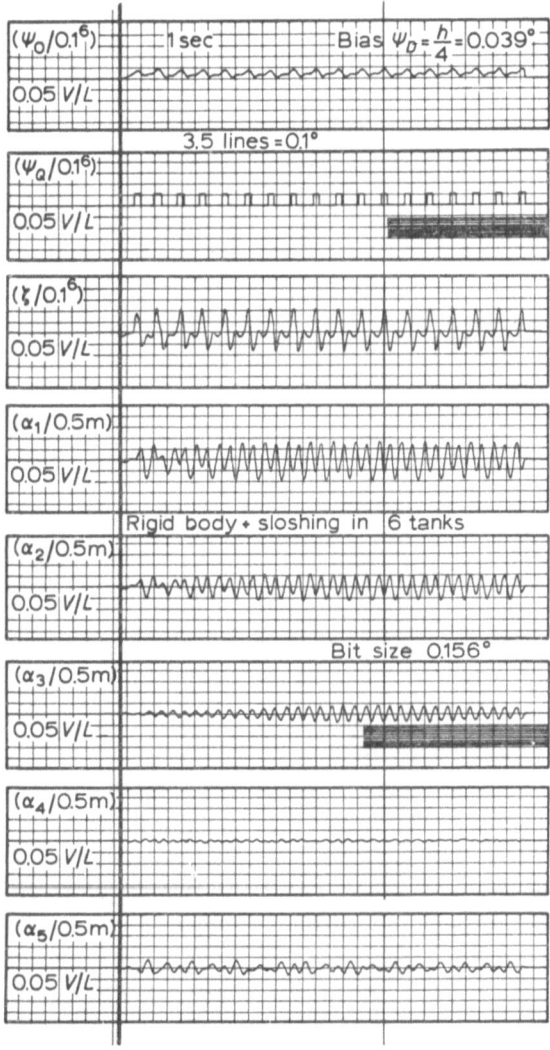

Fig. 10.

to effectively reduce the non-linearity (by reducing bit sizes) or to eliminate or mask it. As an experiment on this latter approach a high frequency dither signal was added to the attitude signal before input to the simulated IGC and it was found that this did indeed have a markedly beneficial effect.

5.12. At this stage of the investigation the first full set of IGC hardware – encoder/computer/decoder – became available and was incorporated into the analogue computer simulation. Because an actual synchro-digital encoder was being used it was also necessary to add a unit to convert the analogue computer attitude angle system into an equivalent synchro output signal before input into the synchro-digital encoder.

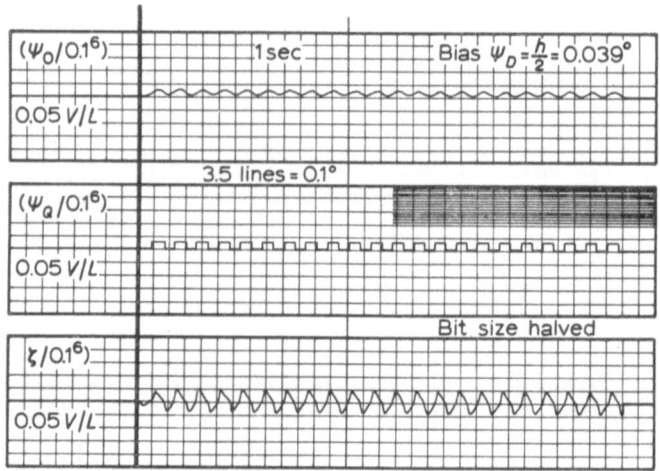

Fig. 11.   Rigid body simulation (no sloshing).

5.13. Once the full equipment had been integrated into the simulation investigation
continued. It was found that significant interactions exciting unacceptable limit
cycles continued to arise and could arise for zero demands. Two factors were found
to be operating that caused difficulties, both having their origin in the synchrodigital
encoder and not related to the output decoder bit size. The first effect arose from the
method of operation of the encoder which was based upon a phase angle difference
measurement which for certain input angle values could lead to steps of 2 bits instead
of 1 bit. The second effect was that the input/output characteristic could vary with
time – probably a temperature effect but one which persisted even after many hours of
continuous running – with the effect that the switching points on the characteristic
could include a switching point at zero input. In Figure 13 the theoretical input/output

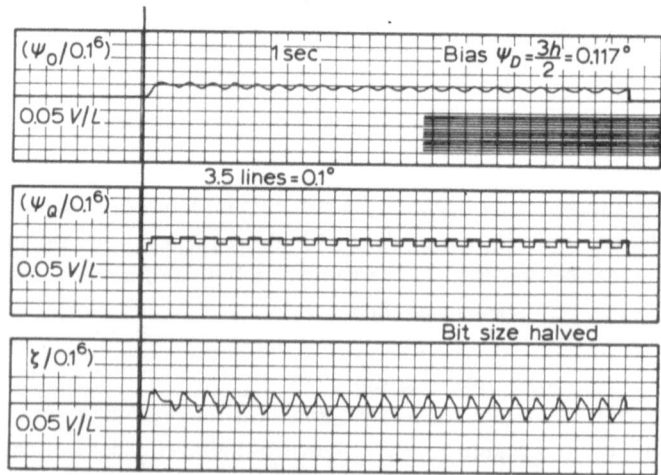

Fig. 12a.

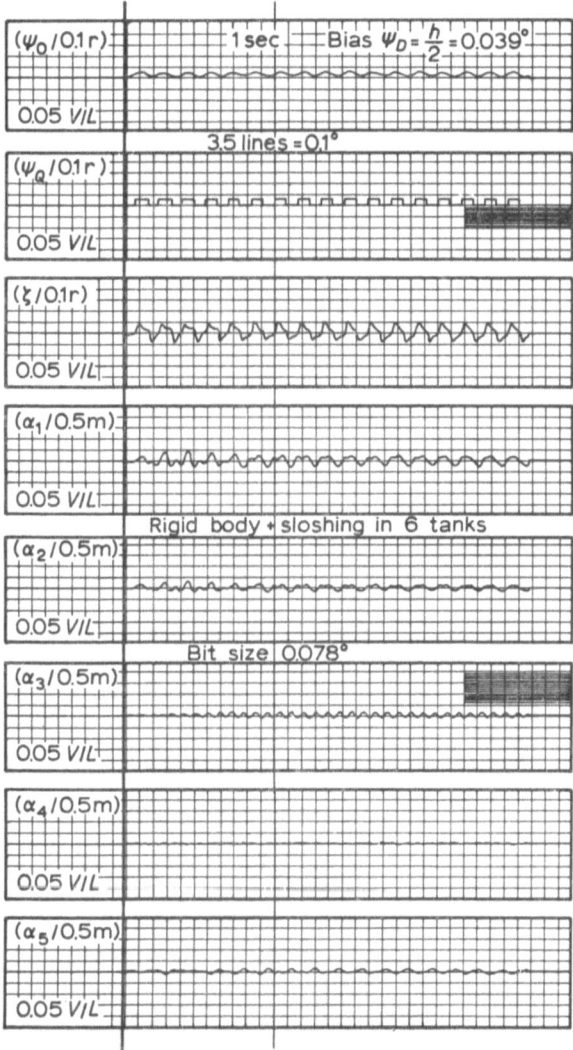

Fig. 12.

characteristic for the nominal encoder (0.087°) and decoder (0.156°) bit sizes is shown together with the range of recorded characteristics. It was found that the slow change with time of the characteristic could be forced by inserting a variable resistance in one of the encoder input lines thus allowing rapid examination of all variant characteristics.

5.14. To overcome the problems and permit use of the IGC as the attitude error source to the first stage autopilot the effects of adding a dither signal was investigated in detail. This could be achieved in practice with minimum difficulty without delay to hardware delivery programmes. All other alternatives involved major equipment modification or re-design and inevitable delays.

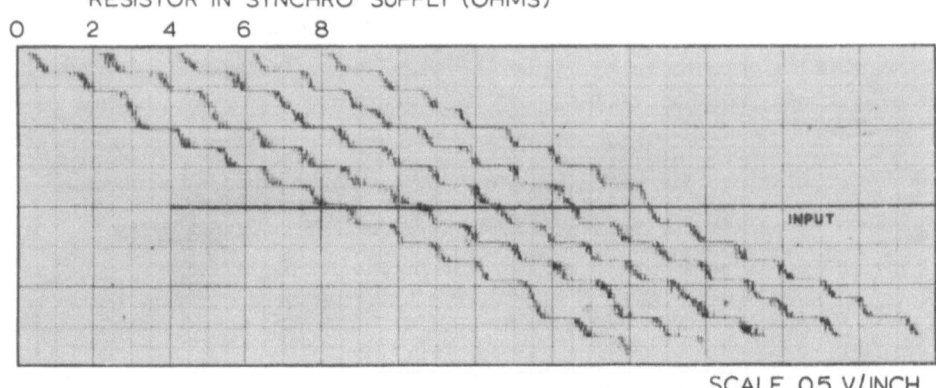

RESISTOR IN SYNCHRO SUPPLY (OHMS)

SCALE 0.5 V/INCH

Fig. 13.   Computer input output characteristics decoder scaled to 10° no dither.

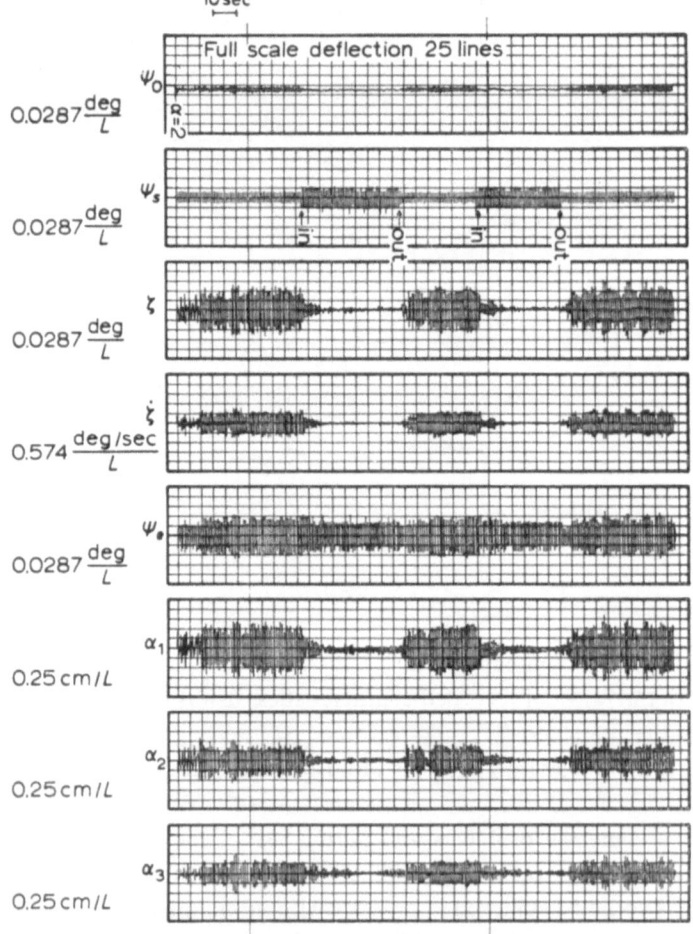

Fig. 14.   Effect of chosen dither signal at 88 sec decoder scaled to 5.537°.

RESISTOR IN SYNCHRO SUPPLY (OHMS)

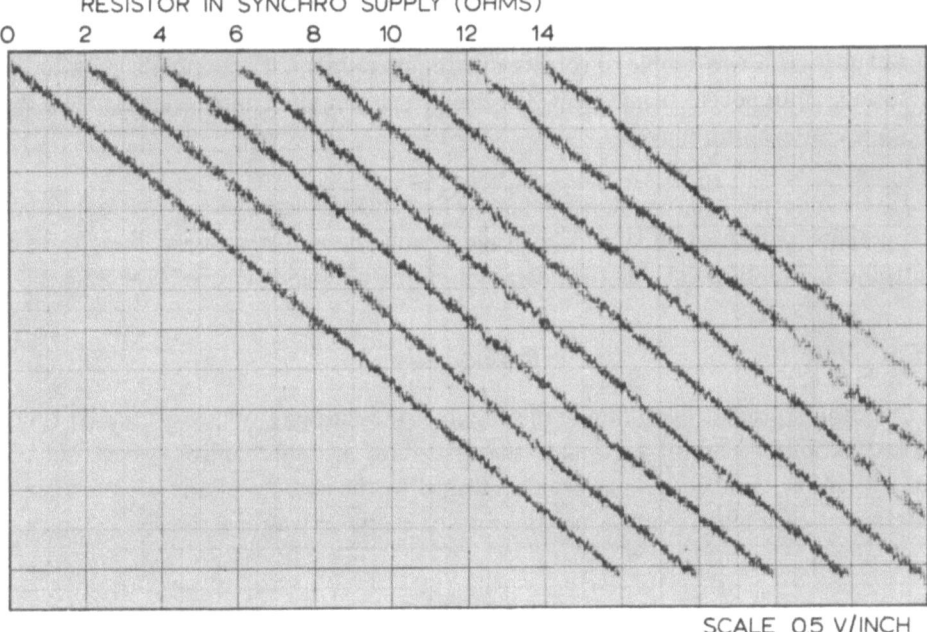

SCALE 0.5 V/INCH

Fig. 15.   Computer input-output characteristics decoder scaled to 5.537° with dither.

5.15.   It had already been found from the simplified IGC simulations that introducing a dither signal appeared satisfactory and detailed studies were commenced to establish:

(a) best waveform – sinusoidal, square and triangular
(b) best amplitude
(c) best frequency
(d) validity under vehicle parameter tolerances.

From these studies a dither signal injected into the synchro-digital convertor of sinusoidal waveform, frequency 112.5 Hz and amplitude $\pm 0.15°$ was found to be satisfactory and the necessary hardware modifications were introduced. Examples of the effect of introducing and removing dither are shown in Figure 14.

5.16.   To improve performance it was decided in parallel to investigate possible methods of achieving the effect of adding dither by changing the computer software processing of the attitude signals. It was found from test runs that a solution by these methods was possibly feasible but shortage of time and of digital computer availability – the other stage designers required the sole set of equipment for their own compatibility tests – meant that this approach could not be fully evaluated before the deadline for deciding whether to modify the hardware to introduce the dither signal.

5.17.   At the time of writing tests are due to start using a set of equipment modified to include dither and prove the validity of this solution for flight in F11. One added feature to the dither solution that has been introduced following the software modification tests is a re-scaling inside the computer to make the input and output bit sizes equal. This was simple to achieve and required only a trivial modification to the first stage autopilot electronics unit. The resulting overall IGC input/output characteristics are shown in Figure 15.

5.18.   It must finally be emphasised that the first stage always flies to a pre-set attitude programme with no Guidance signal added and the above study of the effects of introducing the IGC exclude consideration of any Guidance Loops being added.

6. Conclusion

The steady process of evolution of the design of the Europa I vehicle to Europa II has raised various needs to re-examine the first stage attitude control system but the validity of the basic design has been confirmed by the absence of any need for major alterations even after 7 years since the Europa I vehicle was conceived.

Acknowledgements

The studies described above have involved many colleagues and it is a pleasure to acknowledge contributions made by Messrs. L. Flook, C. Wearmouth, R.D. Mc Queen, K. F. Lofthouse, J. N. Pearson, F. Dove, M. B. Dewe of HSD, by staff of RAE and of the support of Messrs. Elliotts who are responsible for the IGC system.

PART III

PROPELLANTS AND COMBUSTION, I

Chairman: *L. G. Napolitano, Italy*

DECOMPOSITION AND IGNITION OF AMMONIUM PERCHLORATE IN THE PRESENCE OF SnO$_2$-Cr$_2$O$_3$ CATALYSTS

F. SOLYMOSI and T. BANSAGI

Gas Kinetics Research Group of The Hungarian Academy of Sciences, Szeged, Rerrich B.-tér

Abstract. The thermal decomposition and ignition of ammonium perchlorate has been studied in the presence of pure and doped SnO$_2$. Pure SnO$_2$ exerted only a slight effect on the decomposition of AP between 200° and 300 °C and it altered the ignition temperature of AP, only by about 30–40 °C. The effect of SnO$_2$ remained practically the same when it was doped by 1% of WO$_3$ or by 1% of Al$_2$O$_3$. The efficiency of the catalyst was, however, considerably improved when a small amount of Cr$_2$O$_3$ was incorporated into its lattice, and both the rate and the extent of AP decomposition increased with the increase of Cr$_2$O$_3$ content. The activation energy of the catalytic decomposition varied only slightly with the composition of the catalyst and fell in the range of 29–34 kcal. Doping of SnO$_2$ by Cr$_2$O$_3$ markedly lowered, by about 150 °C, the ignition temperature of AP. From the dependence of the ignition delay on the temperature, 15–18 kcal/mole activation energy was calculated.

In order to find out the reason for the marked effect of Cr$_2$O$_3$-doped SnO$_2$ and to elucidate the mechanism of the catalytic decomposition of AP, the investigations were continued in two directions. The first was a kinetic study of the gas-phase decomposition of perchloric acid on the above catalysts. The second was the physical-chemical characterization of the Cr$_2$O$_3$-doped SnO$_2$.

Kinetic data concerning the catalytic decomposition and ignition of AP were discussed in the light of these investigations.

1. Introduction

During recent years there has been much interest in the mechanism of ignition of solid propellants, particularly in the ignition of composite propellants containing ammonium perchlorate (AP) as oxidizer. Since it had been found that AP has a key role in the combustion process which was illustrated by the fact, that AP alone exhibits a number of the burning properties of composite propellants based upon it, for the elucidation of the mechanism of the reactions occurring in composite propellants, the knowledge of the decomposition and burning characteristics of pure AP is an essential prerequisite. Since it had also been observed that the burning properties of the composite propellants can largely be improved by using catalysts, great attention was paid to the evaluation of the role of the catalyst in this reactions.

We have been mainly concerned in our laboratory with the catalytic reactions, and our efforts have been devoted to the study of those properties of added substances which are primarily responsible for the catalysis and the initiation of AP ignition. It was observed that the electric properties of the oxides play an important role in their catalytic influence.

The most effective additives were found to be the *p*-type, in other words the electron hole semiconductors the effect of which sensitively changed with their electron hole concentration [1, 2]. The *n*-type TiO$_2$ proved to be an inactive additive its effect, however, was considerably improved when small amount of Cr$_2$O$_3$ has been incorporated in its surface layer [3].

The primary aim of the present investigation was to study the catalytic effect of

G. A. Partel (ed.), Proceedings of the Second International Conference on Space Engineering. All rights reserved.

pure and doped SnO_2, which has the same crystal and electric structure as that of TiO_2, on the slow and fast decomposition of AP. As in the decomposition of AP, especially at higher temperatures, perchloric acid is formed detailed measurements were carried out on the effect of the same catalysts on the vapour decomposition of perchloric acid, too. As our investigations showed, many oxides catalyse the decomposition of perchloric acid very effectively [4].

2. Experimental Section

A. MATERIALS

AP (BDH), twice recrystallized from water, was used. It was carefully ground to a specific particle size.

SnO_2 was obtained by dissolving of metallic Sn in HNO_3. The substance was dried at 120°C and heated first to 350 for 3 h and to 500°C for 5 h. The doping of SnO_2 with Cr_2O_3 was effected in two ways: (I) SnO_2 was suspended (by means of constant stirring) in a $Cr(NO_3)_3$ solution of a given concentration. After drying at 120°C, the substances were heated at 500°C and sintered at 900°C, both procedures lasting for 5 hours. (II) Cr_2O_3 was measured in the aqueous suspension of SnO_2 and, after a sufficiently long period of stirring the homogeneous suspension was dried and treated as above.

From AP and oxide powders measures in predetermined ratio, homogenized mixtures were prepared and pressed into pellets under 2700 kg/cm^2 pressure.

B. METHODS

The pressure-time measurements of the slow decomposition were made in a conventional vacuum apparatus of constant volume.

The ignition of AP was studied in air at atmospheric pressure. The pellets were dropped into a glass-tube, heated up to required temperature, and the gases evolved before and at the moment of explosion were measured by a gas burette. The vapour phase decomposition of perchloric acid on catalysts was studied in a flow system using nitrogen as carrier gas. The amount of AP used was 0.1 g.

3. Results

A. DECOMPOSITION OF AP

Pure SnO_2 had only a slight effect on the decomposition of AP between 200° and 300°C. The only, noteworthy effect was the decrease in the time lag and the increase in the extent of the decomposition. A greater influence was observed in the high temperature region of AP decomposition, above 300°C where the rate of reaction was doubled.

The effect of SnO_2 remained practically the same when it was doped by 1% of WO_3 (increasing the electron concentration in SnO_2) or by 1% of Al_2O_3 (decreasing the electron concentration in SnO_2).

The efficiency of the catalyst was, however, improved when a small amount of Cr_2O_3 was incorporated into its lattice, and thus both the rate and the extent of AP decomposition increased with the increase of Cr_2O_3 content. In this case detailed kinetic study was performed in the temperature range of 210–240°. Some typical pressure time curves are shown in Figure 1. The kinetic analysis of the pressure-time curves was done by the Prout-Tompkins equation, which resulted in a curve consisting of two straight lines (Figure 2). The first one refers to the acceleration period

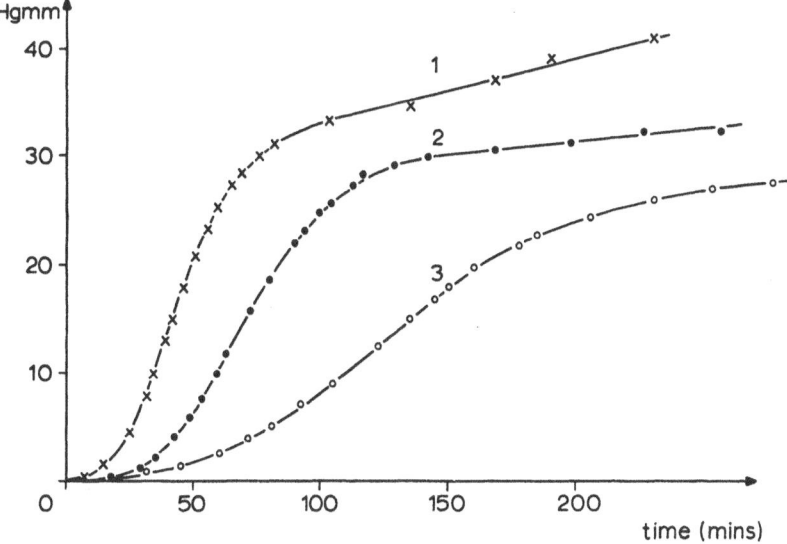

Fig. 1.   Thermal decomposition of AP in the presence of $SnO_2 + 1$ mole % Cr_2O_3 catalyst. (1) 240°C; (2) 235°C; (3) 225°C. AP: SnO_2 mole ratio, 10:1.

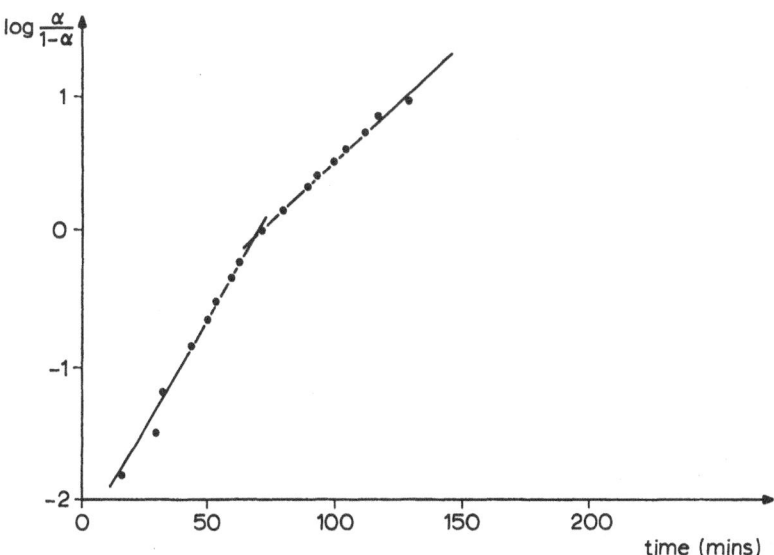

Fig. 2.   Kinetic analysis of the catalytic decomposition of AP by the Prout-Tompkins equation.

and the second one to the decay period of the decomposition. The activation energy of catalytic reaction was calculated by the Arrhenius equation and fell in the range of 29–34 kcal. The composition and the amount of the catalyst had only a slight effect on the value of the activation energy.

B. IGNITION OF AP

The SnO_2 altered only slightly the ignition temperature of AP but doping SnO_2 by Cr_2O_3 considerably influenced the ignition of AP. Detailed measurements were performed with AP–SnO_2 mixtures of 2:1, 4:1 and 10:1 mole ratios. The Cr_2O_3 content of SnO_2 was 0.1, 0.5, 1 and 5 mole%. Figure 3 shows the minimum ignition temperature of the above mixtures as a function of Cr_2O_3 content of SnO_2. Cr_2O_3 is expressed in mole% of SnO_2.

On the basis of these data it can be stated that while pure AP ignites at 430–435 °C and this temperature value decreases only by 30–40 °C using pure SnO_2, the addition of very small amount of Cr_2O_3 to the SnO_2 markedly lowered, by about 150 °C, the minimum ignition temperature of AP. The extent of lowering depended on the mole ratio of AP:SnO_2 mixture and on the Cr_2O_3 content of the SnO_2. The effect of the absolute amount of Cr_2O_3 in SnO_2 is apparent mostly with mixtures containing small amount of catalyst (10:1 mole ratio), where the AP exploded at 320 °C in the presence of 3.8×10^{-4} g Cr_2O_3 incorporated in SnO_2. Taking into account the effect of the same amount of Cr_2O_3 it appears the Cr_2O_3 in the SnO_2 is much more effective than Cr_2O_3 alone.

The activation energy of the reactions leading to the ignition of AP was calculated from the dependence of the ignition delay on temperature and 15–18 kcal was ob-

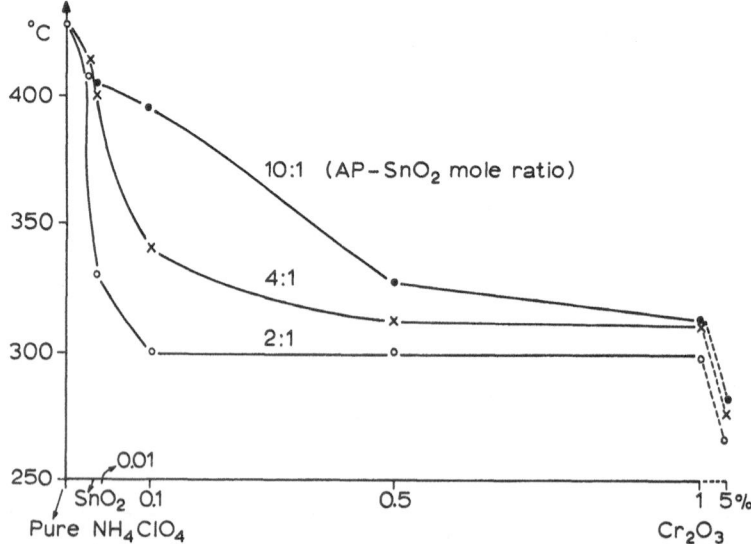

Fig. 3.   Minimum ignition temperature of AP as a function of Cr_2O_3 content of SnO_2 mixed with AP.

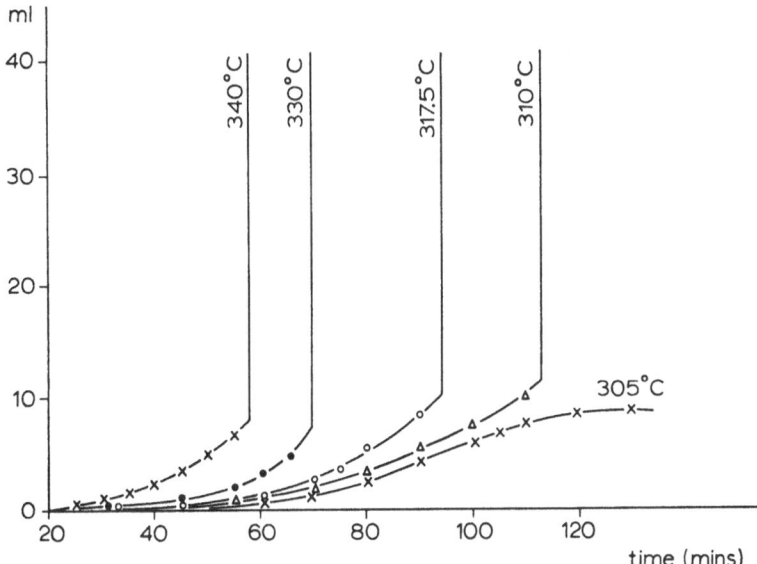

Fig. 4. Induction periods of AP ignition in the presence of $SnO_2 + 1\%$ Cr_2O_3 as a function of temperature. AP–SnO_2 mole ratio; 10:1.

tained. Figure 4 shows the effect of temperature on the ignition delay of AP in the presence of catalyst.

C. DECOMPOSITION OF $HClO_4$ IN VAPOUR PHASE ON THE CATALYSTS

In order to find out the reason for the great effect of Cr_2O_3-doped SnO_2, and to elucidate the mechanism of the catalytic decomposition of AP, investigations were continued in two directions. First, kinetic studies on the catalytic decomposition of perchloric acid were performed. Without going into details we summarize here only the main points of the results concerning this investigation [5]. While SnO_2 is rather inactive in the decomposition and ignition of AP, it catalyses the decomposition of perchloric acid fairly well. In the presence of SnO_2 the decomposition of $HClO_4$ was observed already at 300°C. This temperature was lowered by 50–80°C when SnO_2 was doped by a small amount of Cr_2O_3 ($0.01 - 1$ mole%). The catalytic decomposition was described by the first order equation.

Figure 5 shows the rate constants and the activation energy of the catalytic decomposition of $HClO_4$ as function of the Cr_2O_3 content of the catalyst. The rate of decomposition markedly increased with the amount of Cr_2O_3. The value of activation energy also increased by about 8 kcal/mole.

D. CHARACTERIZATION OF THE CATALYSTS

The second direction in which studies were continued was the characterization of the pure and Cr_2O_3-doped SnO_2. This included the determination of electric conductivity, thermoelectric power, magnetic susceptibility, differential thermal analysis, active oxygen content etc. [6].

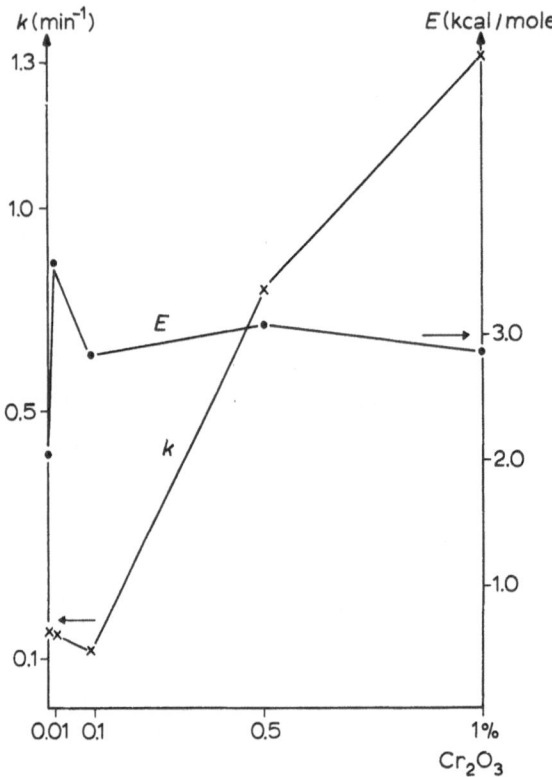

Fig. 5. The first order rate constant and the activation energy of the vapour phase decomposition of HClO₄ as a function of Cr₂O₃ content of SnO₂.

SnO$_2$ heated up to higher temperature has a rutile structure. Its electric conductivity decreases with the partial pressure of oxygen indicating a n-type behaviour. In accordance with this as it is shown on the Figure 6 the lower valence chromium decreases its electric conductivity, both in vacuum and in air. Nevertheless a more detailed study of the mechanism of incorporation of Cr$_2$O$_3$ into SnO$_2$ shows that this system differs in many respects from the SnO$_2$ doped with other tervalent cations. The most important deviation is that during the incorporation of chromium a part of it will be oxidized to higher valence state. In consequence of this oxidation the oxide mixture contains an excess of oxygen, in other words higher valence chromium ions, the amount of which increases with the Cr$_2$O$_3$ content of SnO$_2$. Same behaviour was observed earlier in the case of TiO$_2$–Cr$_2$O$_3$ system with a difference, that the oxidation of chromium ions converted the n-type conductance of the TiO$_2$ to a p-type character [3].

In the present case, however, the conductivity type of the SnO$_2$ has not been changed by doping it with Cr$_2$O$_3$ as is illustrated by the Figure 7. The electric conductivity of both the pure and doped SnO$_2$ decreases with an increase of partial oxygen pressure in agreement with the n-type character. The only exception is the sample containing 5 mole% Cr$_2$O$_3$.

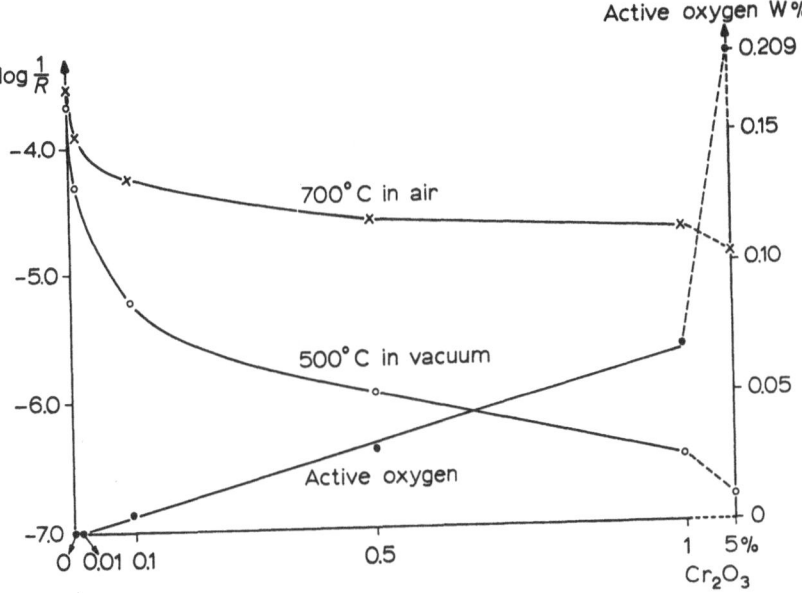

Fig. 6. The electric conductivity and the active oxygen content of SnO₂ + Cr₂O₃ mixture.

Independently from the overall conductivity of SnO_2–Cr_2O_3 mixture it should be expected, that due to the presence of higher valence chromium ions the electron affinity to the catalyst would be much higher than that of pure SnO_2 or SnO_2 containing only tervalent chromium ions.

4. Discussion

Before discussing the kinetic results, we should deal first with the different theories existing for the mechanism of AP decomposition and ignition. Despite several investigations, there are still conflicting views in the literature concerning the mechanism of the uncatalysed and catalysed decomposition of this substance. According to the earlier assumptions AP decomposes by the electron transfer mechanism between 200 and 300°C (activation energy 32 kcal), and by the proton transfer process above 350°C (activation energy 40 kcal [7]).

Recently Davies, Jacobs and Russel-Jones [8] have reported an overall activation energy of 30 kcal/mole in different temperature ranges of the decomposition. Based on this data and some mass spectrometric studies they proposed a proton transfer model for all the temperature ranges of AP decomposition. According to this theory, the role of the catalysts is claimed to be the acceleration of the decomposition of perchloric acid formed in the dissociation process of AP [9]. The mass-spectrometric and some catalytic investigations of Boldyrev et al. [10] seem to support this assumption.

Although the proposed proton transfer model, especially at higher temperatures,

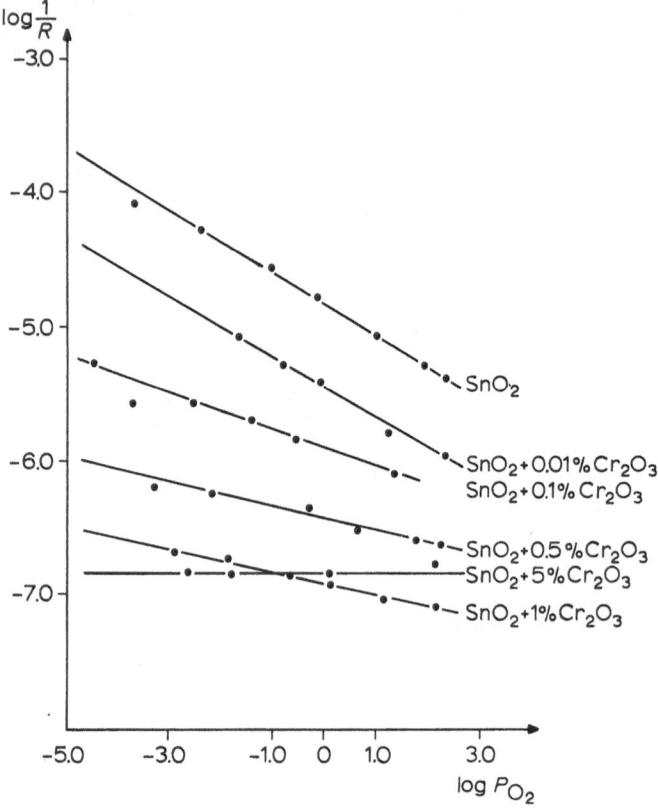

Fig. 7.   The dependence of the electric conductivity of SnO₂+Cr₂O₃ mixture on the
partial pressure of oxygen.

seems very plausible, in our opinion without the detailed knowledge of the catalytic
effect of the solids on the decomposition of perchloric acid, the final decision can not
be made. There are already a number of observations (e.g. the electric behaviour
of AP in the various temperature ranges [11], the effect on point defects on the
stability of AP [11, 12], the results of the product analysis of the catalyzed and
uncatalyzed AP decomposition [13], the marked catalytic effect of certain additives
observed at temperatures as low as 200 °C [14] etc.), indicating that the above unified
theory of the AP decomposition is an oversimplification.

  The n-type SnO_2 turned out to be a very ineffective substance for slow decomposition and the ignition of AP. There was no change observable in its effect even if
small amount of higher or lower valence cations had been incorporated into its lattice.
However a very effective catalyst was obtained when SnO_2 was doped with Cr_2O_3.
On the basis of physical-chemical characterization catalyst it seems likely that the
marked effect of doping with Cr_2O_3 is to be connected partly with the higher valence
chromium ions formed in the surface layer of SnO_2 and partly with high dispersion
of Cr_2O_3 on the surface of SnO_2 carrier.

The low activity of the SnO_2 in the temperature range of 200–300 °C can possibly be explained by the electron transfer mechanism of the AP decomposition. The n-type SnO_2 with high electron density can not participate in the electron transfer reaction between the perchlorate and ammonium ions; in other words it can not promote the formation of perchloric radical

$$ClO_4^- = ClO_4 + \ominus.$$

In accordance with this picture, the catalyst, containing higher valence chromium ion, can easily facilitate the formation of perchlorate radical – due to its high electron affinity – by taking over the electron of perchlorate ion, similarly as was assumed earlier in the case of p-type oxides.

This explanation is in agreement with the recent findings of Rosser et al. [13] who claim that the mechanism of the catalysed decomposition differs profoundly from that of the uncatalysed decomposition, and that the role of the catalyst can be interpreted in terms of electron transfer and the formation of free radicals. In our opinion, however, the above interpretation is valid primarily for the low temperature decomposition of AP but at higher temperatures, above 300 °C, where the dissociation of AP to NH_3 and $HClO_4$ becomes significant the catalytic influence exerted by the oxides on the decomposition of perchloric acid may become dominant. Our kinetic results revealed that while pure SnO_2 accelerated only slightly the decomposition of $HClO_4$, below 300°, this reaction became very fast, however, on the Cr_2O_3 doped SnO_2. On this basis it seems very plausible that, in the presence of this catalyst, the ignition of AP at 300 °C is primarily the result of the marked catalytic influence of the SnO_2–Cr_2O_3 mixture exerted on the decomposition of perchloric acid and of the oxidation of NH_3 by the decomposition products of perchloric acid. The change in the mechanism of AP decomposition is indicated by the change in the activation energy from 29 to 17 kcal. This latter value, determined from the time to ignition at different temperatures, refers to the complex gas-phase reaction. On the basis of these results it seems likely that both the Cr_2O_3 doped TiO_2 and SnO_2 could be an effective burning rate catalysts of AP containing propellants. Experiments on this line are in progress.

References

[1] Solymosi, F. and Krix, N., *J. Catalysis* **1** (1962) 468.
[2] Hermoni, A. and Salmon, A., *8th Symposium (International) on Combustion*, The Williams and Wilkins Co., Baltimore, 1962, p. 656.
[3] Solymosi, F., *Combustion and Flame* **9** (June, 1965) 142.
[4] Solymosi, F., Börcsök, S., and Lázár, E., *Combustion and Flame* **12** (August, 1968) 398.
[5] Solymosi, F., Börcsök, S., and Joó, B., to be published.
[6] Unpublished results.
[7] Hall, A. R. and Pearson, G. S., *Oxidation and Combustion Reviews*, Vol. III (ed. by C. F. H. Tipper), Elsevier Publishing Co., Amsterdam, 1968.
[8] Davies, J. V., Jacobs, P. W. M., and Russel-Jones A., *Trans. Faraday Soc.* **63** (1967) 1737.
[9] Jacobs, P. W. M. and Russel-Jones, A., *11th Symposium (Intern.) on Combustion*, The Combustion Institute, Pittsburgh, 1967, p. 457.

[10] Boldyreva, A. V., Berzrukov, B. N., and Boldyrev, V. V., *Kinetika i Kataliz.* **8** (1967) 299.
[11] Maycock, J. N. and Pai Verneker, V. R., *Proc. Roy. Soc.* **A307** (1968) 305.
[12] Pai Verneker, V. R. and Maycock, J. N., *J. Inorg. Nucl. Chem.* **29** (1967).
[13] Rosser, W. A., Inami, S. H., and Wise, H., *Combustion and Flame* **5** (1968) 427.
[14] Solymosi, F. and Dobó, K., *5th International Symposium on Reactivity of Solids*, Elsevier
      Publishing Co., Amsterdam, 1965, p. 467.
      Solymosi, F. and Raskó, J., *Z. phys. Chem.* N. F. (in press).

ADVANCED CHEMICAL ROCKET PROPELLANTS:
ON THE WAY FROM THEORY TO HARDWARE

R. LO

DVL-Institute for Rocket Propellants, Germany

Abstract. In the first part of the paper the realm of chemical High Energy Propellants is outlined. Theoretical performances of such systems are discussed, comprizing metal-combustion with and without the concept of 'Chemical Heating of Hydrogen'. A survey of the means for the technical realization of such systems is given, including gels, slurries, dust-combustion, etc., with major emphasis, however, upon hybrid and tribrid rocket engines and their peculiarities as they have shown up during the experimental work with such engines at the DVL during recent years.

The second part describes the experimental efforts to get rid of the various shortcomings of metal-combustion. These efforts include the search for new and better materials compatible with the very hot and very corrosive environment of high-energy propellant combustion chambers and the geometrical optimisation of oxidizer-injection, fuel-grains, mixing chambers, etc., along with research on binder materials suitable of improving the combustion efficiency of metallized fuels. At the present state of the art experimental performances closely approaching 100% of theory have been reached with certain hybrid and tribrid systems.

1. Introduction

For still a considerable period of time chemical propellants will provide the major source of energy for lifting space vehicles into their trajectories. Unlike air-breathing propulsion systems, where more than 75% of the exhaust gases are provided by the surrounding air and where, hence, improvements in performance are restricted to changes in the fuel, increases of the performance of rocket propellant systems, even minor ones, can have quite large effects upon payload-capability and liftoff-weight.

Considering the tremendous efforts sometimes necessary to keep payload mass down when there is only a certain amount of energy available for a certain mission, or, on the other hand, considering the giant masses of the rocket-stages necessary to accelerate just a few tons of payload to escape velocity, it is quite obvious that the introduction of propellants with higher energy content than that of the ones available today will be worthwhile.

This paper has two purposes: first, to give a short introduction into the field of high energy propellants to people who might be less familiar with this particular aspect of space technology; second, to show the special demands the engine designer is confronted with by those propellants and how it has been tried to cope with these problems during the recent research done at the DVL.

2. Comparison of High Energy Propellants

There is no general way to compare propellant systems since the desired qualities strongly depend on the requirements of the mission. But, since the thrust available

G. A. Partel (ed.), Proceedings of the Second International Conference on Space Engineering. All rights reserved.

TABLE I

Theoretical performance of HE-Propellants.

$Is, v = $ Specific impulse at $10:0.01$ kp/cm^2 equilibrium expansion into vacuum. $\% = $ Increase of performance as compared with the classical O$_2$/Kerosene system.

System	Is, v	$\%$	System	Is, v	$\%$
O$_2$/Kerosene	374	0	F$_2$, O$_2$/CH$_2$	413	$+10$
O$_2$/H$_2$	469	$+25$	O$_2$/Be	328	-12
F$_2$/H$_2$	493.7	$+32$	O$_2$/Be/H$_2$	564.6	$+51$
F$_2$/Li	456	$+22$	O$_2$/BeH$_2$/H$_2$	559	$+49$
F$_2$/LiH	444.2	$+19$	O$_3$/Be/H$_2$	584	$+56$
F$_2$/Li/H$_2$	520	$+39$	F$_2$, O$_2$/10%LiH, Be/H$_2$	556.5	$+49$
F$_2$/LiH/H$_2$	508.9	$+36$			

per unit mass of propellant is always important, the following table compares propellant systems in terms of specific impulse (see Table I).

Most HEP-combinations have extremely poisoneous products of combustion. This is one of the major reasons that they can be considered for use only in upper stage vehicles.

Therefore, we take vacuum specific impulse at a rather high expansion ratio as the basis of comparison and compare the classical propellant combination LOX/kerosene with performances of HEP-combinations. HEP-combinations can be derived from LOX/kerosene in a formal way by substituting either better oxidizers or better fuels or both. In doing so, two concepts are most important, namely the concept of *oxidizer-fuel pairing* and the concept of the *chemical heating of hydrogene*. Since performance is determined by reaction temperature as well as molecular weight of reaction products, it is advantageous in the case of kerosene which consists of hydrocarbons, to provide fluorine for the combustion of hydrogen, but to keep the oxygen content of the oxidizer high enough to transform the carbon into carbonmonoxide. Indeed, FLOX containing 70% of fluorine and 30% of oxygen, which corresponds to stoichiometric proportions of F to H and O to C, delivers the highest specific impulse with hydrocarbons and is superior to pure oxygen and pure fluorine.

Replacing kerosene by hydrogen leads to the O$_2$/H$_2$ system, the only high energy propellant which is operational today. Consecutively replacing oxygen by fluorine yields the best bipropellant system known, F$_2$/H$_2$ which is not in use yet, though.

In terms of energy per unit mass of reaction mixture there are many combinations with fuels, especially with metals, superior to hydrogen combustion. At the top are beryllium burned with oxygen and lithium burned with fluorine. However, performance of these systems, when used as bipropellants, is not as good as the one of the hydrogen bipropellants, due to the higher molecular weight. To get a lower average, hydrogen has to be added, leading to tripropellant systems which are the best chemical propellant systems known. Beryllium burned with oxygene or ozone used to heat up excess hydrogen yields performances more than 50% above that of LOX/kerosene, while burning of lithium with fluorine give close to 40%. In most of these cases, hydrogen does not take part in the combustion reactions at all. Even more striking, when using Li and Be together, burning them with mixtures of F$_2$ and O$_2$, there are hardly any lithium-oxygen or beryllium-fluorine combustion products and no HF or H_2O appears, when hydrogen is added. This shows the validity of the two concepts mentioned above.

3. Types of HEP-Rocket Engines and HEP-Combustion

The various possibilities for rocket engines with HEP-combustion are listed in Figure 1.

The classical high energy bipropellants O_2/H_2 and F_2/H_2 of course belong to liquid-type rocket engines. Metal combustion normally requires the combustion of solids with the exceptions of using liquid metal-compounds and metal-solutions, which we are not going to discuss since they are of minor interest, and with the exception of using molten lithium. The only other way to use metals in liquid-bi-propellant engines is the application of slurries.

Hybrid engines, where the solid fuel is contained within the combustion chamber, are the most common way of burning solids with liquid oxidizers. Solid lithium is no suitable material for this type of engine since it melts too readily. One has, therefore, to use Li-compounds, the best of which is the hydride. Massive beryllium cannot be used either, on account of its good heat conduction it could not be kept at ignition temperature. A possible way out might be the use of sponge-like porous fuel grains.

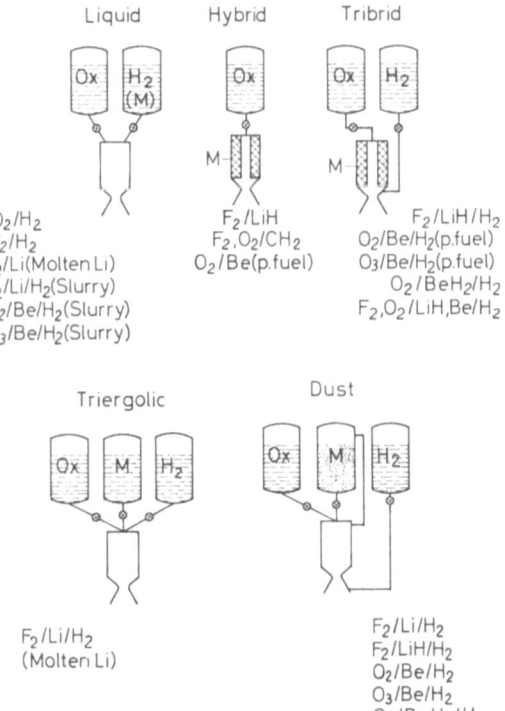

Fig. 1.   Types of HEP-engines.

One of the very few tripropellant systems conceivable with the actual injection of three liquids into the combustion chamber is the $F_2/Li/H_2$ combination. The more general solution, however, is tribrid-combustion with the solid situated within the chamber and the hydrogen being injected additionally. Here again the same restrictions with respect to the choice of the fuel hold as in the case of hybrid combustion.

Metallized solid propellants do not belong to HE-propellants due to the lack of HE-solid oxidizers. Therefore, the only other remaining method of metal combustion is dust combustion. In dust combustion the excess hydrogen is used in the gaseous state as it leaves the regenerative cooling system to transport the powdered solid fuel into the chamber.

There are, as we have seen by now, five types of HEP-engines which differ in terms of propellant feeding. Of these, hybrid, tribrid and dust combustion is being studied at the DVL in Stuttgart and we will now look into some of the results we have got so far.

4. Major Problem Aereas and Recent Achievements at the DVL

With the exception of the O_2/H_2 combination, HE-propellants are not yet in use in actual flight systems.

For this, there are three main reasons:

(a) Poisonous constituents,

(b) Inherently low combustion efficiency,

(c) Corrosive burning.

With maximum tolerable concentrations of 1000 μg per cubicmeter of air for fluorine and 25 μg for beryllium – both concentrations concerning short time exposures – fluorine is certainly a very toxic, beryllium an extremely toxic material.

With respect to test facilities and research, this toxicity of propellants and combustion products is more than a nuisance. It requires complicated safety precautions and has actually delayed tests with these propellants for a considerable period of time. The beryllium facility of the DVL in Lampoldshausen for instance, is still in a pre-experimental stage as far as beryllium is concerned. Considerable trouble has to be taken to test the equipment (scrubber, exhaust gas filters, sampling system etc.) with less harmful substances. In this case aluminum is being used for this purpose.

The second obstacle for the immediate use of HEP-combinations is the inherently low efficiency of metal combustion.

Let us consider a very small scale hybrid combustion chamber of about 5 kp thrust with a shower head for oxidizer injection, a fuel grain in the shape of a hollow cylinder burning at the inner face and a nozzle diameter of about 7 mm. Provided that there are no further means for increasing combustion efficiency, such an engine will deliver only 40–50% of theoretical performance when a metal containing fuel is used. A metal free fuel, such as polyethylene, will deliver 60–80% under the same circumstances, which proves metal combustion to be the main source of difficulties.

Experimental work done in this field has shown that there is no single way to improve the efficiency up to the desired values of 95% or more. This goal can only be achieved by introducing a considerable number of smaller improvements. The means for these improvements as they have shown up at our work at the DVL, are listed in Table II together with an admittedly crude average value of efficiency increase in terms of percent of theoretical performance (mark: not in terms of previous efficiency).

TABLE II

Means for improving combustion efficiency

Comb. eff. influenced by:	%	Remarks	
Injection mode	5	}	
Mixture ratio	5	} interdependant	
Fuel geometry	5	}	
Engine size	10–20	} interdependant	
Afterburner	10–20	} interdependant	
Spoiler plates	20–30	}	
Hydrogen mixing	20–30	}	with I_s increase
Binder addition	5–15	} change of system	—with I_s decrease
Solid oxidizer addition	10–30	}	with I_s decrease

A hollow-cone type of injection proved slightly superior to showerheads which in turn are somewhat better than impinging jet injection. Alltogether, however, injection mode influences mixture ratio and time variation of mixture ratio rather than combustion efficiency.

Mixture ratio itself varies with time since the shape and seize of the burning surface is time dependent in most cases and strongly influenced by injection mode. Aside of that, overoxidation normally leads to more complete metal combustion, although, for obvious reasons, this is not a desirable way of improving efficiency.

Several types of fuel geometry have been tried. Star configurations of the cross section of the inner duct seem to be superior to smooth hollow cylinders, supposedly because they do not lead to stratified flow as much as the latter does. Double cylinders consisting of a fuel-rod inserted into the inner duct of a hollow cylinder lead to constant, time independent surface areas, provided the injection system does not deform the surface. Mixing of propellant vapors is better than in single hollow cylinders.

Larger engines deliver better performance than smaller ones due to the reduction of relative heat losses and due to the changing relation of mixing length to engine dimensions. The values in Table II are related to the performance increase we achieved when increasing the nozzle diameter from 7 to 12 mm.

Another important item is the provision of some length of chamberspace downstream of the fuel grain where after-burning can complete the combustion. Obviously, for watercooled engines the length of this chamber must have an optimum beyond which gain in combustion completion is cancelled by increase of heat losses. Even in regeneratively cooled engines the entropy increase due to increasing heat circulation must have a similar effect.

The most important device for improving combustion efficiency is the spoiler plate. By introducing one or several perforated plates into the stratified stream of combustion gases these are forced to mix. Spoiler plates reduce the required length of the afterburning chamber even more than engine seize does.

There are another three means to influence combustion which essentially require changes in the propellant system, however. First of all, in tribrid systems, the excess hydrogen can be used for mixing the combustion gases. While head end injection of hydrogen leads to very bad combustion of the fuel grain in the resulting flame, injection downstream of the burning channel gives very good results.

Insufficient combustion of the fuel grain can have either of two reasons: being good heat conductors, metals with high melting point remain solid and cannot be brought to or kept at ignition temperature, while metals with low melting point are liquified rather than evaporated and simply melt away. Both shortcomings can be overcome by using powdered metal fuels rather than bulk material. The powder is mixed with a suitable binder, preferably in a castable and curable composition. In the case of high melting points the task of the binder is to transport solid metal particles into the combustion zone while evaporating, in the case of low melting points it ought to form a non-melting skeleton which keeps the melting material together and should reduce the width of the melting layer by reducing the average heat conductivity.

A similar line of reasoning explains the fact that addition of solid oxidizer has great influence upon the efficiency of fuel combustion.

After ignition by the fluid oxidizer, the solid oxidizer causes a much more vigorous surface reaction which vaporizes metal particles. It is important, however, that the solid oxidizer/fuel reaction stops when the main oxidizer is switched off, in order to keep the stop and restart capabilities of hybrid engines. Since binders are fuels of lower energy content and the same holds for solid oxidizers as compared with liquid ones, it is clear that both types of combustion efficiency improvement decrease theoretical performance.

While Table II lists the influences of the various methods in a singled-out way, it is clear that they do not simply add when two or more methods are applied simultaneously.

Figure 2 shows the experimental efficiencies achieved with several hybrid and tribrid propellant systems at the DVL. The values are averaged and presented with somewhat greater accuracy than actually available since some other properties (e.g. chamber pressure, duration of firing) are not quite comparable in some cases.

The efficiencies of the F_2/LiH system where the lowest ones to be achieved experimentally. As expected, there was an increase when a larger chamber was used and a slightly larger one when LiH was burned with binder addition. It was impossible, however, to use spoiler plates with this propellant system. This now brings us to the third one of the shortcomings of metallized HE-propellants: corrosion. With a theoretical combustion temperature of over 4500 K the F_2/LiH system generates a most corrosive environment. Spoiler plates made of ordinary graphite are consumed almost as fast as the fuel while pyrolytic graphite stands the environment for 3 to 5 sec.

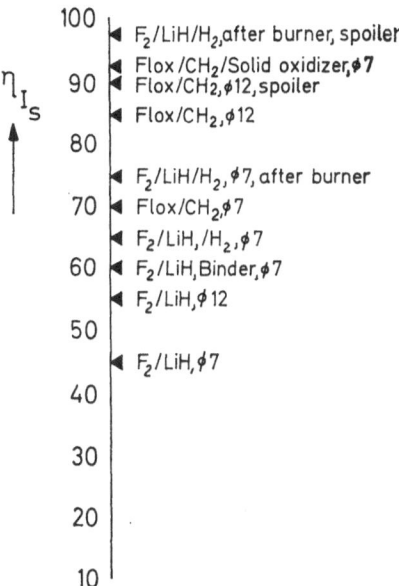

Fig. 2.   Experimental efficiencies of HEP.

Experiments with graphites coated with refractoiy materials proved quite promising, however.

Another aerea of severe coirosion is the nozzle throat, of course. Copper nozzles are severed in a few seconds. Throat inserts out of Tungsten alloys resist up to 15 sec. Here again, nozzles plasma-coated with refractories gave very promising results. So far, however, the firing duration with the F_2/LiH system as a representative of metallized hybrids is limited.

Still another problem, common to all fluorine using hybrids, was what can be called the 'fluorine-torch effect'. During ignition fluorine-rich gases with sharply raising temperature hit all parts of the chamber downstream of the fuel grain. Especially when spoiler plates with narrow holes are built in, the resulting jet can ignite with the chamber, resulting in immediate failure. The problem has been solved, however, by protecting the chamber walls and the convergent part of the nozzle with ablative materials.

Interesting enough, neither one of these problems is encountered with the tribrid $F_2/LiH/H_2$-combination. The excess hydrogen provides a reducing environment and keeps the temperature down.

Figure 2 shows the very satisfactory results. Hydrogen addition to F_2/LiH raised efficiency to 65% (now referring to the theoretical performance of the tripropellant, of course). Optimizing the chamber-length downstream the fuel grain scored another 10%. Finally, efficiency was raised above 95% by spoiler plates.

As has been shown in Table I, the FLOX/polyethylene combination is only very short behind F_2/LiH. Experimentally, its efficiency was far higher. Very little corrosion problems are encountered in this case. Firing duration is arbitrary, since simple copper-nozzles as well as graphite-spoilers stand the environment without damage. Theoretical combustion temperature (4270 K) is not too much below that of F_2/LiH, which proves that metal-vapours are the decisive factor in the corrosiveness of the latter.

Efficiencies of more than 90% have been reached and it is hoped that optimization of chamber-geometry and spoiler plate configuration will push it to a still more satisfactory level.

The hybrid combustion of fuels with a certain amount of solid oxidizer is being done for the sake of studying the influence of the surface reaction upon combustion behaviour. A remarkable 93% efficiency has been reached with mixtures of polyurhane-polyethylene-NH_4NO_3, burned with FLOX. This maximum was achieved with 20% of solid-oxidizer. Similar experiments with polybutadiene-ammonium-perchlorate mixtures proved these to be unextinguishable and had to be abandoned.

Let us now sum up the state of the art of HEP-combustion at the DVL:

(a) In the upper region of HEP-combinations the tribid system F_2/LiH/H_2 has reached a level of satisfactory performances. A still considerable remainder of problems comprises longevity of spoiler plates, constancy of mixture-ratios and thrust variation.

(b) In the lower region of performances the system FLOX/polyethylene is on the point of becoming available for switching from water-cooled to propellant-cooled engines. Some optimizations are presently being studied, but for water-cooled engines no principal technological problems whatsoever are left.

(c) In metal combustion without hydrogen addition, it was possible to surpress combustion instabilities of the system F_2/LiH by binder addition by which the efficiency was rosen as well. Solutions for improving the lifetime of spoiler plates and nozzles are presently being studied.

(d) The triergolic system F_2/LiH/H_2 is being studied with dust-combustion chambers as well. 85% efficiency has been reached without any optimizations aside of mixture ratio. Frequent corrosion or ignition of injection heads is still a problem.

ADVANCED FUELS AND OXIDIZERS*

CLAIR M. BEIGHLEY, WILLIAM R. FISH and ROGER E. ANDERSON

Aerojet-General Corporation, Sacramento, Calif., U.S.A.

Abstract. Guidelines for the search for advanced liquid fuels and oxidizers are developed from theoretical considerations of the reaction principle of rocket propulsion. Advanced oxidizers are based on oxygen and, increasingly, on fluorine. The potential of these elements; their blends; inter-compounds; and compounds with nitrogen, chlorine, noble gases and a few other carrier elements in advanced chemical propulsion are discussed. Fuel research and development are oriented largely toward incorporation of metal additives into liquid fuels to form chemically and mechanically stable heterogeneous gels and emulsions of high heating value and/or density. Factors pertinent to this relatively new area of liquid propellant technology are discussed.

1. Introduction

October 4, 1967 was the decennial of the Space Age (Sputnik I was launched October 4, 1957). Chemical propulsion, more specifically, liquid rocket propulsion, has been the pacing technology which dictated progress and eventually permitted the existing record of achievements in space. The liquid rocket engineer, working with selected sources of liquid propellant energy, has been primarily responsible for the space achievements to date. What does the future hold for the liquid rocket engine and liquid propellants as a source of propulsive energy?

All rocket propulsion systems are based on the reaction principle in which reaction thrust is imparted by the momentum of matter ejected through a nozzle. In a chemical propulsion system such as a liquid propellant rocket, gases generated by chemical reactions are ejected as high velocity gas streams through a nozzle. In essence, propulsive thrust is achieved by converting thermal energy of chemical reactions into directed translational energy of the combustion products.

The primary parameter which indicates the merit of a chemical propellant is the specific impulse, I_s, which is defined as the ratio of the thrust, F, lb, (f), produced to the rate of propellant consumption, w, lb, (m)/sec. Thus,

$$I_s = F/w. \tag{1}$$

An elementary analysis, based on the simplifying assumptions that the generated combustion product gases are ideal and that their heat capacities and composition are independent of temperature and pressure, yields the following expression for specific impulse:

$$I_s = \left\{ \frac{2J}{g} \frac{\gamma R T_c}{(\gamma - 1) M} \left[1 - \left(\frac{P_e}{P_c} \right)^{(\gamma - 1)/\gamma} \right] \right\}^{1/2} \tag{2}$$

G. A. Partel (ed.), Proceedings of the Second International Conference on Space Engineering. All rights reserved.

where the adiabatic coefficient, γ, is defined as

$$\gamma = \left(\frac{\delta \ln P}{\delta \ln \rho}\right)_s \tag{3}$$

and where g = gravitational constant, ft/sec^2, J = mechanical equivalent of heat, erg/cal, P = pressure, lb/sq in; P_c at chamber; P_e at exit plane, M = molecular weight, g/mole, R = universal gas constant, cal/mole K, s = constant entropy, T_c = chamber temperature, K, ϱ = density, g/cc.

(This elementary analysis is presented to emphasize the factors that guide the search for advanced propellants. Equation (2) is no longer used to calculate specific impulse. References 37 and 43 contain more rigorous analyses and the thermodynamic basis for the modern computational methods for obtaining the specific impulse of rocket propellants.) Examination of Equation (2) reveals three variables that are functions of the propellant chemistry, T_c, M and γ. The parameter γ is ca 1.2 for most propellant combinations and cannot be significantly altered. Therefore, the search for higher specific impulse, I_s, must start with propellant combinations which react to yield high temperature combustion products with a low average molecular weight.

In practical applications the weight and size of the tanks which must carry the propellant are important. Therefore, propellants with a high bulk density are preferred and in some cases are essential to minimize these factors.

In a rocket combustion chamber the chemical reaction is essentially adiabatic. This has been expressed by Siegel and Schieler [37] as follows:

$$\text{Reactants}\,(\text{at}\,T) \rightarrow \text{Products}\,(\text{at}\,T_c). \tag{4}$$

Thermodynamically this can be divided into an isothermal step:

$$\text{Reactants}\,(\text{at}\,T) \rightarrow \text{Products}\,(\text{at}\,T) + \Delta H \tag{5}$$

and an adiabatic step

$$\Delta H + \text{Products}\,(\text{at}\,T) \rightarrow \text{Products}\,(\text{at}\,T_c). \tag{6}$$

Because most propellant reactants are endothermic or only slightly exothermic, a high chamber temperature results from selecting propellants which react to yield highly exothermic products. However, from Equation (2) it is apparent that low average molecular weight is just as important as high T_c. Therefore, our search through the periodic table of the elements rapidly narrows to those which yield (a) high exothermicity (negative of the heat of formation per gram of the primary product species), (b) low molecular weight products, (c) high density reactants.

Figure 1 presents a graphical comparison of the exothermicity of the first 40 elements in the periodic table. Compounds formed with oxygen and fluorine have the highest specific exothermicities. Chlorine and nitrogen compounds are considerably lower. Carbides of the light elements, not shown in the graph, are lower than the

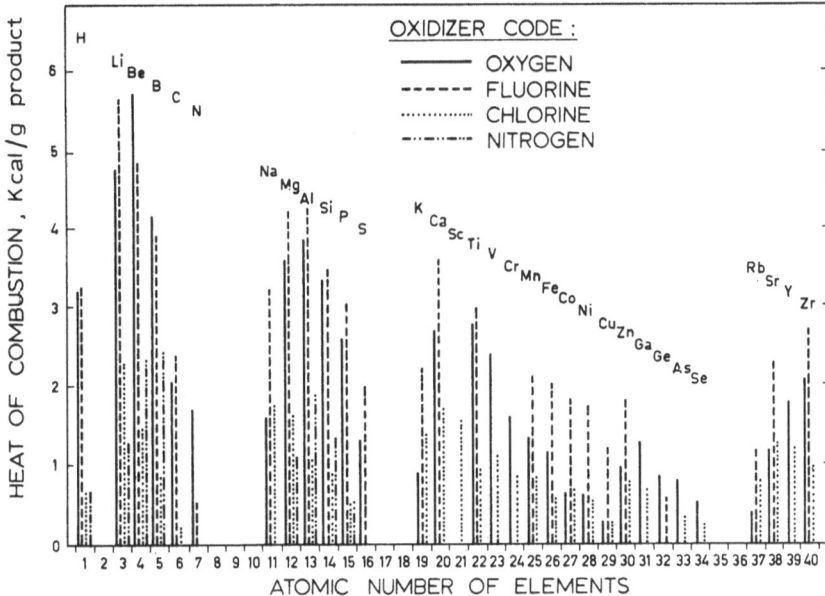

Fig. 1.   Heat of combustion of the low atomic number elements with oxygen, fluorine, chlorine, and nitrogen.

corresponding nitrides. Clearly, both the oxides and fluorides merit serious considerations as desirable propulsion combustion products. Because hydrogen has the lowest atomic weight, propellants rich in hydrogen e.g., N_2H_4 and light metal hydrides, provide high specific impulse. However, hydrogen itself has a low density which is a disadvantage for certain specific applications. Liquid solid mixtures of hydrogen, called slush hydrogen, are being investigated to alleviate the density disadvantage and to increase the heat capacity and lower losses caused by evaporation [6].

Li, Be, and B, metallic elements of the second row of the periodic table, along with hydrogen-rich compounds of these metals, carbon and nitrogen are of primary interest. In the third row of elements, Mg, Al, and Si have some potential. Interest in elements with atomic numbers greater than 16 rapidly decreases because of the high molecular weight of the resulting combustion products. Ti and Zr are potential fuels for systems requiring very high performance on a volumetric basis.

Unfortunately, most of the elements chosen as promising propellant ingredients are metals and hence solids rather than liquids. The endeavor to incorporate solid-phase metals and metal compounds into liquid propellants has led to a whole new area of propellant technology. Heterogeneous fuels are discussed in detail later.

The metals also present an additional problem in that the product oxide, fluoride, or nitride species may be a solid phase at the combustion temperature or condense during expansion through the rocket nozzle. Figure 2 presents a graphical comparison of the phase properties of the primary product species for the elements of interest. The range of typical rocket combustion temperatures and exhaust temperatures

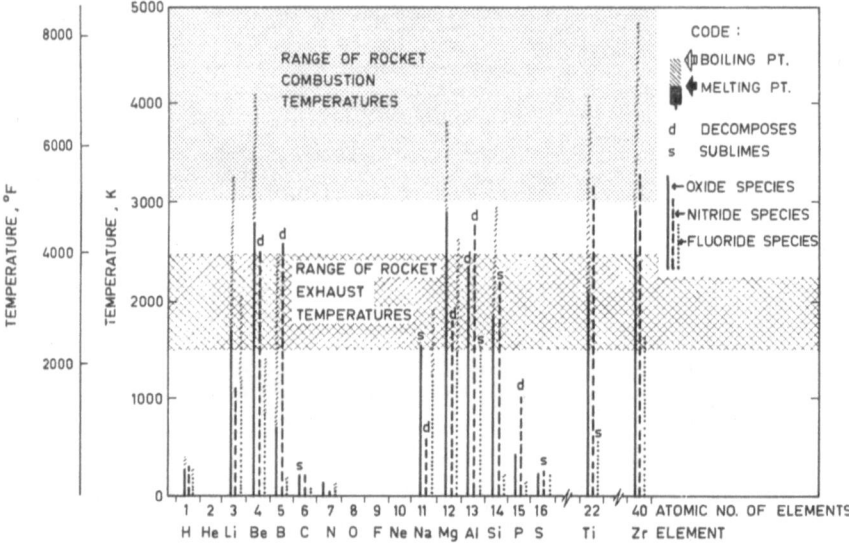

Fig. 2.   Phase properties of the oxides, fluorides, and nitrides of the low atomic number elements.

reveals that in most cases a condensed phase will be present in the exhaust. The energy exchange which occurs during a phase change must be considered. Liquid or solid combustion products cannot expand through the nozzle and must transfer their energy to a working gas during expansion. If the condensed phase grows to large particle diameter, a lag in the transfer of both thermal and kinetic energy occurs, representing a loss in the attainable specific impulse from the system.

An additional consideration is the extent of dissociation of the product species. Rocket combustion temperatures range from 3000 to 5000 K. At these temperatures, many of the product species dissociate, tying up energy in atomic and free radical species. Figure 3 shows the dissociation which occurs in typical product gases at a pressure of 50 atm [37]. The oxygen-oxidized propellant systems generally produce temperatures toward the lower end of the range, 3000 K, where dissociation of species such as CO_2 and H_2O becomes important.

The fluorine-oxidized propellant systems produce temperatures toward the upper end of the range, 4000–5000 K, where the dissociation of species such as H_2, HF, and metallic fluorides must be considered. Pressure has a significant effect on dissociation; lower pressure increases dissociation. A combination of low pressure and high combustion temperature can result in considerable energy being tied up in dissociated species. Conversion of this energy into useful kinetic energy for propulsion presents a design compromise. Rocket nozzles are normally made as short as possible, resulting in very rapid expansion of the product gases. The recombination of the various dissociated product species may not be rapid enough to occur within the nozzle, thus representing a loss in the theoretically attainable specific impulse.

A liquid propellant must satisfy many additional criteria to be considered for a

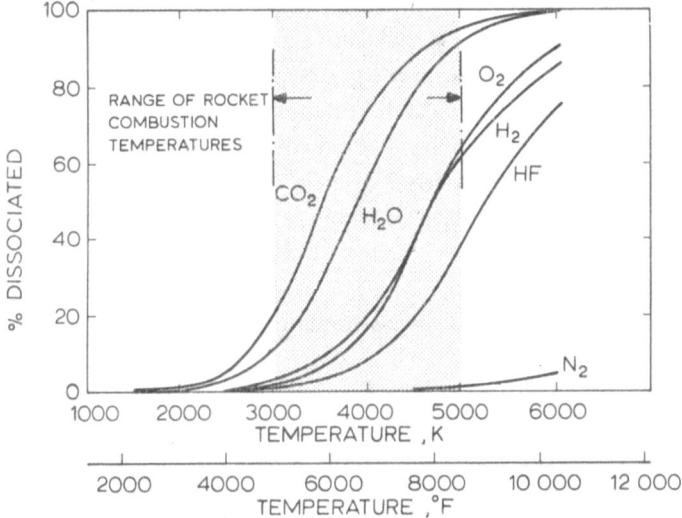

Fig. 3.   Dissociation of typical combustion products at a pressure of 50 atm. [37].

specific application. Rocket engineers have divided propellants roughly into two general types: storable and cryogenic. Storable refers to those which are liquid near standard conditions and can be stored in simple containers for long periods without serious deterioration or loss. Cryogenic refers to propellants which are gaseous at standard conditions and, therefore, must be handled under refrigerated conditions to remain liquid. A detailed discussion of all the criteria which must be satisfied for specific applications has been discussed by Silverman and Constantine [38].

Many compounds have been evaluated as propellants in the past, but only a small percentage of these are currently being used in operational liquid rocket engines. Current operational propellants are based on hydrogen and hydrogen-rich compounds of carbon and nitrogen as fuels and oxygen or oxygen-based oxidizers. Obviously, fuels based on the low atomic number elemental metals, metal hydrides, and organo-metallic compounds and oxidizers based on fluorine and oxygen offer the highest specific impulse potential. They are, therefore, most significant in advanced propellant research and development. Some of the metal-containing fuels such as B_2H_6, B_5H_9, $Al(BH_4)_3$, and $Al(CH_3)_3$ are liquids whose properties are quite well known and documented [25]. The liquid metal compounds are not discussed here because they do not demand radically new fuel technology and because of their limited applicability stemming from a combination of certain disadvantages in cost/availability, toxicity/handling, and logistics/storability. On the other hand, the application of solid metals and metal compounds to liquid systems depends upon new hetero-geneous fuel technology. This technology is the basis for the development of most advanced fuels and is discussed in the last section of this paper.

The advanced oxidizers are based almost entirely on oxygen and fluorine chemistry, and increasingly on fluorine. The utility of various types of oxygen- and fluorine-bearing materials that have been and/or are being considered in rocketry is discussed in detail. This discussion points up the relative usefulness of a broad range of materials in terms of properties and performance and provides insight into the direction future oxidizer research may be most fruitful. For convenience, the discussion of oxidizers is separated into two sections. The first section covers those materials which derive their oxidative capabilities primarily from oxygen. The second section deals with those materials which derive their oxidative capabilities from either fluorine or a combination of fluorine and oxygen.

2. Oxygen-Based Oxidizers

A. OXYGEN AND OZONE

Elemental oxygen exists in two allotropic forms, i.e., as the normal diatomic molecule, O_2, and the more energetic, less stable triatomic molecule ozone, O_3. The normal molecular form in the liquid state has been in use in rocketry for more than 20 years, and, even today, is used in larger quantity than any other oxidizer. Its widespread use is of course readily understandable in view of its high performance capabilities, low cost, and ready availability. Liquid oxygen does, however, present handling problems because of its very low boiling point, sensitivity to shock when contaminated with a variety of organic materials, and its capacity for supporting or intensifying fires. The solution of these problems has been largely responsible for present-day capabilities to handle routinely a broad range of industrially important gases as cryogenic liquids, for the ability to maintain new levels of cleanliness, and for advancements in the prevention and control of fires and explosions.

Ozone has a number of obvious advantages over normal oxygen, including higher performance, boiling point, and density, and the suggestion of its use as an oxidizer for rocket fuels dates back to 1929 [26]. Unfortunately, ozone is both unstable and toxic. Its instability in gaseous and liquid states and in solution in oxygen has been the subject of several studies [1, 2, 9, 30, 32] and presents the major impediment to its practical application to rocketry. Although the presentation of the results of such studies is beyond the scope of this paper, some of the more significant conclusions are: (1) extremely pure gaseous ozone is relatively stable, but its decomposition is accelerated by various agents, and it can decompose explosively; (2) pure liquid ozone detonates at a rate near that of TNT (~ 7000 m/sec), has a critical diameter of less than 1 mm., and is sensitive to contaminants in the oxygen from which it is produced; (3) oxygen–ozone solutions containing up to at least 33 wt% ozone do not detonate in diameters up to at least 38 mm (1.5 in.).

The question of the practical use of ozone in liquid rocket engines is still open, and its resolution will require much more research. On the basis of the existing knowledge it would appear, however, that a relatively dilute solution of ozone could be made a workable advanced oxidizer to replace oxygen. The over-all attractiveness of using ozone to improve rocket capabilities can be judged, however, only after the necessity of such an improvement is established and the alternatives presented by other advanced propellants and engine concepts are carefully considered. A comparison of some pertinent properties of oxygen and ozone is presented in Table I.

B. NITROGEN–OXYGEN COMPOUNDS

When considering nitrogen-oxygen bonded compounds as oxidizers it should be recognized that the nitrogen serves only as a carrier for the oxygen and, therefore, the nitrogen content should be kept small. (Nitrogen itself is important as an oxidizer for the peculiar propellants that depend upon metal nitride (particularly $BN_{(s)}$ and $AlN_{(s)}$) formation for the release of energy; however, in such propellants the introduction of oxygen (or fluorine) is thermodynamically detrimental to metal nitride formation). The nitrogen–oxygen compounds of greatest interest include the nitrogen oxides; nitric acid and nitrates; nitrosyl, nitryl, and nitronium compounds; and nitro organics.

1. Nitrogen Oxides

Among the various nitrogen oxides known (N_2O, NO, N_2O_3, NO_2, N_2O_4, N_2O_5, NO_3, N_2O_7) only the equilibrium mixture of $N_2O_4 \rightleftharpoons 2\ NO_2$, commonly referred to as nitrogen tetroxide, has been utilized extensively in pure form. Nitrous oxide (N_2O) is of little interest to rocketry because of its low boiling point and poor performance capability as an oxidizer. Nitric oxide (NO) and dinitrogen trioxide (N_2O_3) similarly are of little interest as pure components; however, they have been utilized as additives to N_2O_4 to depress its freezing point. NO or N_2O_3 utilized in this manner actually form complex mixtures of nitrogen oxides in accordance with the following reactions:

TABLE I

Properties of Oxygen and Ozone [25]

	Oxygen	Ozone
Freezing (melting) point, °C	−218.9	−192.8
Boiling point, °C	−183.0	−111.9
Critical temperature, °C	−118.8	−12.1
Critical pressure, atm	49.7	54.6
Critical density, g/cc	0.430	0.437
Density at NBP, g/cc	1.140	1.460
Heat capacity, cal/g, °C		
liquid at NBP	0.406	0.354
gas at 25°C (1 atm)	0.220	0.196
Viscosity at NBP, centipoise	0.190	–
Thermal conductivity at NBP, cal/cm sec, °C	0.00050	(0.00056)[a]
Heat of formation gas at 25°C, kcal/g-mole	0	34.2
Maximum theoretical specific impulse, lb(f) sec/lb(m)[b]		
with liquid H₂	391	422

[a] Extrapolated value.
[b] Values calculated by Aerojet-General Corp., assuming shifting equilibrium and optimum expansion from 1000 to 14.7 psia.

$$N_2O_4 \rightleftharpoons 2NO_2 \qquad\qquad (7)$$

$$NO + NO_2 \rightleftharpoons N_2O_3 . \qquad\qquad (8)$$

Such mixtures are commonly referred to as mixed oxides of nitrogen (MON) in the rocket industry and are normally prepared from $NO_{(g)}$ and $N_2O_{4(1)}$ (the equilibrium yields of NO_2 and N_2O_3 are spontaneously achieved upon mixing according to Equations (7) and (8)).

The extensive application of N_2O_4 in pure form to rocketry is a fairly recent development; however, its use as a stabilizer and freezing point depressant for HNO_3 dates back more than 15 years. Mixtures of HNO_3, N_2O_4 and small amounts of water and HF are referred to as inhibited red fuming nitric acid (IRFNA) or without the HF (a corrosion inhibitor) as merely red fuming nitric acid (RFNA).

Dinitrogen pentoxide (N_2O_5) offers a theoretical performance advantage over N_2O_4, but its marked instability presents significant problems as an oxidizer for rockets. Studies of its stability show a rather strong temperature dependence and the stabilizing influence of H_2SO_4, HNO_3, and $HClO_4$ as solvents. While temperature control and/or use of a solvent such as HNO_3 reduces the instability problem, these remedial approaches severely compromise its performance advantage.

Higher oxides of nitrogen have been reported (NO_3, N_2O_6, N_2O_7); however, their isolation has been difficult or impossible because of their rapid decomposition. Such materials are theoretically attractive from a performance standpoint, but their use as oxidizers appears remote at present. The properties of NO, N_2O_3, N_2O_4, and N_2O_5 are presented for comparison in Table II.

2. *Nitric Acid and Nitrates*

The use of nitric acid as a major component in liquid oxidizers dates back to at least World War II when it was used in a mixture with oleum (88wt% white fuming nitric acid and 12 wt% oleum) and was denoted as 'mixed acid'. In later years its use as white fuming nitric acid (WFNA) and inhibited white fuming nitric acid (IWFNA) developed because of its higher performance capabilities in these forms. These acids are fairly pure nitric acid: WFNA contains a maximum of 2 wt% H_2O and 0.5 wt% NO_2; IWFNA additionally contains approximately 0.7 wt% HF to inhibit corrosion. Both acids tend to decompose unless maintained in sealed vessels able to withstand their high equilibrium decomposition pressures (ca 100 atm under temperature and ullage conditions of interest in rocketry). This instability stimulated research which led to the development of IRFNA containing approximately 14 wt% NO_2, 2 wt% H_2O, and 0.7 wt% HF. This composition reduces the equilibrium decomposition pressure to the point that sealed storage vessels are practical and, in addition, depresses the freezing point, increases density, and improves performance capabilities. Nitric acid solutions containing even higher concentrations of NO_2 (up to ca 50 wt%, the limit of solubility at room temperature) have been studied and appear both practical and attractive. These latter compositions are more dense than IRFNA, have low freezing points except near the limit of NO_2 solubility, and have a performance advantage.

TABLE II

Properties of nitrogen oxides

	NO [8, 22, 25]	N$_2$O$_3$ [18, 22]	N$_2$O$_4$ [22, 25, 33, 34, 35]	N$_2$O$_5$ [18, 22]
Freezing (melting) point, °C	−163.6	−102	−11.23	30
Boiling point, °C	−151.8	3.5 (decomp.)	21.15	47 (decomp.)
Critical temperature, °C	−93.0	—	158.0	—
Critical pressure, atm	64.0	—	100.0	—
Critical density, grams/cc	0.52	—	0.56	—
Density, g/cc	1.269 (−152.2°)	1.447 (2°)	1.433 (25°)	1.642 (18°)
Heat capacity, cal/g, °C liquid	0.622 (−152.6°)	—	0.370 (25°)	—
gas	0.238 (25°)	0.206 (25°)	0.201 (25°)	0.213 (25°)
Viscosity (liquid) centipoise	—	—	0.39 (25°)	—
Thermal conductivity (liquid), cal/cm sec, °C	—	—	0.00031 (25°)	—
Heat of formation at 25°C, kcal/g-mole	21.58 (gas)	19.80 (gas)	4.676 (liquid) 2.17 (gas)	2.7 (gas)

The use of these solutions as replacements for WFNA, IWFNA, or IRFNA appears quite likely in a few specialized cases.

A number of other compounds have been considered as additives to HNO_3 to improve its properties and/or performance capilities. Among the more important are N_2O_5, NO_2ClO_4 (nitronium perchlorate), and $HClO_4$. While each of these systems theoretically provides increased performance, their use presents some technical and operational problems. Only careful consideration of the magnitude of these problems vs potential benefits and alternatives can resolve the question of their role as advanced oxidizers.

In general, the salts, esters, and halogen derivatives of nitric acid represent a poor compromise between performance and stability, and their future use in liquid systems appears limited. Several salts or esters of nitric acid have, however, achieved some prominence as oxidizers or monopropellants. The most important are the nitrates of alkali metals, N–H bases, and a few alcohols or polyhydroxy compounds. The nitrates of Li, Na, and K are largely of interest in solid propellant formulations but have been considered to some extent as solutes in liquid propellants (particularly in HNO_3). The low performance of such oxidizers does, however, limit their use to those systems where their other properties are of major interest. The nitric acid salts of N–H bases such as NH_3, N_2H_4, NH_2OH, CH_3NH_2, etc., are similarly of more interest as solid propellant ingredients, but their use as solutes in other liquid propellants has been considered. The sensitivity of these materials, even in solution, limits their applicability to special circumstances. Ethyl and propyl nitrates have been utilized as liquid monopropellants, and glycerol trinitrate and cellulose nitrate are the principal ingredients of double-base solid propellants. The halogen nitrates (NO_3Cl and NO_3F) although intriguing from the standpoint of theoretical performance are so unstable that their use as oxidizers appears very remote.

3. *Nitrosyl, Nitryl, and Nitronium Compounds*

The nitrosyl, nitryl, and nitronium compounds are of only moderate interest as oxidizers for liquid rockets. There has been sufficient interest in them, however, to warrant discussion of some of the more prominent members. Among the nitrosyl and nitryl compounds, NOF and NO_2F are most important. Both are cryogenic, however, and present handling problems not too different from those of liquid oxygen. Theoretically, their performance is intermediate between that of N_2O_4 and O_2. The most important nitronium compounds are the nitrate (N_2O_5) and perchlorate, both of which are solids. The possible use of these materials as solutes in HNO_3 is discussed in the sections on nitrogen oxides and nitric acid and nitrates, respectively.

4. *Nitro-Paraffins*

Nitro organics as a class of compounds are of greater interest in solid than liquid propellants. However, some nitro-paraffins have been utilized in liquid systems. Nitromethane has been used as a monopropellant, while tetranitromethane, hexanitroethane, nitroform, and nitroform salts have been considered as oxidizing components in bipropellant systems. The high melting points of the latter and their sensitivity restrict their use to solutes in lower freezing, insensitive liquids. A solution of 70 wt% $C(NO_2)_4$ in N_2O_4 is an example of such a system and has been proposed as an oxidizer [3].

C. HALOGEN–OXYGEN COMPOUNDS

In halogen–oxygen bonded materials other than those containing fluorine–oxygen bonds it should be recognized that while Cl, Br, and I are strong oxidizers, they are not desirable in rocketry because of their high molecular weights and the strong tendency of their reaction products with rocket fuels to dissociate. For these reasons, the content of Cl, Br, and I must be minimized if the oxidizer is to be of interest to rocketry.

1. Chlorine Oxides

The halogens, like nitrogen, form several oxides. The more important oxides of chlorine include Cl_2O, ClO_2, Cl_2O_6, and Cl_2O_7, but each has a strong tendency to explode. Cl_2O_7 is the most stable of the chlorine oxides, has a low freezing point ($-91.5°C$ [18]), high density, and is of most interest to rocketry. Unfortunately, even Cl_2O_7 is so shock sensitive that its utilization in rocket engines appears unlikely.

2. Perchloric Acid and Perchlorates

Pure perchloric acid is not subject to explosion as is its anhydride, Cl_2O_7. Hence, its consideration in rocketry has been somewhat stronger. From the standpoint of properties, $HClO_4$ is very attractive (f.p., $-112°C$; b.p., $39°C$ at 56 mm Hg; density, 1.764 g/cm^3 at $22°C$ [18]). However, it tends to decompose on standing, and the decomposition products are catalysts which auto-accelerate the reaction. The future of $HClO_4$ in rocketry, therefore, depends upon the ability to control this decomposition. The fact that $HClO_4$-H_2O solutions are much more stable than pure $HClO_4$ is reason to believe that $HClO_4$ may be sufficiently stable in more desirable solvents to be acceptable for rockets.

A number of salts of perchloric acid are stable; in fact, NH_4ClO_4 is one of the most commonly used oxidizers in solid propellants. In general, the perchlorates are of most interest as solid propellant ingredients, but some have been considered as solutes in liquid systems. Those of most importance in tailoring liquids include the light metal perchlorates, perchlorates of N–H bases such as NH_3, and N_2H_4, and nitronium perchlorate. The low boiling material, ClO_4F, sometimes called fluorine perchlorate, is of only theoretical interest because of its extremely explosive nature.

3. Bromine– and Iodine–Oxygen Compounds

The oxidizers containing Br–O and I–O bonds are of little interest to liquid rocketry because of poor performance and properties. The bromine oxides are particularly unstable, and I_2O_5 is the only stable iodine oxide. While I_2O_5 has poor performance potential on a weight basis, its high density (4.799 g/cc at $25°C$, [18]) theoretically predicts high performance on a volumetric basis. Because it is a high melting solid, its use in hybrid or solid propellant systems would be more easily realizable. Some consideration has, however, been given to finding ways to incorporate it into liquid systems.

4. *Peroxides, Superperoxides, and Ozonides*

Among this group of materials, H_2O_2 is the only compound to see actual use as a propellant. Its use in concentrations of 80–85 wt% in water dates back to World War II. It is currently used in approximately 90 wt% concentration in several systems, and considerable effort has been devoted toward the use of material of purity greater than 98%. The stability of H_2O_2 has been studied over many years [37], and a good understanding and control of decomposition have resulted. Although it has not been possible to eliminate decomposition completely, stabilizers and handling techniques have advanced to the point that extended periods of closed storage appear practical, and its application to new fields of liquid propulsion appears possible. The performance of H_2O_2 is not better than that of N_2O_4 with conventional fuels, but with metal-containing fuels it can be significantly superior. The advancements made in metallized fuels in recent years are largely responsible for the renewed interest in H_2O_2. Although well-accepted as a monopropellant, the future of H_2O_2 as an oxidizer will probably depend upon the acceptability of these advanced fuels.

Many metal peroxides, superoxides, organic peroxides, and hydroperoxides are known, but they are not expected to be significant because of their inferiority to H_2O_2 in terms of performance, properties, or stability.

The credibility of the existence of superperoxides of hydrogen has been the subject of considerable controversy in recent years. Ghormley [15] summarized the evidence for H_2O_3 existence, indicating it is unstable above $\sim -60\,°C$, and Czapski et al. [11] found that the half-life of H_2O_3 at 23 °C reaches a maximum of 2 sec in 0.02M (H^+) solution. On thermodynamic grounds, Benson [4] indicates that the alkyl and hydrogen trioxides should be reasonably stable, while the corresponding tetroxides appear unlikely of being produced or isolated above 80–100 K. Skorokhodov et al. (39) report yields of H_2O_4 of $\sim 25\%$ (based on O_3) from O_3 and atomic H reaction in the presence of a film of liquid O_3, and Csejka et al. [10] have determined the heat of decomposition of H_2O_4 at $-196\,°C$, from which they calculate a heat of formation of -27.9 kcal/mole. Thus, the existence of H_2O_3 and H_2O_4 appears established. While these materials are theoretically of interest, their instability creates doubt that they can be used in rocketry.

Alkali metals can form ozonides, and an ozonide is formed when O_3 is bubbled into liquid NH_3. Although such ozonides have theoretical interests, their instability makes their practical use in rocketry also appear remote.

D. NOBLE GAS–OXYGEN COMPOUNDS

Since the discovery in 1962 that the noble gases are not truly chemically inert, propellant chemists became intrigued with the possibility that they could serve as excellent carriers of oxygen (and fluorine) and thus generate a new family of chemical propellants. While the importance of this discovery to chemistry cannot be underestimated, so far it has not led to the preparation of new compounds as significant rocket oxidizers.

Of the noble gas–oxygen compounds, XeO_3 and XeO_4 are known best. Both are unstable and explode easily. Despite their large positive heats of formation (ΔH_f of $XeO_3 = 96$ kcal/mole) [20], the high molecular weight of Xe severely detracts from their theoretical propellant performance potential. Xenic acids and their salts are also known. These materials are more stable than the oxides, but again low performance potential excludes them from serious consideration as oxidizers. Krypton oxides, particularly as kryptic acid and its salts, are known and are theoretically more desirable than corresponding xenon compounds. These materials are, however, less stable than the xenon analogs and still offer poor performance because of relatively low oxygen content and the high molecular weight of krypton. Although the oxides of the light noble gases would be theoretically desirable, the likelihood of their being sufficiently stable for use in rockets is poor.

3. Fluorine-Based Oxidizers

Fluorine is the most energetic oxidizing element and as such is of prime importance in advanced oxidizers. The fluorine-based oxidizers discussed here include elemental fluorine, compounds containing oxygen and fluorine, nitrogen–fluorine compounds, halogen fluorides, and noble gas fluorides.

A. ELEMENTAL FLUORINE

Fluorine is the highest performing, stable oxidizer available to rocketry (ozone theoretically can provide higher performance with some fuels but is highly unstable). Although the performance potential of elemental fluorine has been recognized for many years, some of its properties pose severe problems to rocket application. Basically these problems arise from its low boiling point ($-188.14\,^{\circ}C$), the extreme reactivity and toxicity of it and its reaction products, and the high combustion temperatures it produces. The cryogenic nature of fluorine does not present problems too different from those encountered with oxygen for which solutions are available from modern cryogenic technology. The combination of extreme reactivity of the oxidizer and reaction products and the extremes in the temperatures encountered, however, create problems much more difficult to solve than those encountered with oxygen. These extremes necessitate a thorough knowledge of the capabilities of materials in such adverse environments and careful design to insure that the materials are not exposed to conditions beyond their inherent capabilities. The problem is further compounded because trace contaminants on hardware can initiate a reaction of sufficient intensity to cause burning of primary structures. While the materials problems and their solutions are known, the ability to pre-recognize problems and properly apply the solutions in hardware as complex as a rocket engine becomes the real crux to fluorine usage. Obviously, the high toxicity and cost concomitant with fluorine further impede its application to rocketry. Despite these difficulties, the interest in fluorine and fluorine–oxygen blends (flox) for rocket engines is demonstrated by the fact that a bibliography published in 1964 [5] lists approximately 350 reports which deal with F_2 and F_2-O_2 oxidizers for space applications. It seems safe to assume that fluorine can and will be used in the future if a real need for its superior performance capabilities exists.

B. COMPOUNDS CONTAINING OXYGEN AND FLUORINE

1. Oxygen Fluorides

Four oxygen fluorides are well known (OF_2, O_2F_2, O_3F_2, O_4F_2) and are the subject of an excellent review by Streng [41]. Of these, OF_2 is the most stable and important to rocketry. It is cryogenic and requires handling much like fluorine but does have a sufficiently high boiling point ($-144.8\,°C$) to be considered storable in a closed system in space. Its performance with H_2 is intermediate between that obtainable with O_2 and F_2 but is superior to either with many carbon-containing fuels. The combination of high performance and inspace storability is the strongest basis for its possible future use.

Although the other oxygen fluorides are far less stable than OF_2, O_3F_2 is moderately stable in liquid O_2 when stored in darkness. Despite its low solubility (0.110 wt% at 90 K), undersaturated solutions have been shown to be hypergolic with many fuels nonhypergolic with pure liquid oxygen [24, 41]. This ability to impart hypergolicity is significant to rocketry and could bring about its future use. The properties of the oxygen fluorides are compared in Table III.

2. Fluoroxy Compounds

The desirable oxidizing capabilities of oxygen and fluorine and the energy of the O–F bond make compounds containing the fluoroxy group (–OF) of considerable interest to rocketry. Unfortunately, much of the interest is academic because most of the known fluoroxy compounds (except for OF_2) are either unstable or contain other bonds which drastically reduce their performance potential. Among the better known inorganic fluoroxy compounds, NO_3F and ClO_4F are violently explosive, and the sulfur compounds SF_5OF and FSO_2OF also contain S–F bonds which are undesirable in rocket oxidizers. The organic fluoroxy compound CF_3OF is known and is stable to relatively high temperatures, but the strong C–F bonds degrade performance to the point that it is of no direct practical interest in rocketry. While it seems reasonable that other fluoroxy compounds can and will be prepared, the low strength of the O–F bonds may well limit the number of O–F bonds per molecule and/or necessitate the presence of other strong bonds to such an extent that high performance and stability are not simultaneously possible.

3. Chlorine and Nitrogen Oxygen Fluorides

Most of the more important compounds containing O, F, and N or Cl (i.e., NOF,

TABLE III

Properties of oxygen fluorides [13, 21, 22, 41]

	OF$_2$	O$_2$F$_2$	O$_3$F$_2$	O$_4$F$_2$
Freezing (melting) point, °C	−223.8	−163.5	−189 to −190	> −196 < −183
Boiling point, °C	−144.8(−145.3°)	−57 (decomp.)	−60 (decomp.)	(decomp.)
Critical temperature, °C	−58.0	—	—	—
Critical pressure, atm	48.9	—	—	—
Critical density, g/cc	0.553	—	—	—
Density, grams/cc	1.521(−145°)	1.736(−157°)	1.573(−157°)	—
Heat capacity, cal/g, °C				
Gas at 25°C and 1 atm	0.192	—	—	—
Liquid at −145°C	0.345	—	—	—
Viscosity, centipoise	0.283(−145°C)	—	—	—
Thermal conductivity (liquid), cal/cm sec, °C	0.00061(−183°)	—	—	—

NO_2F, NO_3F, and ClO_4F) have already been discussed. However, one additional compound deserves consideration – i.e., ClO_3F, perchloryl fluoride. In view of its remarkable stability compared with ClO_2F, chloryl fluoride, and ClO_4F, fluorine perchlorate, it is surprising that it was first synthesized in 1952. This compound is more stable than the chlorine oxides and yields some of the performance advantage of fluorine. Unfortunately, its performance gain over more common oxidizers, such as N_2O_4, is small, and its low boiling point ($-46.8\,°C$) creates added handling problems. These factors, plus the toxic nature of its combustion products, have generally caused it to be considered as an additive to other oxidizers rather than as a primary oxidizer.

C. NITROGEN–FLUORINE COMPOUNDS

1. *Nitrogen Fluorides*

Four nitrogen fluorides are quite well known (NF_3, N_2F_4, N_2F_2, and N_3F) but only NF_3 and N_2F_4 are of real interest to rocketry. N_3F is extremely sensitive to shock and light, and N_2F_2, being unsaturated, is less stable and lower in performance than NF_3 or N_2F_4.

Both NF_3 and N_2F_4 are cryogenic, boiling at $-129°$ and $-73\,°C$, respectively. Their high fluorine contents and densities lead to performance capabilities generally intermediate to those obtainable with liquid oxygen and OF_2 but lower than both O_2 and OF_2 with H_2 or high carbon-content fuels such as kerosene. Compared with OF_2 and F_2, NF_3 and N_2F_4 are relatively inert and easier to handle. Like OF_2, both appear capable of closed storage in space. These characteristics plus the fact that they are available in limited quantities at high cost and produce toxic exhaust products would appear to limit their future use largely to the specialized field of in-space propulsion. The properties of NF_3 and N_2F_4 are given in Table IV.

TABLE IV

Properties of NF_3 and N_2F_4 [22, 29]

	NF_3	N_2F_4
Freezing (melting) point, °C	-208.5	-163
Boiling point, °C	-129	-73
Critical temperature, °C	-39.1	36
Critical pressure, atm	44.7	77
Density, g/cc	$1.54(-129°)$	$1.5(-100°)$
Heat capacity (gas) at 25°C, cal/g, °C	0.180	0.182
Heat of formation (gas) at 25°C, kcal/g mole	-30.4 ± 2.0	-2.0 ± 2.5

2. *Fluoroamino Compounds*

A few inorganic derivatives of NF_3 are known (other than the N, O, F compounds previously mentioned) and are commonly referred to as fluoroamines. Two of these difluoroamine (HNF_2) and chlorodifluoroamine ($ClNF_2$), are moderately well characterized. While the physical properties of HNF_2 are desirable (b.p., $-23.6°C$, critical temp., $130°C$) [23], its extreme sensitivity virtually eliminates it from serious consideration as a future rocket oxidizer. $ClNF_2$ is more stable than HNF_2 but its low boiling point ($-67°C$) and low performance potential compared with NF_3 practically negate its usefulness in rocketry.

Many organic N–F compounds are known [19], and the perfluoro derivatives of many different organic nitrogen containing compounds are possible. However, the fluorine in the known compounds is mostly C–F rather than N–F bonded fluorine even though some of the starting materials were rich in nitrogen (e.g., urea). Thus, it appears that organic N–F compounds are sufficiently stable to isolate only when the carbon is highly fluorinated and when the ratio of N–F to C–F bonds is low. Certainly much more of the chemistry of organic N–F compounds must be elucidated before the true relationship between structure and stability can be defined unambiguously, but if the apparent relationship is true, organic fluoroamines with high contents of N–F bonded fluorine will be relatively unstable. It is easily shown on a theoretical basis that only organic fluoroamines having a large content of N–F bonded fluorine are capable of providing a performance advantage over more common oxidizers.

D. HALOGEN FLUORIDES

The halogen fluorides have long been recognized as promising oxidizers. Interest in these compounds is based on their high densities, storability under earth (and perhaps space) conditions, and good performance capabilities. They do, however, suffer some of the same drawbacks as other fluorine-containing oxidizers – i.e., high toxicity and cost and handling problems.

Of the known halogen fluorides, most interest has been paid to the higher fluorides of chlorine (ClF_3 and ClF_5) and, secondarily, to BrF_3 and BrF_5. The exceptionally high density of the iodine fluorides does, however, give them the theoretical capacity to deliver high performance on a volumetric basis and creates some interest in specialized cases. Except for ClF_5, these compounds have been known for many years and are well characterized. Greatest attention has been given to ClF_3, but discovery of ClF_5 in 1963 [40] caused attention to be refocused on this most interesting new member of the halogen fluoride family. The importance of this new oxidizer to rocketry is obvious merely from its theoretical performance potential and reasonably attractive physicochemical properties, and we would expect this material to receive more attention in the future. Some pertinent properties of the halogen fluorides are presented in Table V.

E. NOBLE GAS FLUORIDES

Since the discovery of XeF_4 in September 1962, several fluorides and oxygen fluorides of the three heaviest noble gases have been reported. As previously mentioned, the existence of such compounds intrigued propellant chemists with the possibility of gaining a new family of propellants.

TABLE V

Properties of the halogen fluorides [7, 14, 18, 22]

	ClF_3	ClF_5	BrF_3	BrF_5	IF_5	IF_7
Freezing (melting) point, °C	− 76.32	− 93	8.77	− 62.5	9.43	5.5
Boiling point, °C	11.75	− 12.9	127.6	40.3	100.5	Subl.
Critical temperature, °C	153.5	–	∼ 327	∼ 197	∼ 283	–
Density at 25°C, g/cc	1.81	–	2.797	2.465	3.186	2.8(6.0°)
Heat capacity (gas) at 25°C, cal/g °C	0.165	0.178	0.116	0.139	0.111	0.125
Heat capacity (liquid) cal/gram °C	0.303(5.09°)	–	0.217(22.64°)	–	–	–
Heat of formation (gas) at 25°C, kcal/g mole	− 37.97 ± 0.7	− 57 ± 15	− 61.1 ± 0.7	− 102.5 ± 0.5	− 196.35 ± 1.0	− 224 ± 1.5

Among the noble gas compounds, those of xenon are best known and are the subject of an excellent review by Malm *et al.* [27]. The xenon fluorides (XeF_2, XeF_4, XeF_6) are solids, and compared with the oxides they have a high degree of stability. From the stability standpoint, they are attractive as propellants; however, the high molecular weight of xenon results in low performance and virtually eliminates them from serious consideration. The xenon oxygen fluoride, $XeOF_4$, is also well known and is apparently a stable liquid. This material, like the fluorides, has poor performance potential.

The krypton fluorides, KrF_2 and KrF_4, are moderately stable, but the combination of the relatively high molecular weight of krypton and low fluorine content (< 50 wt%) results in performance capabilities of little real interest. The fluorides of the light noble gases have not been isolated although some spectroscopic evidence for their existence has been found. Theoretically, these materials would be good oxidizers, but present knowledge indicates that they may be too unstable to use as rocket propellants or perhaps even to isolate.

F. CONCLUSIONS REGARDING OXIDIZERS

Much effort has gone into developing advanced oxidizers. While much of the earlier work was devoted to oxygen-based oxidizers, the later work has become increasingly oriented toward fluorine-based oxidizers. In both cases we seem to be reaching the point where the chemical structures required to achieve higher levels of propellant performance are thermodynamically so unstable that they cannot be applied to conventional rocketry. Thus, we need breakthroughs in the stabilization of very energetic materials and in the conversion of this stored chemical energy into propulsive force without allowing the materials to pass through intermediate conditions conducive to the uncontrollable release of the energy.

4. Heterogeneous Fuels

The heterogeneous fuels are a class of liquid propellants that has been considered for propulsion for many years. Over 30 years ago Eugen Sanger suggested the use of suspensions of powdered aluminum in liquid hydrocarbons, and fuels of this type were investigated for air-breathing propulsion systems by the National Advisory Committee for Aeronautics two decades ago. The current interest in heterogeneous fuels began in 1958 and led to significant advances in the technology of these two-phase fuels. This recent work is the subject of this discussion.

A. THEORETICAL BASIS

Heterogeneous fuels are suspensions of finely divided metals or metal compounds in appropriate liquid fuels and represent one approach for using the low molecular weight metals (and certain of their derivatives) that have high heats of combustion with typical rocket oxidizers. The theoretical basis for the interest in metal-containing fuels is discussed in the introduction. Glassman [16] showed that all the metallic elements of molecular weight less than 30 (atomic numbers lower than 15), except sodium, have heats of combustion with oxygen that exceed the heat of combustion of gasoline with oxygen. All these metals, including sodium, have heats of combustion with fluorine that exceed that of hydrazine with fluorine. Of these seven elements, however, lithium, sodium, and silicon are poorly suited for use in heterogeneous fuels because of the reactivity of the first two and the marginal performance potential of the third. Interest has therefore been focused principally on four metals – aluminum, boron, beryllium, and magnesium – and some of their solid hydrides. A discussion of aluminum, beryllium, boron, and lithium as fuels in multicomponent propellants is presented in a paper by Gordon and Lee [17] which includes data from Dobbins [12] on the hydrides of aluminum, beryllium, and lithium. The theoretical specific impulse values of several heterogeneous hydrazine-based fuels oxidized with N_2O_4 are compared in Table VI; the specific impulse of the N_2O_4/N_2H_4 system is included for

TABLE VI

Maximum theoretical specific impulse values of heterogeneous fuels with nitrogen tetroxide

Fuel	Specific impulse[a] lb-sec/lb
$BeH_2 + N_2H_4$	346
$Be + N_2H_4$	327
$AlH_3 + N_2H_4$	318
$Al + N_2H_4$	303
$B + N_2H_4$	296
N_2H_4	292

[a] Pc/Pe, 1000/14.7 psia; shifting equilibrium.

reference. For a discussion of the utilization of heterogeneous fuels in propulsion systems, the reader is referred to the article by Wells [42].

B. MECHANISM FOR DERIVING ENERGY

The mechanism by which propulsive energy is derived from propellant systems containing metals and their compounds is somewhat different from that of conventional liquid propellant systems. For hydrazine and nitrogen tetroxide, for example, their combustion leads to the formation of N_2, H_2O, and H_2 through a relatively simple series of intermediate species:

$$10N_2H_4 + 9O_2 \rightarrow 10N_2 + 18H_2O + 2H_2. \tag{9}$$

In this reaction most of the energy released is derived from the formation of water, and the fuel-oxidizer ratio is adjusted to leave some of the hydrogen unoxidized to achieve an appropriate balance between the release of heat and the molecular weight of the combustion products.

On the other hand, when a metal such as aluminum is added to this system, some energy is derived from the decomposition of the hydrazine, but the principal source is the formation of the metal oxide:

$$8N_2H_4 + 4Al + 3O_2 \rightarrow 8N_2 + 16H_2 + 2Al_2O_3. \tag{10}$$

In this case the hydrazine is the sole source of the working gas, which is expanded by the heat released by the oxidation of aluminum, whereas in the unmetallized system a significant part of the hydrogen from hydrazine decomposition furnishes energy by oxidation as well as part of the working gas in the form of water.

During the combustion of the metallized system in a rocket chamber, it is unavoidable that part of the oxygen is involved first in the oxidation of hydrazine to form N_2 and H_2O, leaving an equivalent amount of the metal unoxidized. A loss of propulsive energy would occur if this situation were to persist throughout the entire length of the combustor and nozzle because the flame temperature would be lower and the average molecular weight of the working gas higher than theoretically possible. The full potential of this type of propellant system can be realized, therefore, only if the conditions in the combustor and nozzle permit the transfer of oxygen from the water to the aluminum and the absorption by the hydrogen of the heat released by this transfer before the products pass the throat of the nozzle. The construction and operation of combustion chambers and nozzles for use with metal-containing propellants must take into account the 'stay time' required for the occurrence of these processes that do not take place in unmetallized systems.

C. UTILIZATION OF MULTICOMPONENT PROPELLANTS

One approach to the utilization of a metallized system such as that involving hydrazine, aluminum, and oxygen is to inject the three components separately into the combustion chamber (tripropellant system). This avoids the problems associated with the suspension of the metal in the fuel or oxidizer (and is therefore not a heterogeneous propellant), but it imposes other problems. A method must be available for the transport and precise metering of the dry powdered metal, and a third tank and associated hardware must be incorporated into the propulsion system. The latter increases the weight and volume of the system and is likely to reduce the reliability-factors that must be weighed against the gain in propulsive energy.

A second approach is to incorporate the metal into a small amount of solid binder (hydrazine is not utilized in a system of this kind) and to burn the metal and the binder with oxygen in a combustion chamber similar to that of a solid rocket motor. A working gas such as hydrogen is then mixed with the hot reaction products in a secondary chamber in which the heat of metal oxidation is transferred to the gas before its entrance into the nozzle. Because this type of system involves both liquids and a solid grain that contains the metal, it is called a tribrid propellant system. It also involves a vessel for containing the metal-containing grain but avoids the transport and metering problems associated with the tripropellant approach. In this case, however, one of the principal problems is the efficient mixing of the hydrogen with the hot metal oxides to effect the high degree of heat transfer desired.

The use of a heterogeneous fuel, in which the metal compound is suspended in a liquid fuel, avoids a third storage vessel because it is used with the oxidizer as in a conventional bipropellant system. The technical problems are then associated only with the stabilization of the suspension, with the rheological properties of the stabilized fuel, and with the reactivity of the suspended solid with its carrier.

D. MECHANICAL STABILIZATION

Four methods may be considered for stabilizing heterogeneous fuels mechanically. The use of the metals in the form of sols is not practical because the powders, when subdivided to the required degree, are expensive and hazardous to handle. If they are allowed to become coated with the metal oxide, to eliminate their pyrophoricity, their metal content is reduced to an unacceptable level. Another approach involves the use of a liquid whose density equals that of the solid phase. Such an approach is also impractical because there are few liquid materials that have appropriate densities, and those that do are inappropriate as fuel components for chemical or thermodynamic reasons. Furthermore, the density equivalence is lost at reduced and elevated temperatures, causing settling or floating of the solid phase.

A third approach is emulsification. Most emulsified commercial products are the oil-in-water type, in which the oil is suspended in the form of small spheres in the water. The oil is the discontinuous or internal phase, and the water is the continuous or external phase. Stabilization of these systems is effected by surface-active compounds that prevent the oil drops from coalescing and by proportioning the two phases so that the lighter phase cannot separate to the top. In applying the emulsion approach to heterogeneous fuel technology water-in-oil types are used because of the nature of the liquid fuels that are desirable as carriers for the solid-phase fuel components. In a sense, they are inverted oil-in-water types of emulsions.

Theoretically, the internal phase of an emulsion can constitute as much as 74% of the total volume without distorting the suspended spheres. To produce emulsions having proportions of internal phase greater than this, the spheres must be distorted to accommodate them in the highly packed condition. The continuous phase, then, is no more than an interstice filler and a barrier to the direct contact of neighboring units of the internal phase. Thermodynamically, this is a highly unstable situation because of the large surface area of the internal phase. There is a tendency for the small volume of external phase to become the internal phase to effect a significant reduction in interfacial surface. If such an inversion can be prevented, however, the internal phase can be used to suspend a large amount of a powdered solid. The solid particles can settle to the bottom of each of the units but no farther, unless the interface between the unit and the continuous phase is broken.

Significant progress has been made in investigating metal-containing fuels stabilized by the emulsion approach. The studies have been confined, of course, to the storable fuels because of the obvious barriers to the emulsification of cryogenic liquids. In addition, the choice of suitable combinations for the phase pairs is limited because not only must they have adequate fuel value and be chemically compatible, they must also be essentially immiscible – e.g., hydrazine and rocket propellant kerosene. Emulsions containing more than 90% by weight of hydrazine dispersed as a phase internal to the kerosene are possible, so that when a powdered light metal is added, the content of combined emulsifier and kerosene is no more than 5%. Although fuel formulations of this type are moderately viscous, they thin as the shear rate is increased and become relatively fluid under the shear conditions of operating propulsion systems. The emulsion approach may ultimately be the favored one for stabilizing heterogeneous fuels, but additional work is required, especially to formulate fuels having proved, long term stability and to establish the storage and use conditions to which they can be subjected without inducing inversion.

The fourth approach to the stabilization of heterogeneous fuels – by gelation of the liquid carrier – has been developed more thoroughly than the emulsification method. Hence, the gelled fuels are discussed in greater detail than emulsified fuels. However, some aspects of heterogeneous fuels are not related to mechanical stabilization; when these are discussed in connection with gelled fuels, it will be apparent that the comments apply also to emulsified fuels.

It is possible to reduce the rate of settling of solids suspended in a liquid by using either a viscous Newtonian liquid or by making a viscous Newtonian liquid by adding a thickener to one of low viscosity. The utility of suspensions made this way would be very brief, for settling would commence immediately upon preparation, the duration of utility depending upon the degree of nonuniformity of distribution that could be tolerated by the application. Their utility in propulsion systems would be limited further because of the magnitude of the pressure differentials required to flow them at high velocities. It is apparent, therefore, that a viscous heterogeneous fuel must satisfy two criteria. The viscosity must be produced by a colloidal structure that has a yield value upon deformation and that is reversible when subjected to shear – i.e., the liquid must be a gel that shear thins. When the yield value is high enough for the colloidal structure to resist the shearing forces of suspended particles acting on the gel under the influence of gravity, the suspended material should remain uniformly distributed indefinitely. If, in addition, the colloidal structure is made up of units loosely united by bonds that are easily broken and reformed, thinning will occur under shear and the gel will recover when the shearing stress is removed – i.e., the system is thixotropic. These, however, are only the minimum requirements of a gelled fuel suitable for use in rocket propulsion. The yield value must be high enough to withstand the shearing forces of particles under many times the force of gravity, depending upon the mission in which the propulsion system will be utilized. The colloidal structure must exhibit sufficient capillarity to hold the liquid phase without bleeding, it must not change appreciably with age either chemically or physically, and these characteristics must be attainable with sufficiently small concentrations of the colloid so that the energy content of the fuel is not adversely affected.

E. TYPES OF GELLING AGENTS

Two general types of gelling agents have been extensively used in the development of heterogeneous fuels – the particulate agents such as silica and acetylene black, and the natural and synthetic hydrophilic polymers. Since they do not involve solubilization in the medium to be gelled, the particulate agents have been used to gel a variety of fuel carriers, including some for which suitable hydrocolloid gelling agents depend largely on electrical charges for the attraction between dispersed particles that hold them in the chain network which constitutes the colloid structure. Consequently, the electrically charged particles are strongly agglomerated in the dry state, and their dispersion in the liquid fuel requires high shear mixers and long mixing times. In addition, relatively large quantities of these inert materials are required to produce gels of adequate yield value and sufficient capillarity to prevent exudation of the carrier. In many cases the gel structure appears to change with age, causing its deterioration and the eventual appearance of supernatant carrier. Although these defects can be remedied to some degree by a hydrocolloid gelling agent in combination with the particulate colloid, the long term storability of such formulations remains to be proved.

F. HYDROCOLLOID GELLING AGENTS

The hydrophilic polymers [28, 31] are generally more effective for gelling the carriers which are most appropriate in heterogeneous fuel formulations. Dozens of such materials, most of which are commercially available, have been evaluated during the past several years for gelling hydrazine, 1,1-dimethylhydrazine, methylhydrazine, other hydrazine derivatives, and mixtures of these materials. They include the natural gums such as guar gum, locust bean gum, and gum arabic; alginates; and gelatin among the natural vegetable and animal products. From the class of modified natural products, various hydroxyalkylcellulose and carboxymethylcellulose products have been evaluated. Poly(vinyl alcohol), poly(vinyl pyrrolidone), sulfonated poly(vinyl toluene), polyacrylamide, poly(acrylic acid) and poly(methylacrylic acid) are examples of the synthetic commercial hydrophilic polymers that have been investigated. Some derivatives of poly(acrylic acid) that are not commercially available have also been studied.

The search for effective gelling agents from among the large number available as commerical products has been undertaken by many research and development laboratories interested in heterogeneous fuels. There has been little agreement, however, with respect to which materials are best suited to the applications, even among those who are investigating the same metal-carrier combinations. This is caused partly by differences in the criteria by which the gelled fuels are evaluated and the absence of standardization of methods and instruments for measuring their properties, but the most important barrier is the competitive spirit that exists among the various industrial, governmental, and research institute laboratories conducting these investigations. At present the gelling agents being used in various fuel formulations based on hydrazine and its derivatives include silica, acetylene black, hydroxyalkylcellulose, alginic acid, and poly(acrylic acid). Certain combinations of these are in use, and in some cases aluminum octonoate is used with alginic acid and with the cellulose derivatives.

G. CHEMICAL STABILITY OF FUELS

Although the principal problem associated with the development of a gelled heterogeneous fuel is the mechanical stabilization of the suspended solids, the solid phase must be chemically compatible with the carrier fuel. The hydrazine-type carriers utilized in these fuels are thermodynamically unstable and decompose to produce gaseous products by mechanisms that are sensitive to catalysts. Special attention must therefore be given to the problem of decomposition because of the large surface areas involved. In many cases it is necessary to investigate the contribution of the solid phase to the gas-producing reactions and to take measures to reduce them to tolerable levels by selecting materials carefully or by treating them by appropriate chemical or physical processes. In most cases, decomposition rates are not large enough to cause significant loss of the propellant itself, but the pressure developed in the fuel tank may reach an intolerable level, and the insoluble gaseous decomposition products are held suspended in the gel structure, causing an undesirable increase in the specific volume of the fuel.

H. IMPORTANCE OF RHEOLOGY

Even though gelling agents are selected on the basis of their capacity (at acceptable concentrations in the fuel) to provide long term mechanical stability of the suspended solids, other factors are of equal importance, and many secondary criteria are used to determine the utility of a heterogeneous fuel in propulsion systems. These factors vary, of course, with the application of the propulsion system. The rheological properties are as important as mechanical stability in view of the influence the flow properties have on the design of propulsion systems. In addition, the effect of temperature on the rheological properties should be as small as possible, and the gel structure should not deteriorate during repeated transfer and other handling operations.

These considerations have resulted in the introduction into liquid propellant research and development laboratories of instruments and techniques not previously used there; further, the production of heterogeneous fuels requires processes and control methods unnecessary in producing conventional propellants. Although some innovation has occurred, for the most part propellant chemists have satisfied their requirements by adopting instruments and methods from various segments of industry involved in the development of commercial products utilizing gel technology. They have found, however, in drawing upon industrial experience, that propellants must be developed to much higher standards of quality and uniformity than ordinary commercial products because the high levels of reliability demanded in missile and rocket propulsion systems cannot be jeopardized by even small changes in the physical and chemical properties of the propellants during storage. The realization of the performance improvements that can be obtained by using heterogeneous fuels in rocket propulsion depends largely on whether or not the propellant chemists can meet these high standards of quality.

188   CLAIR M. BEIGHLEY ET AL.

I. STATUS OF TECHNOLOGY

At present only one heterogeneous fuel, aluminum in gelled hydrazine [42], has been thoroughly developed and evaluated extensively in a static propulsion system. Other metal-containing fuels are being developed, but fuels based on the metal hydrides are still in the research and exploratory development stages. Many problems associated with the development of heterogeneous fuels have been solved, and the obstacles to the development of the more advanced systems are expected to be overcome. The technology of heterogeneous fuels is certainly ahead of the capability of propulsion engineers to use them.

References

[1] Apin, A. Ya., Pshezhetskiy, S. Ya., and Pankratov, A. V., *Zh. Fiz. Khim.* **34** (1960) 1935.
[2] Axworthy, A. F., Jr. and Benson, S. W., *Advan. Chem. Ser.* **21** (1959) 388.
[3] Behrens, H. Z., *Electrochem.* **55** (1951) 425.
[4] Benson, S. W., *J. Am. Chem. Soc.* **86** (1964) 3922.
[5] Cabaniss, J. H., 'Bibliography on Fluorine and Fluorine Oxygen Oxidizers for Space Applications', NASA-TM-X-53149 (Oct. 16, 1964).
[6] *Chem. Eng. News* **43** (Sept. 6, 1965) 41.
[7] 'Chlorine Trifluoride and Other Halogen Fluorides', Allied Chemical Corp., *Tech. Bull.* TA-8532-2.
[8] Cole, L. G., Jet Propulsion Laboratory, *Progr. Rept.* 9-23 (Oct. 18, 1948).
[9] Cook, G. A., Spadinger, E., Kiffer, A. D., and Klumpp, C. V., *Ind. Eng. Chem.* **48** (1956) 736.
[10] Csejka, D. A., Martinez, F., Wojtowicz, J. A., and Zaslowsky, J. A., *J. Phys. Chem.* **68** (1964) 3878.
[11] Czapski, G. and Bielski, B. H. J., *J. Phys. Chem.* **67** (1963) 2180.
[12] Dobbins, T. O., 'Thermodynamics of Rocket Propulsion and Theoretical Evaluation of Some Prototype Propellant Combinations', WADC TR-59-757 (1959).
[13] Fang, F., Allied Chemical Corp., *Tech. Bull.* *65-62* (Nov. 15, 1965).
[14] Gatti, R., Krieger, R. L., Sicre, J. E., Schumacher, H. J., *J. Inorg. Nucl. Chem.* **28** (1966) 655.
[15] Ghormley, J. A., *J. Chem. Phys.* **39** (1963) 3539.
[16] Glassman, I., *Am. Scientist* **53** (1958) 508.
[17] Gordon, L. J., Lee, J. B., *ARS J.* **32** (1962) 600.
[18] *Handbook of Chemistry and Physics*, 46th ed., The Chemical Rubber Co., Cleveland, Ohio, 1965–1966.
[19] Hoffman, C. J. and Neville, R. G., Lockheed Aircraft Corp., *Tech. Rept.* LMSD-703005 (August 1960).
[20] Hyman, H. H., *Science* **141** (1963) 61.
[21] 'Investigations of Space Storable Propellants (OF₂/B₂H₆)', Thiokol Chemical Corp., NASA CR-54741 (Rept. RMD 6039-F) (June 10, 1966).
[22] 'JANAF Thermochemical Data', Dow Chemical Corp., Midland, Mich.
[23] Kennedy, A., Colburn, C. B., *J. Am. Chem. Soc.* **81** (1959) 2906.
[24] Kirschenbaum, A. D., Stokes, C. S., and Grosse, A. V., U.S. Patent 3,170,282 (Feb. 23, 1965).
[25] Kit, B., Evered, D. S., *Rocket Propellant Handbook*, MacMillan, New York, 1960.
[26] Kondratyuk, Yu. V., *Zavoyevaniye Mezhplanetnykh Prostranstr* [The Conquest of Interplanetary Space], Novosibirsk, 1929.
[27] Malm, J. G., Selig, H., Jortner, J., and Rice, S. A., *Chem. Rev.* **65** (1965) 199.
[28] *Natural Plant Hydrocolloids*, ACS Monograph 11, Reinhold, New York, 1954.
[29] 'Nitrogen Trifluoride, Tetrafluorohydrazine', Stauffer Chemical Co., *Tech. Bull.* (Feb. 1960).
[30] Paushkin, Ya. M., *Khimya Reacktivnykh Topliv* [The Chemistry of Reaction Fuels], Izdatel'stvo Akademii Nauk SSSR, Moscow, 1962.
[31] *Physical Functions of Hydrocolloids*, ACS Monograph 25, Reinhold, New York, 1960.
[32] Platz, G. M. and Hersh, C. K., *Ind. Eng. Chem.* **48** (1956) 742.

[33] Reamer, H. H., Sage, B. H., *Ind. Eng. Chem.* **44** (1952) 185.

[34] Richter, G. N. and Sage, B. H., *Chem. Eng. Data Series* **2** (1) (1957) 61.

[35] Richter, G. N., Reamer, H. H., and Sage, B. H., *Ind. Eng. Chem.* **45** (1953) 2117.

[36] Schumb, W. C., Satterfield, C. N., and Wentworth, R. L., *Hydrogen Peroxide*, Reinhold, New York, 1955.

[37] Siegel, B., Schieler, L., *Energetics of Propellant Chemistry*, Wiley, New York, 1964.

[38] Silverman, J., Constantine, M. T., *Advan. Chem. Ser.* **88** (1969).

[39] Skorokhodov, I. I., Nekrasov, L. I., Kobozev, N. I., *Zh. Fiz. Khim.* **38** (1964) 2198.

[40] Smith, D. F., *Science* **141** (1963) 1039.

[41] Streng, A. G., *Chem. Rev.* **63** (1963) 607.

[42] Wells, W. W., *Space/Aeronautics* **45** (1966) 76.

[43] Wilkins, R. L., *Theoretical Evaluation of Chemical Propellants*, Prentice-Hall, Englewood Cliffs, N.J., 1963.

EFFECT OF ADDITIVES ON THE COMBUSTION OF HYDRAZINE/NITROGEN TETROXIDE

FORREST S. FORBES and ROBERT A. BIGGERS

*Air Force Rocket Propulsion Laboratory, U.S. Air Force Systems Command,
Edwards, Calif., U.S.A.*

Abstract. The effect of additives on the combustion stability of the hydrazine/nitrogen tetroxide propellant combination was investigated using experimentally measured droplet burning rates. Initially the burning rate and flame structure of neat hydrazine droplets were established to provide a baseline for additive effects and result comparison. Neat hydrazine burning was found to vary linearly with the droplet diameter and a hydrazine decomposition flame was observed to exist very close to the droplet surface. This decomposition flame causes a large temperature gradient at the droplet surface resulting in a much higher burning rate than found with thermally stable fuels. Additives selected on the basis of modifying the hydrazine decomposition flame were found to be effective in modifying the hydrazine/nitrogen tetroxide burning rate; however, if additive distillation or accumulation occurred, erratic burning resulted. This disruptive burning led to the testing of additives to produce droplet shattering. It was found that either $\frac{1}{2}\%$ tetranitromethane or 5% hydrazine nitrate is very effective for shattering hydrazine droplets burning in N_2O_4 vapor and thus may be used in a controlled way as an atomization agent. A series of engine firings was accomplished to demonstrate the effect of an additive on the combustion instability of a $25\#$ thrust laboratory engine using hydrazine and nitrogen tetroxide. At a normal chamber pressure of 300ψ the addition of hydrazine nitrate improved the stability of the engine by lowering the 'critical' chamber pressure required for stable operation.

1. Introduction

The literature abounds with references of studies to eliminate combustion instability in liquid rocket engines. While most of the efforts centered around the elucidation of the liquid oxygen/hydrocarbon propellant combination, data show that very few bi-propellants are free of combustion instability. With the current emphasis on non-cryogenic systems, hydrazine and hydrazine derivatives are receiving prime consideration as rocket fuels. Unfortunately, the design criteria for stable engine combustion has not been well established.

Combustion dynamics research to eliminate instability has ranged from empirical testing of large thrust engines to theoretical studies of unsteady flow phenomena. Various techniques have been used to alleviate combustion instability such as baffles, acoustical liners, and changing injector patterns and coarseness.

The Priem vaporization controlled combustion model provides a better understanding of the physical-chemical parameters affecting the combustion efficiency [1]. This model postulates that a low relative velocity between droplet and gas exists in a region near the injector face (Figure 1). Since the vaporization rate is a function of convective heat transfer and Reynolds number the burning rates are predicted to be low. It is also in this region that pressure perturbations have a profound effect on the vaporization rate and thus on combustion stability [2]. When the bipropellants are N_2O_4 and hydrazine, fuel vaporization may often control because of its lower vapor pressure. As pressure increases during a transient pressure pulse, fuel vaporization

G. A. Partel (ed.), Proceedings of the Second International Conference on Space Engineering. All rights reserved.

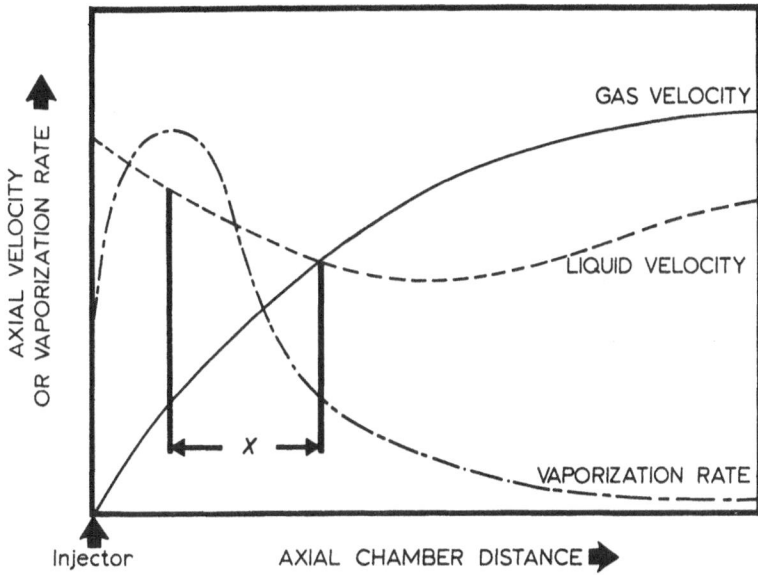

Fig. 1. Typical velocity and vaporization rate profiles.

decreases, and after the pressure wave passes, vapors flash off and combust in the oxidizer-rich environment. The Priem model has been found to correlate well with experimental results for a large number of propellants; however, the predicted combustion efficiencies for the hydrazine type fuels are below the experimental values. Hydrazine burning in nitrogen tetroxide undergoes decomposition prior to combustion resulting in 'two flames' [3]. By substituting experimentally obtained burning (decomposition) rate data into the Priem vaporization controlled model, good correlation between the predicted and measured combustion efficiency of hydrazine fueled rocket engines is obtained [4] (Figure 2).

The Priem-Guentert nonlinear instability model was extended by Dynamic Science Corporation to determine stability maps and scaling criteria for actual engines [5]. A related experimental program was conducted by the Air Force Rocket Propulsion Laboratory [6] using this instability model to provide experimental correlation between thrust chamber design parameters of propellant injection velocity, propellant droplet size, propellant flow rate and chamber pressure, and combustion stability.

Rocket engine stability is very sensitive to changes of burning rate in the low relative velocity region near the injector face [7]. Hydrazine decomposition controls the burning rate within this region and is the major source of heat in the droplet near the injector. It was postulated that additives to hydrazine could drastically affect the burning (decomposition) rate and the resulting combustion stability. Using the prediction methods [5] these additives should effectively change the relative distance (X in Figure 1) between the point of greatest evaporation and minimum ΔV to provide operation within the stable region (Figure 3).

Based upon the vaporization/decomposition stability models, an experimental

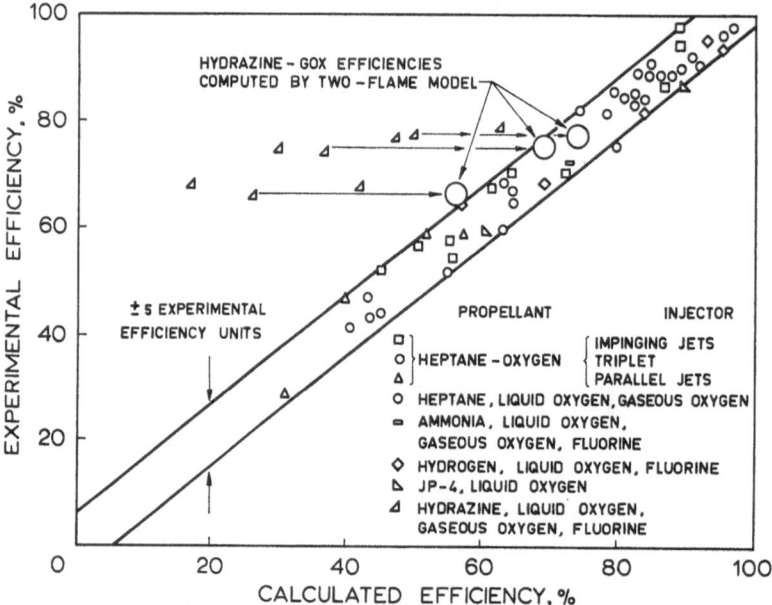

Fig. 2.   Comparison of experimental and calculated efficiencies.

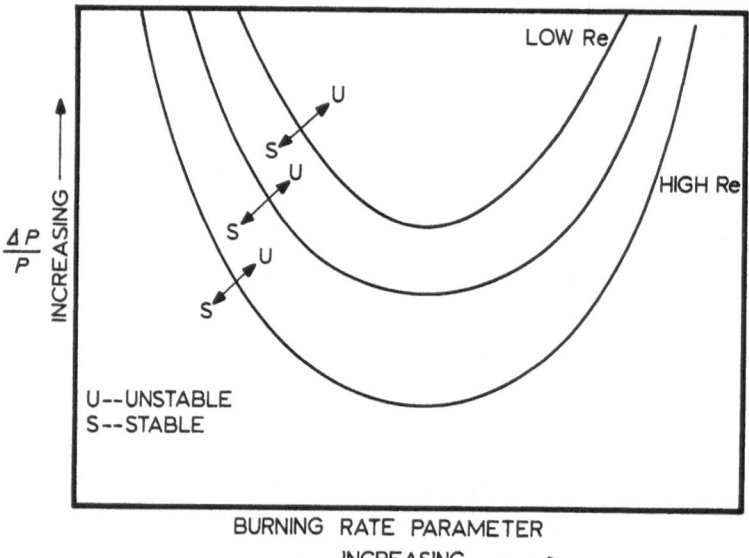

Fig. 3.   Typical stability limit curves.

program was undertaken to measure the effect of various additives on the burning rate of hydrazine droplets burning in nitrogen tetroxide vapor. An additional objective was to demonstrate the ability of selected additives to extend the stability limits of a small laboratory thruster.

2. Experimental Program

The initial approach was to experimentally measure hydrazine droplet burning rates with and without additives in an N_2O_4 atmosphere. Droplet burning was selected because it provided a means for studying individual droplet behavior. This behavior could then be related to the combustion chamber burning rate profiles and instability sensitivity in the steady-state and instability computer programs. The suspended droplet avoids the problems associated with the porous sphere technique such as large surface area and excessive heat transfer and fixed droplet size. Promising additives were subsequently evaluated in a 25-lb rocket chamber.

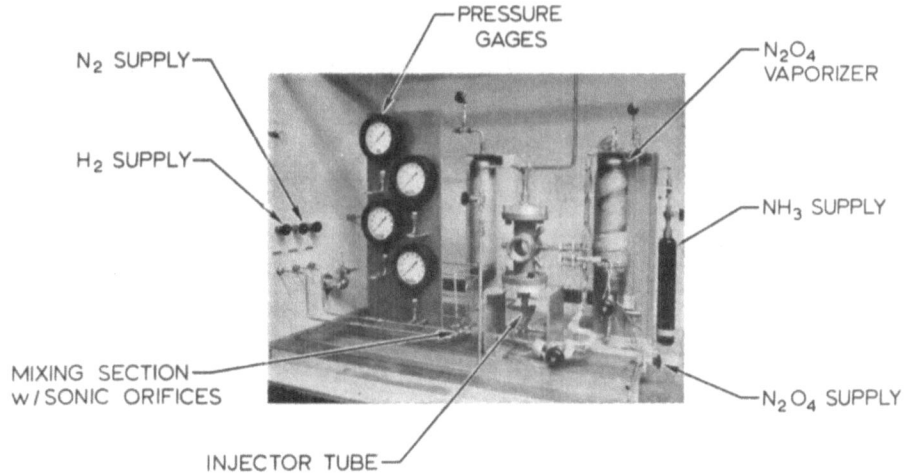

Fig. 4.   Droplet burner apparatus.

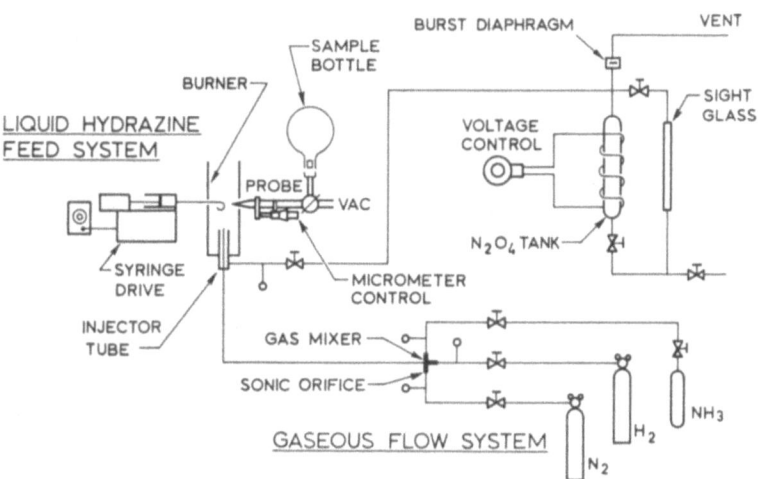

Fig. 5.   Suspended droplet burner.

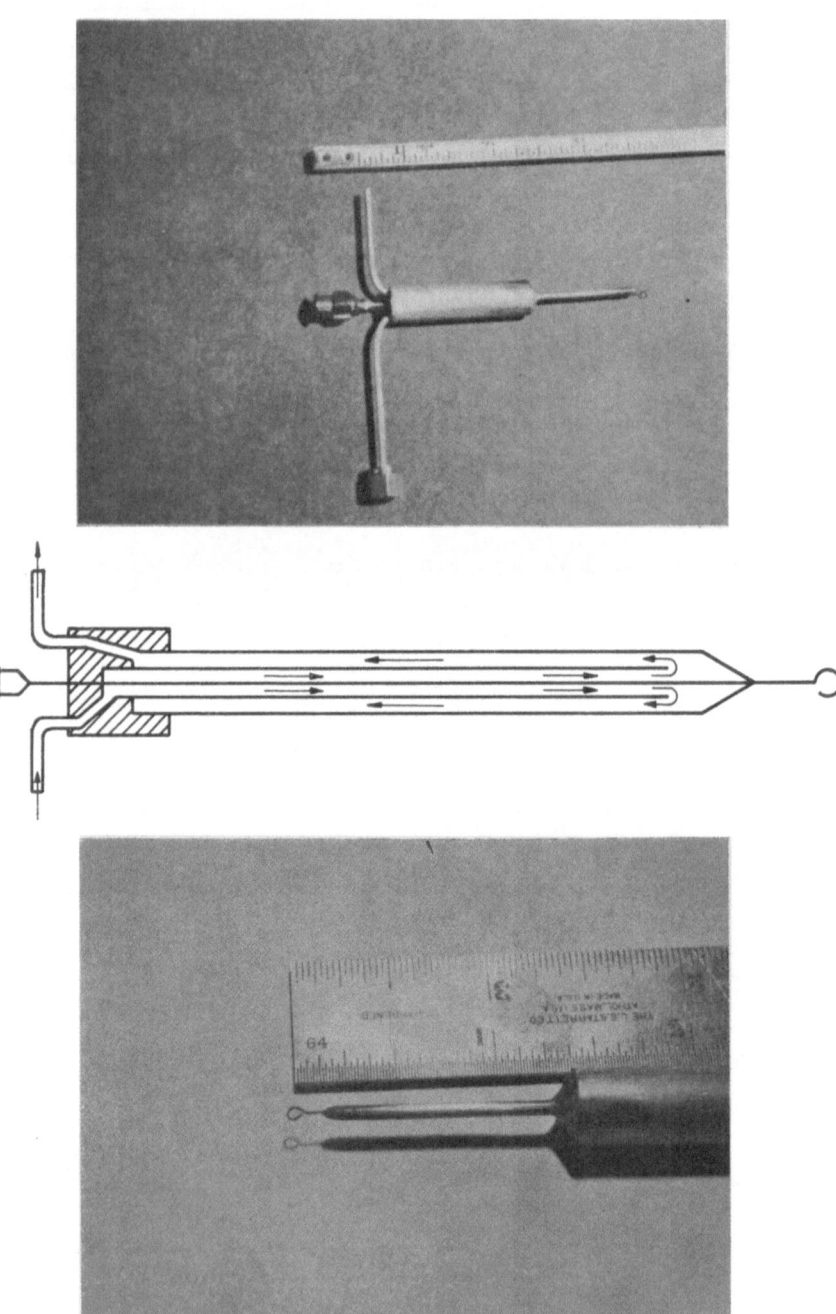

Fig. 6.   Water-cooled needle.

A. DROPLET BURNER APPARATUS

Droplet burning tests were conducted with the apparatus depicted in Figure 4 and shown schematically in Figure 5. The heart of the apparatus is the burner tube in which a water-cooled hypodermic needle was used to suspend the fuel droplet. N_2O_4 gas at various temperatures and concentrations was fed into the burner tube through an injector tube located in the base plate. Windows were built into the burner tube for visual and photographic observation. Oxidizer flow rates were measured with calibrated sonic orifices and were controlled by regulating upstream pressures. The liquid N_2O_4 supply bottle was heated electrically to vaporize and dissociate the N_2O_4 and provide gas for the oxidizing atmosphere.

The fuel droplet suspension needle (Figure 6) consists of a 0.0015-in. I.D. stainless steel tube jacketed with two major concentric tubes through which cooling water is circulated. This arrangement was selected to prevent boiling and thermal decomposition of the hydrazine within the needle. Fuel was fed from a syringe into the suspension needle with a variable-speed syringe drive. The droplet diameter range of 0.06-in. to 0.12-in. was maintained by proper adjustment of the syringe drive.

Experimental runs were started by flowing hydrazine through the water-cooled suspension needle, and then initiating the N_2O_4 gas flow. The variable-speed syringe drive was adjusted until the droplet diameter became constant. Burning rates are calculated from the measured syringe drive rate. Droplet diameter, flame thickness, and flame positions are measured from color photographs (Figure 7) showing the

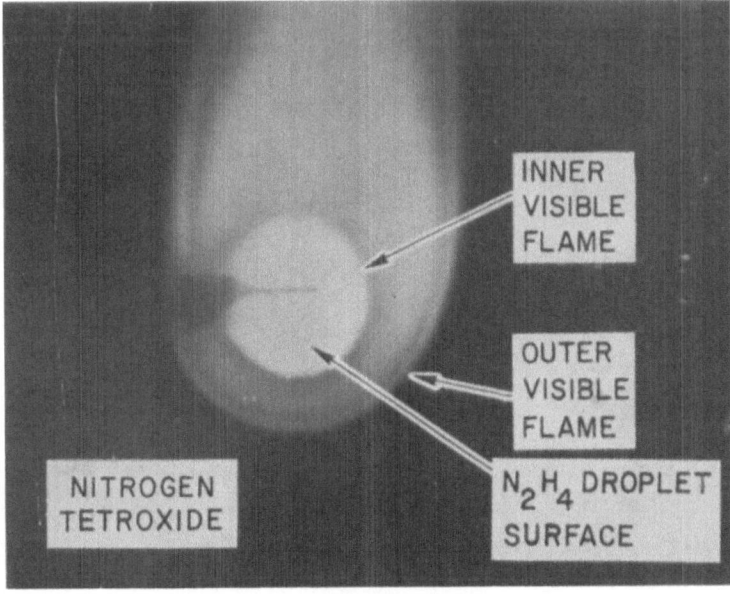

Fig. 7. Hydrazine/nitrogen tetroxide flame.

thinning of the decomposition flame as compared to the oxidation flame at the diffusion boundary layer.

B. 25-LB LABORATORY THRUSTER

The rocket engine used in this program is depicted in Figures 8 and 9. The chamber pressure of 300 psig was chosen as a value typical of those in operational engines. Severe erosion and burnout due to localized overheating from unstable engine operation was experienced with the original stainless steel nozzle. This was alleviated by silver-soldering a nozzle machined from a copper block to the stainless steel chamber. All experimental runs were 3.25-sec duration, which was more than adequate for achieving steady-state conditions. The O/F ratio was varied over the range of 1.05 to 1.45, with most runs near the stoichiometric value of 1.2. Since propellant temperatures drastically affected engine stability, the temperatures at the injector manifold of the 25-lb thruster were varied from 9° to 52°C.

Chamber pressure was monitored, and pressure fluctuations were used as the criteria for stability. During stable operation with neat hydrazine, the chamber pressure oscillated with a frequency of about 450 Hz, and with an amplitude of 8 psi or less. In the unstable mode, the pressure fluctuations were generally 15 to 20 psi with a frequency of about 4500 Hz.

C. PROPELLANTS AND ADDITIVES

The hydrazine and nitrogen tetroxide used in this program were of the propellant grade. All of the additives evaluated were chemically pure grade materials, and are

Fig. 8.   Stability-rating test stand for hydrazine/nitrogen tetroxide engine.

Fig. 9.   25-lb hydrazine/nitrogen tetroxide engine.

listed, with their properties, in Table I. Commercial quality nitrogen gas was used for
pressurization and as a diluent.

3. Results

The data from the hydrazine/additive burn-rate studies are presented in Table I and
are plotted in Figure 10. Effects of each additive are discussed below:

(a) Water. Because of its low boiling point, the principal influence was found to be
dilution of the decomposition and oxidization flame with the effect being more
pronounced in the lower temperature decomposition flame.

(b) Hydrazine nitrate. This additive caused erratic burning. Thus it was difficult
to adjust the flow rate to maintain a constant droplet diameter; the burn-rate data are
only approximate. The basic phenomenon was one of droplet shattering.

(c) Fluorobenzene. Brilliant white flashes were observed in the droplet flame. It was
postulated that these flashes were caused by periodic flashing of the fluorobenzene in
the droplet due to its relatively low boiling point.

(d) Urea. Increased droplet burning rate was noted, although the erratic burning
caused some measurement errors. Tiny burning particles (shooting stars) were
randomly ejected from the hydrazine droplet surface. An increase in the vaporization
rate of the droplet was probably due to agitation by the ejected particles.

(e) Dimethylsulfoxide. Stable droplet burning could not be achieved because of
gas bubbles formed within the droplet. This material apparently reduced the decomposition combustion kinetics.

(f) Phenylhydrazine. Smooth burning occurred and the burning rate was increased

TABLE I

Additive properties and effects

Additive	wt %	Formula	Boiling point °C	Melting point °C	N_2H_4/N_2O_4 droplet test [a]	Postulated mechanism
Water	10	H_2O	100	—	0.818	Diluent
Hydrazine nitrate	2	$N_2H_5NO_3$	—	—	1.44	NO_2 radical source and shattering agent
Fluorobenzene	1	C_6H_5F	84.9	—	1.09	H atom scavenger
Urea	1	$NH_2CO\ NH_2$	—	132.7	1.14	NH_2 radical source
Dimethylsulfoxide	1	$(CH_3)_2\ SO$	100	—	0.887	CH_3 radical source
Phenylhydrazine	1	$C_6H_5NH.NH_2$	243	—	1.07	NH, NH_2 radical source
Nitromethane	1	CH_3NO_2	101	—	—	NO_2 radical source
MMH	1	$CH_3(N_2H_3)$	89.5	—	1.00	CH_3 radical source
	2	$CH_3(N_2H_3)$	89.5	—	0.90	CH_3 radical source
	10	$CH_3(N_2H_3)$	89.5	—	0.745	CH_3 radical source
	10	$CH_3(N_2H_3)$	89.5	—	0.745	CH_3 radical source
UDMH	1	$(CH_3)_2\ N_2H_2$	63	—	1.00	CH_3 radical source
	2	$(CH_3)_2\ N_2H_2$	63	—	0.90	CH_3 radical source
	5	$(CH_3)_2\ N_2H_2$	63	—	0.80	CH_3 radical source
Tetranitromethane	0.25	$C\ (NO_2)_4$	125.7	—	—	Shattering agent
	0.5	$C\ (NO_2)_4$	125.7	—	—	Shattering agent

[a] $\dfrac{\text{Burn Rate of Fuel with Additive}}{\text{Burn Rate of Neat Fuel}}$

Propellants used: Hydrazine (N_2H_4)   B.P. 113.0°C   MIL-P-26536B

Nitrogen tetroxide (N_2O_4)   B.P. 21.2°C   MIL-P-26539B

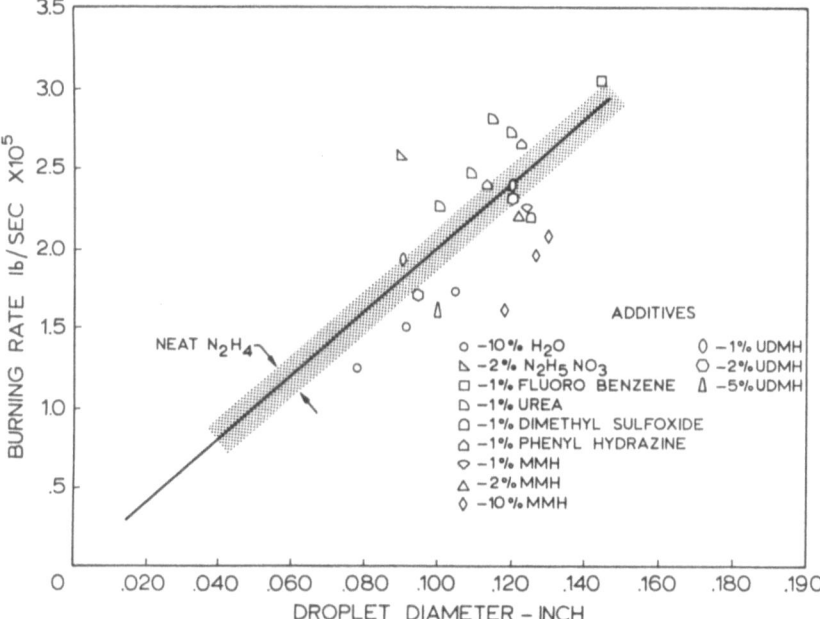

Fig. 10.   Effect of additives on the N_2H_4/N_2O_4 droplet burning rate.

very slightly. Although this additive is a radical producer, its low vapor pressure rendered it relatively ineffective.

(g) Nitromethane. This additive reacts with hydrazine and continuously evolves gas. Droplet burning was very violent with ejection of tiny burning particles. The vigorous bubbling of the droplet precluded measurement of its diameter. The basic change was from droplet burning to droplet shattering; effective as an atomization agent.

(h) Monomethylhydrazine (MMH). All mixtures with hydrazine burned smoothly but at a significantly lower rate than neat hydrazine. Since 10% MMH produced a greater decrease in burning rate than the addition of 10% water, and MMH being a fuel, a strong kinetic effect was indicated.

(i) Unsymmetrical-dimethylhydrazine (UDMH). A reduction in burning rates was also noted with this additive. Mixtures containing more than 5% UDMH caused erratic burning although neat UDMH burned smoothly. The erratic burning was attributed to distillation effects.

(j) Tetranitromethane (TNM). This material was chosen deliberately to produce droplet shattering, which it did in a violent manner. With 0.25 to 0.5% TNM, the droplet would burn smoothly for a few seconds, progressively turning darker in color, then explode.

A. PRESSURE EFFECTS

Tests were also conducted to determine the effect of pressure on the hydrazine droplet burning rate in the N_2O_4 environment. The results are shown in Figure 11. At pressures up to 3 atm, the burning rate appears to be linear. The tests were repeated with 10 percent water added to the hydrazine. First-order reactions were indicated; thus the additive effect should also be apparent at the higher pressure encountered in a rocket chamber.

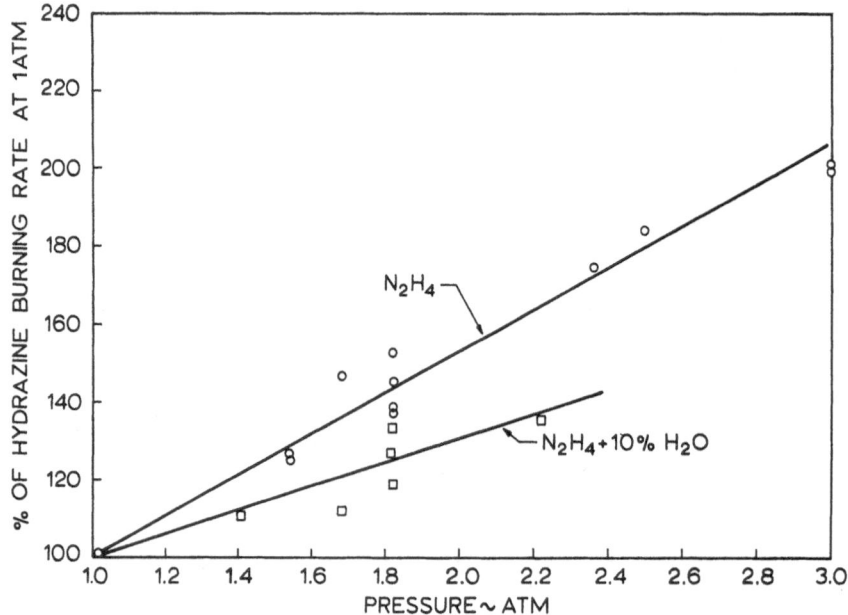

Fig. 11.   Effect of pressure on the N_2H_4/N_2O_4 droplet burning rate.

B. ENGINE APPLICATION

The droplet burning tests showed that several additives were very effective in shattering the hydrazine droplet. Atomization of the fuel would greatly increase the vaporization rate (burning rate) and thus alter the combustion profile. As shown in Figure 3, an engine operating within the unstable regime can be stabilized by shifting the burning rate. Additives should have the most pronounced effect along the upper portions of the curve. Also, an increase in burning rate could have an adverse effect if one is operating along the left portion of the curve, thus the steady-state combustion characteristics of a chamber must be analyzed before one can successfully hope to predict relative additive effects. Hydrazine nitrate effected excellent shattering of the hydrazine drops, thus it was chosen for the 25-lb thrust chamber studies. The hydrazine nitrate was prepared *in situ* by adding ammonium nitrate to the hydrazine and drawing off the resulting ammonia under vacuum.

The engine test data taken at 3.0 sec into the run are summarized in Table II and are plotted in Figure 12. Since increasing chamber pressure can increase combustion stability [6], this parameter was used in determining the effectiveness of the hydrazine. The pressure at which an unstable run became stable was tabularized in Table III

TABLE II
Engine stability tests

Run	ṁ$_{tot}$ (#/sec)	O/F	T_{ox} (C)	T_f (C)	T_{inj} (C)	P_c (Psia)	ΔP_c Ampl. (Psia)	ηc^*	Stable?	Fuel
7/29-1	0.0980	1.12	19	16	76	315	2.7	95.0	Yes	
7/29-2	0.0962	1.18	15	16	92	289	5.3	89.3	Yes	
7/29-3	0.0937	1.26	15	15	99	255	13.3	81.0	No	
7/29-4	0.0940	1.25	15	15	94	275	13.3	87.7	No	
7/39-5	0.0966	1.19	15	14	107	272	14.6	84.3	No	
7/29-6	–	–	17	14	94	255	20.0	–	No	
7/29-7	0.0930	1.45	14	12	88	297	2.7	90.3	Yes	
7/29-8	0.0956	1.05	13	12	52	307	4.0	96.4	Yes	
7/29-9	0.0917	1.26	12	8	75	283	12.0	93.0	No	Neat N$_2$H$_4$
7/29-10	0.0945	1.21	12	9	74	299	2.7	96.1	Yes	
7/29-11	0.0949	1.20	13	9	52	305	2.7	97.7	Yes	
7/30-1	0.0969	1.24	35	35	152	305	2.7	96.2	Yes	
7/30-2	0.0969	1.24	34	32	147	310	5.3	98.0	Yes	
7/30-3	0.0940	1.33	34	32	155	297	4.0	97.2	Yes	
7/30-4	0.0969	1.20	49	42	164	305	5.3	97.0	Yes	
7/30-5	0.0977	1.20	39	36	150	319	2.7	100.3	Yes	
7/30-6	0.0959	1.23	41	35	155	305	2.7	98.6	Yes	
7/30-7	0.0959	1.23	50	43	154	297	2.7	95.6	Yes	
7/30-8	0.0979	1.18	52	44	160	313	2.7	100.2	Yes	
7/30-9	0.0948	1.05	72	55	160	297	2.7	96.6	Yes	

(Table II continued)

7/31-1	0.0913	1.33	22	18	—	275	26.7	95.4	No
7/31-2	0.0942	1.23	16	18	—	295	2.7	97.8	Yes
7/31-3	0.0934	1.27	20	24	135	297	5.3	100.0	Yes
7/31-5	—	—	40	16	114	227	13.3	—	No
7/31-6	0.0942	1.17	31	27	114	309	4.0	104.0	Yes
7/31-7	0.0940	1.14	28	—	117	295	4.0	100.0	Yes
7/31-8	0.0904	1.24	30	—	117	272	13.3	96.2	No

Neat Hydrazine

8/1-1	0.0969	1.27	19	15	120	288	6.6	95.9	Yes
8/1-2	0.0964	1.25	18	14	117	297	6.6	99.8	Yes
8/1-3	0.0976	1.20	—	—	114	297	6.6	98.5	Yes
8/1-4	0.0973	1.19	26	—	104	297	6.6	95.5	Yes
8/1-5	0.0974	1.16	58	—	109	289	6.6	92.6	Yes
8/1-6	0.0974	1.16	37	—	109	297	6.6	95.0	Yes
8/1-7	0.0991	1.20	25	12	114	297	6.6	97.9	Yes
8/1-8	0.0985	1.19	12	19	112	299	5.3	99.2	Yes
8/1-9	0.0957	1.27	11	16	117	289	6.6	99.5	Yes

1 wt% $N_2H_5NO_3$ in N_2H_4

8/2-1	0.0982	1.27	11	18	106	291	8.0	97.9	Yes
8/2-2	0.0964	1.23	12	18	106	291	6.6	99.5	Yes
8/2-3	0.0963	1.18	9	16	103	271	6.6	93.0	Yes

8/5-1	0.0798	1.25	36	26	—	213	13.3	88.5	No
8/5-2	0.0969	1.19	16	15	—	276	2.7	94.0	Yes

2 wt% $N_2H_5NO_3$ in N_2H_4

8/14-1	0.0956	1.27	—	22	98	283	2.7	97.0	Yes
8/14-2	0.0861	1.28	—	22	98	243	16.0	92.0	No
8/14-3	0.0881	1.28	—	22	106	248	13.3	92.0	No
8/14-4	0.0927	1.20	—	22	106	283	6.6	99.5	Yes
8/14-5	0.0938	1.22	—	22	98	289	2.7	100.2	Yes
8/14-6	0.0930	1.20	—	22	104	283	4.0	99.5	Yes

5 wt% $N_2H_5NO_3$ in N_2H_4

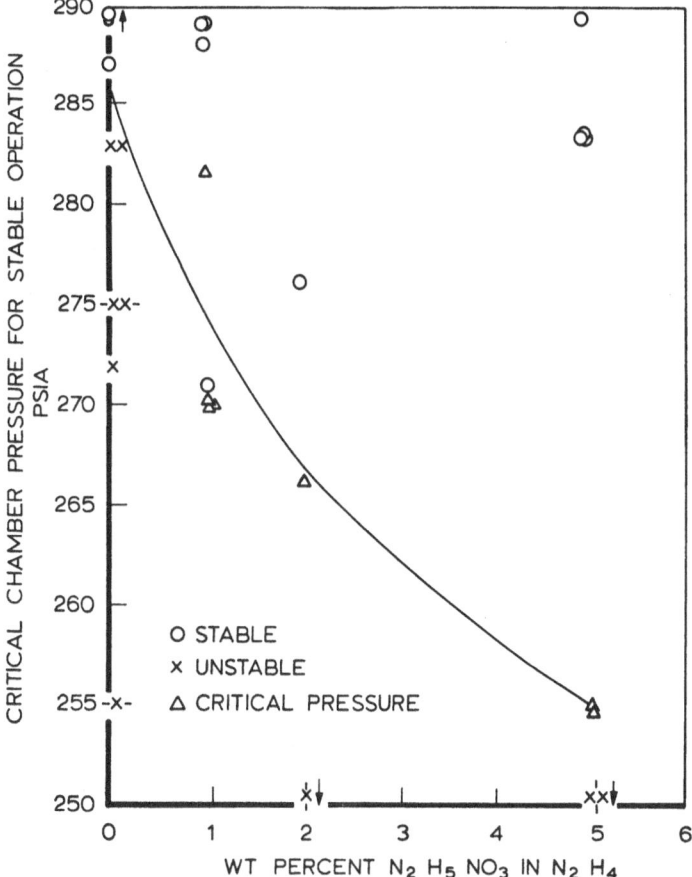

Fig. 12.   Effect of additive on laboratory engine stability limits.

and plotted as the critical pressure in Figure 12. A curve approximating these points is shown. Also plotted are the chamber pressure of runs that remained unstable, and those that were stable initially. In general, stable runs fall above the curve and the unstable runs fall below the curve. Thus, hydrazine nitrate has a definite stabilizing effect on this small thruster.

The Dynamic Science steady-state-combustion computer program was modified to include the measured additive effects of both distillation and shattering of statistical drop groups on propellant burning rate. Comparison of various additive effects upon an actual rocket engine combustion profile can now be made.

4. Conclusions

From this study it was found that additives to hydrazine burning in N_2O_4 vapor can have a profound effect on the burning rate. Changes in the peak burning rates of up to 20% were achieved with less than 2 wt% additive.

TABLE III

Operating conditions just prior to performance shift of runs which were unstable at start up

Run	Time of shift (sec)	$\dot{m}_{tot}$ (#/sec)	O/F	T_{ox} (°C)	T_f (°C)	T_{inj} (°C)	P_c (Psia)	Fuel	Engine
7/29-1	2.53	0.0981	1.11	18	18	101	291		
7/29-7	2.97	0.0929	1.08	14	11	122	283		
7/29-8	1.97	0.0957	1.16	12	7	68	283	Neat N_2H_4	
7/29-10	1.75	0.0944	1.20	12	9	60	286		
7/29-11	1.63	0.0947	1.20	13	10	78	288		
7/30-3	1.42	0.0944	1.33	38	41	161	280		
8/1-1	1.05	0.0965	1.27	15	15	88	270		
8/1-2	1.71	0.0962	1.24	15	14	104	275		
8/1-4	2.12	0.0974	1.19	28	–	99	294		300 psia
8/1-5	1.34	0.0946	1.04	65	–	68	230		
8/1-6	2.50	0.0971	1.17	33	–	104	297	1 wt% $N_2H_5NO_3$ in N_2H_4	
8/1-7	1.40	0.0987	1.19	15	22	91	283		
8 1-9	1.77	0.0952	1.27	11	17	104	270		
8/2-1	0.91	0.0982	1.26	28	18	96	270		
8/2-2	1.88	0.0971	1.23	11	19	94	283		
8/2-3	2.77	0.0961	1.20	9	18	104	270		
8/5-2	1.83	0.0965	0.94	16	21	195	267	2 wt% $N_2H_5NO_3$ in N_2H_4	
8/14-4	1.28	0.0931	1.20	23	26	98	255	5 wt% $N_2H_5NO_3$ in N_2H_4	
8/14-6	0.90	0.0948	1.22	24	26	86	255		

A number of additives were studied in the suspended drop burning apparatus and their effects on burning rate were measured. The effects of MMH, UDMH, and water were given a more extensive treatment. Tetranitromethane and hydrazine nitrate were found especially effective as shattering agents. Water, dimethylsulfoxide, MMH and UDMH suppressed the burning rate.

Hydrazine nitrate improved the stability of the 25-lb laboratory thruster operating at a nominal chamber pressure of 300 psi. The 'critical' chamber pressure, the minimum pressure at which stable operation was achieved, was the criteria used for determining additive effectiveness.

Additional studies are required to correlate additive effects with other parameters affecting combustion stability. Propellant flow rates and temperatures affect stability; these parameters were not adequately investigated in this program. Further refinement of the DSC modified Priem model to include these and other basic parameters is necessary before one can analytically choose the best additive and its optimum concentration.

Acknowledgement

The authors are appreciative of the analytical studies and model refinement accomplished by Dynamic Science Corporation. The efforts of Dr. B. P. Breen are especially noteworthy.

References

[1] Priem, R. J. and Heidmann, M. F., 'Propellant Vaporization as a Design Criterion for Rocket-Engine Combustion Chambers', NASA TR R-67 (1960).
[2] Priem, R. J. and Guentert, D. C., 'Combustion Instability Limits Determined by a Nonlinear Theory and a One-Dimensional Model', TN D-1409, Oct. 1962, NASA.
[3] Lawver, B. R., 'Some Observations on the Combustion of N_2H_4 Droplets', *AIAA* 4, No. 4 (April 1966) 659.
[4] Breen, B. P. and Beltran, M. R., 'Steady-state Droplet Combustion with Decomposition: Hydrazine/NTO', presented at AIChE 61st National Meeting, Houston, Texas, Feb. 19–23, 1967.
[5] Beltran, M. R., Breen, B. P., Kosvic, T. C., Sanders, C. F., Hoffman, R. J. and Wright, W. O., 'Liquid Rocket Engine Combustion Instability Studies', AFRPL-TR-66-125, 1 July 1966, Dynamic Science Corp., Monrovia, California.
[6] Abbe, C. J., McLaughlin, Weiss, R. R., 'Influence of Storable Propellant Liquid Rocket Design Parameters on Combustion Instability', *Spacecraft* 5, No. 5 (1968) 588.
[7] Beltran, M. R. and Frankel, N. A., 'Prediction of Instability Zones in Liquid Rocket Engines', *AIAA* 3, No. 3 (March 1965) 516–518.

EFFECTS OF PROPELLANT INJECTION ON
RESONANT COMBUSTION

W. HARTUNG

Deutsche Forschungsanstalt für Luft- und Raumfahrt, Trauen ü. Soltau, Germany

Abstract. A research program has been conducted with the objective of investigating the effect of injector-induced mass flux and mixture ratio-distribution characteristics on combustion chamber volume, combustion performance, and stability. The program included both analytical and experimental efforts to give correlations between the fluid dynamic phenomena of the injected propellants and the behavior of the resonant combustion.

Data are presented for various impinging injector types and two propellant combinations (one hypergolic and one inactive combination).

The effects of hydraulic flow characteristics of the injector elements on the combustion performance and the combustion stability were established by giving theories for the various combustion processes.

Mass-flux distribution was observed to have a significant effect on the necessary chamber length. The correlation between mixture ratio distribution of the cold-flow studies and combustion efficiency is not quite as clear. There are some differences, especially with hypergolic propellants due to reaction kinetics between the hydraulic criteria for optimum mixture ratio distribution and the criteria for obtaining the best combustion efficiency with a given injection element.

The effects of specific spray characteristics on the combustion stability were seen to be related to the specific injector type and the propellant combination.

The result of extended experimental studies on the two-on-one element show a clear effect of momentum and orifice area ratios of the impinging streams.

Definitions

(1) Mixing factor:

$$E_m = 100 \left[1 - \left(\frac{\sum_0^{n'} \dot{m}_x (\Omega_0 - \Omega_x)}{\Omega_0 \cdot m_{\text{ges}}} + \frac{\sum_0^{n''} \dot{m}_x (\Omega_0 - \Omega_x)}{(\Omega_0 - 1) \cdot \dot{m}_{\text{ges}}} \right) \right].$$

(2) Mass flux standard deviation:

$$\delta = \left[\sum_0^n \left(\frac{\dot{m}_x}{\dot{m}_{\text{ges}}} - \frac{1}{N} \right)^2 \right]^{1/2}.$$

(3) Mixture ratio:

(a) $\qquad \Omega = \dfrac{1}{\dfrac{\dot{m}_1}{\dot{m}_2} \cdot \dfrac{\varrho_2}{\varrho_1} + 1} \cdot 100 \quad [\%]$

(b) $\qquad r = \dfrac{\dot{m}_{\text{Ox}}}{\dot{m}_{\text{Br}}}.$

G. A. Partel (ed.), Proceedings of the Second International Conference on Space Engineering. All rights reserved.

Nomenclature

A = area (orifice or nozzle throat)
D = orifice diameter
$C^* = p_c \cdot A_t / \dot{m}_{ges}$ characteristic velocity
J = momentum of a jet
$\dot{m}$ = mass flux per second
n' = number of samples with $\Omega_x < \Omega_0$
n'' = number of samples with $\Omega_x > \Omega_0$
p_c = chamber pressure
L_{zyl} = cylindrical length of the chamber
w = jet velocity
ϱ = density
ξ_b = C^*_{eff} / C^*_{th} combustion efficiency
Ω_0 = overall or nominal mixture ratio
Ω_x = local mixture ratio

SUBSCRIPTS

Ox   oxidizer
Br   fuel
th   theoretical
eff   effective
ges   total
x local point in a chamber cross-section.

1. Introduction

The performance of fuel combustion in rockets depends on one side on the chemical and physical properties of the fuel combination and on the other on the hydrodynamic characteristics of the injection elements.

A research program has been conducted with the objective of investigating the effects of injector induced mass flux and mixture ratio distribution on combustion chamber volume, combustion performance, and stability.

The operating characteristics of thrust chambers, as combustion efficiency, chemical and thermal load on the walls, and stability, depend mostly on the injection elements.

The evolution of optimized injector design criteria will require a clear understanding of those injection parameters which control the combustion process. For most propellant combinations, three physical processes are critical to the achievement of efficient propellant combustion:

(1) Mass and mixture distribution (mixing)
(2) atomization and
(3) vaporization.

The manner in which these three processes are completed depends almost entirely on the design characteristics of the injector and the properties of the propellants. Thus, the eventual determination of injector design criteria leading to predictably high combustion performance must result from correlations of atomization and distribution characteristics with combustion efficiency and injector mechanical design parameters.

Separate from other phenomena, high-combustion efficiency in volume-limited thrust chambers will occur when the initial, local mixture ratio distribution is at or near the target chamber mixture ratio. This implies that the injector should provide a spray field having an uniform mixture ratio over the entire flow cross section. Any departures from this initial uniformity must be corrected by turbulent mixing.

2. Cold Flow Tests

Cold-flow studies on single-injector elements of various types were conducted to define the major variables affecting the distribution of mixture ratio and mass distribution. The propellants were simulated by dyed water, so that influences of the density, immiscibility, surface tension, and viscosity were eliminated.

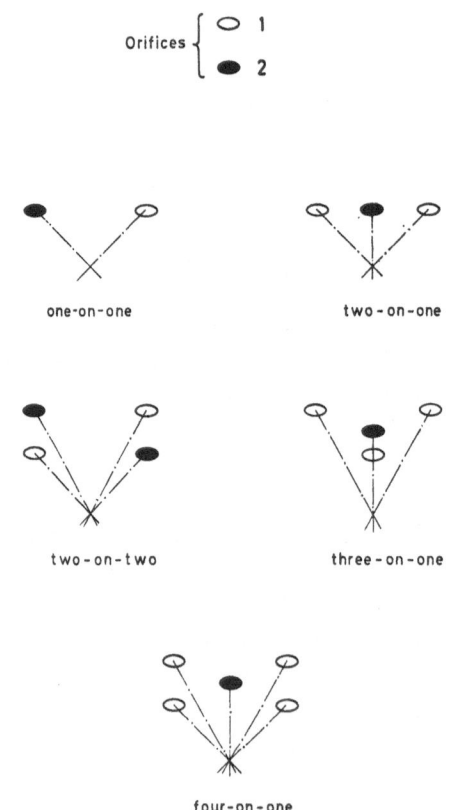

Fig. 1.   Schematic representation of various kinds of symmetrical impinging-jet injector elements.

The mixture ratio distribution may be characterized by a so-called mixing factor E_m, which is essentially a summation of the mass-weighted differences between the local O/F values and the over-all, or nominal value. The uniformity of the mass flux distribution was characterized using an expression which is known as the standard deviation of an arithmetical distribution.

Based on the mixing factor E_m criteria for optimum mixing have been determined

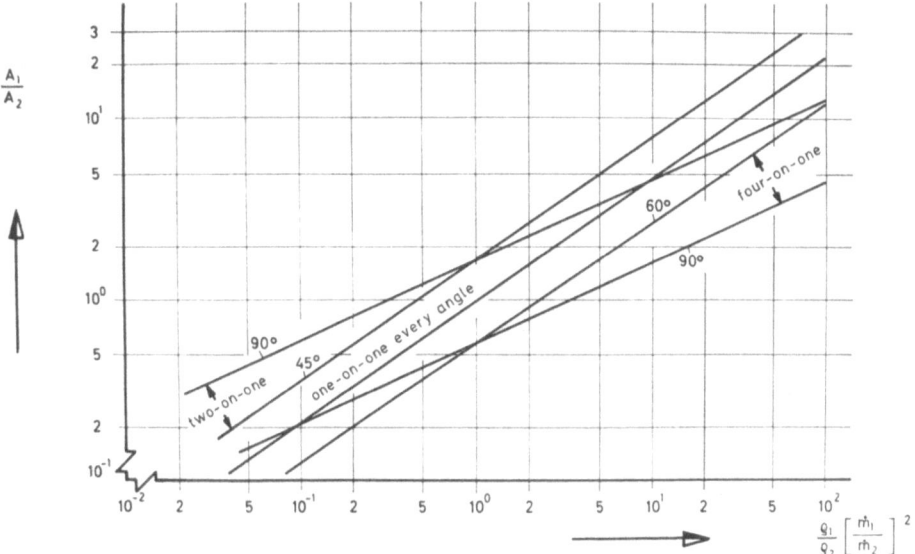

Fig. 2.   Optimum area ratios of various impinging elements with respect to mixing factors versus mixture ratio.

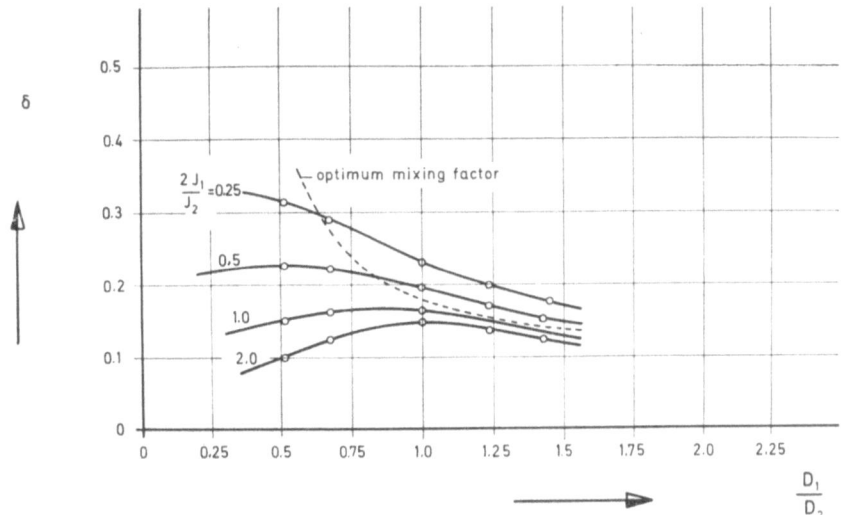

Fig. 3.   Mass flux standard deviation of two-on-one impinging elements versus jet diameter ratio

for doublet, triplet and quintuplet impinging-jet elements (Figure 1). The results were typically obtained with symmetrical single elements, in order to eliminate the effects of interactions with sprays produced by adjacent elements, which can augment liquid-liquid mixing in multi-element injectors.

Figure 2 shows a correlation between the area ratios A_1/A_2 of several element-types and the nominal mixture ratio of the propellants for which the mixing factor E_m will reach a maximum. The included impingement angle has a clear influence on the mixture-ratio distribution of the element-types with a central nozzle.

Figure 3 shows the standard deviation of the mass flux for 'two-on-one' elements versus the diameter ratio of the jets. These hydrodynamic properties of the injection elements provide the basis of relating injector design parameters to the associated performance characteristics.

3. Hot Firing Results

Differences between cold flow tests and hot firing data can be expected due to influences of vaporization (especially when the heat of vaporization of the propellant-components is different) and the subsequent combustion.

The influences of the various hydraulic parameters on steady state combustion may be determined by analytical calculation or theories, as follows:

3.1. MASS FLUX STANDARD DEVIATION

Gross maldistribution of propellant mass is characterized by small radial movement of the propellant mass and leads to droplet agglomeration. Analytical calculations indicated that radial movements of the droplets have a great effect on the length of the combustion chamber.

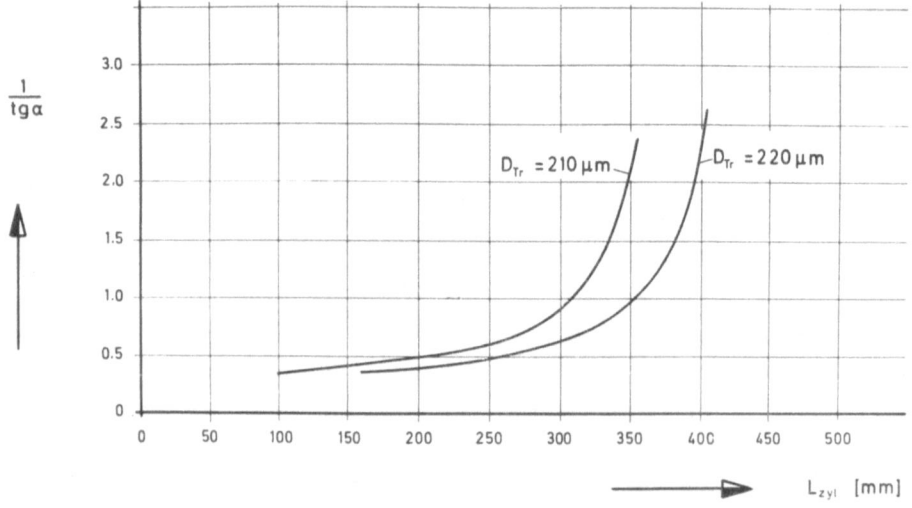

Fig. 4.   Effect of injection angle on droplet vaporization.

Figure 4 shows the characteristic velocity plotted against the chamber length for several diameter ratios of the triplet elements. The tests were conducted with N_2O_4 and JP_4 as propellants. The table at the right side on the picture gives the standard deviations for each diameter ratio of the elements. The table shows that with increasing mass flux standard deviation the chamber length increases too, if the characteristic velocity remains constant.

Figure 5 indicates the effect of the mass flux standard deviation on the length of the chamber for two propellant combinations. With the hypergolic propellant, the injection elements proved to be very sensitive against changes in the over-all-mixture

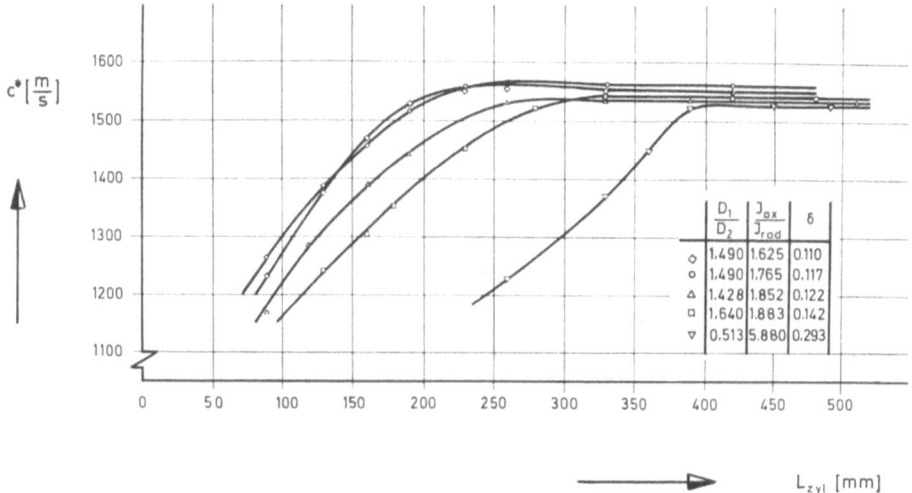

Fig. 5. Effect of chamber length on characteristic velocity for various jet diameter ratios of the triplet element.

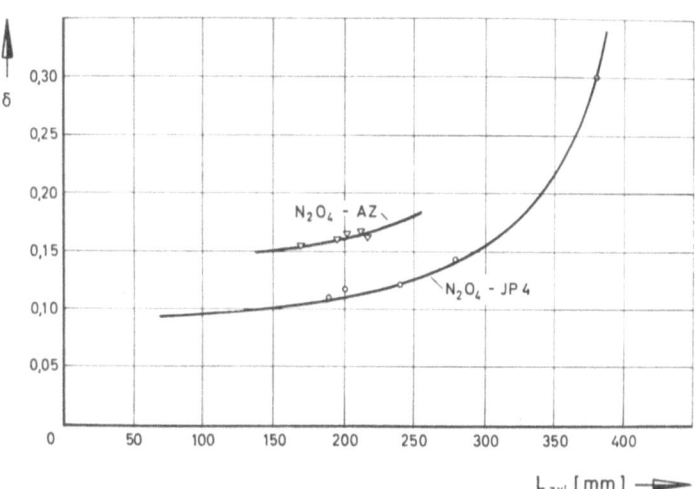

Fig. 6. Effect of mass flux standard deviation on the chamber length.

ratio, therefore it was not possible to make the tests in the same large region of mass flux distributions as with the non-hypergolic propellants.

Figure 6 shows the effects of injection angle for single droplets on the chamber length on the basis of an analytical calculation. The curves are very similar to those of the former figure.

3.2. MIXTURE RATIO DISTRIBUTION

Several models have been established to characterize the influence of O/F distribution on performance. The following have worked very successfully in many trials.

In this analysis, the flow is separated into stream tubes of various O/F ratios based on the injector mass distribution. The fluids issuing from the injector into each stream tube are presumed to mix, combust, and expand through the nozzle without mixing or interacting with any adjacent stream tube. The geometrical shape of a given stream tube is determined by the dimensions of a local area over which O/F can be considered constant.

The calculation based on this model presents losses of 2% to 6% of the characteristic velocity at a given over-all-mixture ratio for most of the elements.

According to the theoretical model, hot-firing performance results should show a clear correlation with the mixing factor E_m of the cold flow spray characteristics. But in the case of highly reactive hypergolic propellants the energy release creates a gas film at the interface of the impinging point which tends to decrease the combustion efficiency, due to jet separation.

Figure 7 shows the combustion efficiency versus the over-all-mixture ratio for triplet elements with various jet diameter ratios and for two different propellant combinations. The performance curves for N_2O_4–Aerozin at the right side show

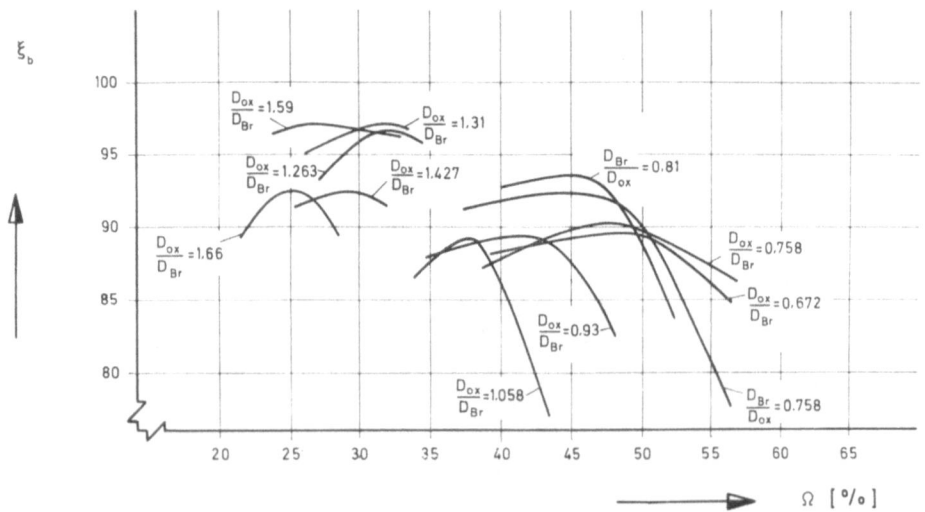

Fig. 7. Combustion efficiency of triplet impinging elements with various diameter ratios plotted against mixture ratio.

significant lower efficiencies than those for the non-hypergolic combination N_2O_4–JP_4.

The efficiency of the latter one corresponds to the region given by the analytical calculation. The mixture ratio is defined as the weight of component 2 divided by the whole propellant mass and is calculated for the density ratio 1.

In Figure 8 the optimum area ratios with respect to combustion efficiency are plotted against a standardized mixture ratio raised to the second power. A comparison with the hydrodynamic data show significant differences between cold flow tests and

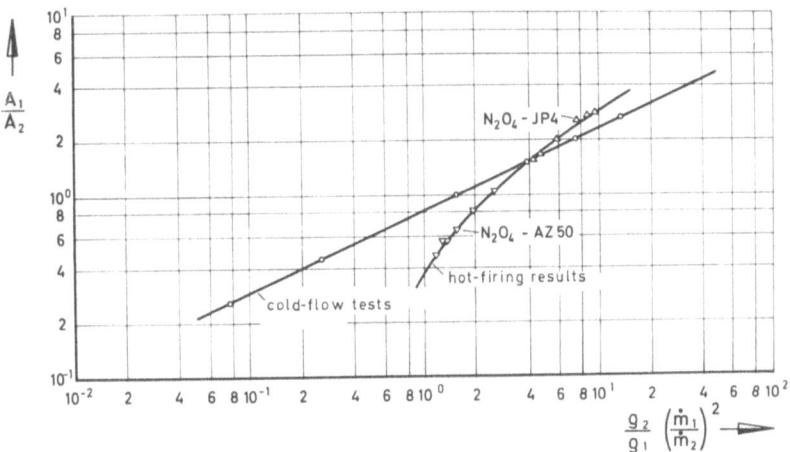

Fig. 8.   Comparison of cold flow tests and hot-firing results of triplet impinging elements.

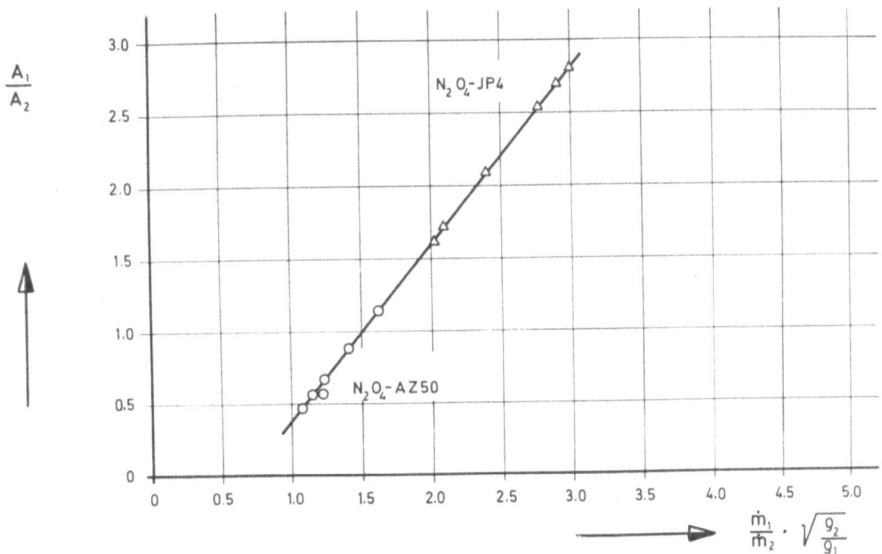

Fig. 9.   Optimum area ratios of triplet impinging elements with respect to combustion efficiency versus mixture ratio.

hot firing results. But surprisingly a common curve was found to exist for both pro-pellant combinations. Here we have to realize that the standardized mixture ratio is defined as weight of component 1 divided by component 2. The ratio of the components is corrected by the square root of the reciprocal ratios of the densities, due to an influence of the momentum ratio of the jets.

Figure 9 gives the relationship between injector design parameters and performance characteristics of the triplet injection elements. The optimum area ratios of the jets with respect to combustion efficiency are plotted against the total mixture ratio. It is evident now, that the area ratio of optimum performance elements is only a function of the standardized mixture ratio.

3.3. COMBUSTION OSCILLATIONS

One of the objectives of the research program was to find out a correlation between the hydraulic characteristics of the injector elements and combustion chamber stability. Combustion instability was arbitrarily defined as periodic pressure fluctuations of the combustion gases at more than 800 cps. At these frequencies the noise of the motor is a sharp whistling or screaming. Due to chamber and injection head geometry the oscillations are of a longitudinal mode.

Systematic investigations on several types of impinging elements indicated that all of the types tend to augment high frequency oscillations.

In the case of invariable geometry of the chamber and constant over-all chamber pressure the experimental results show clear effects of hydraulic characteristics of the impinging elements on the region of mixture ratio where high frequency instability will occur.

Figure 10 shows in the plane of the jet velocity ratio of the propellants and the

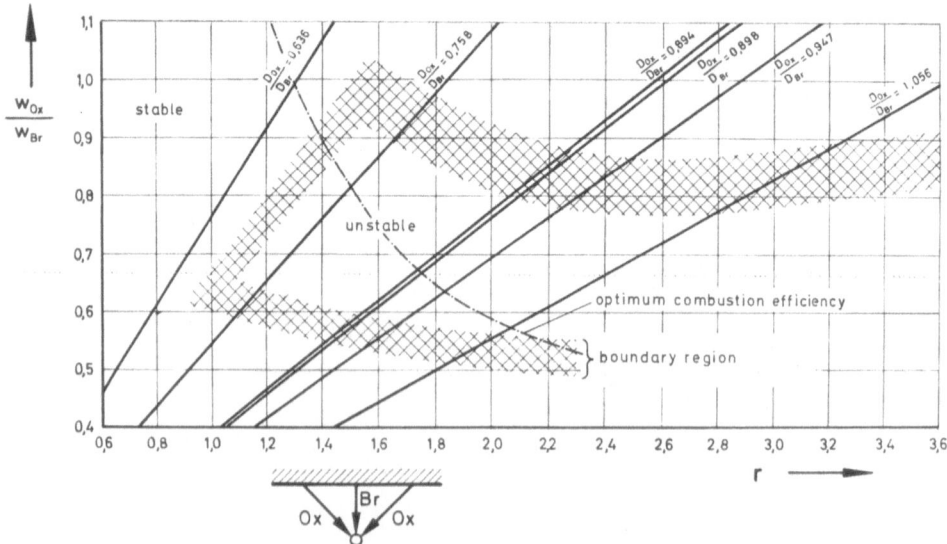

Fig. 10.   Region of high frequency instability in the plane velocity ratio – mixture ratio.

over-all mixture ratio the region of instability of the triplet impinging elements with
the oxidizer flowing through the outer orifices. The dashed line marks the zone of
optimum combustion efficiency.

In the region of the jet diameter ratios 0.76 and 0.64 the combustion instability is
ending suddenly. All elements with jet diameter ratios smaller than about 0.7 seem to
be stable. They also indicate a relative high combustion efficiency.

Figure 11 is showing the same for triplet elements with the fuel flowing through the

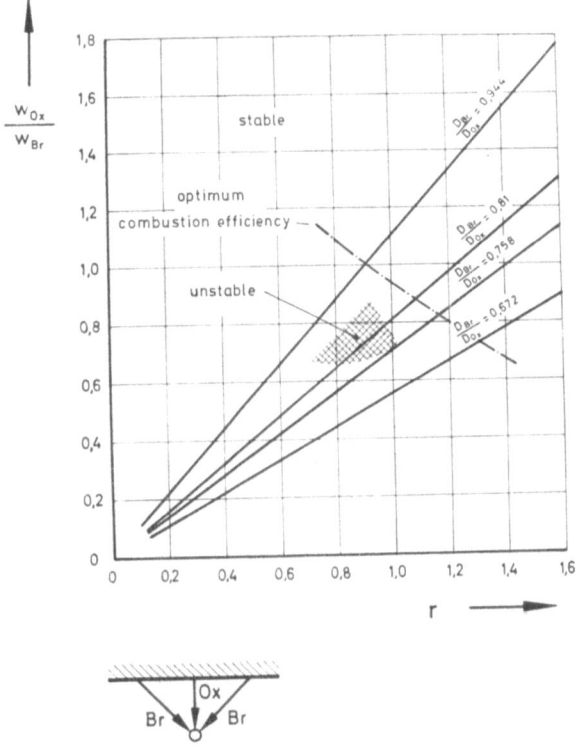

Fig. 11.   Region of high frequency instability in the plane velocity ratio – mixture ratio.

outer orifices. This case looks much better, because a slight instability could be found
only in the region of the jet diameter ratio 0.8. But it is well aside of the point of
optimum combustion performance. The application of this type of elements is somewhat limited, because the optimum characteristic velocity lies fairly in the fuel rich.

For comparison some tests were made with several other impinging jet configurations as the two-on-two, three-on-one, and four-on-one elements.

Figure 12 shows in the plane of jet velocity ratio and mixture ratio the boundaries
of stable burning for two-on-two elements. The region of instability becomes wider
with decreasing mixture ratio on the other side it ends at a diameter ratio of about 1.6.
But principally the field, where combustion oscillations occur, looks very similar to
that of the triplet (two-on-one) element.

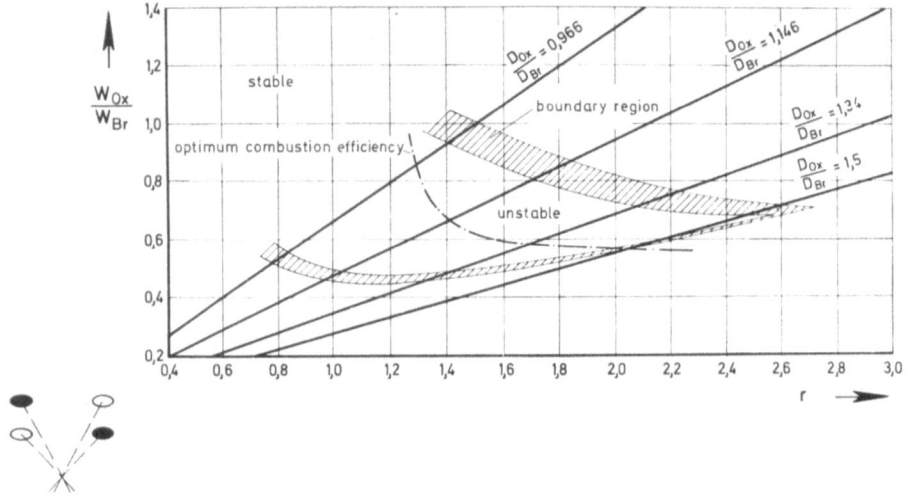

Fig. 12.   Region of high frequency instability in the plane velocity ratio – mixture ratio.

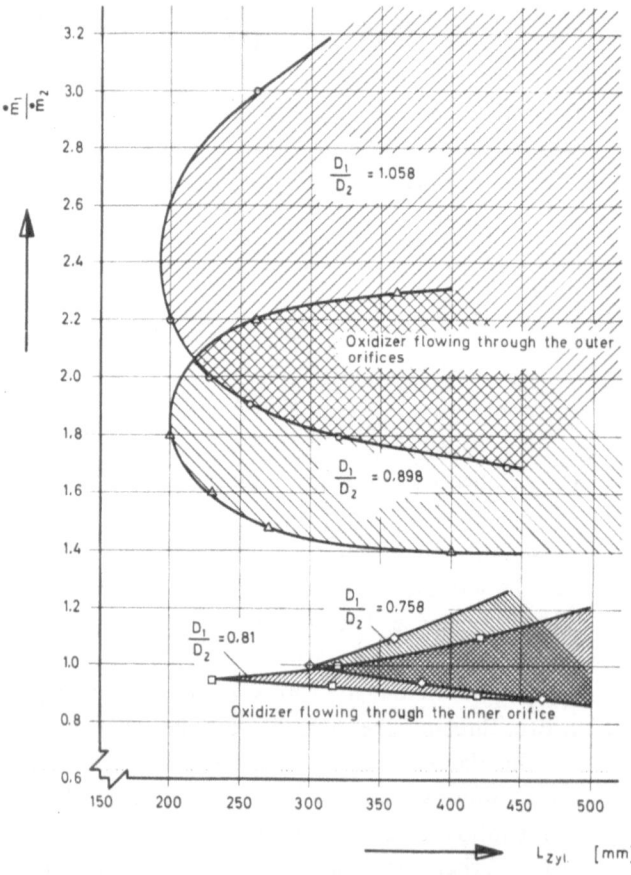

Fig. 13.   Zone of stability in the plane mixture ratio-chamber length.

Figure 13 indicates a clear effect of the cylindrical part of the chamber on the range of mixture ratio, where high frequency oscillations will occur. Increasing length of the chamber makes the combustion more sensitive to instable burning.

Figure 14 shows that over-all chamber pressure has a similar effect. Increasing chamber pressure enlarges the field of unstable burning with regard to mixture ratio.

It is believed, that the following mechanism causes the high frequency oscillations:

(1) Due to arrangement of the single element in the middle of the injection head, a strong back-flow is induced.

(2) In combination with some special jet diameter ratios and mixture ratios hypergolic propellants create a gas film at the interface of the impinging point which may separate the oxidizer from the fuel.

Parts of the unreacted propellants join the back flow and went in the proximity of the injector head, there the components will be mixed.

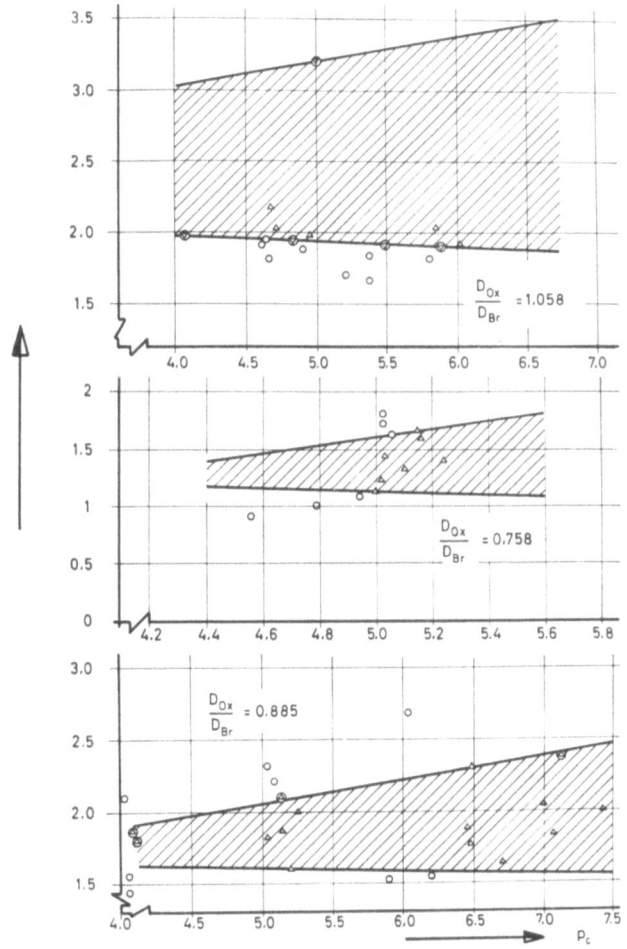

Fig. 14.   Effect of the chamber pressure on the high frequency oscillations.

If the mixture ratio of the back flow propellants is near the stoichiometric mixture ratio, sudden ignition occurs. The shock-wave travels down the chamber and then returns to the injection head after being repelled near the nozzle throat. If a sufficient quantity of unreacted propellants have been collected until now, ignition will occur again and the releasing energy will augment vibration.

The frequency of the oscillation is determined only by the time the shock-wave needs for completing the hole cycle described above. This corresponds to a frequency of about 2000 cps with the chamber that was used.

4. Conclusions

The experimental results of cold flow spray and hot-firing in this report illustrate that variations in combustion performance can be related to variations in spray parameters. A strong influence on performance is shown by mixture ratio uniformity factor. But effects of vaporization and reaction kinetics are visible too. The mass flux standard deviation shows clear effects on the cylindrical chamber length.

All types of impinging element configurations tend to augment high frequency combustion oscillations at certain environmental conditions. The hypothetical model shows the reasons for the dangerous type of oscillations and it may help to avoid them by proper design of the injection head.

References

[1] Rupe, J. H., 'A Correlation Between the Dynamic Properties of a Pair of Impinging Streams and the Uniformity of Mixture-Ratio Distribution in the Resulting Spray', Jet Propulsion Lab., Progress Rep. No. 20–209, Calif., March 1956.
[2] Hartung, W., 'Der Einfluß des Einspritzvorganges auf das Betriebsverhalten kleiner Flüssigkeits-raketen, Teil 1: Hydrodynamische Gesetzmäßigkeiten bei der Treibstoffeinspritzung durch Prall-strahlen', DLR FB 68-17.
[3] Fisher, R. A. et al., 'Study of Droplet Effects on Steady-State Combustion, Vol. I: Measured Spray Parameter Analysis and Performance Correlation', Rocketdyne, Canoga Park, Calif., Technical Report AF PRL-TR-66-152-Vol. I, August 1966.

ANALYTICAL AND EXPERIMENTAL INVESTIGATIONS OF AEROTHERMOCHEMICAL PROCESSES IN LIQUID PROPELLANT ROCKET MOTORS

T. PAUL TORDA

Illinois Institute of Technology, Chicago, Ill., U.S.A.

Abstract. Rapid increase in rocket performance level in the last 20 years was plagued by combustion instability. Investigators devoted considerable effort to solving this problem but no valid analysis exists for predicting stable operation of a new design. To fill this need, a non-linear approach was developed of liquid rocket design which is also useful for the investigation of oscillatory (unstable) combustion.

First, a brief history is given of the linear or linearized analytical approaches used by other investigators to the solution of combustion instability starting from the 'time lag' concept suggested by von Kármán and developed by Summerfield, Tsien, Crocco, Barrère, and others. Then, the non-linear analysis is developed which is useful for both the design of liquid rocket chambers (steady state) and for the investigation of combustion instability (non-steady state). The model used allows incorporation of the most advanced information of the aerothermochemical processes involved as they are obtained from available experimental and analytical investigations.

Starting with the basic equations of aerothermochemistry, as written by von Kármán in 1955, the analysis is developed correlating the design parameters of pressure, temperature, gas velocity and cross-sectional area variation. The analysis incorporates the liquid phase (injector spray and droplet histories), the liquid-vapor transition (evaporation of propellants), as well as the energy release histories. It is shown that all the parameters can be determined and their individual influences evaluated and it is established that additional information is needed on the evaporation and shattering of drops in high-pressure environment at elevated temperatures. The set of simultaneous equations is solved by numerical methods.

Next, the non-steady, non-linear study of longitudinal combustion instability is presented and an extension to three-dimensional analysis is proposed.

For both the steady-state analysis (design) and the non-steady one (instability), realistic chamber pressures and chamber dimensions are used and a 100000 lb liquid oxygen-liquid hydrogen rocket motor is chosen. The steady-state analysis shows the important correlation of the design parameters. The result of the non-steady analysis are illustrated by the effects of realistic disturbances (both step-functions and oscillatory disturbances) which are introduced at various times and at various spatial positions. The obtained non-steady behavior of the rocket chamber agrees well with measured oscillatory behavior of actual rocket motors.

Finally, recent experimental results are presented of liquid propellant droplet evaporation under high-pressure (up to several times critical pressure) and high-temperature conditions. It is shown that the existing models for drop evaporation lack reality and that a new non-steady analysis is needed for the correlation of drop vaporization under supercritical conditions. The experimental apparatus and methodology are presented which were used in the experiments. The time-histories of drop temperature and drop radius are shown and heat and mass transfer coefficients are derived for use in rocket design. Drop shattering and spray breakup are discussed also both at atmospheric and at high-pressure conditions.

A 5-min film is shown relating to droplet evaporation and spray and drop shattering.

List of Symbols

A	chamber cross-sectional area, amplitude in forcing function
A_i	injector area

G. A. Partel (ed.), Proceedings of the Second International Conference on Space Engineering. All rights reserved.

A_t	throat area
B	amplitude in forcing function
C_D	drag coefficient
C_d	discharge coefficient
C_l	specific heat of liquid
C_p	specific heat at constant pressure
D	drop diameter
D_{ABf}	binary diffusion coefficient at film temperature
e	internal energy
F	thrust
f	forcing function
g	gravitational constant
h	mean heat transfer coefficient
I_s	specific impulse $[F/\dot{w} = V_e/g]$
k	thermal conductivity
k_{xm}	mass transfer coefficient
L^*	characteristic chamber length $[V_c/A_t]$
$\dot{m}$	mass rate of flow propellant
M	momentum addition to gas per unit of time and unit of distance
M_l	mass of liquid drop
N_{Nu}	Nusselt number $[hD/k]$
N_{Pr}	Prandtl number $[\mu C_p/k]$
N_{Re}	Reynolds number $[D(u-V)\varrho/\mu]$
N_{Sc}	Schmidt number $[\mu_{mf}/\varrho_{mf} D_{ABf}]$
N_{Sh}	Sherwood number $[k_{xm}D/\varrho_{mf} D_{ABf}]$
p	pressure
p_c	chamber pressure
p_i	injector pressure
Q	energy addition to gas per unit of time and unit of distance
Q_l	heat transferred to liquid
Q_v	heat transferred to drop
r	drop radius
t	time
t_0	initial time
T	temperature
T_l	temperature of liquid
u	gas velocity
V	drop velocity
V_c	chamber volume
V_e	exhaust velocity
V_l	drop volume
$\dot{w}$	propellant weight flow rate
W	evaporation rate from drop

W_e	Weber number $[\varrho_l D(u-V)|u-V|/\sigma_l]$
x	axial distance
x_0	initial distance
Δ	difference
λ	latent heat of liquid, factor in forcing function
μ	viscosity
μ_{mf}	mixture viscosity at film temperature
ϱ	density
ϱ_g	gas density
ϱ_l	liquid density
ϱ_{mf}	mixture density evaluated at film temperature
σ_l	surface tension of liquid
ψ	mass addition to gas per unit of time and unit of distance
ω, ω'	frequencies in forcing function.

1. Introduction

In reviewing research work on liquid propellant rockets, I will restrict myself to American investigation. I do not wish to imply that no important work has been carried out in other countries, nor do I wish to detract from the importance and quality of work carried out elsewhere, but American developments are important enough to merit independent treatment. Right from the start I want to commit myself by stating that I am a firm believer of the nonlinear approach in investigating the aerothermochemical phenomena in liquid propellant rocket motors: these highly complex processes can be approached by nonlinear methods only. Proposing this twenty years ago in America generated strong opposition, but in the intervening years sufficient evidence accumulated so that some of the most vehement proponents of linear analyses have been converted to nonlinear analysis.

The history of modern rocketry in America starts with Goddard in the first decade of this century, but practical application of rockets was introduced by von Kármán and his students in the 1940-ies. As the thrust level of liquid propellant rocket motors was increased, unstable combustion appeared often resulting in destruction of the rockets. The unstable (oscillatory) combustion may be of low, intermediate, or high frequency and of varying amplitude. Both low and intermediate frequency oscillatory combustion are coupled with the feed-line system and are well understood and usually easily remedied. High frequency combustion oscillations are decoupled from the feedline and much experimental and analytical work has been devoted to clarification of the interaction of the various physical and chemical phenomena which cause high frequency oscillatory combustion.

The analytical efforts dealing with oscillatory combustion are either attempts at formulating an analytical framework which may be useful for a priori determination of the stability of a new design; or they consist of scaling analyses: once a stable rocket motor has been developed (usually by judicious trial and error methods), the analysis should enable designers to scale up this rocket to higher thrust levels. This would have obvious economic advantages (lower development costs, etc.). The latter efforts are based on dimensional analyses and have led to a great number of scaling laws which have been unsuccessful to date.

The first attempts at formulation of an analytical model for the investigation of the aerothermochemical phenomena in liquid propellant rocket motors were based on an integral concept suggested by von Kármán. He proposed use of the time lag between introduction of the propellants into the rocket chamber and the energy release through combustion [1]. Von Kármán postulated that the time lag is a constant characteristic for each rocket motor (thus it depends on the type of propellants, chamber pressure, temperature and chamber volume). Crocco et al. [2] modified the time lag concept by assuming that it has a constant and a pressure sensitive part. Barrère et al. [3] further refined the time lag concept by introducing a third, chemical kinetics dependent, part. All these analyses using the various time lag models originated in the early 1950-ies and use linearized forms of the aerothermodynamic equations for the gas phase phenomena. These equations are applied to simplified models of aerothermochemistry and both the linearization and use of simplified models prevented successful application of the various theories to the design of stable rocket motors.

It was again von Kármán who, in 1955, presented the fundamental equations of aerothermochemistry [4]. He also discussed similarity parameters available which might lead to scaling laws. While similarity analyses have not been successful, analytical efforts based on the fundamental equations of aerothermochemistry have been the most promising in the development of rational design procedures for liquid rocket motors and also for the analyses of nonsteady combustion phenomena in such engines. In 1956, author presented a new nonlinear approach [5] based on the fundamental equations and incorporating realistic models for liquid and gas phase preparation, combustion and gas dynamics as well as heat balance in liquid propellant rocket motors. Another important feature of this new approach was pointing out for the first time the important influence of chamber geometry (cross-sectional area variation) on efficient operation and stability. These efforts resulted in presentation of two papers, one devoted to nonsteady aerothermochemical analyses of combustion instability [6] in 1962, and the other [7], in 1963, to the rational design of liquid propellant rocket motors.

By the early 1960-ies, the proposed nonlinear approach has been followed by at least one group [8] and a steady state analysis of a liquid propellant rocket motor with constant cross-sectional chamber area was published by Lambiris et al. in 1963 [9]. The interesting contribution of this paper is a comprehensive discussion of drop evaporation models existing at that time. Although Lambiris pointed out that the important influence of high pressure and high temperature on the evaporation of sprays and droplets has been recognized by Torda [10] and Spalding [11], he and his co-authors came to the conclusion that an elapsed time has to be defined from injection to breakup for each injector design and that after breakup, the droplet size is uniform and of 50 μ magnitude.

Besides the chamber geometry of the rocket motor and the liquid jet and drop breakup, the preparation of the combustible mixture has strong influence on performance. This preparation occurs in two steps: the phase change (evaporation of liquid propellants) and the mixing of the fuel and oxidizer vapors. Near the injector, the injected liquid jets entrain hot combustion gases which help heating up the jet and promote phase change (from liquid to vapor). This entrainment process also promotes the breakup of liquid jets and drops, since there is a velocity difference between the injected liquid and the entrained gas (the injection process in itself induces oscillations in the liquid streams). In addition to the dynamic effects, shear forces may also become important at some significant Reynolds number. The relationship between the dynamic forces acting on the drop and its surface tension is expressed by the Weber number. Thus, the drops will break up either when the Weber number is critical, or when a certain critical relationship between Weber and Reynolds numbers is reached. The influence of the various modes of stream and drop breakup on the gas dynamics of the rocket motor are discussed in some detail in References 6, 7 and 9.

2. Discussion

A. NONLINEAR ANALYSIS

The basic features of the nonlinear analysis will be presented next. It should be remembered that the particular model used in this analysis serves as an example only and, though it is realistic, it has shortcomings, since some of the physical and chemical phenomena are still not known. However, the analysis is capable of incorporating any new information whenever such becomes available.

Both the steady and nonsteady analyses are governed by the same equations and these will be written in their more general form.

The equations of motion used for the gaseous phase are obtained neglecting gaseous heat conduction, viscosity, and interdiffusion of chemical species. Variations in properties across the chamber are neglected so that one-dimensional flow may be considered. It is assumed that an amount of mass, Ψ, momentum, M, and energy, Q, is added to the gas in each unit time and distance. This represents the coupling from liquid to gaseous phase. The mass addition is the mass added by liquid evaporation. The momentum term includes the momentum addition of the evaporating liquid and the momentum loss from the drag of the liquid droplets on the gas. The energy term includes energy addition from the kinetic energy of the evaporating liquid and the heat of combustion, and energy loss from the drag of the liquid phase, the heat transferred from the gas to the liquid phase, and heat transferred to the chamber walls. In addition, terms representing recirculation patterns may be added to the coupling terms.

Under these conventions the conservation equations become

$$\frac{\partial}{\partial x}(\varrho u A) + \frac{\partial}{\partial t}(\varrho A) = \Psi \tag{1}$$

$$\frac{\partial}{\partial x}(\varrho u^2 A) + \frac{\partial}{\partial t}(\varrho u A) = M + p\frac{\partial A}{\partial x} - \frac{\partial}{\partial x}(pA) \tag{2}$$

$$\frac{\partial}{\partial x}(\varrho A e u) + \frac{\partial}{\partial x}\left(\varrho A \frac{u^3}{2}\right) + u^2 \frac{\partial}{\partial x}(\varrho A u) + \frac{\partial}{\partial t}(\varrho A e) + \frac{\partial}{\partial t}\left(\varrho A \frac{u^2}{2}\right) = Q. \tag{3}$$

Equations (1), (2) and (3) are the one-dimensional, nonsteady equations and are applied to the flow of the combustion gases. A global equation of state is used (e.g., in the form of a density weighted ideal gas or other appropriate form)

$$p = p(p, T). \tag{4}$$

The mass, momentum and energy terms are evaluated next.

The mass addition is the mass added by evaporation of the liquid propellants. Although it is realized that the evaporation model used is inadequate since it does not account for high pressure and high temperature effects (the gaseous environment in which the liquids evaporate is well above the critical point of either the fuel or the oxidizer), for lack of a better one, the Ranz and Marshall [12] correlation will be used to account for heat transfer to and mass transfer from the liquid drops. Naturally, the drop size distribution and their physical location has to be taken into account after the liquid sprays have broken up into drops.

The Ranz and Marshall correlations yield for heat transfer

$$N_{Nu} = 2 + 0.6(N_{Re})^{1/2}(N_{Pr})^{1/3} \tag{5}$$

and for mass transfer

$$N_{Sh} = 2 + 0.6(N_{Re})^{1/2}(N_{Sc})^{1/3}. \tag{6}$$

The temperature change of the liquid drop is computed from

$$\frac{dT_l}{dt} = \frac{dQ_l/dt}{M_l C_l}. \tag{7}$$

The heat transferred to the liquid is determined by

$$\frac{dQ_v}{dt} - \lambda\frac{dW}{dt} = \frac{dQ_l}{dt}. \tag{8}$$

The drop diameter history is determined from

$$\frac{dD}{dt} = \frac{V_l\dfrac{d\varrho_l}{dT_l}\dfrac{dT_l}{dt} - W}{\dfrac{\pi}{2}\varrho_l D^2}. \tag{9}$$

The drop velocity is computed from

$$\frac{dV}{dt} = -\tfrac{3}{8}C_D\frac{\varrho_g}{\varrho_l}\frac{(u - V)^2}{r}. \tag{10}$$

It is generally accepted that the drag coefficient is given by

$$C_D = \frac{27}{N_{Re}^{0.84}} \tag{11}$$

where

$$N_{Re} = \frac{D(u - V)\varrho_c}{u}. \tag{12}$$

When a critical value of the Weber number is reached, the drops are assumed to break up into smaller size droplets during a finite time element.

The enumerated equations allow the evaluation of Ψ and M, and part of Q. The portions of Q not yet accounted for are due to the heat of combustion and due to heat transfer to the walls. The combustion process may be calculated from the relevant chemical kinetics appropriate for the particular fuel-oxidizer combination used. The heat transfer to the wall may be obtained from empirical relationships, e.g., References 13 and 14.

The set of equations govern the interrelationship of the parameters and are subject to boundary conditions appropriate to each particular problem. Both design problems and oscillatory combustion problems may be solved by the set of equations and for design problems (steady state), all $\partial/\partial t$ terms are zero. The steady state solution is necessary also as initial condition for each nonsteady problem. First, the results of a steady state (design) problem will be presented. Then a nonsteady analysis will be discussed.

B. DESIGN PROCEDURE FOR LIQUID PROPELLANT ROCKET MOTORS

The examples presented are based on selected design parameters. The influences of the variation of the significant aerothermochemical parameters are explored. The basic design parameters are:

> thrust: 100 000 lb
> propellants: liquid hydrogen and oxygen
> mixture ratio: O/F=4 (mass)
> design chamber pressure: 1000 psi
> design $L^* = 110$ in.
> specific impulse: $I_s = 400$ sec
> propellant mass flow: 309 lb/sec
> chamber area at injector: 1.67 ft^2.

The upstream boundary conditions are defined by assumption of a showerhead injector for the propellants and an appropriate drop size distribution is assumed at a distance downstream from the injector where the injected propellant is broken into droplets due to liquid stream instability and due to dynamic interaction with the recirculating hot gases [15]. It was further assumed that the droplets will shatter at a critical Weber number and that this shattering occurs within a small distance traveled (approximately 5% chamber length).

The heat transfer is calculated both to the liquid propellants and to the walls of the rocket chamber. For the heat transfer to the propellants, the Ingebo–Priem–Heidman method [16, 15] is used, while the heat transfer through the chamber walls is calculated according to the empirical methods of Bartz [13] and Witte [14].

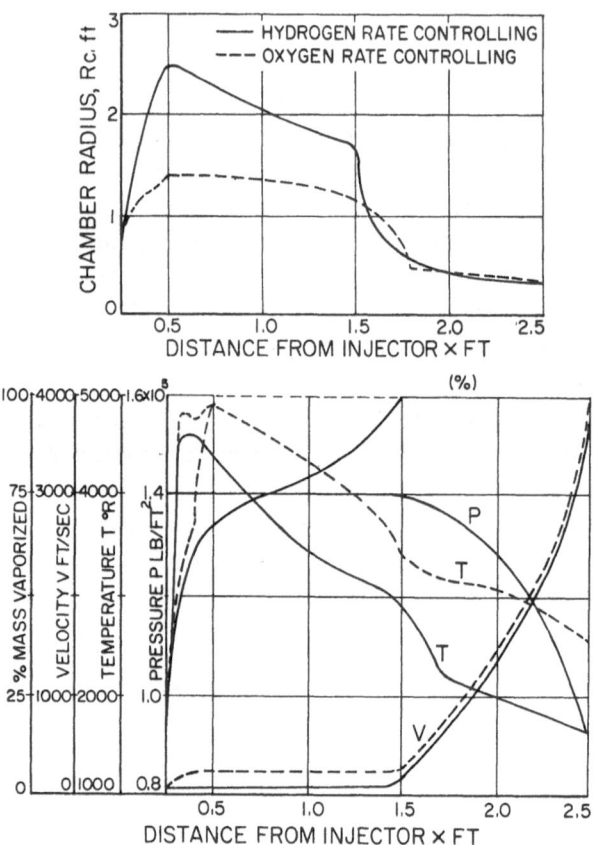

Fig. 1.   Results with assumed pressure variation.

For the propellant combination selected (liquid hydrogen–liquid oxygen), it is safe to assume that the rate controlling process is evaporation. This means that it is assumed that immediately upon evaporation the vapors react chemically to form the product gases. Therefore, if sufficient hydrogen and oxygen vapors are present at any point in the chamber, the product gases will be at equilibrium conditions at the local pressure and temperature. Two limiting cases have been calculated assuming that the evaporation takes place in an atmosphere of combustion products (steam plus excess hydrogen). In one case the evaporation of the hydrogen was computed using the mass transfer relation, while that of the oxygen was controlled by the imposed 4:1 oxygen hydrogen mass ratio. In the second case it was assumed that the oxygen evaporation was governed by the mass transfer relations and the hydrogen evaporation by the imposed 4:1 ratio. The reaction is considered instantaneous despite the diluting effects of the atmosphere. For both cases, the oxidizer-fuel mass ratio was arbitrarily set equal to that assumed for the overall combustion process ($O_2/H_2 = 4$).

Besides the assumption of either H_2 atmosphere or O_2 atmosphere, the variation of three parameters was investigated as they affect design. First, the chamber pressure variation was prescribed, in the second case, Mach number variation was assumed, and finally, the variation of the chamber cross-sectional area was prescribed.

Figure 1 shows the design results for prescribed pressure variation. The most radical consequence of the particular pressure variation is the resulting chamber shape, particularly when hydrogen evaporation is rate controlling. The large difference is due to the fact that oxygen evaporates faster than hydrogen since the oxygen drops shatter earlier. These computations confirm the important effect of drop evaporation, valid information on this is not available and is badly needed. Naturally, the influence is also marked in the rate of evaporation on the axial variation of the other parameters (temperature, velocity, etc.).

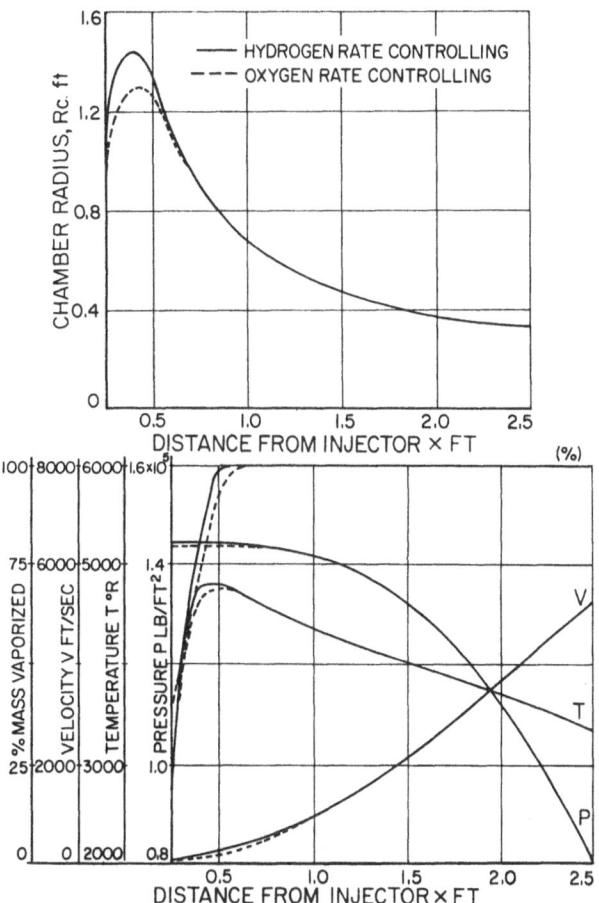

Fig. 2.   Results with assumed Mach number variation.

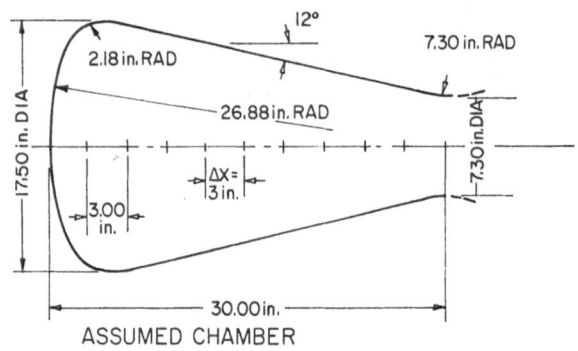

ASSUMED CHAMBER

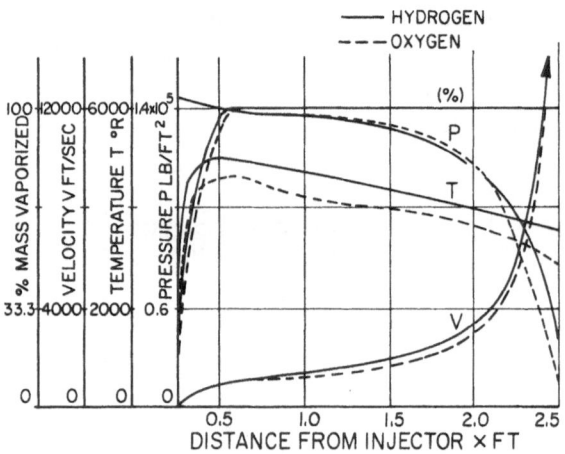

Fig. 3.   Results with assumed chamber area variation.

When the velocity variation along the chamber axis is prescribed, the influence in the difference between the hydrogen and oxygen evaporation rates is not too pronounced as may be seen in Figure 2. Since it is reasonable to assume that the average velocity of the combustion gases is small near the injector, the chamber area variation resulting from these computations shows the need for a large local chamber volume near the injector which may be decreased rapidly towards the nozzle.

Assumption of a 'pear' shaped combustion chamber (which resembles modern rocket motors) results in temperature, pressure and velocity variations shown in Figure 3. Again, the evaporation rate of fuel and oxidizer does not seem to influence the variation of parameters. The gas velocity seems to have a steeper slope near the injector than in the previous two cases.

The choice in the three cases treated aimed at bracketing parametric behavior and no optimum design was attempted at this time. However, some general observations are in order regarding the results from these exploratory investigations.

In all cases early vaporization and, therefore, early energy release occurs. (Even in the most extreme case where constant chamber pressure is assumed, energy release is completed in the first sixty per cent of the chamber length and at twenty per cent of the chamber length approximately seventy per cent of the energy is released.) This indicates that the chamber is too long and it could be made lighter by reducing its length without loss of efficiency. As a matter of fact, in a shorter chamber, the product gas velocity increase would be more rapid so that the critical Weber number would be reached earlier and total energy release achieved closer to the injector.

It may be that the critical Weber number is chosen too low in the present report. Though the value used is generally accepted, even larger values would not change the presented results significantly, since the Weber number increases rapidly with increase of gas velocity.

Another point worth mentioning is the size of the initial chamber area. This is dictated by the injector design (showerhead in the present case) and is too large for all three cases. A reduction of this initial area would result in higher velocity of the product gases and their greater directional stability. Divergent-convergent chambers and novel types of propellant injection may make rocket motors lighter, more efficient and more stable.

The steady state analysis presented illustrates the degree of control designers have over the variation of parameters in liquid propellant rocket motors.

C. NONLINEAR THEORY OF OSCILLATORY COMBUSTION

Use of the basic equations in their time dependent form allow the evaluation of the rocket chamber response to arbitrary disturbances. Thus, it may be evaluated whether a pressure pulse will be attenuated or not by the coupled aerothermochemical processes. Physically, the disturbances may be caused by concentrations of unburned fuel oxidizer vapor mixtures which are ignited releasing locally higher than average energy, thus, generating a pressure and temperature pulse. If such a pulse occurs close to the injector face, it will disturb the injection pattern and may be reinforced to generate more concentration of local energy release and finally lead to shock generation. Of course, these disturbances are all three-dimensional in nature and a one-dimensional analysis cannot account for them in detail. However, for longitudinal oscillations, the one-dimensional nonsteady analysis is valid. Later in this paper, an extension to three-dimensional nonsteady analysis will be proposed.

The nonsteady upstream boundary conditions define the injected mass of liquid. Usually, the time dependent phenomena in the chamber are decoupled from the propellant manifold by proper design of the injector (e.g., sufficiently large Δp across the injector plate). Even so, the injected mass is time dependent if pressure oscillations are present in the rocket chamber. This may be expressed as

$$\dot{m}(t) = f\left(p_i, p_c, A_i, C_d\right)$$

and may be obtained empirically for each injector design.

The liquid jet and drop shattering will be time dependent also, since all liquid streams oscillate on emission from an orifice and the shattering is enhanced if the injected jets impinge upon each other. The drop shattering has to be taken into account in an appropriate manner depending on the particular shattering and evaporation model used in the analysis.

If the aim of the analysis is the determination of occurrence of oscillatory combustion and its amplification, then it is sufficient to determine whether or not the pressure waves in the chamber will steepen into shock waves. Then, when shock waves appear anywhere in the chamber, the analysis may be terminated.

The energy release occurs after the fuel and oxidizer evaporated and the vapors thoroughly mixed. The energy release then depends on the local as well as on the overall fuel oxidizer ratio.

The reaction kinetics is not completely understood for many fuel oxidizer combinations. If it may be assumed that the reaction is not rate controlling, a simple chemical reaction may be assumed which goes to completion yielding products. In other cases, it is more appropriate to assume chemical equilibrium for the gases, and for other propellants it may be important to permit non-equilibrium gas composition and use an appropriate set of kinetic equations for the reaction which are to be solved simultaneously with the aerothermodynamic set. A note of caution is important here. In the difference equations, to insure numerical stability, often very small time intervals have to be used. It is important to check whether the small time intervals make it necessary to consider intermediate steps in chemical kinetics even for very fast reactions.

The already complex phenomena in liquid propellant rocket motors are further complicated by a number of damping mechanisms which include energy and momentum loss to the boundary layer as well as the liquid streams and droplets, and also losses due to chemical reactions. Thus, it seemed justified to explore the dynamic behavior of liquid propellant rocket motors by treating the simplest mathematical model, the one-dimensional longitudinal instabilities before extending them to multidimensional oscillations.

The numerical solution of the simultaneous equations was carried out using the method of characteristics in the particular form as discussed by Courant et al. [17] who proved convergence and stability for the particular method used. (The method allows use of square mesh grids which greatly simplifies the computer logic.)*

The rocket engine considered was defined as follows:
        thrust: 100 000 lb.
        chamber pressure: 600 psi
        propellants: liquid hydrogen and liquid oxygen
        overall weight ratio of oxidizer to fuel: 4
        chamber geometry: 30 in. from injector to throat with a 20 in. long
                          straight section of 61.5 in. diameter and converging
                          to a 38.9 in. diameter throat.

Mass and energy release curves are similar to those of Reference 9.

Several disturbances were introduced in the chamber as forcing functions of the

* A Univac 1105 was used for the computations.

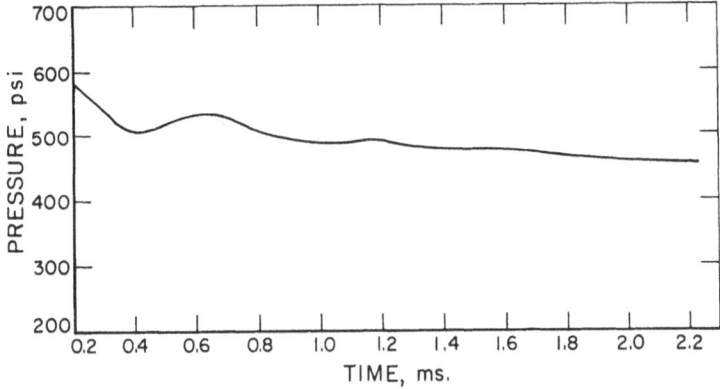

Fig. 4.   Pressure as a function of time at $x = 16''$ with no oscillatory forcing function.

form

$$f(x, t) = 1 + A \sin \omega (t - t_0) \cos \lambda \{x - [x_0 + B \sin \omega' (t - t_0)]\} \qquad (13)$$

Various frequency and amplitude forcing functions were applied to pressure and to velocity at various chamber locations and the reaction of the chamber calculated.

The first disturbance introduced was a pressure wave without oscillatory component. The resulting pressure response as a function of time at 16 in. from the injector plate is shown in Figure 4. After about two full chamber oscillations the pressure wave is substantially damped. The period is somewhat longer than the time required for a wave to travel the length of the chamber at the speed of sound. The particle velocities corresponding to a small amplitude disturbance at 8 in. from the injector are shown in Figure 5. The amplitude of the oscillations is larger than those for the pressure and the period of oscillations is about 1 msec corresponding to a wavelength of two chamber lengths. Apparently, the region of high mass release near the injector acts as a reflecting region. The case presented in Figures 6 and 7 shows the pressures and velocities as functions of time for a case where instability is generated by an oscillating forcing function. The chamber is relatively stable for about 2.9 msec when the disturbance drives the chamber unstable.

The presented results show that the nonlinear chamber response may be computed. This method should aid designers in determining what kinds of disturbances will drive the chamber unstable and what other kinds will be damped.

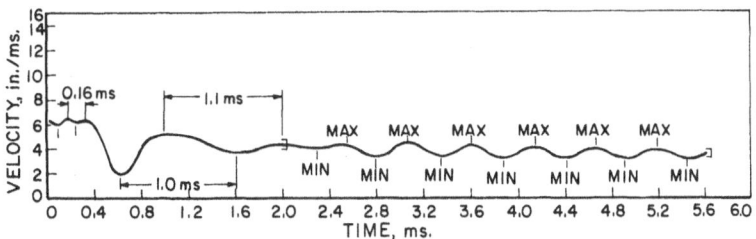

Fig. 5.   Particle velocity as a function of time at $x = 8''$.

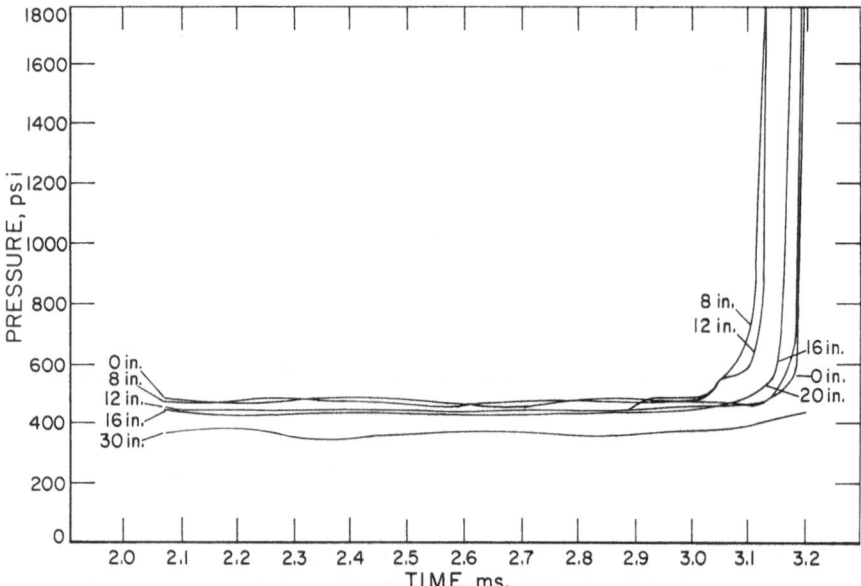

Fig. 6.   Pressure as a function of time at various distances with strong oscillation
and quadratic feedback.

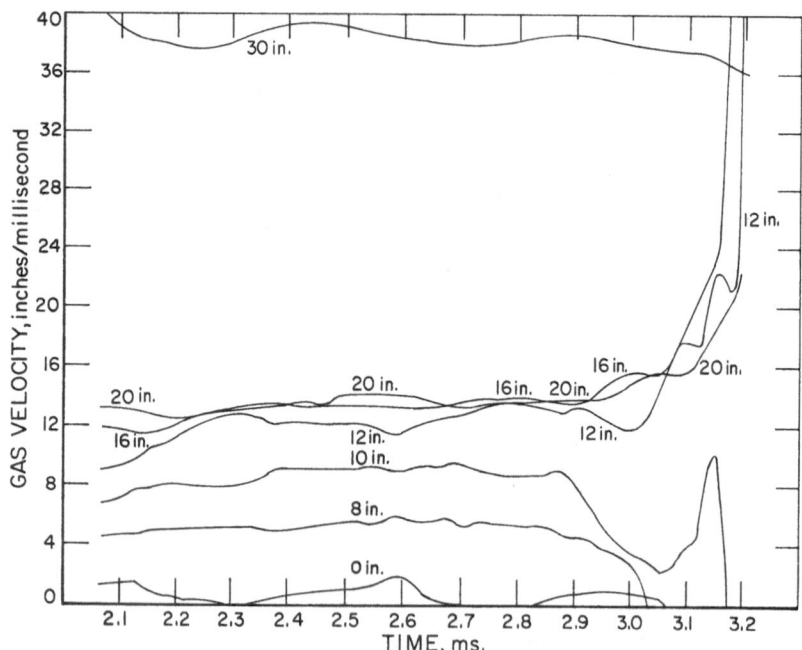

Fig. 7.   Gas velocity as a function of time at various distances with strong oscillation
and quadratic feedback.

The spatial distribution in radial and tangential directions of non-uniformities in flow parameters as well as liquid and evaporated propellants and reaction centers cannot be accounted for by the one-dimensional model discussed. Several attempts have been made to extend analyses to two- and three-dimensional models but, to date, no successful analysis has been developed. Most of the recent attempts at two- and three-dimensional analyses are based on the Priem-Guentert analysis [18] who used a nonlinear theory and a one-dimensional model. The flow is essentially in the axial direction, but variations and disturbances are considered in the tangential direction. Thus, average properties and parameters are used in the axial and radial directions and it cannot be expected that their results be applicable to actual rocket motors where three-dimensional effects are important. Priem and Guentert used two models for energy release: The one assumed that evaporation of the propellants is rate controlling, while the second assumed that the chemical reaction is rate controlling. For large rockets, the evaporation model showed sensitivity to pressure disturbances, while for small rockets the reaction rate model exhibited pressure sensitivity. None of their analyses showed sensitivity to velocity disturbances. However, experimental evidence shows [19] that mixing and velocity fluctuations (particularly near the injector) have a gross effect on combustion stability.

It is proposed that an analysis may be developed based on a segmented combustion chamber model since it is accepted that local disturbances trigger combustion instability. Each segment of Δx length is treated as a two-dimensional chamber where radial and tangential variations are explored. The disturbances are introduced either as variations in flow parameters or as local energy release above the average value. Once a disturbance is introduced in one of the segments, its history is investigated as it propagates in the axial direction. Since the governing differential equations have to be solved by numerical methods, such an analysis may be carried out with the aid of a sufficiently large digital computer. While such an analysis would take diffusion of the disturbance into account, it would allow also consideration of mixing which has not been treated successfully to date.

D. NEW RESULTS IN LIQUID DROP EVAPORATION

Possibly the last unexplored phenomena are liquid jet and droplet breakup and evaporation in high pressure and high temperature environments. Even the most advanced recent analyses use an oversimplified drop evaporation model based on experimental results obtained at low temperatures and low pressures. The validity of the model at high chamber pressures is questionable since there is lack of understanding concerning the evaporation processes in a supercritical gas environment (high chamber pressures). Torda [10, 20] indicated that liquid propellant droplets evaporating in a supercritical environment may remain in a transient heating state while evaporating. This is a deviation from the accepted 'normal' evaporation process.*
He concluded that a thorough investigation of the mechanism of droplet evaporation

* The 'normal' process is defined as droplet evaporation at a constant wet-bulb temperature; this is generally accepted in the literature.

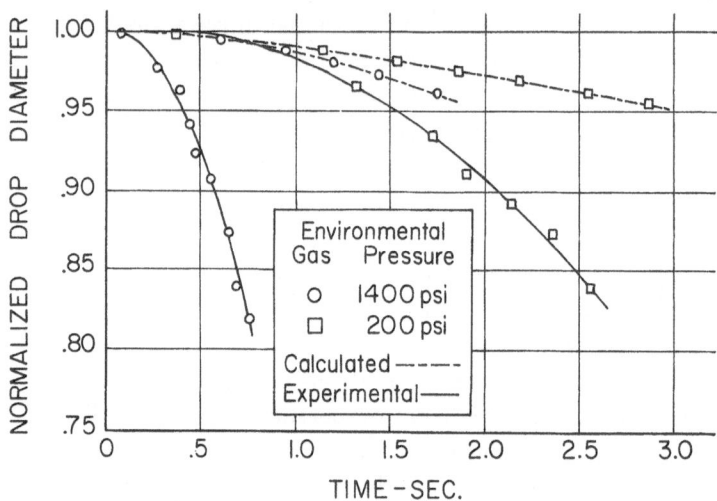

Fig. 8.   Experimental and calculated radius-time histories for N-pentane drops.

at elevated pressures and temperatures is a necessary requirement for a more valid rocket combustion model for design and instability studies. Recent investigations [21] have shown that the generally accepted evaporation models yield erroneous results.

The experiments of Reference 21 were carried out with a single droplet, suspended from a thermocouple, evaporating in a quiescent environment whose temperature and pressure approach, are equal to, or exceed the critical temperature and pressure of the evaporating fuel droplet. The time histories of the temperature and size of the droplet were recorded. The experimentally obtained droplet evaporation histories were compared with existing analytical methods used for prediction of droplet evaporation histories.

Figure 8 shows the calculated and experimental radius-time histories for the 200 psi and the 1400 psi runs. Since there is no agreement between the analytical and the experimental results, an energy balance was made on the vaporizing drop. The purpose of the energy balance was two-fold: (1) to determine the effects of radiation and of heat conduction through the thermocouple on the evaporating drop, and (2) to calculate a realistic design heat transfer coefficient which includes mass transfer.

The first objective of the energy balance analysis indicates whether or not the quasi-steady evaporation model with natural convection is applicable, since the evaporation model does not account for droplet heating due to radiation and heat conduction through the thermocouple.* If the energy contributions from radiation and/or heat conduction through the thermocouple are negligible compared to the energy requirements for latent heat of evaporation and superheat, then the discrepancy between the calculated and experimental radius-time histories cannot be explained in terms of the

---

* The energy contributions for droplet evaporation from radiation and heat conduction through the thermocouple affect the experimental radius-time histories of the evaporating droplet. These effects are included in the experimental radius-time histories, but the radiation and heat conduction effects on the evaporating droplet are not included in the analytical model.

TABLE I

Results of the energy balance analysis

Run 63-7	$P_T = 1400$ psi		$T_e = 850°R$					
sec	h	q_T	$\dfrac{q_{th}}{4\pi r^2}$	q_{rad}	q_a	q_{SH}	$\%q_{th}$	$\%q_{rad}$
	B/ft²-sec-°F	B/ft²-sec	B/ft²-sec	B/ft²-sec	B/ft²-sec	B/ft²-sec		
0.25	.033	8.91	.112	.1905	6.75	2.16	1.6%	2.8%
0.50	.0476	10.87	.105	.1745	7.07	3.80	1.5%	2.5%
1.00	.0885	15.915	.1515	.1026	9.455	6.46	1.6%	1.08%

Run 63-6	$P_T = 200$ psi		$T_e = 850°R$					
sec	h	q_T	$\dfrac{q_{th}}{4\pi r^2}$	q_{rad}	q_a	q_{SH}	$\%q_{th}$	$\%q_{rad}$
	B/ft²-sec-°F	B/ft²-sec	B/ft²-sec	B/ft²-sec	B/ft²-sec	B/ft²-sec		
1.00	.0088	2.710	.058	.1960	1.948	0.762	3.0%	10.0%
1.50	.01325	3.630	.057	.1890	2.180	1.450	2.6%	8.7%
2.00	.01210	3.280	.063	.1862	2.730	1.550	2.3%	6.8%

effects of radiation and of heat conduction through the thermocouple on the evaporation of the droplet.

The second objective of the energy balance analysis is meaningful only if the effects of radiation and of heat conduction through the thermocouple are negligible compared to the energy requirements for latent heat of evaporation and superheat. Then, the calculated heat transfer coefficients can be used as design heat transfer coefficients for evaporating droplets.

The results of the energy balance analysis are shown in Table I and it may be seen that the energy contributions from radiation and from heat conduction through the thermocouple are small compared to the energy requirements for evaporation and superheat. Thus, the discrepancy between the calculated and experimental radius-time histories cannot be explained in terms of the effects of radiation and of heat conduction through the thermocouple on the evaporation of the droplet. The following conclusions are drawn from the energy balance analysis: (1) the quasi-steady evaporation theory [12, 22] is not valid for droplet evaporation at high temperatures and high pressures (supercritical environmental conditions), and (2) a non-steady evaporation theory using the energy, momentum, species, and global continuity is required to solve this complex problem.

3. Summary

Analytical and experimental work on steady state and nonsteady behavior of liquid propellant rockets has been reviewed. In particular, the importance of nonlinear analyses has been pointed out. The examples cited show that analytical tools are available for liquid propellant rocket chamber design and that certain modes of oscillatory combustion are amenable to analysis. The extension of nonsteady analyses to multidimensional ones is proposed. Finally, recent work is discussed on drop and spray evaporation at high temperatures and high pressures.

References

[1] The time lag concept was suggested by Dr. Theodore von Kármán and was used by Martin Summerfield, 'A Theory of Unstable Combustion in Liquid Propellant Rocket Motors', *J. Am. Rocket Soc.* **21** (1951), and by H. S. Tsien, 'Servo-Stabilization of Combustion in Rocket Motors', *J. Am. Rocket Soc.* **22** (1952).

[2] Crocco, L. and Cheng, S. I., 'Theory of Combustion Instability in Liquid Propellant Rocket Motors', *AGARDograph* No. 8, Butterworth Scientific Publications, London (1956).

[3] Barrère, M. and Moutet, A., 'Inflammation et Allumage dans les Moteurs-fusées à Propagols Liquides', in *Agard – Selected Combustion Problems*, Vol. II, Butterworth Scientific Publications, London, 1956.

[4] Kármán, Th. von, 'Fundamental Equations in Aerothermochemistry', in *Agard – Selected Combustion Problems*, Vol. II, Butterworth Scientific Publications, London, 1956.

[5] Torda, T. Paul, 'Analysis of Chemical Kinetic and Aerothermodynamic Processes in Liquid Propellant Rocket Motors', presented at the Agard Combustion Panel Meeting, Oslo, 1956.

[6] Torda, T. Paul and Schmidt, L. A., 'Aerothermochemical Analysis of One-Dimensional Unsteady Flow Phenomena in Liquid Rocket Chambers', presented at the XIII International Astronautical Congress, Varna, 1962. Also published as 'One-Dimensional Unsteady Aerothermochemical Analysis of Combustion Instability in Liquid Rocket Engines', *Pyrodynamics*, **1** (1964).

[7] Torda, T. Paul, Busenberg, S. N., Kaufmann, J. C., and Steinke, R., 'Rational Design Procedures for Liquid Propellant Rocket Motors', in *Proceedings Fifth International Symposium on Space Technology and Science*, AGNE Corporation, Tokyo, 1963.

[8] Rocket Research Group of the Rocketdyne, A Division of the North American Aviation Corporation, under the leadership of Dr. R. S. Levine.

[9] Lambiris, S., Combs, L. P., and Levine, R. S., 'Stable Combustion Processes in Liquid Propellant Rocket Engines', in *Combustion and Propulsion – Fifth Agard Colloquim* (A Pergamon Press Book), The Macmillan Company, New York, 1963.

[10] Torda, T. Paul, 'Aerothermochemistry of Liquid Propellant Rocket Combustion Chambers', Polytechnic Institute of Brooklyn TR-5-R85OX-713428, Brooklyn, N.Y., October, 1958. Also, 'Aerothermochemistry of Jet Propulsion-Liquid Propellant Rocket Motors', Preprint No. 8, Symposium on Thermodynamics of Jet and Rocket Propulsion, Fourteenth National Meeting,, American Institute of Chemical Engineers, Kansas City, Missouri, May, 1959.

[11] Spalding, D. B., 'A One-Dimensional Theory of Liquid-Fuel Rocket Combustion', *Aeronautical Research Council A.R.C.* 20, London, 1958. Also, *J. Am. Rocket Soc.* **29**, No. 11 (1959).

[12] Ranz, W. E. and Marshall, W. R., 'Evaporation from Drops', *Chem. Eng. Progr.* **48**, No. 3, and No. 4 (1952).

[13] Bartz, D. R., 'A Simple Equation for the Rapid Estimation of Rocket Nozzle Convective Heat Transfer Coefficients', *Jet Propulsion* **21** (1957).

[14] Witte, A. B. and Harper, E. Y., 'Experimental Investigation of Heat Transfer Rates in Rocket Thrust Chambers', *AIAA J.*, **1**, No. 2 (1963).

[15] Priem, R. J. and Heidmann, M. F., 'Propellant Vaporization as a Design Criterion for Rocket-Engine Combustion Chambers', NASA TR R-67, 1960.

[16] Ingebo, R. D., 'Vaporization Rates and Heat Transfer Coefficients for Pure Liquid Drops', NACA TN 2368, 1951.

[17] Courant, R., Isaacson, E., and Rees, M., 'On the Solution of Non-Linear Hyperbolic Differential Equations by Finite Differences', *Comm. Pure Appl. Math.* **5** (1952).

[18] Priem, R. J. and Guentert, D. C., 'Combustion Instability Limits Determined by a Non-Linear Theory and a One-Dimensional Model', NASA TN D-1409 (1962).

[19] Levine, R. S., 'Development Problems in Large Liquid Rocket Engines', in *Combustion and Propulsion – Third Agard Colloquium*, Pergamon Press, New York, 1958.

[20] Torda, T. Paul, 'Combustion Instability of Liquid Propellant Rocket Engines – Notes on the State of the Art and Proposed Areas of Investigations', presented to AFAOR, January, 1962. Armour Research Foundation TM D-29, January, 1962.

[21] Torda, T. Paul and Matlosz, R. L., 'Investigation of Liquid Propellants in High Pressure and High Temperature Gaseous Environments', presented at the 19th Congress of the International Astronautical Federation, IAF Paper P 62, October, 1968.

[22] Bird, R. B., Stewart, W. E., and Lightfoot, E. N., *Transport Phenomena*, John Wiley and Sons, Inc., 1960.

Bibliography

1. Damkohler, G., 'Einflüsse der Strömung, Diffusion und des Wärmeüberganges auf die Leistung von Reaktionsöfen', *Z. der Elektrochem.* **42** (1936).

2. Keenan, J. H. and Keyes, F. G., *Thermodynamic Properties of Steam*, J. Wiley & Sons, New York, 1936.

3. Buddenberg, J. W. and Wilke, C. R., 'Calculation of Gas Mixture Viscosities', *Ind. Eng. Chem.* **41**, No. 7, (July, 1949) 1345.

4. Maxwell, J. B., *Data Book on Hydrocarbons*, D. Van Nostrand Co. Inc., 1950.

5. Shapiro, A. H., *The Dynamics and Thermodynamics of Compressible Fluid Flow*, Vols. I, II, The Ronald Press, New York, 1953.

6. El Wakil, M. M., Uyehara, O. A., and Myers, P. S., 'A Theoretical Investigation of the Heating-up Period of Injected Fuel Droplets Vaporizing in Air', NACA TN 3179, 1954.

7. Miesse, C. C., 'Ballistics of an Evaporating Droplet', *Jet Propulsion* (July–Aug., 1954).

8. Ross, C. C. and Datner, P., 'Combustion Instability in Liquid Propellant Rocket Motors – A Survey', in *AGARD Selected Combustion Problems*, Butterworths, London, 1954.

9. Barrère, M. and Bernard, J. J., 'Etude théorique des instabilitiés de basse fréquence', O.N.E.R.A. No. 79 (1955).

10. Barrère, M., Moutet, A., and Sarrat, P., 'Etude expérimentale des instabilitiés dans un moteur-fusée', O.N.E.R.A. (1955).
11. El Wakil, M. M., Priem, R. J., Brikowski, H. J., Myers, P. S., and Uyehara, O. A., 'Experimental and Calculated Temperature and Mass Histories of Vaporizing Fuel Drops', NACA TN 3409, 1955.
12. Hilsenrath, J., Beckett, C. W., Benedict, W. S., Fano, L., Hoge, H. J., Masi, J. F., Nuttall, R. L., Touloukian, Y. S., and Wolley, H. W., 'Tables of Thermal Properties of Gases', *Natl. Bur. Std. (U.S.), Circ.* 465 (1955).
13. Miesse, C. C., 'Correlation of Experimental Data on the Disintegration of Liquid Jets', *Ind. Eng. Chem.*, 1955.
14. Priem, R. J., *Vaporization of Fuel Drops Including the Heating-up Period*, Ph.D. Thesis, University of Wisconsin, 1955.
15. Wilke, C. R. and Lee, C. Y., 'Estimation of Diffusion Coefficients of Gases and Vapors', *Ind. Eng. Chem.* **47**, No. 6 (June, 1955) 1253.
16. Crocco, L., 'Considerations on the Problem of Scaling Rocket Motors', in *Selected Combustion Problems, II, AGARD*, Butterworths Scientific Publications, London, 1956.
17. Penner, S. S., 'On the Development of Rational Scaling Procedures for Liquid-Fuel Rocket Engines', presented during the Fall Meeting of the ARS, Buffalo, N.Y., 1956.
18. Ross, C. C., 'Scaling of Liquid Fuel Rocket Combustion Chambers', in *Selected Combustion Problems, II, AGARD*, Butterworths Scientific Publications, London, 1956.
19. Priem, R. J., Heidmann, M. F., and Humphrey, J. C., 'A Study of Sprays Formed by Two Impinging Jets', NACA TN 3835, 1957.
20. Torda, T. Paul, Burstein, S. Z., and Gegenwarth, R. E., 'Non-Linear Method for Combustion Instability Analysis of Liquid Propellant Rocket Motors', ARS Preprint No. 556–57.
21. Torda, T. Paul, *Notes on Problems in Combustion Instability of Rocket Motors, Aeron. Eng. Rev.*, **16**, No. 11 (Nov., 1957).
22. Zemansky, M. W., *Heat and Thermodynamics*, 4th ed., McGraw-Hill Book Company, New York, 1957.
23. Bittker, D. A., 'An Analytical Study of Turbulent and Mixing in Rocket Combustion', NACA TN 4321, 1958.
24. Bittker, D. A. and Brokaw, R. S., 'An Estimate of Chemical Space Heat Rates in Gas-Phase Combustion with Applications to Rocket Propellants', Paper No. 824–59, American Rocket Society, 1959.
25. Reba, I. and Brosilow, C., 'Combustion Instability: Liquid Stream and Droplet Behavior, Part III: The Response of Liquid Jets to Large Amplitude Sonic Oscillations', WADC Technical Report 59–720.
26. Schuyler, F. L., 'Combustion Instability: Liquid Stream and Droplet Behavior, Part I: Experimental and Theoretical Analysis of Evaporating Droplets', WADC-TR 59–720, 1959.
27. Torda, T. Paul, Lewkowski, Z., and Schuyler, F. L., 'Convective Heat Transfer Through Three-Dimensional Non-Steady Boundary Layers with Fluid Injection', Technical Note PRL-TN-59-4, Contract NAW-6594, Polytechnic Institute of Brooklyn, May, 1959.
28. Bird, R. B., Stewart, W. E., and Lightfoot, E. N., *Transport Phenomena*, John Wiley & Sons, Inc., 1960.
29. Lewis, B. and von Elbe, G., *Combustion, Flames and Explosions of Gases*, Academic Press, N.Y., 1961.
30. Priem, R. J. and Morrell, G., 'Application of Similarity Parameters for Correlating High-Frequency Instability Behavior of Liquid Propellant Combustors', Preprint ARS 1721–61.
31. *Handbook of Chemistry and Physics*, 35th ed., Chemical Rubber Publishing Co.,, 1961.
32. Wieber, P. E., 'Calculated Temperature Histories of Vaporizing Droplets to the Critical Point', *AIAA J.* **1**, No. 12 (1963) 2764.
33. Combs, L. P., 'Calculated Propellant Droplet Heating Under F-1 Combustion Chamber Conditions', Rocketdyne RR 64-25, June 1964.
34. Brzustowski, T. A., 'Chemical and Physical Limits on Vapor-Phase Diffusion Flames of Droplets', *Can. J. Chem. Eng.* (Feb., 1965).
35. Hersch, M., 'A Mixing Model for Rocket Engine Combustion', NASA TN D-2881, June 1965.
36. Reid, R. C., and Sherwood, T. K., *The Properties of Gases and Liquids: Their Estimates and Correlation*, 2nd ed., McGraw-Hill Book Company, New York, 1966.
37. Campbell, D. T., 'Combustion Instability Analysis at High Chamber Pressures', AFRPL-TR-67-222, August, 1967.

PART IV

PROPELLANTS AND COMBUSTION, II

Chairman: N. Takagi, Japan

INVESTIGATIONS ON PROPELLANT VAPORIZATION AND GAS MIXTURE FORMATION IN A LIQUID ROCKET MOTOR USING NONHYPERGOLIC PROPELLANTS

H. TWARDY

Deutsche Forschungsanstalt für Luft- und Raumfahrt, Trauen ü. Soltau, Germany

Abstract. Theoretical considerations are made on the vaporization of a mixture of hydrazine and oxygen droplets in a combustion chamber. Purpose of the calculations is to become acquainted with the influence of different parameters on the vaporization of a single droplet as well as droplet groups on their way through the combustion chamber. In this, particularly the injection direction of droplets with respect to the axis of combustion chamber as a parameter is considered together with the droplets diameter and the gradient of gas velocity of the combustion gas.

It is shown that it is not sufficient for an estimation of the droplet vaporization to know only the respective timely vaporization rate at the different points of the combustion chamber. The respective residence time of droplets must also be considered.

Usually, the mixture ratio distribution of an injection head is determined by means of cold-flow tests, collecting the fluid components at several points of a plane. In this, only the local gross masses of propellant can be determined. Since in vaporizing, however, always only shares of mass of this local gross mass vaporize, and that with oxydizer and propellant vaporizing besides with different velocity, other mixture ratios must be expected above the length of combustion chamber than those determined from cold-flow tests. Besides, the influence of the combustion-chamber gas flow on the variation of direction of the droplet movement must be added, what is also not considered in cold-flow tests.

Due to the shorter vaporization length of the oxygen droplets, an oversupply of oxygen exists in the region near the injection head of the combustion chamber, which downstream continually becomes lower; that is, the mixture ratio will decrease downstream. In consequence of the accumulation of mass of liquid propellant during injection for instance in the middle axial zone of the combustion chamber, also larger gas masses are to be expected at these points after vaporization and/or combustion; with a view to the mass equilibrium these will also flow off in radial direction and influence the mixture ratio distribution.

Concludingly, investigations are reported in which measurements on the temperature distribution of the combustion gas of a model rocket motor have begun. The results are expected to give further information on the mixing process in a liquid rocket motor.

Nomenclature

c_p $\left[\dfrac{J}{kg \cdot grd}\right]$ specific heat of the environmental gas of the droplet

c_w $[\cdot / \cdot]$ drag coefficient of droplet within the gas flow

D $[m]$ droplet diameter

H $\left[\dfrac{J}{kg}\right]$ heating power of fuel

$\dot{M}$ $\left[\dfrac{kg}{sec}\right]$ mass vaporized by a droplet group during the time $\varDelta t$

G. A. Partel (ed.), Proceedings of the Second International Conference on Space Engineering. All rights reserved.

M^* $\left[\dfrac{\text{kg}}{\text{sec}\cdot\text{m}}\right]$ mass vaporized by a droplet group during the unit of axial path length

$\dfrac{dm}{dt}$ $\left[\dfrac{\text{kg}}{\text{sec}}\right]$ vaporization rate of a droplet

$\dot{n}$ $\left[\dfrac{1}{\text{sec}}\right]$ number of droplets per second within a group

r [m]   radial distance from impingement axis

Δr [m]   distance between adjacent droplet paths

r_v $\left[\dfrac{\text{J}}{\text{kg}}\right]$ heat of vaporization

T_1 [K]   temperature of droplet

T_c [K]   gas temperature within the combustor

t [sec]   residence time of a droplet within the combustor

u $\left[\dfrac{\text{m}}{\text{sec}}\right]$ axial velocity vector of the droplet

x [m]   axial distance from injector face

Y_∞ $\left[\dfrac{\text{kmol}}{\text{kmol}}\right]$ mole ratio of oxidizer mass/gas mass in the environment of the droplet

λ $\left[\dfrac{\text{J}}{\text{m}\cdot\text{sec}\cdot\text{grd}}\right]$ heat conduction of gas in the environment of the droplet

θ_s $\left[\dfrac{\text{kg}}{\text{kg}}\right]$ stoichiometric ratio of oxidizer/fuel

Ω $\left[\dfrac{\text{kg}}{\text{kg}}\right]$ mixture ratio of oxidizer/(oxidizer + fuel)

Re [·/.]   Reynolds number

Nu [·/.]   Nusselt number.

1. Introduction

Within the scope of a research program, we are concerned with investigations on the interaction of liquid propellant injection and mixture formation of combustion gases in the combustion chamber of a liquid rocket engine. An estimation of the thermodynamic performance of the systems is only possible, if the local mixture ratio distribution of oxidizer and fuel above the injector face is taken into account. A nonuniform local mixture ratio distribution within the combustion chamber has a disadvantageous effect on the combustion efficiency, as is well-known from other papers [1]. The valuation of the mixture ratio distribution provided by the injector is usually made by means of cold-flow hydraulic studies [2]. In this, however, the influences of the liquid propellant vaporization in the combustion chamber and the turbulent diffusion are not considered.

We, therefore, are concerned with theoretical investigations on liquid propellant vaporization in a combustion chamber, particularly with respect to mixture formation. By means of tests, where the temperature distribution in the gas of the combustor of a liquid rocket engine is measured, we will further get data on the influence of turbulence on the mixing process.

2. Theoretical Model

For the considerations, a theoretical model was taken as a base that on the one hand is simple enough for the surveyability of the results obtained with it, and on the other hand is of general validity for complex injectors. We, therefore, chose impinging jets, where the spray pattern of the droplet mixture, which in reality is tridimensional, is considered as two-dimensional, see Figure 1. At the impingement of fuel and oxidizer

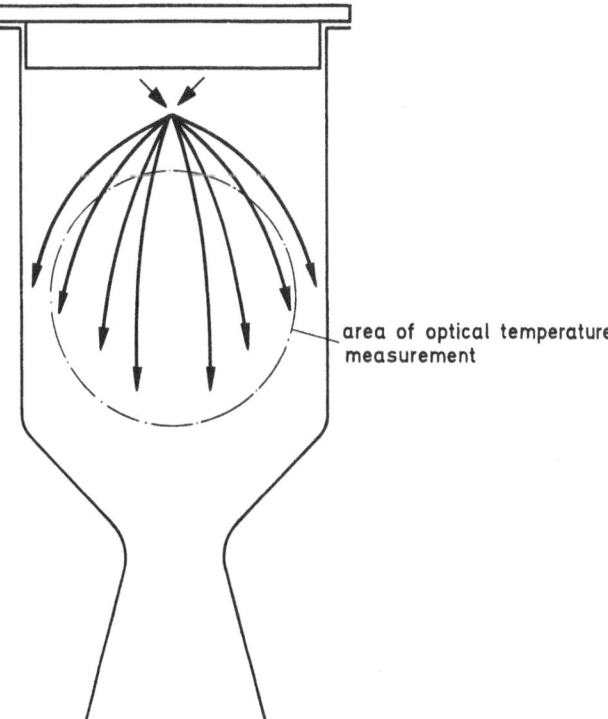

area of optical temperature measurement

Fig. 1.   Propellant distribution in the combustion chamber after the injection by impinging jets.

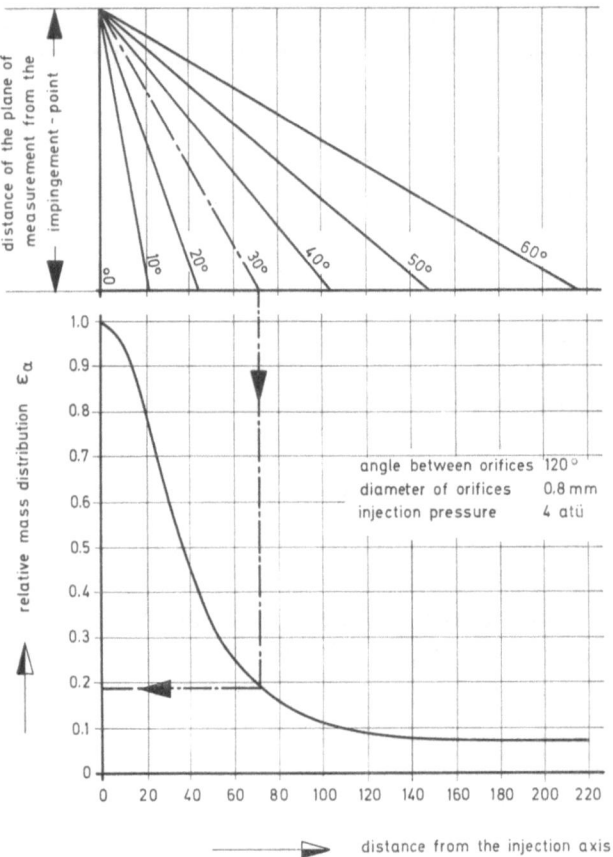

Fig. 2.   Relative mass distribution in the spray zone of impinging jets.

flow, the droplet formation shall start in the immediate vicinity of the impingement point. It is assumed that each droplet only consists of one of the two propellant ingredients and all droplets are of the same size. Since in such injectors a cumulation of mass of the injected propellant occurs in the environment of the injection axis, this is also considered in the model, Figure 2. The velocity variation of the combustion gas, which results due to the vaporized propellant mass and the cross section size of the combustion chamber, is already advanced for the calculations, Figure 3.

Considerations can, therefore, be made independently of a definite size of the combustion chamber.

3. Vaporization within the Combustion Chamber

Up to now, already a variety of calculations have been made with respect to droplet vaporization in liquid rocket combustion chambers. These calculations, however, are in most cases concerned with the determination of the required combustion chamber length, where the *overall* propellant mass has vaporized. For a determination of the *locally* vaporizing propellant masses, however, the analytical tracking of the individual droplets on their way through the combustion chamber is required. It is therefore necessary to take into account the axial motion as well as the radial motion of all droplets from the center to the chamber wall, Figure 4.

The well-known equations for a calculation of the droplet vaporization, Figure 5, assume that the droplet is surrounded by a diffusion flame. For this purpose, it must be in an atmosphere of constantly sufficient oxidizer and/or fuel. The vaporization rate

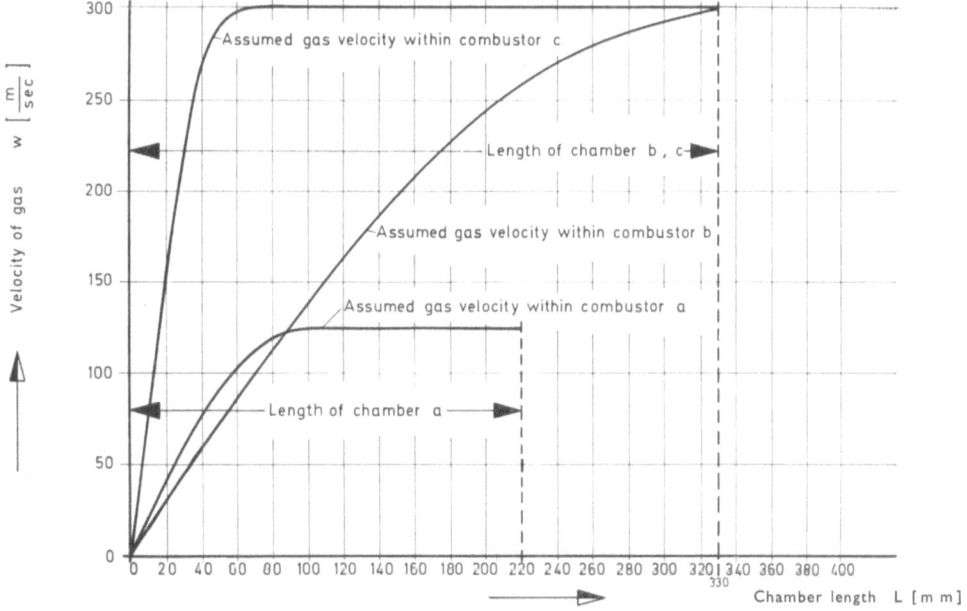

Fig. 3.   Plot of theoretical gas velocity for calculation.

is determined by heat transfer from the environment to the droplet surface and by diffusion. The chemical reaction mechanism runs much faster than the vaporization process, so that the combustion of a droplet is determined by all processes influencing the vaporization [3]. Since the total vaporized mass could react with the surrounding gas mass in a constant stoichiometric ratio, a constant temperature in the diffusion flame would result on these assumptions [4]. The combustion of a droplet mixture can, however, not be understood as the sum of all droplets burning with a diffusion flame, as the assumption of a constantly sufficient oxydator and/or fuel mass in the droplet environment is not always satisfied. The vaporized mass can be considerably higher than the reactive mass. As is well-known, a droplet vaporizes also in a hot neutral gas atmosphere without any reaction. Such surplus masses then have the possibility to mix at another place of the combustion chamber with oxidizer and/or fuel mass which are available there and to burn as premixed gases. The combustion of a droplet mixture may, therefore, be imagined as a heterogenous combustion, that is one portion of the vaporized masses burns in the diffusion flames possibly surrounding the droplets, whereas the rest mixes outside the diffusion flames and burns there.

Since the vaporization shall be the rate controlling process at the propellant preparation in the combustion chamber, it must be presupposed that fuel and oxidizer have no chemical affinity to each other in the liquid phase, that is they are not allowed to react hypergolically.

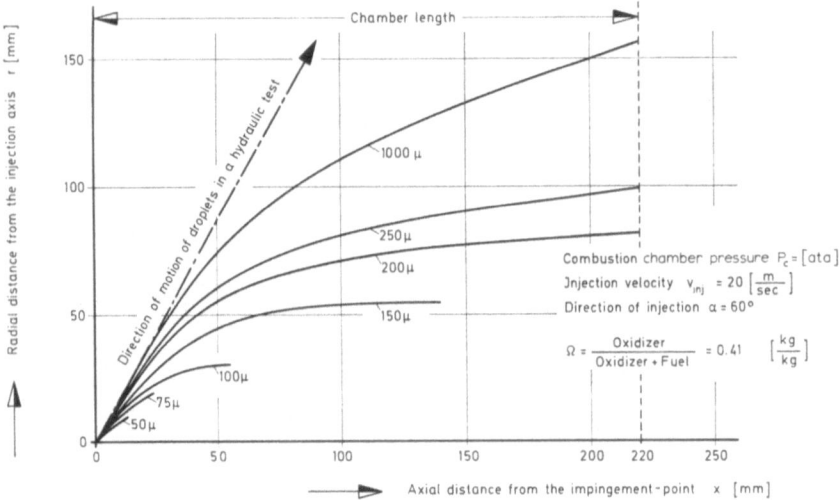

Fig. 4. Paths of motion of burning hydrazine droplets in the gas flow of the combustion chamber.

$$c_w = 27 \cdot Re^{-0.84} \qquad\qquad 0 < Re < 80$$

$$c_w = 0.271 \cdot Re^{0.217} \qquad\quad 80 < Re < 10^4$$

$$c_w = 2 \qquad\qquad\qquad\quad Re > 10^4$$

$$\frac{dm}{dt} = 2 \cdot \pi \cdot D \cdot \frac{\lambda}{c_p} \cdot \ln\left[\frac{H}{r_v} \cdot \frac{Y_\infty}{\theta_s} + 1 + \frac{c_p}{r_v}(T_c - T_1)\right] \cdot \frac{Nu}{2} \quad \left[\frac{kg}{sec}\right]$$

$$\frac{Nu}{2} = 1 + 0.276 \cdot Re^{1/2} \cdot Pr^{1/3}$$

Fig. 5. Droplet drag coefficient and vaporization equation.

4. Theoretical Results and Discussion

For an estimation of the vaporization process in a spray zone, the knowledge of the vaporization rate dm/dt of the droplets has proved not to be sufficient, as will still be shown with an example. The mass vaporized by a droplet or a droplet group during the unit of axial path length Δx is essentially more informative. Here, the respective droplet velocity is also considered. It is defined:

$$M^* = \frac{dm}{dt}\,\dot{n}\,\frac{\Delta t}{\Delta x} = \frac{dm}{dt}\,\dot{n}\,\frac{1}{u}\left[\frac{kg}{sec \cdot m}\right].$$

$\dot{n}$ represents the number of droplets per second within a droplet group. The features for a droplet group are: (1) all droplets are of equal diameter when injected, (2) droplets contain the same liquid, and (3) motion direction and velocity of all droplets are equal at the injection face.

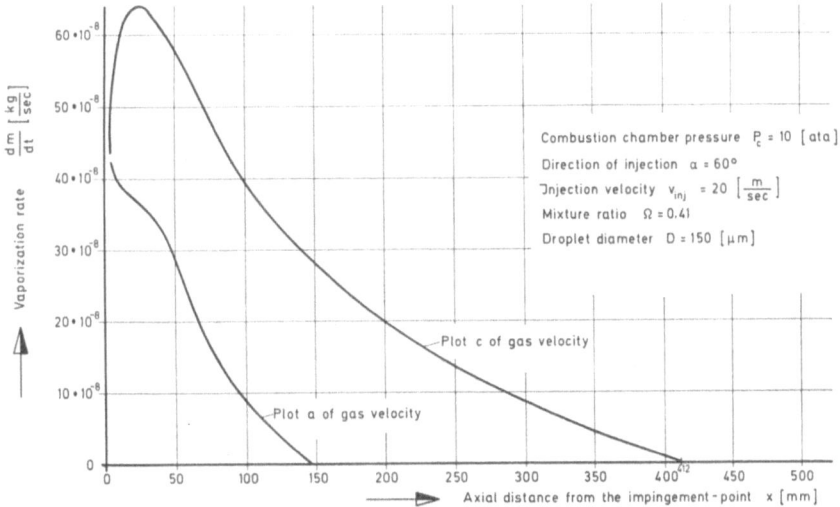

Fig. 6.   Vaporization rate of burning hydrazine droplets with differing plot of the combustion chamber gas velocity.

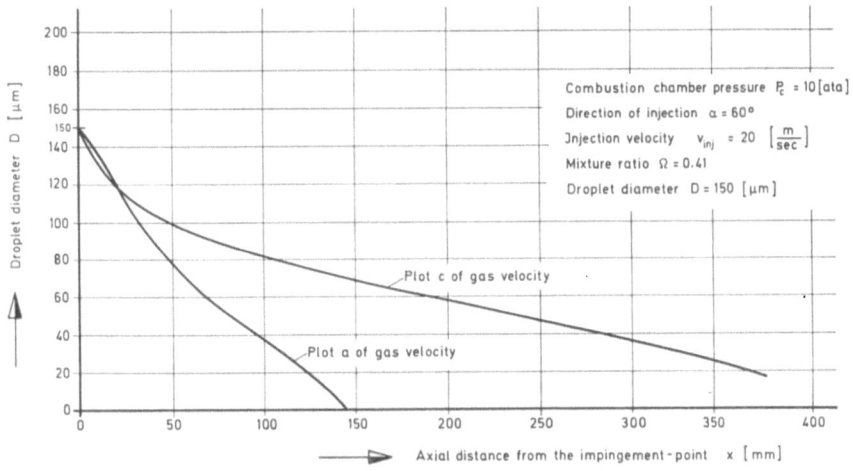

Fig. 7.   Reduction of the diameter of burning hydrazine droplets with differing plot of the combustion chamber gas velocity.

In order to demonstrate the advantage of this definition with an example, the development of the vaporization rate dm/dt was plotted versus the combustion chamber length for two different velocity variations of combustion gas, Figure 6. Though the vaporization rate is very much higher in one case than in the other one, the droplet diameter versus the combustion chamber length will be reduced considerably more slowly in case of the higher vaporization rate, Figure 7. At first, this appears to be contradictory. Plotting the mass $M*$ vaporized per axial element of path, however, Figure 8, shows that the curve with the lower vaporization rate in the beginning attains even higher values of $M*$ in consequence of the lower droplet velocity; therefore, a considerably smaller combustion chamber length is necessary in this case for a complete vaporization.

It has been investigated, in which area of a spray zone of oxygen and hydrazine droplets of the same size a vaporization takes place, Figure 9, if the spray zone is superposed by the combustion gas flow. Thereby, the droplets experience a drift off their original direction of motion. The shape of the areas in which oxygen and hydrazine vaporize is formed by a connection of the most outside droplet path of a component with the terminal points of all droplets paths lying inside. It is noticeable that all the oxygen vaporizes already in the upper part. In the lower part, therefore, only hydrazine vaporizes. A combustion in this part is only possible, because the surplus of oxygen from the upper area which is rich in oxygen comes downstream.

The vaporization mass $M*$ can be stated for every point of the spray zone in the combustion chamber, Figure 10. Particularly when taking into account the non-uniform mass distribution due to the injection process, as in this figure, a survey is obtained on the magnitude of the locally vaporized mass.

For calculating the mixture ratio distribution of vaporized fuel and oxidizer masses,

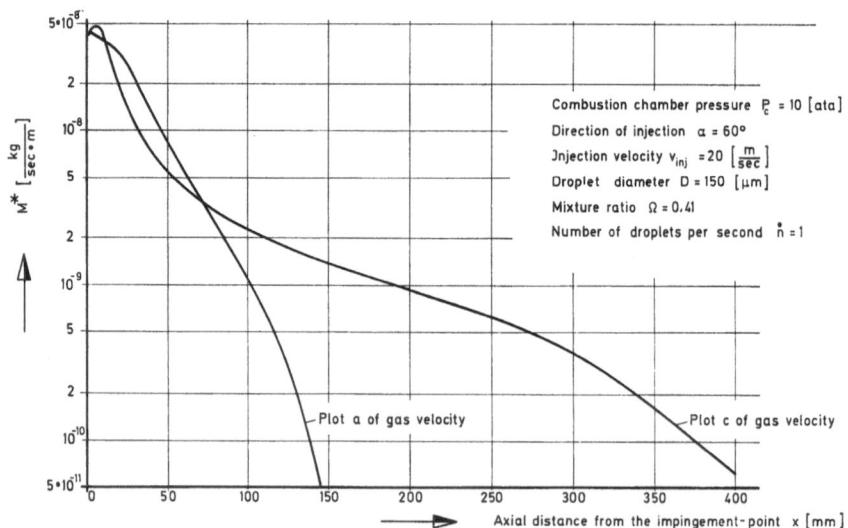

Fig. 8.   Vaporization of burning hydrazine droplets with differing plot of the combustion chamber gas velocity.

the term *density of vaporized mass* is defined, Figure 11. This is the mass vaporized per unit of time and of area in the plane spray zone. With this, the mixture ratios of the locally vaporizing propellant can be determined, Figure 12. The diagram was plotted on the assumption that the mixture ratio Ω_{fl} of the liquid phase was constant everywhere during the injection. The mixture ratio of the locally vaporizing propellant ingredients, however, decreases downstream, since the oxidizer vaporizes already in the upper region of the combustor for the most part. The irregular plot of the curves

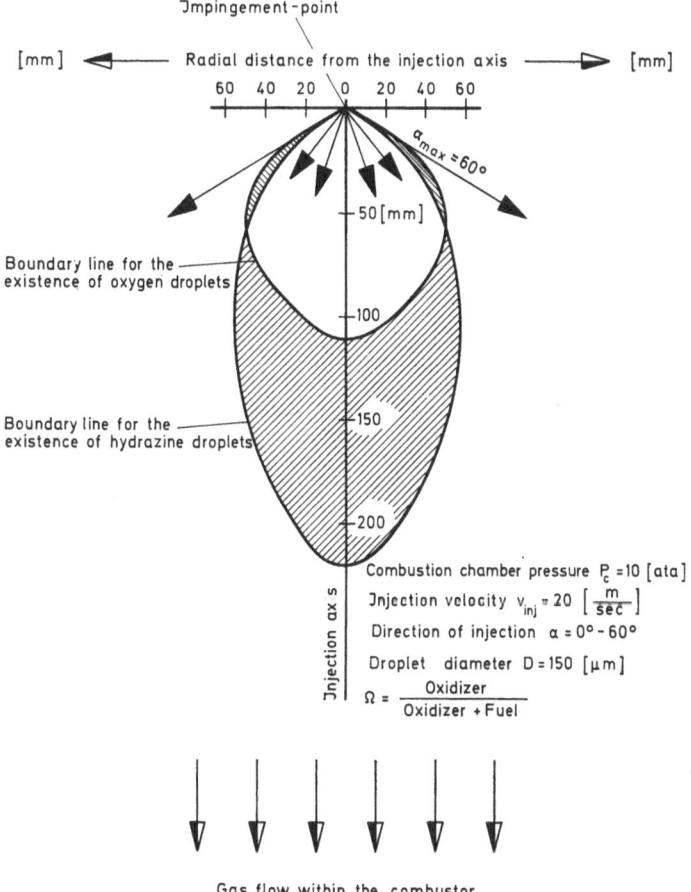

Fig. 9.   Spray zone of vaporizing oxygen droplets and burning hydrazine droplets in the gas flow
of the combustion chamber.

showing the values 1 and/or 0 occurs at the locations, where only oxydator and/or fuel droplets appear (compare Figure 9) and because a mass equilibrium between the stream tubes has not been considered here. Adding up all the masses vaporized in a stream tube – combustion shall hereby not take place – yields, after finished vaporization, mixture ratios in the part of the combustion chamber lying downstream, which in places essentially deviate from that of the liquid phase, Figure 13. The devi-

ations increase in the direction of the pheriphery of the spray zone, that is to the combustion chamber wall.

   In the sectioning of the spray zone into stream tubes of the same size it has not been
considered that the mass distribution differs very much. In order to induce a mass
equilibrium, one part of the gas masses will flow to the chamber wall, that is to

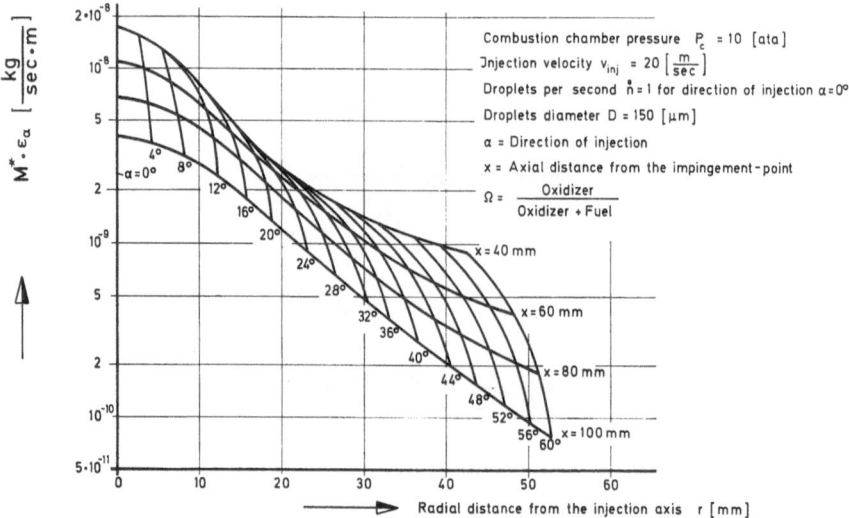

Fig. 10.   Droplet vaporization of hydrazine depending on the injection direction into the combustion
chamber taking into account a mass distribution measured in a cold-flow hydraulic analysis.

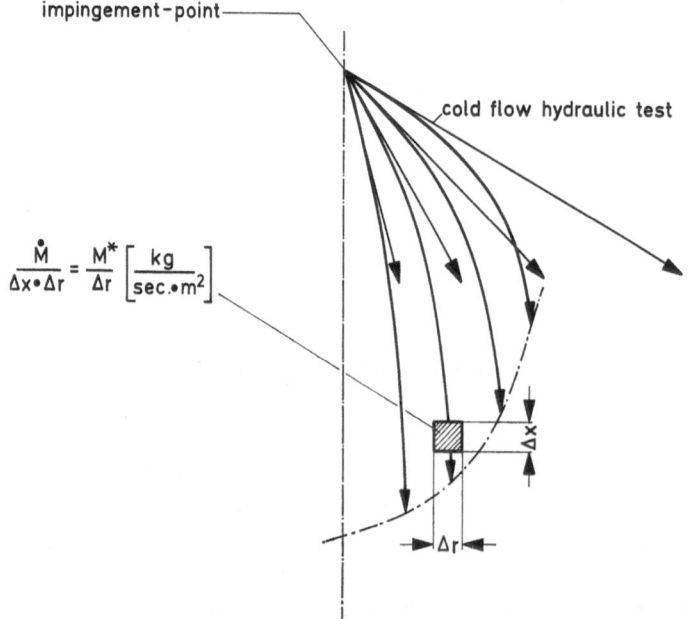

Fig. 11.   Definition of the density of vaporized mass.

produce the so-called 'cross flow' [1, 5, 6]. Hereby, shiftings in the mixture ratio distribution will occur. Since the combustion gases will take part in this as well as the not yet reacted gas masses, the relations for a calculation can not easily be overlooked. A statement of the mixture ratio distribution actually effective in case on the combustion is, therefore, not possible up to now. Thus, investigations are made with the aim of determining the mixture ratio distribution during the combustion process. This is made by measuring the local gas temperature in the combustion chamber of a liquid rocket engine by means of an optical measuring method.

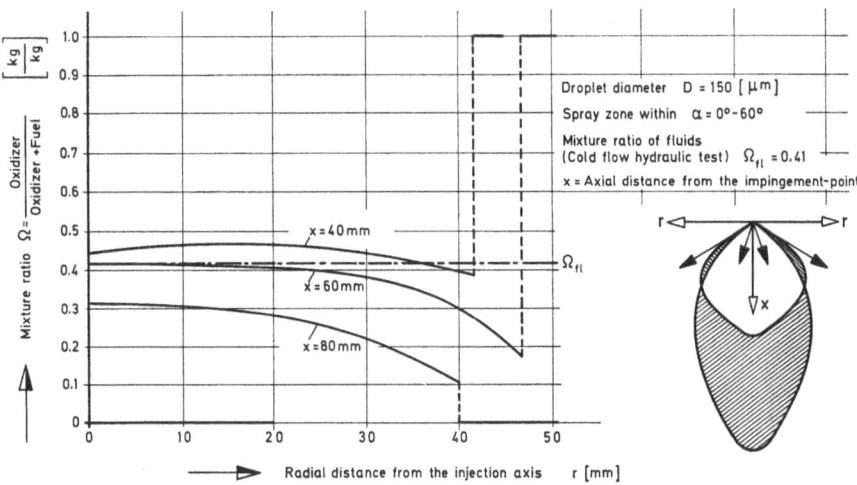

Fig. 12.   Mixture ratios from the local densities of vaporized mass in a spray zone of oxygen and hydrazine droplets with constant fluid mixture ratio distribution.

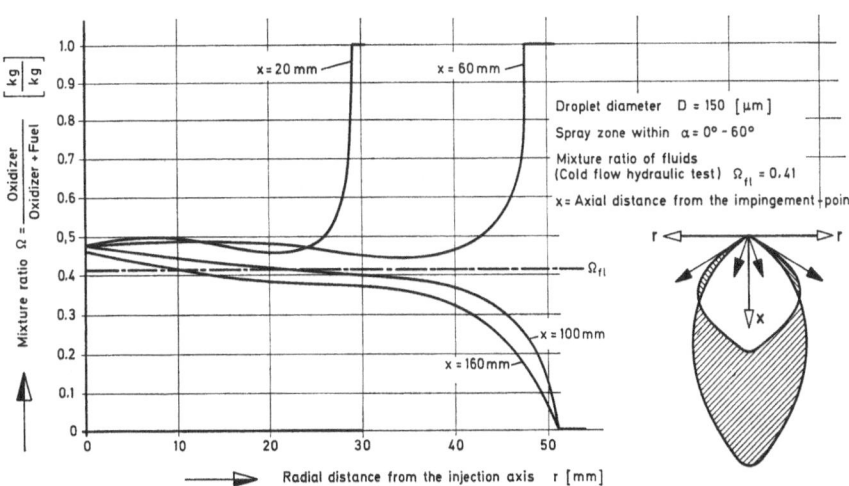

Fig. 13.   Mixture ratios of the totally vaporized masses in a spray zone of constant fluid mixture ratio distribution of oxygen and hydrazine droplets.

5. Apparatus and Procedure of Experiments

5.1. COMBUSTION CHAMBER

For a clear analysis of the results, a quasi one-dimensional combustor is used, Figure 14, with a wall distance of only 15 mm. In the width, the wall distance is 100 mm. In consequence of the low thickness of gas layer, the problems which occur in measuring thick layers with different temperatures do not exist here. In our case, only the colder boundary layer on the inner side of the quartz glass windows the combustor was equipped with has to be considered. As injection device a triplet is used, the outermost orifices of which are arranged with an angle of 60° to each other; the orifice diameter is 0.6 mm.

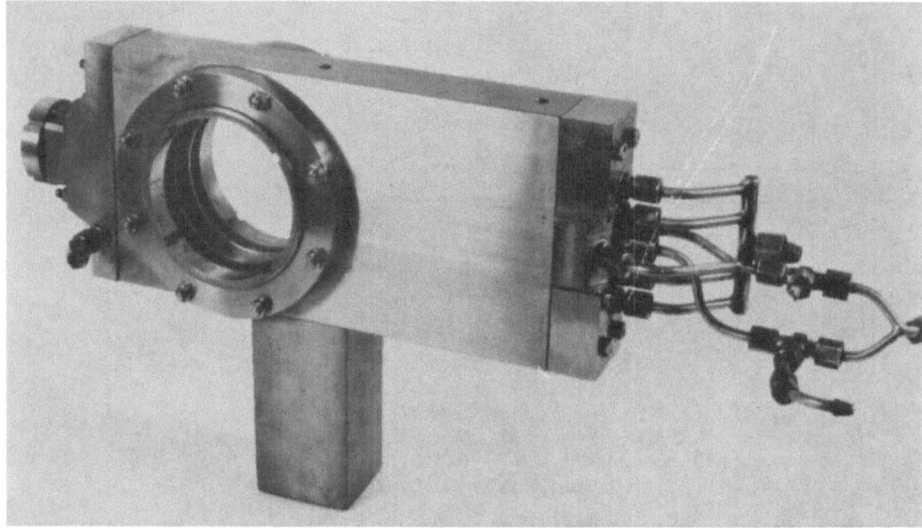

Fig. 14.   One-dimensional combustion chamber for optical temperature measurement.

The injection pressures of the outer orifices and the inner orifices are adjusted so that the base of the spray fan takes a very narrow elliptic shape and nearly adapts to the form of the cross section of the combustor. By this, an impact of the droplets on the wall of the combustor is avoided, since otherwise the value of the results would be affected. Ignition of the non-hypergolic propellant combination of hydrazine and oxygen is made by means of a hypergolic lead with N_2O_4. All the fuel is injected into the combustor by means of the triplet, whereas the oxygen uniformly distributed over the injector face is injected as gas. The theoretical velocity of flow of the gas near the injector head is very low, it is only approximately 2 [m/sec].

It shall be attained by this type of fuel injection that a droplet mixture only consisting of fuel burns in an oxidizing atmosphere. Different mixture ratio distribution in the combustion chamber then occurs due to a nonuniform distribution and/or vaporization of fuel.

5.2. OPTICAL TEMPERATURE MEASUREMENTS

The measuring method operates according to the sodium D-line reversal method, Figures 15 and 16. For this, 0.5% sodiumtrichloracetate is solved in the hydrazine.

In case of the line reversal method, use is made of the physical fact that with equal brightness of a resonance line emitted by a flame and of a reference radiation transmitted by a flame the sought temperature equals the blackbody temperature of the reference source. A high-pressure xenon lamp XBO 450 W is used as a reference source. The brightness balance is made periodically by means of a combination of a motionless and a rotating polarizing filter. This arrangement – for the first time used by Rössler [7] – allows the measuring of short-time aperiodic processes. It is the disadvantage of this arrangement, however, that the maximum spectral transmission with parallel position of the planes of polarization theoretically amounts to only 50%.

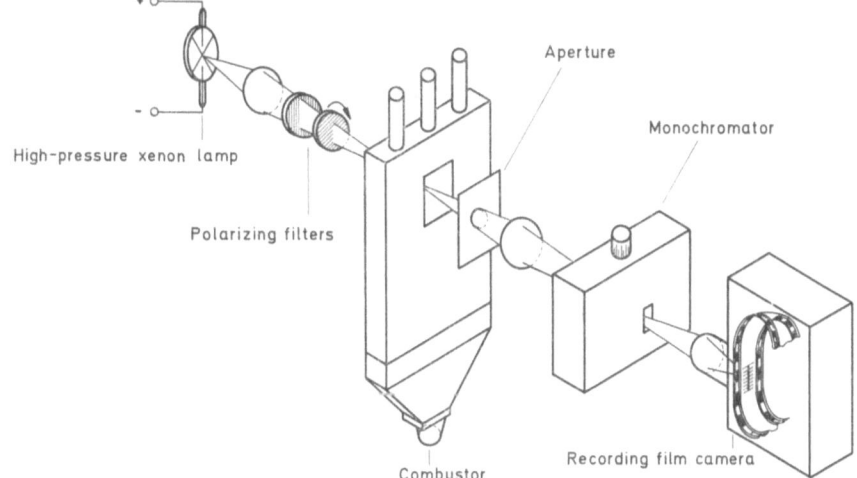

Fig. 15.   Plan of the measuring installation for optical temperature measurement according to the line reversal method.

In consequence of reflection and absorption, the value sinks still further in the application; in this case to 31.8% for yellow light.

The maximum measurable flame temperature thus is considerably below the blackbody temperature of the reference source. It amounts to approximately 3250 K; for a sufficiently accurate evaluation, the highest flame temperature must, however, by approximately 130 K remain below the theoretical limit. For measuring higher temperatures – the maximum temperature of the propellant combination amounts to, for instance, 3160 K – a mechanical modulation device is used with a maximum degree of transmission of, nevertheless, 80.3%. This operates according to the jalousie shutter principle differing only in that the shutter leaves make full turns. Sodium line radiation and modulated reference source radiation are recorded on a continually running photographic recording film, from which the position of the reversal points with the appertaining brightness balance can be gathered. The evaluation for a determination of the flame temperatures is, however, very time-consuming, so that in future a recording of reference source radiation and sodium line radiation shall be made by means of photomultipliers. A resolution of the turbulent temperature variations is

Fig. 16.   Combustion chamber and measuring installation in the test cell.

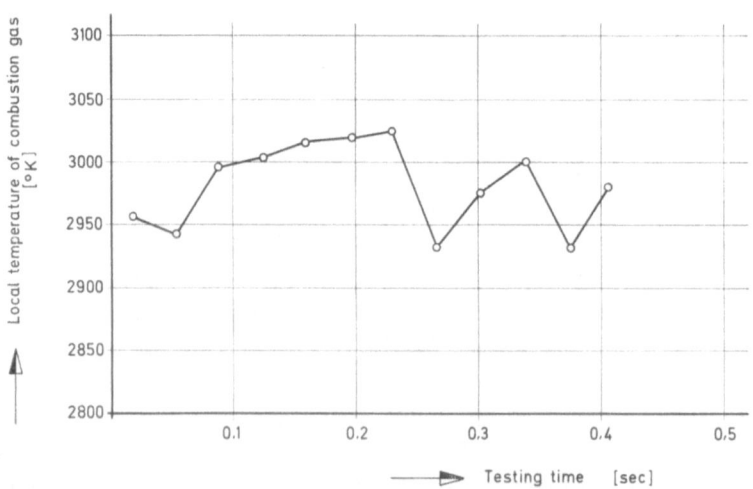

Fig. 17.   Example of a temperature plot variable with time in turbulent combustion chamber gas flow.

made with a sequence of measurement of 28 per second, Figure 17. The error of measurement varies between 2.7% and 4% according to the filter position at which the line reversal takes place.

6. Conclusions

Theoretical considerations are made on the vaporization of a mixture of hydrazine and oxygen droplets in a combustion chamber. For that purpose it is not sufficient only to know the timely vaporization rate at different points of the combustor. The residence time of droplets must be also considered. Furthermore the influence of gas flow within the chamber on the direction of droplet motion must be taken into account.

Usually, the mixture ratio distribution of an injection head is determined by cold-flow hydraulic tests. In this, fluid mixture ratio distribution of the local gross masses of propellant is determined. During vaporization, however, only shares of the local gross masses vaporize. Due to usually shorter vaporization length of the oxidizer droplets, an oversupply of oxidizer exists in the region near the injection head of the combustor, which downstream continually diminishes; that is, the mixture ratio will decrease downstream. Mass accumulation of liquid propellant during injection for instance in the middle axial zone of the combustor will create 'cross flow' of gas masses, which will influence mixture ratio distribution again.

Measurements of the temperature distribution of the combustion gas inside a model liquid propellant engine have begun. The results are expected to give further information on the mixing process in a liquid rocket motor.

References

[1] Valentine, R. S., Dean, L. E., and Pieper, J. L., 'An Improved Method for Rocket Performance Prediction', *J. Spacecraft Rockets* 3, No. 9 (Sept. 1966).
[2] Hartung, W., 'Der Einfluß des Einspritzvorganges auf das Betriebsverhalten kleiner Flüssigkeits-raketen', Teil I, DLR-Forschungsbericht 68–17, 1968.
[3] Lewis, J. D., *5th AGARD Colloquium*, 1963, p. 143.
[4] Williams, F. A., *Combustion Theory*, 1965, p. 270.
[5] Twardy, H., 'Berechnungsverfahren zur Ermittlung der Mischungsverhältnisverteilung in einer Raketenbrennkammer unter Berücksichtigung von Verdampfung und Verbrennung', DLR-Forschungsbericht 65–32, 1965, p. 39.
[6] Buschulte, W. and Twardy, H., 'Theoretische Untersuchungen zur Treibstoffverdampfung und Mischungsbildung in einer Flüssigkeits-Raketenbrennkammer', DLR-Forschungsbericht 66-83, 1966, p. 23.
[7] Rössler, F., 'Temperaturmessung nach der Methode der Linienumkehrung bei kurzzeitigen aperiodischen Vorgängen', *Z. angew. Phys.*, IV, No. 1 (1952).

PHYSICAL AND CHEMICAL KINETICS OF Al AND B FLAMES

(An Extended Abstract)

S. N. B. MURTHY

Dept. of Mechanical and Aerospace Engineering,
University of Massachusetts, Amherst, Mass., U.S.A.

Abstract. The reaction mechanisms of Al and B particles will be reviewed from the point of view of (a) the nature of the combustion process and the controlling mechanism, (b) the particulates in the products of combustion, and (c) the implications of different experimental techniques adopted in these studies. In determining the overall effectiveness of the energizing of propellants by the addition of metal particles, the heat transfer, drag and relaxation coefficients play a major role. Recent data pertaining to these will be presented with the primary emphasis on the combustion zone.

In conducting further experimental and theoretical studies, it will be shown that the most meaningful results are likely to be obtained in the combustion zone, especially under oscillatory combustion conditions.

In that connection, the theoretical results obtained from studies conducted on the longitudinal mode acoustic wave propagation through the preignition and combustion zones of metal particles will be compared with the results obtained for the case of a weak shock wave propagating through the same zones. This comparison reveals the difference in the amplification of a linear vs. a non-linear disturbance. Some of the important parameters examined in this study are (i) the steady-state combustion laws, (ii) the influence of the inertial terms or the effect of flow Mach number, (iii) the change in the physical state of the particles, and (iv) the size and void fraction of the particles. Very extensively calculated numerical results are presented for the amplification factors both in the linear as well as in the non-linear case.

The principal results may be summarized in terms of contributions to the resolution of a series of controversies pertaining to (a) the nature of the reaction mechanism under steady and non-steady conditions, (b) the formation of particulate products from the particulate reactants, (c) the dependence of the amplification of the waves on the directivity of the waves with respect to the flow direction, and (d) the experimental techniques.

1. Introduction

In the study of oscillatory processes in combustion involving discrete solid particles, it is difficult to postulate a simpler system than the one presented schematically in Figure 1 when the coupling between the gas dynamic and chemical reaction processes is to be retained in any significant measure. The system has two gaseous reactants (which will be referred to as the primary reactants) and the solid phase; Al, Be or B particles may be injected as a distinct stream as in Figure 1a or 1b. In the former case the particles are supposed to arrive in the chamber on completion of the gas-gas homogeneous combustion. The simplest of the subregions is obviously the region (A1).

The methane–oxygen combination has been chosen as the primary combustion system in view of possible application in high Mach number flight vehicles. The postulated model can have liquid instead of gaseous primary reactants and also a slurry of particles. The practical application of such a system, with the strong emphasis on flow inertial terms, is in ramjets and ducted rocket motors and also in MHD power generation devices. The objective generally is to obtain propellants which are considerably more energy rich than the usual propellants. The high density and the

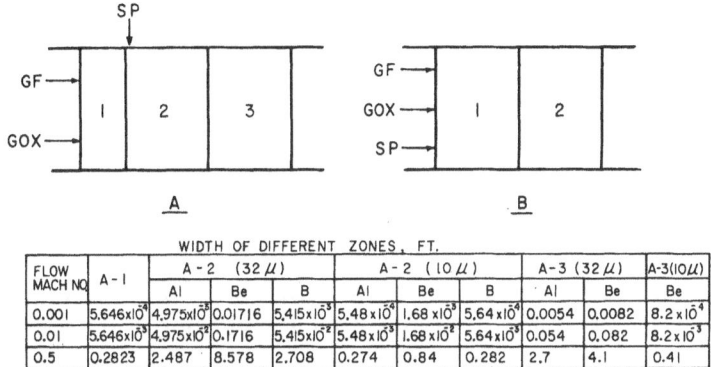

FLOW MACH NO	A-1	A-2 (32 μ)			A-2 (10 μ)			A-3 (32 μ)		A-3(10 μ)
		Al	Be	B	Al	Be	B	Al	Be	Be
0.001	5.646×10^{4}	4.975×10^{3}	0.01716	5.415×10^{3}	5.48×10^{4}	1.68×10^{5}	5.64×10^{4}	0.0054	0.0082	8.2×10^{4}
0.01	5.646×10^{3}	4.975×10^{2}	0.1716	5.415×10^{2}	5.48×10^{3}	1.68×10^{4}	5.64×10^{3}	0.054	0.082	8.2×10^{3}
0.5	0.2823	2.487	8.578	2.708	0.274	0.84	0.282	2.7	4.1	0.41

WIDTH OF DIFFERENT ZONES, FT.

Fig. 1.   Reaction zone models.

high heats of reaction with oxygen have made aluminum, beryllium and boron attractive as additives to the primary propellants.

The model is also useful in gaining a further understanding of the manner in which metal particles assist in the suppression of combustion pressure oscillations in solid propellant rocket motors.

While the combustion process may be intrinsically unsteady, the problem of the growth of an induced disturbance, such as an acoustic or weak shock wave, may be posed separately. The most direct aspect of the problem is the determination of the change in the disturbance and the effect of the disturbance during its passage through the combustion zone along and counter to the flow.

It is well recognized that the phenomena of combustion in such a system, involving as it does complex, interconnected, physical and chemical features, is highly non-linear whether the process is non-steady or not. When oscillations are present, the balance of gains and losses will dictate the growth of the amplitude of fluctuations until, in general, the oscillations grow to a limiting amplitude. The importance of the role played by the particles, from the stage of being one of the reactants to the stage of particulate products, in determining the growth of the oscillations (magnitude, rate and eventual limiting value, if any) and the overall energy release is the reason for this study. It is apparent how an understanding of the reaction mechanism in such heterogeneous systems is essentially incomplete until the non-steady state kinetics are understood.

While the model under consideration is simple, a number of uncertainties arise as one proceeds to attempt the calculation of even progressive waves. Those uncertainties may be grouped broadly under the following: the energy release rate during the different processes; the details of the process occurring in the reaction and the pre-reaction zones; the change in the physical structure of the particles; the transport processes between the gas and the particles; and the nature of the postulated instability.

2. The Energy Release Rate

Perhaps the most serious question in the postulated model is the implication of the well-defined subsections and the overall boundaries of the combustion zone. This is the Berke-Schumann model but it has been shown that such a model is the first outer solution in an asymptotic representation of the flow field. However, the question remains whether the oscillatory processes during the non-steady state operation will not progressively affect the widths and even the location of the combustion zone. It can be shown that the displacement of the boundary due to the heat transfer during the oscillatory process is indeed very small until a situation arises when it becomes questionable whether the integrity of the flame zone itself will have any meaning under those conditions. It is also interesting in this connection that the dynamic and thermal characteristic times for Al are about 25 and 12 times and for Be about 18 and 30 times respectively the time for the combustion of methane; the response of the metal particles should be even less critical.

The second problem of importance is the interaction of the two phases in such subregions as (B1) and (B2). The region (B2) is particularly complicated in view of the facts: (a) should the gas-gas combustion be not complete it is necessary to have a means of dividing the available oxygen between the methane and the metal particle and furthermore of establishing reaction rates for each in the presence of the other and (b) in the case of boron, for instance, it is necessary to consider the endothermic reaction in the formation of BO as well as the subsequent exothermic reaction in the formation of B_2O_3. Hence, in the case of this region, it seems possible at this stage only to perform calculations with the total energy release in the region as a parameter.

3. The Particle Kinetics

In regard to the combustion of the particles, some general assumptions may be made as follows: (i) the particles undergo heating up to the ignition temperature and the width of the subsection for heating may be calculated on the basis of heat transfer from the gas to the particulate phase (dictated by the relative motion of the particles with respect to the gas and the absolute velocity of the particles) and (ii) the reaction zone and the pre-ignition zone may be clearly demarcated for an assigned size of particles and given environmental conditions in terms of temperature boundaries.

The dominant mode of heat transfer to the particles is through conduction unless the particles are larger than 100 μ in diameter when corrections have to be made for convective heat transfer.

Experimental evidence seems to indicate that in the case of non-fragmentary burning of Al and Be particles, the vapour phase diffusion model of Bruzustowski and Glassman is valid. On the other hand the experimental results of Macek may be utilized to relate the burning rate to the partial pressure of oxygen in the mixture and the mean diameter of the particle. The experimental data for different partial pressure values can be converted into equivalent mixture pressure values.

The reaction mechanism of boron particles has been demonstrated to be different from that of Al and Be particles; the high melting point and the low vapour pressure of boron seem to dictate the reaction. The initial step during steady state reaction is surface reaction unless the particles are very fine. The rate limiting process in the steady state reaction seems to be diffusion of oxygen through the oxide vapour. Again this latter may be a function of the particle size and, for fine particles the chemical kinetic equations may dictate the final reaction.

The basic parameter in the study of particulates in a gas–solid system is the size and number density distributions, their instantaneous values and the rate of change of the values. The particle sizes are described by such relations as the mass median diameter, the volume mean diameter and the generalized mean diameter based upon the appropriate moments. However, there is no substantial evidence for the proper mean diameter to be employed except for data from stationary flames, or at the completion of combustion as in the exit plane of rocket motor nozzles or rocket motor configurations without nozzles. The view taken in the calculations presented here is that calculations may be performed in the different subsections of the flame zone for different mean sizes without in fact assigning causes for such changes and the results may be employed in assessing experimental data where the mean particle size may change. In other words, for example, if the particle size in region (A2) is $32\ \mu$ and in region (A3) is $10\ \mu$, one can find the overall change in a pressure disturbance by using the appropriate sizes in the different zones.

Regarding the momentum and energy transfer between the particles and the gas, certain characteristic times are postulated. At the same time, on the basis of the low number density of heavy particles in the present analysis, it is taken that the product (frequency) × (characteristic time) is very much less than one so that the steady state drag coefficients are not in error when applied to essentially non-steady processes.

4. The Nature of the Postulated Instability

In the present analysis, only longitudinal, non-shifting, uncoupled disturbances are considered. However, both acoustic as well as weak shock type disturbances are studied and the results compared.

In regard to the acoustic disturbances, the analysis presented here is restricted to high and discreet frequency disturbances. The weak shock wave theory, it may be stated, is not truly a nonlinear wave theory. In the quasi-linear theory, the shock wave is affected by the disturbances it creates in finite, small regions.

An important factor in the propagation of the waves, acoustic or the weak shock type, is the direction of propagation of the wave relative to the fluid velocity. It is extremely important to realize that the fluid velocity affects the propagation of the wave in a significant measure and, furthermore, that the velocity disturbance is not truly a wave-like process but is dictated by wave diffusion. This can be established analytically.

5. The Computed Results for Acoustic Waves

The range of parameters for which calculations have been performed is as follows:

          mean combustion pressure   10 atm
          methane-oxygen mixture ratio   0.5 to 2.0
          particle size   10 to 32 μ
          particle mass fraction   1 to 2%
          flow Mach number   0.001 to 0.5
          frequence of oscillations   0.5 to 3 kcs.

Some typical results are given in Tables I and II and in Figure 2.

As an example, at Mach 0.5, a wave of nominal frequency 2.5 kcs on passing as a Left Running wave (LRW, opposed to flow) through a flame zone with Al particles may undergo an attenuation of 0.90493 if the size of the particles is taken to be

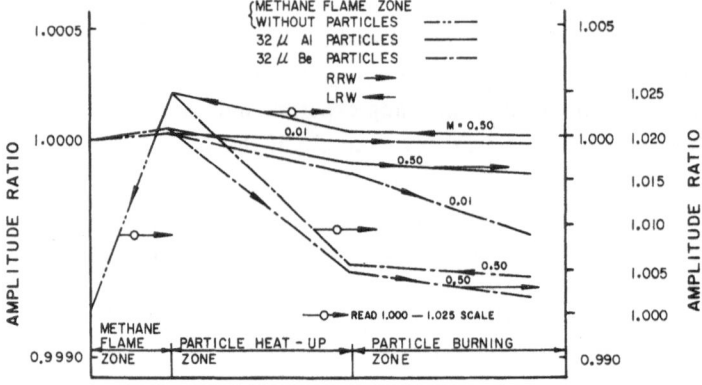

Fig. 2.   Amplitude ratio of the acoustic wave across various zones.

32 microns in the heat up zone but 10 microns in the burning zone; on the other hand, the wave undergoes an amplification of 1.01528 if the particle size is held constant at 32 microns.

6. The Computed Results for Weak Shock Waves

The range of parameters for which shock wave calculations have been performed is as follows:

          mean combustion pressure   10 atm
          methane-oxygen mixture ratio   1.0
          particle size   10 and 32 μ
          particle mass fraction   1 and 2%
          flow Mach number   0.001 to 0.5
          shock Mach number   1.01

The results of the calculations have been presented in Figure 3.

TABLE I
Widths of sub-zones, Ft

Flow M	A-1	A-2		A-2		A-2	A-3
		Al (32 μ)	Al (10 μ)	Be (32 μ)	Be (10 μ)	B (32 μ)	Be (32 μ)
0.001	5.646×10^{-4}	4.975×10^{-3}	5.48×10^{-4}	0.0176	1.68×10^{-3}	5.45×10^{-3}	0.0082
0.01	5.646×10^{-3}	4.975×10^{-2}	5.48×10^{-3}	0.176	1.68×10^{-2}	5.415×10^{-2}	0.082
0.5	0.2823	2.487	0.274	8.578	0.84	2.708	4.1

TABLE II
Amplification factor

M	0.1	0.5	0.5	2.5
ω, kcs	0.5	2.5	0.5	2.5
Al. Flame				
Amp. Ratio RRW	0.99815	0.99921	0.98514	0.99675
LRW	0.99822	0.99925	0.98832	1.01528
Be Flame				
Amp. Ratio RRW	0.9940	0.99626	0.94958	0.98517
LRW	0.99394	0.99571	0.9564	0.98449

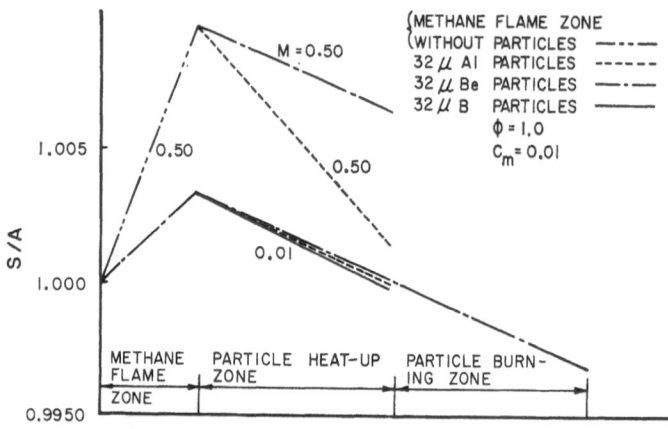

Fig. 3.   Comparison of shock wave and acoustic wave theories. S – Shock wave theory; actual
property distribution. A – Acoustic wave theory; mean properties. $\omega = 2.5$ kcs.

7. Discussion

In the analysis of heterogeneous combustion involving two phase flows with a discreet solid phase, a serious problem is the determination and the incorporation of the changes in the discrete phase into the flow process under investigation. It is clear that any method of correlating the size distribution is not independent of the processes occurring in the system. The results of the present analysis are encouraging in regard to the experimental determination of the size distribution in the reasonably large widths of the heat up and combustion zones and also in regard to obtaining a check on the measured distribution by its effect on induced wave motion.

In the homogeneous combustion zone, the energy release (or the equivalence ratio) affects the wave in the intuitively predictable fashion. The effect of the inertial terms (or the flow Mach number) on acoustic waves is to decrease the amplification of the RRW (right running wave, in the direction of flow) and to increase it for the LRW. The weak shock wave on the other hand undergoes an increase in the amplification in the flow direction with increasing Mach number.

In the nonhomogeneous heat up zone, for the same mass fraction and particle size, Be is more effective in attenuating acoustic waves than B which in turn is more effective than Al. On the other hand, the weak shock wave under the same conditions shows an amplification, B being able to reduce the amplification more than Al or Be. This difference in the two theories points up the need for a proper nonlinear wave theory and a detailed analysis of the structure of the shock wave in the nonhomogeneous zone. The inertial effects of gas motion are again significant.

Finally, in the nonhomogeneous reaction zone, results are included only for Al and Be, wherein it will be observed that for Al combustion, the gas phase temperature in fact drops when the methane-oxygen combustion is complete. In general the conflicting effects of the presence of a discreet phase and of heat release are apparent in both the acoustic and weak shock wave theories; in the former, the frequency of oscillation is of significance. For the weak shock waves, the attenuation of the wave even in the presence of energy release shows the importance of the transfer processes. Furthermore, there arises the question of the section at which the attenuation begins between the heat up zone and the reaction zone. The flow inertial terms once again cause a difference between the LRW and the RRW.

References

[1] Markstein, G. H., *Nonsteady Flame Propagation*, The Macmillan Company, New York, 1964.
[2] Price, E. W., *AIAA J.* **7**, 1 (1969) 153.
[3] Summerfield, M. and Krier, H., AIAA Paper No. 69–178.
[4] Fendell, F. E. and Chung, P. M., Aerospace Corp. Rep. No. TD6-469 (55240-10)-1, 1965.
[5] Bakhman, N. N. and Kundrashkov, Yu. A., NASA-TT-F-8436, June 1963.
[6] William G. C., Hottel, H. C. and Morgan, A. C., 'The Combustion of Methane in a Jet Mixed Reactor', *Twelfth Symposium (Int.) on Combustion*, 1968.
[7] Kuehl, D. K., *AIAA J.* **3**, 12 (1965) 2239.
[8] Friedman, R. and Macek, A., 'Combustion Studies of Single Al Particles', *Ninth Symposium (Int.) on Combustion*, 1963.
[9] Kuehl, D. K. and Zwillenberg, M. L., AIAA Paper No. 68–49A.
[10] Macek, A. and Semple, J. M., 'Experimental Burning Rates and Combustion Mechanisms of Single Beryllium Particles', *Twelfth Symposium (Int.) on Combustion*, 1968.
[11] Brzustowski, T. A. and Glassman, I., 'Vapor Phase Diffusion Flames in the Combustion of Magnesium and Aluminum: I, Analytical Developments', *Progress in Astronautics and Aeronautics*.
[12] Macek, A., 'Fundamentals of Combustion of Single Aluminum Particles', *Eleventh Symposium (Int.) on Combustion*, The Combustion Institute, 1967, p. 203.
[13] Talley, C. P., 'The Combustion of Elemental Boron', *Progress in Astronautics and Rocketry*, Vol. 1, Academic Press, 1960, p. 279.
[14] Schadow, K., AIAA Paper No. 68–634.
[15] Sehgal, R., Jet Propulsion Lab. TR 32-238, 1962.
[16] Cheung, H. and Cohen, N. S., *AIAA J.* **3** (1965) 250.
[17] Dobbins, R. A. and Strand, L. D., AIAA Paper No. 69–146.
[18] Colucci, S. E. and Adams, J. M., Aerojet-General Corp. Rep. SRO20, November 1965.
[19] Crowe, C. T., *et al.*, UTC 2128-QT3, United Tech. Center, October 1965.
[20] Povinelli, L. A., AIAA Paper No. 67–104.
[21] Crowe, C. T. and Willoughby, D. G., AIAA Paper No. 66–639.
[22] Hoglund, R. F., *ARS J.* **32** (1962) 662.
[23] Carlson, D. J. and Hoglund, R. F., *AIAA J.* **2**, 11 (1964) 1980.
[24] Marshall, R. L. *et al.*, 'An Experimental Study of the Drag Coefficient of Burning Aluminum Droplets', presented at the 3rd ICRPG Combustion Conference, October 17–21, 1966.
[25] Fuchs, N. A., *The Mechanics of Aerosols*, The Macmillan Company, New York, 1964, p. 84.
[26] Kovasznay, L. S. G., *J. Aeron. Sci.*, **20** (1953) 657.
[27] Salant, R. F. and Toong, T.-Y., 'Some Fundamentals of Combustion Instability', *Oxidation Combust. Rev.* **2** (1967) 185.
[28] Culick, F. E. C., Tech. Rept. 480, MIT Aerophysics Lab., Cambridge, Massachusetts, 1961.
[29] Murthy, S. N. B. and Osborn, J. R., *J. Acoust. Soc. Am.* **37**, 5 (1965) 872.
[30] Murthy, S. N. B., NAL Tech. Note TN-PR-1-65, 1965.
[31] Lehmann, G. M. and Murthy, S. N. B., *J. Spacecraft* **2**, 5 (1965) 828.
[32] Sirignano, W. A. and Crocco, L., *AIAA J.* **2** (1964) 1285.
[33] Chisnel, R. F., *Proc. Roy. Soc. (London)* **232**, 1190 (1961) 350.
[34] Rudinger, G., 'Shock Wave and Flame Interactions', in *Combustion and Propulsion (Third AGARD Colloquium)*, Butterworth's Scientific Publications, London, 1958.
[35] Lehmer, D. H., *J. Ass. Computing Machinery* **8** (1961) 2.

INVESTIGATION OF BORON COMBUSTION FOR ITS APPLICATION IN AIR AUGMENTED ROCKETS*

K. SCHADOW

Deutsche Forschungsanstalt für Luft- und Raumfahrt, Trauen ü. Soltau, Germany

Abstract. During the past years different types of air-augmented rockets has been receiving increased emphasis. This presentation is related with the ducted type of air-augmented rocket for missions of long distance at low altitude. In the ducted rocket, ram-air is introduced into a secondary chamber where subsonic turbulent mixing and burning will occur with the fuel-rich exhaust of a primary chamber. A second nozzle is provided for acceleration of the product gases to supersonic velocity. Propellants with boron are attractive for air-augmentation because of their high density and high heat of reaction with oxygen. From the standpoint of volumetric energy release it is desirable to maximize the boron content in the propellant of the primary chamber (up to 70 % by weight). However, the theoretical gain in specific impulse can, of course, only be attained with efficient boron combustion.

In a short review of specific literature it will be shown that the conditions for boron-particle ignition and efficient combustion are poor (in comparison to aluminium particles, for example) and not yet well understood. Therefore combustion studies of propellants heavily loaded with boron are necessary in order of developing air-augmented rockets with high combustion efficiency.

In the following results of an experimental research program will be presented which has been performed at the U.S. Naval Weapons Center to study the combustion behavior of boron propellants. An experimental apparatus will be described which permits the observation of the boron-combustion zone in the secondary chamber through quartz windows. During the tests the combustion efficiency can be controlled by c^*-efficiency measurements and particle-sampling probes. The primary rocket motor is burning gaseous hydrogen-gaseous oxygen-boron mixture, so that the boron-particle temperature in the primary rocket can be easily varied by changing the hydrogen/oxygen mixture ratio.

By increasing the primary chamber temperature from 750 K to 2000 K there is a considerable change in the boron-combustion behavior (demonstrated by color cinephotography) and an increase of the overall reaction efficiency from 0.85 to 0.95. The same effect of increased combustion efficiency can be demonstrated by boron-particle sampling in the exhaust of the secondary chamber. These results show that the initial boron-particle temperature in the primary rocket should exceed a certain temperature before the particle makes contact with the air in the secondary chamber.

Considering a propellant with oxidizer, fuel, and boron, the temperature of the boron particles in the primary chamber has to be established by the oxidizer-fuel matrix if one assume that the boron does not start to burn violently in the primary motor (this assumption is well established for propellant heavily loaded with boron). Therefore the primary combustion temperature (computed with no boron reacting) becomes an important criterion for determining propellant compositions for air-augmented application. This requirement of high primary chamber temperature, of course, has to be optimized with other propellant and motor conditions. So an increase in primary chamber temperature is followed by a decrease in specific impulse. Results of a theoretical program will be shown in which the thermodynamic data of different propellants are computed in order to optimize the requirements of high primary chamber temperature and high specific impulse.

1. Introduction

The development of air-breathing rockets is closely related with combustion of propellants heavily loaded with boron. Considering a typical air-breathing rocket

* The experimental part of this work was performed under an unclassified portion of the Naval Ordnance Systems Command ORD TASK ORD-033-123/200/F 009-06-05 at the Naval Weapons Center, China Lake, Calif., U.S.A., and under Fellowship of Bundesministerium für wissenschaftliche Forschung, Germany.

(the ducted rocket) these propellants react in a primary chamber producing a fuel-rich exhaust. These products mix and burn with the ram-air in a secondary chamber. A second nozzle is provided for acceleration of the product gases to supersonic velocity.

In the first part this paper will discuss the ignition and combustion of single boron particles. Following this review results of combustion studies with a fuel-rich propellant system and air will be presented. In these tests actual propellants for air-breathing application have been simulated by burning hydrogen-oxygen-boron mixtures in the primary chamber. On the base of thermodynamic calculations the results of these experiments will be used for demonstrating possibilities for improving the combustion characteristics of propellants heavily loaded with boron. This will be the subject of the third part of this paper.

2. General Consideration on Boron Combustion

2.1. HEATS OF COMBUSTION OF BORON

The use of metals in air-breathing rockets is very attractive because of their high density and high heats of combustion with oxygen. Various metals have been considered as fuel for conventional rocket motors and for air-breathing rockets. A summary of the metals of interest and their heats of combustion (see Table I) will illustrate the advantage and disadvantage of specific propellant ingredients for both types of rockets. On the base of kcal/gram fuel + oxidizer (which term is used in conventional rocketry) boron appears to be already very attractive. Compared with beryllium and aluminum, two well-known metals in conventional rocket motors, beryllium is superior to aluminum and boron, but boron shows a higher heat of combustion than aluminum. There are some reasons that aluminum instead of boron has received by far the widest use in conventional solid rocketry (beryllium has not been used for its well known aggressivity).

Cohen [2] summarized the disadvantages of using boron in conventional rockets. One important point he mentioned was that the ignition and combustion character-

TABLE I

Heats of combustion [1]

Fuel	Combustion product	kcal/ gram fuel	kcal/ gram fuel + oxyd.
H_2	H_2O (1)	33.9	4.23
C	CO_2	7.83	2.14
Li	Li_2O	10.25	4.79
Be	BeO	16.3	5.86
Mg	MgO	5.9	3.56
B	B_2O_3	14.1	4.43
Al	Al_2O_3	7.43	3.94
Si	SiO_2	7.5	4.00
Zr	ZrO_2	2.85	2.11
Ti	Ti_2O_3	3.79	2.65

istics of boron are not sufficient for achieving complete combustion in the rocket motor.

After briefly discussing the use of boron in conventional rockets in the following the application of boron in air-breathing rockets will be demonstrated. In air-breathing rockets the ram-air is the major oxidant, so that oxidizer can be considered as 'free' available. Therefore, looking again on the heats of combustion for different metals in Table I the comparison has to be done on the base of kcal/gram fuel. Now the data show that boron is superior to aluminum by a considerable factor almost reaching the value of beryllium. Although the heat of combustion does not tell the entire story for propellant evaluations these data show that boron is the most attractive metal for air-breathing application if the air/fuel ratio is high to fulfill the requirements of the relatively high stoichiometry of the boron-air reaction. A typical air/fuel ratio is about 12 or higher with 50 to 70% boron in the propellant of the primary rocket.

In the following special consideration will be given to the ignition and burning characteristics of boron particles in these motors. It is the main object of this paper to discuss the problems on the base of combustion studies in ducted rockets and calculated thermodynamic data. But prior to these investigations of fuel-rich systems, it may be instructive to review briefly the ignition and combustion characteristics of single boron particles.

2.2. COMBUSTION OF SINGLE METAL PARTICLES

The ignition and combustion characteristics of metals depends upon the physical properties of the metal and its oxide. A tabulation of these properties is given in Table II. The classification of the metal in four groups is made by Glassman and Brzustowski [3]. A summary is given in [2].

Boron is classified in the group of nonvolatile metal (high metal boiling point) – volatile oxide (relatively low oxide boiling point). According to the Glassman Criterion [4], which has been established from energy considerations, the combustion will occur by a heterogeneous surface reaction. It will be demonstrated that this surface oxidation process is relatively slow compared to the vapor-phase burning. Vapor-phase combustion takes place if the boiling point of the metal is below that of its oxide. This is true for aluminum. As well known aluminum shows sufficient combustion efficiency in conventional solid propellant rocketry.

It seems to be instructive to discuss the boron particle combustion characteristics by comparing the typical criteria of a vapor-phase combustion (aluminum) and of a surface oxidation process (boron). This will be done using two pictures made by Prentice [5]. In these tests boron and aluminum particles of 400μ particle size are ignited while in free-fall by a xenon lamp and are permitted to fall down a glass pipe where the reaction can be quenched on a glass plate. The specimens are examined on the microscope. Two pictures of these investigations are shown in Figure 1.

The flame structure can be derived from these pictures by looking at the particle and the surrounding quenched combustion products (oxides). The aluminum particle

TABLE II

Properties of metals and their oxides [3]

Metal			Oxide	
MP K		BP K	MP K	BP K
1. Volatile metal				
Li	454	1620	1700	3200[a]
Mg	923	1381	3075	3350[a]
2. Nonvolatile metal-insoluble oxide				
Al	932	2740	2318	3800[a]
Be	1556	2750	2823	4123[a]
Si	1685	3582	1883	3000[b]
3. Nonvolatile metal-soluble oxide				
Ti	1950	3550	2128	4100[a]
Zr	2125	4650	2960	5200[a]
4. Nonvolatile metal-volatile oxide				
B	2300	3950	723	2520[b]

[a] $BP_{METAL} < BP_{OXIDE}$ Vapor-phase combustion [4]
[b] $BP_{METAL} > BP_{OXIDE}$ Surface reaction [4]

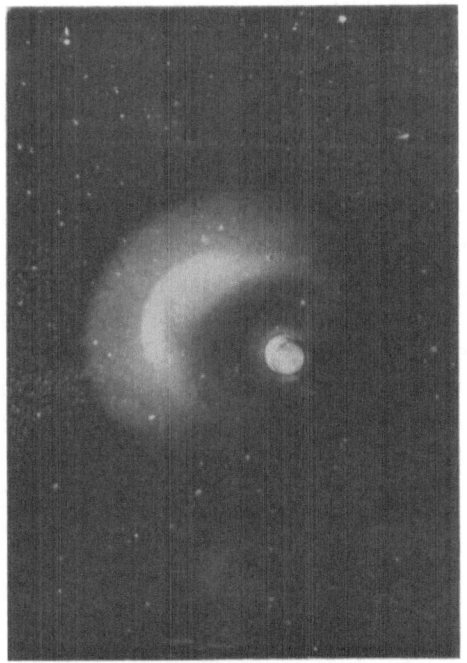

Aluminium

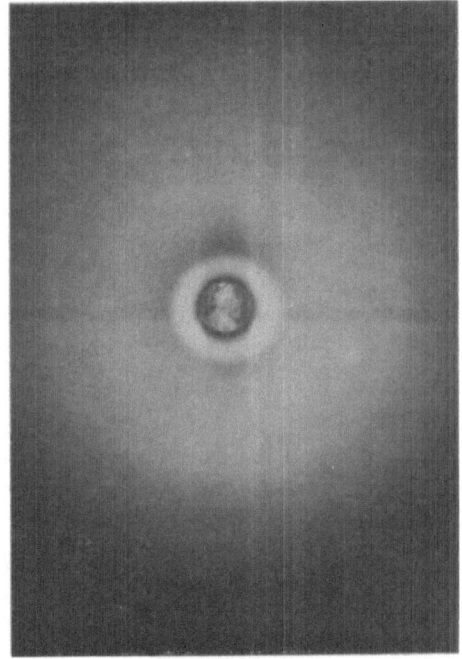

Boron

Fig. 1.   400 μ aluminium and 400 μ boron particle burning in air. Quenched on a glass plate [5].

is covered on one side by a accumulation of surface oxide, so that the combustion zone is associated to the other side with the exposed metal particle surface. The flame structure is a typical vapor-phase combustion zone with its combustion center in a distinct distance from the metal surface. The aluminum vapor diffuses into the oxidizing environment and the combustion forms a smoke of very fine oxide particles. The reaction is limited by the vaporization rate of the aluminum.

Looking at the boron particle there is no oxide on the metal particle. The surface reaction can be realized by the large amount of oxide surrounding the particle very closely. The reaction in this phase is rate limited by the diffusion of the oxygen through the oxide vapor to the surface. From these two different combustion characteristics two different combustion rates result. In the case of vapor-phase combustion (aluminum) the combustion rate is about proportional to the square of the particle diameter, in the case of surface reaction the combustion is only proportional to the particle diameter. Depending on the particle size and the nature of the environment to which the particle is exposed combustion rates of 10–100 μ per msec have been measured for aluminum. The reaction rate of boron is not known exactly; 0.2–2 μ per msec seems to be a good value for demonstrating the relatively slow reaction rate compared with aluminum.

The reaction rate of boron is strongly dependent on the particle temperature. This temperature dependency is important for examining the rocket motor tests in Section 3.

Although there is little known about this problem, it will be discussed using a schematic diagram with the scaleless combustion rate of boron vs. particle temperature (Figure 2). On the temperature axis the important physical properties of the metal and its oxide are marked. It should be again emphasized that the curve of the rate of reaction is strictly schematic because very little information has been given in the literature.

Data of [6] show that clean boron particles react spontaneously with air at temperatures of ca. 400 K. But, according to [7] only a slow surface oxidation takes place below the oxide melting point (720 K). By increasing the temperature to the region of the metal melting point (2300 K) and oxide boiling point (2500 K), a sharp increase in boron oxidation rate was noticed by the same author. The same observation was made by Prentice [8] in preliminary tests when burning single boron particles in air. Color pictures indicated that the temperature of the particle rose relatively slowly after the particle was exposed to a thermal energy source; this slow surface oxidation phase was finally followed by vigorous particle combustion. Considering the relatively short residence times associated with rockets this sharp increase in reaction rate should be termed with 'ignition'. The rate of combustion rises further with increasing temperature. The nature of the combustion should remain a surface oxidation up to high temperatures, but it is expected that a vapor-phase combustion can be reached at higher particle temperatures. The rate limiting processes in the three mentioned regions are shown in Figure 2.

From the schematic curve in Figure 2 it can be realized that the burning time for

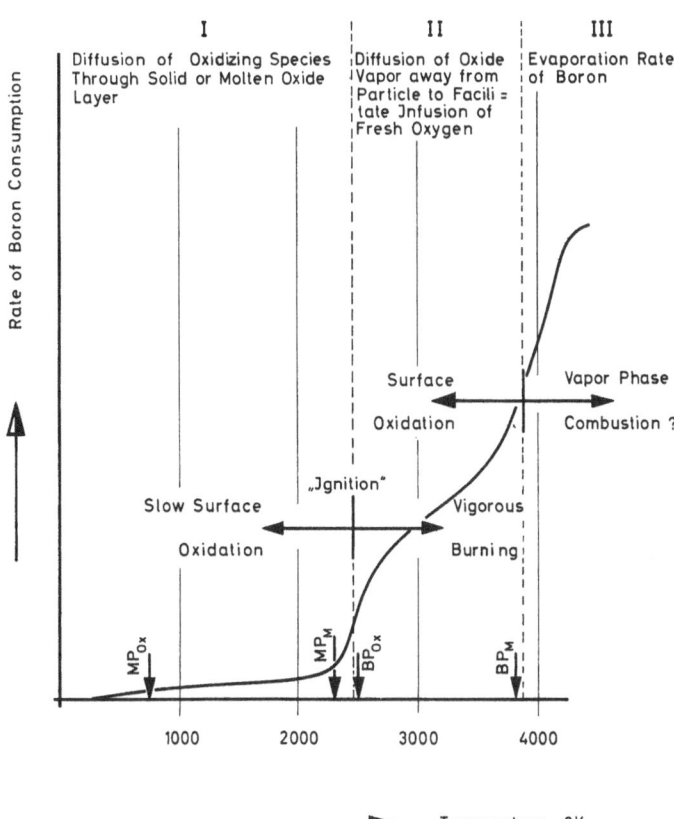

Fig. 2. Schematic diagram of rate of boron combustion in air vs. temperature (single particle).

complete combustion of the boron particle is strongly dependent on the original particle temperature. The dependency is unknown, but is a very important value for the development of boron propellants, because in a ducted rocket the boron particle temperature at the primary nozzle exit (which is the starting point of the boron–air reaction) is dependent on the propellant composition in the primary rocket. The influence of this temperature on the boron reaction process in the secondary chamber has been studied at the Naval Weapons Center and the results will be presented in following Section 3. Before discussing the combustion of fuel-rich propellant systems in ducted rockets it should be emphasized that the combustion characteristics of single particles in air and of propellant systems heavily loaded with metal particles in ducted rockets show significant differences:

(1) The particle size of the investigated single particles is about 400 μ instead of 20 μ to 1 μ particle size of the metal products used in propellants.

(2) The combustion of a group of particles is more intensive than of a single particle, as the heat radiation loss of a single particle to its environment is larger than that of a particle cloud ('cooperative effect' [9]).

(3) The mentioned combustion studies of single particle have been done in air. The reaction kinetics will change with changing chemical environment composition. It is expected that there is an increase in combustion rate of boron in the presence of water, fluorine, chlorine etc.

3. Combustion Studies with Propellants Heavily Loaded with Boron

A detailed description of these investigations is presented in [10]. In this paper a survey of the test apparatus and the results concerning the primary chamber temperature are given.

3.1. TEST APPARATUS

A schematic diagram of the combustion tunnel is shown in Figure 3. The air coming from a heater which burns liquid alcohol directly with the high pressure air enters the tunnel from the left side through a choked throat, which makes the air flow independent from the pressure in the tunnel as long as the ratio of the pressure before and after the throat is larger than about two. In the tunnel the air is divided into two streams which pass the primary rocket on the top and on the bottom through two rectangular channels. Then the air enters the secondary combustion chamber where it mixes with the exhaust of a rocket. This reaction zone can be optically observed through four pairs of quartz windows mounted on opposite sides of the tunnel walls. The reaction products leave the combustion tunnel through a throat with 75 mm diameter. The characteristic length of the secondary combustion chamber (L^*) based

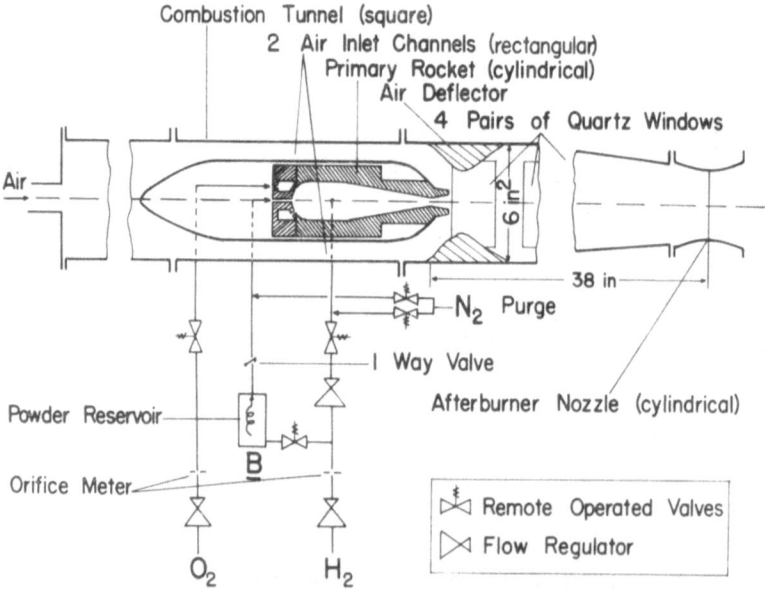

Fig. 3.   Schematic diagram of combustion tunnel with primary rocket.

on the length from primary to secondary nozzle is 4.3 m; the corresponding residence time based on a temperature of 2500 K is 4 msec.

The control devices of the propellant feed system for the primary chamber are described in the figure. The boron powder (Kawecki, with an average particle size of less than 1 μ) is transported to the burner by entraining it in the H_2-gas by means of a pneumatic transport system. The average boron powder flow rate is determined by weighing the powder reservoir before and after each test. The copper nozzle has a 12.2 mm throat diameter, and an expansion ratio of 2.68. The characteristic length (L^*) of the rocket is 2.1 m; the corresponding residence time based on a temperature of 2500 K is 1.9 msec.

Following measurements have been made:

(1) Recording of the afterburning behavior with a Hycam camera on Anscochrome D/200 color film at 2500 frames/sec.

(2) Schlieren pictures of the reaction in secondary chamber.

(3) Particle sampling by dropping a glass plate fastened on a metal block in free fall through the exhaust gases of the secondary chamber.

(4) Combustion efficiency measurement (c^*) based on the static pressure in the secondary chamber.

In the following the results of tests will be presented in which the primary chamber temperature was varied by changing the oxygen/hydrogen/mixture ratio.

Prior to showing the results two remarks have to be made for the following discussion.

3.2. NONEQUILIBRIUM COMBUSTION TEMPERATURE

The thermodynamic equilibrium flame temperature is an important value for comparing energy potentials of the different reaction processes for rocket motor processes in absence of metal particles. In a system with metal particles, where the physical and kinetic conditions are such that the metal ignition requirements are not being reached, it is more meaningful to consider the metal in the fuel-oxidizer combustion matrix as an inert heat sink. The temperature produced by the nonmetallic fuel-oxidizer reaction determines whether or not the metal will melt or ignite in the combustion zone environment. This specific combustion temperature calculated with the metal considered as a heat sink only is called the nonequilibrium combustion temperature T^* [11].

3.3. COLOR PICTURES EVALUATION

In using a color picture for information about the boron combustion behavior it is necessary to determine the different colors which correspond to specific processes of the hydrogen-oxygen-boron reaction.

If one studies the produced combustion species, only the molecular bands of the boron oxides and the continuous body radiation from the boron particles have to be considered. If the boron particles leave the primary rocket at very low temperature the changing color scale is visible during the heat-up time of the particles in the secondary chamber. These colors recorded by photography can be roughly compared with the colors of a black body at different temperatures. The true temperature of the particle is only slightly higher than its color temperature, as determined by the comparison with the black body color.

The characteristic molecular emission of the boron oxides are the green bands of the species BO_2, which is formed by the reaction $2\,B_2O_3 + O_2 \rightleftharpoons 4\,BO_2$. Therefore, the green color is a good indicator of boron combustion, if there is an excess of oxygen and a sufficient environment temperature to fulfill the conditions for forming BO_2 from B_2O_3. The gaseous species B_2O_3 are colorless while condensed B_2O_3 is white. The B_2O_3 should be in the gaseous phase in the afterburner flame if the temperature is above 2326 K (B_2O_3 boiling point) and condensed boron oxide should be noticed as a white vapor in the exhaust and on cooler parts of the test sections.

3.4. PRIMARY CHAMBER TEMPERATURE VARIATION

The investigations have been started with tests run without air flowing through the tunnel. In these tests the percentage of the boron in the propellant was about 40%, while the oxygen/hydrogen mixture ratio was changed from about 0.6 to 4.3 simultaneously increasing the theoretical primary chamber temperature T^* from 800 K to 2600 K. With the expansion of the primary combustion gases to the secondary chamber pressure there is a temperature decrease. Assuming a thermal equilibrium between the expanding gases and the boron particles the boron particle temperature at the primary nozzle exit can be calculated. These calculated data have been compared with the experimental particle temperature, which has been determined from the colors recorded by photography. Under the described conditions only moderate accuracy for the calculated and semiquantitative temperature determination can be expected. The color temperature was slightly higher than the calculated temperature [10].

These tests show that a temperature region from about 500 K to 1500 K for the particle at the primary nozzle can be established. Remembering the discussion of single particle burning this temperature would be the original particle temperature before starting the combustion in air. The influence of this temperature on the combustion behavior of the boron particles in the ingested air will be discussed in the following figures.

It should be mentioned at this point that there is a considerable amount of gaseous hydrogen in the fuel-rich exhaust of the primary rocket. The reaction of these fuel species with the air will not be discussed separately from the boron-air combustion process in this paper. For information on the hydrogen-air reaction in connection with the boron combustion see [10].

Figure 4 shows the boron combustion behavior at two different primary chamber temperatures. In both tests the mass flow rates of hydrogen, boron, and air were kept almost constant, while the oxygen mass flow rate in Test 1 was higher than in Test 13; thus, the nonequilibrium chamber temperature was increased from about 750 K to about 2000 K. With increasing oxygen mass flow the propellant weight increases, so that the characteristic parameters related to the propellant weight (air

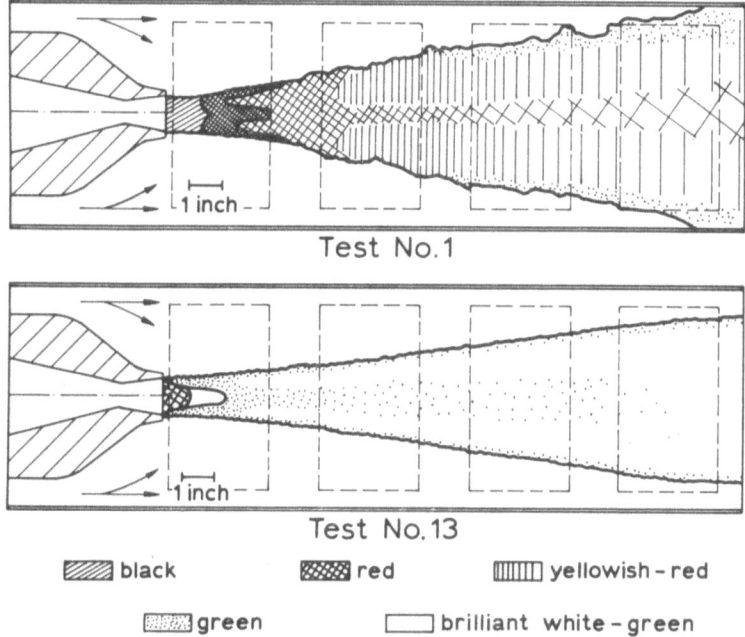

Fig. 4. Boron combustion behavior at different primary combustion temperatures. Test 1: $T^* = 750\,K$, test 13: $T^* = 2000\,K$. (For further test data see Table III.)

to propellant ratio, per cent boron in propellant) changed, although the air and boron mass flow rates were constant. For specific test conditions see Table III. As discussed earlier the particles in Test 1 left the primary rocket showing no radiation. As the particle temperature increased, the entire heat color scale from dark red to bright yellow was observed over all the visible reaction zone. In the third and fourth windows green color indicated boron combustion on the outer boundaries of the reaction plume. This showed that the energy generation from the hydrogen-air reaction and from the slow surface oxidation of boron went into heating the boron particles. Boron combustion only occurred on the jet boundaries near the hydrogen-oxygen combustion zone and in regions with large oxygen excess. Where these conditions were not established (plume center) the color temperature of the particles was relatively low.

The described boron reaction behavior, indicating very poor combustion, changed entirely at higher chamber temperature. The particles were expelled from the rocket with red radiation and the boron combustion started immediately when the exhaust was mixed with the air. The green mixing and combustion zone spread very fast to the center of the jet, so that after a mixing length of about $1.5D_e$ the entire reaction zone was green. At the outer regions of the jet this green changed to brillant white-green with the intensity increasing with increasing mixing length.

The improvement of the boron combustion behavior with primary chamber temperature has been confirmed by particle sampling probes behind the afterburner nozzle. The probes (25 mm × 75 mm) showed both the quenched boron particles and con-

TABLE III

Test Conditions

Test No.	Total propellant flow $\dot{m}$ kg/sec	% Boron in propellant	Oxygen/ hydrogen ratio by wt.	Air flow m_A kg/sec	Air temp T_A K	Air/propellant ratio by wt.	Primary chamber psia	Secondary pressure psia
1	0.087	55.2	0.61	0.59	280	6.7	95	26.5
13	0.135	32.6	2.34	0.48	280	3.6	320	29.4
17	0.105	41.9	1.16	0.548	280	5.5	167	28.1
18	0.109	33	1.82	0.55	280	5.0	263	28.7

densed B_2O_3. The density of the boron particle accumulation changed at different locations on the glass plate depending on their position in the exhaust jet. The density was the greatest in the jet center. Two microscope magnifications of these areas are shown in Figure 5 comparing the results for Test 1 and Test 13. The pictures were made with transmitted light showing the B_2O_3 as white areas and the quenched particles as dark. The picture indicated that the amount of boron burned was significantly higher at higher primary chamber temperature.

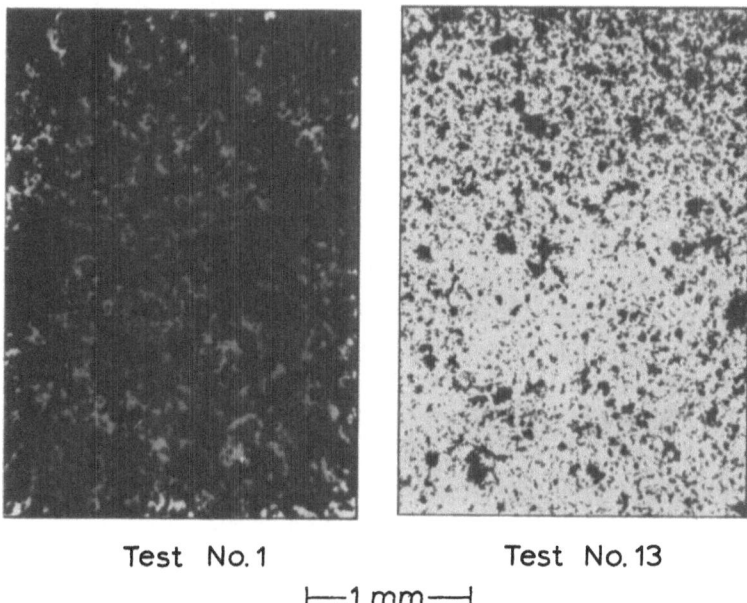

Test No. 1   Test No. 13

⊢—1 mm—⊣

Fig. 5.   Particle sampling probes at different primary chamber combustion temperatures. Test 1: $T^* = 750\,K$, test 13: $T^* = 2000\,K$. (For further data see Table III.)

The qualitative but instructive results of Tests 1 and 13 have been confirmed by c^*-efficiency determinations. Figure 6 shows the influence of the chamber temperature on the experimental c^*-efficiency for four tests, in which the oxygen mass flow was increased from Test 1 to Test 13 keeping the mass flow of boron, hydrogen and air almost constant. The c^*-efficiency, including the hydrogen and boron combustion, as well as the mixing process of the air with reaction products can be improved from about 0.85 to about 0.94 by increasing the primary nonequilibrium chamber temperature (0% boron reacted) from about 750 K to about 2000 K.

4. Thermodynamic Calculations

It has been demonstrated that the temperature of the boron particles in the exhaust from the primary rocket has a significant influence on the combustion of boron in the ingested air. To achieve highly efficient combustion the particle temperature

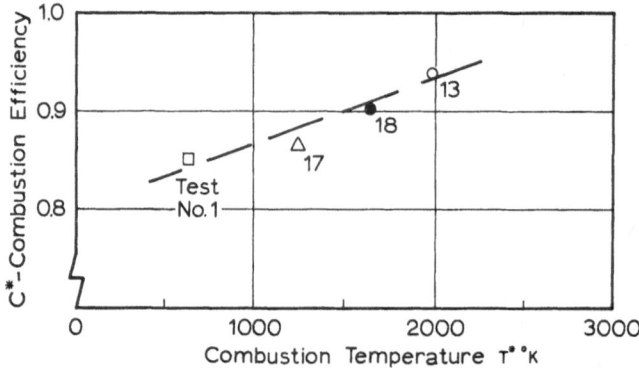

Fig. 6.   Correlation of c^*-efficiency with primary rocket combustion temperature.
(For test data see Table 5.)

should be as high as possible. Therefore, the combustion temperature of the non-metallic propellant should be as high as possible. This requirement, of course, has to be optimized with other propellant and motor conditions in an air-augmented rocket: (1) good production behavior of heavily metallized propellants, (2) efficient, sustained burning, (3) positive expulsion characteristics from the primary rocket, (4) nonagglomeration of condensed-phase particles, and (5) high metal loading for maximum specific density impulse.

There are difficulties in transferring the results of the investigated hydrogen-oxygen-boron-air system to typical propellants for air-breathing application. A comparison should consider both thermodynamics and reaction kinetics. A comparison based on thermodynamic data (for example, the nonequilibrium primary temperature) is possible, but the reaction kinetics aspects, especially the boron oxidation rate as a function of the temperature and the surrounding gas conditions, are not well enough understood to make exact comparisons for typical propellants for air-breathing application using the results of this paper. Therefore the following discussion has to be considered more as a speculation than as a result, and the data have to be viewed with the necessary limitations.

Two typical propellants for air-augmentation applications are solid propellants with 50% boron and liquid propellants (slurries) with 70% boron in combination with an interhalogen oxidizer. The calculated nonequilibrium primary chamber temperatures of these propellants are compared in Figure 7 with those calculated for the hydrogen-oxygen-boron system. The nonequilibrium temperature of this combination has been calculated for an oxygen-hydrogen mixture ratio of 2 (where high efficiency boron combustion has been demonstrated) as a function of the boron concentration in the propellant. For this propellant combination the temperature dropped to about 1250 K as the boron weight was increased to 70%. In comparison to the 50% boron solid propellant and the 70% boron liquid propellant system, the nonequilibrium combustion temperature for the hydrogen-oxygen system was 500° to 700° higher.

On the basis of the calculated thermodynamic data the 70% boron system does

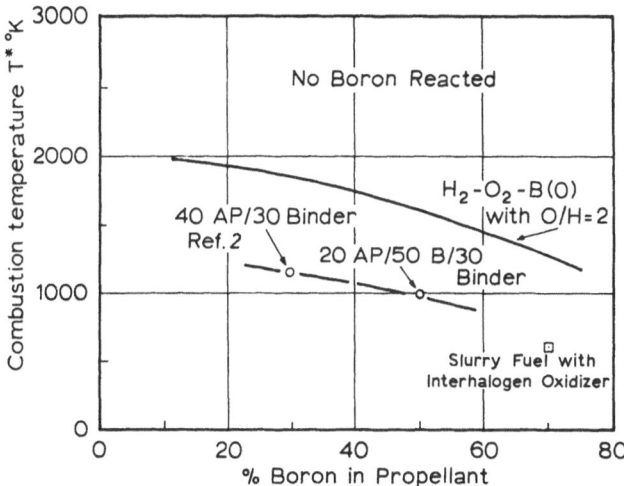

Fig. 7.   Primary chamber T^* calculations for different propellants with boron.

appear to fall well short of achieving T^* levels, in the primary chamber, for which high efficiency combustion has been demonstrated with the hydrogen-oxygen-boron system. The solid propellant system with 50% boron is comparable (on the base of the calculated thermodynamic data for the primary nonequilibrium chamber temperature) with the conditions of Test 1 in this paper ($T^*_{\text{PRIMARY}} \sim 1000\,\text{K}$). It was demonstrated that the combustion efficiency of this test could have been increased by a considerable factor by increasing the primary chamber nonequilibrium temperature.

Looking for possibilities of increasing the primary nonequilibrium chamber temperature it should be realized that this has to be done without decreasing the characteristic velocity of the propellant-air reaction and without loosing the requirements for processing these propellants. In this paper one possibility of increasing the primary chamber temperature will be discussed. Thermodynamic calculations of propellants with hydrogen and oxygen have been made in which aluminum was used additional to the boron. In these calculations it is assumed that the aluminum particles burn completely with the oxygen in the primary rocket while the boron is considered as a heat sink. In the different investigated propellant compositions the hydrogen concentration (simulating a constant binder mass fraction in a solid propellant) is kept constant, while the oxygen concentration is changed with varying aluminum content to fulfill the requirements of stoichiometry for the aluminum-oxygen reaction. The air to propellant ratio is 1200.

The results of these calculations concerning the primary nonequilibrium chamber temperature and the characteristic velocity are given in Figure 8. It can be seen that the primary temperature can be increased from 912 K to 1560 K without decreasing the characteristic velocity (1396 m/sec instead of 1398 m/sec) when a propellant with 65 B/20 H_2/15 O_2 is changed to a composition of 50 B/15 Al/20 H_2/15 O_2. The primary

278   K. SCHADOW

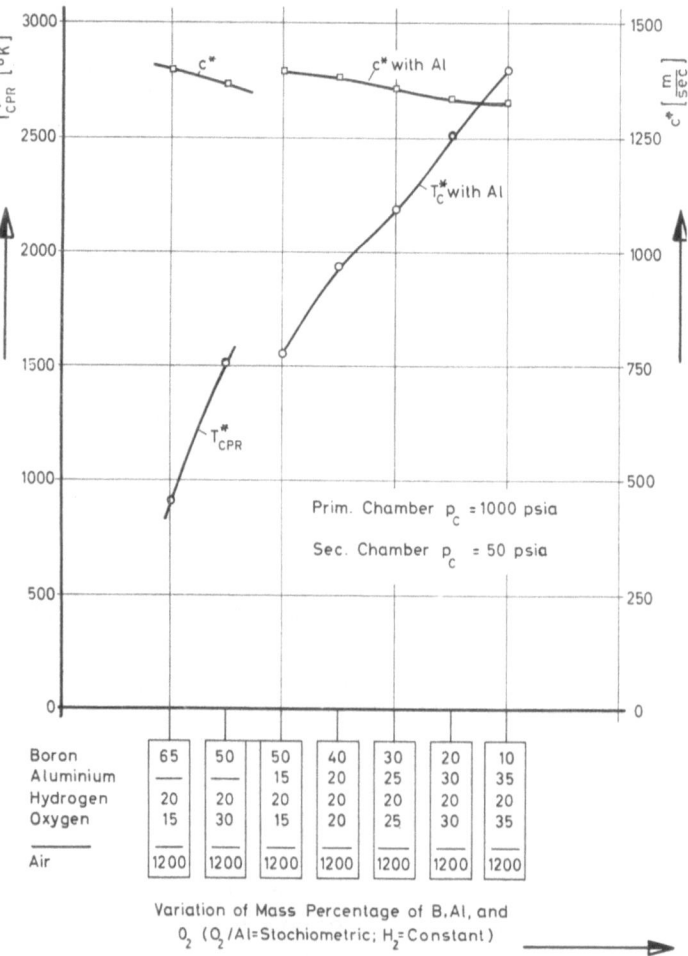

Fig. 8. Primary chamber temperature T^* and characteristic velocity c^* for different propellants with B, Al, H₂, O₂, and air (equilibrium composition).

temperature can be further increased by additional aluminum concentration, but the drop in c^* is significant.

These calculations have shown one way for improving the combustion behavior of boron systems in air-breathing rockets by increasing the primary chamber temperature. An experimental verification of the calculations has to be made.

Acknowledgements

The author wishes to express his appreciation for the help he received during the investigations in the Aerothermochemistry Division of the Naval Weapons Center, and especially for the helpful discussions with J. E. Crump and Mr. Jack Prentice.

The support of Mr. A. Ballew for running the experiments is gratefully acknowledged. The test facility was planned and designed by Mr. R. Slates (deceased).

The author wishes also to acknowledge the help of Mr. A. Langemeyer during the thermodynamic calculations.

References

[1] Grosse, A. V., Conway, J. B., 'Combustion of Metals in Oxygen', *Ind. Eng. Chem.* **50**, No. 4, (April 1958).

[2] Cohen, N. S., 'Combustion Considerations in Fuel-Rich Solid and Hybrid Propellant System in Air-Breathing Propulsion', AIAA Sixth Aerospace Sciences Meeting, New York, January 22–24, 1968.

[3] Brzustowski, T. A. and Glassman, I., 'Spectroscopic Investigation of Metal Combustion', in *ARS Progress in Astronautics and Rocketry: Heterogenous Combustion*, Vol. XV, Academic Press, New York, 1964.

[4] Glassman, I., 'Combustion of Metals-Physical Considerations', in *ARS Progress in Astronautics and Rocketry: Solid Propellant Rocket Research*, Vol. I, Academic Press, New York, 1960.

[5] Prentice, J. L., 'On the Combustion of Single Aluminum Particles', *Combustion and Flame* **9** (1965) 209 and unpublished report on 'Boron Particle Combustion'.

[6] Anderson, R., private communication, CETEC, Mountain View, Calif., June 1968.

[7] Talley, C. P., Woods, H. P. and Popkin, G., 'Combustion of Elemental Boron', Progress Reports on Contract NOnr-1883 (00), 1960–1962 Texaco Experiment Inc., Richmond, Va.

[8] Prentice, J., private communication, Naval Weapons Center, China Lake, Calif., May 1968.

[9] Gordon, D. A., 'Combustion Characteristics of Metal Particles', in *ARS Progress in Astronautics and Rocketry: Solid Propellant Rocket Research*, Vol. I, Academic Press, New York, 1960.

[10] Schadow, K., 'Experimental Investigation of Boron Combustion in Air-Augmented Rockets', AIAA 4th Propulsion Joint Specialist Conference, Cleveland, Ohio, June 1968 – *AIAA Paper* No. 68–634.

[11] Baumgartner, W. E., 'Combustion of Thermogens in Solid Propellants', presented at the First AIAA/ICRPG Solid Propulsion Conference, Washington, D.C., July 1966.

PROPELLANT EXPULSION TECHNIQUES

EDWARD E. STEIN and PETER A. VAN SPLINTER

Air Force Rocket Propulsion Laboratory, Edwards, Calif., U.S.A.

Abstract. A positive gas-free source of propellant must be supplied to the rocket engine of a space vehicle if successful operation is to be achieved during zero or adverse acceleration. Various techniques have been used to accomplish this, and a description of each is presented along with its advantages and limitations. Typical areas of application are also identified.

Items such as elastomeric and plastic bladders, metal bellows and diaphragms, ullage orientation rockets, surface tension devices, and electric fields are included in the discussion.

Tests conducted on some of these devices, including those accomplished on zero-gravity aircraft, are described and the results are presented.

1. Introduction

The successful operation of vehicles in space has been due in part to the efforts of propulsion system designers in developing methods of feeding liquid propellants to the rocket engines under conditions of low or zero gravity. During these conditions, the propellants in the vehicle tankage may migrate to a minimum energy configuration where surface tension forces predominate or they may become oriented in such a way that the tank outlet is completely uncovered. The engines require a gas free supply of propellant at the right pressure to achieve a successful start and various techniques have been explored to assure a positive propellant supply. Several of the more important methods are described in the remainder of the paper, along with their advantages and limitations.

2. Expulsion Devices

A. ULLAGE ROCKETS

Ullage orientation rockets have been used in several applications to settle space vehicle propellants prior to starting the main rocket engine. Solid rocket motors or stored gas attitude control thrusters were provided to produce the required vehicle acceleration. Tests conducted by personnel from the Rocket Propulsion Laboratory in 1962 determined the amount of acceleration and the duration it must be applied to obtain a successful start. These tests were conducted on a Boeing 707-type aircraft flying the zero gravity Keplerian trajectory (Figure 1). Zero gravity test durations of from 5 to 10 sec could be achieved. Figure 2 shows the schematic of the test capsule. Acceleration forces were applied to the capsule after it had reached zero g conditions floating in the aircraft cabin. An expulsion of fluid from one tank to the other was then initiated and pictures were taken of a transparent section in the feed line to determine if any gas was ingested. It was found that an acceleration of approximately 0.1 G for the period of time required for the tank to move a distance equal to its length would insure gas-free propellant flow from one tank to the other. On a space vehicle this acceleration can be provided to start the main engine by either small

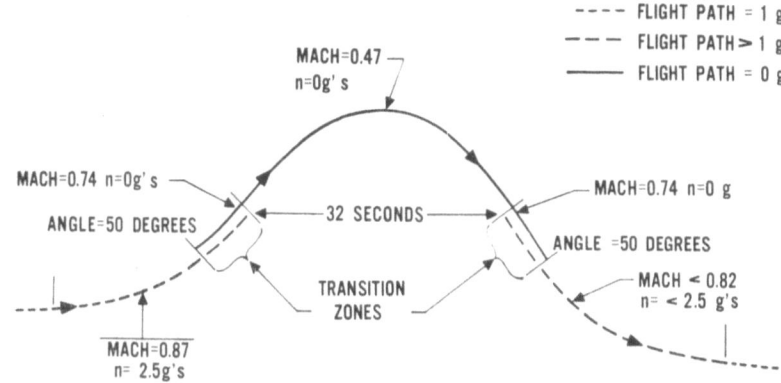

Fig. 1. Keplerian trajectory.

solid rockets using one for each start or by small liquid thrust chambers connected to the attitude control positive expulsion propellant feed system.

The main advantage of ullage orientation rockets is their proven ability to orient either storable or cryogenic propellants in large space vehicles prior to main engine start. Their disadvantages include complexity and cost. Ullage rockets are used on the Saturn S-4 stage to orient the liquid oxygen and liquid hydrogen before main engine start and they were also used on the Agena target vehicle in the Gemini program.

B. PLASTIC AND ELASTOMERIC BLADDERS

Positive expulsion of the vehicle propellants can also be accomplished using plastic or rubber bladders. These have been used extensively in the past on a variety of space

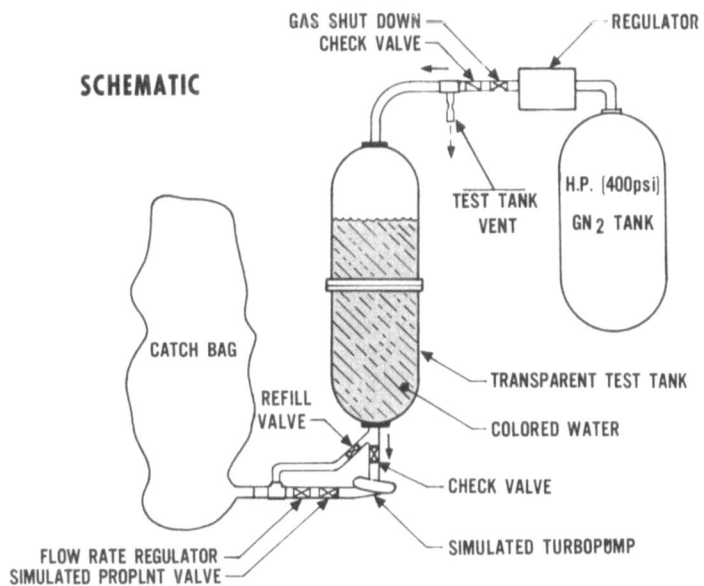

Fig. 2. Zero gravity test capsule.

vehicles in both main and attitude control propulsion systems. They can be fabricated for tank shapes ranging from spheres and cylinders to toroids and sizes from less than a gallon to several hundred gallons. A typical bladder and tank configuration is shown in Figure 3.

Teflon has seen wide use primarily due to its good compatibility with most storable fuels and oxidizers. Difficulty was experienced with this material early in its development due to pinhole leaks caused by three-corner folds during the expulsion process. Laminating the higher molecular weight TFE teflon with the lower molecular weight FEP teflon has resulted in much improved material. The bladder can be cycled for checkout and repeated expulsions over the temperature range associated with currently used storable propellants. Teflon bladders do not provide an impermeable barrier between the propellant and pressurant gas. Propellant vapors and pressurant gas will intermix, especially under conditions of long-term storage. The care and quality control required in manufacturing has made these bladders somewhat expensive. They are being used in the Apollo reaction control system, in the Surveyor Spacecraft and in various other space and launch vehicles.

C. RUBBER BLADDERS

Rubber bladders are less susceptible to the three-corner fold problem than teflon and are, therefore, capable of a greater cycle life and more reliable operation. Rubber materials however, are not compatible with some of the currently used rocket propellants and are therefore incapable of expelling these propellants. Butyl rubber and ethylene propylene rubber have been used successfully with amine fuels and have performed very well. These materials are also somewhat permeable to the pressurant gases and propellant vapors, so impermeable versions have been developed by laminating a thin metal foil into the bladder. A recently developed new material, carboxy nitroso rubber, shows promise of being compatible with N_2O_4. If subsequent development of the material into bladders is successful it may be possible to obtain elastomeric expulsion bladders for the currently used storable propellants, N_2O_4 and amine fuels that have a near unlimited cycle life. In addition to the advantage mentioned above, rubber bladders are rugged, provide good expulsion efficiency and are comparatively easy and inexpensive to fabricate. They have been used in the propellant tanks of space vehicles having hydrazine mono-propellant attitude control systems.

D. METALLIC DIAPHRAGMS AND BLADDERS

In applications where only a single or one shot expulsion is required, metallic expulsion devices have been used successfully. Various sizes and shapes have been explored for their compatibility with nearly all propellants. Being one shot devices they are not capable of an operational checkout. However, their inherent reliability has precluded the need for this. Soft aluminum has been used predominantly as the material of construction providing good formability and propellant compatibility. Its use also results in low actuation force. Some of the designs used for these devices are shown in Figures 4 through 8. In Figure 4 the piston rolls the diaphragm inside of itself to expel the propellant. In Figure 5 the diaphragm reverses through itself. In Figure 6 it expands outward and in Figure 7 it collapses inward. Figure 8 shows a typical metal bellows. All of these devices provide a positive barrier between the pressurant gases and the propellant and with the exception of the metal bellows are

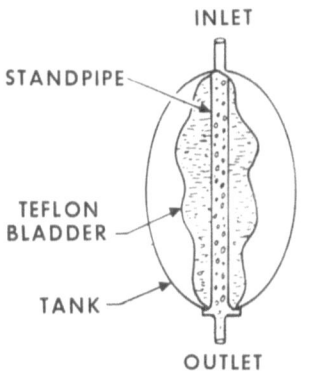

Fig. 3. Plastic expulsion bladder.

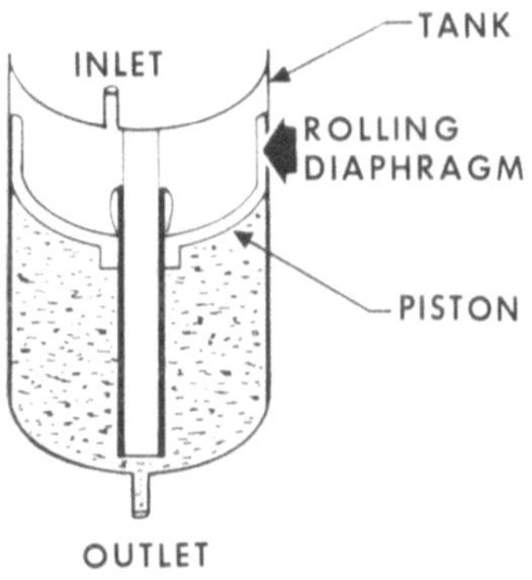

Fig. 4. Rolling diaphragm.

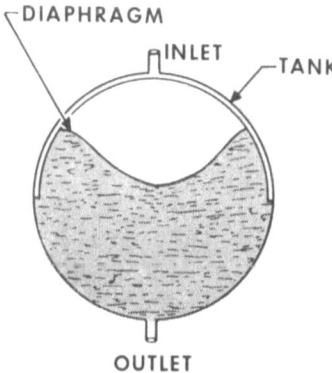

Fig. 5. Reversing diaphragm.

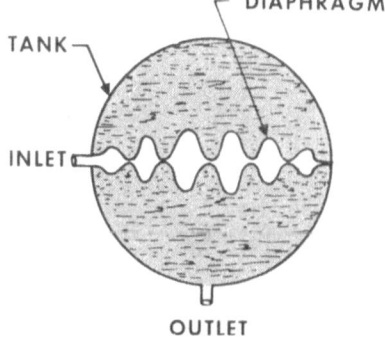

Fig. 6. Convoluted diaphragm.

limited to one expulsion cycle. While these devices have not been used as extensively as non-metallic bladders, their use in the future seems assured due to their ability to separate the propellant from the pressurant and their compatibility with the more reactive propellants. These devices are expensive to develop; however, their production costs are comparable with non-metallic bladders. An additional advantage is their inherent ability to eliminate propellant slosh.

E. SURFACE TENSION DEVICES

The surface tension characteristics of the propellant combined with very fine mesh screens can provide the capability of orienting propellants during space flight. A containment device made of the fine mesh screen, when covered with a film of the propellant, will prevent the passage of gas and permit free flow of the propellant.

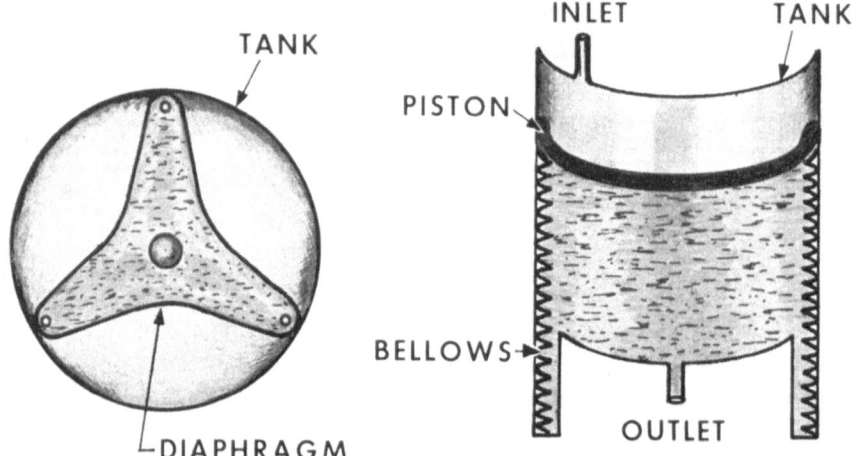

Fig. 7.   Collapsing diaphragm.   Fig. 8.   Metal bellows.

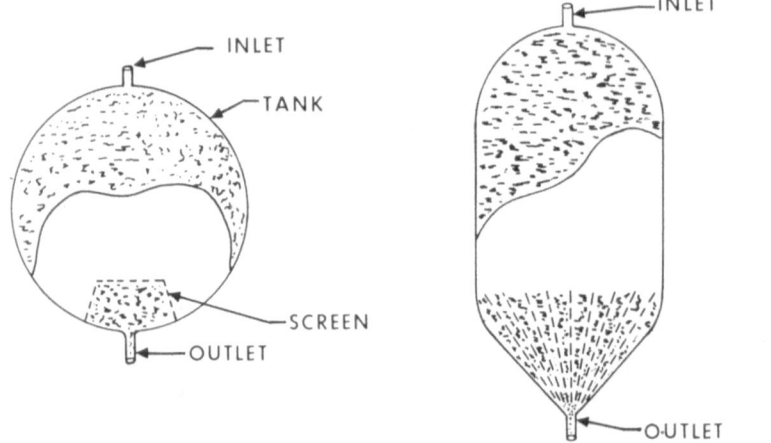

Fig. 9.   Surface tension device.   Fig. 10.   Dielectrophoretic device.

Figure 9 shows a typical screen device in a propellant tank. It is designed to contain enough propellant to start the engine and orient the remaining propellant in the tank. The ability of these devices to function as designed has been demonstrated in both ground and flight tests and their use is being explored for several applications. They have the advantage of being completely passive, are capable of 100% expulsion efficiency, are low in both weight and cost. They do not however, separate the pressurant from the propellant and will not prevent slosh.

F. DIELECTROPHORETIC EXPULSION

Orientation of fluids can be accomplished by using the polarization forces induced in the fluid through the application of an electric field. Proper orientation of the electrodes in the tank, combined with a propellant having a high dielectric strength will result in orientation of the propellant at the tank outlet. Figure 10 shows a typical configuration. The system is attractive for use with cryogenic propellants due to their high dielectric strength properties. However, the use of high voltage is required with the resulting problems of power supplies and tank feed-throughs. The feasibility of the system has been demonstrated on the ground and its eventual use in space will depend upon the development of practical hardware for flight vehicles.

3. Conclusions

Several methods of orienting and expelling propellants from space vehicle tankage have been described, along with their advantages and factors that affect their use. These methods which have been developed since the early days of space flight now provide the designer with several options in selecting an expulsion system for his particular application.

References

[1] Ward Putt, J., 'Experience with Teflon Positive Expulsion Bladders for the Surveyor Vernier Propulsion System', Joint AIAA/Aerospace Corporation Symposium, May 1968.
[2] Petriello, J. V., 'Modifications in Fluorocarbon Bladder Structures in Improving Flexibility and Impermeability', Joint AIAA/Aerospace Corporation Symposium, May 1968.
[3] Wendt, D. A., 'High Expansion Metal Bellows with Non-Welded Convolution Edges for Propellant Expulsion Systems', Joint AIAA/Aerospace Corporation Symposium, May 1968.
[4] DiPeri, L. J., 'A Resume of the Management of Liquid Gas Interface Using Surface Tension Technology', Joint AIAA/Aerospace Corporation Symposium, May 1968.
[5] Clodfelter, R. G. and Lewis, R. C., 'Fluid Studies in a Zero Gravity Environment', ASD Tech. Note 61–84, June 1961.
[6] Melcher, J. R., Hurwitz, M., Fax, R. G., and Blutt, J. R., 'Dielectrophoretic Liquid Expulsion', Joint AIAA/Aerospace Corporation Symposium, May 1968.

INVESTIGATION OF AN INJECTOR CONCEPT FOR THROTTLEABLE HIGH-ENERGY ROCKET ENGINES

W. BUSCHULTE

Deutsche Forschungsanstalt für Luft- und Raumfahrt, Trauen ü. Soltau, Germany

Abstract. A new injection system of simple construction and suitable for developing a concept for a throttleable rocket engine will be described. The system is applicable for propellants with one component in gaseous phase; this would be the case using hydrogen as a fuel. In this system the oxidizer is injected into the combustion chamber in radial direction through a central injection element with numerous holes. The gaseous fuel is injected in axial direction through a porous plate covering almost the entire area of the injection head; by this a good mixing characteristic is achieved. A simple method for throttling this injection system will be discussed, the throttling can be easily achieved by partly covering the injection holes for the liquid oxidizer in the central element.

In order to demonstrate the working conditions of this system in different thrust levels test results of rocket firings with gaseous hydrogen and liquid oxygen in the 100 pound and less thrust region will be presented. In these tests injection elements for LOX with different number of holes (corresponding to different thrust levels) are used. The data of the combustion efficiency and heat transport to the chamber wall will be discussed. The results will show that the injection processes of the oxidizer have a considerable influence on the combustion efficiency and heat transport. Up to this time, there is only little known about physical conditions of the mixing processes occurring in this system with a cryogenic jet in a cross-flowing gas field. This process will be investigated in a two-dimensional, transparent model chamber burning a single liquid oxygen jet in a gaseous hydrogen field. The test apparatus and its working conditions will be discussed.

1. Introduction

In connection with spacecraft and spaceprobe rocket propulsion the demand for rocket engines with thrust-throttling ranges greater than 10:1 is more and more raised. Besides this it can be foreseen that such types of engines will also be needed for propulsion systems combined from rocket and air breathing engines, such as ram-rockets.

Assuming that in future cryogenic high energy propellants will be increasingly used, such a throttleable engine should fulfil the following four conditions:

(a) an acceptable energy-conversion efficiency over the whole throttling range

(b) show stable combustion characteristics and no excessive heating of the engine walls

(c) give a concentrated flow of the cryogenic liquid to avoid long cool-down phases for the engine start period

(d) have mechanically simple and inexpensive throttling devices.

This being the task, the resulting design is described in the following part.

2. Principle of an Injector Head for a Throttleable High Energy Rocket

Thinking of high energy propellants means in most cases that hydrogen is the fuel. Due to its excellent cooling properties it will also be used as the coolant in a regenerative cooling system of the engine, so that the fuel leaves the cooling passages in the

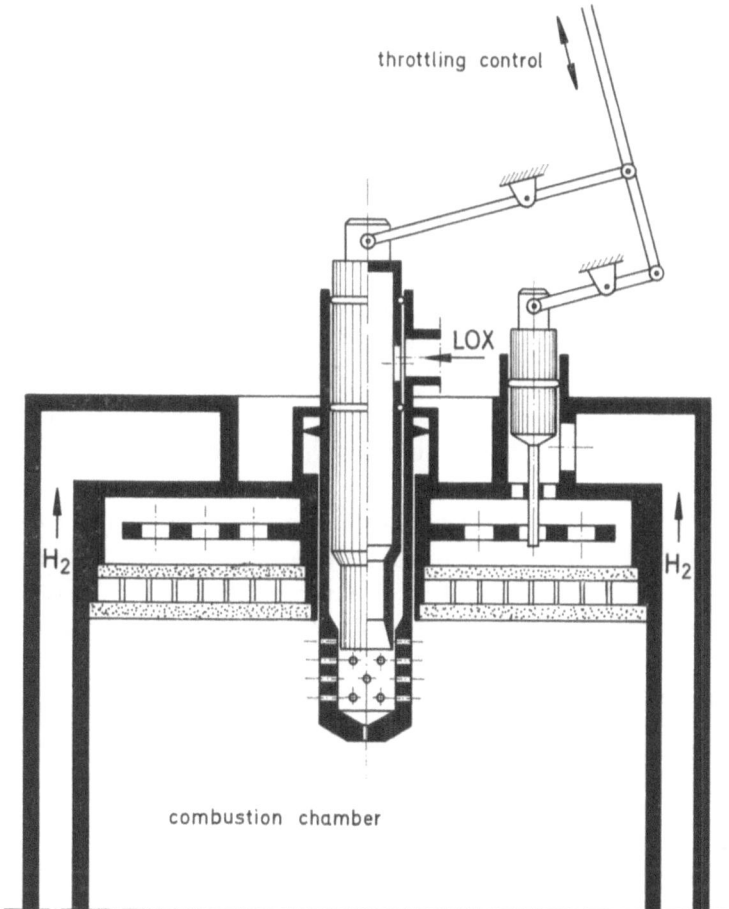

Fig. 1. Schematic drawing of the injector head of a throttleable H_2/O_2 rocket engine.

gaseous state. In this case therefore it turns out to be sensitive to leave the gaseous fuel enter the combustion chamber equally through a porous plate of sintered metal or mesh wire over almost the whole cross-section of the chamber head end (Figure 1). Thus we get a full fuel stream parallel and concentric to the chamber axis.

The throttling and metering of a gaseous propellant component is common technique and shows no problems.

As far as the impartment of the oxidizer into the combustion chamber is concerned the choice fell on a single injector element, which is located in the center of the injector head, emerges into the combustion chamber and shows several rows of injector bores or slots, leaving a central passage and pointing rectangular to the axis towards the chamber wall.

As is shown in the schematic drawing of Figure 1 one possible design of the throttling device consists of a moveable pintle in the control passage which covers and closes a greater or less number of the injection bores and has a pressure throttling device at the inlet of the oxidizer element to maintain the injection pressure constant at all flow rates of the throttling range.

The pressure throttling devices for the oxidizer and the fuel must be coordinated in the thrust control device such that the designed mixture ratio of the propellant components is achieved at all throttling stages.

The variation of the total injector cross-section area for the liquid oxidizer component is a fundamental need for all throttleable rocket engines of large throttling range, because the change of injection pressure can only be applied for narrow throttling. If the injection pressure deviates too strong from the design figure to lower values the propellant jets or sheets are not enough disintegrated so that the engine sooner or later falls into very rough combustion or even low frequency instability. If on the other hand the injection pressure is raised much above the appropriate design value, droplet velocity becomes so high, that an increasing part of the propellant is likely to leave the combustion chamber unburnt. The engine then delivers a bad combustion efficiency.

The radial pointing of the oxidizer injector streams here is applied to achieve a mixing of the propellant components in a crossflow-field. As the fuel mass flow is constant across the cross-section of the chamber, the main task for the research program is to learn about the necessary conditions and design means that are to be applied for achieving an oxidizer mass distribution by droplet formation and vaporisation which forms an optimal mixture ratio distribution across the reaction chamber.

3. Experimental Investigations with a Small Rocket Engine

After having developed the working principle, the following program is divided into several steps in order to avoid being faced with all problems at one time. The first step then was to study the ignition, combustion efficiency, heat transfer and general behaviour of a rocket engine, when burnt at different thrust levels of the throttling range.

This part of the program has been performed with a small water-cooled rocket engine with a nominal thrust of about 60 kp at 10 kp/cm² chamber pressure. The engine burns liquid oxygen and gaseous hydrogen.

Figure 2 shows the test installation shortly before an engine firing.

For simplicity reasons the engine ignition was here performed by an electrically initiated commercial primary squib. The electrical ignition signal is coordinated by a sequencer with the main propellant valves.

During the test program it turned out that due to the special injector head design the time between oxidizer valve opening and fully liquid injection into the combustion chamber was between 100 and 200 msec.

For firing at the different thrust levels five different oxidizer injector elements are used having 40, 30, 20, 10 and 6 injection bores of 0.6 mm diameter.

Besides this three types of fuel injector plates are used (Figure 3). One is a plane plate and two have copper splash rings of different depth.

Figure 4 shows a photograph of the firing rocket engine.

Fig. 2.   Test stand for LOX-hydrogen throttleable rocket engine.

In Figure 5 the measured experimental characteristic velocities are shown for the investigated throttling range, characterised by the chamber pressure variation. A general slow decrease in efficiency can be stated, when the chamber pressure is lowered. At design pressure the combustion efficiency reaches normal values. But the most remarkable result here is that especially in the lower pressure region improvements can be achieved by applying devices which alter the oxidizer distribution in the chamber. As the mechanism of oxidizer distribution for this design situation is not enough understood, future efforts are to be concentrated on this point.

Certain explanations of the development of the results could already be tried, but should be postponed until a more detailed view has been taken into the mechanism.

In Figure 5 it is indicated that the test results at the different steps are obtained over a certain range of mixture ratios.

In that test program the cooling heat flux has been measured too. As mentioned before the rocket engine was intensively watercooled. As can be expected the heat flux increased, when the mixture ratio was raised. But for this case it is of more interest to learn about the development of the cooling behaviour over the throttling range. The experimental results are shown in Figure 6.

The experimental and theoretical curves correspond to a mixture ratio $r = 4$.

The heat capacity of the coolant has been calculated assuming heating of the coolant of 250 K and using the theoretical characteristic velocity and a mixture ratio of 4. This line is looked upon as the upper limit of heat flux into the coolant.

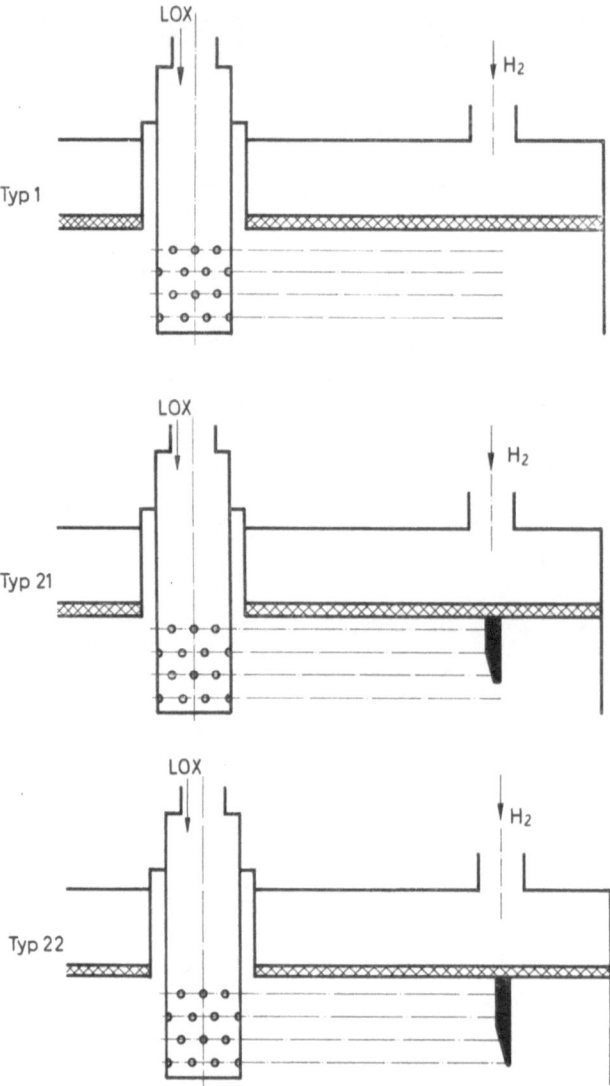

Fig. 3.   Types of fuel injector plate used in test program.

The first overall view on the diagram shows that over the whole range of thrust variation the cooling capacity of the hydrogen flow is sufficiently high to absorb the heat flux coming in through the wall.

In detail it can be seen that the injector plates with splash rings on their frontside show markedly less heat transfer in the upper thrust region compared to the plain plate. But towards lower thrust levels the heating of the walls decreases more slowly with these injector plates. The type 22 where all oxidizer jets impinge on the splash ring shows the highest values of heat transfer – but also the best efficiency – at low thrust which means a good but fairly hot mixture ratio at the wall also.

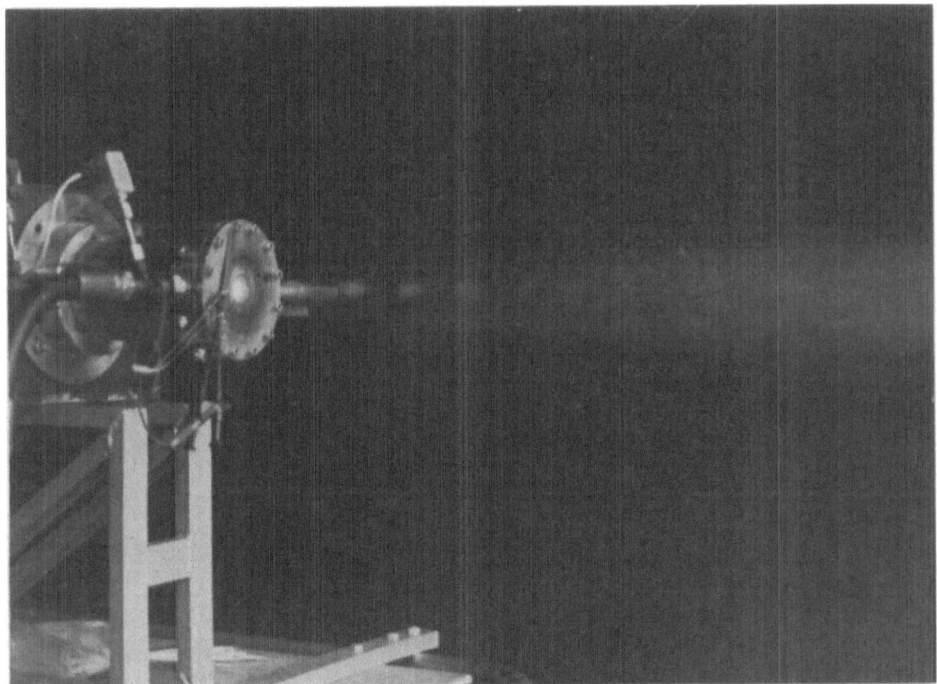

Fig. 4.   Firing of the rocket motor.

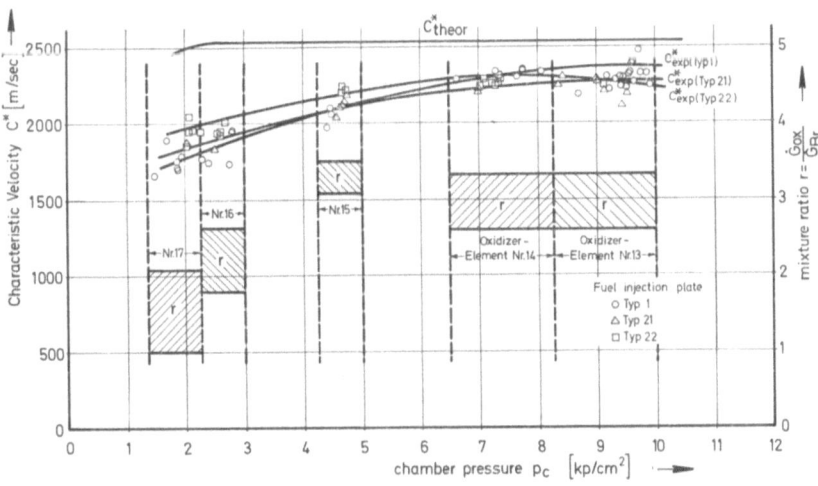

Fig. 5.   Characteristic velocity C^* versus chamber pressure p_c for three different injector plates (combustion efficiency in thrust throttling).

As to the general behaviour of the test engine the following experiences have been observed:

The ignition of the engine in about 250 test firings was successful and performed well at all thrust levels without changing the timing in the ignition sequence.

The combustion was very even and smooth at all thrust levels. Only at very fuel-rich mixture ratios where the oxidizer injection pressure drop became very small it

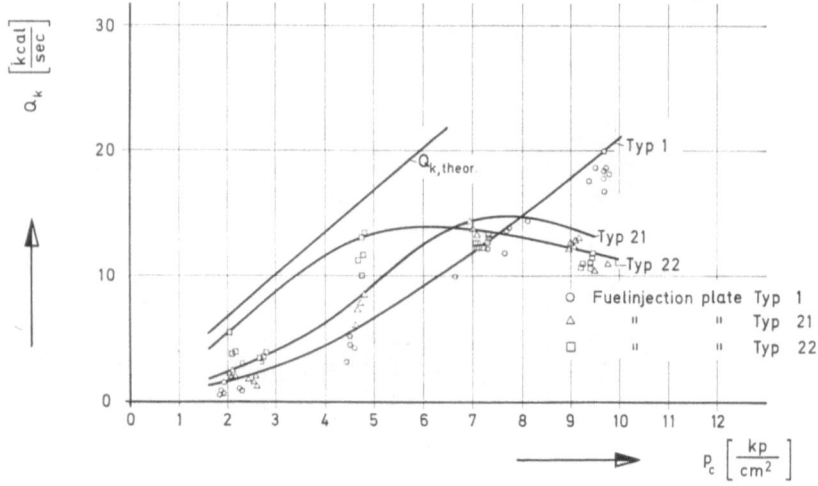

Fig. 6.   Heat flux into the coolant against throttling range.

tended to fall into low frequency instability. But for normal application this is not of interest.

Various materials were used for the oxidizer injector elements. Copper turned out to be the best as it showed no signs of overheating or oxidation at the outer surface.

4. Throttleable Oxidizer Element for Functional Tests

When the test firings with the experimental motor gave these encouraging results, the design of a first functional apparatus for the throttling of the oxidizer flow in the injector element was started.

First it must be achieved that the sealing between the bore wall of the element and the moveable pintle is so tight that no significant leakage flow may pass through the covered injection holes. Because if the oxidizer leakage flow is too high, it leaves the injector holes at very low pressure drop – and therefore with low velocity – so that it is to be feared that combustion takes place at the element surface and the element would be burned locally.

The main difficulties to meet the aim arise from the high difference between normal ambient temperature and the temperature of liquid oxygen. Therefore material shrinking was looked for to support the sealing effects at low temperatures.

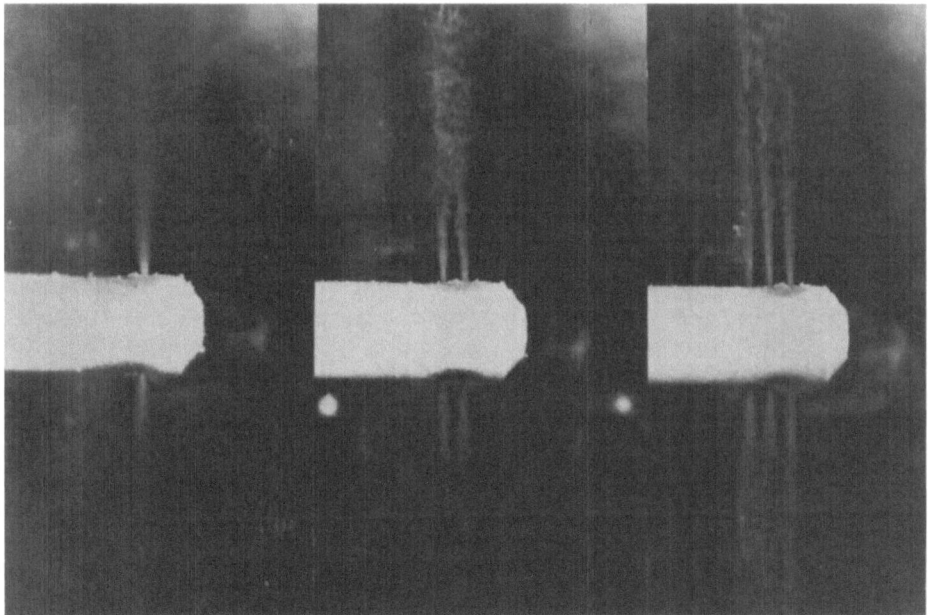

Fig. 7. Photograph sequence of throttling of an oxidizer element.

In Figure 7, where for various throttling steps liquid air is fed through the functional test element into the open air, it can be seen that this part of the concept can technically be realized by meeting the basic demand.

For the case that a higher leaking rate should arise the possibility still exists to lead a channel from the rear side of the seal to a central bore in the front end of the oxidizer element where a liquid oxidizer jet sucks off the evaporated leakage mass.

The second test object was to see which kind of jet is formed when the pintle closes only a part of the injection hole inlet area.

This is also important with respect to the danger of burning the element.

As Figure 7 shows such a jet stemming of a partially covered injector hole has a normal injection velocity. It does not attach again to the wall during its passages through the injector bore. Also it shows a downward turn in its mass flow direction. This fact is important when splash rings are applied. Whether an engine operation is possible with a part of the injector holes are partially closed cannot be said at this time. It has to be tested later in the rocket motor.

5. Research Program on Combustion in the Cross Flow-Field

Following the experiences of the motor test program it is felt that more research effort must be invested with the disintegration of a jet into droplets, the droplet evaporation and burning, when one or more jets are injected into a cross flowing stream of gaseous fuel.

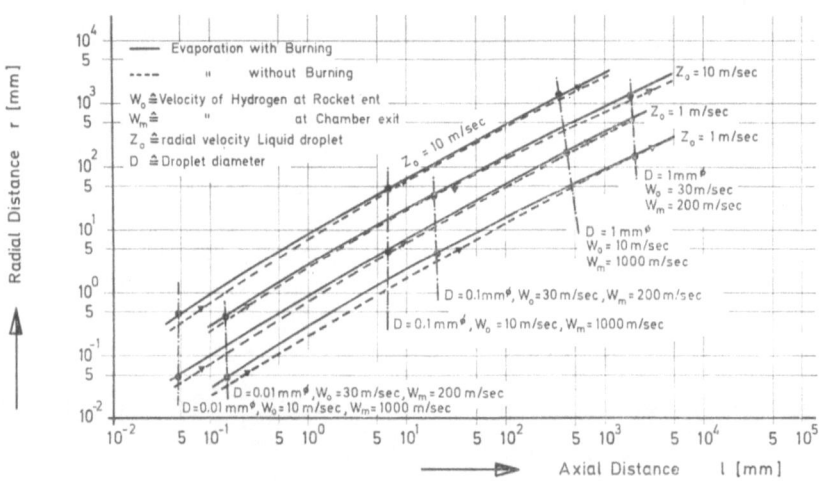

Fig. 8.   Evaporation and burning of droplets of liquid oxygen in a hydrogen gas flow.

A theoretical study has been made about the evaporation and the path of droplets
of different size and under various ambient conditions.

Figure 8 shows all those geometrical points of droplets where they have reached
99% evaporation. The graph demonstrates clearly the effects of the variation of some

Fig. 9.   Test arrangement with window chamber for crossflow combustion studies.

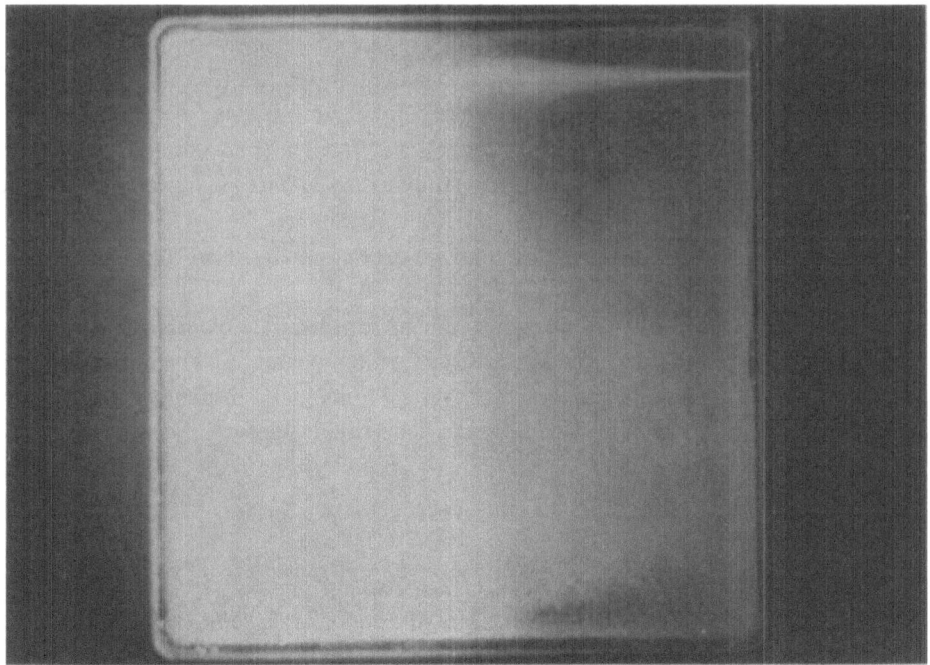

Fig. 10.   Burning of single LOX-jet in the window-chamber.

parameters on increasing the axial and radical distance of the droplets end point. The most significant effect comes from the droplet diameter.

To perform this program a twodimensional combustion chamber with transparent windows on each side and the necessary feed system has been fabricated and is being tested now. Figure 9 shows the arrangement of the testing apparatus.

First it is intended to observe and investigate the disintegration and oxidizer mass distribution of a single jet along the jet axis and its rear side flowfield. The same program will follow for the splash plate arrangement. Then the situation has to be studied, when in the direction of the gas flow several jets follow one after the other so that the combustion of the streamup jet influences the evaporation and combustion process of the following jets.

While in the first two cases gas velocity, jet velocity, bore diameter and length and for the splash plate impingement distances and angle are the main variables, the third case then brings additional geometric parameters into the picture concerning the location of the various injection bores to each other.

Additionally it might also be necessary to study the situation with a partially closed injector bore.

Overall it will be a lengthy program but of fundamental importance. To give an impression of the combustion of such a jet, Figure 10 shows the model chamber in operation. The liquid jet and its combustion zone can clearly be seen.

6. Summary

An injector head concept, which is especially suitable for throttleable rocket engines for cryogenic high-energy propellants, has been tested in a small rocket motor. The achieved combustion efficiencies and heat transfer rate show encouraging results and possibilities for improvements by further study of the system. The general behaviour of the test motor is good.

Preliminary tests with throttling mechanism for the oxidizer element at low temperatures have so far been positive also.

The disintegration and distribution mechanism for the liquid oxidizer in a crossflow-field is specially be studied in a twodimensional window-chamber. Various parameters, geometric jet arrangements and design features are included in this program. This is for improving the knowledge for designing such rocket motors.

References

[1] Lysdale, C. A., 'Investigation of Space Rendezvous Propulsion System Requirements', Vol. I, Aerojet General Corp., Azusa, California, U.S.A., 1963.

[2] Johnson, R. J., Boyd, B. R. and Smith, T. H., 'Application of the Mira 150 A Variable Thrust Rocket Engine to Manned Lunar Exploration Flying System', 3rd AIAA Prop. Joint Spec. Conferenec, Washington, U.S.A., 1967.

[3] Carey, L., 'Dual Mode 100:1 Thrust Modulation Rocket Engine', 3rd AIAA Prop. Joint Spec. Conference, Washington, U.S.A., 1967.

[4] Pinker, R. A. and Herbert, M. V., 'The Pressure Loss Associated with Compressible Flow Through Square-Mesh Wire Gauges', NGTE-R 281, National Gas Turbine Establishment, Great Britain, 1966.

[5] Hersch, M., 'Combined Effect of Construction Ratio and Chamber Pressure on the Performance of a Gaseous Hydrogen-Liquid Oxygen Combuster for a Given Propellant Weight Flow and Oxidant-Fuel Ratio', NASA TN D-129, 1961.

[6] 'ELDO Future Program Preliminary Study of Liquid Hydrogen-Liquid Oxygen Stages', SEREB, Courbevoie, France, 1964.

[7] Dardare, I., 'Preliminary Study of a Liquid Oxygen–Hydrogen Rocket Motor of 6 Tons Thrust' SEPR, Villejuif, France, 1965.

[8] Dardare, I., 'Rocket Motors Using Liquid Oxygen and Hydrogen', *Air and Cosmos* **96** (1965) 26–30.

[9] Combs, L. P. and Schuman, R., 'Steady State Rocket Combustion of Gaseous Hydrogen and Liquid Oxygen', Part II, Rocketdyne, Canoga Park, U.S.A., 1965.

[10] Bailey, C. R., 'A Preliminary Investigation of Oxidizer Rich Oxygen-Hydrogen Combustion Characteristics', NASA MSFC, Huntsville, U.S.A., 1966.

[11] Hannum, Scott, 'The Effect of Several Injector Face Baffle Configurations on Screech in a 20.000-Pound Thrust Hydrogen-Oxygen Rocket', NASA Lewis Research Center, Cleveland, U.S.A., 1966.

[12] Wanhainen, J. P., Parish, H. C. and Conrad, E. W., 'Effect of Propellant Injection Velocity on Screech in a 20.000-Pound Hydrogen-Oxygen Rocket Engine', NASA Lewis Research Center, Cleveland, U.S.A., 1966.

[13] Tomazic, W. A., Conrad, E. W. and Godwin, T. W., 'M-1 Injector Development-Philosophy and Implementation', 3rd AIAA Prop. Joint Spec. Conference, Washington, U.S.A., 1967.

[14] Hoare, F. E., Jackson, L. C. and Kurti, N., *Experimental Cryophysics*, Butterworths, London, 1961.

[15] Timmerhaus, K. D., *Advances in Cryogenic Engineering*, Vol. 7, Plenum Press, New York, 1962.

[16] Scott, R. B., *Cryogenic Engineering*, D. van Nostrand, Princeton, 1962.

[17] Scott, R. B., *Technology and Uses of Liquid Hydrogen*, Pergamon Press, Oxford, 1964.

[18] Weil, L. and Perroud, P., *Liquid Hydrogen*, International Institute of Refrigeration, Pergamon Press, Oxford, 1965.

[19] Zeleznik, F. I. and Gordon, S., 'General IBM 704 or 7090 Computer Program for Computation of Chemical Equilibrium Compositions, Rocket Performance, Chapman-Jouguet Detonations', NASA TN D-1454, 1962.

[20] Barrère, M., Jaumotte, A., Fraeijs de Veubeke, B., and Vandenkerckhove, J., *Raketenantriebe*, Elsevier Publishing Company, Amsterdam, 1961.

[21] Buschulte, W., 'Untersuchungen zu einem System zur Schubregelung von hochenergetischen Raketentriebwerken', DLR-FB 68-57, 1968.

[22] Buschulte, W., 'Schubregelung bei Flüssigkeitsraketen', *DFL-Mitteilungen*, Nr. 8 (1968) 372–375.

[23] Hartung, W., 'Der Einfluß des Einspritzvorganges auf das Betriebsverhalten kleiner Flüssigkeitsraketen, Teil 1: Hydrodynamische Gesetzmäßigkeiten bei der Treibstoffeinspritzung durch Prallstrahlelemente'.

[24] Twardy, H. 'Berechnungsverfahren zur Ermittlung der Mischungsverhältnisverteilung in einer Raketenbrennkammer unter Berücksichtigung von Verbrennung und Verdampfung'.

[25] Buschulte, W. and Twardy, H., 'Theoretische Untersuchungen zur Treibstoffverdampfung und Mischungsbildung in einer Flüssigkeitsraketenbrennkammer', DLR-FB 66-33, 1966.

PART V

PROPULSION, I

Chairman: A. Dadieu, West Germany

DETERMINATION OF THE POSITION
OF THE SUPERSONIC FLOW CLOSING SHOCK WAVE IN THE
CHANNELS OF AN AIR-BREATHING JET ENGINE

G. I. PETROV

Academy of Sciences, U.S.S.R.

Abstract. The following problems are planned to be discussed in the paper:
(1) Static stability of the closing shock in the diffuser and the nozzle.
(2) Interaction of the shock wave with the turbulent and boundary layers.
(3) Limiting pressure ratios for a bridgelike shock.

The main problem in designing and studying of the air-breathing jet engine channels, where the transition from the supersonic to the subsonic flow takes place, is the determination of the boundary which separates the supersonic flow from the flow with subsonic velocity. This boundary is complex combination of shock waves and regions with large velocity gradients at the boundaries of the streams (Figure 1).

We shall name as the trailing shock wave the shock wave of a complex form, the position and shape of which depend upon the pressure at the end of the channel where the flow is completely subsonic. Statically stable position of the trailing shock wave determines: the intake stagnation recovery pressure in the channel of the supersonic inlet, nozzle thrust which operates in the counterpressure mode, etc.

In experiments, carried out in 1948–1950 by me and my colleagues, it was determined that for all cases of shock wave interaction with the turbulent boundary layer, for the development of the shock in front of an obstacle, for the reflection of the shock from the wall or for the flow at the nozzle (Figure 2), there exists a limiting

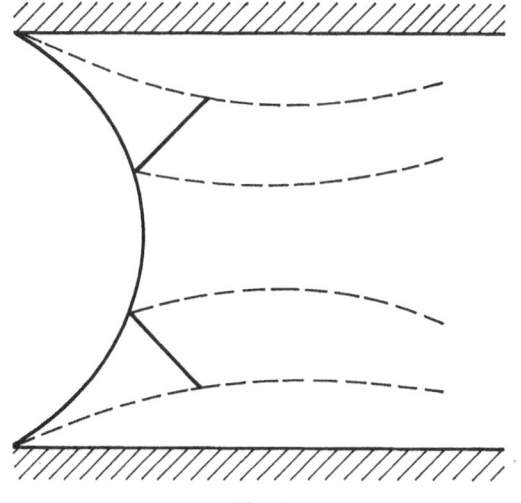

Fig. 1

ratio $K=P_2/P_1$ of the pressure behind the shock to the pressure in front of it, which determines the intensity of the shock, adjacent to the wall with a boundary layer. This ratio for the turbulent layer (Figure 3) is a function only of the mach number.

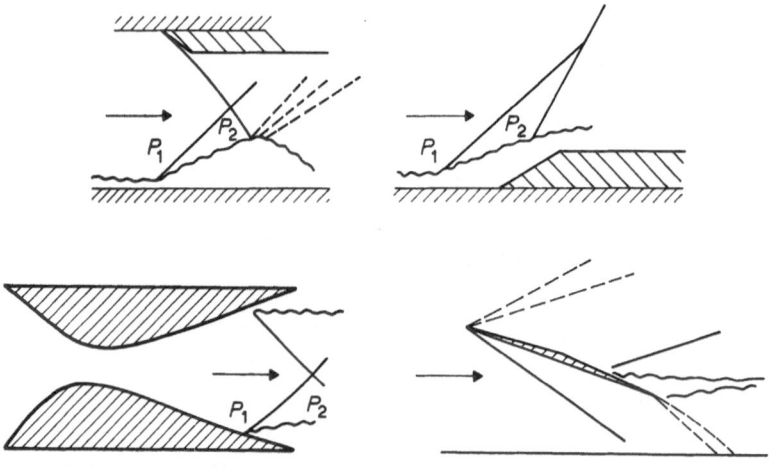

Fig. 2

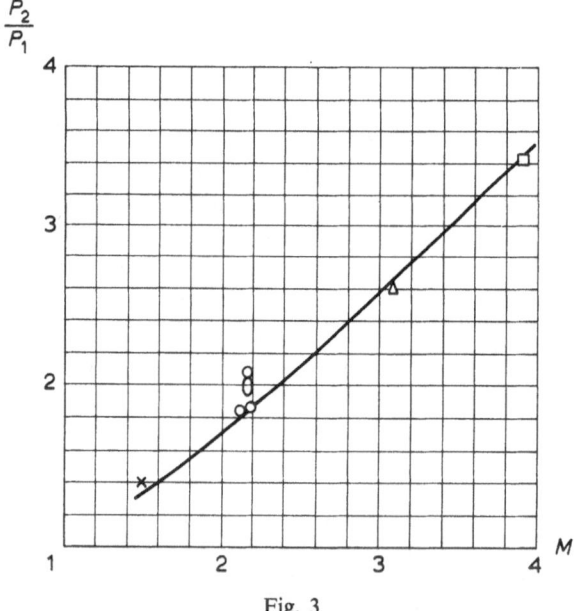

Fig. 3

Let P_0 denote the stagnation pressure of a given flow

P_k – static pressure in the chamber or the end of the channel;

P_1 – pressure at the wall in front of the trailing shock;

P_2 – pressure at the wall behind the trailing shock.

Then

$$\frac{P_0}{P_k} = \frac{P_0}{P_1}\frac{P_1}{P_2}\frac{P_2}{P_3} = K_0 K K_k$$

K_0 – is easily calculated over the supersonic part of the channel in any point of it.

K – is determined from Figure 3 for M number at the wall of the channel.

K_k – in some cases may be assumed to equal 1 or not dependent on the position of the trailing shock.

In other cases additional pressure recovery for particular channel shapes should be determined experimentally.

For a given geometric channel shape the P_0/P_k curve can be plotted for each channel wall.

Each position of the trailing shock has a definite corresponding value of P_0/P_k.

It is obvious that the increase in pressure, P_k by throttling the exit, or by increasing the heat flow to the chamber of a ram-jet, shall result in the displacement of the trailing shock towards the flow. If the calculated value of P_0/P_k for this displacement can increase up to a required level, a further increase of P_0/P_k for random pressure changes is possible. The position of the trailing shock will be statically stable and the channel can operate at a given regime. For the opposite case a break-off of the flow will take place and there will be a transfer to a pumping regime in the ram-jet inlet or the locking of the wind-tunnel.

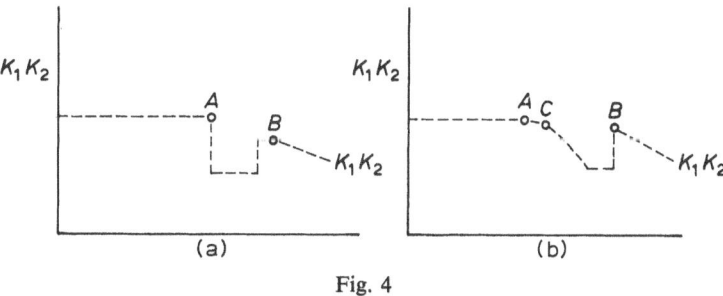

Fig. 4

An example of such a calculation for an inlet of an air-breathing jet is presented in Figure 4. Here point B represents the stable position of the trailing shock for the smooth contour of the inlet channel, what ensures a large pressure recovery coefficient in comparison with the a contour, for which the stable position can be only in point C. The position of the operating point is determined by the necessary reserve $\Delta P < a$ (I – the pressure oscillation amplitude, which are always present in the region of the trailing shock), and also by the necessary reserve for the control system. This simple method is also applicable for calculating the nozzle thrust of a jet engine, operating in a mode with counterpressure. The ratio K_2 may be considered also as a pressure limit ratio in the shock wave, the exceeding of which shall cause development of a sufficiently extended break-off region of the turbulent boundary layer and, consequently, shall lead to a substantial change of flow form in the supersonic part of the channel.

The dependence of P_2/P_1 on the Mach number, displayed in Figure 3, is determined experimentally and approximately from the bias of the front shock, which, of course, is not strictly linear and for this approximation the noted dependence for the turbulent boundary layer does not depend on Re number and the thermic capacity ratio.

The trailing shock wave in the majority of cases has a shape, which is close to a shape called in English language publications as the 'bridge wave' (Figure 5). A shock of this form develops in a jet, flowing from a nozzle when the pressure in the surrounding medium is higher than at the end of the nozzle.

For a decreasing pressure ratio P_a/P_H the central shock increases. This shape can be calculated approximately for a plane as well as an axis-symmetrical case, but the essential point is that for both cases the limit ratio P_a/P_H can be determined, which cannot be exceeded when this form of shock exists.

This ratio is determined through a simple consideration, that the stream which is developed behind the central shock cannot be completely subsonic, because the

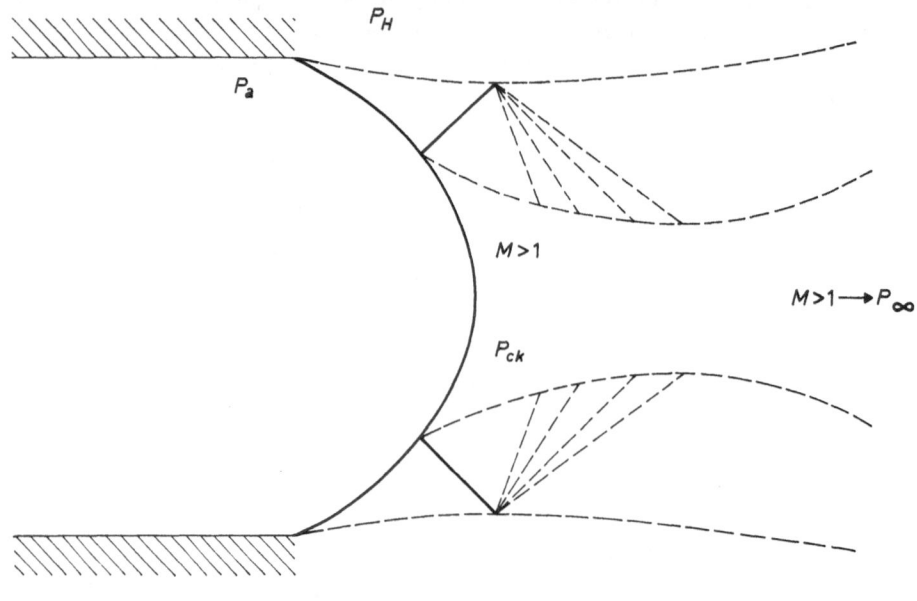

Fig. 5

pressure P cannot be smaller than P_H. Therefore, pressure ratio behind the shock at the axis of the flow (Figure 5)

$$\frac{P_{ck}}{P_H} \geqslant \frac{P_{ck}}{P_{crit}}$$

P_{crit} – pressure in point $V = a_{kp}$.

P_{ck}/P_H – is easily calculated through Mach number and in front of the central shock.

Thus the limit ratio as well as coefficient K_2 determines the position of the channels' trailing shock, if the boundary layer is drawn-off in that part of the channel where the trailing shock is possible.

For a stream flowing from a nozzle into a medium with pressure lower than the pressure on the edge of the nozzle $P_a > P_H$ the stream obtains a shape as shown in Figure 6, in other words, the flow overexpands and the static pressure on the axis of the stream $P_c < P_H$.

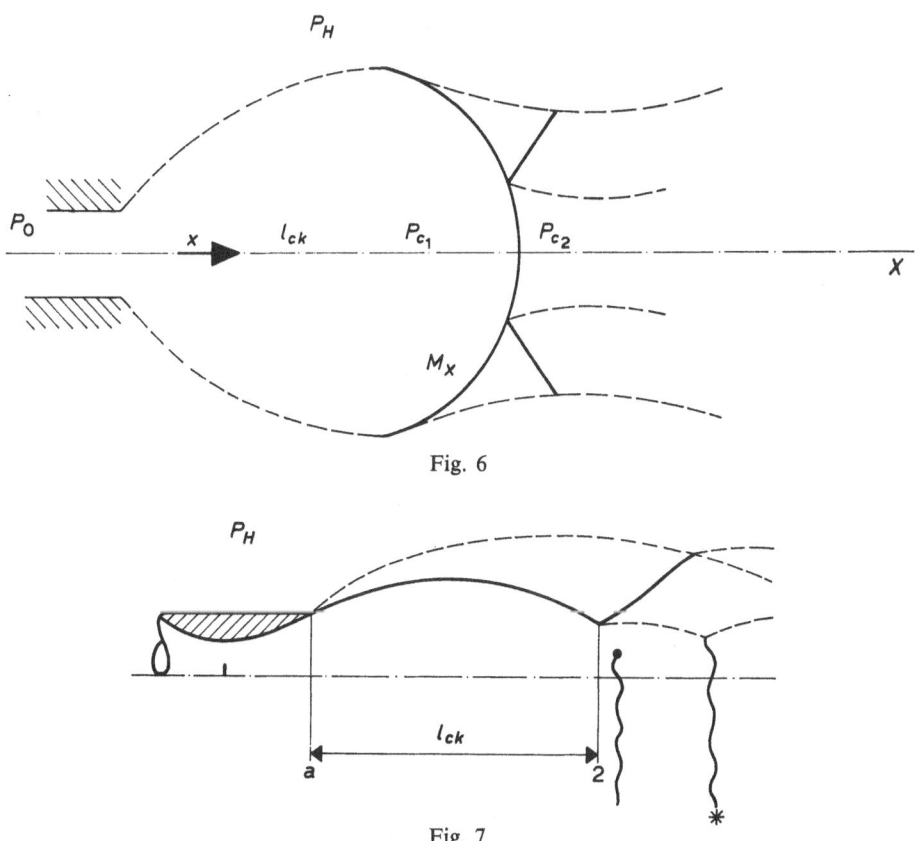

Fig. 6

Fig. 7

The position of the central shock can be determined by calculating the Mach number on the axis of the stream $M_x = f(x)$ through the use of the characteristics method, and assuming

$$\frac{P_0}{P_H} = \frac{P_0}{P_c}\frac{P_0}{P_H}$$

where:

P_0 – the pressure of adiabatic braking;

P_1 – pressure in front of the shock;

P_2 – pressure behind the shock;

P_H – pressure in the medium.

If P_0/P_H is known in accordance with the abovementioned, then (Figure 7)

$$\frac{P_c}{P_0} = \left(\frac{k-1}{2}\right) k/(k-1),$$

in other words, the limit pressure ratio, P_0/P_k is easily determined and, consequently, M_c and l_c.

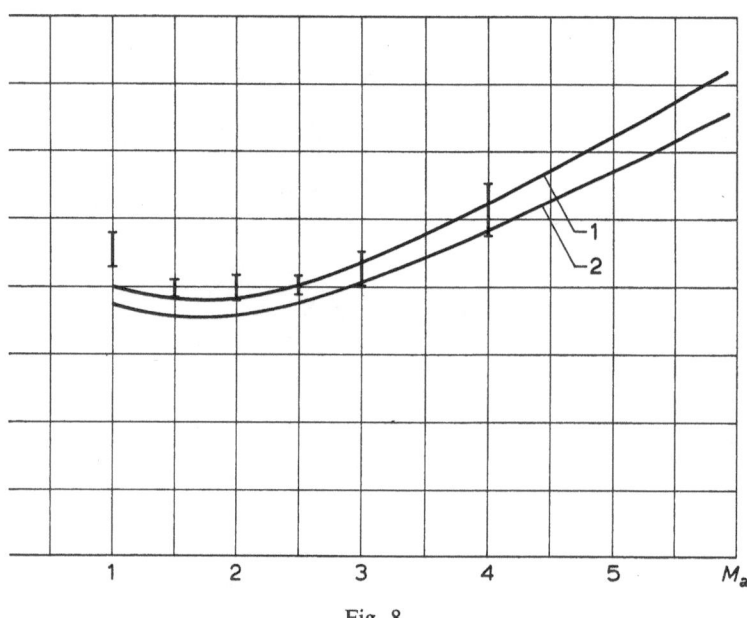

Fig. 8

An example of such a calculation and a comparison with an experiment, conducted by M. Ydilovich, are presented on Figure 8. Here $n = P_a/P_h$.

Thus, quite simple considerations, based on experimentally established facts, permit solution of many problems of complex flow forms related to break-off of the boundary layer, which are impossible to solve only on the basis of the ideal liquid or the boundary layer theories.

REDUCTION IN COST OF THE FIRST STAGE, 'RECOVERABLE' FIRST STAGES, AND FIRST STAGES 'TO BE THROWN AWAY'

A. ANGELONI

Snia Viscosa, Division BPD, Italy

Abstract. Assuming that, by given increases in weight and cost, it is possible to recover the first stage of space-launch vehicles, on firings successive to the first one only the costs of propellant supply and of overhauling will bear.

The cost of a single firing decreases therefore proportionally to the number of scheduled firings.

Even in the case of a first stage to be 'thrown away', the cost per firing decreases according to the number of scheduled firings, provided it is possible to arrange for a mass production.

The problem in question is taken into consideration for different values of the main parameters, and mention is given to solutions to be proposed for both cases.

One can foresee that space activity, after the first landing on the Moon, will enter into a new expansion phase.

In order to analyse the means that will be suitable in the near future for this expansion phase, it is necessary to make some forecasts about the general features this activity will assume.

The future demands that are to be expected concern:

(1) orbital laboratories;

(2) traffic with moon;

(3) exploration of planets.

The orbital laboratories might be bases orbiting near Earth and capable of giving hospitality to the scientists who will attend to observations and researches about Earth and universe and to the technicians who will attend to:

(1) maintenance of equipment for terrestrial services (telecommunications, navigation and so forth);

(2) assembly and refuelling of the space vehicles which will perform transport service to and from the Moon scientific bases;

(3) assembly, refuelling and maintenance of the spacecrafts assigned to the exploration of planets;

(4) space traffic communication, control and rescue services.

If this is the general feature toward which the space activity development will be directed in the near future, it stands to reason that a transport service between terrestrial and orbiting bases will have to be organized.

We will see therefore frequent departures from terrestrial bases of vehicles which will carry to orbiting bases: equipments, propellants, spare parts, means of subsistence and personnel.

These vehicles will have to be reliable and cheap.

American and European Organizations proposed for this end several years ago,

G. A. Partel (ed.), Proceedings of the Second International Conference on Space Engineering All rights reserved.

the use of recoverable vehicles, capable of taking-off from Earth and of re-entering into the atmosphere and landing.

Vehicles as the Aerojet Astroplane, The Douglas Astro; the Nexus, the Rombus and the European Raumstransporter, all of them being rather complex and requiring a long and costly development.

We will consider here a more simple and immediate, even if partial, solution of the problem, considering, in the first place, that the first stage of a three stage space vector constitutes the most of the total weight and, secondly, that the performance requirements of the first stage are less critical in comparison with those of the other stages.

The use of an expendable solid propellant first stage is considered.

In order to determine, though in a quite rough way, size and weight of this Earth-orbit vehicle, we assume that the space base will have a minimum starting crew of ten men; this crew will increase exponentially up to 150 men in ten years.

We assume besides, again with a rough estimate, that each orbitating man requires 60 ton/year of material consisting of means of subsistence, building materials, tools and equipment, permanent equipment and installation, transport means for shifting crews.

Figure 1 shows the weight that has to be put into orbit each year for the first ten years, in conformity with these assumptions.

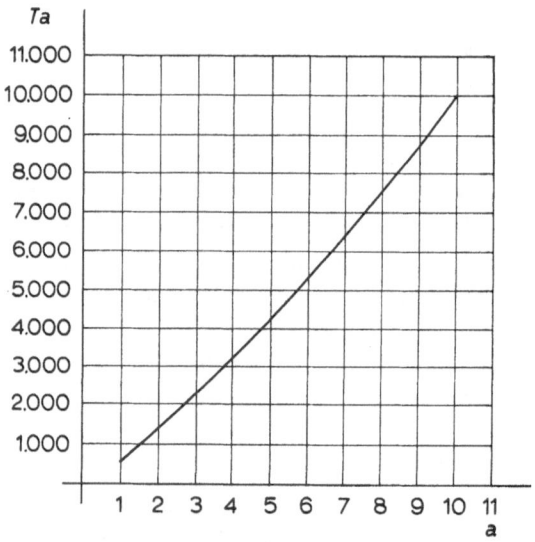

Fig. 1.   Orbiting weight versus time. $Ta = $ Ton; $a = $ year.

Carrying on this rough estimate, we assume that the solid propellant first stage consists of a cluster of seven boosters.

Each booster consists of five segments.

The solid propellant segmented motors technique is well known and tested.

Each segment (Figure 2) consists of a cylindrical steel envelope containing a solid

propellant grain with cylindrical port and head not inhibited surfaces; during combustion the front surfaces compensate the progressivity of the radial surface.

Assuming $K_i = 100*$ and initial burning area = final burning area, we obtain:

$$l_0 = 1.75D.$$

The grain volume is about D^3 and the port diameter is $d_0 = \frac{1}{2}D$.

As these data have not been optimized, the filling ratio is rather low (0.75) and
could be easily improved.

Assuming a mass ratio of 4 at the first stage cut off so that respective speed and
altitude values could be of the same order as those of to-day's vehicles, and a first
stage structural ratio (inert weight divided by propellant weight) of 0.20, which is
allowed by present technology, one gets:

$$
\begin{aligned}
W_1 \quad &= 60\ W_{cu} & W_1 \quad &= \text{1st stage weight} \\
W_{2+3} &= 6\ W_{cu} & W_{2+3} &= \text{2nd and 3rd stages weight} \\
W_t \quad &= 67\ W_{cu} & W_t \quad &= \text{total weight} \\
& & W_{cu} &= \text{payload.}
\end{aligned}
$$

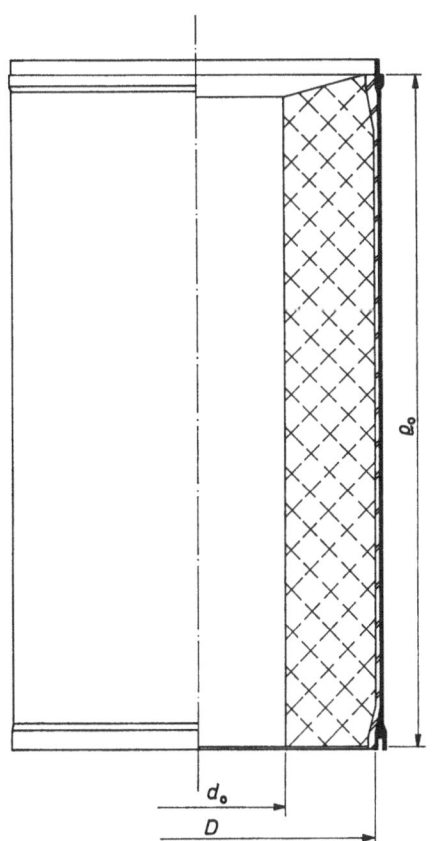

Fig. 2.   Segment.

* K_i = burning area/port area.

The general shape of our vehicle is given in Figure 3.

Let us consider now the cost of the activity based on the assumed system.

Suppose that the cost inclusive of rough materials, labour, equipment amortization and overheads is 5 Million Lit per ton with reference to ten units per year.

As the cost of qualified rough materials is about 1 Million Lit per ton, we assume that this is the value which the inclusive cost will approach for a large number of units.

Figure 4 shows then the cost per ton as a function of the units number.

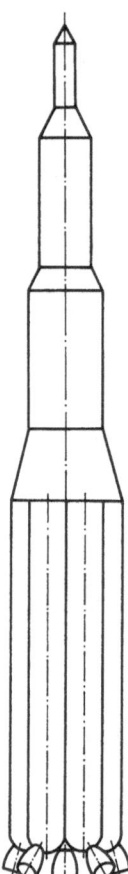

Fig. 3.   Configuration sketch.

Let us compare now the costs of the first stage of three vehicles having the same shape and designed to put into orbit payloads respectively of 25, 50 and 75 ton, which will be used for the above assumed ten years service.

Table I shows the main data of these three vehicles.

The result is given in Table II where the expenditures and therefore the production lots are considered to be annual.

A fixed expenditure for study, system setting up and preproduction should be added to the figures given in the table.

TABLE 1

First stage general data

Payload ton	Weight 1st stage ton	Weight 2nd and 3rd ton	Total weight at the start ton	Weight of 1 segment ton	Weight propellent 1 segment ton	ϕSegment m	L Booster m
25	1500	150	1675	42.8	35.6	2.76	24.2
50	3000	300	3350	85.7	71	~3.5	~30
75	4500	450	5025	128.6	107	~4	~35

A. ANGELONI

TABLE II
Annual expenditures estimate

Year	Payload = 25 ton				Payload = 50 ton				Payload = 75 ton				
	Launches No.	Segment No.	Cost Lit × 10⁶	Total cost Lit × 10⁶	Launches No.	Segment No.	Cost Lit × 10⁶	Total cost Lit × 10⁶	Launches No.	Segment No.	Cost Lit × 10⁶	Total cost Lit × 10⁶	
1	24	840	2.32	83400	12	420	2.6	93600	8	280	2.8	100800	600
2	56	1960	2	167770	28	700	2.4	144000	19	665	2.45	209475	1400
3	98	3430	1.8	264250	50	1750	2.05	307500	33	1150	2.2	325280	2440
4	130	4550	1.75	340795	65	2270	1.95	379500	44	1540	2.12	419800	3260
5	170	5950	1.62	415095	85	2970	1.87	476000	56	1960	2	504000	4250
6	216	7560	1.56	504770	110	3850	1.79	591000	72	2520	1.9	615600	5400
7	258	9030	1.52	587460	130	4550	1.75	682000	86	3020	1.85	718300	6450
8	320	11200	1.45	695010	152	5320	1.7	775200	101	3540	1.79	814700	7560
9	350	12250	1.4	734000	175	6120	1.62	849800	117	4100	1.76	927700	8760
10	400	14000	1.37	820900	200	7000	1.6	960000	134	4690	1.74	1049000	10000

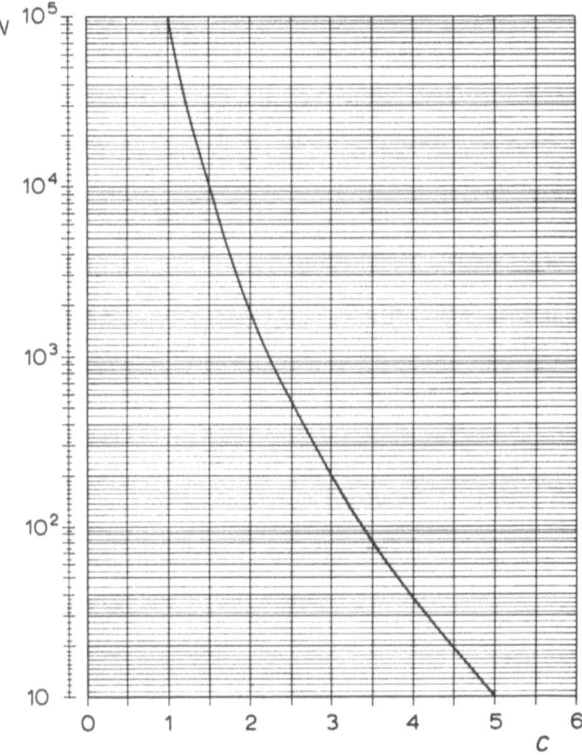

Fig. 4.   Cost versus quantity of produced units. N = Number of produced units; C = Cost,
Italian lires × 10^6/ton.

These figures show that, in the best case, that is in the tenth service year of a 25 ton payload vehicle, the incidence of the first stage cost on a kg in orbit is about 80 000 Lit, i.e. about 65 $/1b.

These values are higher than those evaluated for recoverable vehicles but it must be taken into account that the proposed system will require a much lower development cost.

In conclusion, as it could be expected, this very short investigation shows that our system, even if producing less brilliant performances and is probably more expensive in the long run, has the advantage of not requiring a huge initial investment for its development: therefore its cost falls more easily within the range of the world space research typical appropriation of funds.

References

[1] Koelle, H. H. and Rutland, C. H., 'Toward a Reusable Earth–Moon Transportation System', *Astron. Aeron.* (January 1964).
[2] Koelle, H. H., *Handbook of Astronautical Engineering*.
[3] Lambrecht, J., Überlegung zu einem europäischen Wiederverwendbaren trägersystem, *Raumfahrtforsch.* (April–June 1965).

LOW COST LAUNCH- AND RE-ENTRY VEHICLES –
A REQUIREMENT FOR FUTURE SPACE FLIGHT

B. H. GOETHERT

The University of Tennessee Space Institute, Tullahoma, Tenn., U.S.A.

Abstract. The paper will present a survey of the technological status and the economy of current space boosters and re-entry vehicles. Despite large reductions of launch costs, the specific cost per pound in orbit is still so high that only a few large-scale flights per year can be supported even by the economically strongest countries. It is pointed out that low-cost space launch- and re-entry vehicles are the key for extended space flight of the future. Several concepts for re-useable ferry-type vehicles offer for the future most attractive possibilities for making 'space flight traffic' approach the economy and reliability of 'atmospheric flight traffic'.

A brief survey of the technological state of the art will be presented. Various approaches to the low cost boost- and re-entry vehicle will be compared and the most significant technical problems for their realization identified. Particular emphasis will be given to the relative merits of rocket type boosters and air-breathing type boosters, to be used for expandable and re-useable vehicles.

1. Introduction

The last decade has seen a tremendous progress in space flight. Beginning with small grapefruit-sized payloads launched into low Earth-orbits, the technology has advanced to deep-space probes, to large 50 ton payloads, three-men space capsules circling not only the Earth but venturing out toward the Moon, and capable of maneuvering in space, rendezvous, docking, etc.

Examples are the recent Apollo flights of the US Space Program. During the historic Apollo 8 space flight, human eyes have seen for the first time in history, the surface of the Moon, from close distance, unaided by optical instruments.

Indeed, advancements beyond prior expectations for this short period of time!

In the United States, the launch cost per pound payload in low Earth-orbit has been reduced from $ 3000 for vehicles as the Thor, to $ 800 for vehicles as the Atlas, and to $ 400 and $ 500 for the Saturn-5. As impressive as these cost reductions are, their significance has been somewhat obscured by the associated increases of the launcher size.

While an Atlas launch vehicle would cost approximately $ 8 million, the cost of a Saturn-5 vehicle is in excess of $ 120 million. Adding to the launch vehicle the cost of the payload and the operational cost for launch, re-entry and recovery, a single launch in the Saturn-5 class will require funds of between $ 400 million and $ 500 million. With such a high cost per launch, even most generous space budgets are not expected to support more than limited space programs, that is only a few launches of boosters in the Saturn-5 class per year. Such number of vehicles will be too small to achieve appreciable cost reductions from mass production or systems maturity.

Enthusiastic public support for adequate space budgets can in the long run, be counted on only if *utilization of space* assumes a strong place beside exploration of space. If this situation is not achieved, funding for the space program can scarcely

G. A. Partel (ed.), Proceedings of the Second International Conference on Space Engineering. All rights reserved.

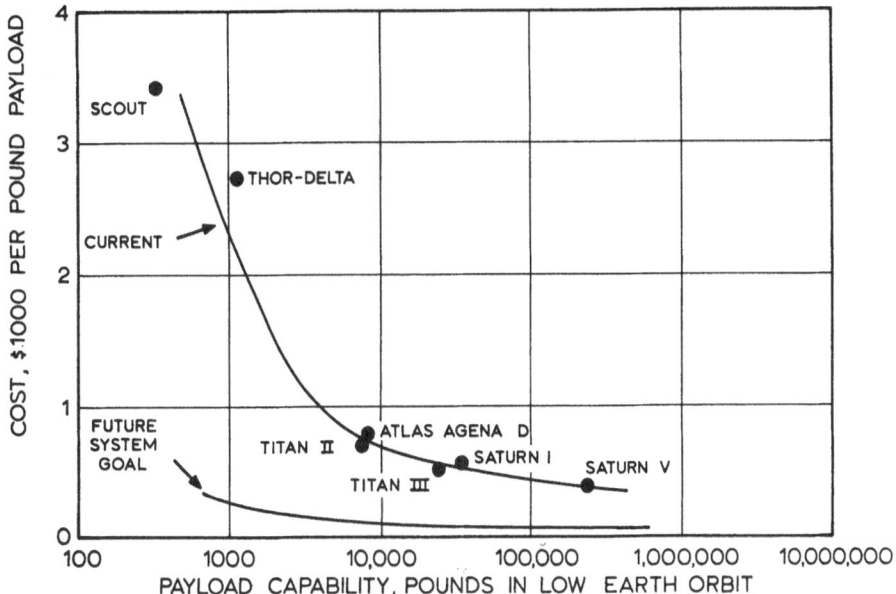

Fig. 1.   Launch vehicle cost trends.

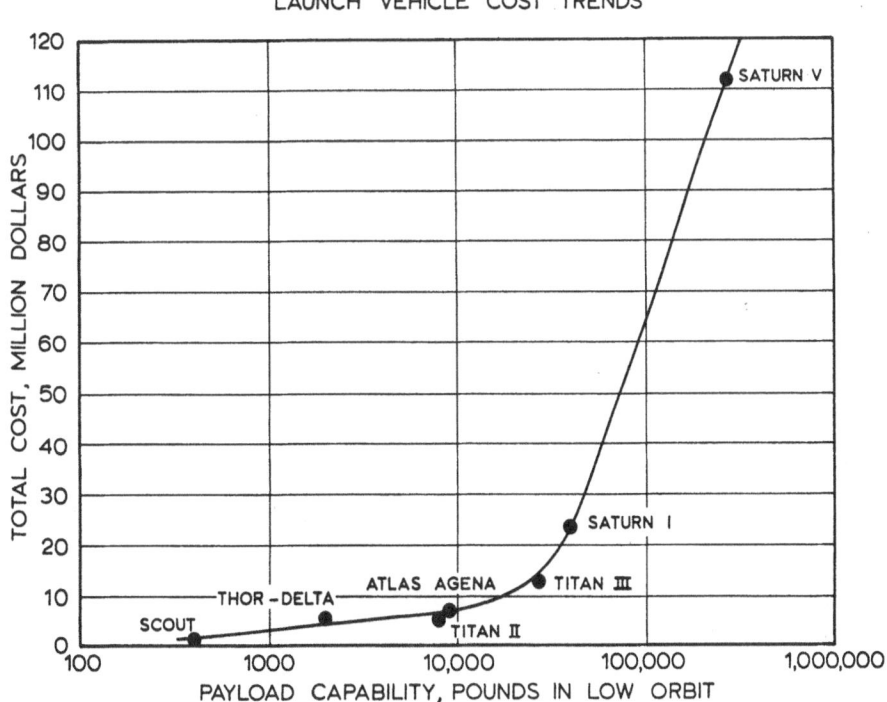

Fig. 2.   Launch vehicle cost trends.

be expected to increase greatly or even to maintain its current high level, but must count on being reduced, for instance, to the considerably more modest level of research in astronomy.

Looking into the future, it may be firmly assumed that a considerable portion of future space activities will be directed toward *applications* of space technology and commercial utilization of space, that is, to areas affecting directly the everyday life of each human being and the entire economy on Earth, such as communication satellites, navigation satellites, etc., are doing already today.

In a vigorous space program, with emphasis on 'space-utilization', the cost of the launchers must be drastically reduced. This need becomes particularly pressing when in the near future large space stations are assembled and manned for long-time operation in Earth-orbit. Their operation, and hence their usefulness, may be severely curtailed by the high cost of the first generation type space launch vehicles, as the only vehicles available to provide their continuous logistic support.

It can no longer be afforded to dump the launch vehicle into the ocean, after only one-time-use of costly $ 100 million hardware.

Also, cost reductions by means of component recovery with parachutes or similar devices, will be no more than a short-time expediency. Economical fully re-useable ferry-type vehicles are needed, having *airline-like* reliability and operational capabilities similar to our aircraft, that is, capable of taking off and landing from space-ports, similar to our regular airports, depositing the payload, men as well as goods, at low-orbit space stations and returning safely to Earth.

Only with such multi-purpose ferry vehicles, the era of 'space utilization' instead of mere 'space exploration' can in earnest begin.

Current technology is advanced to such an extent that systems of the re-useable ferry-type are technologically feasible. With appropriate support for research, proto-type design, prototype and final configuration testing, the available scientific and technological base can be systematically extended to provide the required detail design data.

The launch cost of such vehicles is estimated to go down from $ 500 per pound to $ 50 per pound, and will even approach $ 10 per pound in the future.

The tremendous impact of such low launch and recovery costs on the volume of space traffic is hard to imagine.

Many missions, which are utterly unthinkable at today's high costs of launch, will become attractive and economical.

2. Operational Characteristics of Space-Ferry Vehicles

The operating characteristics for commercial Earth–space–Earth traffic should include the following key elements:

A. COST

The cost of transporting one pound of payload into Earth orbit should approach $ 10, as feasibility studies have shown to be achievable as a logical extension of the existing technological base. Since the return flight to Earth is essentially free, the round trip cost per passenger from Earth to space and back to Earth comes to $ 2000, instead of the presently required one quarter of a million dollars, including recovery.

B. FLEXIBILITY

The space vehicle should be capable of taking off and landing at the discretion of the pilot, on numerous conventional airports, distributed over the entire globe. It should be capable of serving multi-purpose missions for both passenger and freight transportation.

C. ENVIRONMENT

Passenger traffic should provide environmental and physical conditions to which an average traveller can readily adapt himself, without the need for a special training program. Consequently, provisions for horizontal takeoff and landing, shirt-sleeve cabin atmosphere, and acceleration and deceleration in the direction of flight of not much more than 1 g are extremely desirable, and probably indispensable for airline-like service.

D. RELIABILITY

Reliability and safety of operation must equal that of accepted transportation means on Earth, e.g., that of airline traffic. This requirement includes the need for 'fail-safe' vehicles, aimed at the capability of terminating the flight at almost any time without excessive danger to either the space travellers or the ground base population.

3. Design Concepts of Space-Ferry Vehicles

During the last decade, essentially all elements of engineering were pushed to their limit, in order to achieve orbital flight at all, largely disregarding the economical aspects. Starting from the current technological base, all areas of space vehicle engineering will undoubtedly continue to advance, in a usual evolutionary manner, and thus contribute to more efficient designs of space vehicles and reduction of cost.

In this process, the following areas need particular attention, to name only a few:

In the aerodynamics field – high Mach number phenomena, such as friction, heat flux, flow separation, etc.; conceptual design evaluation of vehicles with high 'lift-over-drag' ratios for both subsonic, supersonic and hypersonic flight.

In the propulsion field – improvements of chemical rockets, both in the areas of propellants and rocket design.

In the field of materials – improved high-temperature materials, capable of resisting the environmental conditions at high speed flight in the atmosphere with long stay times. Ablating materials to be avoided due to their degrading effect on vehicle performance; major emphasis on regeneratively or radiation cooled materials and structures.

Besides such evolutionary advances, however, there are some key elements of design concept which are expected to produce major *improvement jumps* in performance and economy of space ferry operations:

(1) *Fully re-useable vehicles* – with low refurbishing costs, and airplane-like operation.

(2) *Medium-size payload* – to allow many flights per vehicle, and achieve design and constructive maturity.

(3) *Airbreathing propulsion for major portions of the launch* – to make use of the oxygen in the air, instead of carrying it on board, and thus reduce greatly the take-off weight of the vehicle.

A. FULLY RE-USEABLE VEHICLES

The pay-offs of re-useability of a space-launch vehicle are so obvious that the *next generation of launchers* can hardly be imagined without having this feature.

Neglecting the costs for refurbishing and the greater complexity of re-useable vehicles, 5 re-uses would reduce the cost per pound in orbit to $ 100/lb. and 50 re-uses to $ 10/lb. instead of the currently required $ 500/lb. in orbit. Assuming refurbishing costs after each flight as a maximum of 10% of the cost of the original vehicles, a 50 re-use vehicle could still provide a launch service for $ 60/lb. Assuming a refurbishing cost of 1%, which is still much higher than for today's airplanes, the cost per pound in orbit would be no more than $ 15.

It is also to be noted that the assumed 50 re-uses for each vehicle is not a definite limit at all. There is no reason that it cannot be pushed to 100 or above.

Even, when the high complexity of a re-useable vehicle is accounted for by higher cost of the vehicle, say twice as high as an expendable vehicle, the costs are nevertheless attractively low.

B. MEDIUM-SIZE PAYLOAD

With the present system of expendable boosters, the opportunities for *systematic post-flight inspection and subsequent use of the inspection results for vehicle improvement is practically non-existent.* Furthermore, the check-out man hours for each individual vehicle, have remained nearly at the same high initial level (100 000 man hours, corresponding to $ 1 million) as the examples of two space boosters indicate.

The 're-useability' feature of the proposed space ferry vehicle provides the opportunity to inspect the vehicle after each flight, and thus to systematically correct deficiencies due to low design loads, etc. Consequently, each individual flight contributes to the development of an improved vehicle line. The significance of this point is clearly visible from statistical data on a typical aircraft. The delivery check-out for each of the first five aircraft of the same type required approximately 80 000 man hours, estimated to cost approximately $ 800 000. These check-out man hours and costs were reduced to approximately 4000 man hours of $ 40 000 per airplane at the 100th delivery. At larger numbers of vehicles beyond 100, the check-out costs continue to decrease significantly.

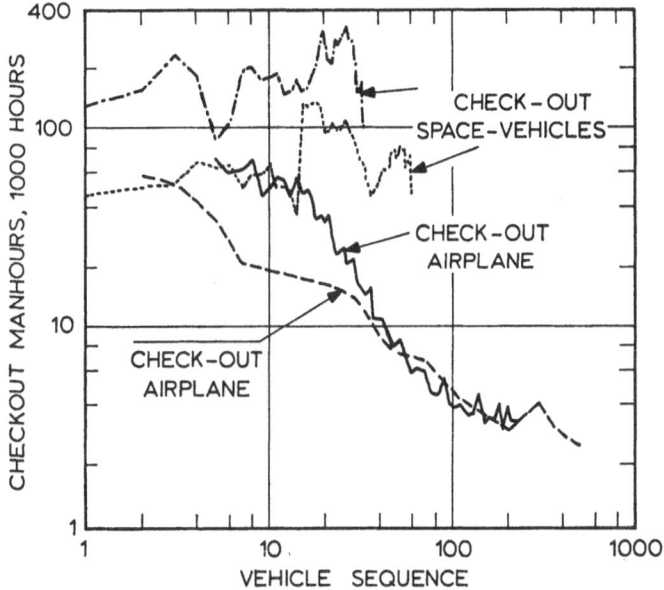

Fig. 3.   Vehicle checkout time.

Hence, a large number of flights with the same vehicle hardware has been vital for achieving low costs of airplane transportation.

Observing this experience, it is apparent that neither the extremely small nor the extremely large payload transporters are conducive to large trip numbers. The transportation needs are met best by intermediate size vehicles which are large enough to handle a sizable economical load, but on the other hand, are small enough to provide a high frequency of trips to and from diversified stations.

A standard re-useable space-ferry vehicle with very low refurbishment cost, in the intermediate payload size of about 50 000 pounds is promising for future space missions.

C. AIRBREATHING PROPULSION

The performance of chemical rockets has been extensively improved in the last decades. It appears however, that except for a major breakthrough in this field, the performance of chemical rockets is now approaching asymptotically an upper limit.

The fact that *rockets* are indispensable for operations in space, does not exclude using *airbreathing* propulsion for space launchers during the atmospheric part of the trajectory.

One of the most promising performance improvements of the future is associated with the application of airbreathing systems, particularly in the lower stages of the next generation of space launch vehicles. The *supersonic-combustion ramjet* with its low specific fuel consumptions, varying from $\frac{1}{8}$ to $\frac{1}{4}$ of that of our best rocket engines, in the speed range from supersonic to hypersonic Mach numbers of approximately 14 is the key to spectacular performance improvements.

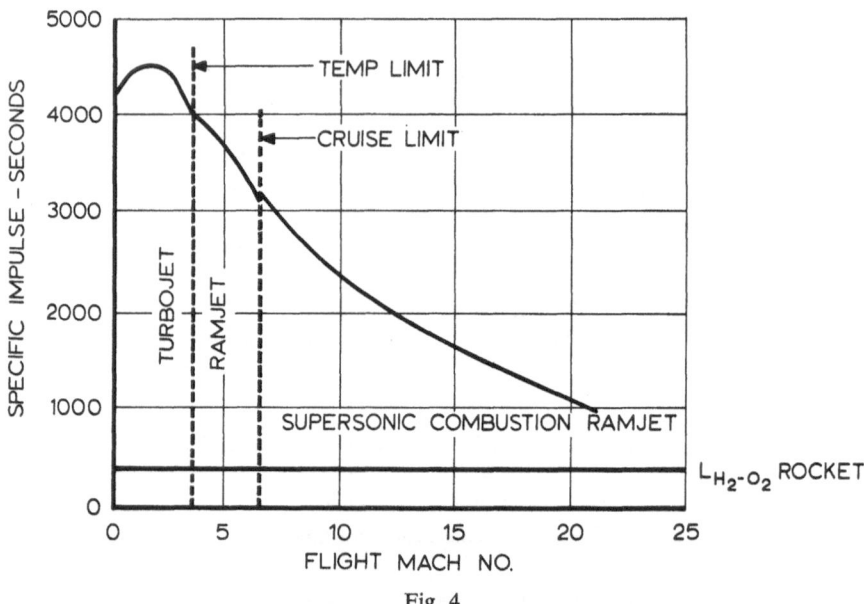

Fig. 4

Realizing that the propellants on board a conventional rocket-powered launch vehicle represent approximately 85% of the total launch weight, and the payload is usually less than 5%, it is evident that the quoted reductions of fuel consumption will result in major improvements of the payload fraction and thus of the launch cost.

On the other hand, the structure of the air-breathers will become heavier due to larger dimensions and the necessity of flying longer in the high heat-flux environment within the atmosphere. The superior vehicle performance achievable with airbreathers over rockets, depends critically on the skill of the designers in arriving at an optimum balance between propellant reduction and structural weight increases. Various design studies have been conducted to determine the relative performance of launch vehicles with advanced chemical rockets and airbreathers.

In the studies discussed above, the first stage was assumed to be powered by air-breathers, while the second stage was assumed to use advanced chemical rockets.

Such estimates indicate consistently for *equal takeoff weight* of approximately 1 million pounds that:

> First stage airbreathers, particularly SCRAMJETS, offer a two to three-fold increase of the payload in comparison to chemical rocket systems.

In summary, re-useable launchers of the future having an airbreathing first stage and a rocket second stage, and a high staging Mach number of 10 or above, offer great promise for a low-cost, airline-like transportation system between Earth and low orbit.

4. Space-Ferry Vehicle Design

Numerous design studies have been conducted to examine closely the performance and recovery potential of space-ferry vehicles with airbreathing propulsion. Promising designs tended to be more closely associated with airplane design than with conventional rocket vehicle design. This latter fact is evident in the layout of a re-useable two-stage ferry vehicle for service between the Earth and low orbit, by the John Stack Group of Fairchild-Hiller, USA. This ferry vehicle exhibits many features of the hypersonic cruise vehicle, designed by the same engineering group.

Both vehicles have horizontal takeoff, low acceleration, high L/D values of the aerodynamic configuration, and efficient integration of propulsion system and airframe.

5. Space Ferry Vehicle Operations of the Future

The ferry vehicles will not only serve as boosters to space, or only as return vehicles from space, but similarly to conventional airplanes, serve both missions with the same vehicle. They will provide much less costly transportation to and from space than is possible today. For example, trips between Earth and low orbits could approach *round trip fares as low as $ 10 per pound* provided that technological possibilities are fully exploited.

When payloads of 45 000 pounds each are launched into orbit, in intervals of somewhat less than *one launch* per month, with the current *high-cost expendable* launchers at $ 500 per pound or whether the same payloads are transported by an *advanced ferry vehicle*, into orbit and return to Earth, at a rate of *two launches per week* at $ 50

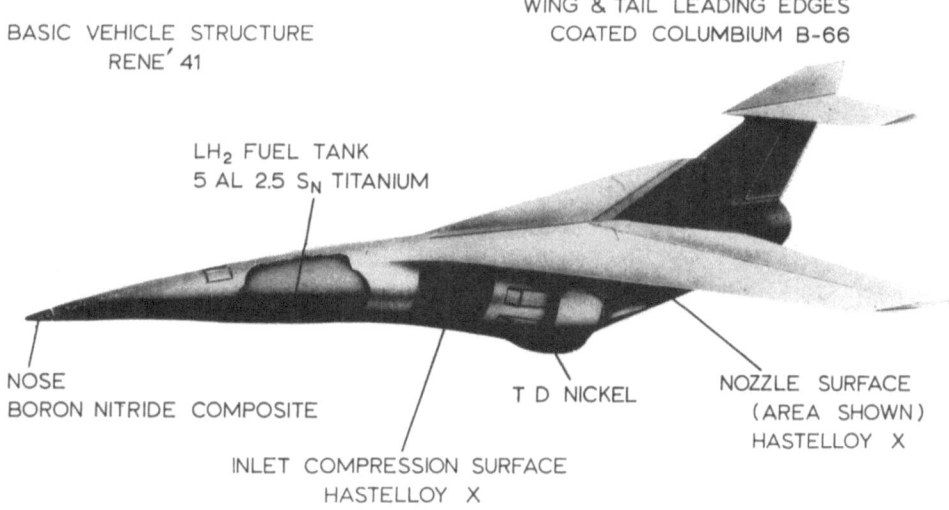

Fig. 5.   Mach 12 hypersonic transport (high wing) materials.

per pound, the *total* launch costs over a period of 10 years are estimated to be *the same*; that is, about $ 230 million per year.

When *3 flights a day* with the same individual payload are conducted, over the same 10 year period at $ 10 per pound, the total cost will only double, that is $ 500 million per year.

This conclusion is naturally valid only under the assumption that future high-density space traffic will develop for the medium payload range.

At the threshold of airline-like low fare space traffic, the planners need to think of space as no longer being a hostile environment, but as being an environment which provides *numerous unique features and opportunities, unmatchable on the ground.* Weightlessness, ultrahard vacuum, extremely low temperature heat-sink, etc., represent some of the parameters which can be established on the ground only at very high cost, or in some cases, cannot be duplicated at all.

If the development of advanced launch and re-entry vehicles is not to be delayed to other decades, but if such vehicles are to be operational in the early 1980 period, we must start now without delay, to refine and verify the required technological base, since a system development, beginning with the conceptional design until operation, requires a period of approximately 10 years.

The aforementioned space-ferry systems are technologically feasible. However, the time of their realization depends upon the availability of competent research and engineering teams, of both industrial and academic background, to guide the design, and finally, upon the *determination of the people* who want to extend the human aspirations into space.

COMPARISON OF A BOOSTER-RAM ROCKET COMBINATION WITH A TWO-STAGE ROCKET FOR AN ACCELERATION MISSION

H. GAWLIK and E. RIESTER

Institut für Strahlantriebe DFL, Braunschweig, Germany

Abstract. In recent time spacecraft launchers always have been rockets. In order to save costs some proposals have been made to make the launcher reusable. In this case weight is increasing. For these launchers often air-breathing engines are proposed because of their less propellant weight, but structure weight increases again. There have been made many calculations, and the optimum depends on many parameters and assumptions, for example on the number of launches needed.

In this paper a launcher is investigated being a combination of a rocket booster with a second-stage ram rocket, but the launcher shall not be reusable. So the ram rocket must be built simple, inexpensive and in an extreme lightweight construction. This is possible, because combustion time is short, less than $\frac{1}{2}$ min, as will be shown. In the calculations a fixed total starting mass and a fixed specific total energy at propellant cut-off of the ram rocket are assumed, and the engine is optimized to minimum structure and propellant mass of the whole launcher (booster and ram rocket), that means to a maximum so-called 'payload' of the launcher. Calculations are compared to those with a single-stage and with a two-stage rocket launcher. To simplify calculations, always a vertical climb is assumed for the launcher, that means worst conditions for the ramjet.

The result of the calculation is a gain in 'payload' of the launcher up to 35 p.c., if the second-stage rocket of the launcher is replaced by a ram rocket. But as the ranges in which the ram rocket has sufficient performance are rather narrow, parameters of the engine must be optimized. To get into orbit, upper-rocket stages must be added.

Notations

LATIN SYMBOLS

A	$[\text{m}^2]$	cross-sectional area of air flow in the ramjet
a	$[\text{m/sec}]$	exit velocity of rocket
b	$[\text{m/sec}^2]$	missile acceleration without gravity
g	$[\text{m/sec}^2]$	acceleration due to gravity
h	$[\text{m}]$	altitude
k	$\left[\dfrac{\text{kg sec}}{\text{m}}\right]$	mass flow of rocket
m	$\left[\dfrac{\text{kg sec}^2}{\text{m}}\right]$	missile mass
m_s	$\left[\dfrac{\text{kg sec}^2}{\text{m}}\right]$	ramjet structure mass
t	$[\text{sec}]$	time
w	$[\text{m/sec}]$	velocity

GREEK SYMBOLS

α $[-]$ ratio of rocket propellant mass $k t_e/m_0$

β $[-]$ interior rocket size factor $ak'/m_s g$

δ $\left[\dfrac{\mathrm{m}^3}{\mathrm{kg\ sec}^2}\right]$ light weight construction factor A_3/m_s

SUPERSCRIPTS

without: regime of the first acceleration stage (booster)
dash: regime of the second acceleration stage
star: after density decrease in ramjet combustion chamber.

SUBSCRIPTS

0   at ignition time of the stage
e at propellant cut-off of the stage.

RAMJET SUBSCRIPTS

1   cowl lips
3   entrance of the combustion chamber after fuel addition
5   nozzle throat.

1. Introduction

In recent time spacecraft launchers always have been rockets. In order to save costs some proposals have been made to make the launcher re-usable. In this case weight is increasing. For these launchers often airbreathing engines are proposed because of their less propellant weight, but structure weight increases again. There have been made many calculations, and the optimum depends on many parameters and assumptions, for example on the number of launches needed. There have been many discussions according to these proposals, but as nobody likes to spend money for these developments without knowing how many launches will be made in future, all proposals did not become more than projects.

The matter of this paper is another step towards using airbreathing engines with launchers: The launcher may not be re-usable, too. In this case its utilisation does not depend from the number of launches. The proposed launcher will be a combination of a rocket booster with a second stage ram rocket. The booster may accelerate up to velocities of somewhat more than the velocity of sound. The ram rocket serves as a second acceleration stage up to Mach numbers of 4 to 6 and altitudes up to 30 km, a regime in which conservative ramjets can work. Of course the combination rocket booster with ram rocket (we will call it launcher and the whole orbit vehical may be called missile) is not sufficient to launch a payload into orbit but must be completed by upper rocket stages. But we may limit the postulations for our so called launcher to a fixed sum of potential and kinetic energy related to the total starting mass.

Object of the following investigations is to find a launcher with maximum payload for a given total starting mass and to compare it with pure rocket launchers fulfilling the same postulations.

Originally the following calculations have been made[*] for an altitude test missile. So the postulated sum of energies – related to the total starting mass – has been chosen rather low with 10^5 m, a vertical climb is assumed always and the assumed acceleration values are somewhat higher than usual in launching of space vehicles. But the results can be useful for orbit launchers, too: as the vertical climb gives the worst conditions for a ramjet, it will work better in an inclined flight, and the postulated sum of energies may be enlarged if flight gets a horizontal component. If it can be shown that the ram rocket launcher will be superior under the chosen conditions, it will be better than pure rockets in usual climb, too. The need of more stages to get into orbit will be a disadvantage of all airbreathing engines.

Now at first a booster as first stage will be regarded. Secondly for comparison a second stage rocket will be added to the booster fulfilling the same postulations as given before for the ram rocket. Thirdly some details of ramjet calculation are given. Furthermore some combination parameters for ramjet and rocket will be defined, and a family of climbs will be calculated. The results of the ram rocket calculation will be compared with the results of calculating a two-stage pure rocket. The regimes in which the ram rocket is superior will be shown. A typical ram rocket will be chosen to show the variation of its characteristic data during the flight.

2. Booster Performance

The booster is assumed to be a rocket. It has the task to accelerate the missile to a velocity at which the second stage drive can work. For comparison the booster may be the same both for a second stage pure rocket and for a ram rocket. Therefore the assumptions for the booster are not so very important. In this case the booster may have solid propellant with constant mass flow, constant specific momentum and constant thrust. The exit velocity may be 2000 m/sec, and a structure mass part of 10 p.c. of the rocket may be assumed to be realistic.

There are two free parameters of the booster: Firstly the starting acceleration, mostly without consideration of gravity. This may be expressed as the booster thrust related to the total starting mass:

$$b_0 = \frac{ak}{m_0}.$$

Secondly the booster propellant related to the total starting mass

$$\alpha = \frac{kt_e}{m_0}.$$

[*] In the Institut für Strahltriebe der Deutschen Forschungsanstalt für Luft- und Raumfahrt e.V. (DFL), Braunschweig, directed by Prof. Dr.-Ing. habil. O. Lutz.

To calculate what the booster can do is rather simple. Figure 1 shows the results of the calculation. Each point in the field is characterized by an acceleration b_0 and by a relative propellant value α, and at both axes we can read the altitude and the velocity at booster propellant cut-off. This velocity and altitude may be used as initial conditions for the second stage.

In Figure 1 all curves end at a dotted line. At this line the booster can already fulfil the postulations for both the first and the second stage, and the second stage therefore can be omitted. Another limit is not shown in Figure 1: acceleration must be larger than one or the missile will not climb.

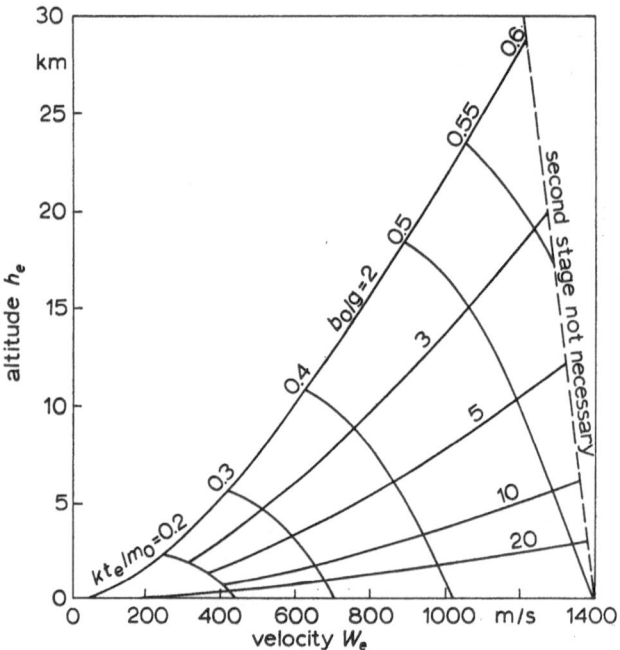

Fig. 1.   Altitude and velocity at booster propellant cut-off depending on booster size and booster initial acceleration.

3. Second Stage Pure Rocket

The second stage will find its initial conditions by the booster propellant cut-off, the parameters again are booster initial acceleration b_0 and booster propellant α. End condition for the second stage is the postulated sum of potential and kinetic energy at its propellant cut-off. There is one free parameter left for the second stage. In the case of a ram rocket this parameter will be used to get an optimum ram rocket. In the case of a second stage pure rocket the free parameter at first will be used for the additional postulation that the propellant is fully consumed at an altitude of 30 km, that means that second stage pure rocket acceleration is comparable to a ram rocket acceleration.

For calculating the second stage a pure rocket again may be assumed to be a solid propellant rocket with the same data as for the booster rocket: constant massflow, constant exit velocity (2000 m/sec), and relative structure mass 10 p.c. of the rocket mass. Calculations have been made by an electronic computer. The results are pointed out in Figure 2. On the horizontal axes the relative booster propellant is plotted which

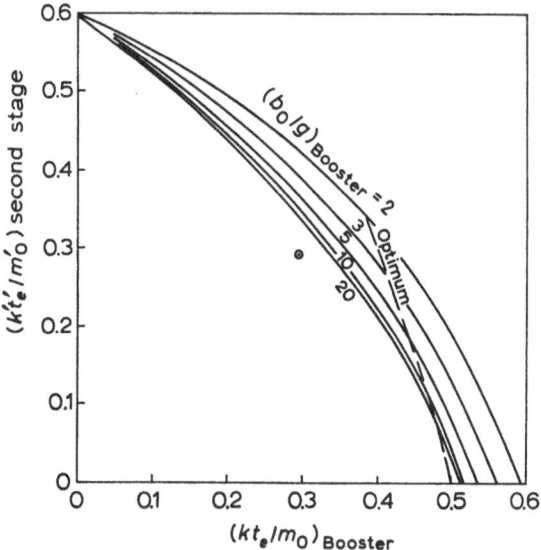

Fig. 2. Propellant mass of the second stage rocket as a function of propellant mass and acceleration of the booster. ○ Optimum two stage rocket with infinite acceleration in both stages.

is – because of the little structure mass – nearly the ratio of booster mass to the total starting mass. On the vertical axes the same is plotted for the second stage rocket but related to the total mass at second stage ignition. We can see that a big booster needs only a little second stage rocket, and with little booster the second stage rocket becomes big. There are solutions without second stage rocket – depending on booster acceleration – and without booster, too.

Furthermore one single point is plotted in Figure 2: This is a theoretical optimum point for infinite acceleration in both rocket stages. In this case both relative rocket masses – related to the total mass – are the same at ignition of their stages. It is to be seen that the point is not very far from the curves.

4. Combination of Two Pure Rocket Stages

Figure 3 shows how the two rockets are working together. It may be mentioned again, that this investigation on a two-stage rocket shall be done only to enable us to compare a ram rocket with a pure rocket. This picture will be discussed very briefly. Main result is that the whole acceleration period of both stages takes between about 30 sec at high booster acceleration and 1 minute at low booster acceleration.

Figure 4 is of some more interest. It shows at the vertical axes the so-called payload of the two-stage rocket, that means the whole missile mass except structure and fuel of the two stages, related to the total missile starting mass. The curves have an optimum at rather high booster masses, and the payload at the optimum is not far from the value of the right limiting curve, at which the booster alone is sufficient to

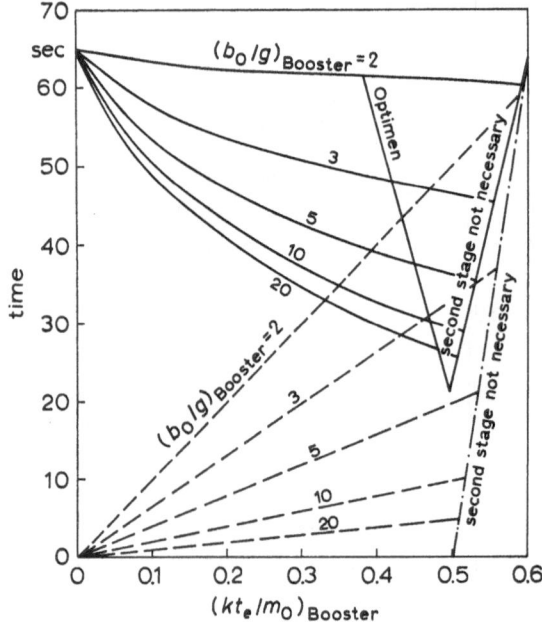

Fig. 3.   Time after starting as a function of booster propellant mass ratio and acceleration. Both stages are rockets. Dotted lines: Booster only. Full lines: Booster and second stage rocket combined.

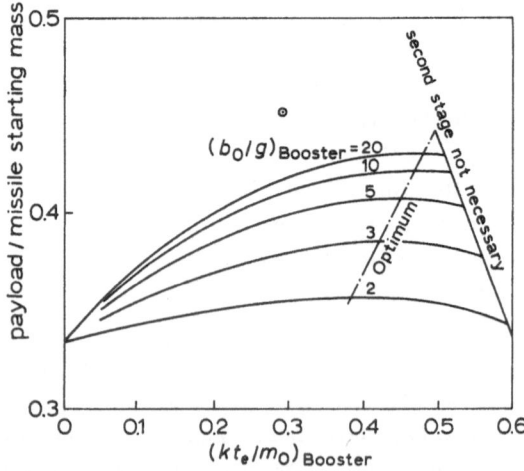

Fig. 4.   Payload as a function of booster propellant mass and acceleration. Both stages are rockets.
○ Optimum two stage rockets with infinite acceleration in both stages.

do the task of both stages. It is, therefore, for the postulated conditions not very useful to divide the rocket into two stages.

When the 'second stage not necessary'-line and the 'optimum'-line are meeting, booster acceleration becomes infinite. Additionally, there is a single point – as mentioned already with Figure 2 – plotted at a booster propellant of about 0.3. This point means an optimum two-stage rocket with infinite acceleration at each stage. We can see that both the meeting point and the single point do not exceed the curve with 20 g booster acceleration very much although there are extreme conditions calculated. So the 20 g-curve later on will be used for comparison with ram rockets.

5. Ramjet Performance

In literature mostly a large number of ramjets are calculated, and the results are presented in form of pictures and tables. There is also a large number of off-design calculations. But if anybody is looking for a special powerplant and the same powerplant will run in large ranges of velocity and altitude while some of the parameters are connected by a special mission and cannot be chosen free it is rather difficult to use literature data, if the control-ranges of the plant will be rather narrow. So in the chapter a special computing program is made for a ramjet for necessary conditions in climbing up.

Of course there must be made some simplifying assumptions or the program will become complicated. First of all a separation is made between velocity and altitude. This is possible if the static air temperature during in flight – connected with the velocity of sound – is taken constant. This assumption may be admitted because operation begins in some kilometers of altitude and ends beneath 30 km. Medium temperature is about 240 K, the velocity of sound is about 310 m/sec. This assumption allows to calculate the influence of the velocity of flight independent from altitude while real thrust and specific fuel consumption is proportional to the density, which is only a function of altitude.

Another main assumption is concerned to the ratio of specific heats. In order to simplify the computing program there are only two values of the ratio used: 1.4 for flight and inlet conditions and 1.3 for the conditions of combustion chamber exit and nozzle. The value of 1.3 belongs to a temperature of about 2000 K, this being a medium static temperature of combustion chamber exit and nozzle. As can be shown the influence of the specific heats on the payload of the two stages is only very small.

In the combustion chamber the heat addition will be as high as possible to get a high thrust coefficient. At rather low flight Mach numbers the heat addition is limited by reaching velocity of sound in the nozzle. If flight Mach number is increasing temperature becomes too high for the nozzle material. In this regime total temperature is limited to 2400 K. This is a rather high value and can only be submitted because the mission time of the ramjet is rather short. Another limit for heat addition is given by combustion: Air mass flow must be higher than stoichiometric. This limitation has never been reached during computing and is therefore only of theoretical interest.

What happens if heat addition is reduced from maximum possible addition to reach velocity of sound in the nozzle to a lower addition caused by the limit of 2400 K in nozzle total temperature? In all cases being computed nozzle exit pressure is low enough to get the velocity of sound in the nozzle. Therefore density in the combustion chamber is going down to fulfil the continuity equation, and there will be a further pressure loss at the diffusor exit together with an increase of velocity in combustion chamber entry.

The areas of diffusor entry, of combustion chamber, of nozzle throat, and of nozzle exit are fixed for each powerplant. The area of nozzle exit always equals the area of the combustion chamber. The area of the diffusor throat is controlled, a second control is that of fuel consumption. But a family of ramjets with different areas of diffusor entry and nozzle throat are calculated. Some of them are shown in Figure 5.

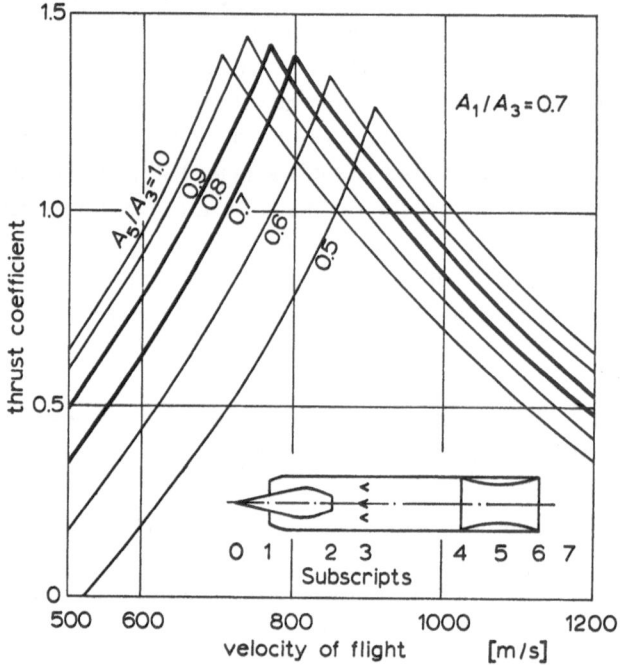

Fig. 5. Ramjet performance. $A_1/A_3 = 0.7$.

6. Ram Rocket Parameters

In the section before we have got the thrust coefficient and the specific fuel consumption as a function of velocity of flight for a family of ramjets. At first one or two special plants with fixed geometry – except diffusor throat area – must be chosen. The plants having a good thrust coefficient over the whole velocity range needed (see Figure 5) are as follows:

1. $A_1/A_3 = 0.7$ $A_5/A_3 = 0.8$
2. $A_1/A_3 = 0.8$ $A_5/A_3 = 0.8$.

A_1 is the area of the cowl lips, A_3 is the area of the combustion chamber and of the nozzle exit, A_5 is the area of the nozzle throat.

Now some size-parameters must be found. The size of a rocket may be described in the first order by its propellant mass. That is not true for a ramjet. The ramjet size must be characterized by its main cross sectional area and by its fuel mass. But it is difficult so to get suitable dimensionless parameters in this way. According to the fuel we may think of the short running time we had seen in Figure 3 and therefore neglect the fuel consumption as a parameter – of course not in mass calculation. According to the cross sectional area we may regard the ratio of it to the ramjet structure mass $\delta = A_3/m_s$ as being a measure for the light weight construction of the ramjet. The area A_3 of the combustion chamber is proportional to the square of the linear size of the ramjet while its structure mass at first seems to be proportional to the third of the linear size. But in practice length and wall thickness do not rise proportional to the linear size so that in a good approach δ can be regarded independent from ramjet size. The value of δ will be as high as possible. If structure mass of the rocket was assumed to be only 11 p.c. of propellant mass a ramjet will be equivalent with a factor $\delta = 0.02 \text{ m}^3/\text{kg sec}^2$, that means 500 kp weight per m². $\delta = 0.02$ is a rather high value. Perhaps it may be necessary to build the shroud of plastics while part of the centrebody may be a casting for the payload and therefore

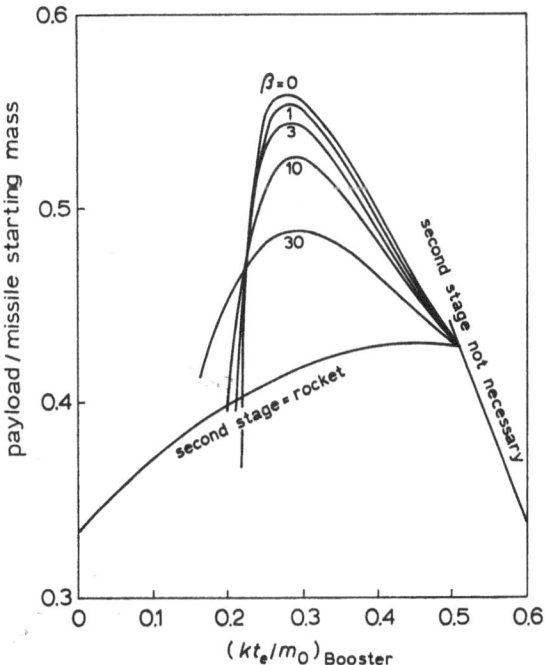

Fig. 6.   Payload of a missile with booster-ram rocket drive.

$(b_0/g)_{\text{Booster}} = 20$ $\delta = A_3/m_s = 0.02 \text{ m}^3/\text{kg sec}^2$
$A_1/A_3 = 0.7$ $A_5/A_3 = 0.8$

– see the former definition of payload – regarded to be 'payload' itself. For the calculation δ is a fixed value.

Now we can define a good parameter for the ramjet size: m_s/m'_0 is the ratio of ramjet structure mass to the missile mass at the ignition of the second stage. This ratio is a constant value for one mission, but in an optimisation the value connected with the highest payload must be found.

Another difficult problem is to find a parameter for the size of the interior rocket as a part of the ram rocket. The parameter must have a constant value for one mission. Here it is proposed to use $\beta = ak'/m_s g$, that means the ratio of the thrust of the interior rocket to ramjet structure weight. For a pure ramjet β is zero. For a pure second stage rocket β must be infinite, but in this case the used computing program is not convergent. Values of β of 100 or more are not suitable.

7. Results of Calculation, Comparison with Equivalent Rocket

In the calculation a number of details must be considered. Integrations must be done numerical, and there are some solutions possible only by trial and error. It will not be carried out here. But some of the results may be shown on the following pictures.

In Figure 6 it can be seen that payload increases if the second stage rocket is

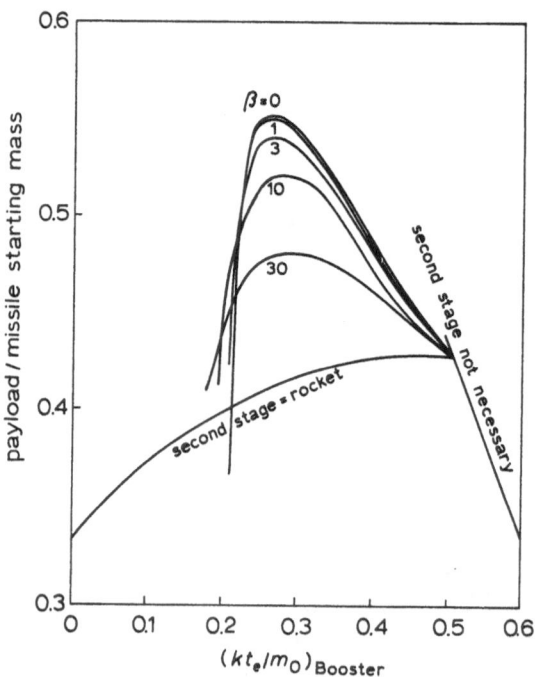

Fig. 7. Payload of a missile booster-ram rocket drive.

$(b_0/g)_{\text{Booster}} = 20$ $\qquad \delta = A_3/m_s = 0.02 \text{ m}^3/\text{kg sec}^2$

$A_1/A_3 = 0.8$ $\qquad\qquad A_5/A_3 = 0.8$.

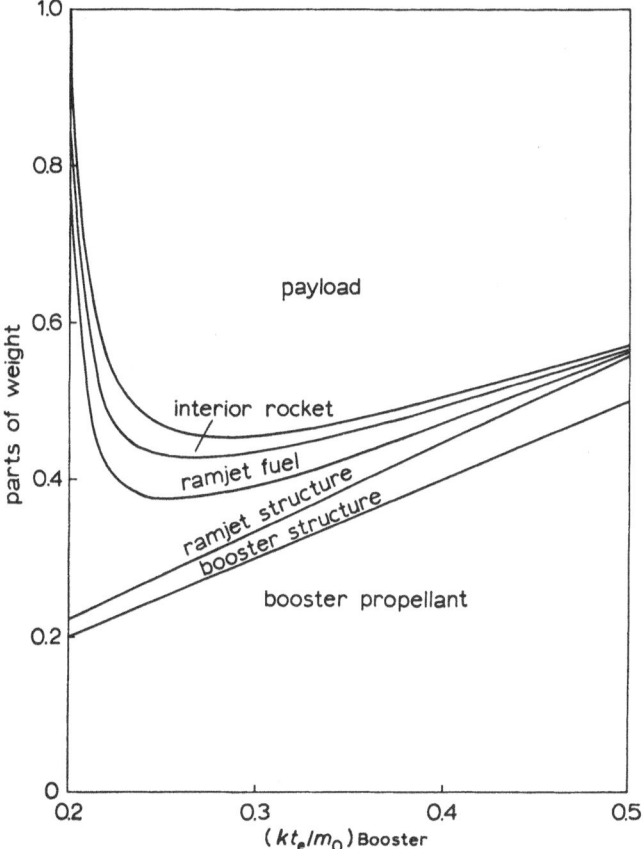

Fig. 8.   Parts of weight as a function of booster size.
$(b_0/g)_{\text{Booster}} = 20$ $A_1/A_3 = 0.8$ $A_5/A_3 = 0.8$.

$$\beta = \frac{ak'}{m_s g} = 3 \qquad \delta = A_3/m_s = 0.02 \text{ m}^3/\text{kg sec}^2$$

replaced by a ramjet or ram rocket. The lower curve relates to a pure rocket and is transferred from figure 4 for booster acceleration of 20 g. This is also true for the right limiting line on which pure booster is sufficient to fulfil mission postulations. The upper curves relate to ram rockets as second stage drive for different sizes of the interior rocket. The top curve shows the performance of the pure ramjet ($\beta=0$). If the size of the interior rocket becomes larger, maximum of the curve goes down but the range of possible boosters becomes larger. That means that it is best to have a pure ramjet without any interior rocket. But there are some other features, i.e. ignition and combustion in large altitudes which make the interior rocket useful. So in practice β-values of 1 to 10 will be used. In all there may be a gain in payload of 25 to 35 p.c. by a ram rocket.

Figure 6 relates to a special type of ramjet with a diffusor inlet area ratio of $A_1/A_3 = 0.7$ and a nozzle throat area ratio of $A_5/A_3 = 0.8$. In Figure 7 the same curves

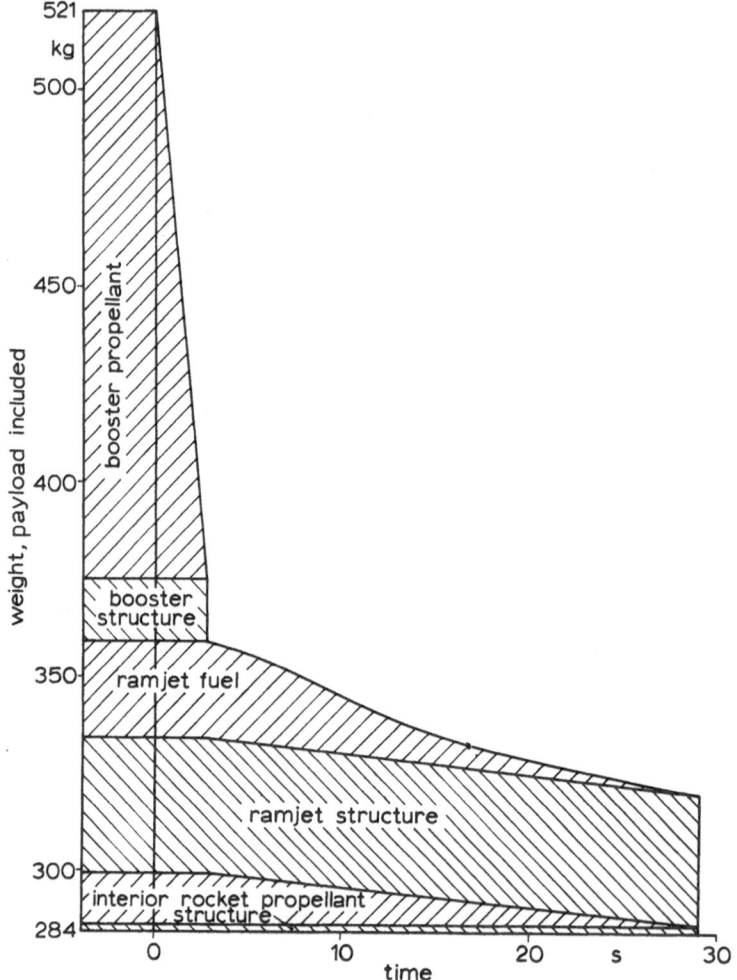

Fig. 9.   Example engine: Weights of components of the missile as a function of time.

$$h_e' + \frac{w_e'^2}{2g} = 10^5 \frac{\text{mkp}}{\text{kg}} \qquad\qquad A_1/A_3 = 0.8$$

$$\alpha = \left(\frac{k \cdot t_e}{m_0}\right)_{\text{Booster}} = 0.28 \qquad A_5/A_3 = 0.8$$

$$\delta = \frac{A_3}{m_s} = 0.02 \frac{\text{m}^3}{\text{kg sec}^2} \qquad \beta = \frac{ak'}{m_s g} = 3.0$$

$$\frac{m_s}{m_0'} = 0.09833$$

are shown for a larger mass flow across the ramjet ($A_1/A_3=0.8$). Optimum payload here is nearly the same as before but the booster must be some larger. If here is a ram rocket taken instead of a pure rocket payload shows a little increase in comparison with the same curve of Figure 6.

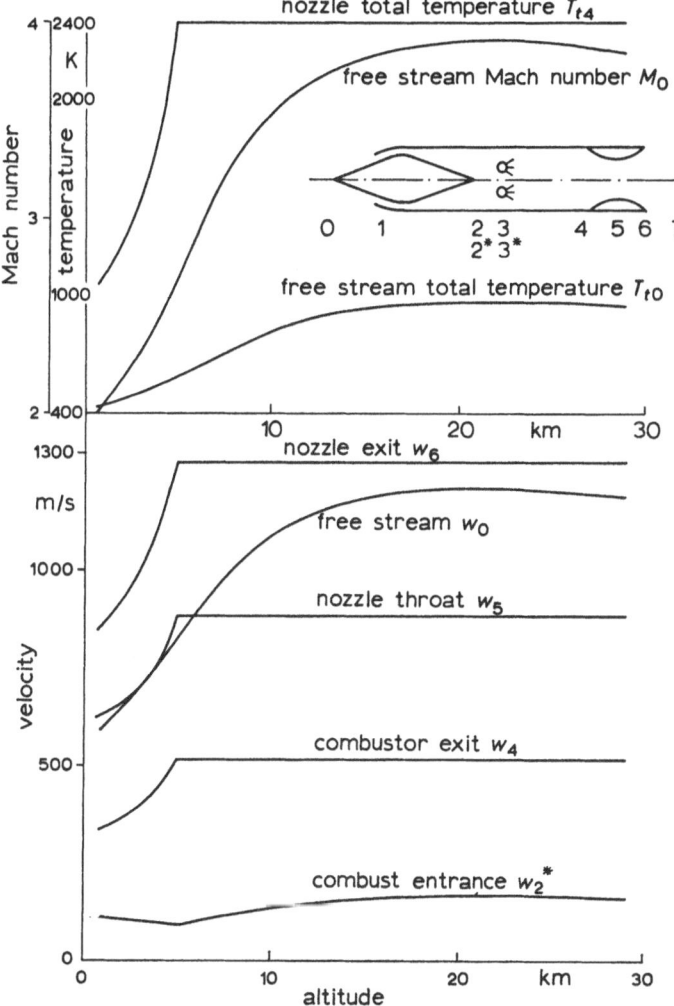

Fig. 10.   Example engine continued: Temperatures and velocities. Fixed datas see Figure 9.

In Figure 8 the different weights of the missile parts are shown for a ram rocket with $\beta = 3$, depending from booster size. We learn from this picture that ramjet fuel is only a little part of ramjet total weight if the booster is not too large.

Figures 9, 10, and 11 are related to a typical engine for the postulated mission (sum of potential and kinetic energy $= 10^5$ m). Booster propellant weight is about 130 kg, booster structure weight 14 kg. Booster thrust is 10000 kp, acceleration at starting 20 g. At booster propellant cut-off that means after somewhat less than 3 seconds the missile has got a velocity of 626 m/sec at an altitude of nearly 850 m. The second acceleration stage drive consists of a ramjet and an interior rocket with 14 kg propellant weight, 1.5 kg structure weight, 104 kp thrust and a flow rate of 0.5 kg/sec. The ramjet has a diameter of 30 cm, a structure weight of 35 kg and a fuel weight of

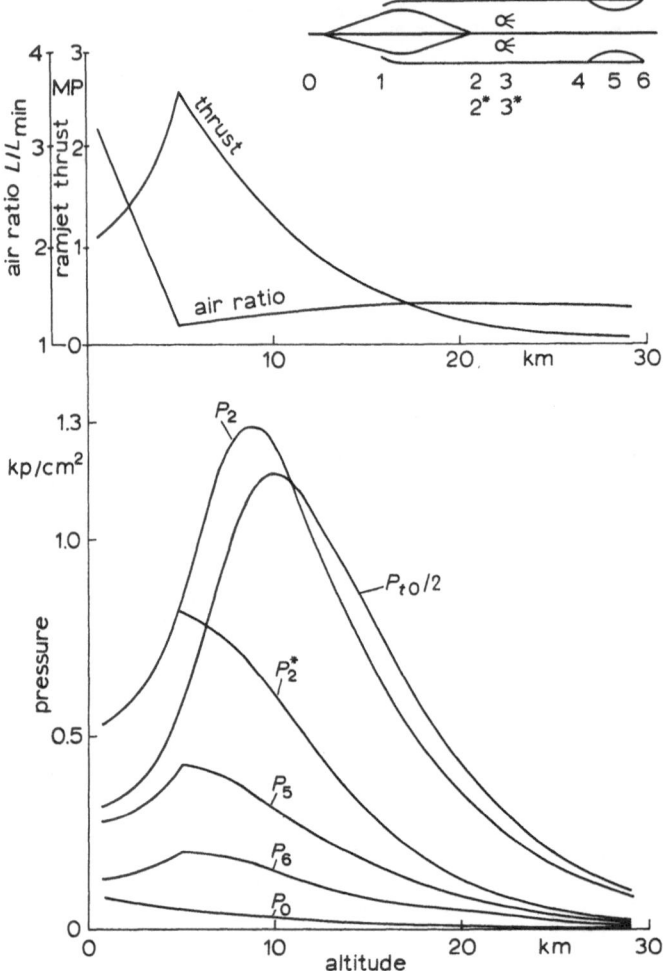

Fig. 11.   Example engine continued: Thrust, air ratio and pressures. Fixed datas see Figure 9.

25 kg. The starting weight of the whole missile is 521 kg, the payload being 284 kg.

In Figure 9 weight variations of the components of the typical engine are shown as a function of time. $\beta = 3$ means that the thrust of the interior rocket is three times the structure weight of the ramjet. As we can see the fuel weight of the ramjet is not so important in comparison with the structure weight.

In Figures 10 and 11, we can see some detail data for the ramjet during flight. At an altitude of somewhat more than 4 km the limiting total temperature of 2400 K is reached in the combustion chamber. Therefore at this point all curves for values of combustion chamber exit or nozzle have an angle. In larger heights density in the combustion chamber is decreasing to keep velocity of sound in nozzle throat. So velocity at combustion chamber entry becomes rather high.

8. Conclusions

It was shown that there may be a gain in payload of 25 to 35 p.c. if in missile acceleration the second stage rocket is replaced by a ram rocket or ramjet. But as the ranges in which the ram rocket has sufficient performance are rather narrow, parameters of the plant must be optimized. Diffusor throat and fuel consumption of the ramjet must be controlled. To get into orbit, upper stages must be added.

References

[1] Schmidt, E., *Thermodynamik*, Springer-Verlag Berlin/Göttingen/Heidelberg, 1960.

[2] Sänger-Bredt, J., *Bemessungstafeln für Staustrahlantriebe*, Part 1 and 2: *Flugtechnik*, Ernst von Olnhausen, Stuttgart, 1959.

[3] Oswatitsch, E., *Gasdynamik*, Springer-Verlag, Vienna 1952.

[4] Lutz, O., 'Diagrammdarstellungen der Vorgänge in Brennkammern und Staustrahltriebwerken', *Yearbook 1955 of the WGL*, pp. 252–265.

[5] Krause, E., 'Entwurfsprobleme supersonischer Einlaufdiffusoren', *DVL-Bericht* No. 198 (June 1962).

[6] Mascitti, V., 'Charts of additive Drag Coefficient and Mass-Flow Ratio for Inlets Utilizing Right Circular Cones at Zero Angle of Attack', NASA-TN-D 3434, May 1966.

[7] Trommsdorff, W., 'Einlauf-Diffusoren im Überschall', *Luftfahrttechnik* 6, No. 12 (1960), 361–368.

[8] Triebnigg, H. and Schmidt, G., 'Zur rechnerischen Erfassung der Einlaufverluste bei Gasturbinen und Staustrahltriebwerken, insbesondere in Überschallflug', *Luftfahrttechnik* 6, No. 12 (1960) 369–374.

[9] Söffker, E., 'Beitrag zur Frage des Lufteinlasses bei Strahltriebwerken für Überschallgeschwindigkeiten', *Jahrbuch 1957 der WGL*, pp. 108–117.

[10] Söffker, E., 'Untersuchungen an einem Überschall-Einlaßdiffusor mit stabilisierendem Ring'. *Z. Flugwelt* 8, Part 2 (1960), 33–44.

[11] Gibbings, J., 'Pressure Measurements on Three Open Nose Air Intakes at Transonic and Supersonic Speeds, with an Analysis of their Drag Characteristics', *Aero Research Council Current Papers* No. 544 (1961).

[12] *U.S. Standard Atmosphere*, U.S. Government Printing Office, Washington 25, D.C., 1962.

[13] Hermann, R. R., 'Supersonic Inlet Diffusors', Minneapolis-Honeywell Regulator Co., Minneapolis, Minnesota (1956).

[14] Trommsdorff, W., 'Untersuchungen an Triebwerkseinläufen', DLR-FB 66–34.

[15] Fabri, J., 'Air Intake Problems in Supersonic Propulsion', *AGARDograph* No. 27, Pergamon Press (1958).

HYBRID MOTORS IN SOUNDING ROCKETS
TECHNICAL AND ECONOMICAL ADVANTAGES

LENNART PERSSON

Svenska Flygmotor AB, Trollhättan, Sweden

Abstract. Conventional sounding rockets use solid propellant motors for their propulsion. The introduction of hybrid motors will lead to considerable decreases in overall cost. The transportation, handling and storage procedures are greatly simplified due to the fact that a hybrid motor is not considered as explosive matter. A hybrid motor may also be designed to stand a severe environment without resorting to conditioning arrangements.

There are also a number of interesting technical possibilities which can be realized using hybrids. These include engine cut-off and re-start and thrust programming.

Svenska Flygmotor AB has started a development program aiming at producing a sounding rocket with a hybrid capable of lifting 20 kg to an altitude of 175 km. The main features of this project are discussed in the report with regard to the above remarks.

1. Introduction

Hybrid rocket motors have been studied actively only during the last ten years. Although a full understanding of the combustion process is not yet achieved several motor projects are now under development. The fact that a motor with such a limited professional background as the hybrid motor can compete with well-established solid or liquid propellant systems shows that the inherent qualities of the hybrid system must offer considerable advantages.

What are these qualities? Some of the features of a hybrid motor system have been listed below. These are

Safe handling

Thrust programme capability

Capability to meet difficult environment requirements

High performance

Low production costs.

Depending on the particular application (missile, space booster etc.) the different features may seem more or less attractive. The competitive situation (hybrid versus conventional motors) may therefore turn out more or less satisfactory from the hybrid motor point of view. For one particular application however, that of a sounding rocket, the hybrid motor offers a superior choice in all respects.

With this in mind Svenska Flygmotor AB has recently started the development of a medium sounding rocket. This rocket called SR-1 uses a prepackaged hybrid motor as the propulsion unit. The liquid oxidizer (nitric acid) is expelled by nitrogen gas from a high pressure tank in a blow-down mode. The solid fuel is a Flygmotor developed amino-plastic.

In the following the arguments which have led to the selection of hybrid motors as the preferred choice for sounding rocket propulsion will be detailed.

G. A. Partel (ed.), Proceedings of the Second International Conference on Space Engineering. All rights reserved.

2. Safe Handling

A hybrid motor, based e.g. on a liquid oxidizer and a solid fuel, will in general not be classified as explosive matter. Such is the case for instance for the propellant combination studied by Svenska Flygmotor AB. This rocket motor can be *transported* with no other restrictions than those which apply to transportation of conventional chemicals. The very special precautions which must be taken when transporting solid propellant motors (classified as explosives) are thus not required. This results in turn in considerable savings in transportation costs.

Solid propellant motors are subject to *storage* restrictions again as a result of the classification as an explosive. The total amount of propellant which may be housed in a certain store is then prescribed. No such restrictions apply to hybrid motors.

The *handling* of a hybrid motor during launch preparations will also gain from the reduction in safety precautions which may be allowed. There are now restrictions at the launch ranges as to e.g. the number of rockets which can be worked on in the same locality. Also work on the launch pad with a hybrid motor rocket will not interfere with other activities. At some ranges the use of radar is now prohibited when staff is attending the rocket in the final launch preparations.

3. Thrust Programme Capability

The hybrid motor lends itself very well to thrust programming. The reason for this is that there are several parameters that govern the internal ballistics of the motor like fuel grain dimensions and burning rate, injector performance, oxidizer mass flow rate etc. These facts can be made use of advantageously already at the design stage. It is therefore a rather simple task to design a hybrid motor to fit to a given envelope.

Variations in maximum altitude can be achieved by *partial tanking* of the rocket motor. This is of interest for instance when it is desired to obtain maximum flight time at a lower altitude.

Thrust *monitoring* in flight can be performed by variation of the oxidizer mass flow. Modulation of the rocket motor thrust by ratios of the order of 8:1 has been demonstrated by e.g. Svenska Flygmotor AB.

Impact dispersion of sounding rockets can be minimized by coasting through the 'jet-stream' part of the atmosphere. From this point of view *engine cut-off* and *re-start* is a very interesting possibility. This can be achieved with hybrids as again has been demonstrated.

Conventional methods for obtaining flight termination in the case of rocket malfunctioning are now based on destruction of the rocket using explosives. From the safety point of view such methods have certain disadvantages; for instance the explosive contained in the sounding rocket always represents a safety risk during handling and launch preparations. An interesting alternative to this technique for obtaining flight termination is offered by the hybrid motor through the possibility of engine cut-off. In the Flygmotor project this is realized by simply venting the nitrogen tank which then leads to combustion and thrust termination. This method has the added advantage that the rocket may be recovered intact (using parachute) which then allows examination of e.g. engine parts.

4. Environment

The environment a sounding rocket meets before and during flight constitutes a vital part of the rocket motor design constraints. Shock and vibration spectrum during transportation and flight and temperature limits encountered during storage, at the launcher and during flight form important requirements.

Solid propellant motors are particularly sensitive to shock and vibration. The formation of cracks in the solid propellant grain frequently leads to explosive rupture of the motor casing. A fair amount of the motor development costs is usually consumed in the efforts to meet the shock and vibration requirements. The situation for the hybrid motor is much less critical. A crack in the solid grain does not produce any effects of significant consequence to the motor.

A similar situation exists vis-à-vis thermal environment. The temperatures which a solid propellant motor may be exposed to affect the burning rate directly. Also, the thermal stresses produced may lead to crack formation with effects similar to those produced by vibration and shock. Again, extensive and expensive development may produce solid propellant grains which meet the given requirements. Very often however the budget for the development of a sounding rocket is limited and other ways of solving the problem have to be looked for. This has led to that a number of the sounding rockets now in use need conditioning arrangements during storage and launch preparations. This in turn means that the actual cost which has to be paid in order to meet temperature requirements is paid no longer by the rocket motor manufacturer as part of the development cost but instead by the user.

Temperature requirements (say storage and use between $-50\,°C$ and $+50\,°C$) may again easily be met with hybrid motors. The solid fuel which has physical properties similar to those of a conventional plastic is not at all as sensitive to thermal stresses as for instance a composite solid propellant which has a very complex mechanical structure. The bond between the solid propellant and its insulation is another source of problems; again tests have shown that hybrid motors are much less susceptible to faulty insulation.

The liquid oxidizer which has been selected (nitric acid) for the sounding rocket SR-1, is also used in prepackaged liquid motor systems produced by Flygmotor. The temperature limits for these systems are $-50\,°C$ to $+65\,°C$.

In short it appears that a hybrid motor is ideally suited to fit environmental requirements for sounding rockets.

5. Performance

Theoretical performance of hybrid propellant combinations are usually in the same class as those of liquid propellants. For the combination Flygmotor has selected

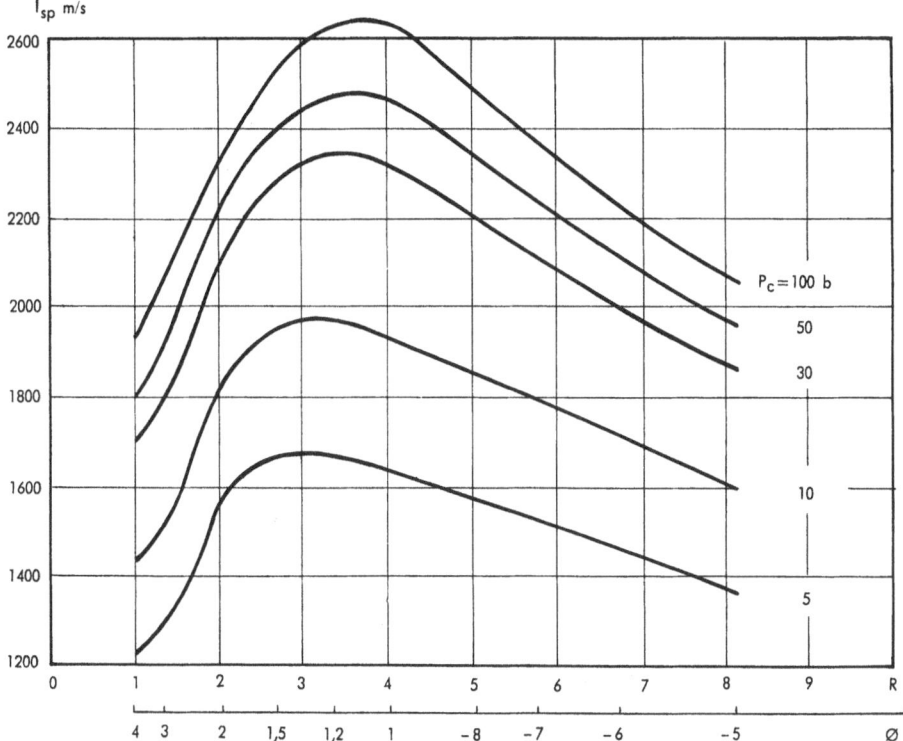

Fig. 1. SAGAFORM-HNO₃: Specific impulse – mixture ratio. Shifting equilibrium; $P_c = 1$ bar.

theoretical specific impulse at 70 bars and expansion to 1 bar is 2550 Ns/kg. (Compare Figure 1.)

6. Flygmotor Sounding Rocket SR-1

As shown a number of arguments in favor of the hybrid motor as a sounding rocket motor can be listed. One single sounding rocket may however not be able to make full use of all the advantages noted. In the case of the Flygmotor sounding rocket SR-1 the following main arguments were used in the selection of the propulsion unit.

Handling: The fact that no handling restrictions apply to the hybrid motor – it does not contain explosive matter – will certainly produce substantial reductions in the total launching costs. The hybrid motor has a 'civilian' touch compared to the solid propellant motor.

Trajectory control: Thrust termination can be accomplished simply by reducing the oxidizer tank pressure. There is thus no need to resort to e.g. explosive devices for destructive purposes.

Environment: Substantial reductions in development cost will result from the fact that the hybrid motor so easily meets difficult environment requirements.

The points above were considered of sufficient weight to choose a hybrid motor in favor of a prepackaged liquid motor although the company has considerably more experience with the latter type of engine. It was also thought that the hybrid motor production costs will be lower than those of the liquid motor.

A major problem in present sounding rocket launchings from European inland ranges is the impact dispersion. Recent ESRANGE launchings have resulted in impacts outside the range in a considerable number of cases. The prominent cause of impact dispersion are the wind gusts at low altitudes. The resulting forces produce a deviation in the initial flight direction. To neutralize this the Flygmotor sounding rocket is equipped with an Attitude Control System which will keep the rocket axis along the selected flight direction during motor burning time.

Main data for the SR-1 sounding rocket are given in the following table which lists dimensions, performance figures, and data on the motor system.

Rocket length	5 m
diameter	0.25 m
launch weight	200 kg
propellant weight	140 kg
pay-load	20 kg
Apogee altitude	175 kg
Altitude at end of burning time	16 km
Time to apogee	202 sec
Maximum velocity	1909 m/sec ($M = 6.5$)
Maximum acceleration	25 g (at end of burning time)
Motor thrust	14.7 kN
Motor burning time	20 sec
Motor total impulse	294 kNs

Figure 2 shows the relations between pay-load and apogee altitude.

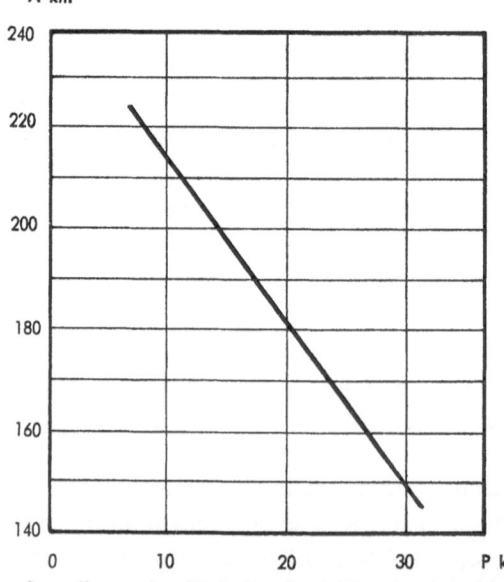

Fig. 2.   Sounding rocket SR-1: Pay-load (P) – Apogee altitude (A).

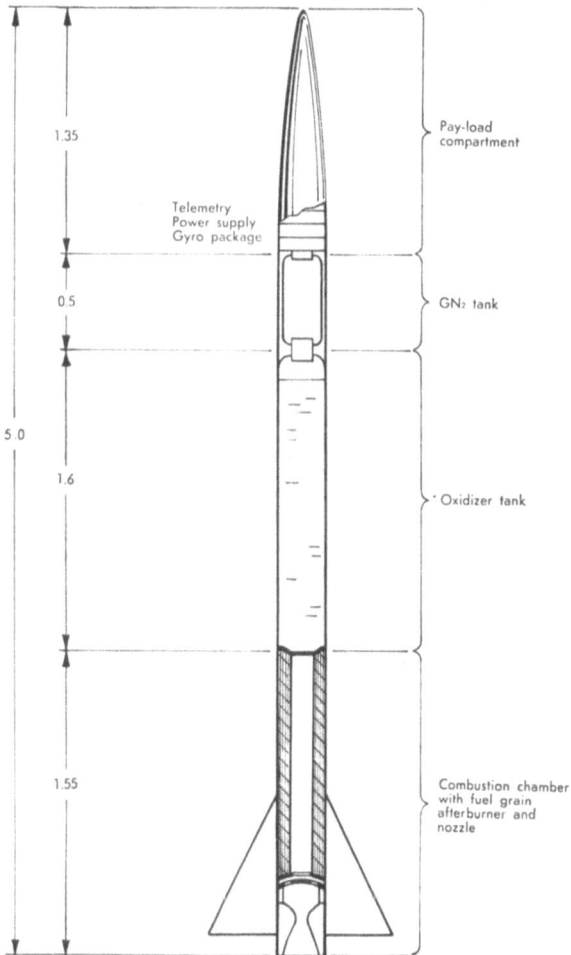

Fig. 3. Flygmotor sounding rocket SR-1.

The SR-1 sounding rocket system consists of the following subsystems
    Structure including fins
    Propulsion unit
    Attitude Control System
    Pay-load.
A lay-out of the rocket is shown in Figure 3.

The structure will be manufactured in aluminum. The engine casing forms a part of the free bearing structure.

The propulsion unit consists of a pressure vessel for gaseous nitrogen, a reduction nozzle with a burst disk, nitric acid tank, injector, combustion chamber with the fuel and the rocket motor nozzle. The secondary injection valves and orifices are located around the supersonic part of the nozzle.

The rocket motor is prepackaged which means that the liquid propellant is contained in a closed tank which is filled at manufacture. There are thus no tanking procedures at the range. Present Flygmotor prepackaged systems have a storage life of six years.

The attitude control system (ACS) is based on one attitude gyro and a differentiating device which provide control signals to a secondary injection thrust vector control system. The medium injected in the nozzle will be nitric acid taken from the oxidizer tank.

The pay-load compartment will house a standard telemetry unit (partly for housekeeping), power supply and gyro package in addition to the scientific experiment. A total volume of approximately 50 litres is available in the compartment.

Operation of the system starts by the rupture of the burst disc in the transfer tube which connects the nitrogen tank with the oxidizer tank. Gas from the high-pressure GN_2 tank (210 bars) then pressurises the nitric acid to a pressure of 40 bars. After the rupture of another burst disc the oxidizer flows into the combustion chamber where

Fig. 4. Experimental sounding rocket HR 3-F.

a reaction starts spontaneously with the solid fuel. The high-temperature combustion gases (3100 K) then generate thrust when passing the rocket nozzle and the rocket consequently starts moving.

The rocket will be launched from a simple rail launcher. It is intended to adapt the rocket to already existing launchers. As the ACS starts operating immediately the length of the launcher may be very short.

7. Flygmotor Hybrid Motor Experience

Hybrid motor technology has been studied at Flygmotor since 1962. The first phase of the work covered development of a solid propellant hypergolic with the selected oxidizer (nitric acid). Ignition phenomena, combustion mechanism and tank pressurization technique where covered during this first phase. In 1965 a sufficient background existed for the undertaking of the development of a complete rocket system. In the fall of that year two small sounding rockets (Figure 4) were launched from a military

Fig. 5.   Recovered rocket.

test range. The experiments proved a full success. Parachute recovery functioned satisfactorily as shown in the Figure 5 where a helicopter is shown carrying one of the rockets back after impact in water.

Further development of the solid fuel gas dominated the continued studies. Improvements in propellant grain strength and in storage life times combined with increased burning rates have been the major achievements. The present fuel now has a performance in these respects which is fully satisfactory for use in e.g. military missile projects.

Work on the sounding rocket SR-1 started a year ago with a pre-project study. This phase has now been completed and the actual development programme has started. Anticipated development time including first flight tests is two years. The programme is at the moment a company-sponsored venture.

EVOLUTION OF REACTION CONTROL ENGINES
FOR U.S. MANNED SPACECRAFT

DWIGHT A. MOBERG

The Marquardt Corp., Van Nuys, Calif., U.S.A.

Abstract. Recent manned-spacecraft achievements have highlighted the importance of small re-action-control rocket engines in the successful accomplishment of such missions. Successive generations of U.S. manned spacecraft – Mercury, Gemini, and Apollo – have utilized reaction-control engines of increasing sophistication and capability. As missions have become longer and more complicated, a corresponding improvement has been required in engine characteristics such as operating life, minimum impulse bit, duty cycle flexibility, pulse performance, and reliability.

In this paper, the reaction-control engines used on Mercury, Gemini, and Apollo are described, and significant design characteristics and performance parameters are compared. In-flight experience is also discussed, to the extent of data availability.

1. Introduction

Five to ten years ago, most attention in the aerospace propulsion field was focused on the development of very high thrust engines that were needed to boost larger payloads into orbit. Recently, with manned orbital flight becoming almost routine, increasing emphasis is being placed on maneuverability of the spacecraft, and the propulsion devices employed for this purpose. The purpose of this paper is to review the evolution of the reaction control engines (thrustors) that have been used to date on U.S. Manned Spacecraft – Mercury, Gemini and Apollo. This review will include a discussion of the different types of system installations used, engine design features, critical performance parameters, and a summary of flight test experience.

2. Discussion

A. MISSION REQUIREMENTS

The driving force for improvement of reaction control engine capability has been the development of spacecraft of increasing weight, for missions of increasing complexity and duration. The trend in spacecraft weight is shown in Figure 1. In slightly less than 8 years, the launched weight has grown by a factor of more than 30. The complexity of missions flown has similarly increased, as indicated in Figure 2. The Mercury missions basically required only that the reaction control system orient the spacecraft during a brief stay in earth orbit, and then provide attitude control during reentry. The Gemini and Apollo programs, however, have been directed at longer duration missions requiring the reaction control system to perform numerous functions, including rendezvous and docking maneuvers. The Apollo mission also requires reaction control activities associated with the lunar landing itself. A measure of the increased utilization of reaction control engines is the total impulse available from the Mercury, Gemini, and Apollo reaction control systems (Figure 3). In this case, there has been a

G. A. Partel (ed.), Proceedings of the Second International Conference on Space Engineering. All rights reserved.

U. S. MANNED SPACECRAFT

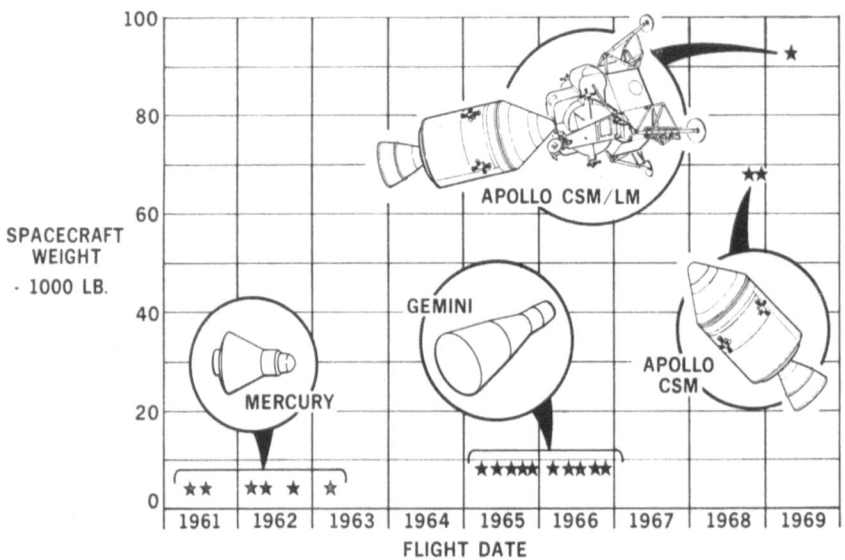

Fig. 1.

REACTION CONTROL SYSTEM MANEUVERS

	MERCURY	GEMINI	APOLLO
• SEPARATION FROM BOOSTER			X
• ABORT MANEUVER			X
• ULLAGE POSITIONING		X	X
• RATE DAMPING	X	X	X
• SETUP ORBIT ATTITUDE	X	X'	X
• IN-PLANE ORBIT CHANGE		X	X
• ORBITAL PLANE CHANGE		X	X
• ON-ORBIT ATTITUDE HOLD	X	X	X
• RENDEZVOUS MANEUVER		X	X
• PASSIVE THERMAL CONTROL			X
• NAVIGATION HOLD			X
• MID-COURSE CORRECTION			X
• ATTITUDE CONTROL FOR LUNAR LANDING/TAKEOFF			X
• TRANSLATE OVER LUNAR SURFACE			X
• SEPARATION FOR REENTRY		X	X
• RETRO ATTITUDE HOLD	X	X	X
• REENTRY POSITIONING	X	X	X
• REENTRY RATE DAMPING	X	X	X

Fig. 2.

REACTION CONTROL SYSTEM TOTAL IMPULSE

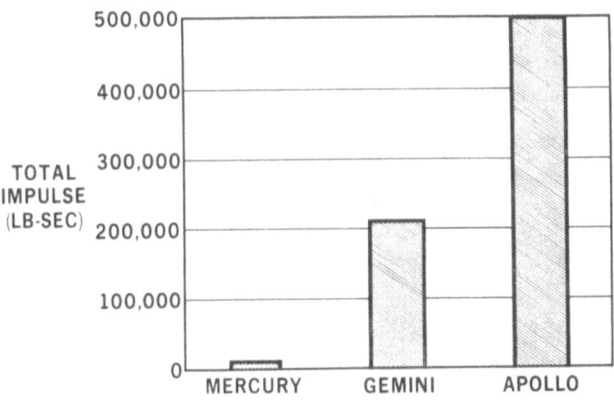

Fig. 3.

growth in capability by a factor of 70. This growth has been due not only by an increase in the weight of RCS propellant carried, as suggested by the increase in overall spacecraft weight, but is also due to the use of more energetic propellants and the use of reaction control engines that are more efficient in the pulsing mode of operation.

B. SYSTEM INSTALLATIONS

The reaction control system installations on the various spacecraft are illustrated in Figures 4 to 8. For the Mercury spacecraft, shown in Figures 4a and 4b, eighteen reaction control engines were used. Of these, six were 1 lb thrust, four were 6 lb thrust, and eight were 24 lb thrust. These were grouped into two systems, superimposed on the same spacecraft structure, with each system providing for rotational control about the pitch, roll, and yaw axes. One of the groups of thrustors (the 12 thrustor Automatic System shown in Figure 4A) could be controlled by use of either the ASCS (Automatic Stabilization and Control System) or the Fly-By Wire mode, with direct, control by the astronaut. In both modes, the thrustors were operated by electrical actuation of solenoid valves.

The other group of thrustors (the 6 thrustor Manual System shown in Figure 4b) was operated directly by the astronaut through pushrod linkages between his control stick and mechanical throttle valves upstream of each thrustor. Four 24 lb thrustors and two 6 lb thrustors were connected in this manner, and thrust was varied by proportional movement of the control stick. On certain of the Mercury flights, another control mode, RSCS (Rate Stabilization and Control System), employed the Manual System to provide auxiliary damping as a backup for the ASCS, which normally controlled this function.

For the Gemini spacecraft, reaction control was achieved through the use of two independent systems; the OAMS (Orbit Attitude and Maneuver System), and the

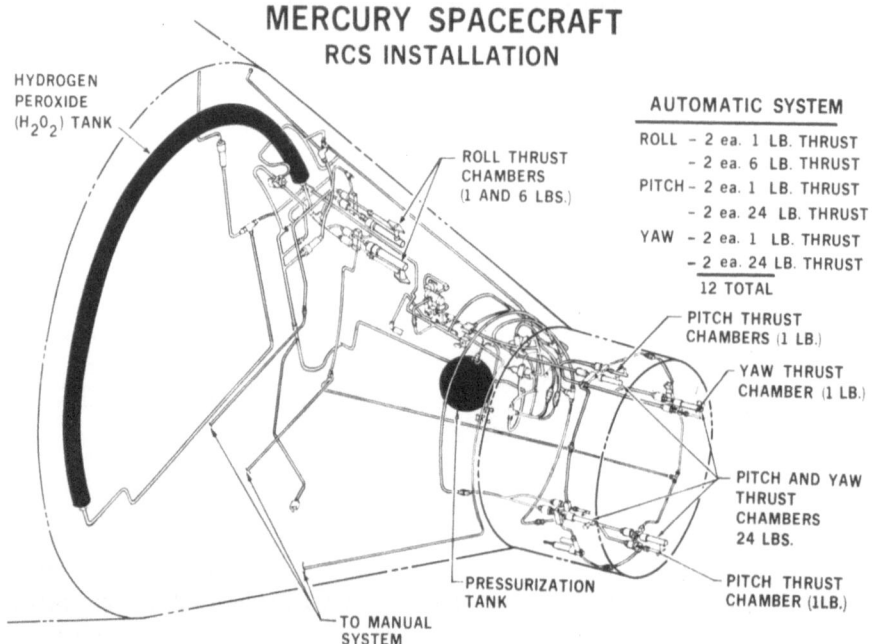

Fig. 4a.

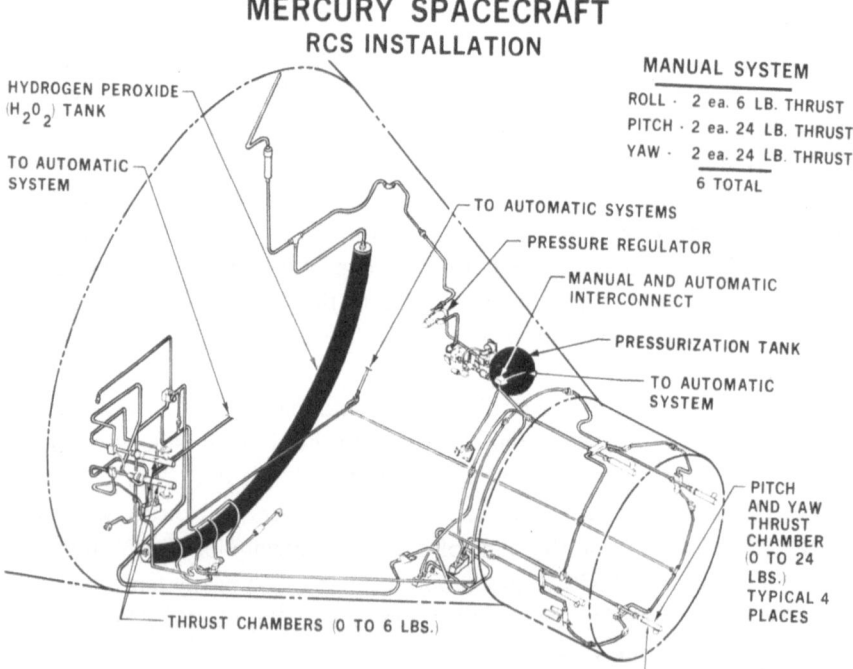

Fig. 4b.

RCS (Reentry Control System). As shown in Figure 5, the OAMS engines were flush mounted around the periphery of the spacecraft Adapter Section, and the RCS engines were similarly mounted around the Reentry Section. The OAMS system provided for spacecraft rotation and translation while in orbit. Rotational control was provided by eight 23.0 lb thrust engines, while translation maneuvers were accomplished by two 79.0 lb thrust engines and six 94.5 lb thrust engines. The RCS, used for rotational control during reentry, employed sixteen 23.5 lb thrust engines, connected in two completely redundant systems. All Gemini thrustors were operated by electrical control. Each thrustor propellant valve was equipped with two independent solenoid windings, providing for control circuit redundancy. Both OAMS and RCS engines used the hypergolic propellant combinations of nitrogen tetroxide oxidizer and monomethyl hydrazine fuel.

The reaction control system installations on the Apollo spacecraft are presented in Figures 6, 7, and 8. As shown in Figure 6, the astronaut's Command Module is maneuvered through the use of twelve, flushmounted, 93 lb thrust engines, which provide rotational control around the pitch, roll, and yaw axes. These engines are connected in redundant systems of six engines each, with either system capable of the required reentry control maneuvers. As in Gemini, all of the thrustor propellant valves on Apollo are equipped with two electrical solenoids, and the astronaut can elect control through either circuit. In one case, control is accomplished by the SCS (Stabilization Control System) and in the other directly by the astronaut's control stick.

GEMINI SPACECRAFT
OAMS/RCS INSTALLATION

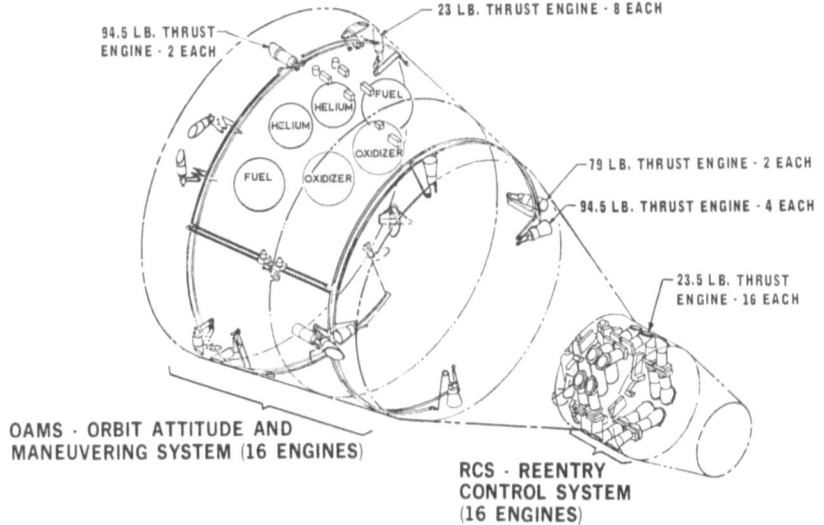

Fig. 5.

APOLLO COMMAND MODULE
RCS INSTALLATION

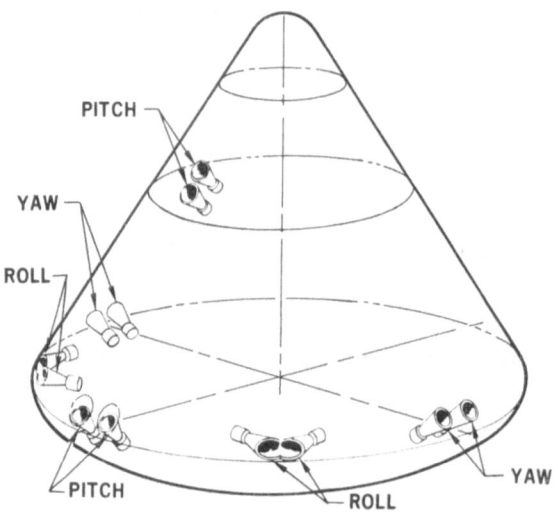

93 LB. THRUST ENGINES · 12 EACH

Fig. 6.

APOLLO SERVICE MODULE
RCS INSTALLATION

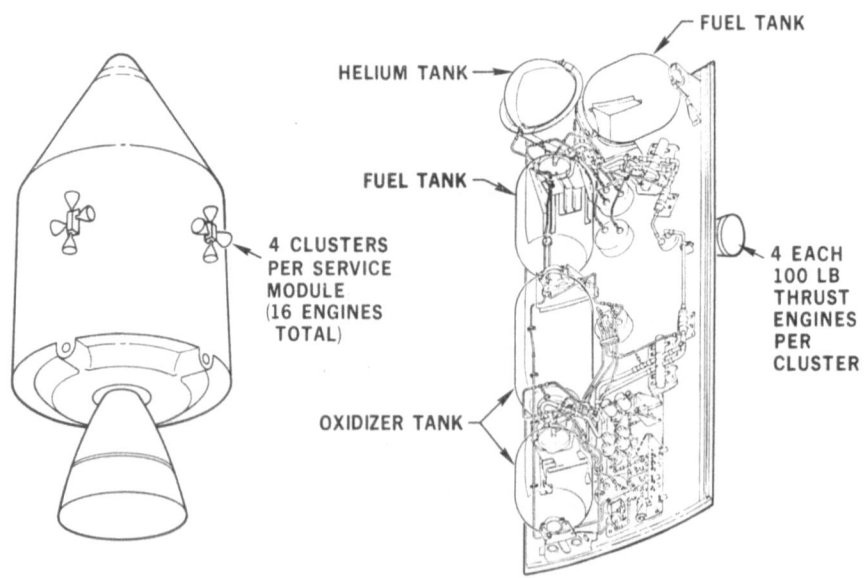

Fig. 7.

APOLLO LUNAR MODULE
RCS INSTALLATION

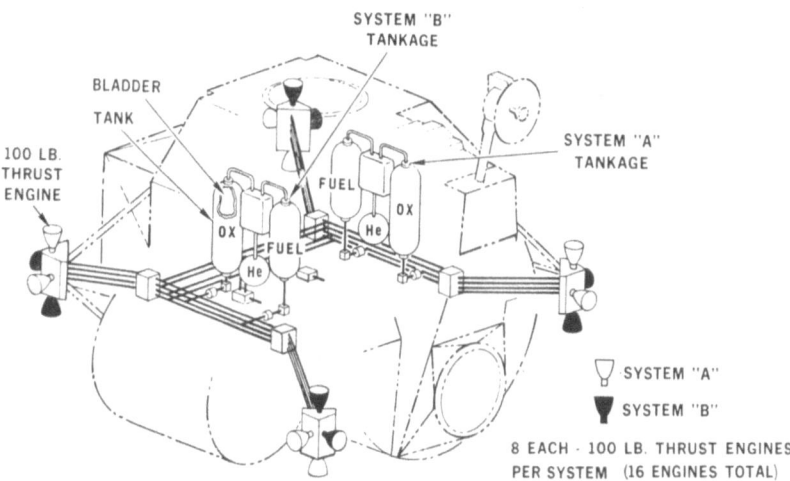

Fig. 8.

The Service Module Reaction Control System is illustrated in Figure 7. Sixteen 100 lb thrust engines are mounted externally in clusters of four, at 90° intervals around the perimeter of the spacecraft. Each cluster has its own propellant supply and pressurization system, mounted on a panel which can be removed from the spacecraft for ground maintenance. For both Command and Service Modules, the reaction control engines employ the nitrogen tetroxide/monomethyl hydrazine propellant combination.

Figure 8 shows the reaction control system used for the Apollo Lunar Module. Again, sixteen engines are employed in clusters of four, in this case mounted on booms at 90° intervals around the spacecraft. Propellant lines lead along each boom to redundant tankage within the spacecraft. Two engines from each cluster are connected to one of the two independent systems, of eight engines each. An interconnect is provided between the RCS tankage and the tankage for the Ascent Engine. Propellants used are nitrogen tetroxide and a 50/50 blend of hydrazine and unsymmetrical dimethyl hydrazine.

C. REACTION CONTROL ENGINE DESCRIPTIONS

The Mercury thrustors, illustrated in Figure 9, provided thrust through the decomposition of 90% hydrogen peroxide. Each thrustor consisted of a propellant valve, heat barrier, metering orifice, and a stainless steel thrust chamber. The radiation-cooled thrust chamber enclosed a distribution disc, catalyst screen bed, plenum chamber, and exhaust nozzle. The peroxide decomposition products, at 1380 °F, were expanded through a 15:1 area ratio nozzle. The thrustor components were scaled up as required to provide 1, 6, and 24 lb of thrust.

MERCURY THRUSTORS

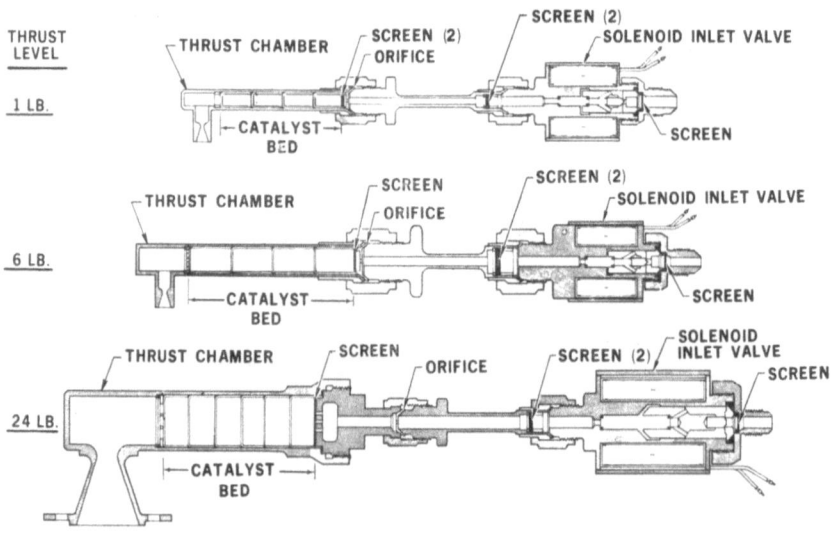

Fig. 9.

A significantly different type of thrustor, shown in Figure 10, was developed for the Gemini program. Aerodynamic considerations required that a submerged thrustor installation be employed. In order to limit thrustor exterior wall temperature, so that surrounding spacecraft equipment would not be overheated, an ablative material was used for the thrust chamber wall. Heat is absorbed by the chamber wall as the ablative material chars and releases gaseous products back into the exhaust stream. Chamber life is limited by the rate at which the ablative material chars. Thus, the chamber weight is strongly related to the required operating life of the thrustor. At the chamber throat, a silicon carbide insert is used to provide increased erosion resistance. The thrust chamber is encased in a thin-walled stainless steel jacket, which, with a scarfed exit nozzle provides for flush mounting with the spacecraft skin. Each thrust chamber is equipped with two solenoid-operated valves which control propellant flow to the injector. For the 23.5 lb thrust RCS engine and the 23.0 lb thrust OAMS engine, these are fast response valves. Together with the small injector manifold volume employed, this allows for the very short pulse widths which are required for precision attitude control.

Like the Gemini thrust chambers, the Apollo Command Module thrustor, shown in Figure 11, is ablatively cooled. The design approach employed for this 93 lb thrust bipropellant engine was generally quite similar to that used for Gemini. Fast response propellant valves were employed to provide for the pulse mode operation needed for precise orientation during the reentry maneuver. A relatively short life was required for this engine, since the command module RCS is activated only for this final phase of the mission. These thrustors also provide a backup capability for deorbit Delta V.

A substantially different design approach was used for the reaction control engines

GEMINI THRUSTORS

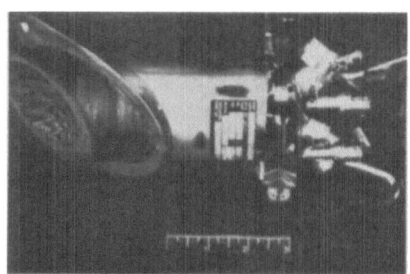

RCS - 23.5 LB. THRUST -16 EACH

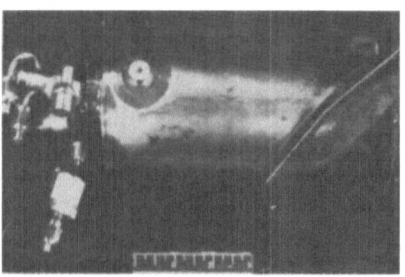

OAMS - 23.0 LB. THRUST - 8 EACH

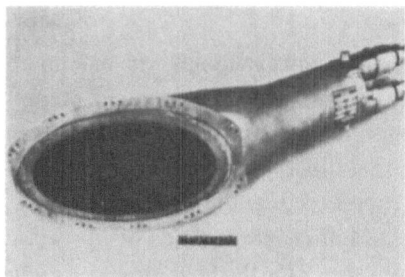

OAMS - 79 LB. THRUST - 2 EACH

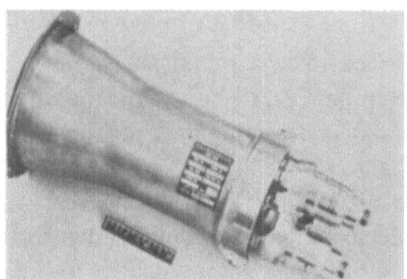

OAMS - 95 LB. THRUST - 6 EACH

Fig. 10.

APOLLO COMMAND MODULE THRUSTOR

100 LB. THRUST - 12 EACH

Fig. 11.

for the Apollo Service Module and Lunar Module (Figure 12). Since the mission allowed for mounting of this engine external to the spacecraft, a radiation-cooled thrust chamber could be employed. A lightweight molybdenum chamber with an oxidation resistant disilicide coating was developed for this purpose. Although this chamber can be operated at wall temperatures as high as 3100°F, an injector design was selected which would operate the chamber at 2000°F, resulting in effectively unlimited chamber life. Fast response propellant valves were also developed for this engine to provide short pulse width for precision attitude control of the Service Module and Lunar Module, in addition to translation maneuvers requiring long burn times. Dual solenoid windings were used on the valves to allow for control of the engines by either the SCS or the astronaut. The weight of this highly flexible engine is only 5 lbs compared to the 8 to 9 lb weight of the Command Module or Gemini engines of similar thrust level.

D. ENGINE OPERATING CHARACTERISTICS

One of the more significant characteristics of a reaction control engine is the maximum operating life, or burn time, for which it can be expected to function reliably. Representative operating lives for the Mercury, Gemini, and Apollo engines are shown in Figure 13. Each of these engine designs is limited in a different way, relative to maximum operating life. The Mercury thrustor's catalyst bed gradually becomes less effective in decomposing the hydrogen peroxide, and after 900 seconds of operating life, the thrust produced is 9% less than its original value. For the Gemini and Apollo

APOLLO
SERVICE MODULE / LUNAR MODULE THRUSTOR

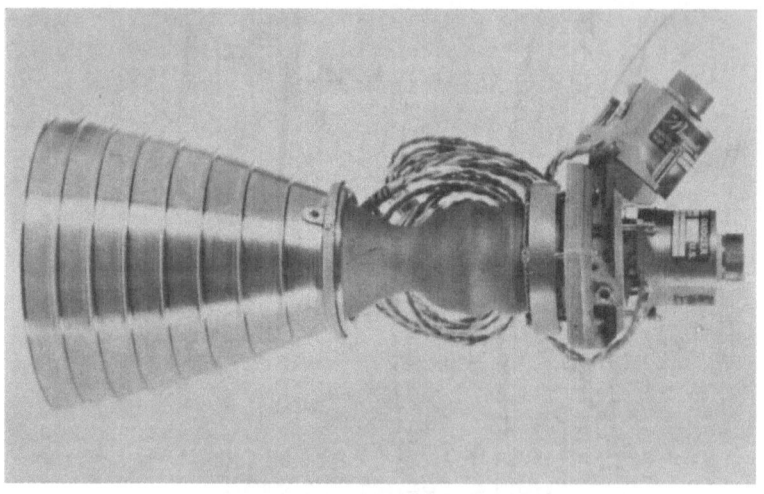

100 LB. THRUST · 16 EACH

Fig. 12.

THRUSTOR OPERATING LIFE

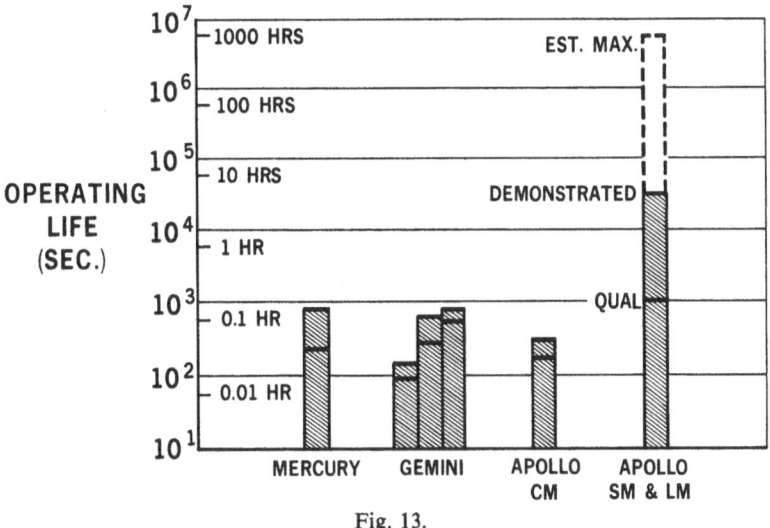

Fig. 13.

Command Module engine designs, operating life is limited due to charring through of the chamber wall and/or eventual erosion of the throat insert. Certain of the Gemini thrustors were operated at very fuel-rich mixture ratios in order to extend the allowable burn time. The radiation-cooled engine used for the Service Module and Lunar Module systems, however, has an expected operating life of approximately 1000 hours, or four orders of magnitude greater than the ablative engine designs. Since the maximum combustion chamber temperature for this design is only 2000°F, degradation of chamber coating takes place very slowly. Thus, excess margin is available for more extensive future missions.

Another vital characteristic of the reaction control engine is the ability to make many starts in a space environment. For the brief Mercury mission, less than a thousand starts were required of the thrustor. As shown in Figure 14, thousands of starts are now required of present day engines. Further increases in this requirement can be anticipated in direct proportion to mission life. A potential limiting factor will be wearout of the fast response propellant valves in which the wearing and sealing surfaces must function properly during long-term exposure to corrosive propellants.

The trend toward use of fast response reaction control engines is shown in Figure 15. The order of magnitude reduction in response time from that of the Mercury thrustors was due primarily to the switch in propellants, since the relatively slow peroxide decomposition process was replaced by the near-instantaneous hypergolic ignition of the bipropellant systems. However, improvements in solenoid valve design and care in minimizing injector manifold volume have also contributed to faster thrustor

THRUSTOR CYCLE LIFE

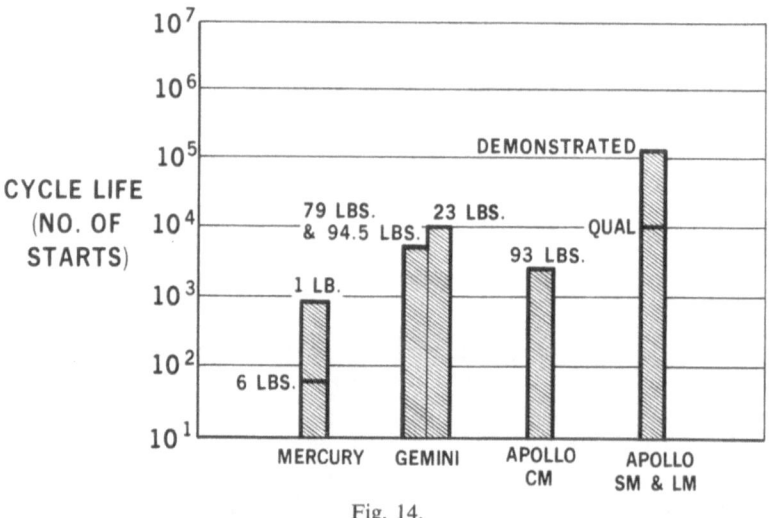

Fig. 14.

Fig. 14.

THRUSTOR RESPONSE
TIME FROM ELECTRICAL SIGNAL TO 90% THRUST

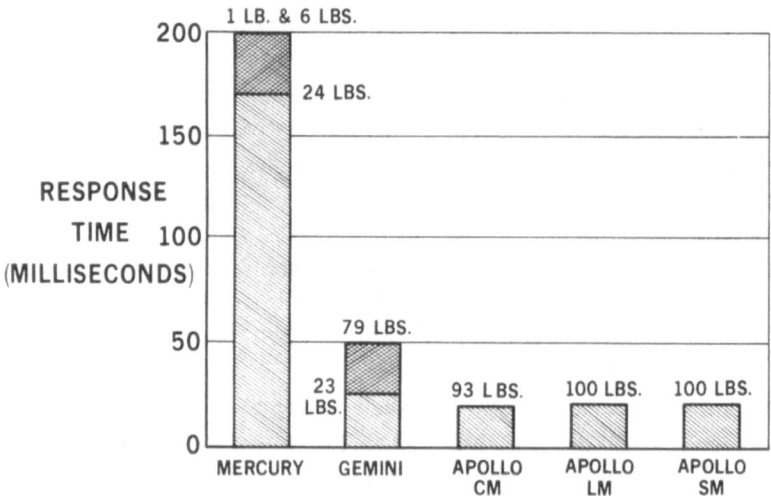

Fig. 15.

response. Because of such improvements, the 100-lb thrust engines on the Service Module and Lunar Module can be used for both the steady and pulsing requirements of the mission, eliminating the need for separate special purpose engines.

The change to a more energetic propellant system for Gemini and Apollo was made to provide a substantial increase in reaction control engine specific impulse, in order that the weight of RCS propellant carried for these more ambitious missions not be prohibitive. Figure 16 shows the substantial gain that was achieved in delivered specific impulse as a result of this change. In terms of steady state performance, the improvement has been on the order of 70 percent. The increase in performance at short pulse widths has been even more substantial. During operation at 100 milliseconds, performance of the Mercury thrustors was strongly dependent on duty cycle. When the catalyst bed was cold, performance was poor, but improved with repeated pulsing, which raised the catalyst bed temperature and insured rapid decomposition of the peroxide. At very short pulse widths, no thrust was produced on the first few pulses commanded. For the hypergolic bipropellant reaction control engines, however, pulsing performance has been shown to be essentially independent of duty cycle. The performance levels realized depend principally on the minimization of injector manifold volume, use of fast response injector valves, and employment of low L^* combustion chambers. By applying these techniques, delivered pulse specific impulse for bipropellant engines operating can be held at a level which is typically greater than the delivered steady state specific impulse for peroxide monopropellant thrustors.

E. MANNED FLIGHT TEST RESULTS

A total of 19 manned spaceflights (6 Mercury, 10 Gemini, 3 Apollo) have now been made by the United States. Performance of the reaction control systems, and engines, used on these flights is summarized in Figure 17. Some degree of problem with the

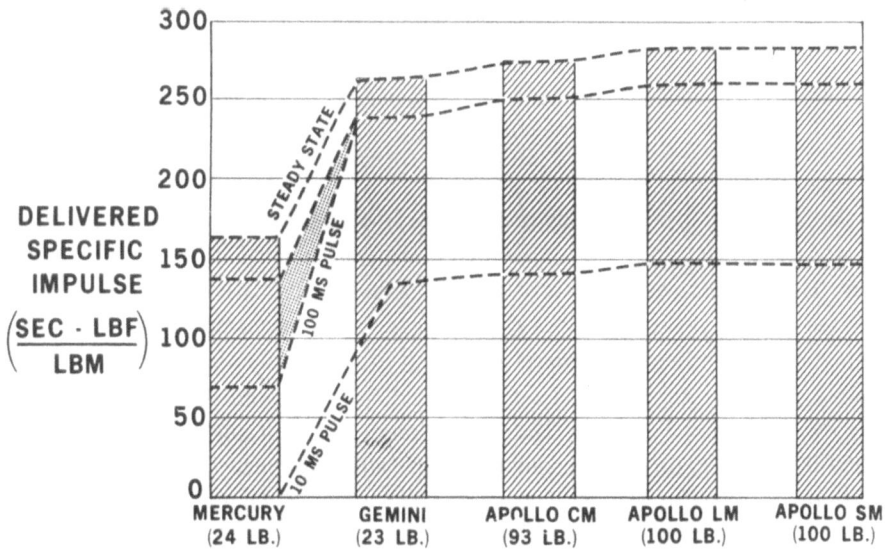

Fig. 16.

MANNED FLIGHT TEST EXPERIENCE
REACTION CONTROL SYSTEMS

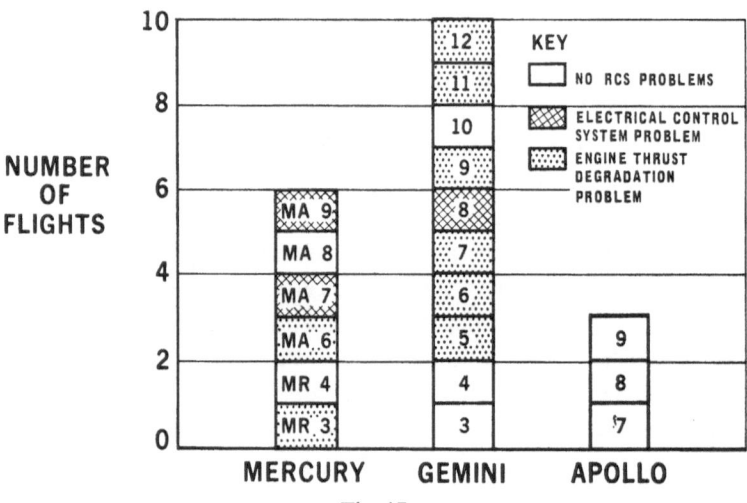

Fig. 17.

overall reaction control system has been experienced in 11 of these flights. Three of these problems were caused by malfunction of electrical components or circuitry in the control system for the thrustors. These problems were as follows:

Flight	Problem
Mercury MA-7	Pitch horizon scanner malfunctioned resulting in improper ASCS operation.
Mercury MA-9	Short circuit at autopilot power plug disabled ASCS.
Gemini 8	Short Circuit in control circuitry caused an OAMS 23 lb thrustor to fire continuously; eventually caused early termination of mission.

In eight of the remaining missions, problems were experienced in varying degrees with the reaction control engines themselves. In all cases, the problem was a degradation of the thrust produced by one or more of the engines in the reaction control system. Troubleshooting of the thrustor problems on the early Mercury flights was facilitated by direct examination of the hardware from the recovered spacecraft. Such examination showed small particles lodged at critical points in the thrust chamber assembly. Some of these particles were found to be broken pieces of the catalyst screen. Minor design changes to the screen design and metering orifice location were successful in eliminating this problem on the later Mercury flights. Thrust degradation also occurred to varying degrees during 6 of the 10 Gemini flights. As was the case for Mercury, the spacecraft pilot was able to make use of redundant thrustors, or systems, to successfully complete the mission. The Gemini thrust degradation problems were believed to have been the result of propellant flow reduction due to contamination blockage of small thrustor orifices. Based on Mercury and Gemini experience in this regard, improvements in engine and propellant supply system design and in overall cleanliness procedures have been made for the Apollo program. Initial Apollo flight tests have shown 'as planned' reaction control engine performance, indicating that the corrective action taken has been effective in preventing such problems.

3. Conclusions

(1) Increases in spacecraft weight and in the complexity and duration of missions to be flown have required the development of reaction control engines of correspondingly increased capability.

(2) Both monopropellant and hypergolic bipropellant reaction control engines have been successfully developed and flight tested. Combustion chamber cooling has been accomplished by both ablative and radiation cooling techniques. Fast response propellant valves, suitable for long-term service in corrosive propellants, have been developed to permit operation at very short pulse widths.

(3) For Mercury and Gemini, performance and operating characteristics of the reaction control engines were adequate for the planned missions. In the case of Apollo, the engine developed for the Service Module and Lunar Module application possesses considerable margin for use in longer missions.

(4) Flight test results, relative to reaction control system and engine performance have typically demonstrated accomplishment of overall mission objective, although in several cases mission success was enabled because of system redundancy, or astronaut corrective action. Improvements in contamination control techniques, and in engine and system design approach, have been required to avoid reduction of propellant flow and subsequent thrust degradation because of clogging of critical orifices.

References

[1] Purser, P. E., Faget, M. A., and Smith, N. F., *Manned Spacecraft: Engineering and Operation*, Fairchild Publications, Inc., New York 1964.
[2] Kleinknecht, K. S., Bland, W. M., *et al.*, 'Mercury Project Summary, Including Results of the Fourth Manned Orbital Flight, May 15 and 16 1963', NASA Report SP-45, Office of Scientific and Technical Information, National Aeronautics and Space Administration, Washington, D.C., October 1963.
[3] Malik, P. W. and Souris, G. A., 'Project Gemini, A Technical Summary', NASA Contractor Report CR-1106, National Aeronautics and Space Administration, Washington, D.C., June 1968.
[4] Grimwood, J. M., 'Project Mercury, A Chronology', NASA Report SP-4001. Office of Scientific and Technical Information, National Aeronautics and Space Administration, Washington, D.C. 1963.
[5] Gatland, K., *Manned Spacecraft*, The Macmillan Company, New York, 1967.

PART VI

PROPULSION, II

Chairman: H. G. S. Murty, India

THE MORPHOLOGICAL CONTINUUM IN
SOLID PROPELLANT GRAIN DESIGN*

JOHN S. BILLHEIMER

Aerojet-General Corp., Sacramento, Calif., U.S.A.

and

F. R. WAGNER

University of Utah, Salt Lake City, Utah, U.S.A.

Abstract. Grain configurations for solid propellant rockets are classified by relative web thickness and mean vector direction of burning surface into a topological continuum. This ranges from the thin web dendrite (web equal to $\frac{1}{8}$ of charge radius and entirely in the cross-section plane) thru the wagon-wheel- and star-perforated grains ($\frac{1}{4}$ to $\frac{1}{2}$ web range and partial use of end effects in burning surface area control) to the slotted, conocyl, and finocyl grains (web 0.6–0.8 of radius and burning front partially in the axial direction). These geometrical principles relate to the mission by the ratio of thrust-to-duration squared (F/t^2) which requires a dendrite grain for $F/t^2 \approx 3000$ lbf/sec^2 and a slot or finocyl for $F/t^2 \approx 30$ lbf/sec^2. This effect is counterbalanced by the range of burning rates available. Burning rate, relative web thickness, chamber pressure, length-to-diameter ratio, and volumetric loading affect F/t^2 attainable in a descending significance. The prevailing style of grain design in any era, although optimized mathematically within itself, depends more on technological breakthroughs in materials and propellant properties, than on factors of ballistic performance. Grain design is primarily a graphic subject. There are two aspects: performance attributes and description of the grain configuration.

1. Preface

A. FUNCTIONAL CHARACTERISTICS IN GRAIN DESIGN

Certain definitions are essential to the analysis of solid propellant grains. It is said the solid surface *regresses*, or the flow space *progresses*, as the charge burns. The grain may be shaped so that the burning surface area remains constant (*neutral-burning*), increases with time (*progressive-burning*), or decreases with time (*regressive-burning*). In a star-perforated grain, this depends on the *fillet angle* between the attachment of the wedge or protrusion to the case-bonded *web*, which is the minimum burning distance from the *port space* or flow channel to the chamber wall. After the web is burnt out, the regression of the burning surface recedes along the chamber wall. This portion of the propellant is called the *attached sliver*, and its rapidly decreasing burning surface results in a *tail-off* in the pressure-time curve. Neutrality, progressivity, and regressivity are zero, positive and negative values of the first derivative of the performance curve respectively.

Figure 1 shows a concave performance curve (positive second derivative) which amplifies the significance of *terminal acceleration* and *erosive-burning*. This is typical of a star-perforated grain with web greater than the *critical burning distance* (y^*). This results in adverse initial and final pressure peaks unless corrected by an (1) outside

---

* Prepared for the NASA Design Criteria Monograph program under Contract NAS3–11179.

round to give deficient initial surface area to allow for erosive-burning, and (2) *end-effects* to compensate for terminal progressivity.

Figure 2 defines a desirable convex performance curve (negative second derivative), as obtainable in anchor, slot, and multipropellant grain designs. This favors absorption of the erosive burning peak and minimizes terminal acceleration.

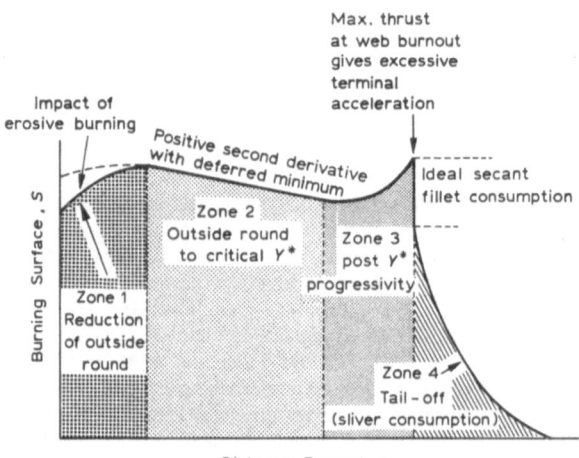

Fig. 1.   Schematic for two-dimensional star-perforated grain with web $> Y^*$.

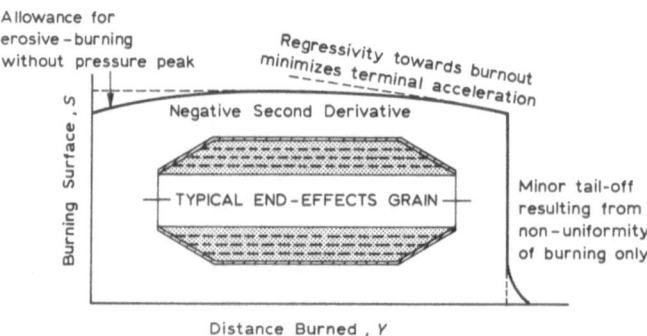

Fig. 2.   Advantage of convex performance curve from bipropellant, slotted cylinder, or retractive-burning in head effects.

Figure 3 shows progressive, neutral, and regressive thrust-time curves. For a linear surface area – distance burnt curve, the exponential relation of burning rate to chamber pressure results in a slight curvature [the K-ratio transformation –

$$P/P_0 = (S/S_0)^{1/1-n}].$$

To evaluate the effect of performance curve shape on performance of the missile, this figure defines a 'configuration efficiency'. This is the ratio of actual delivered thrust under the performance curve until action time, divided by that obtainable in an ideal

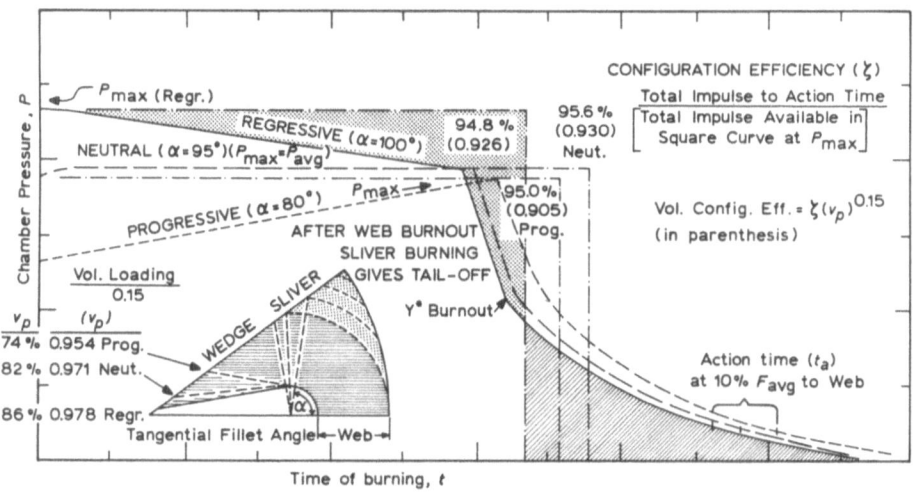

Fig. 3.   Configuration efficiency for progressive, neutral, and regressive grain designs.

square curve at the maximum operating pressure permitted by the chamber (i.e., the maximum occurring anywhere in the curve). This coefficient incorporates the effect of both sliver and neutrality, and is thus more explicit than the peak-to-average pressure ratio often used.

These characteristics will be used to arrange the various configurations in a logical morphological procession in accordance with the web thickness, from the two-dimensional dendrite-wagonwheel-star set to the three-dimensional slotted cylinder and conocyl grains. In the first section the nature of the burning surface mechanism will be introduced, and the logic of burning of various surface contours will be established.

2. Introduction

A. GEOMETRICAL DESIGN OF THE SOLID PROPELLANT CHARGE FOR CONTROLLED THRUST PROGRAMS

The reliability and short development time of the solid propellant rocket arise from the precise knowledge of the burning surface area and burning rate due to the geometrical burning of the charge. Penner [40] states that a liquid propellant rocket requires extensive empirical testing to establish performance of the injector spray pattern and the combustion efficiency. The solid propellant combustion mechanism on the other hand occurs in a mixing zone having dimensions comparable to the diameter of an oxidizer particle, i.e., 10–100 μ in thickness. The magnitude of burning rate cannot be calculated theoretically from the composition, but must be measured by strand burners or small motors [1]. The granular diffusion flame (GDF) theory as illustrated in Figure 4 from Steintz et al. [51] permits interpretation of the effect of particle size, burning rate additives, chamber pressure, and transverse velocity on the thickness of the reaction zones and thus the magnitude of the steady-state regression

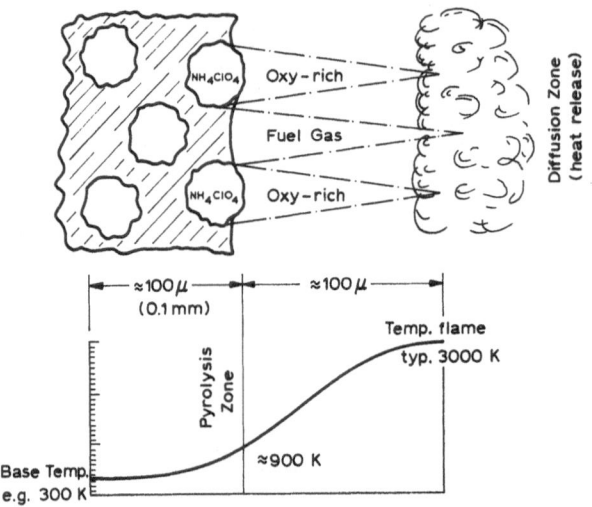

Fig. 4.   Schematic of burning surface showing mechanism related to size of oxidizer particle.

(burning rate). Miller [34] applies the GDF theory to the interpretation of erosive burning.

Since a micro depression would be less effective in heat adsorption and a micro protrusion more effective, the system is self-correcting for minor variations in composition and the macro surface regresses as a simple normal projection of the exposed burning surface, much in the manner of chemical milling, making solid propellant charge-design a simple matter of geometrical analysis. To obtain optimum thrust programming, taking account of the changing density of air as described by Miele [33], or to balance between sufficient initial thrust for flight stability and not so much later to cause excessive dynamic pressure or terminal acceleration as described by Vandenkerckhove [57], geometrical shapes are selected having the necessary burn-back sequence (Figure 5). Since this grain design problem can be rigorously solved mathematically, it can be coupled with trajectory analysis and optimized by computer. The critical reactions occur in a boundary zone of a few hundred microns thickness. Thus the ballistic design of a solid propellant rocket is virtually dimensionless and rapid scale-up becomes only a process engineering and not a technical design problem. Thus Thackwell [55] notes that there is a greater tendency to develop custom-designed solid rockets for particular applications, in terms of envelope, thrust, and impulse levels for optimum mission, while the longer development time and empirical scaling of the liquid rocket relegates it to fewer, general-purpose-type propulsion units. It is for this reason that the competitive design of solid propellant rockets has become very highly automated [7]. 'Factory-sealed thrust programming' or exact shaping of the charge to the optimum burning sequence is thus the source of the solid propellant reliability, although this limits command variation in flight unless hybrid or controllable motors are used [24].

What is a solid rocket charge configuration? Figure 6 shows a cross-section of the

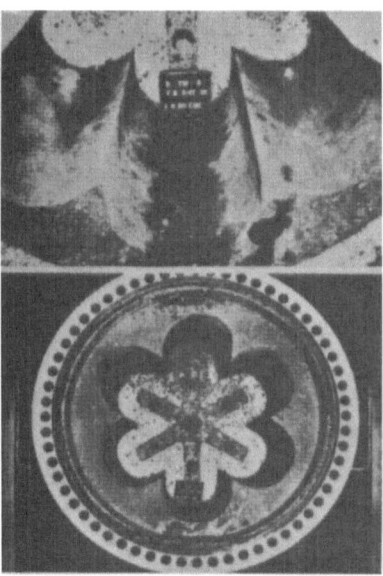

Fig. 5.   Precision of surface reproduction in stop-fire of star-perforated grain.

Fig. 6.   Segment of 100 in. solid booster enroute to assembly.

propellant charge for the first 100-inch segmented solid rocket enroute to assembly.*
Figure 8 shows a photoelastic test specimen in the manner of Durelli [14–16] and
Ordahl and Williams [38]. The generalization of photoelastic stress concentrations as a
function of web thickness, wedge angle, and fillet radius has been developed by Fourney
[19]. It is obvious that the structural analyst would desire small web thicknesses, wide
wedge angles, and large fillet radii to minimize these stress concentration areas. Why
then does a ballistician develop a complex charge configuration such as this shown?

* 100TW-1, fired by Aerojet-General Corporation at Sacramento, California, on 3 June 1961,
showing predicted and actual performance curve in Figure 7 [75].

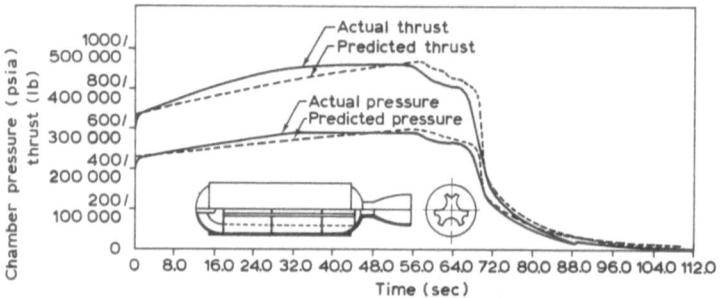

Fig. 7.   100-FW-1 solid rocket motor, predicted performance curve.

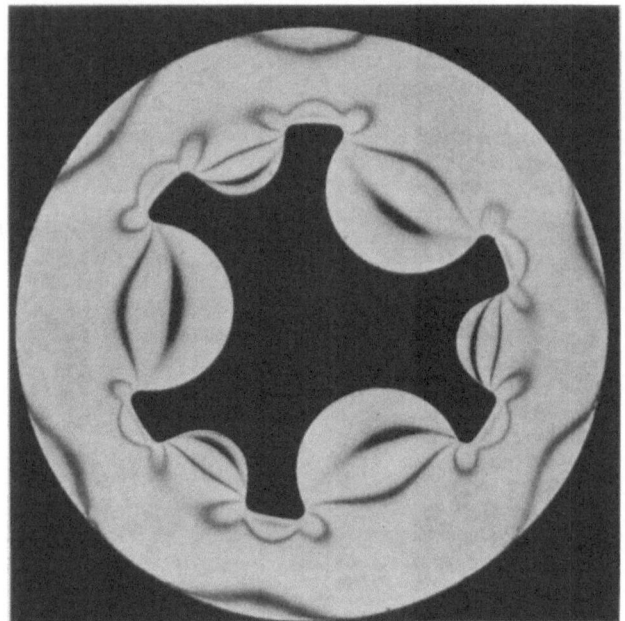

Fig. 8.   Photoelastic test specimen of forked cloverleaf grain design.

B. HISTORY OF GRAIN DESIGN AND PRINCIPLES TO BE SOUGHT

Emile Sarran (1875) showed gun performance related to the shape of the grain. Grain design for solid propellant rockets derived from the extruded perforated-cylinder of cannon powder (whence the term grain design) to the complex case-bonded configuration of today. The World War II development of the artillery rocket is based heavily on the work of the British. This work was reviewed by Wheeler [64], and includes the introduction of the star-perforated charge by Poole [43]. The subsequent United States development is reviewed by Shafer [48] and by Bartley and Mills [5].

Actually the history of rockets goes back many hundreds of years, based on tamped saltpeter, sulfur, and carbon. With no concept of thermodynamic calculations, the formulations were empirically arrived at and jealously guarded. Neither were internal pressure measurements or the concept of nozzling available. The last tamp charge was made of clay, which formed a refractory orifice for the propellant gas, and 'by tightening the port the friction imparted by the gases onto the propellant surface and thus the rocket case, increased the efficiency'. Thus neither internal pressure nor the physics of reaction had been applied to rockets, and the Congrieve rocket of the 18th century, as noted by Sokol'skii [50] was primarily noted for its disruption of cavalry rather than its precision of bombardment. When quantitative metallurgy permitted muzzled cannon and later rifled guns, the rocket fell out of favor in military operations. This held until mobility and stand-off distance resumed importance in World War II. By this time, the thermo-dynamics of the gun had been worked out, and it was logical that the new rocket science should adopt its terminology from interior ballistics of guns (whence grain design), which distinguishes gas dynamics in the combustion chamber from the exterior ballistics or dynamics of flight. It is also of interest that liquid propellant rockets did not evolve from guns, but rather from extension of turbine engines, and have a more aerodynamic vocabulary. This produces interesting problems in communication when designing contemporary hybrid and air-augmented rockets, depending upon whether the conversants have a liquid or solid rocket background.

Turning solely to post World War II solid rockets, the evolution of nomenclature and concept will be traced thru the end-burning, star-perforated, envelope optimization, multi-propellant, three-dimensional, monocoque chamber, segmented and conocyl, and the ultimate ICBM charge-design periods. At all times attention will be on the direction of the burning front, the extent of chamber wall protected from flame, the support of the propellant by the chamber wall, and the ratio of web thickness to charge radius, or the relative web thickness. This latter quantity will be shown to migrate systematically from $\frac{1}{6}$ with the dendrite grain burning entirely in the cross-section plane and with total case-bonded chamber protection, to the order of 0.8 for the segmented and conocyl grains which burn in partly an axial coordinate and depend heavily on programmed exposure of an ablative insulation to control the burning surface area. Recognition of position in this morphological spectrum is the key to effective selection of grain design and realization of whether gaps in design capability exist. The tradeoff of this web thickness geometrical characteristic with the mission parameters expressed in the ratio of thrust to duration squared (F/t^2) is also the key to optimization of propellant as described by Dudley *et al.* [13].

3. Evolution of Nomenclature and Concept in Grain Design

A. END-BURNING (RESTRICTED), TUBULAR, AND EXTERNAL-BURNING CHARGES

A brief history of grain design will assist in orienting the names and analysis styles which dominate today's design practice. It will be noted that mass ratio, structural integrity, and ballistic precision have formed the goals by which each step in charge design was motivated.

In World War II solid propellant grains were classified as end- or cigarette-burning in which length controlled the burning time, or for shorter duration motors radial burning in which the web was determined by (1) clusters of external burning rodular elements (bazooka charge) or (2) an internal-external burning tube with raised spacers (triform or quadriform), or (3) an internal burning cylinder with coned end to slightly increase the burning area [30, 80]. Ten years later the classic Princeton University Press compendium on Jet Propulsion Engines [5] and the corresponding John Wiley and Sons text on Space Technology [48] showed these former designs plus (a) the star perforated, (b) the slotted cylinder, (c) the rod and tube concentric, and (d) the multi-perforated integral grain. These designs have also been reviewed by Barrere [4] and Shapiro [49]. What were the factors that contributed to this development?

Rocket charges took their nomenclature from gun interior ballistics, so Kelley, McClure, and Rosser [29] discussed the regressiveness of the burning surface for "one 15-inch powder grain with lateral holes compared to three 5-inch powder grains without holes", the latter being shown more progressive by mathematical formulae. Similarly, Davis and Mills [12] and Clautice [10] were concerned with the packing of multiple cylindrical grains in a cylindrical chamber. This contrasts to the Rodman multiperforated grain of artillery use (Figure 9). Gen. Rodman (1857) introduced the term 'progressivity' in describing the surface area change of a grain with burning. This change is called the 'shape factor' or form function by Hayes [25], Hunt [26], and the Naval Powder Plant [91].

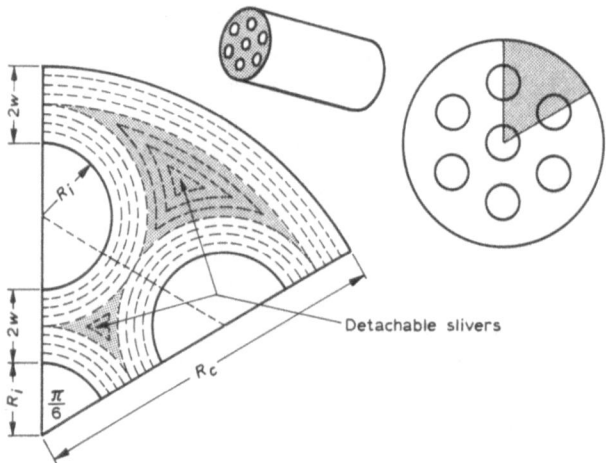

Fig. 9. Rodman internal–external burning seven perforation grain, $R_c = 3R_i + 4w$; $u = 2\pi[7/6(R_i + y) + (R_c - y)]$ $(0 < y \leqslant w)$; $du/dy = 7/6 - 1 = +1/6$ (slight progressivity); Loading $= 1 - 7\pi R_i^2/\pi R_c^2 = 1 - 7(R_i/R_c)^2$.

The optimization of flow space between the inner and outer burning surface of a single-perforated multigrain charge was noted in a Rohm and Hass report in 1954 [92]. The design of internal-burning and multiperforated charges, including specifically the star- and anchor-perforated designs, was studied by Aerojet in 1950 using graphical design and temperature cycling tests. In 1956, Aerojet [67] compared the concentric

Rod and Shell to an annular perforated grain (Figure 10) and favored the one-piece, cartridge loaded charge [22]. Nicholson [37] integrated the surface area functions of the tapered bore star-perforated grain (Figure 11) for optimum in satisfying total surface area neutrality. The star-perforated grain is typified by a web thickness equal to $\frac{1}{4}$ to $\frac{1}{3}$ of the charge radius. Problems in temperature cycling of the multiperforated grain, however, and in support for multitubular charge-assemblies led Aerojet in 1968 to use an internal-burning case-bonded '12-point star' (Figure 12), actually composed of 6 major and 6 minor rays, the prototype of the dendrite perforation [74]. A multi-slab grain was also tested in this period by Allegany Ballistics Laboratory using various combinations of inhibitors and propellants to control the shape of the perfor-mance curve [76]. Orlov [39] describes a wide variety of external-burning and multi-perforated ordnance charges. Astrodyne [78–79] considered a triform and a slab-type

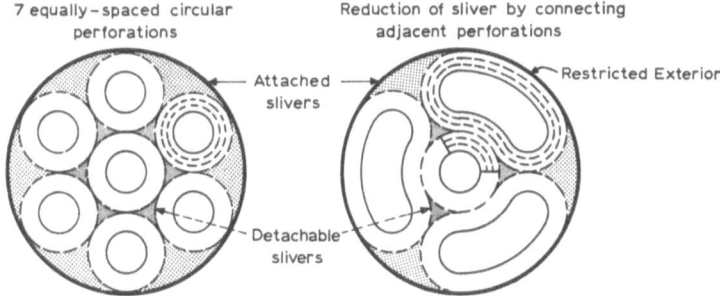

Fig. 10.   Case-bounded multiperforated grain.

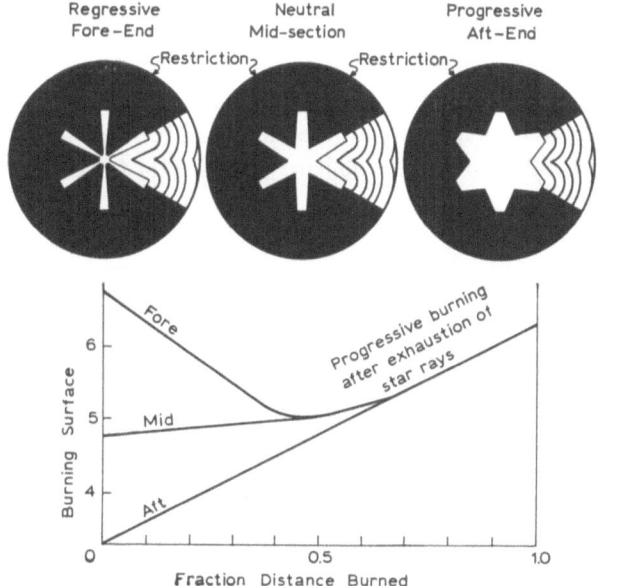

Fig. 11.   Balance of forward regressivity and aft progressivity in tapered star with web $= Y^*$.

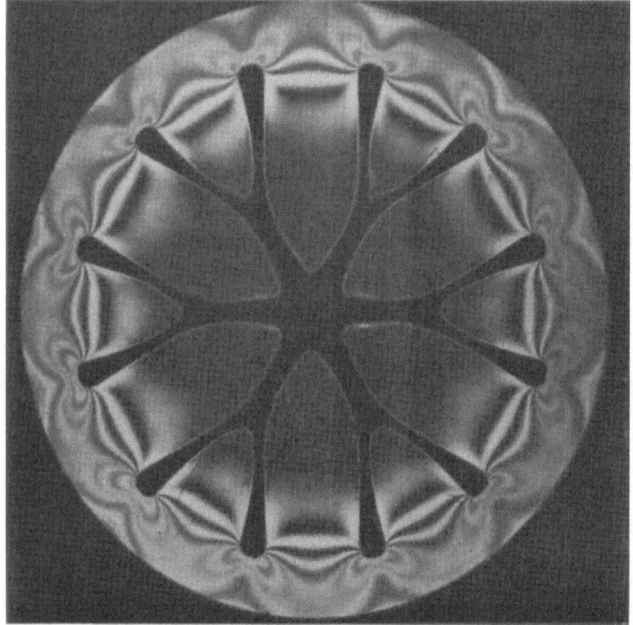

Fig. 12. 'Twelve-pointed star' or dendrite grain design.

charge assembly to obtain maximum burning surface area. The dendrite and multislab design are typified by web thicknesses of the order of $\frac{1}{6}$ of the charge radius.

B. INTRODUCTION OF THE STAR-PERFORATED GRAIN AND COMPUTER ANALYSIS

Meanwhile at Rohm and Haas, Stone [53, 95] reduced the design of an internal-burning star-perforated grain to mathematical form, in terms of number of star points, the web, the progressivity ratio, the sliver fraction, and the loading fraction (Figure 13), and design graphs were prepared with the aid of an IBM-650 computer. A similar study was made by Piasecki and Robillard at JPL [41] for the star configuration

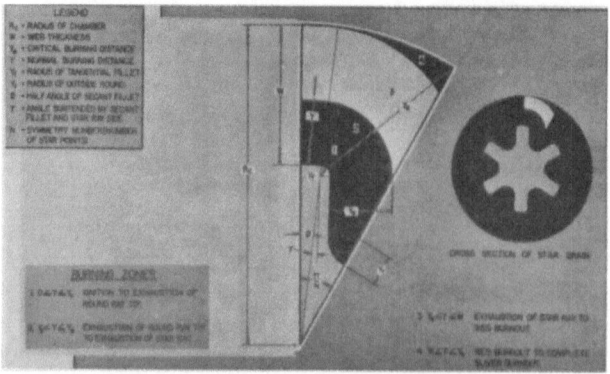

Fig. 13. Symbols in periphery equations.

which documented a method of calculating the pressure-time and thrust-time relationship. Stone [53, 95] produced similar charts extending the sharp-pointed star to a broad-pointed form (known as the HR design) and introducing the wagonwheel and stator configuration (Figures 14).

The stator design is also described by NOTS [90]. Vandenkerckhove [56, 58] and

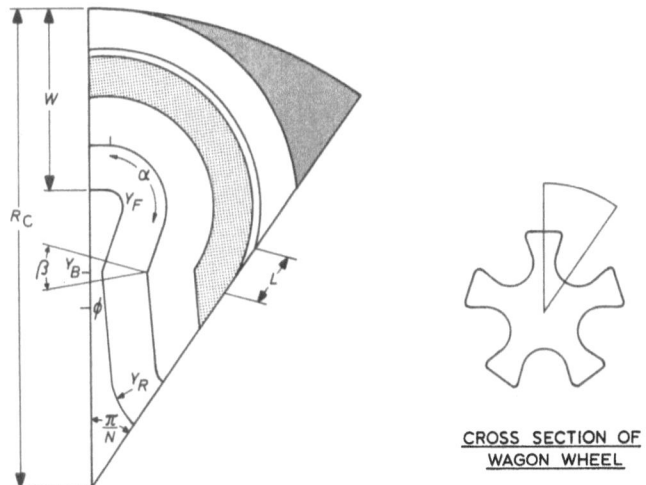

CROSS SECTION OF
WAGON WHEEL

Fig. 14. Wagon wheel configuration. Independent wagon wheel parameters: R_C – outer grain radius; N – symmetry number; W – Web; ϕ – secant fillet half angle; α – tangent fillet angle; β – break round angle; Y_F – tangent fillet radius; – Y_B – break round radius; Y_R – outside round radius; L – ray side length.

Godai [23] developed analyses for star and wagonwheel designs. Kasky at the Ordnance Missile Laboratories produced a Grain Design Handbook in 1956 [28] containing design parameters for five types of solid propellant grain-configurations obtained by digital computer solution, and ARGMA issued a further manual by Lumpkin in 1959 [31]. This included plots of the web thickness corresponding to neutrality as a function of volumetric loading factor for various values of the symmetry number. Billheimer [6] and Kakima [27] established the outside round (Figure 6) as an antidote for erosive-burning in large solid boosters. By modifying the wagonwheel so that the protrusion burned out decidedly before the case-bonded web, Ritchey [44] and Grand and Barney [21], or by introducing an outer ring of slow-burning propellant [72], a dual thrust effect is obtained. Vogel [61] assembled a morphological classification of grain designs in 1956.

The slotted tube design (Figure 15) was programmed by Stone of Rohm and Haas in 1959 for a Royal-McBee LGP-30 Computer [52, 96–99]. The effect of insulation is shown in Figure 16. The burning in the slot requires auxiliary insulation in proportion to the time of exposure to protect the chamber from propellant flame while operating at full pressure. The star-perforated charge, on the other hand, uses only

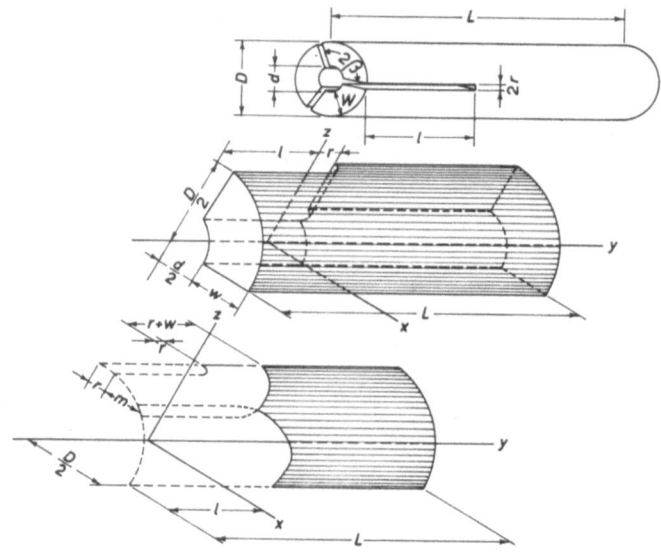

Fig. 15.   The aft-ended slotted cylinder charge with three 120° slots for half length.

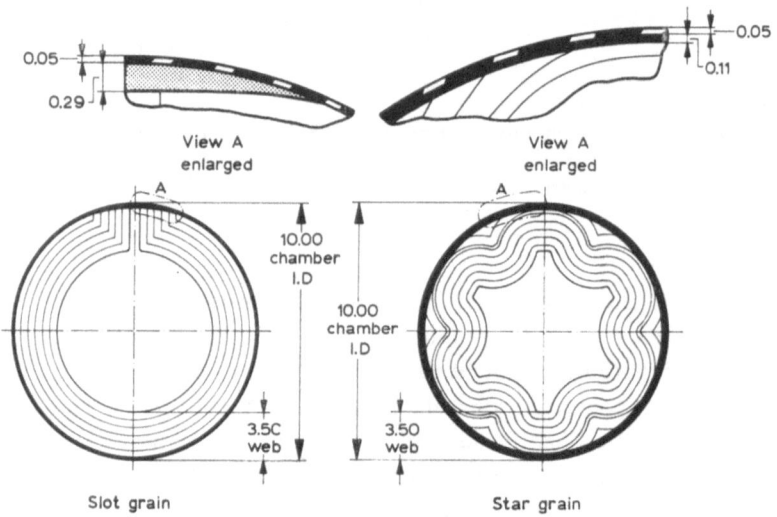

Fig. 16.   Built-up insulation for slot compared to uniform liner for star.

a minimal propellant-to-chamber liner, since the protection by the propellant web is maintained until the rapid pressure decay or tail-off commences.

A generalized slotted-grain solution using Legendre elliptical integrals is obtained [96]. This slotted design has a typical web thickness of 0.6–0.8 of the charge radius. Aerojet made a comprehensive investigation of internal-burning grain configurations in 1959 [69] showing the stepwise integration of the burning area to accurately predict the actual firing curve and comparing the stress analysis of typical star and slot configurations. The generalized star-wagonwheel-dendrite grain design program on the IBM-704 computer was described by Billheimer in 1959 [6] including mass flow effects and head-loading calculations. This led to a system for prediction of erosive burning and tailoff and a system of key points as shown in Figure 17. This relates changes in the shape of the performance curve to geometrical properties of the charge [8], showing the systematic relation of configuration efficiency and volumetric loading factor to the web thickness in the sequence dendrite-wagonwheel-star-slot by Rossini [46]. Whetstone, Threewit and Billheimer [65] established a rigorous degrees-of-freedom analysis for the star-wagonwheel-dendrite system, and this was extended to a morphological transformation incorporating the anchor design [66]. A similar computer program was developed by Vellacott at Thiokol/Redstone including the 'broken-back forked wagonwheel' which incorporates the conventional star-wagonwheel-dendrite with up to 21 degrees of freedom [59, 60]. The thin-web grains, when used in couple with high burning rate propellants, obtain minimum time to target for interception rockets, as characterized by Dudley *et al.* [13] for high F/t^2 values.

C. OPTIMIZATION WITH RESPECT TO MOTOR WEIGHT, TRAJECTORY, AND GRAIN STRESS

This computer capability in grain design opened the door for more exact optimization of the entire motor design. Previously Wall [62] had attempted graphical representation of the optimization relations between propellant characteristics, geometrical characteristics, and mission parameters. Struble and Black [54] followed by Nead [36], added inert component weight formulas to develop a rational process of design to reduce total motor weight employing isometric surface representation. These contributions led to the Waco, Texas, 1962 ARS Symposium on Solid Rocket Optimization [77] and subsequently to the formation of the ICRPG Working Group on Design Automation [20].

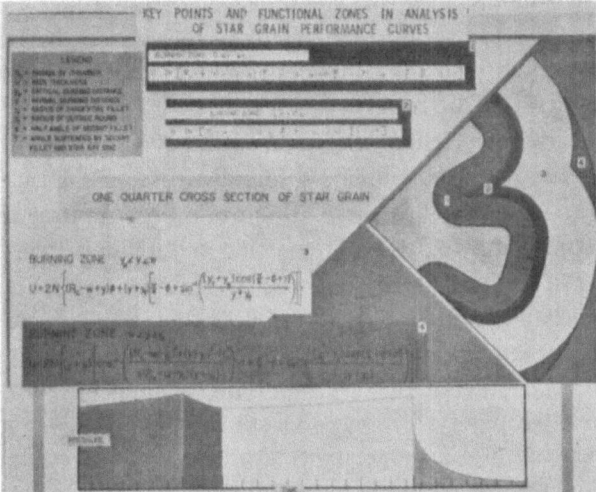

Fig. 17. Key points and functional zones in analysis of star grain performance curves.

During this period of analytical development of the ballistic design of propellant charges, Durelli [14] tested photoelastic models of the original sharp fillet star design, and showed the benefit of circular and elliptical fillets in reducing the stress concentration. Filleting was noted however to increase a typical sliver loss of 3.75% by an additional 1.4% in a normal case [15]. A 'bent ellipse' was recommended as the ideal solution of the star-point stress problem [16]. Thirty-seven new possible grain configurations had been studied. Ordahl and Williams in 1957 [38] made a photoelastic stress analysis on a family of thick-walled cylinders with symmetrical internal slots of various widths, depths and fillet shapes. Aerojet [72] reported several methods of stress analysis on a wagonwheel dual thrust motor in 1959 and photoelastic tests have been used by Thiokol to modify the stress concentration in a fillet design [102].

D. SURFACE REGRESSION, THE DIRECTRIX, AND MULTIPROPELLANT CHARGES

Both the graphical and computerized analysis of burning in a grain is based on the equal regression principle of Piobert, which was confirmed by Stone at Rohm and Haas [93] with interrupted burning of interior angles, exterior angles, interior arcs, and exterior arcs (Figure 18). These tests demonstrated that for an interior angle the sides remain parallel to the respective original sides while at the point of an angle an arc is formed with its center at the vertex and its radius equal to the distance burnt. For an exterior angle, on the other hand, the burning layers remain parallel to the original sides, and the identity of the angle is retained. Epstein [17] generalized these principles by considering the directrix of the burning surface, the evolute of the directrix being invariant. Thus a cusp convex towards the gas phase remains a cusp, while a concave one becomes an arc of a circle.

With grain designs being thought of mathematically, it became logical to seek neutrality without the severe stress concentration and sliver problems of the star-perforated grain.

The most logical solution of the constant thrust, internal-burning grain is continuous variation of burning rate inverse to the radius of the burning surface. A sawtooth performance curve could be obtained from concentric cylinders of propellant of the appropriate burning rates [70]. An alternative developed at JPL is a conical boundary between fast and slow propellant in a cylinder with tapered port, so that the burning time of all cross-sections is constant [86]. Within the cross-section itself Fey *et al.* [18] and Morey [35] at ABL provided fast burning rate propellant in the sliver region to salvage this propellant in a shorter time period and minimize tailoff. Billheimer at Aerojet [71] used an elliptical interface between an initial layer of fast and an outer layer of slow propellant, so adjusted that the lines of burning refracted after crossing into the slower propellant and met the chamber wall on a radius (Figure 19).

Solution for the exact interface contour to assure simultaneous burnout of all paths from the initial fast-burning surface assures a sliverless and nearly neutral grain, with loading factor increasing and progressivity decreasing as the symmetry number is increased, much as in a star-perforated grain (Figure 20). This design was reduced to practice by Podell [42] using a symmetry-of-four bi-propellant sliverless charge design

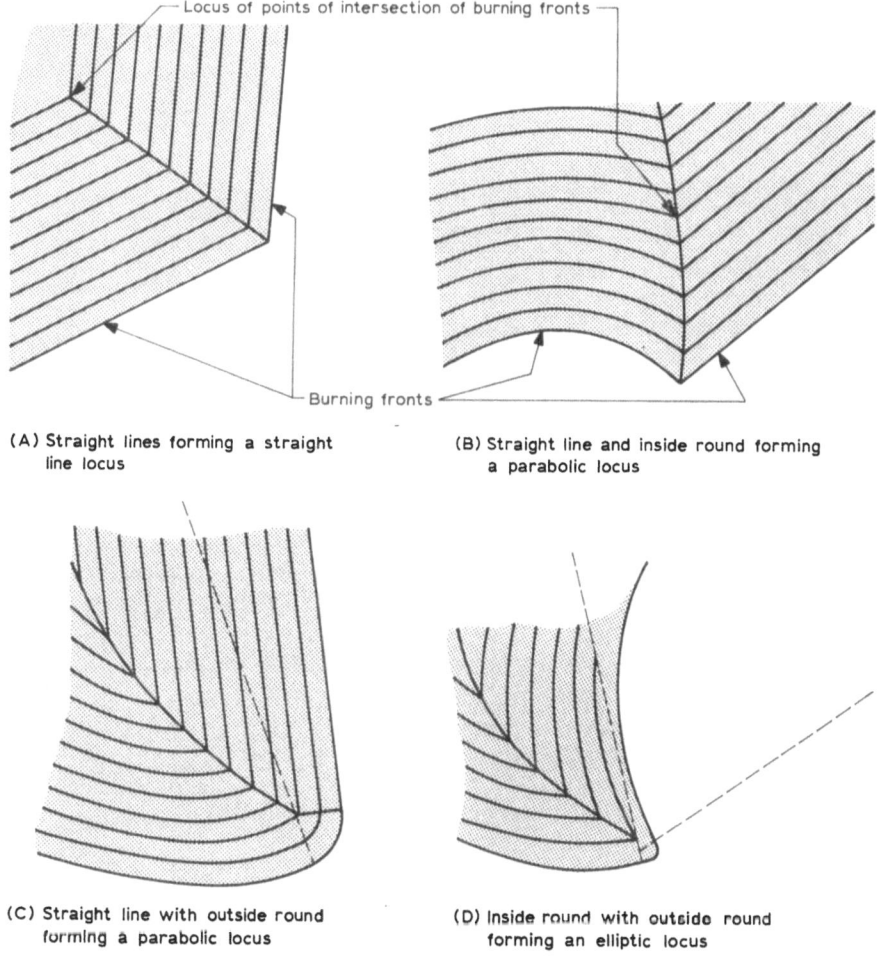

(A) Straight lines forming a straight line locus

(B) Straight line and inside round forming a parabolic locus

(C) Straight line with outside round forming a parabolic locus

(D) Inside round with outside round forming an elliptic locus

Fig. 18.   Loci of intersection points of various burning front contours.

for ICBM application at Aerojet (Figure 21), and similar designs were also established by Rogers at Thiokol [45, 101]. This grain has the feature, in addition to being sliverless, that the fillet form is substantially elliptical, corresponding to stress minimization without sliver penalty, and that the resulting performance curve is convex (compared to a single propellant star of comparable web which produces a concave performance curve). The convex curves permit absorption of ignition and erosive-burning early effects without pressure peak, and allows a slight regressivity as web burnout is approached to ease terminal acceleration values [73].

The axial interface bipropellant charge was further developed by JPL using two concentric frustums of cones [85]. Parlaying this concept into a circular bore, conical port design, employing three propellants, a conceptual design for minimum stress concentration and minimum axial pressure drop is obtained yielding neutral burning and sliverless performance. If a burning rate spread of 10-fold is obtainable without

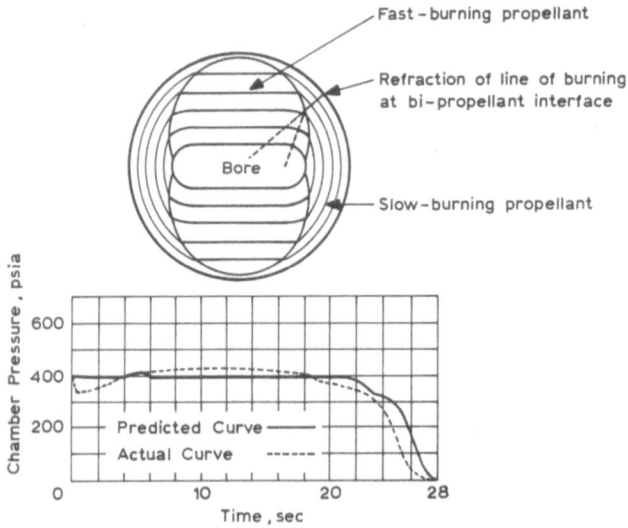

Fig. 19.   Demonstration of oval-bore bipropellant sliverless charge design.

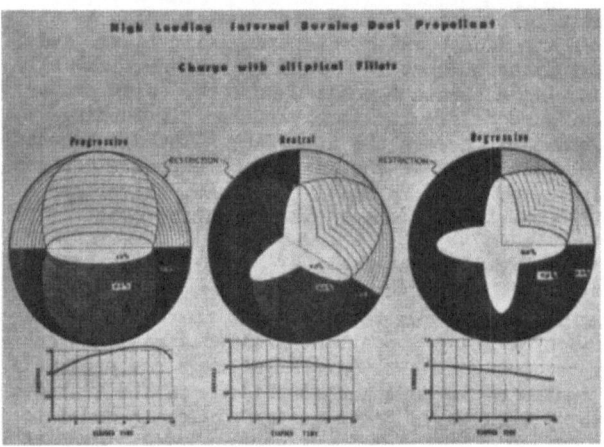

Fig. 20.   High loading internal burning dual propellant charge with elliptical fillets.

undue sacrifice in specific impulse, the surfaces can be modified to obtain an initially end-burning, ultimately radial-burning charge giving 100% volumetric loading. This technique was explored at ERDE [82], CARDE [81] and by Bacon and Braun at WADC [2] to produce a 100% volumetric loading charge based on a very fast burning axial core, a medium rate transition layer, and an ultimate burnout in a radial cross-section while achieving a constant total mass flow. With proper casting no difficulty is experienced in burning reproducibly across the interface. The subsequent development of fore-end insulation exposure designs has however made the multi-propellant sliverless charge obsolete on the basis of casting cost.

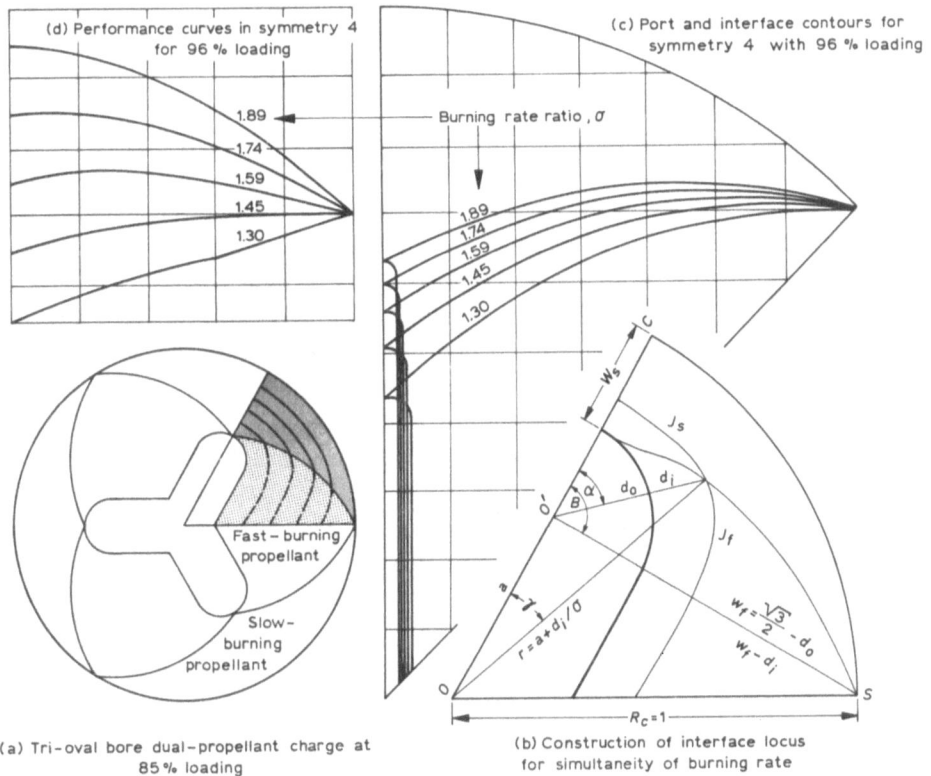

(d) Performance curves in symmetry 4 for 96 % loading

1.89
1.74
1.59
1.45
1.30

Burning rate ratio, σ

1.89
1.74
1.59
1.45
1.30

(c) Port and interface contours for symmetry 4 with 96 % loading

Fast – burning propellant

Slow – burning propellant

(a) Tri-oval bore dual-propellant charge at 85 % loading

(b) Construction of interface locus for simultaneity of burning rate

Fig. 21.   Analytical construction of bipropellant sliverless charge.

E. SUMMARY AND THE STATE OF OPTIMIZATION, 1960–1965

This completes the survey of nomenclature and origins of solid propellant grain-design practice for the period 1943–1962. An attempt has been made to indicate the time and conditions of the introduction of each basic grain-design style, in order to clarify overlapping and ambiguous nomenclature and establish a workable description system to assist in design comparisons. Equivalent names have been documented wherever possible.

The era is summarized at the Motor Design Symposium of the 1963 Seattle ICRPG Solid Propulsion Meeting with the 'Report on the Working Group for Design Automation' [20], the 'Automatic Design Selection and Optimatization Program for Solid Rocket System' [47], and the 'Integrated Solid Rocket Design Procedure Utilizing Computers' [9]. These papers establish that, within the scope of two-dimensional cylindrical grain design with end effects, and in the objective of maximization of neutrality and protection of the chamber wall, the morphological state of grain design is fully analytical, and subject to complete optimization in a stable (convergent) computer program linked to appropriate component weight and mission evaluation programs.

The above sequence represents the completion of the transition from packed bed combustion of propellant grains for gun-type charges (in which the shape factor was secondary to empirical flame spread in determining performance); through the era of fixed-position, multipiece charges; to integral multi-perforated grains – still cartridge-loaded; and finally to cast-in-case internal burning charges for which regorous mathematical relations of charge shape to ballistic performance curve have been developed. Appropriate erosive-burning and flow-with-mass-addition relations are coupled with the geometrical surface regression analysis in these comprehensive computer programs. The fetish of achieving neutrality of performance curve by case-supported integral castings with a minimum of exposure of chamber wall until all propellant is consumed reached an ultimate in first the two-dimensional, and finally the three-dimensional multipropellant charges where the refraction of the line of burning in layers of successively slower burning propellant compensates for the geometrical growth of the surface area from the initial bore space to approach the entire chamber interior simultaneously at burnout.

The sliverless, internal-burning grain problem has been solved, and the reliability of the method established, but the fabrication cost is a handicap.

F. TRANSITION TO THE MONOCOQUE CHAMBER

During this period the combustion chamber had progressed from a cylinder with separate head attachments, and cartridge loading; to integral fore-head closure with aft-end casting and cores tapered for aft extraction in agreement with the aftward enlargement of flow space to minimize erosive-burning.

However as the fiberglass chamber with integral heads and a nozzle opening less than the maximum breadth of the core became predominant for the ballistic missile systems, it became necessary to replace the rigid extractable core with collapsible cores. The extreme case of this is the NASA spherical motor [88], using a fusible core of low melting alloy (Figure 22). Similar results were obtained experimentally with frangible cores such as styrofoam, which could be left in place and were effectively collapsed by the igniter brisance [32]. Other spherical grains employ also a submerged nozzle, with propellant cast outside the nozzle entrance section [83].

In the more conventional designs, the slotted propellant charge, which originally had the slots in the aftward portion of the charge for ease of extraction, now adopted the lower gas velocity regions of the forward end of the chamber for preferred exposure, and the optimum design employs radial slots arranged as required in the forward section, with the gas exiting through a circular bore to the nozzle. Figure 23 shows the relative insulation requirement between fore and aft positions in a slotted grain. This technique reduces the erosive requirement on the liner, and when coupled with advances in sacrificial (ablative) insulation technology, has made conscious use of programmed exposure of the forward chamber wall a principal degree of freedom in ballistic design of propellant charges, whereas it had been an apologetic approximation of end effects in the prior two-dimensional case-bonded propellant era.

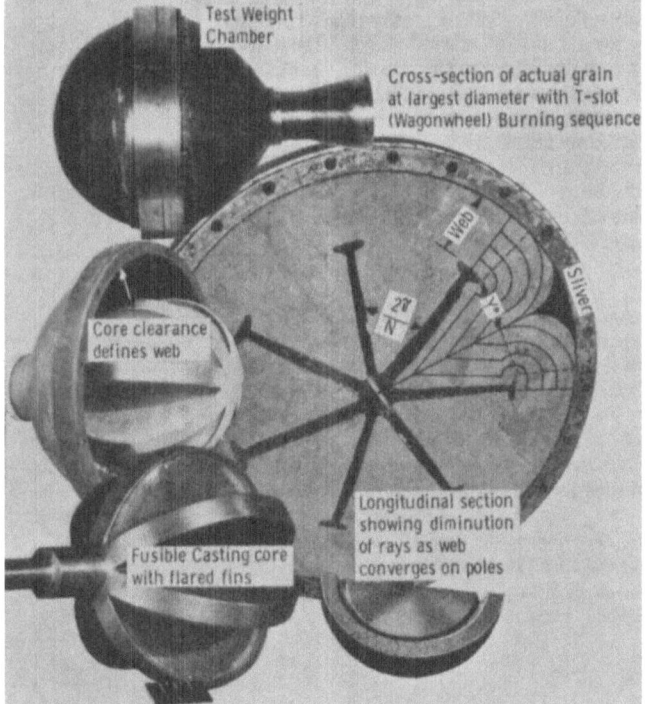

Fig. 22.   Construction of melon-slice spherical propellant charge.

G. THE SEGMENTED CYLINDER AND CONOCYL DESIGN

The segmented rocket motor [87, 89], although substituting mobility for maximum performance as a design objective, also incidentally established the practicality of insulating sections of chamber wall in direct proportion to their time of exposure. This discovery returned grain design to a cylindrical three-dimensional concept with attention on the shape of the confining surface. The interior bore geometry of the segmented motor has essentially no degrees of freedom, except for the frequency of axial cuts (number of segment faces), although the forward section may be made a star-perforated charge as regressive trim for the normally progressive pressure-time curve.

The logical progression of collapsible core technology and preformed sacrificial insulation placement in burning surface area control led to the change of foreward and axial slots to a conical slot formed by inserting a forward-end conical propellant element into a coned surface of a basic cylinder with circular bore. This design is termed the cone-in-cylinder or conocyl [11]. Two opposed conical-burning surfaces are formed, resulting in a funnel-shaped flow space, and strictly speaking the design should be named the 'funnel-perforated grain' for consistency with the two-dimensional perforated grain nomenclature which precedes it (Figure 24). Nevertheless the inherent simplicity and fabrication convenience of the grain has given it rapid acceptance and the name conocyl appears well established.

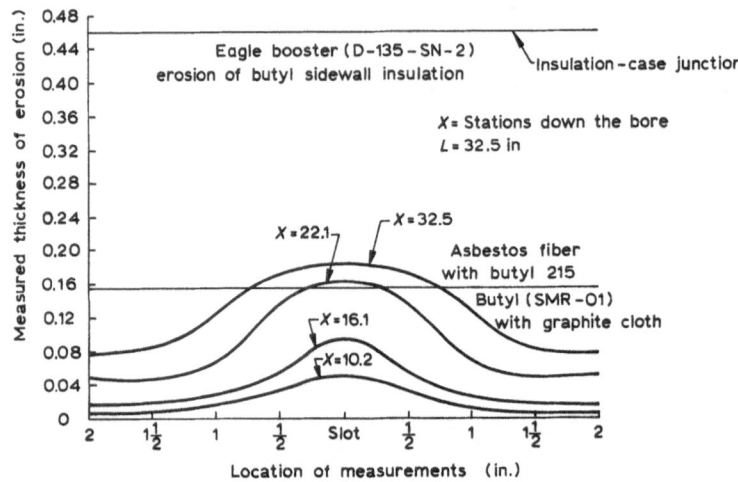

Fig. 23.   Compensatory insulation for burnback of slot faces as a function of position from the fore-end.

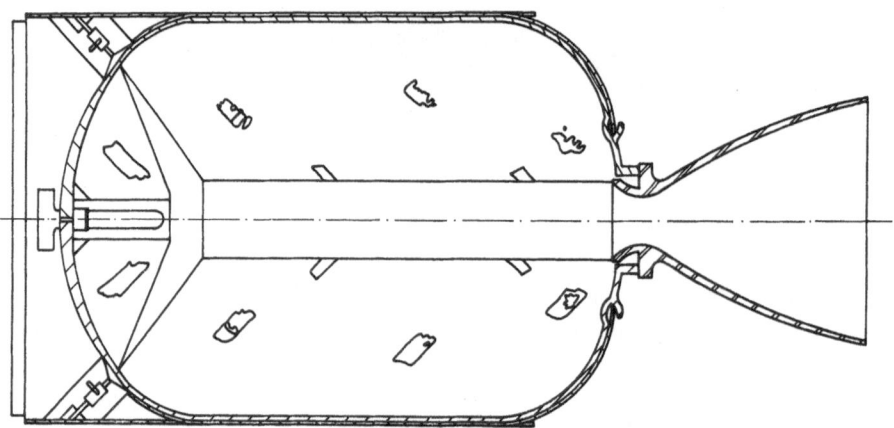

Fig. 24.   The *cone*-in-*cyl*inder or conocyl grain design.

This design is not to be confused with the 'cone and cylinder design' employed in 1959 by the British Ministry of Aviation at ERDE [3] and by Warren [63], who coned the free-burning *aft* end of a circularly perforated cylinder (externally restricted) to increase slightly the burning surface area, but experienced difficulty in the undesirable exposure of aft-end chamber wall where the erosive heat transfer duty was greatest. This is readily understood in contemporary high mass ratio standards.

It should be noted that morphologically the conocyl design is the three-dimensional analog of the star-perforated from two-dimensions, in that the cone is the most regressive-burning three-dimensional body possible (analogous to the wedges in the case-bonded star construction), and the singly perforated cylinder is the most progressive element short of an internal-burning sphere, which has no conceivable gas exit. The conocyl design is thus the fundamental opposition of a conical regressive surface and a cylindrical progressive surface, and is subject to optimization in terms of the web thickness, cone angle, space between the cone and the matching conical surface of the end of the cylinder, and the orientation of these features with respect to the contour of the head closure. A simple circular slot in the cross-section plane, approximately separating the cylindrical from the head closure portion of the grain, will yield a satisfactory performance curve in certain length-to-diameter ratios [100]. Whether this is simply a one-segment-and-end-piece segmented grain, or a limiting case of the conocyl, is significant only in understanding the degrees of freedom of the design. Actually the conocyl's simplicity belies its lack of flexibility, and shows its extreme dependence on chamber envelope. Emphasis on the conical confining surface recalls the earliest three-dimensional 'end-effects' design, the Aerojet cone-ended 15KS-1000 JATO (Figure 25), although the cartridge loaded restricted grain has today been replaced by case-bonded cast-in-place charges.

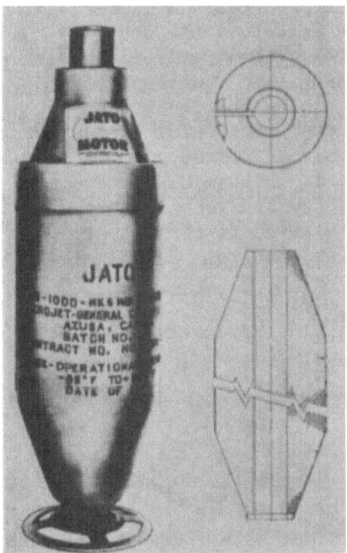

Fig. 25.   The cartridge-loaded cone-ended JATO motor.

Eventually axial slots reappeared on the conical face of the cylindrical element, to give the cross-section in this region a star-perforated behavior, resulting in a 'finocyl' name by analogy for the fins which modify the basic conocyl idea. The design is also known as 'star-in-a-pocket' or 'winged slot' [84, 103]. This concept is the fundamental grain of the advanced ICBM systems, and replaces with collapsible core and planned sacrificial insulation exposure, the bipropellant grain as a minimum sliver, convex performance curve, relatively long burning time, high web charge configuration. It will be noted that the mean direction of the burning front is no longer contained in a cross-sectional plane, as with earlier low-web designs, but is now partially in an axial direction, permitting effective web thicknesses approaching or exceeding the radius of the chamber. This couples with the use of low burning rate propellants to obtain minimum terminal acceleration for low F/t^2 ballistic missile application as characterized by Dudley, Veit, and Billheimer [13].

Virtually no analytical grain design literature has appeared in this period, since the designs are essentially made to order for specific applications, and apply only to large missile programs of limited competition. Such designs are difficult to generalize outside of mission specification and the chamber envelope, compared to the dimensionless treatment adopted for the two-dimensional designs. In fact, structural analysis, core technology, and the liner and insulation placement all become critical aspects of charge design, often overrriding the purely ballistic considerations. Grain design has become integrated by necessity, but with pragmatic considerations not amenable to computer expression. It would appear that the topological science of deployment of propellant surfaces to satisfy various restraints: support continuity, near constancy of total burning surface area, and minimization of exposure of confining surface envelope, which had been essentially solved in the two-dimensional domain, remains a challenge and requires close collaboration with chamber insulation designers and propellant structural analysts in the three-dimensional monocoque chamber domain.

Analytical studies have developed the basic characteristics of the star-perforated grain, shown the introduction of secant and tangent fillets, and the evolution of wagonwheel- and dendrite-perforated grains. From this there is established a first and second law of grain design (neutrality of first derivative, and convexity or negativity of the second derivative for optimum trajectory). Subsequently comparison to the slotted grain which uses systematic chamber wall exposure as a burning surface area control device implies a third law relating to chamber protection. It may be suggested that a fourth law pertaining to principles of structural support of the grain would be desirable, but this is not quantitatively established at the present state of the art.

4. Conclusion – Points of Evolution and Resultant Principles

The evolution of solid propellant grain design has been shown to consist of five principal stages:

(1) Cartridge-loaded, partially restricted tubular, end-burning or multiperforated charges.

(2) The analytically-designed star-, wagonwheel-, dendrite-, and anchor-perforated two-dimensional, singly-connected grain configurations.

(3) The multipropellant sliverless charge using refraction of the line of burning to avoid sliver and obtain approximately simultaneous burnout with constant mass flow rate.

(4) The slotted and segmented grain, which depends on exposure of the chamber wall to control burning-surface area, but is primarily concerned with fabrication logistics.

(5) The monocoque chamber, with collapsible or consumable core, having a cir-

cular bore main section and configuration detail (slots, wedges, stars) in the forward stagnation area only.

Each of these periods depended on a change of technology for their existence: from the ordnance heritage of the cartridge-loading period, the chamber wall protection of the internal-burning (star-perforated) grain with case-bonded propellants, the elimination of sliver and total respect to chamber wall protection at any process cost in the multipropellant systems, to the dependency on functionally placed ablative insulation to permit the chamber wall to serve as a burning surface area control device, and finally to the collapsible or consumable core permitting placement of control surfaces in the forward stagnation region and use of a low stress cylindrical bore, maximum web main section for maximum mass ratio.

These designs can be organized by two principles. One is the sequence of relative web thickness, whereby the functional use of the configuration along the scale from very high-thrust, short-duration anti-missile to relatively low-thrust, long-duration ballistic missiles is represented by a sequence of characteristic web thicknesses from $\frac{1}{8}$ (dendrite), $\frac{1}{4}$ (wagonwheel), $\frac{1}{3}$ (two-dimensional star), $\frac{1}{2}$ (star-with-end-effects), 0.7 (longitudinal slot), 0.8 (conocyl/finocyl), >1.0 (end-burner). The second is that the mean vector of burning surface migrates systematically from normal to a radius in the cross-sectional plane, to along a radius in the cross-sectional plane, to along the flow axis of the charge as the web thickness ranges from $\frac{1}{8}$ to 0.8 of the radius, respectively (Figure 26).

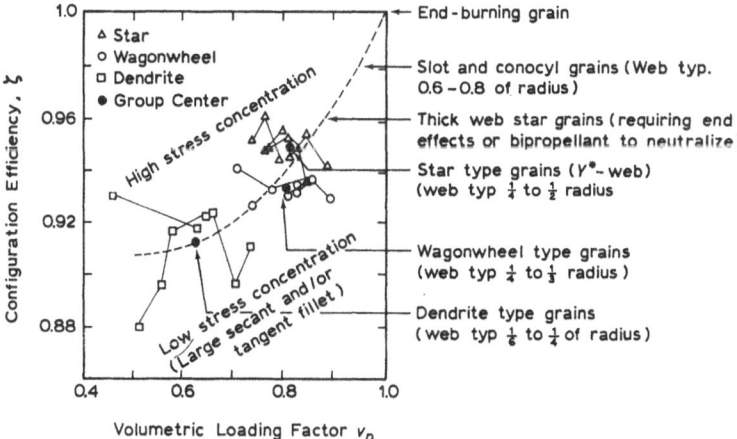

Fig. 26.   Configuration efficiency as a function of volumetric loading factor for grain designs of various typical web thickness.

Within a given configuration, four 'laws' apply:

(1) The first derivative of burning surface area with respect to distance burnt should be approximately zero (neutral burning) for the main portion of the performance curve.

(2) The second derivative thereof should be negative to allow absorbance of ignition

and erosive-burning initial effects without an undue pressure peak, and to allow muting of thrust as the propellant is exhausted to limit terminal acceleration.

(3) Web thicknesses exceeding the critical burning distance (y^*), which favors high volumetric loading, oppose the second derivative principle while preserving chamber protection in a single propellant. The fabrication alternatives of multi-propellant, or slotting to expose chamber wall, favor the negativity of the second derivative.

(4) Support of the propellant grain by the chamber wall (and stiffening of the chamber by the propellant) are desirable for structural integrity of the charge, but are fundamentally opposed to the prior principles of internal burning at a constant mass rate to expose minimum chamber wall until all propellant is consumed. Quantitative expression of the structural criteria in a manner suitable to directing a design search for optimum configuration is not yet available.

5. Summary

The grain design or charge configuration of a solid propellant rocket has been shown to form a logical sequence from two-dimensional design of relatively low web thickness, burning entirely in the cross-section plane (dendrite, wagonwheel, star) thru thick-webbed stars, multipropellant, and slotted grains to a family of partially axially burning charges depending on programmed exposure of the chamber wall and having segmented or conical burning faces intercepted by radial growth from a central perforation. Many different types of charge configuration have appeared for solid rockets, including multi-perforated, concentric tubes, cone-ended, and many with no attempt at functional nomenclature but only a serial designation. However all designs when expressed in terms of their relative web thickness, neutrality and sliver, chamber wall exposure and method of support, fall into a logical progression from thin-web radial to thick-web semi-axial burning characteristics, and thus allows mating of grain configuration with burning rate for particular missions, and the estimation of difficulty in meeting arbitrary mission and envelope restraints with available propellants. This factor is expressed in the ratio F/t^2, which represents the span from interceptor ($F/t^2 \approx 3000 \text{ lbf/sec}^2$) to ICBM ($F/t^2 \approx 30 \text{ lbf/sec}^2$).

The mathematical precision of grain design, based on the self-adjusting feature of the granular diffusion flame mechanism of burning, permits predicting performance from analytical geometry of the charge independent of size. This minimizes scale-up cost and permits custom design of propellant charges for optimum ballistic satisfaction. Account is taken of thrust-to-weight ratio, atmospheric density and dynamic pressure, terminal acceleration, and the envelope and pressure of the combustion chamber, as well as the internal design features of web thickness, symmetry, volumetric loading factor, method of support, and chamber wall protection. Structural aspects of the grain are of as great importance as the ballistic, and frequently in opposition thereto. They have not however been developed to the level of morphological relation that is employed in the ballistic solution. This is now a most fertile area of mathematical study.

Glossary of Solid Propellant Grain Design

There is universal recognition of the star (étoile) grain* (Poole, Robillard, Barrere, Shapiro, Godai) with simple circular fillet

звездообразный заряд со округлеными углами
zvezdoobraznyi zaryad so okruglenymi uglami
star-shaped charge with rounded corners

although it may be noted as a

модифицированая форма заряда с каналом звездообразного сечения
modifitsirovanaya forma zaryada s kanalom zvezdoobraznogo secheniya
modification of the form of the charge with canal of star-shaped section.

Similarly wagonwheel (Stone) designates a star-perforation with two distinct linear segments defining the protrusion boundary (voiture-roue, вагонное колесо – vagonnoe koleso), although care must be taken that mere broadening of the attachment fillet and rounding of the protrusion tip (Ka kimi) does not meet the wagonwheel criteria. The forked wagonwheel (Vellacot, Barrere), is cited also by Shapiro as

модифицированая форма заряда 'вагонное колесо' древовидный
To minimize compound terms and emphasize the characteristic of branched flow space, dendrite drevovidnyi is recommended (Billheimer).

The refraction of burning rate between grains of differing burning rates (Billheimer, Rogers) or "fusees à propergol solide utilisant deux poudres ayant des vitesses de combustion differentes" (Barrere) is described as

схема заряда из двух одновременно горящих топлив с различными скоростями горения
skhema zaryada iz dvukh odnovremenno goryashchikh topliv s razlichnymi skorostyami goreniya
scheme of charge with two simultaneous burning fuels with different rates of burning.

In three-dimensional charges, the slotted cylinder (Stone) is noted by Barrere and by Shapiro

щелевой заряд
shchelevoi zaryad
slotted charge

* Since internal-burning grains (French: side-burning to distinguish from end-burning) are generally named by their port or channel (каналом – kanalom) shape (формч – formu), "-perforated" (-percement, продырявленный) is implied in mention of star, wagonwheel, anchor, etc., grains. These are not to be confused with external-burning (triform, cruciform) ordnance charges.

as well as the coned cylinder (Badrick) or cylindroconical (Barrere)

трубчатый заряд с компенсационным конусом
trubchatyi zaryad s kompensatsionnym konusom
tubular charge with compensating cone

which is to be distinguished from the cone-in-cylinder or 'conocyl' (Billheimer) in which the cone is opposite to and mated with a flared end of the cylindrical section. End-burning (cigarette) grains are designated

стержневой заряд горящий с торца
sterzhnevoi zaryad goryashchii s tortsa
rod charge burning at face

The internal-burning shell is noted by Shapiro as

трубчатый заряд с внешней бронировкой
trubchatyi zaryad s vneshnei bronirovkoi
tubular charge with external sheeting

while concentric tubes (multi-tubular charge) becomes simply

телескопический заряд
teleskopicheskii zaryad
telescopic charge

It is concluded that the geometrical forms of solid propellant grain design are universally recognized and that the nomenclature is literal or at least similar in concept in each instance. The performance curve

характеристика прогрессивности
kharakteristika progressivnosti
characteristic progressiveness

is related to the number of rays,

(числа лучей – chisla luchei)

and the web thickness or

толщины горящего свода
tolshchiny goryashchego svoda
thickness of burning arch*

---

* The Russian concept of 'arch' for minimum burning distance suggests original application to the internal-burning star. This is more logical than the English/French 'web thickness', which derives from multi-perforated artillery grains, and is ambiguous between physical web (distance between opposite free surfaces) and burning web – distance to mid or restrictive surface. Nonetheless, the use of 'web' is well established.

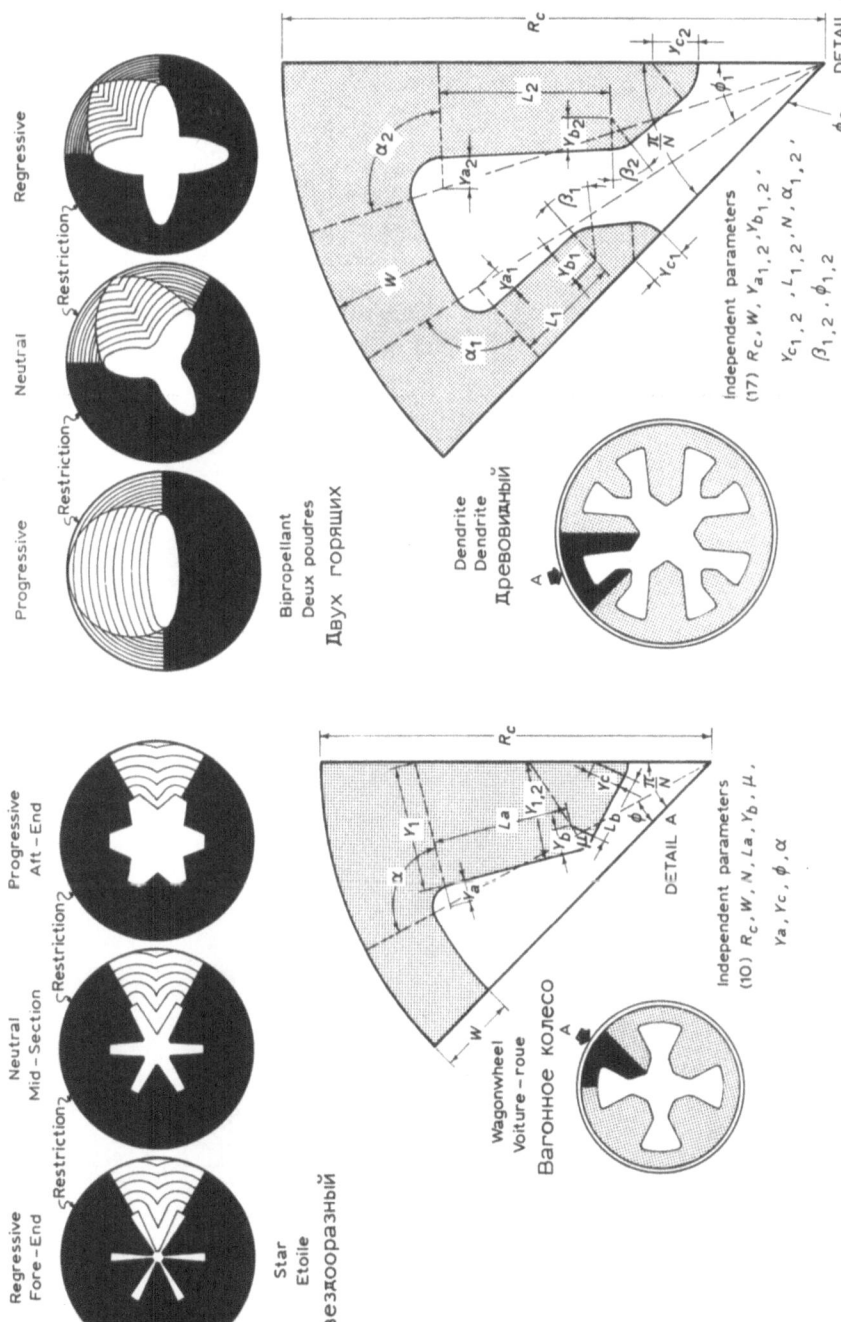

Fig. 27. Case-protecting/case-supported grains with all burning fronts outbound (attached sliver or simultaneous burnout).

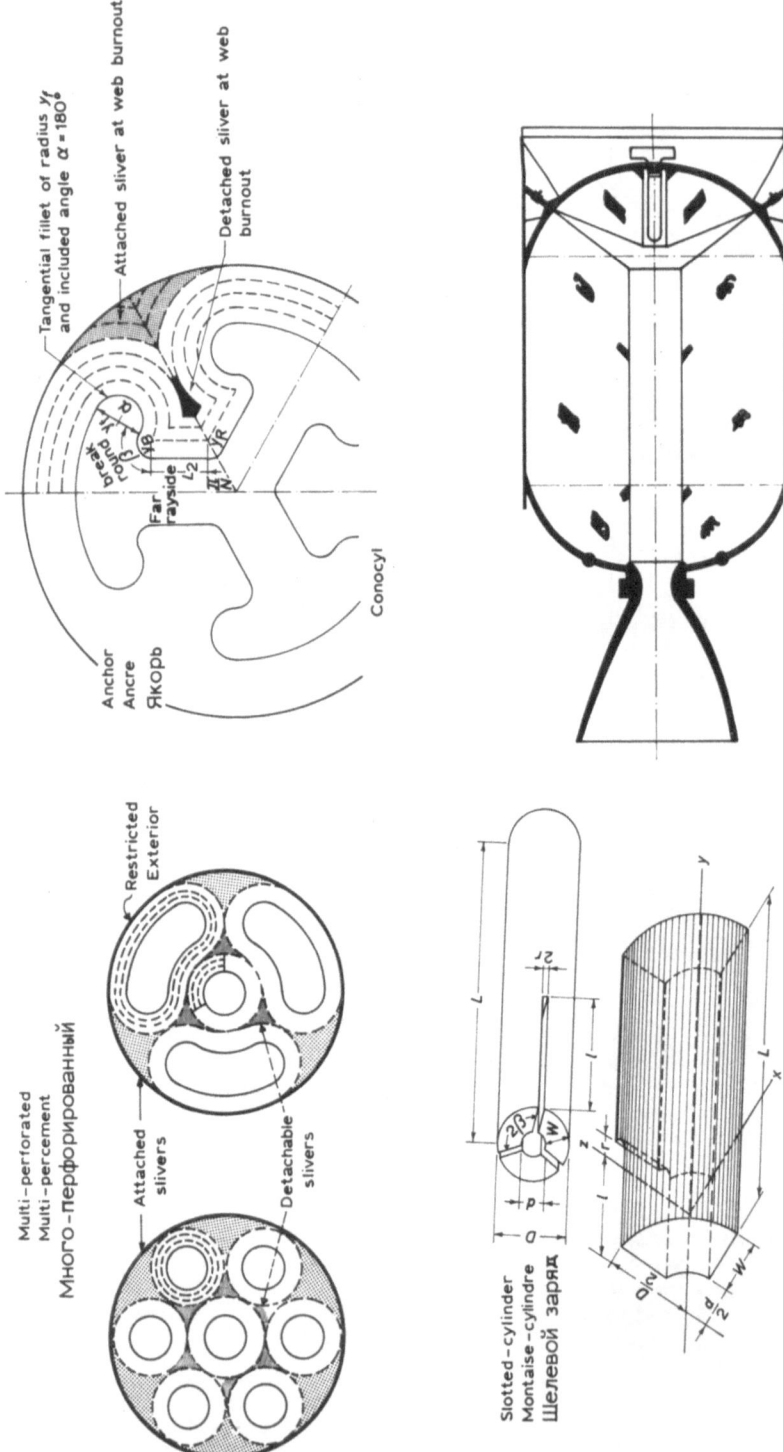

Fig. 28. Outward and inward or axial burning grains with either detachable sliver or chamber exposure before web burnthru.

yielding the familiar three zones of burning

три фазы горения заряда
tri fazy goreniya zaryada
three phases of burning

of the star grain. This incorporates the essential classification of grain design (Figures 27 and 28).

References*

[1] Anderson, S. E., 'Ballistic Scale-Up of NF Propellants; I. 80-1bm Motor Demonstration Firing (U)', Rohm and Haas, Huntsville, Report no. S-94, 17 March 1966.

[2] Bacon, W. S. and Braun, J. V., 'High Mass Ratio by the Use of Multiple Propellant Grains', *Bull. of the 15th Meeting of the JANAF Solid Propellant Group*, Vol. I, June 1959, pp. 165–192 (*0127, #0925)

[3] Badrick, C. M. and Vernon, J. H. C., 'Small Rocket Motors for Ballistic Assessment: The Geometry of the Cone and Cylinder Charge', Explosives Research and Development Establishment, Tech Memo No. 9/M/58, December 1958. (*0024)

[4] Barrere, M., Jaumotte, A., Fraeys de Veubeke, B., and Vandenkerckhove, J., *Rocket Propulsion*, Elsevier Publ. Co., Amsterdam, 1959; *Ibid.*, *La propulsion par fusées*, Dunod, Paris, 1957.

[5] Bartley, C. E. and Mills, M. M., 'Solid Propellant Rockets', in *Jet Propulsion Engines* (ed. by O. E. Lancaster), Princeton University Press, 1958, pp. 521–624. (#0004)

[6] Billheimer, J. S., 'Case Bonded Grain Design for High Loading, Long Duration Solid Propellant Motor', *Bull. of the 15th JANAF Solid Propellant Group*, Vol. I, June 1959, pp. 231–245. (*0129)

[7] Billheimer, J. S., 'Optimization and Design Simulation in Solid Rocket Design', ICRPG/AIAA 3rd Solid Propulsion Conference, Atlantic City, June 4–6, 1968, AIAA preprint 68-488.

[8] Billheimer, J. S., 'Use of the Computer in Direct Grain Design for Erosive Burning, Sliver, Neutrality, and Tail-Off Considerations', *Bull. of the 16th Meeting of the JANAF Solid Propellant Group*, Vol. V, June 1960, pp. 211–252. (*0042, #0594)

[9] Billheimer, J. S., and Wiegand, J. H., 'Integrated Solid Rocket Design Procedure Utilizing Computers', *Bull. of the Interagency Solid Propulsion Meeting, Seattle*, Vol. III, July 1963, pp. 11–36.

[10] Clautice, A. W., 'A Study of Loading of Cylindrical Shape Grains', Ballistics Research Laboratory, Memorandum Report No. 783, May 1954. (*0018)

[11] Crooks, J. R., 'Skybolt Propulsion System', *Bull. of the 18th Meeting of the JANAF-ARPA-NASA Solid Propellant Group*, Vol. 1, 5–7 June 1962, pp. 113–123. (#1044)

[12] Davis, L. Jr. and Mills, C. D., Jr., 'Dependence of the Mass of Propellant in a Rocket Motor on the Web Thickness and Motor Dimensions', OSRD-1319, A63, cit/jac4, March 1943. (*0108)

[13] Dudley, D. P., Veit, P. W., and Billheimer, J. S., 'The Man-Computer Link in Solid-Propellant Rocket Preliminary Design and Optimization', ICRPG/AIAA 3rd Solid Propulsion Conference Atlantic City, June 4–6, 1968, AIAA preprint 68-489.

[14] Durelli, A. J., 'Investigation of Distribution of Stresses in Propellant Grains', Armour Research Foundation, Report Numbers 1–7, 13 September 1953. (*0098)

* Acknowledgement is made to two comprehensive bibliographies in grain design:

(*XXXX) referes to citations in Thiokol Chemical Corp., Elkton, 'A Research Study to Advance the State-of-the-Art of Solid Propellant Grain Design; Bibliography and Abstracts Report', an attachment to Quarterly Progress Report No. 5, TCC-E E-181-60 (Addend. Conf.), Contract AF 33 (616)-6530, by E. D. Cohen (28 October 1960).

(#XXXX) refers to citations in Wagner, R. F., 'Bibliography on Internal Ballistics and Grain Design of Solid Propellant Rocket Motors', Univ. of Utah, UTEC-ME-67-080, Contract NAS3-11179 (November 1967).

[15] Durelli, A. J., Report Numbers 8–14, 4 January 1954. (*0091)

[16] Durelli, A. J., Final Report of 26 May 1954. (*0089)

[17] Epstein, L. I., 'The Design of Cylindrical Propellant Grains', *Jet Propulsion* **26**, No. 9, 757–759. (*0123, #0987

[18] Fey, R. S., Angelus, T. A., Sherman, J. N., and Skurzynski, E., 'New Principles of Propellant Grain Design', *Bull. of the 12th Meeting of the JANAF Solid Propellant Group*, May 1956, pp. 175–190. (*0133)

[19] Fourney, M. E. and Parmerter, R. R., 'Parametric Study of Rocket Grain Configurations by Photoelastic Analysis', Mathematical Science Corp., Seattle, Report No. 65-29-12, AFRPL-TR-66-52 (Mar. 66), Contract AF 04(611)-10529, 113 pp.

[20] Gale, H. W., 'Report on the Working Group for Design Automation', *Bull. of the Interagency Solid Propulsion Meeting, July 1963, Seattle*, Vol. III, pp. 1–10.

[21] Grand, H. R. and Barney, J. D., 'Dual Thrust Solid Propellant Rocket', *Bull. of the 9th Meeting of the JANAF Solid Propellant Group*, May 1953, pp. 95–106. (*0135)

[22] Geckler, R. D., Noland, R. L., Roberts, E. R., and Rogers, W. L., 'Internal Burning Grains and Related Components for Solid-Propellant Rocket Motors,' Aerojet Report 445 (Final), 23 June 1950, 252 pp. (U)

[23] Godai, Tomifumi, 'Grain Configuration and Length-Diameter Ratio of Solid Propellant Rockets', *First Symposium (International) on Rockets and Astronautics*, Tokyo, 1959, pp. 87–93.

[24] Gustavson, John, 'Small Sounding Rocket Propulsion Systems', *XIth International Astronautical Congress, Stockholm, 1960*, Springer-Verlag, 1961, pp. 53–62.

[25] Hayes, T. J., *Elements of Ordnance*, John Wiley and Sons, N.Y., 1958, p. 11.

[26] Hunt, F. R. W. (ed.), *Internal Ballistics*, Ministry of Supply, His Majesty's Stationery Office, London, 1951, Chapter IV, 'The Form Function', pp. 40–52.

[27] Kakimi, Tsuneo, 'On the Optimum Design of Three-Stage Solid Propellant Rockets with Special Reference to Their First-Stage', *First Symposium (International) on Rockets and Astronautics*, Tokyo, 1959, pp. 94–106.

[28] Kaskey, B., *Grain Design Handbook*, Report 3413, Rocket Development Laboratory, Ordnance Missile Laboratory, Redstone Arsenal, 8 June 1956.

[29] Kelley, B., McClure, R. T., and Rosser, J. B., 'A Less Regressive Design for Powder Grains', OSRD-2069, A75M, November 1943. (*0107)

[30] Kershner, R. B., 'Interior Ballistics of Rockets', in *Rocket Fundamentals*, OSRD, George Washington University, 1954, Chap. 3, pp. 39–68.

[31] Lumpkin, H. K., *Mathematical Approach to Solid Propellant Grain Design*, Army Rocket and Guided Missile Agency, Ordnance Missile Laboratories Division, ARGMA TN 1G5N, 21 December 1959.

[32] Manfred, R. K., Wilkes, B. F., and Brown, J. M., 'Design and Development of Solid-Rocket Combustible Mandrels (U)', Aerojet-General Corp., Sacramento, Rept. 0630-81Q-2 and 0630-81Q-3 (16 May–15 Nov. 1962), Conf. (#0758)

[33] Miele, Angelo, 'Optimum Burning Program as Related to Aerodynamic Heating for a Missile Traversing the Earth's Atmosphere', *VIIIth International Astronautical Congress*, Barcelona, 1957, pp 257–277.

[34] Miller, E., 'Erosive Burning of Composite Solid Propellants', *Combustion and Flame* **10** (December 1966) 330–336. (#0375)

[35] Morey, L. E., 'Dual Propellant Grain Configurations', *Bull. of the 13th Meeting of the JANAF Solid Propellant Group*, June 1957, pp. 829–839. (*0136)

[36] Nead, D. M., 'Normalized Equations for Rocket Motor Design', *Bull. of the 12th Meeting of the JANAF Solid Propellant Group*, May 1956, pp 131–143. (*0132)

[37] Nicholson, A. H. and Wilsten, D. B., 'Derivation of the Surface Integral and Neutrality Control Function for the Axially Tapered Internal-Burning Star Solid Propellant Charge', Aerojet-General Report No. 1785/89-2, Appendix A, 15 November 1956.

[38] Ordahl, D. D. and Williams, M. L., 'Preliminary Photoelastic Design Data for Stresses in Rocket Grains', *Jet Propulsion* (June 1957). (*0016, #0805)

[39] Orlov, B. V., and Mazing, G. Yu., 'Thermodynamic and Ballistic Design Fundamentals of Solid-Propellant Rocket Engines', Translated by Foreign Technology Div., AFSC-RTD-65-191 (16 June 1966), AD 645 793, pp 124–129,

[40] Penner, S. S., 'The Role of Combustion Research in Rocket Propulsion', *XIIth Congress of the International Astronautical Federation*, Washington, D.C., 1961.

[41] Piasecki, L., and Robillard, G., 'Generalized Design Equations for an Internal Burning Star-Configuration Solid Propellant Charge and Method of Calculating Pressure Time and Thrust Relationships', Jet Propulsion Laboratory, Memorandum No. 20-135, September 18, 1956. (*0021, #0900)

[42] Podell, H. L., 'The Practical Application of a Bi-Propellant Grain System in Large Solid Rocket Engines', *Bull. of the 16th Meeting of the JANAF Solid Propellant Group*, Vol. II, June 1960. (*0038, #0922)

[43] Poole, H. J., unpublished reports of the British Ministry of Supply, 1937-1940.

[44] Ritchey, H. W., 'Thrust Programming of Solid Propellant Boosters by Choice of Propellant Configuration and Composition', *Jet Propulsion*, **25**, No. 10 (Oct. 1955).

[45] Rogers, K. H., 'Mathematical Design of a Sliverless Rocket Engine', ARS Solid Propellant Rocket Conference, Salt Lake City, Feb. 1-3, 1961, ARS preprint 1616-61.

[46] Rossini, R. A., Billheimer, J. S., and Threewit, T. R., 'Configuration Efficiency: a New Measure of Ballistic Quality for a Grain Design', *ARS J.* (Dec. 1961) 1761-1766; ARS Solid Propellant Rocket Conference, Salt Lake City, Feb. 1-3, 1961.

[47] Rossini, R. A., and Threewit, T. R., 'An Automatic Design – Selection and Optimization Program for Solid-Rocket Systems', *Bull. of the Interagency Solid Propulsion Meeting, July 1963, Seattle*, Vol. III, pp. 37-59.

[48] Shafer, J. I., 'Solid Rocket Propulsion', Chapter 16 in *Space Technology* (ed. by Howard S. Seifert), John Wiley and Sons, N.Y., 1959, pp. 16-04 to 16-09.

[49] Shapiro, Ya. M., Mazing, G. Yu., Prudnikov, N. E., *Teorya raketnogo dvigatelya na tverdom toplive* [Theory of Solid Fuel Rocket Motors], Boennoe Izdatel'stvo, Ministerstva Oboroni, SSSR, Moscow, 1966.

[50] Sokol'skii, V. N., *Russian Solid Fuel Rockets*, Acad. Sciences USSR, Inst. History of Science and Engineering. Translated from Russian by S. G. Kozlov, Israel Program for Scientific Translations, Jerusalem, 1967.

[51] Steinz, J. A., Stang, P. L., and Summerfield, M., 'The Burning Mechanism of Ammonium Perchlorate-Based Composite Solid Propellants', AIAA 4th Propulsion Joint Specialist Conference, Cleveland, June 10-14, 1968, AIAA preprint 38-658, 23 pp.

[52] Stone, M. W., 'Slotted Tube Grain Design and Some Practical Modifications and Use by Grain Designers', *ARS J.* (Jan 1960), 1055-1060. (*0111)

[53] *Ibid.*, 'A Practical Mathematical Approach to Grain Design', *Jet Propulsion* **28**, No. 4 (April 1958). (*0017, #0906)

[54] Struble, R. A., and Black, H. D., 'A Generalized Closed Form for Burnt Velocity and the Effect of Drag on Rocket Design', *Bull. of the 12th Meeting of the JANAF Solid Propellant Group*, Vol. II, White Oak, 7-9 May 1956, pp. 145-64 (U).

[55] Thackwell, H. L., Jr., 'The Application of Solid Propellants to Space Flight Vehicles', *Xth International Astronautical Congress, London, 1959*, Springer-Verlag, 1960, pp. 155-170.

[56] Vandenkerckhove, J. A., 'Internal Burning Star and Wagon-Wheel Designs for Solid Propellant Grains', Université libre de Bruxelles, USAF-ARDC Contract S61(052)58-13, 1958.

[57] Vandenkerckhove, J. A., 'Note on the Optimum Design of Solid Propellant Power-Plants for Missiles Systems Engineering', *IXth International Astronautical Congress, Amsterdam, 1958*, Springer-Verlag, 1959, pp. 149-167.

[58] Vandenkerckhove, J. A., 'Recent Advances in Solid Propellant Grain Design', *ARS J.* (July 1959) 483-491.

[59] Vellacott, R. J., 'Design Study of Solid Propellant Configurations', Thiokol Chemical Corp., Redstone Div., Huntsville, Final Rept. 28-61, U-A-61-28A, 299 pp. (28 July 1961), Conf.

[60] Vellacott, R. J., 'A Computer Program for Solid Propellant Rocket Motor Design and Ballistic Analysis', ARS preprint 2315-62 (1962).

[61] Vogel, J. M., 'A Quasi Morphological Approach to the Geometry of Charges for Solid Propellant Rockets', *Jet Propulsion* **26**, No. 2 (Feb. 1956).

[62] Wall, R. W., 'Designing Solid Propellant Rocket Engines for Optimum Ballistic Performance by Use of Graphical Solutions', *Bull. of the 12th Meeting of the JANAF Solid Propellant Group*, May 1956, pp 111-129. (*0131)

[63] Warren, F. A., 'A Survey of Properties of Solid Propellants for Gas Generator Applications

(U)', *Bull. of the 20th Interagency Solid Propulsion Meeting*, Vol. IV, July 1964, pp. 291–310. (#0116)

[64] Wheeler, W. H., 'Rocket Development', *Nature*, **158** (Oct 5, 1946) 464–469.

[65] Whetstone, A. E., Threewit, T. R., and Billheimer, J. S., 'Basic Grain Design and the 564 Interior Ballistics Computer Program', Aerojet-General Corp., Sacramento, STM-143, 10 June 1961, Contract DA-04-200-506-ORD-1120, 142 pp.

[66] Whetstone, A. E., Threewit, T. R., and Rossini, R. A., 'Intermediate Grain Design and the ACP 564B Interior Ballistics Computer Program, Aerojet-General Corp., Sacramento, STM-148, 20 Feb. 1962, Contract AF 04(611)-6358, 72 pp.

[67] Aerojet-General Corporation, Azusa, 'Development of Solid Propellant Rocket Engines for Long Range Missiles', Report No. 1785-89-1, 17 Aug. 1956. (*0076)

[68] Aerojet-General Corporation, Azusa, 'Development of Solid Propellant Rocket Engines for Long-Range Missiles', Report No. 1785/89-2, 15 Nov. 1956. (*0101)

[69] Aerojet-General Corporation, Azusa, 'Investigation of Internal Burning Grain Configuration', PR No. 5-985/710, 985/711, 3 Jun 1959. (*0105)

[70] Aerojet-General Corporation, Azusa, 'Research and Development of Large High Performance Solid Propellant Rocket Engine', QPR No. 1314-1, Jan. 1957. (*0060)

[71] Aerojet-General Corporation, Azusa, 'Research and Development of Solid Propellants for Large High Performance Rockets,' QPR No. 1314-3, July 1957. (*0068)

[72] Aerojet-General Corporation, Sacramento, 'Development of Dual Thrust Rocket Motors for the Hawk Missile', Report No. 1914-Q-11, 15 Aug. 1959. (*0034)

[73] Aerojet-General Corporation, Sacramento, 'Development of Large Solid Propellant Rocket Motors', Report No. 0162-01M-3, 20 Nov. 1958. (*0074)

[74] Aerojet-General Corporation, Sacramento, 'Final Development and Prequalification Report: 1.5-KS-12000 Solid Propellant Motor', Report 0139-01-1 through 8, 10 Nov. 1958. (*0057)

[75] Aerojet-General Corporation, Sacramento, 'Large Solid Rocket Motor Program, TW-1 Motor Firing Report', Report No. 0434-0158-T3 (8 Aug 1961) and '... 100 FW-1 Motor Firing Report', Report No. 0434-01S11-T4 (16 Oct. 1961).

[76] Allegany Ballistics Laboratory, 'Status of Development Project', ABL-QPR 10, 15 Feb. 1959. (*0069)

[77] American Rocket Society, Solid Rocket Symposium, Waco, Texas, 1962, 'Solid Rocket Optimization Session', ARS preprints 62-2315 to 2319.

[78] Astrodyne, Inc., 'Development and Qualification of High Performance Case-Bonded, Solid Propellant Rocket Motors, Report No. 760-1-59, 2KS-10, 650 (1 Apr. –30 June 1959). (*0028)

[79] Astrodyne, Inc., 'Development of Composite Propellant Booster Unit', Report 726-1-58. (*0053)

[80] California Institute of Technology, 'Fundamentals of Solid Propellant Rockets', Chapter VII in *Jet Propulsion*, prep. for the Air Technical Service Command, 1946, pp. 169-171.

[81] Canadian Armament Research and Development Establishment, 'Design of Conical Conduit Bi and Tri-Propellant Charges', CARDE Tech Memo 237/59, May 1959. (*0035)

[82] Explosive Research and Development Establishment, 'Progressive Charges with Constant Thrust: Concentric Cylinder and Double Cone Types', Tech Memo No. 6/M/55, Oct. 1955. (*0083)

[83] Grand Central Rocket Company, 'Development of a Propulsion System for the Nike Zeus Missile', Report 1, Jan.–Dec. 1956. (*0064)

[84] Hercules Powder Co., Chemical Propulsion Div., Bacchus, 'Minuteman Stage III, Weapon System 133-A (U)', QPR MCS-51 (Apr.–June 1960), Conf.

[85] Jet Propulsion Laboratory, 'Development of Composite Propellants in Rocket Motors', Research Summary No. 5, 15 Oct. 1959. (*0031)

[85] Jet Propulsion Laboratory, 'Dual Propellant Design Studies', CBS No. 53, 15 Jun 1956. (*0046)

[87] Lockheed Propulsion Co., Redlands, '156-Inch Diameter Motor Liquid Injection TVC Program (U)', QPR 1 (30 June 65), 699-Q-1, RPL-TR-65-148, and QPR 2 (3 Sep. 65), LPC 699-Q-2, RPL-TR-65-212, Conf. (#1090)

[88] National Advisory Committee for Aeronautics, 'Analytical and Experimental Studies of Spherical Solid Propellant Rocket Motors', RML 57G12A, 16 Aug. 1957. (*0066)

[89] National Aeronautics and Space Administration, Washington, D.C., 'Feasibility Demon-

stration of Large Solid-Propellant Motors (U)', Nov. 1962, 70 pp., Conf. (#0980)
[90] Naval Ordnance Test Station, 'Solid Propellant Progress', Tech Program Report 201, 10 Nov. 1957, NOTS 1880. (*0051)
[91] Naval Powder Plant, Indian Head, 'Form Functions for Use in Interior Ballistics and Closed Chamber Calculations', Memo Report No. 3, 15 Feb. 1951. (*0099)
[92] Rohm and Haas Co., Redstone Div., 'Quarterly Progress Report on Interior Ballistics', 10 May 1954. (*0085)
[93] Rohm and Haas Co., Redstone Div., Report No. P-55-7, 10 May 1955. (*0079)
[94] Rohm and Haas Co., Redstone Div., Report No. P-55-15, 10 Aug. 1955. (*0081)
[95] Rohm and Haas Co., Redstone Div., Report No. P-56-1, 10 Feb. 1956. (*0043)
[96] Rohm and Haas Co., Redstone Div., Report No. P-59-7, Jan. 1959–Apr. 1959. (*0008)
[97] Rohm and Haas Co., Redstone Div., Report No. P-59-13, 14 Apr. 1960. (*0119)
[98] Rohm and Haas Co., Redstone Div., Report No. 59-19, July 1959–Oct. 1959. (*0010)
[99] Rohm and Haas Co., Redstone Div., 'Use a New Propellant Composition in the Pershing Missile', Report No. S-23, 29 Dec. 1959. (*0004)
[100] Thiokol Chemical Corp., Redstone Div., Huntsville, 'Pershing Propulsion System Development Program (U)', Progress Repts. C-A-62-101A, -116A, -142A (21 Nov. 1961–20 Feb. 1962), Conf. (#1020)
[101] Thiokol Chemical Corp., Wasatch Div., Brigham City, 'Minuteman Development of Large, High Performance Solid Propellant Rocket Engines', QPR No. 1, AF 33(600)-36514, 24 Feb–31 May 1958. (*0030)
[101] Thiokol Chemical Corp., Wasatch Div., Brigham City, 'XM-38 Solid Propellant Rocket Motor, Vol. II: SN-73 (Goose) Guided Missile (weapon System 123A)', TU-68-5-59, June 1959. (*0029)
[103] Thiokol Chemical Corp., Wasatch Div., Brigham City, '156 Inch Fiberglass LITVC Motor Program (U)', AFRPL-TR-65-192, Vol. I and II, QPR 1 (15 May–30 Aug. 1965), and AFRPL-TR-66-19,(Vol. I and II, QPR 2 (Sep.-Nov. 1965), Conf. (#1091)

UNE FAMILLE DE PROPULSEURS BI-ERGOLS
À POUSSÉES COMPRISES ENTRE 40 ET 1 kp

Technologie et possibilités d'application
pour le Satellite 'Symphonie'

H. STROBL et G. MUNDING

Messerschmitt-Bölkow, Allemagne

Abstract. Bi-propellant rocket propulsion engines with combustion chamber temperatures exceeding 3000 °C are cooled in such a manner that one propellant component is passed through a cooling jacket prior to injection. This cooling method is highly expedient for greater units. For smaller engines, however, it is problematic, since the cooling capacity of the smaller quantities of liquid available is no longer sufficient to ensure regenerative cooling of the total surface of the combustion chamber. A practical limit is set at thrusts of about 50 kp. The cooling problems in connection with smaller thrusts must be solved in a more unconventional manner by applying film, capacitive, ablation, and radiation cooling, or relevant cooling method combinations.

Messerschmitt-Bölkow was confronted with this specific task when manufacturing a 40 kp-engine for the hypergolic propellants N_2O_4 and AZ50 for the third stage of the ELDO-A launcher. This task was solved by the development of an injection procedure permitting combined engine cooling (film, regenerative, and radiation cooling) as well as high combustion efficiency despite of commercially available materials used (V2A, nimonic, titanium).

The 40 kp-engine is ready for series production since 1964. About 100 engines of this type were subjected to more than 1000 ground and altitude tests with a total test time of more than 100 hours. This total test time includes an uninterrupted permanent test of 3 hours duration without cut-off necessity. This engine is equipped with an electrically operated pneumatic propellant cut-off unit and excells – as to the required mission of the ELDO-A launcher – by a proven total reliability of 97,8 % (confidence level for 6 min mission of ELDO-A 90 %).

It was a matter of fact to try out whether reasonably scaled down injection systems would also be of satisfactory operation. The following engines were subject to gradual diminution:

The 8 kp-engine was established in 4 specimens in 1965. The engine was ground-tested in 58 tests with a total test time of $2\frac{1}{2}$ hours by means of test-bay integrated propellant cut-off units.

The 5 kp-engine established in 3 specimens in 1965 underwent 43 tests with a total test time of 3 hours.

The 3 kp-engine established in 12 specimens in 1966 was subjected to ground testing. A total of 39 permanent tests of up to 30 min duration, and 5 pulsation tests of up to 15 Hz, were carried out by means of test-bay integrated propellant cut-off units. A total of about 6450 ignitions were initiated within a total test time of about 6 hours.

The 1 kp-engine was started with in 1967. A great variety of engine versions (more than 30) were hitherto subjected to 100 tests with 750 ignitions. The tests were carried out under ground and altitude conditions. The total test time is more than 7 hours. The longest uninterrupted test lasted 30 min.

The bi-propellant engines mentioned excell by high specific impulse (between 285 and 300 sec), practically unlimited lifetime and constant thrust.

The engines of this family offering clear advantages against other conceptions could – due to their specific characteristics and due to the propellant storageability – be proposed for the communication satellite 'Symphonie'.

Several 1 kp-engines are provided to serve for the orbit correction system of the satellite because of advantages not offered by hydrazine monopropellant engines coming into question for this project. These advantages are:

(1) Higher specific impulse – hence weight advantages,
(2) Faster thrust build-up and thrust decay – hence better pulsability for more exact corrections,
(3) Hypergolic ignition of propellants – hence no catalyst.

A 40 kp-engine operating with the same propellants was proposed for apogee propulsion pur-

G. A. Partel (ed.), Proceedings of the Second International Conference on Space Engineering. All rights reserved.

poses. In comparison to a solid propellant engine possible for such a mission, the above 40 kp-engine shows the following advantages:

(1) Higher specific impulse – hence weight advantages,

(2) Low thrust level for a propulsion time of 18 min (instead of 23 sec with a solid propellant engine) – hence low acceleration; gain of weight in satellite construction,

(3) Possibility for cut-off and reignition – hence possibility of more exact injection,

(4) Possibility to test flight-ready propulsion units prior to launch – propulsion unit not wasted as is usual in case of solid propellant engines – hence better conditions for reliability determination and lower development costs.

1. Introduction

La Société Messerschmitt-Bölkow travaille depuis 1957 au développement de propulseurs à ergols liquides, et depuis 1960 entre autre et avec un intérêt particulier aux petits propulseurs à moyenne énergie utilisés pour les contrôles d'attitude et d'orbite.

Il existe dans ce domaine pour les propulseurs chimiques deux solutions:

– propulseurs à monergols avec décomposition par catalysateur,

– propulseurs bi-ergols de préférence avec des carburants hypergoliques (figure 1).

Ils sont tous les deux développés chez Messerschmitt-Bölkow. Tandis que les propulseurs mono-ergols sont encore au stade de développement, les propulseurs bi-ergols sont plus avancés et quelques-uns sont déjà qualifiés.

Pour définir les raisons qui ont poussé à préférer les propulseurs bi-ergols pour le

Fig. 1.   Groupe de propulseurs de petite taille de Messerschmitt-Bölkow.

satellite expérimental de télécommunications on a établi des comparaisons entre les deux solutions représentées au tableau I d'après le point de vue de la Maison.

On peut voir d'après le tableau I que les propulseurs bi-ergols présentent, étant donné l'impulsion spécifique plus élevée, des avantages de masse pour une impulsion totale plus élevée, que leur pulsabilité est bien meilleure et que la gamme de température du stockage des ergols est bien plus avantageuse.

C'est pourquoi cet exposé traitera, en rapport avec Symphonie, uniquement ces propulseurs bi-ergols utilisant les ergols Aerozine 50 et le tétroxyde d'azote.

C'est à l'intérieur de la chambre qu'a lieu à 3000°C une combustion effective des ergols. C'est pourquoi le refroidissement pose un grand problème: un refroidissement régénératif complet de la chambre de combustion n'est plus possible pour les propulseurs dont la poussée est inférieure à 50 kp, la capacité de refroidissement de la quantité d'ergols n'étant plus suffisante.

Notre équipe chargée du développement à Lampoldshausen trouva une solution avantageuse à ce problème: un procédé d'injection fut développé qui permet le refroidissement des propulseurs par film, régénération et rayonnement et avec lequel il est tout de même possible, en utilisant des matériaux courants sur le marché et en évitant l'emploi de matériaux d'ablation d'obtenir un rendement de combustion allant jusqu'à 98%. La durée de vie des propulseurs pour une poussée restant constante est ainsi pratiquement illimitée.

Le système d'injection fonctionne de telle sorte qu'un film d'oxydant homogène se propageant le long de la paroi de la chambre de combustion est consommé sous l'effet du carburant pulvérisé centralement, l'ergol en s'évaporant maintient la chambre

TABLEAU I

	Hydrazine propuls. monerg.	N₂O₄/AZ50 propuls. bi-ergols
Masse du système d'après une impulsion totale comprise entre 4000 et 8000 kp	100%	70–80%
Impulsion spec. eff. dans le vide	225 sec	290 sec
Pulsabilité	10 Hz en fonction de la température de départ	30 Hz
Catalysateur	oui	non
Température à la paroi de la chambre de combustion	700°C sans isolation 150°C avec isolation	120°C
Stockabilité des ergols	applicable pour la mission du satellite	applicable pour la mission du satellite
Température limite du stockage	a partir de 0°C	a partir de −6°C
Fiabilité des propulseurs prévision de succès pour plusieurs années	env. 0.98	env. 0.98
Stade de développement	construction en série aux USA et au stade de modèle de développement en Europe	types testés de nombreuses fois en Europe
Pièces non trouvables sur le marché en Europe	catalysateur	aucune

de combustion sur toute sa surface à une basse température. La partie du col de la chambre de combustion est refroidie par régénération par le carburant. La tuyère de détente est refroidie par rayonnement.

Ce système est employé pour toute une famille de propulseurs dont la poussée est de 40, 8, 5, 3 et 1 kp (Figure 2).

2. Développement conduisant aux propulseurs bi-ergol de petite taille

Etant donné les expériences positives faites par l'équipe de Lampoldshausen lors du développement du moteur vernier 40 kp, on a essayé de fabriquer en plusieurs étapes successives des propulseurs de plus en plus petits.

Le propulseur 40 kp est un moteur vernier conçu pour le 3ème étage de la fusée ELDO-A dont le développement commencé en 1964 conduisit à une fabrication en série. Avec un nombre de propulseurs supérieur à 100, on a réalisé 5200 allumages pendant plus de 1000 essais au sol et en altitude, y compris aussi des allumages faits sous vide poussé pendant 100 heures d'essai, ainsi qu'également un essai continu pendant plus de 3 h avec un propulseur sans arrêt sans nécessité d'interruption. Ce propulseur comprenant des valves d'arrêt incorporées a, pour la mission requise de la fusée ELDO-A, une fiabilité prouvée de 97.8% avec une affirmation de 90% pour un temps de fonctionnement de 6 min.

Le propulseur 8 kp a été construit en 1965 en 4 exemplaires et éprouvé au sol

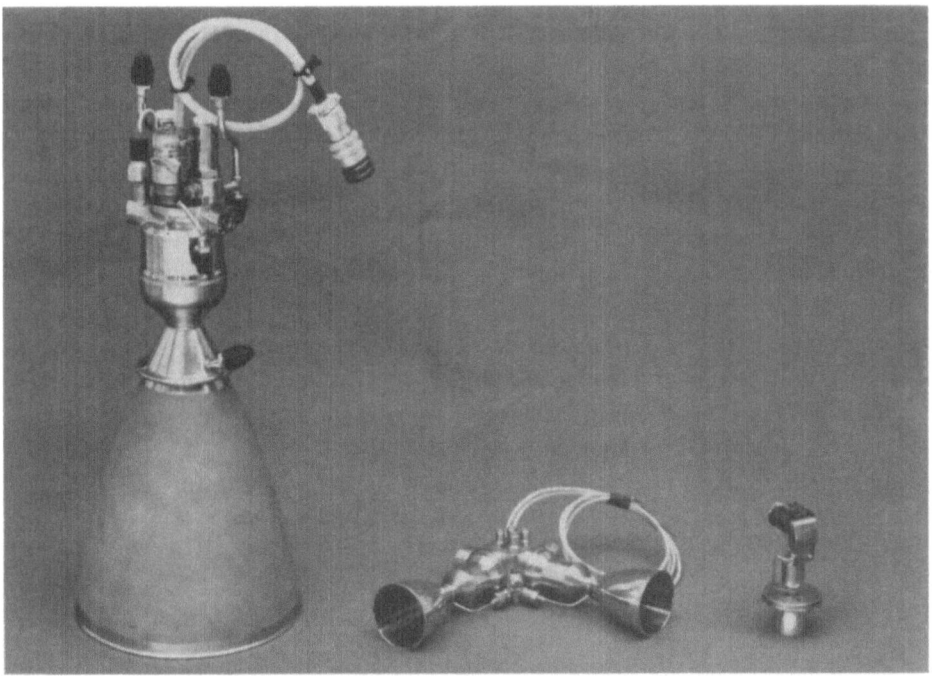

Fig. 2.   Propulseurs bi-ergols à 40, 3 et 1 kp de poussée.

Fig. 3.   Maquette du satellite 'Symphonie'.

pendant 58 essais d'une durée de deux heures et demie avec des valves d'arrêt appartenant aux bancs d'essai et non au propulseur.

Le propulseur 5 kp a été également construit en 1965 au nombre seulement de 3 exemplaires qui furent éprouvés pendant 3 h au cours de 43 essais.

Le propulseur 3 kp a été construit en 1966 en douze exemplaires et éprouvé au sol. Avec des valves d'ergols appartenant aux bancs d'essai ont été réalisés 39 essais continus durant jusqu'à 30 min et 5 essais d'impulsion allant jusqu'à 15 pulsations/sec. 6450 allumages ont eu lieu pendant une durée totale d'environ 6 h.

Le propulseur 1 kp ne vit le début de son développement qu'en 1967. 20 propulseurs avec 35 systèmes d'injection furent éprouvés au sol et en altitude pendant environ 7 h au cours de 100 essais comprenant 750 allumages ainsi qu'un fonctionnement interrompu de 30 min. Ce propulseur est conçu pour une correction d'orbite de satellites de grande dimension et de longue mission.

Nous ne décrirons ici en détails que le propulseur 40 kp et le propulseur 1 kp, ces deux propulseurs étant utilisés pour le satellite de télécommunications expérimental franco-allemand 'Symphonie'.

3. Le propulseur 40 kp (Figure 4)

Ce propulseur présente le système d'injection mentionné antérieurement. Avec la chambre de combustion, les valves forment un ensemble intégré incorporable. Toutes

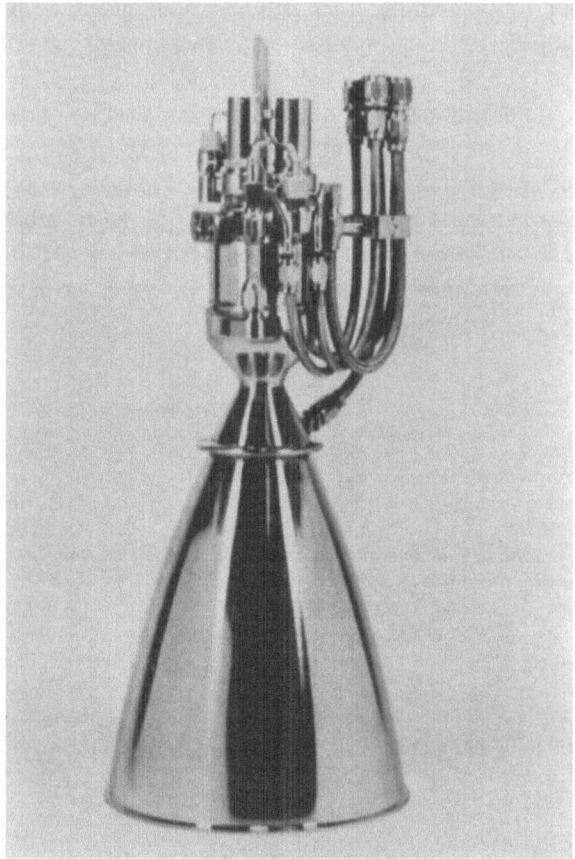

Fig. 4.   Propulseur 40 kp.

les pièces servant au fonctionnement, telles que la tête d'injection, la chambre de combustion, la tuyère de détente et la chemise réfrigérante sont soudées entre elles par jet d'électrons ou par soudure d'argonarc.

Les valves d'arrêt d'ergols sont vissées dans la tête d'injection, l'étanchéité étant réalisée au moyen de joints teflon. Elles sont commandées indirectement, c'est-à-dire une soupape de précommande électro-magnétique également intégrée commande le gaz (helium) de pressurisation pour l'ouverture et la fermeture simultanée des 2 valves d'ergols pour le comburant et le carburant. La consommation de courant est nulle aussi bien dans la position fermée que dans la position ouverte. Seul le changement de position des valves entraîne des coups de courant de 100 msec à 28 V et 750 mA pour l'ouverture et 320 mA pour la fermeture.

L'étanchéité des soupapes est obtenue par la combinaison métal–téflon. Lors d'essais à longue durée, on constata un taux de fuite de 10^{-5} torr l/sec par siège. La fiabilité totale du propulseur 40 kp est à présent pour un fonctionnement d'une durée de 6 min pour le 3ème étage de la fusée ELDO-A de 97.8% avec une affirmation de 90%. Cette fiabilité totale comprend les fiabilités d'allumage, de fonctionnement et de fin de combustion. Des essais réalisés jusqu'en janvier 1969 sur 50 propulseurs pendant plus de 58 h ont donné cette fiabilité totale.

Le tableau II résume les données techniques du propulseur 40 kp tel qu'il est utilisé dans Symphonie.

La figure 5 montre la répartition des températures le long de la périphérie extérieure du propulseur. D'après cette figure on peut voir que la chambre de combustion proprement dite, et que le col de la tuyère ne dépassent pas 120°C. La tuyère de détente refroidie par rayonnement atteint 800°C, ce qui n'est pas une température trop élevée pour le matériau Nimonic (figure 6).

TABLEAU II

Poussée nominale dans le vide	40 kp
Oxydant	N_2O_4 (tetroxyde d'azote)
Combustible	AZ 50 Aerozine
Pression nominale de chambre de combustion	7 at
Rapport de mélange Ox/comb	1.7
Rapport de mélange admissible	1.6 à 2.2
Impulsion spécifique sous vide	300 sec $\pm$ 2 sec
Débit d'ergols	0.1335 kg/sec
	($= 0.0841\,Ox + 0.0494\,comb$)
Rapport de détente	1000:1
Pression d'injection	11.5 at Ox et 13 at comb
Température admissible des ergols stockés	$-6°C$ à $+60°C$
Masse du propulseur, valves d'arrêt inclues	1.8 kg
Longueur hors-tout	392 mm
Diamètre hors-tout (à l'extrémité de la tuyère)	150 mm
Stade de développement	série

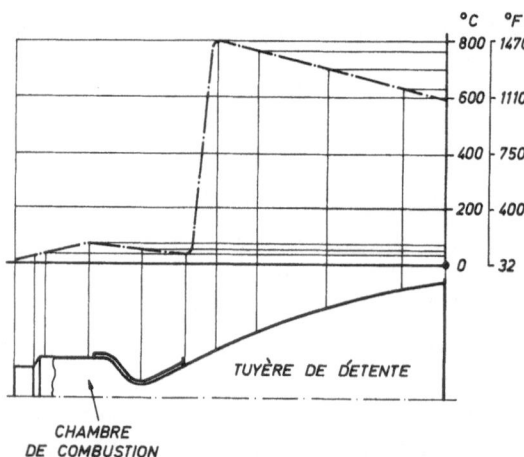

Fig. 5. Profil de la température le long de la chambre de combustion du propulseur 40 kp.

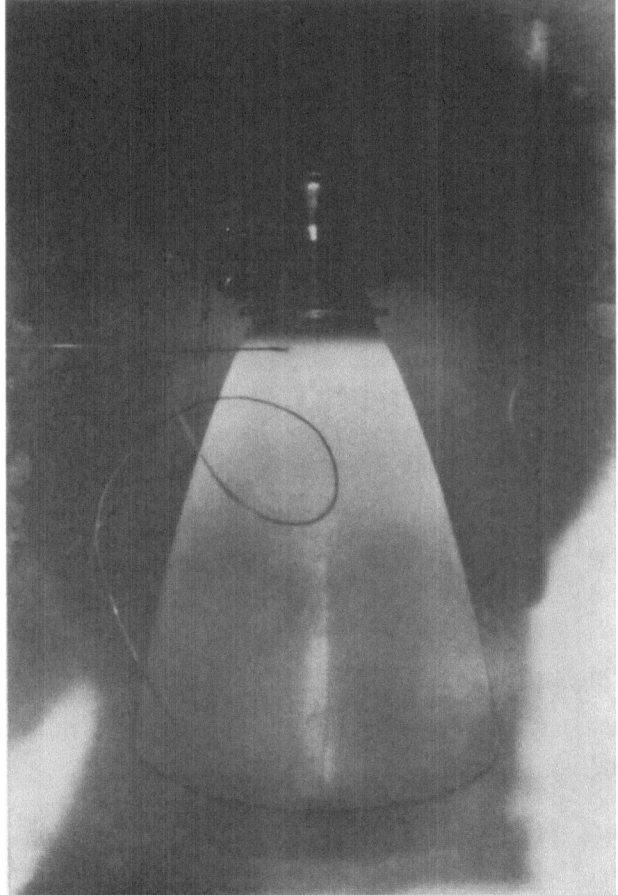

Fig. 6.   Profil incandescent de la tuyère en fonctionnement.

Ce propulseur 40 kp est prévu pour la propulsion d'apogée dans le satellite Symphonie. Un seul propulseur forme avec un système de réservoirs un sous-système facile à intégrer dont le but est de donner à une masse au départ du satellite de 360 kg un incrément de vitesse de 1473 m/sec. La trajectoire de transfert elliptique est transformée de ce fait en une trajectoire circulaire géostationnaire. La masse finale se monte à 200 kg sans la structure du moteur d'apogée.

La tableau III indique les données les plus importantes.

La figure 7 montre un schéma de propulsion. Le propulseur est relié au réservoir d'ergols par un dispositif de poussée. Le réservoir en titane est lui-même divisé en deux parties au moyen d'une paroi intermédiaire pour les deux composants liquides. Pour des raisons de température on a monté le réservoir d'hélium soumis à une pression de 400 at pour l'alimentation des gaz de pressurisation dans la partie réservée au carburant. Pour éviter une fuite éventuelle du gaz on a placé sur le réservoir principal une soupape de surpression pour 14 at, pression à laquelle le réservoir et

TABLEAU III

Masse d'ergols stockée (AZ 50/N_2O_4)	144 kg
Gaz de pressurisation (He)	0.5 kg
Structure du moteur d'apogée	15.5 kg
Masse du satellite	200 kg
Masse au départ (orbite de transfert)	360 kg
Poussée	40 kp
Durée de combustion	1060 sec
Accélération initiale	1.09 m/sec^2
Accélération finale	1.80 m/sec^2

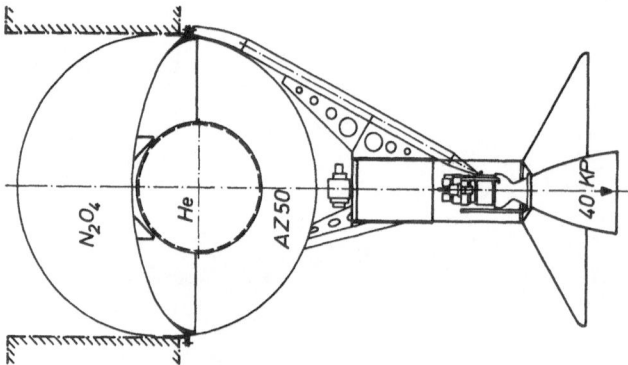

Fig. 7.   Schéma du moteur d'apogée.

la paroi intermédiaire ont été conçus pour résister en toute sécurité. Les accessoires tels que réducteur de pression, valves pyrotechniques et valves anti-retour sont incorporés à proximité du propulseur. Un bouclier thermique protège la surface du satellite contre un rayonnement de la tuyère. Le remplissage et éventuellement la vidange se font au moyen d'un appareil de remplissage indépendant par des raccords se trouvant à la périphérie du satellite. Des tuyaux d'aspiration allant jusqu'aux endroits les plus bas servent à faire la vidange. Une séparation mécanique entre gaz et liquide est superflue, étant donné qu'ils sont séparés par des forces de rotation dues à un spin d'environ 120 t/min. C'est pourquoi les endroits où les ergols sortent du réservoir se trouvent au point le plus éloigné du centre. L'aménagement décrit du système de propulsion liquide pour l'injection du satellite Symphonie dans un orbite géostationnaire présente des avantages, comparé aux moteurs à poudre d'une force de poussée allant jusqu'à 1.7 et 1.8 tonnes et une combustion plus courte durant 22 et 23 sec.

Ce sont des avantages que l'on exploitera d'une part en général et d'autre part dans le cas de Symphonie:

– Une impulsion spécifique élevée entraîne un gain de charge utile.
– Les risques de développement sont minimes, étant donné que le propulseur lui-même est déjà qualifié.
– La fiabilité qui peut être obtenue est plus élevée pour la période de développement

donnée, les équipements de vol pouvant être éprouvés au sol avant chaque fonctionnement.

- Une précision d'injection plus grande, le moteur pouvant être arrêté et reallumé.
- Une structure plus légère du satellite car des efforts mécaniques moindres se manifestent, étant donné une poussée plus faible.
- Moins de problèmes thermiques, une éjection ou une isolation thermique importante n'étant pas nécessaire.

Une augmentation de consommation en ergols en raison d'une longue durée de combustion (18 min), la poussée ne se faisant pas continuellement tangentiellement à la trajectoire, n'est que très peu sensible. Une étude a montré que l'influence pendant les 18 min ne se manifeste que par $\frac{1}{10}$ sec.

5. Propulseur 1 kp (voir figure 8)

Un autre propulseur de la famille des propulseurs bi-ergols est utilisé pour Symphonie: C'est le propulseur 1 kp prévu pour le système de correction d'orbite.

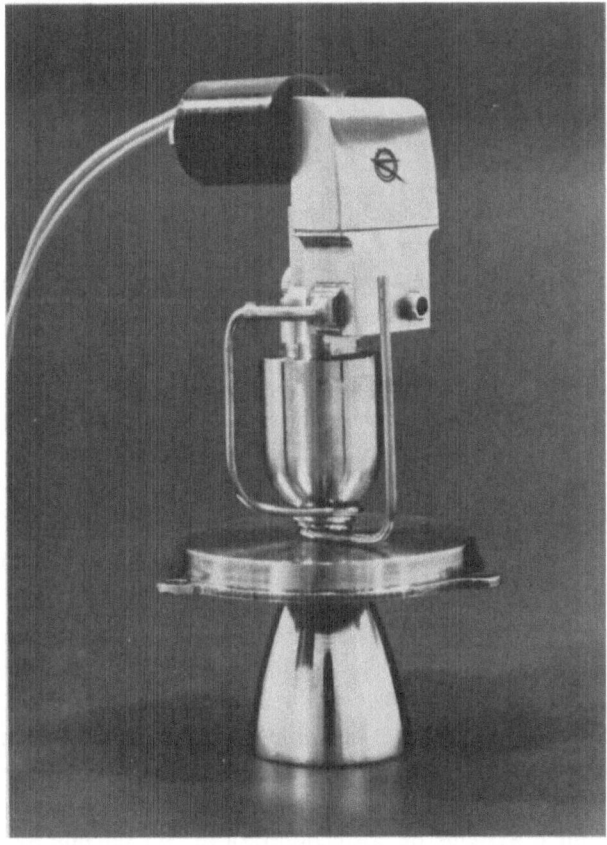

Fig. 8.   Propulseur 1 kp.

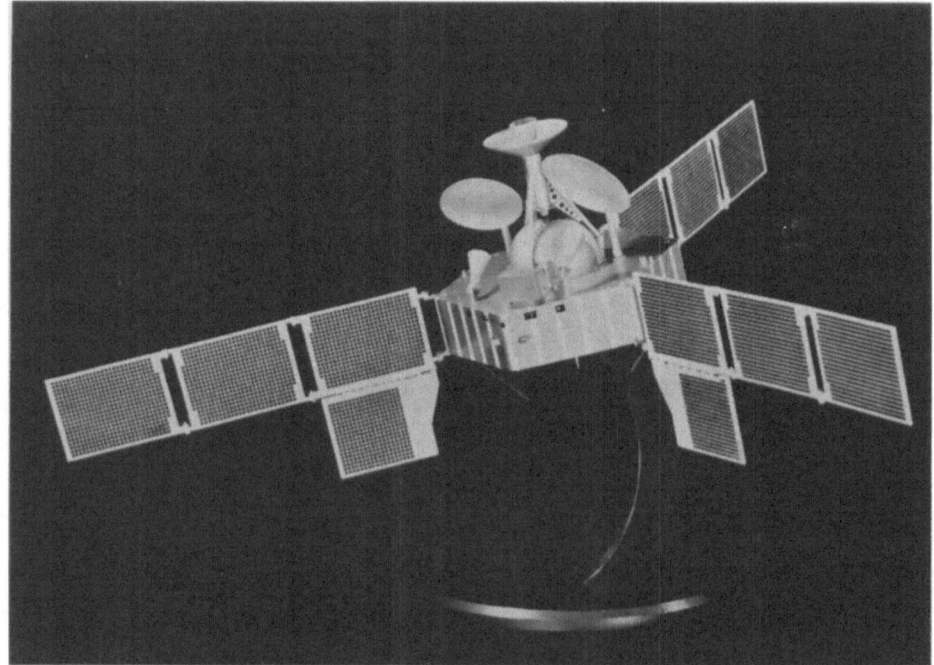

Fig. 9.   Maquette du satellite 'Symphonie'.

La chambre de combustion du propulseur 1 kp ressemble dans sa construction et sa fonction à celle du propulseur 40 kp. Elle possède également un refroidissement interne et régénératif. La tuyère de détente est refroidie par rayonnement. Pendant le fonctionnement la paroi de la chambre de combustion ne dépasse pas une température de 120 °C et la tuyère de détente une température de 800 °C.

La pression nominale de la chambre de combustion pour une poussée de 1 kp se monte à 7 at et la surpression d'injection à 3 at. Le propulseur peut fonctionner dans une gamme de poussée comprise entre 0.8 et 1.3 kp, la pression du réservoir diminuant progressivement. Cette propriété est très importante lors de l'utilisation de cette diminution progressive de pression des réservoirs dans le satellite.

La valve d'arrêt intégrée, spécialement développée par Messerschmitt-Bölkow, se trouve immédiatement au-dessus du système d'injection, sa commande n'est pas électro-pneumatique mais directement électrique. Son principe de fonctionnement se base sur un seul aimant qui ouvre ou ferme en même temps les deux valves. Sans qu'aucun courant ne passe, ces deux valves restent fermées sous l'action d'un ressort.

Elle présente ainsi que la courte chambre de combustion de bonnes conditions pour obtenir des nombres de pulsations élevés.

Le tableau IV indique les données techniques les plus importantes.

On a prévu pour le système de correction d'orbite du satellite Symphonie 7 propulseurs 1 kp.

TABLEAU IV

Poussée nominale dans le vide	1 kp
Pression nominale de la chambre de combustion	7 at
Rapport de mélange Ox/comb	1.6
Impulsion spécifique dans le vide	285 sec
Pulsabilité	30 Hz
Débit d'ergols	3.5 g/sec
Masse du propulseur avec valves d'arrêt inclues	180 g
Longueur hors-tout	120 mm
Diamètre hors-tout	45 mm
Stade de développement: modèles testés de nombreuses fois par essais au sol et en altitude.	

Des corrections ouest et nord-sud sont ainsi possibles pour compenser les perturbations de trajectoire en raison de l'ellipticité de la terre, de l'attraction du soleil et de la lune et de la pression de rayonnement, ainsi que des erreurs d'injection. Dans le cas de corrections N–S, c'est-à-dire d'inclinaison, la pulsabilité des propulseurs permet que la direction de la poussée totale passe par le centre de gravité du satellite.

Pour le temps de mission de 5 ans, les exigences d'impulsion pour le système à gaz chaud avec ses propulseurs se monte à environ 6700 kpsec. Les propulseurs peuvent être actionnés indépendamment l'un de l'autre pour un fonctionnement continu ou un fonctionnement d'impulsion.

Comme le montre le schéma suivant, le système de propulsion est, pour des raisons de redondance, subdivisé en deux systèmes entièrement indépendants l'un de l'autre (Figure 10).

Pour fournir les conditions nécessaires à une fiabilité élevée on fera diminuer

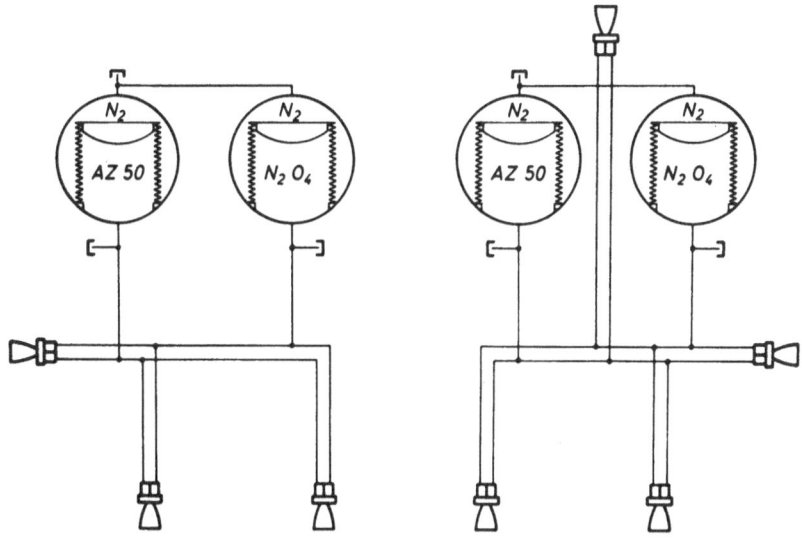

Fig. 10   Schéma de propulsion du système à gaz chaud.

progressivement la pression des réservoirs. Ceci est donc un aménagement sans réducteur.

A cause de la non-pesanteur pendant la mission, une séparation mécanique de l'azote et du liquide devra se faire dans les réservoirs. On prévoit pour cela des soufflets.

Les 4 réservoirs d'ergols en Titane pour une masse totale d'ergols de 23.4 kg (19.9l) ont le même volume, ce qui donne le rapport de mélange mentionné de 1.6 en raison des masses spécifiques des composants d'ergols différentes. Afin d'éviter lors de la vidange progressive des réservoirs un déplacement sensible du centre de gravité, sont incorporés deux par deux, face à face à de différentes distances de l'axe central du satellite. La masse totale du système à gaz de correction d'orbite est de 27.3 kg.

DESIGN OF THE APOLLO SERVICE MODULE ROCKET ENGINE FOR MANNED OPERATION

A. L. FELDMAN and DAN DAVID

Aerojet General Corporation, Sacramento, Calif., U.S.A.

Abstract. The Apollo spacecraft service module engine utilizes a liquid rocket engine generating 20000 lb of thrust. Presented is a summary of the techniques employed to ensure maximum pilot safety. Man-rating techniques described include extensive use of redundancy on critical moving parts and the use of unusual design margins on other parts. Particular emphasis is placed on the redundancy developed for the bi-propellant valve and the gimbal actuators.

1. Introduction

The Apollo service module provides the primary propulsion system, electric power supply, and part of the basic life support apparatus for the Apollo spacecraft command module. The service propulsion system (SPS) rocket engine is employed for all man-euvers requiring major spacecraft velocity changes subsequent to insertion of the vehicle into translunar flight by the S-IVB third stage of the Saturn V launch vehicle. It is used for midcourse corrections, lunar orbit injection, lunar mapping and rendez-vous maneuvers, and transearth injection.

This engine is utilized also in the event it is necessary to abort the spacecraft after completing the initial launch to Earth-orbit phase. Firing durations planned for the engine are varied. A typical lunar firing duty cycle is shown in Table I; the engine firings which may be required to abort the spacecraft during a lunar mission are shown in Table II.

TABLE I

Lunar orbit mapping mission

Type of maneuver	Firing duration = seconds)
Translunar midcourse correction	17
Translunar midcourse correction	7
Translunar midcourse correction	7
Lunar orbit injection	380
Mapping and rendezvous maneuvers	40 (10 3-sec firings)
(total of twenty maneuvers)	(10 1-sec firings)
Transearth injection	140
Transearth midcourse correction	5
Transearth midcourse correction	2
Transearth midcourse correction	2
Total No. starts	28
Total duration	600 sec

G. A. Partel (ed.), Proceedings of the Second International Conference on Space Engineering. All rights reserved.

TABLE II

Abort of lunar mission

Type of maneuver	Firing duration (seconds)
Translunar midcourse correction	9
Translunar midcourse correction	1
Maximum abort maneuver	586
Transearth midcourse correction	3
Transearth midcourse correction	1
Total No. Starts	5
Total duration	600 sec

There is no backup to the SPS engine for the critical firings, of a lunar mission. It must be, without question, of a basically reliable and proven design.

2. Engine Description and Design Approach

Figure 1 shows the SPS engine, which basically consists of a bipropellant valve, injector, combustion chamber, nozzle extension, and gimbal actuators. It is over 13 ft in height and has a maximum diameter of 8 ft. It is designed to operate for at least 750 sec, producing a vacuum thrust of 20 500 lb, and is capable of 36 separate starts during a mission. Earth-storable hypergolic propellants, nitrogen tetroxide (oxidizer), and a 50–50 mixture of hydrazine and unsymmetrical dimethylhydrazine (fuel) are used at a mixture of 1.6:1 (oxidizer to fuel) and at a nominal chamber pressure of 99 psia. Basic engine operating parameters are shown in Table III.

The high reliability and safety requirements for a man-rated engine were considered and planned into each design and development activity undertaken. Principal elements assuring the required reliability are (1) simplicity in concept; (2) maximized redundancy; (3) additional design margins for nonredundant parts; and (4) comprehensive testing and quality assurance programs.

A. BIPROPELLANT VALVE – MAXIMIZED REDUNDANCY

The element of simplicity was achieved by selection of a pressurized propellant feed system and the use of storable hypergolic propellants, thus alleviating requirements for a propellant turbopump and an ignition system. Examples of the applications of redundancy and unusual design margins as applied to the SPS engine follow. Two redundant components are discussed – the bipropellant valve, which is fully redundant, and the partially-redundant gimbal actuator. Three components with unusual design margins are presented – the injector, combustion chamber, and nozzle extension. At the initiation of the program, it was recognized that critical failure modes would include failure of the thrust chamber valves to open, failure of the valves to close, and failure of the valves to seal. To minimize these possibilities, the valve was designed in a doubly-redundant, series and parallel arrangement. A ball design was selected to

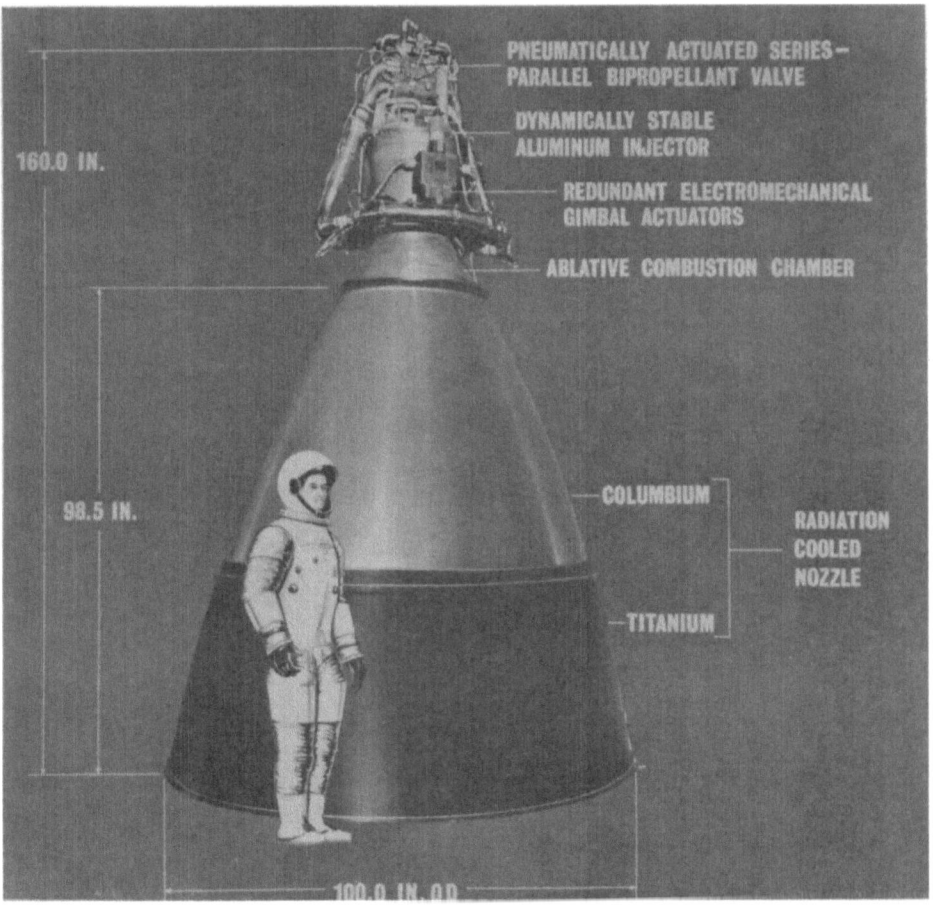

Fig. 1.   Apollo service propulsion system engine.

minimize the pressure drop of the valve system. The propellant flow system is shown in the schematic of Figure 2. Propellant delivered from the pressure-fed tanks is passed through articulated lines around the gimbal mount. As shown in the Figure 2 schematic, each propellant feed line terminates in a Y connection to supply the two parallel legs of the thrust chamber valve. The parallel arrangement of the valve passage is shown in the cutaway photograph of Figure 3. Two ball seals in series are then employed to seal the propellant from the injector; with this arrangement, the valve is actually four bipropellant valves in a single housing. The valve as installed on the engine is shown in Figure 4.

The valve actuation system is shown schematically in Figure 5. Nitrogen gas is stored in the gas-storage bottle at a pressure of 2500 psia. Prior to providing a firing signal, the stored gas shut-off valve is energized open, thereby allowing high pressure nitrogen gas to flow through the regulator and up to the two 3-way pilot valves. A firing signal will actuate the 3-way valves open allowing regulated nitrogen gas to

TABLE III

Engine operating data

Parameter	Nominal Value	Remarks
Thrust, lb	$20\,500 \pm 1\,\%$	Min of 18 500 and max of 25 000
Mixture ratio, O/F	1.6 ± 0.02	Capable of operation from 1.4 to 1.8
Oxidizer	N_2O_4	
Fuel	50% UDMH-50% N_2H_4	
Chamber pressure, psia	99 ± 1	Capable of operation from 91 to 124
Specific impulse, seconds	313	Minimum of 311
Propellant temperature, °F	70	Capable of operation with oxidizer 30–110 fuel 40–120
Total duration, seconds	750	Max continuous operation of 610 sec, minimum of 0.5 sec.
Restart capability	36	Limited only by actuation gas supply
Valve operation	Dual bore	Capable of operation on either single bore
Engine compartment temperature, °F	30 to 130	

flow into the actuation cylinders, thereby moving the piston outboard against a closing spring. Movement of the piston shaft (rack), which is meshed with a gear, rotates each fuel and oxidizer ball assembly open. The functional unit (4 per bipropellant valve) is shown in Figure 6. Shutdown is achieved by interrupting the firing signal, thereby causing the 3-way valve to close and simultaneously vent the gas from the actuation cylinder. This allows the spring to force the piston to its inboard position and hence rotate the ball pair to its closed position.

Each of the two pneumatic actuation systems has a separate and independent electrical harness. The bipropellant valve system was designed to be versatile in regard to modes of operation. The engine may be operated on either a single actuation system or on both systems simultaneously; it also may be started on one system and the redundant system opened seconds later. Shutdown can be achieved in a similar operating mode. This concept of redundancy, also utilized within the various elements of construction of the valve, is illustrated in Figure 7. Shown is one of the two propellant flow systems; the identical parallel redundant system is not shown, although it is incorporated within the same housing. The added redundancy in series is shown by the two seals on each ball. All ball drive shafts are provided with dual series shaft seals as well as shaft cover dual seals as shown in the gear cavity section. All four oxidizer inboard shaft covers (leakage-retention cavities) are manifolded together and routed overboard; the fuel circuit is similarly manifolded. Instrumentation is incorporated that provides a continuous ball pair assembly position indication to the astro-

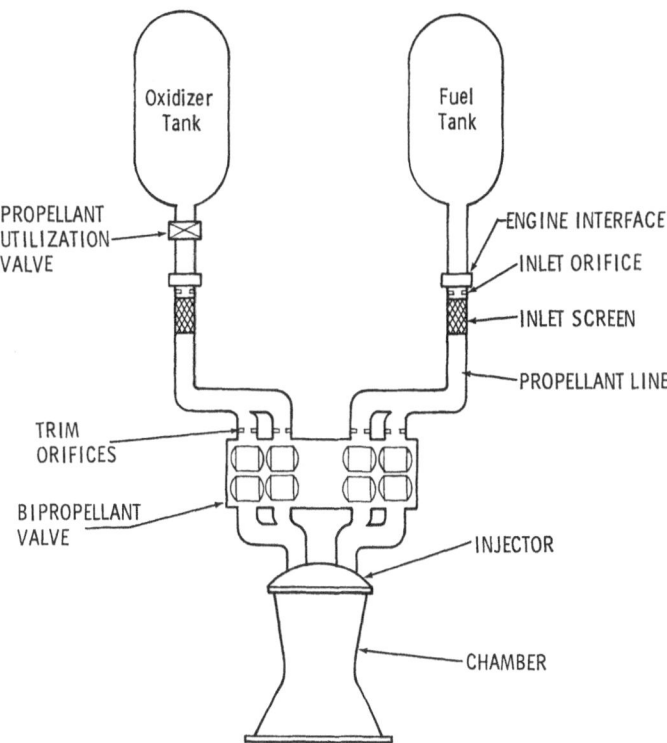

Fig. 2.   Propellant flow system.

nauts. This is accomplished by a linear potentiometer attached to each actuator (as shown in Figure 6). Redundancy also is provided within this component in that each potentiometer housing has two separate wiper assemblies with separate and redundant electrical windings and circuits.

Extensive tests have been conducted during bi-propellant valve development to demonstrate operation at both nominal and off-design conditions. Acceptance of the valve was further accomplished by qualification tests at the component level and by engine testing under both sea level and simulated-altitude conditions. These firings were conducted at the extreme limits of temperature, pressure, and propellant flow rates anticipated for planned space missions, and the bipropellant valves were utilized for a total number of firings over three times that planned for any mission. Dry cycle testing (opening and closing the valve to simulate dry system checkout functional tests) was accomplished for a total of approximately three times the number of normal mission valve operations. The wear-out effect of engine firings was simulated by opening and closing the valve under flowing propellant conditions for a total of 1500 cycles, compared to the maximum possible mission cycles of 36 engine starts. In addition, numerous valve flow tests were conducted at rated flow rates and at 160°F, which are the most severe valve flow conditions. Vibration testing was conducted at power levels approximately 10 times higher than experienced during normal engine firing.

Fig. 3. Bipropellant valve cutaway photograph.

A space mission was simulated by periodically operating a valve that was exposed to a hard vacuum for a 14-day period, the length of the longest-scheduled space mission. Adequacy of the valve for its intended thermal environment was demonstrated by functional and leak testing of the valve at its lowest temperature, which results from a long-duration coast period, and at its highest temperature, the result of heat soakback from chamber to valve after engine firings. The valve design was also subjected to all types of decontamination and purging operations planned to be utilized in engine test and spacecraft handling operations; the absence of adverse effect was verified by post-procedural functional and leak tests. Compatibility of the valve with the propellants was demonstrated by contact exposure of a valve to propellants under pressure for a 30-day period; pretest and posttest functional and leak tests verified the absence of any negative effects of the exposure. Valve bores, actuators, and pneumatic tanks were subjected to their respective design burst pressures without failure. In addition, all components of the pneumatic actuation system were subjected to the full range of expected temperature, pressure, vibration, and cycling environments during the course of the valve development program.

B. GIMBAL ACTUATORS – PARTIAL REDUNDANCY

The two electro-mechanical gimbal actuators provide the SPS engine with the ability to gimbal $\pm 4.5°$ at a minimum actuator linear drive rate of 1.57 in./sec. in both

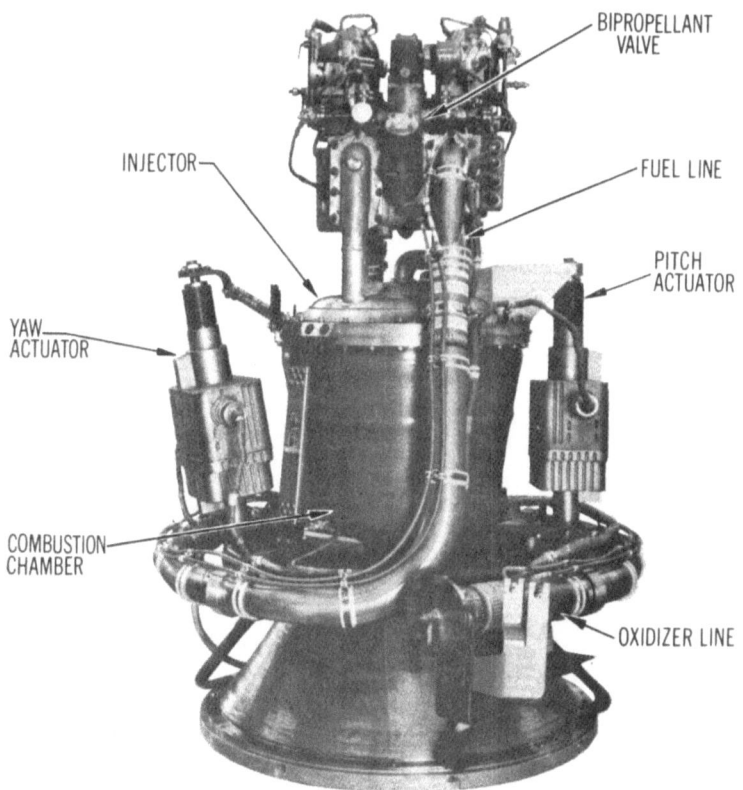

Fig. 4. SPS engine components relationship.

the pitch and yaw planes. The actuators are shown assembled to the engine in Figure 4. Redundancy is achieved in each actuator by providing two separate functional channels of drive and feedback elements; either channel is capable of satisfying all mission objectives. The mechanical elements of the actuator are shown in Figure 8. Each drive system consists of a direct current electric drive motor and two counter-rotating magnetic control clutches. By applying differential current to one clutch or the other, the actuator can be made to extend or retract on command and at a rate of motion proportional to the signal amplitude. The torque from the drive clutch transmits rotational power to the jack-screw drive nut, which in turn converts the rotary power input to linear motion. The coupling between the nut and screw is provided by recirculating ball bearings riding in a parallel double race to distribute the load. Each channel also has separate rate and position feedback transducers that function independently and provide data to the spacecraft servo system.

Redundancy of the electrical circuit is achieved by providing two electrical connectors on each actuator to isolate the power input, control signals, and feedback for each channel. The jack-screw assembly, thrust bearings, and mechanical linkage are

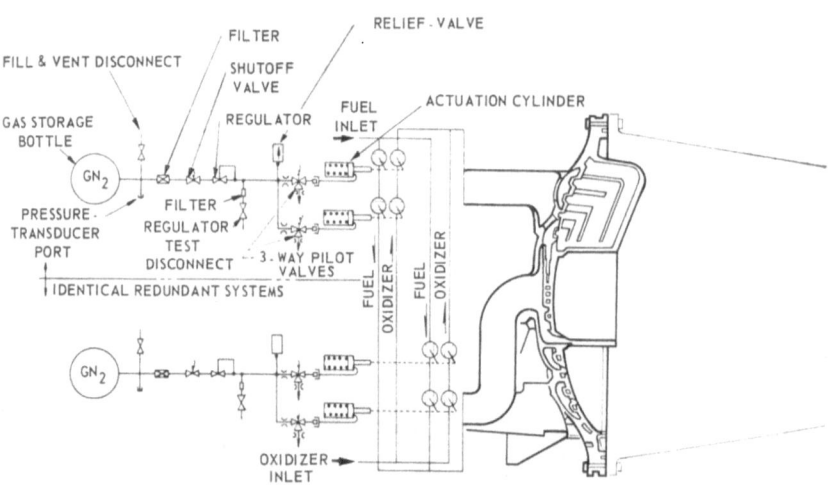

Fig. 5.   Bipropellant valve control system schematic.

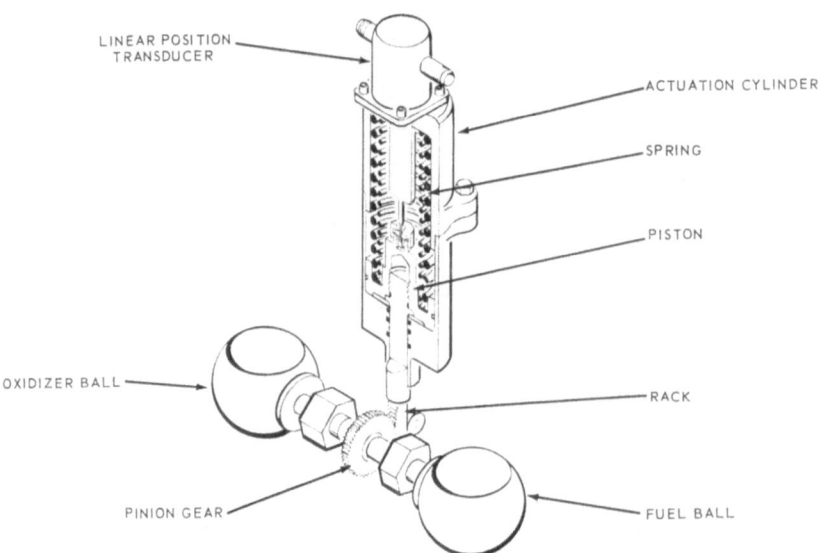

Fig. 6.   Bipropellant valve actuator and gear system.

common to both channels and are the basic non-redundant parts of the actuator assembly. To assure the required reliability in these areas, the system has been designed to withstand a linear force of 9000 lb, whereas the maximum emergency operating force is 1000 lb. The entire actuator mechanism described is assembled into a hermetically-sealed sheet metal container. A positive pressure of 3.5 psia is maintained within the container to protect the internal components from particle contamination as well as the wearing effects of prolonged vacuum.

Through a series of development and qualification tests, the gimbal actuator has

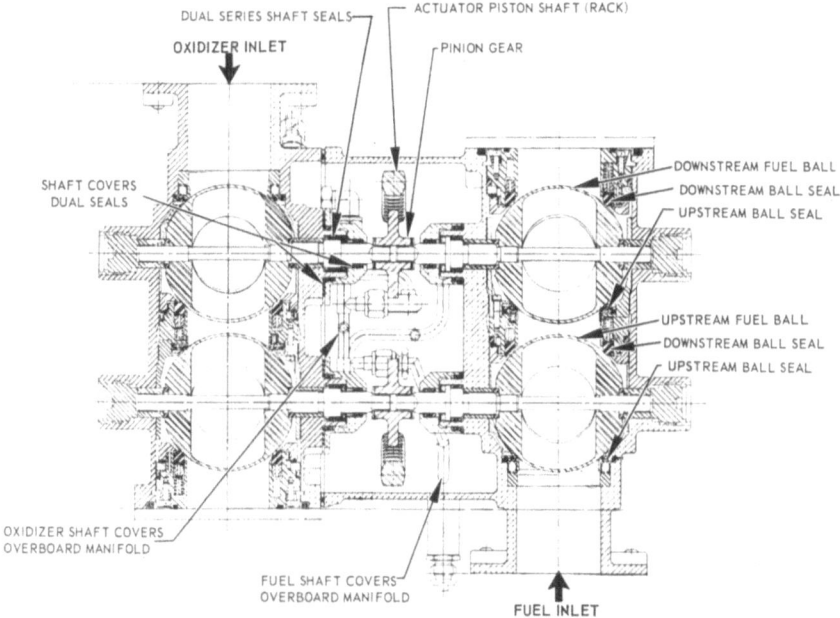

Fig. 7.   Bipropellant valve cross section.

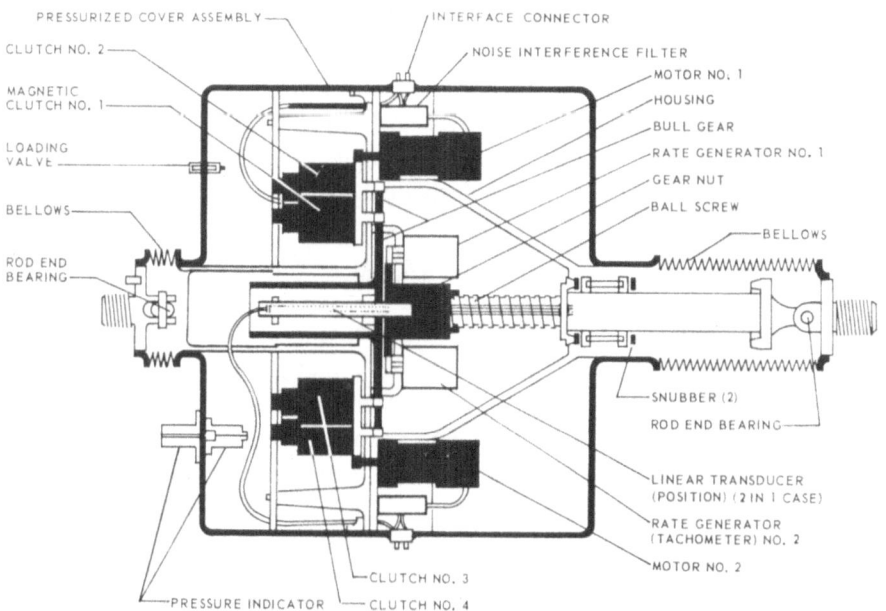

Fig. 8.   SPS engine gimbal actuator.

demonstrated that it can perform as commanded under extreme environmental and engine-operating parameters. Reliability has been proven by repeating these tests on each item of test hardware until the life of the item has been established or has been proven to be far beyond requirements. The tests conducted have been in three main categories – actuator components, actuator assembly, and engine firings with actuators installed.

The component tests designed to prove the ability of the assembled actuator to pass qualification requirements. In addition, they identified failure modes under life and environmental conditions not possible in actuator assembly tests. As an example, clutch slip rings and brushes were tested for 5 h at 10^{-6} torr vacuum, whereas actual flight usage in vacuum would not exceed 10 min, and then only if the sealed actuator cover was punctured. Likewise, gear trains have been run up to 50 h under maximum load, whereas the maximum duty cycle in this condition would be 10 min. Similar tests have been conducted on all functional actuator components.

The actuator assembly tests demonstrated that the actuator could perform at environmental temperatures from 0 to 140°F and at random vibration levels of up to 0.1 g^2/cps. In addition, each of the 15 life-cycle tests completed on each of four clutches in both the pitch and yaw actuators was equal to a lunar mission at maximum engine offset of 450 lb. The maximum force capability of the actuators was measured periodically throughout the life cycle testing, which proved that they still could meet the 1000 lb force requirement that may be required in the event of a spacecraft malfunction condition. Other cycle tests were conducted with the actuator interior at a 10^{-6} torr vacuum condition. Two life-cycle test series per clutch were successfully completed, proving that a container leak will not affect performance.

Simulated flight mission engine-firing tests with actuators installed and engine gimballing throughout the firing repeatedly demonstrated engine and actuator compatibility. Redundancy and the ability of the actuator to switch channels while maintaining gimbal control was verified during the flight of Apollo 7; with engine firing, the channels were switched without perturbation of the spacecraft.

3. Design Margins

Design reliability for the non-redundant components of the engine assembly is achieved by providing a significant performance margin over maximum mission requirements. In the case of the three most critical non-redundant components, a single injector has operated for over 9000 sec and been subjected to 480 starts; a single ablative combustion chamber has demonstrated a duration life of 2392 sec; and a nozzle extension demonstrated a total firing duration of 333 sec while being subjected to 280 starts, as compared to a mission flight requirement of only 600 sec and 36 starts for these components.

A. INJECTOR

The injector is a dish-shaped configuration with an unlike doublet pattern of 15 coaxial circular channels. Construction details are shown in Figure 9. Fuel is distributed to the outer channels through the regeneratively-cooled central hub and five equally-spaced radial baffles. Oxidizer is admitted through annular manifolds on the back of the injector. The injector is fabricated from aluminum alloy 5083, which is a high-strength, nonheat-treatable, high-magnesium alloy. Each baffle and hub assembly is installed to the injector face as one piece; the face rings and manifold covers are welded into the injector body by the electron beam process. There are no welds common to both propellants. Fabrication quality for a reliable component is assured by the non-destructive X-ray inspection of each weld.

Dynamic combustion stability has been demonstrated under all flight conditions, as well as off-design conditions. The injector has been pulsed with explosive charge devices under conditions of high and low propellant temperatures; high and low mixture ratio and chamber pressure; operation with and without helium saturated propellant; and during start-transient as well as steady-state engine operation. The injector has never been driven unstable under any of these conditions, and the perturbation is damped well within the required 0.040 sec. Satisfactory injector operation has been demonstrated during engine system testing under the following induced malfunction modes: oxidizer depletion, fuel depletion, feed pressure decay, and the bipropellant valve approximately 50% open. Injectors have consistently demonstrated satisfactory operation for durations and restarts of over three times those required on any flight mission.

B. COMBUSTION CHAMBER

The combustion chamber for the engine also is a non-redundant component; however,

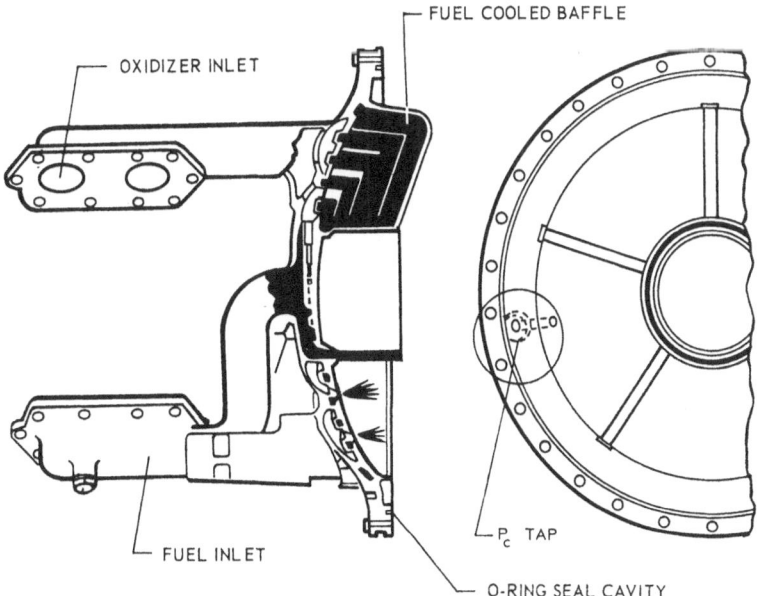

Fig. 9. SPS engine injector.

it incorporates redundant design features in its construction (chamber is shown in Figure 10). A rugged chamber is assured by the proper utilization of ablative materials in its construction. The ablative liner is fabricated with an oriented silica fabric tape. Longitudinal and hoop strength is provided by an overwrap of phenolic impregnated glass roving. Of utmost importance in any ablative chamber, in addition to excellent ablative characteristics controlled by the liner material and a compatible injector, is construction durability and reliability. The Apollo chamber does not rely solely on bond strength for the various laminates or for flange attachment, but incorporates mechanical locking features for added redundancy and increased reliability. As shown in Figure 10, the movement of the liner aft is prevented by the mechanical 'steps' provided in the asbestos overwrap. The asbestos is, in turn, mechanically locked to the metal flange by the use of a metal Z-ring; finally, added mechanical locking is

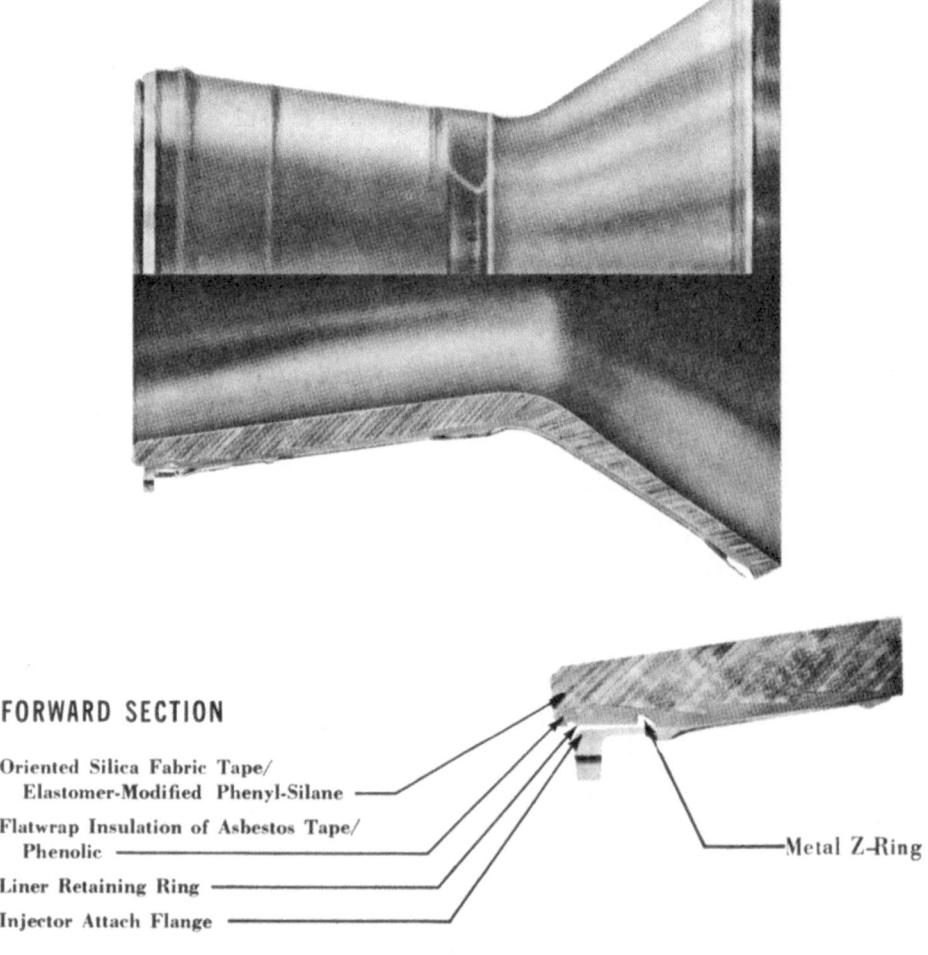

FORWARD SECTION

Oriented Silica Fabric Tape/
   Elastomer-Modified Phenyl-Silane

Flatwrap Insulation of Asbestos Tape/
   Phenolic

Liner Retaining Ring

Injector Attach Flange

Metal Z-Ring

Fig. 10.   SPS engine combustion chamber.

provided for the overall assembly by extending the 'fingers' of the metallic flange well over the chamber tapered section.

The chamber has demonstrated satisfactory operation for durations and conditions in excess of any flight requirements. Compatibility of the ablative material with duty cycles more severe than any anticipated flight duty cycle were also satisfactorily demonstrated. These demonstration test firings utilized a 'heat pump' duty cycle, which was planned with re-firings at peak chamber temperatures during coast periods to continually drive more combustion heat into the chamber. Change in throat diameter has consistently been demonstrated to be well within a narrow range of 0 to 2% increase for durations of over 900 sec under all combinations of operating conditions. Two chambers have been tested to destruction to determine the maximum accumulated operating time before chamber burn-through. One of these chambers fired for 2392 sec and the other for 1956 sec. As a normal life usage for various engine and injector development tests, chambers are consistently fired for durations of approximately 1400 sec.

C. NOZZLE EXTENSION

The nozzle extension for the SPS engine is radiation cooled and attaches to the ablative combustion chamber at the 6:1 expansion ratio and continues out to an exit area ratio of 62.5:1. The construction details of the nozzle are shown in Figure 11. Temperatures at the 6:1 area ratio are in the range of 2000 °F, thus requiring a material with adequate structural properties at high temperatures. As this temperature is in excess of the desired temperature for use of titanium alloy, a columbium alloy is used. To maintain minimum weight, the thickness of the columbium is reduced at the 20:1 area ratio from 0.030 in. to 0.020 in. At an area ratio of 40:1, the gas temperatures are low enough (approximately 1500 °F) to utilize titanium that extends to the exit. The columbium and titanium are joined by a resistance seam-weld technique. Two circumferential stiffeners are utilized to suppress bell-mode vibrations and are located at the exit and 40:1 sections. The nozzle flange is attached to the nozzle with René 41 (a high-strength, heat-and-corrosion-resistant, nickel base alloy) bolts; a metallic seal provides a high-temperature, leak-free joint. An oxidation-protective coating is applied to the columbium portion of the nozzle to protect it from becoming brittle from exposure to combustion gases. An emissive coating is applied to the exterior surface of the titanium section to enhance radiative cooling.

The alloys, coatings, and fabrication techniques utilized in the nozzle have all undergone thorough laboratory evaluation tests. Numerous oxidation protective coatings were screened prior to final selection. When a coating was selected, extensive laboratory tests were designed to thoroughly evaluate it under simulated-environment conditions. Coated columbium samples were temperature-cycled with liquid nitrogen spray (-320 °F) and then immediately heated to 2000 °F. No coating failures were ever observed during these tests. Flexure cycling tests were satisfactorily conducted with bend radii impossible to achieve in an actual firing environment. Included in the qualification test program for the engine were special qualification tests of sample nozzle material. Final nozzle qualification was accomplished at simulated altitude conditions, where complete engine assemblies were fired with flight nozzles. These tests were conducted at nominal values and at extremes of operational environments (i.e., mixture ratio, chamber pressure, long duration firing, repeated pulse firings, and spraying of liquid nitrogen on the nozzle prior to and during the start transient).

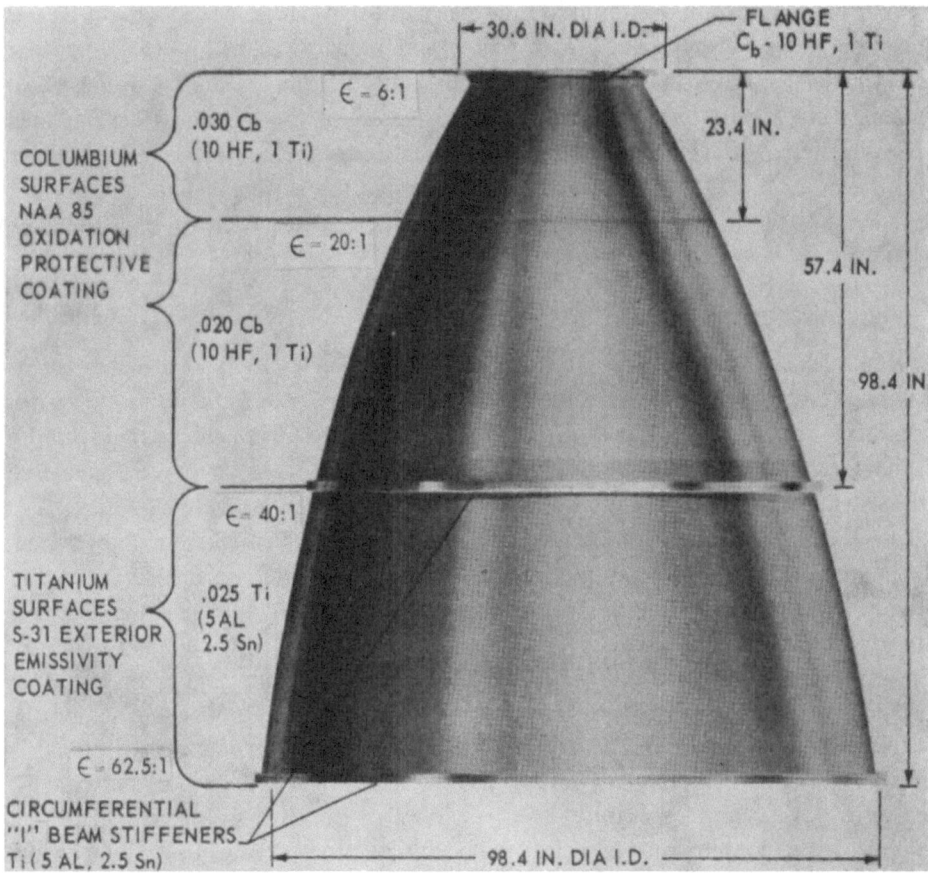

Fig. 11.   SPS engine nozzle extension.

4. Conclusion

Five major components of the engine have been described illustrating different facets of the design and testing approaches used in man rating the Apollo SPS engine. These techniques, along with an extensive engine system test program at sea level and simulated altitude and an effective change control, have resulted in a continued growth in confidence as the flight program has progressed.

Acknowledgement

The authors gratefully acknowledge the efforts of Mr. R. E. Field of North American Rockwell Corporation, Downey, California, under whose direction the SPS engine was developed. The North American Rockwell Space Division, Downey, California, is producing the Apollo Spacecraft command and service modules for the National Aeronautics and Space Administration Space Fight Center, Houston, Texas.

SOME PROBLEMS OF THE DYNAMICS OF A SPACE VEHICLE
WITH TANKS PARTIALLY FILLED WITH LIQUID

V. M. SHASHIN

U.S.S.R.

During a space flight conditions on board the space vehicle approach zero-g conditions. As a result, many physical processes, in which mass forces are of considerable importance, differ from those in the Earth. In particular, surface tension forces become especially effective on the liquid free surface. The liquid partially filling the space vehicle tanks, tends to assume the position corresponding to the energy minimum of the gas–liquid–solid system. Equilibrium forms of the liquid free surface affected by gravity and surface tension forces may be determined by [1]:

$$2H - B^*z - \lambda = 0, \tag{1}$$

where

$H = \frac{1}{2}(R_1^{-1} + R_2^{-1}) =$ surface mean curvature;

$R_1, R_2 =$ the principal radii of curvature at any point;

$B^* = ng\varrho l^2/\sigma =$ Bond number which is the ratio of gravity forces to surface tension forces (a dimensionless parameter);

$z =$ dimensionless free surface ordinate (z-axis is parallel to gravity forces);

$\lambda =$ a constant.

The surface to be determined and a tank wall are to form a given angle of wetting γ. The liquid volume bounded by this free surface and the tank walls must be equal to a given quantity. Thus, the equilibrium form of the surface depends on the following three parameters: a Bond number, an angle of wetting, and a given liquid volume (or a tank filling factor equivalent to it).

The solution of Equation (1) in its general case is for the present an unsolvable problem since the mean curvature is expressed by a complex nonlinear differential operator in partial derivatives of the second order. With large Bond numbers the theory of boundary layer may be successfully used to determine the surfaces. With $B^* = 0$ the surfaces of constant mean curvatures are obtained. The problem is considerably simplified if tanks are bodies of revolution and the gravity force acts along the axis of revolution. Thus (1) takes the form of an ordinary differential equation of the second order which it is possible to solve using some numerical method, for example, the Runge-Kutta method. In case of absolute weightlessness with $B^* = 0$ the liquid surface becomes spherical. It is necessary to point out, however, that specific parameters of this surface are to be determined from the system minimum energy conditions. Figure 1 shows the gas–liquid system typical configurations of in a spherical tank with different filling factors.

G. A. Partel (ed.), Proceedings of the Second International Conference on Space Engineering. All rights reserved.

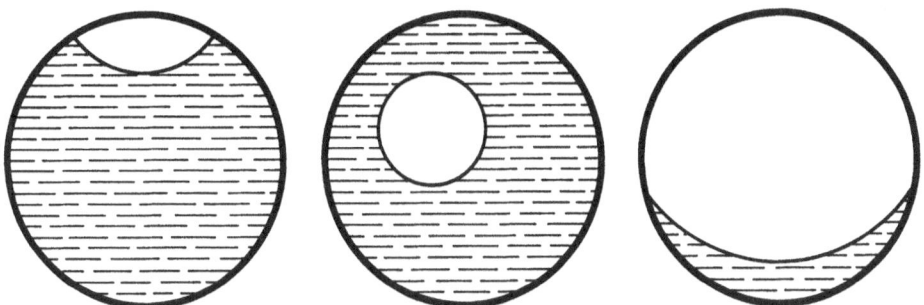

Fig. 1.   Location of zero-g liquid in a spherical tank.

During the flights of space vehicles their sustainer motors would be, as a rule, ignited many times. To ensure the engine safe ignition as well as to investigate the space vehicle dynamics at the ignition time it is necessary to study the response of the liquid free surface to a variation in mass forces with time. The stepped variation of a mass force is most typical. A sudden transition from a substantial weight to a negligible one corresponds to an engine cut-off; a sudden transition from a negligible weight to a substantial one corresponds to an engine ignition.

Liquid is considered to be perfect and the motion as beginning from rest. These assumptions allow us to introduce the speed potential φ into our discussion and formulate the problem as follows [2]:

In the volume occupied by the liquid:

$$\Delta \varphi = 0. \tag{2}$$

On the free surface $z - f(x, y, t) = \psi = 0$

$$\frac{\partial f}{\partial t} = \nabla \varphi \, \nabla \psi \tag{3}$$

$$\frac{\partial \varphi}{\partial t} + \tfrac{1}{2}(\nabla \varphi)^2 + n(t) \frac{B}{1+B}(z - h) + \frac{1}{1+B} 2H = 0. \tag{4}$$

On solid walls

$$\frac{\partial \varphi}{\partial \nu} = 0. \tag{5}$$

On the line of intersection of the free surface and the solid wall:

$$\frac{\nabla \psi \, \nabla F}{|\nabla \psi| \, |\nabla F|} = \cos \gamma. \tag{6}$$

Initial data:

$$\text{with } t = 0 \quad z = f_0; \quad \varphi = \varphi_0 \tag{7}$$

428   V. M. SHASHIN

where

  F = the equation of the tank solid wall;
  h = liquid level with no surface tension;
  B = Bond number at Earth conditions;
  $n(t)$ = the law of variation of acceleration with time;
  v = normal to the solid wall.

Of all possible varieties of hydrodynamic problems we shall discuss only those in which it is possible to neglect surface tension forces.

Furthermore, only the liquid axially symmetric motion will be considered. In other words, we shall discuss Cauchy-Poisson nonlinear axially symmetric problem with an initial surface form corresponding to its equilibrium position with due regard for surface tension forces.

The solution of the problem may be found by a well-known technique of harmonic functions [2]. Let us seek for the potential φ as

$$\varphi = \sum_{n=1}^{\infty} \tau_n(t)\, \Phi(k_n, q_1, q_2) \tag{8}$$

where

  $\tau_n(t)$ = time factors;
  $\Phi(k_n, q_1, q_2)$ = harmonic functions satisfying the condition for nonleakage on the solid walls;
  k_n = roots of an equation corresponding to the condition for nonleakage;
  q_1, q_2 = independent variables (coordinates).

With the tank formed, for example, by two coaxial cylinders, we have

$$\varphi = \sum_{n=1}^{\infty} \tau_n(t)\, [c_1\, J_0(k_n r) + c_2 N_0(k_n r)],$$

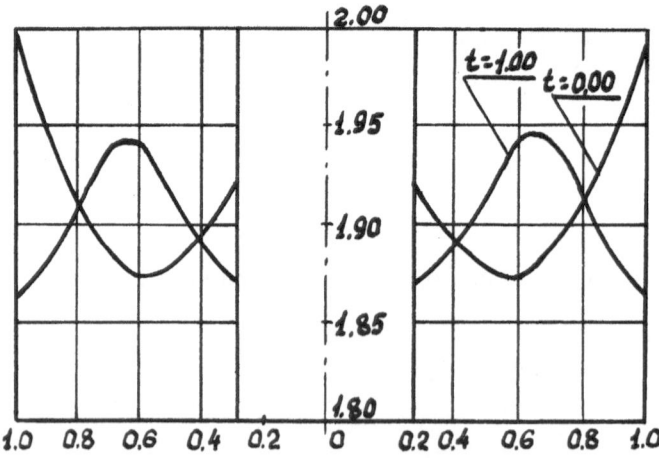

Fig. 2. Liquid dynamics in an annular tank ($B^* = 0$, $\gamma = 45°$, $F = 0.3$; $n = -1$).

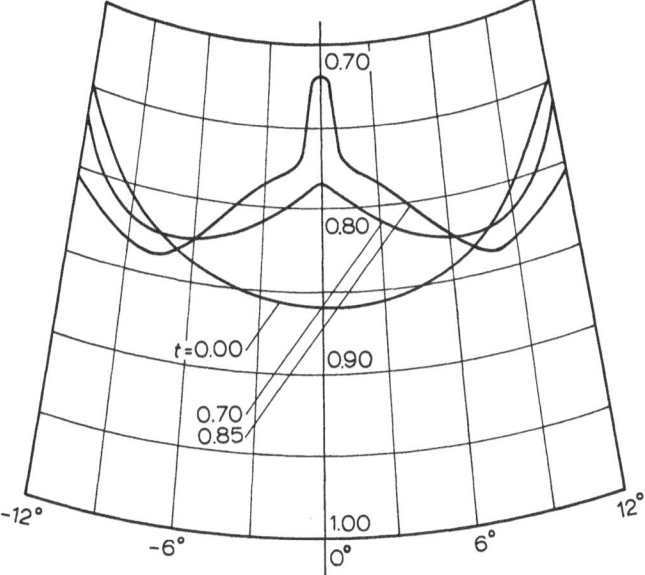

Fig. 3.   Liquid splash in a conical tank ($B^* = 1$, $\gamma = 5°$; $n = -1$).

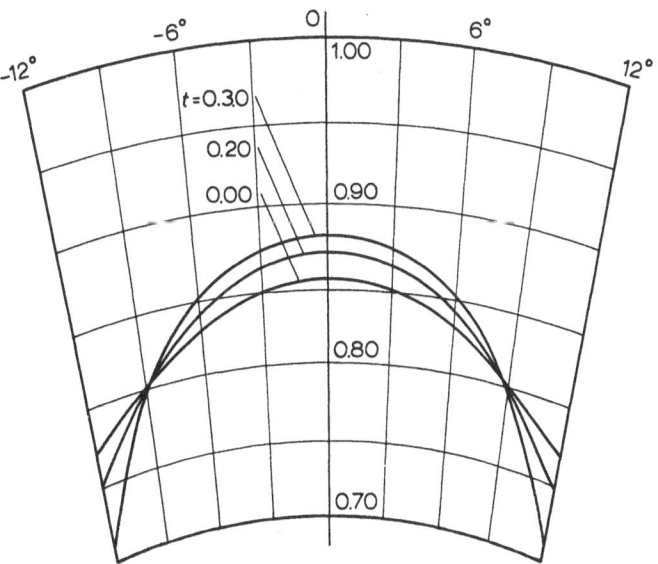

Fig. 4.   Liquid spreading in a conical tank ($B^* = 0$, $\gamma = 30°$; $n = +1$).

where $J_0(k_n r)$ and $N_0(k_n r)$ are Bessel and Neumann functions respectively. For the cone, we have

$$\varphi = \sum_{s=1}^{\infty} \tau_{v(s)}(t) \left(r^{v(s)} + \frac{v(s)}{v(s) + 1} r^{-v(s)-1} \right) P_{v(s)}(\cos \theta),$$

where $P_{v(s)}(\cos\theta)$ are Legendre functions of the first type. For the sphere with insignificant filling, we have

$$\varphi = \sum_{n=1}^{\infty} \tau_n(t) \left(r^n + \frac{n}{n+1} r^{-n-1} \right) P_n(\cos\theta),$$

where $P_n(\cos\theta)$ are Legendre polynomials.

Coordinates of the free surface and the $\tau_n(t)$ function are determined by solving integro-differential equations resulting from boundary conditions (2) and (3). The solution is found numerically. In practice the series employed converge rather rapidly.

Figures 2, 3, 4 and 5 present successive forms of the free surface in a tank formed by two coaxial cylinders, in a conical tank, and in a spherical one. One may observe either a powerful splash in the centre of the tank (resembling a cumulative effect) or the liquid spreading about the tank walls, depending on the direction of acceleration.

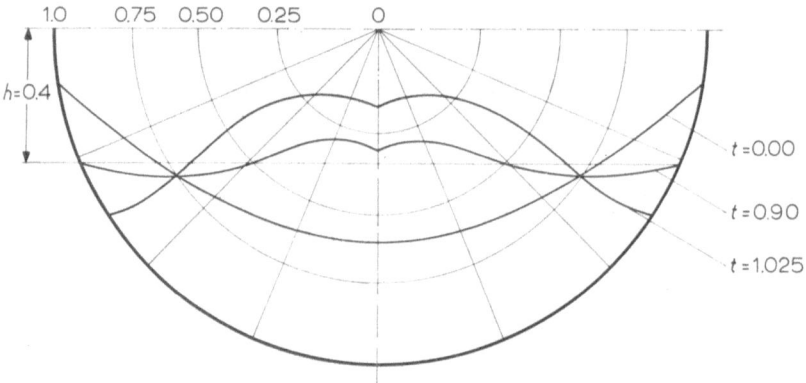

Fig. 5.   Liquid splash in a spherical tank ($B^* = 0$, $\gamma = 30°$, $n = -1$).

Theoretical results were compared with experimental data obtained with the help of a zero-gravity tower. Figure 6 shows typical photographs of the liquid surface shapes in a cylindrical tank during the transition from a negligible weight to a substantial one.

Circles in Figure 7 represent a quantitative treatment of similar photographs for a speed of splash in the centre of a conical tank.

Theoretical values are shown by a solid line. The agreement between theoretical and experimental results is seen to be quite acceptable.

It should be emphasized that the proposed method, when properly modified, may be used for solving more general problems, namely if surface-tension forces and/or three-dimensional character of the motion are considered.

Thus, the technique presented makes it possible to determine motion of the perfect free-surface liquid in a time-dependent mass-force field and may serve as the basis for solving quite a number of design and construction problems.

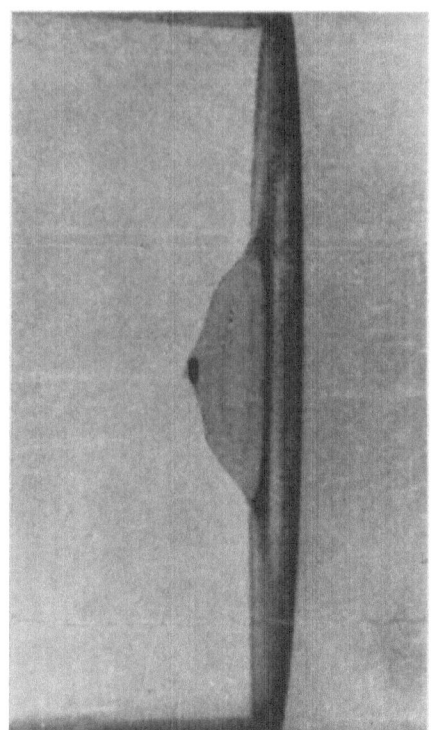

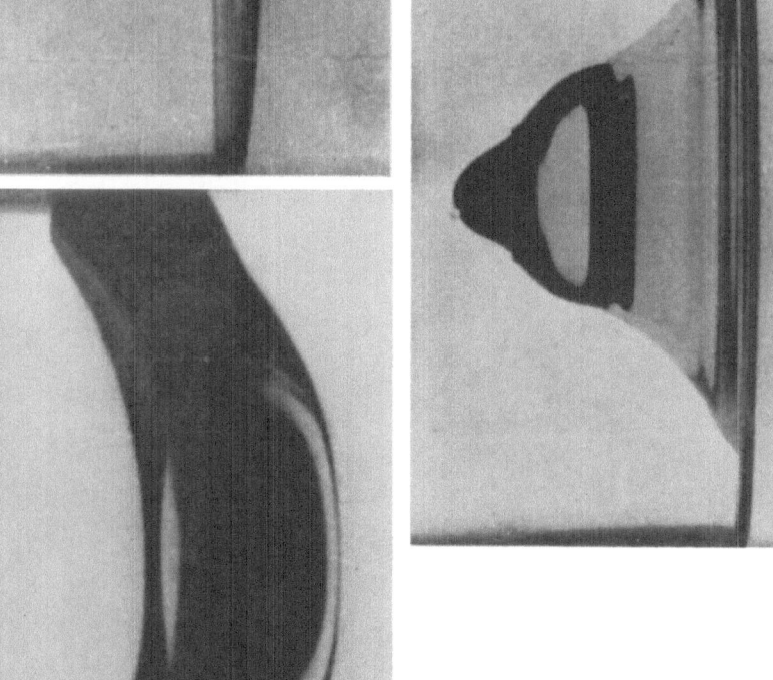

Fig. 6. Liquid surface shapes in a cylindrical tank.

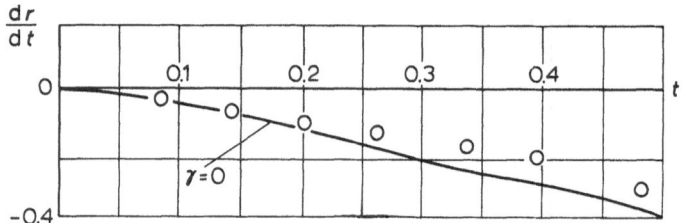

Fig. 7.   Speed of splash in the centre of a conical tank.

References

[1] Moiseev N. N., and Rumjantsev, V. V. *Dynamics of a Body with Liquid Tanks*, Moscow, 1965.
[2] Shashin, V. M. 'Dynamics of Liquid of a Conical Tank During the Transition from a Negligible Weight to a Substantial One', *USSR Acad. Sci. News* (Fluid Mechanics Series) Nr. 6 (1968).

PART VII

AUXILIARY POWER SYSTEMS

Chairman: M. Douat, France

COMPACT ZrH REACTOR DEVELOPMENT STATUS AND REACTOR THERMOELECTRIC SPACE POWER SYSTEMS*

R. F. WILSON

Atomics International Division of North-American Rockwell Corp., U.S.A.

and

W. L. KITTERMAN

Space Nuclear Systems Division, U.S. Atomic Energy Commission, U.S.A.

Abstract. The zirconium hydride reactor has been under development by the Atomic Energy Commission since 1955 as a power source for space vehicle application. To date, over 35000 hours of successful reactor operation, on five different reactors, has been achieved. The zirconium hydride reactor concept is characterized by the use of a metal hydride (ZrH) moderator material uniformly alloyed with highly enriched uranium. These thermal neutron spectrum reactors are small in size (about 18″ by 24″ including reflector), easy to control, highly reliable, and can provide power levels exceeding 1 MWt with 1300 °F (570 °C) liquid metal coolant temperatures. The reactors are characterized by reflector control. This reactor system is actively being considered in the U.S.A. for use with Mercury Rankine power conversion machinery and thermoelectric power conversion systems. One reactor of this type, with a thermoelectric power conversion system, was space-flight tested in 1965.

The ground test of a flight-configured zirconium-hydride reactor, designed to produce 600 kWt at 1300 °F coolant temperatures for greater than 10000 hours was initiated in December 1968. This reactor, designated S8DR, is the final confirmatory ground test and embodies all of the development information obtained over the last 10 years. This reactor is composed of 211 0.56″ diameter zirconium-hydride uranium fuel rods within a 9″ diameter by 20″ high stainless steel core vessel. The reflector is beryllium metal and contains 6 semi-cylindrical control drums for reactor control. The drum bearings, drum motor actuators, and all other reflector components are flight-type. The reactor is currently operating successfully towards achieving its design objectives. Current planning calls for the ground test to be run for a minimum of 10000 hours. The first 4 months of reactor operation will be more fully described at the conference.

This zirconium-hydride reactor is currently being seriously studied as the heat source for a 25 kWe thermoelectric space power system for use on a large manned space station. Studies of this system show that it can meet the performance requirements of the application. The power conversion equipment is composed of a series of compact tubular lead telluride thermoelectric modules. The overall system weighs less than 20000 lb. including manned 4π shielding; requires 1280 sq. ft of radiator area and has an overall diameter and length of 11.2 ft by 49.7 ft respectively. This power system offers many overall performance advantages to the space station compared to conventional solar cells. The selection of this type of powerplant for use on a manned space mission will introduce a new era of space power and provide the experience for larger, higher performance, future reactor space power systems.

1. Introduction

Over ten years ago the U.S. Atomic Energy Commission recognized the need for a small compact reactor heat source for application to space missions. Initial objectives of this program were to develop a reactor, which could be combined with various types of power conversion equipment (e.g., mercury Rankine, thermoelectric, gas Brayton, and/or organic Rankine) to produce electrical power in the few kilowatt

---

* Work performed under AEC Contract No. AT(04–3)–701.

G. A. Partel (ed.), Proceedings of the Second International Conference on Space Engineering. All rights reserved.

range. The Atomic Energy Commission has continuously pursued this development, and in the early sixties increased the objectives of the reactor development program to provide systems in the tens of kilowatts electrical power range.

Very early in the program the initial objective was the selection of a reactor-type on which development could proceed with a reasonable level of confidence in success. The important parameters for the reactor were a small compact size, hence minimizing shield weight; a temperature capability suitable for efficient conversion of the thermal energy to electrical power; simplicity and hence high reliability; and a reactor-materials technology which would not unduly extend the state of the art. At this early date, a conservation of fissionable material was also a consideration, although currently with somewhat larger quantities of enriched U^{235} available, this is no longer a necessary objective. Based on these objectives the early reactor engineering work rapidly focused on a system using liquid metal coolants; hydrogen moderation to achieve a small compact size; a thermal neutron spectrum for ease of control; and reflector control systems which require no penetrations or seals in the liquid metal circuit. This reactor concept was simple and was felt to result in high reliability. These features were incorporated in a reactor which uses zirconium hydride as the hydrogen moderator material, NaK (a eutectic mixture of sodium and potassium) as the coolant, and a beryllium reflector. Control is achieved by varying the neutron leakage from the reactor with movable semi-cylindrical control drums in the reflector. The early operational temperature goal was 1200°F (650°C) coolant outlet temperatures which permitted the use of conventional austenitic steels or nickel alloys for containment. Zirconium hydride was chosen as the neutron moderator material because at the operating temperatures of interest its hydrogen density is essentially equal to that of cold water, and the dissociation pressure is generally less than one atmosphere. The low dissociation pressure rendered tractable the problem of retaining the hydrogen in the reactor core necessary for long system life. This high hydrogen density also results in small reactor cores, varying between 8 in. to 9 in. (20.3 to 22.8 cm) in diameter and 12 in. to 16 in. (30.5 to 40.6 cm) in length. The zirconium hydride reactor concept has been under continuous development for over ten years, and the program has achieved some significant milestones. These will be discussed later. At the present time in the U.S.A., programs in space exploration and utilization are on the threshold of requiring the power capabilities and performance that only reactor systems can provide.

The reactor development work has included the construction and operation of five reactors of the zirconium hydride type. The second prototype power-producing reactor, designated as SNAP 2DR was operated in 1961–1962 and successfully produced 50 kWt at the designed 1200°F (650°C) outlet temperature. The reactor was operated for 10 500 h, and was then shut down for examination of the fuel and associated components.

As a result of the early successful reactor operations, the SNAP 10A program was initiated by the AEC. This program, for the first time, combined a thermoelectric power conversion system with a nuclear reactor to produce useful electrical energy.

The major objective of the SNAP 10A program was to produce a flight prototype unit which could be tested in orbit. The resultant system (shown in Figure 1) produced over 500 W electrical power at a reactor rating of approximately 35 kWt and 1050°F (567°C) coolant outlet temperature. The system weighed 960 lb, including shielding for electronic type payloads, and required a radiator area of about 62 sq. ft. The SNAP 10A program culminated in the launch and flight test of a system in April 1965. The system operated successfully for 43 days in orbit before being inadvertently shut down due to a failure in the voltage regulation equipment on the spacecraft. While the space test unit was shut down after 43 days, a companion system tested in a space simulation chamber on the ground was operated in an unattended fashion for a total of 10 005 h. During this 10 005 h run no manual control was exercised and the system

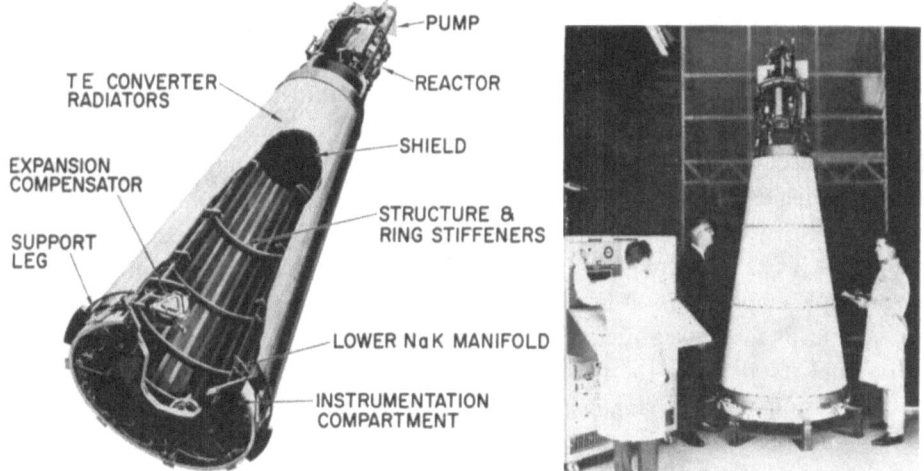

Fig. 1. SNAP 10A, world's first orbital test of a reactor thermoelectric space power system. Launched April 1965.

operated in a mode identical to that as if it were in orbit. Examination of the reactor, liquid metal circuits, and thermoelectric power conversion system after shut down confirmed that the system could have operated for several additional years. This SNAP 10A program achieved several historic firsts.

(a) The first operatio of a reactor in space, the operation of which confirmed the performance predictability from ground test data and further demonstrated that space was an ideal environment for reactor operation.

(b) Demonstrated the relative ease with which reactor power plants may be integrated and used with typical unmanned payloads and launch vehicles.

(c) Demonstrated the reliability and performance to be expected from this type of zirconium hydride reactor-thermoelectric space power system.

This early low powered reactor operating experience demonstrated that the early goals set for the zirconium hydride reactor could, in fact, be achieved with confidence. The early objectives were increased to include a reactor rated at 600 kWt at 1300°F

(706 °C) coolant outlet temperatures. Coupling a reactor of this size with a mercury Rankine power conversion system (PCS) could produce approximately 35 kWe in first generation units and as much as 50 kWt in improved versions. The zirconium hydride reactor-mercury Rankine PCS system was designated SNAP 8. Work was initiated on the reactor in 1960 and the first experimental reactor (designated S8ER) was placed on test in 1963. While this reactor had a slightly larger core than the earlier, smaller power models, its concept was identical. The S8ER reactor was operated for 12 000 h, including one continuous run of 5 000 h at rated power and temperature. While operation of the reactor was completely successful, examination of the fuel following shutdown indicated some failures in the fuel element cladding material. Subsequent development work, which is now completed, was directed toward eliminating the cause of these cladding cracks.

The zirconium hydride reactor program has, to date, achieved over 35 000 successful reactor operating hours (this reactor operation is summarized in Figure 2). The program has produced the first successful reactor-thermoelectric space power plant and has demonstrated a capability for the system to be integrated with space payloads and successfully and safely launched. This past work has likewise demonstrated the performance of the various liquid metal components and the inherent reliability to be expected from the zirconium hydride reactor with static thermoelectric power conversion. The zirconium hydride reactor program is now at the point in time where the principle emphasis is on final reactor demonstration and the application of this extensive technology to future space missions. This paper will principally discuss the operational results from the final reactor qualification test (designated S8DR) and the application potential for a large 25 kWe reactor thermoelectric system on an orbiting manned space station.

2. S8DR Reactor and Operation

The S8DR reactor incorporates the technology of the last 8 years of the zirconium hydride reactor development program. The reactor is designed as a final demonstration test of the high powered, high temperature (>300 kWt, 1300 °F) hydride concept and has design features which would permit its direct application to unmanned space power systems. Unlike the earlier S8ER reactor, the S8DR incorporates flight configured reflector components (for example, bearings, actuators, switches) and in addition incorporates improvements in the fuel to eliminate the earlier observed clad cracking and to further reduce hydrogen loss. The S8DR was originally designed to operate at 600 kWt, 1300 °F (706 °C) liquid metal coolant outlet temperature with a design lifetime of 10 000 h. Based on latest fuel element test data, it is expected that the S8DR life capability more nearly approximates 16 000 to 20 000 h. The reactor is designed for and capable of fully automatic control.

A cutaway of the reactor indicating its general design features and a picture of the completed unit is shown in Figure 3. The core consists of 211 cylindrical rods of uranium zirconium alloy, hydrided to an equivalent hydrogen density of 6.1×10^{22}

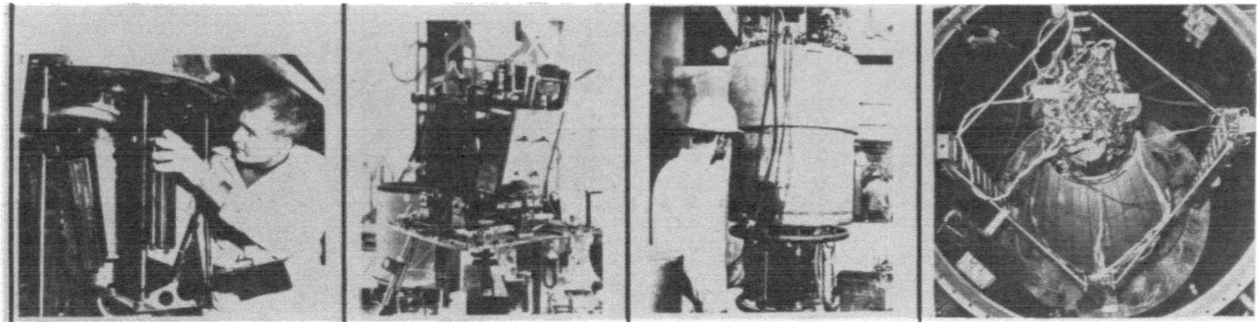

	SNAP EXPERIMENTAL REACTOR (SER)	SNAP DEVELOPMENTAL REACTOR (SDR)	SNAP 8 EXPERIMENTAL REACTOR (S8ER)	SNAP 10A FLIGHT SYSTEM (FS-3)	(FS-4)
CRITICAL	SEPTEMBER 1959	APRIL 1961	MAY 1963	JANUARY 1965	APRIL 1965
SHUTDOWN	DECEMBER 1960	DECEMBER 1962	APRIL 1965	MARCH 1966	MAY 1965
THERMAL POWER	50 kwt	65 kwt	600 kwt	38 kwt	43 kwt
THERMAL ENERGY	225,000 kwt-hr	273,000 kwt-hr	5.1×10^6 kwt-hr	382,944 kwt-hr	41,000 kwt-hr
ELECTRIC POWER	—	—	—	402 watts	560 watts
ELECTRIC ENERGY	—	—	—	4028 kw-hr	574 kw-hr
TIME AT POWER AND TEMPERATURE	1800 hr AT 1200°F 3500 hr ABOVE 900°F	2800 hr AT 1200°F 7700 hr ABOVE 900°F	1 yr AT 1300°F 400 TO 600 kwt	10,005 hr (417 days)	43 days

Fig. 2. Zirconium hydride reactor test experience. Over 35 000 h of successful power operation achieved through 1968.

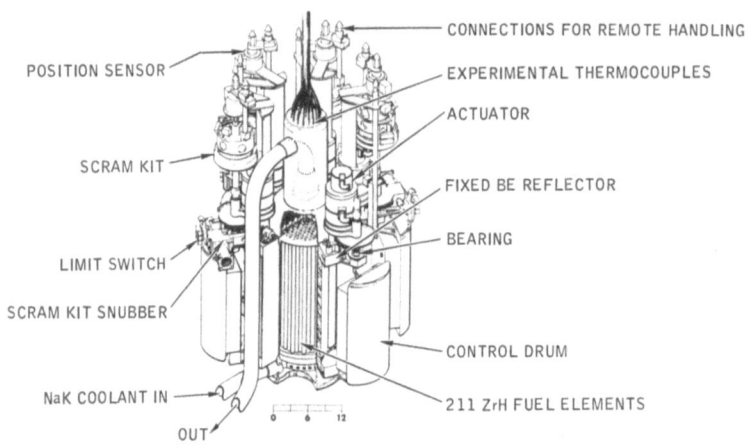

Fig. 3.   SNAP 8DR reactor design. The reactor is rated at 600 kWt at 1300°F liquid metal coolant outlet temperature.

atoms/cc. Each rod contains 10.5 wt% enriched U^{235}. The zirconium rod is approximately 0.53 in. (1.35 cm) in diameter and 16.8 in. (42.8 cm) in length, and is clad with 0.010 in. (0.25 mm) thick Hastelloy N. The hydrogen retention capabilities of the cladding tube are enhanced by the presence of a proprietary hydrogen barrier bonded to the inside surface of the tube. The reactor vessel is austenitic stainless steel approximately 9.4 in. (23.9 cm) in diameter by 32.2 in. (81.8 cm) in length. The liquid metal coolant (eutectic NaK) enters through a nozzle in the bottom of the vessel at approximately 1100°F (596°C) and exits at the top at 1300°F (706°C). Provision is made, for the purposes of the ground test, for monitoring individual fuel channel outlet temperatures through the incorporation of thermocouples located in the coolant exit region. Surrounding the reactor core vessel is a beryllium reflector approximately 5 in. (12.7 cm) thick which is dark anodized on its surface to enhance radiative heat transfer. Control of the reactor is effected through rotary motion of six semi-cylindrical control drums of beryllium. Each drum can be rotated individually through its own stepper motor, and moves on its own set of high temperature bearings. The location of the control drums determines the neutron leakage from the core. The control components have been successfully developed to withstand the intense radiation fields and operate at temperatures up to 1000°F (537°C) in a hard vacuum environment. For the purpose of the ground test only, the reactor is equipped with quick acting shutdown devices called 'scram kits' which are located above the reactor as indicated in Figure 4. With the control drums rotated to their most reactive position (full-in), the overall reactor size is approximately 20 in. (50.8 cm) in diameter by 36½ in. (92.7 cm) in length and weighs 950 lb. complete. As part of the development effort leading to the S8DR test, individual components have been separately subjected to the temperatures and vacuum environments expected during the test. These components have successfully completed over 10 000 hours of operation and the test program is continuing.

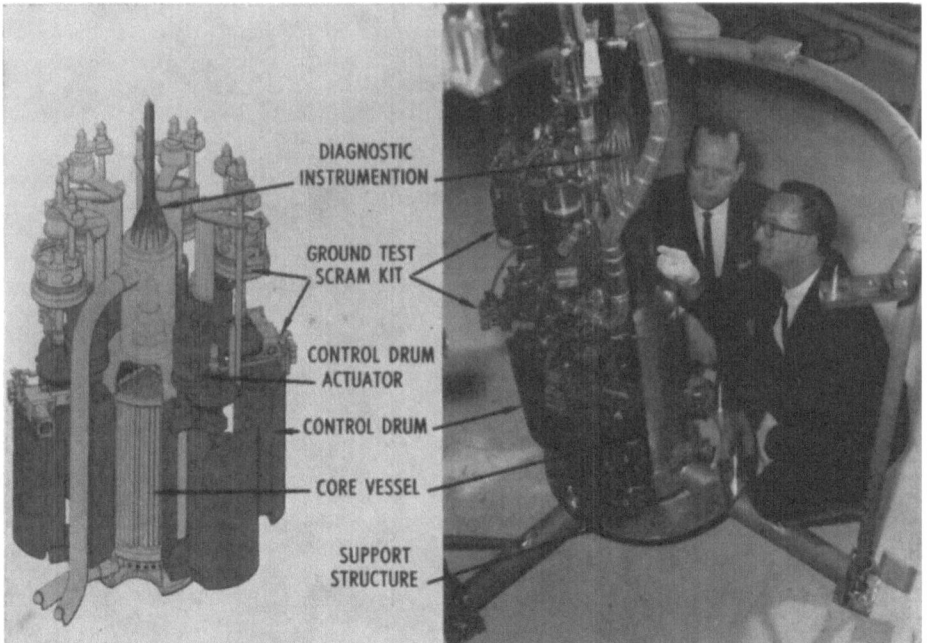

DIAGNOSTIC
INSTRUMENTION

GROUND TEST
SCRAM KIT

CONTROL DRUM
ACTUATOR

CONTROL DRUM

CORE VESSEL

SUPPORT
STRUCTURE

Fig. 4. SNAP 8DR reactor. The photograph shows the reactor with only one-half of the reflector installed.

The test facility and configuration is shown in Figure 5. The reactor is installed within a large vacuum vessel 72 in. (183 cm) in diameter by 177 in. (450 cm) in length which is located within a shielded pit. Since there is no power conversion system included as part of the test, the reactor heat is discharged to the air through primary and intermediate loops. During operation the vacuum vessel is maintained at a pressure of 10^{-6} torr by two large (35 in. (89 cm) in diameter) oil diffusion pumps. Operation of the reactor is controlled remotely from the control room and incorporates provisions for completely automatic operation. The facility has provisions for emergency power and all other auxiliary systems required for a ground test of this nature. The facility further incorporates extensive remote handling facilities located in the vault area above the vacuum vessel such that at the conclusion of the test program the entire reactor may be remotely disassembled, inspected, and packaged for shipping to eventual disposal. During testing the facility can be operated by a staff of only four persons.

The test program for S8DR is designed to explore its full operating capabilities, including both transient and steady state operation. The test program is further designed to demonstrate the environmental and life capability of the reactor. In recognition of these objectives, the test program outlined in Figure 6 is being followed. Physics and transient tests on the reactor were completed in December of 1968 and the reactor brought to full power for the start of the first 500 h, 600 kWt run in mid January of 1969. Both the 600 kWt and 1 MWt operation conditions have been

Fig. 5.   Test facility in which the SNAP 8DR test is being conducted. The reactor is located in a
vacuum tank within the shielded vault as indicated on the drawing.

completed, and the reactor is now operating at its designed condition of 600 kWt
and 1300 °F (706 °C) coolant outlet temperature; power conditions which will be
maintained throughout the endurance phase. While the figure indicates a test program
of approximately 10 000 h, it is hopeful that the reactor can be operated to its full
inherent lifetime.

The early transient testing of the reactor was designed to simulate the occurrences
expected from the startup of a complete mercury Rankine-reactor space power
system. This type of startup normally occurs in two distinct phases:

(a) Bringing the reactor from the subcritical (shutdown) condition up to full
temperature and approximately 50–100 kWt while the Power Conversion System
(PCS) is inoperative.

(b) The simulation of mercury injection during which the reactor temperature is
maintained close to the 1300 °F (706 °C) design point while the reactor power is
increased to match the heat extraction from the mercury boiler.

Figure 6 also shows experimental results obtained and compares them with the
analytical predictions. The simulation of the mercury injection phase does not dupli-
cate in all respects that which will occur in an actual space power system since the
transfer functions associated with the facility heat transfer systems are not identical
to those of the space system, but the tests do confirm that the startup transient will

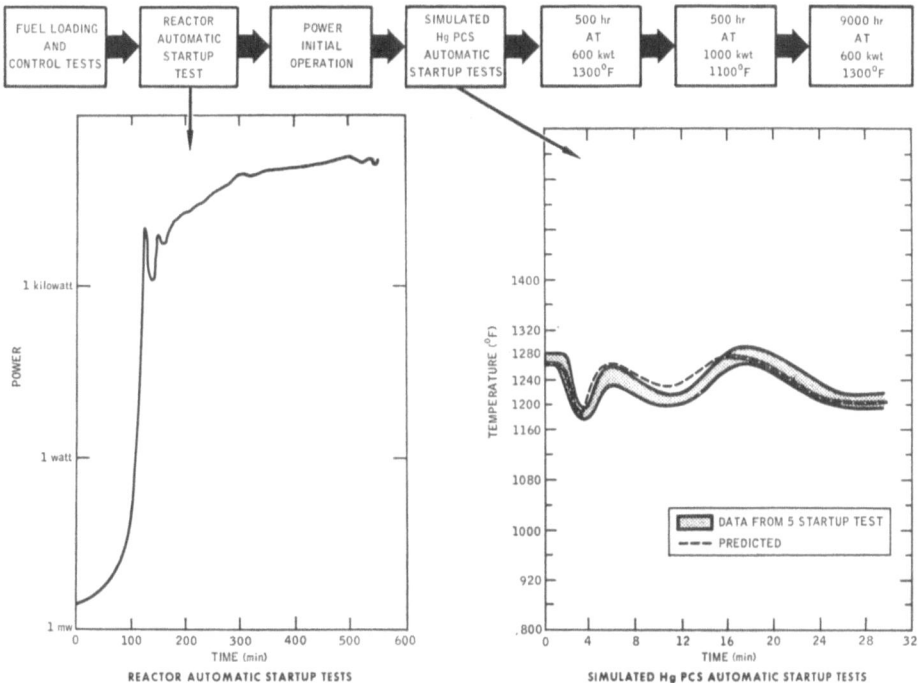

Fig. 6.   SNAP 8DR reactor test program. Testing through the 1000 kWt phase is complete. The curves indicate experimental results for two of the transient tests.

not unduly stress the reactor components and the controller can be programmed for full automatic control of the startup. During this test phase the plant behaved predictably. The reactor-mercury Rankine PCS can be brought from the shutdown condition to power in a time period of about 6 to 8 h.

The reactor is currently running at 600 kWt, 1300°F (706°C) and over 1000 h have been accumulated as of March 30, 1969. The operation at 600 kWt and 1 MWt indicates the reactor is meeting all of its design objectives. Temperatures on the various reflector components are slightly lower than predicted and no failures have occurred. The operation is also confirming the solution of the fuel element clad cracking observed on the previous S8ER reactor test. Very successful operation allows the next step in the program – the engineering, testing, and use of large complete reactor thermoelectric space power systems – to proceed with a high degree of confidence.

3. A 25 kWe Reactor Thermoelectric System for a Manned Space Station

The U.S. Atomic Energy Commission, in cooperation with the National Aeronautics and Space Administration, has been actively studying the utilization of a large reactor-thermoelectric power system on an orbiting manned space laboratory. It should be noted that the space station is not, at this time, an officially approved mission.

Nevertheless, studies of this type are very useful to identify potential problem areas and explore (analytically) alternate solutions to these identified problems.

The system heat source is the basic zirconium hydride type reactor, and the thermo-electric power conversion system utilizes compact tubular modules of lead telluride. In the selection of this system for detailed evaluation on the space station application; important considerations included the following.

The well-developed basic technology of the heat source and power conversion system.

The availability of the system on a time schedule consistent with the assumed space station flight schedule in the mid-1970's.

The past experience and excellent performance of the smaller SNAP 10A system which was flight tested in 1965.

The high technical confidence that the reactor-thermoelectric system could provide the required power levels and the required reliability.

The integration flexibility of the system and its growth potential.

As the study of the application progressed, several other potential advantages of the reactor-thermoelectric system, over that of competitive type systems, were developed and are discussed later.

The space station concept requires system and subsystem lifetimes of two to five years in an earth orbit of approximately 300 nautical miles (556 km). The six to nine man crews would be rotated on a 90 to 180 day cycle. Specific requirements assumed for the space station study include:

A 1975 launch date.

An unmanned launch using the existing two-stage Saturn V standard United States launch vehicle.

A nominal 250–300 mile orbit.

A design lifetime of two years with a five year goal.

An operational zero gravity mode either manned or unmanned.

The space station is designed to serve as a general purpose space laboratory and as the focal point for a variety of scientific experiments. These experiments include both those of an astronomical type as well as those devoted to earth resources. Some of the experiments would be integrated directly into the main space station, while others were conceived as small satellites in parallel orbits to the space station but perhaps 2 to 20 miles away. These small satellite experiments would be placed into orbit and retrieved back to the main space station by a small manned vehicle operating from the space station. Manned spacecraft would affect the resupply of the station and serve as crew exchange transporters.

Figure 7 is an artist's conception of the zirconium hydride reactor-thermoelectric system integrated into an operational space station. The reactor power plant is located on the very upper end of the figure with the upper cylindrical section being the heat rejection radiator. A detailed evaluation of the power requirements for the station was made, which indicated a requirement varying between 17 and 25 kWe. A significant fraction of this power was required for the closed cycle environmental control

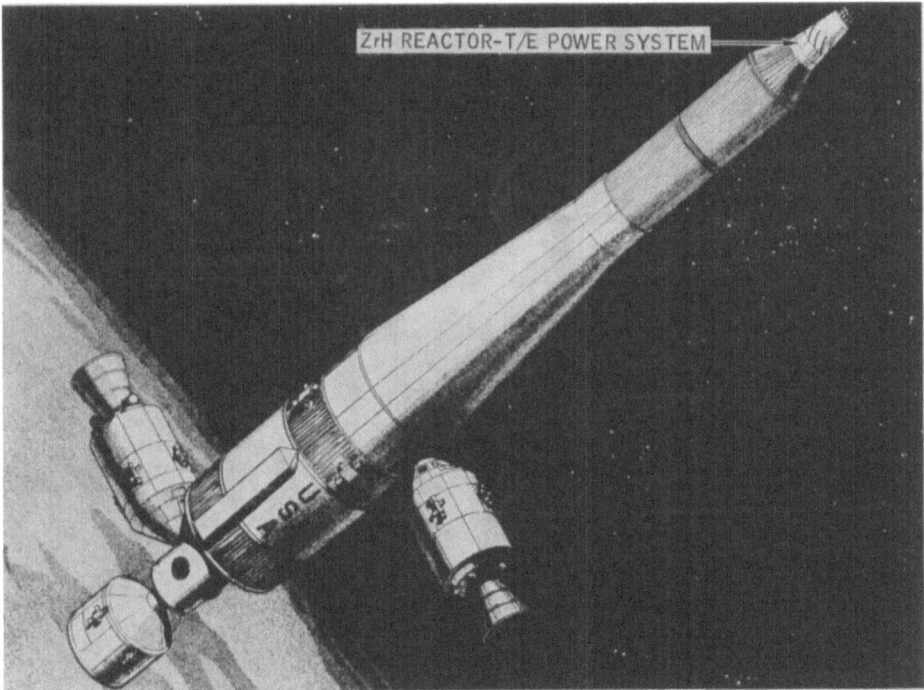

Fig. 7.   25 kWe reactor thermoelectric space power system for a manned space station application.

subsystems. The upper power level (i.e., 25 kWe) was used for the power plant design with the expectation that power growth requirements, as the program matured, would require this level. The total power system has a gross weight of 20 000 lb and a requirement for 1280 ft² of radiator. The weight of the power system includes the weight of a shaped 4π steradian manned type shield, which accounts for approximately 60% of the total system weight. The overall power system is approximately 11 ft (3.35 m) in diameter and 50 ft (15.25 m) in length.

The reactor is a direct outgrowth of the S8DR which has been modified to operate within a 4π shield geometry. The modification is one in which the control drums are changed to more effectively regulate the power level when surrounded by the radiation shield. The nominal operating point of the reactor is approximately 580 kWt at 1250 °F (676 °C) outlet coolant temperature.

The overall design of the power system is shown in Figure 8. The detailed arrangement of the reactor, shield, and power conversion equipment is shown in Figure 9. The reactor is totally surrounded by the shield but the bulk of the shielding is located between the reactor and the space station. The system pumps and power conversion equipment are located in a gallery within the shield region. The gallery is required because the primary coolant is radioactive, and therefore, must be shielded from the space station. The gallery region is about 24 in. (61 cm) high and 62.5 in. (159 cm) in diameter. Waste heat from the power conversion system is transported to the radiator

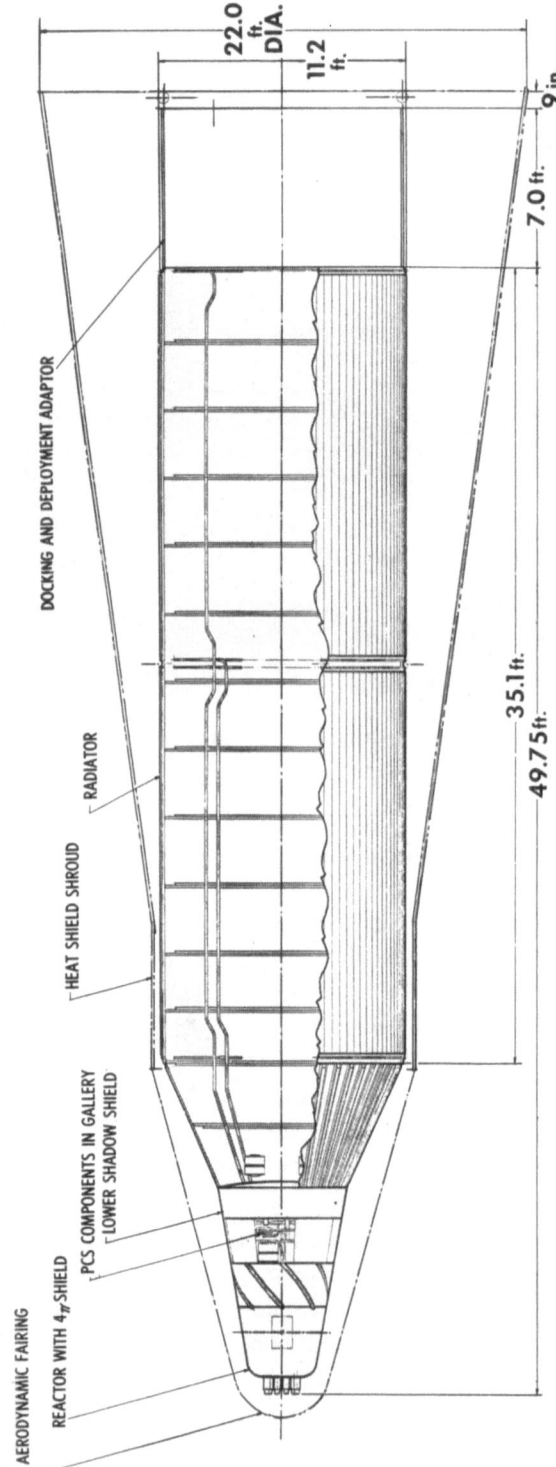

22.0 ft. DIA.

11.2 ft.

9 in.

7.0 ft.

35.1 ft.

49.75 ft.

DOCKING AND DEPLOYMENT ADAPTOR

RADIATOR

HEAT SHIELD SHROUD

PCS COMPONENTS IN GALLERY
LOWER SHADOW SHIELD

AERODYNAMIC FAIRING

REACTOR WITH 4π SHIELD

- URANIUM ZIRCONIUM HYDRIDE, NaK-COOLED REACTOR
- LEAD TELLURIDE THERMOELECTRIC CONVERTER
- POWER OUTPUT   25kwe
- SYSTEM WEIGHT   19,880 lbs (795 lbs/kwe)
- EFFECTIVE RADIATOR AREA 1,280 ft² (51 ft²/kwe)

Fig. 8.   Design of 25 kWe reactor plant for a space station application.

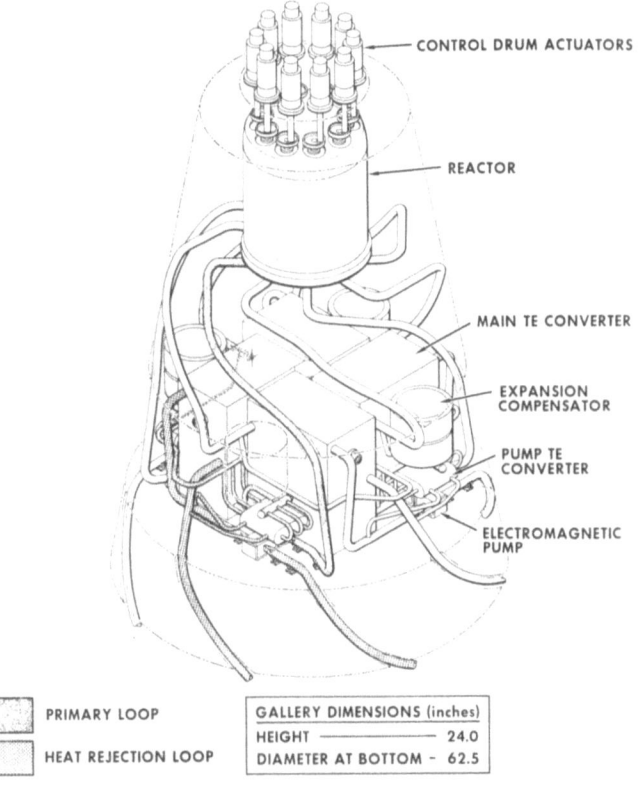

Fig. 9. Reactor, shield, and gallery arrangement for 25 kWe space power system (shielding is shown in phantom).

through four independent loops; hence the power conversion equipment is joined in a common primary loop but each quadrant has an individual and isolated secondary coolant heat rejection loop. The thermoelectric power conversion system is composed of a matrix of individual tubular modules, each module capable of producing 262 W (electric) at design temperature. Twenty-four of these modules are hydraulically connected into a power producing quadrant, producing 6.3 kWe. The arrangement of the modules within the quadrant is shown in Figure 10. The liquid metal coolant is circulated from the reactor to the central tube (hot side) of the power conversion units, and from the outer surface (cold side) of the power conversion units to the radiator, by electromagnetic pumps. These pumps have the advantage of requiring no moving parts for their operation. A few of the key parameters and operating conditions of the power system are summarized in Table I.

The shield design is interesting in that it incorporates the 'split shield concept' in which the radioactive primary liquid metal system transfers its heat to the power conversion system in the gallery region between the two shield halves. The shield materials are lithium hydride and tungsten within a stainless steel container. The integration of the reactor thermoelectric system into the space station is largely

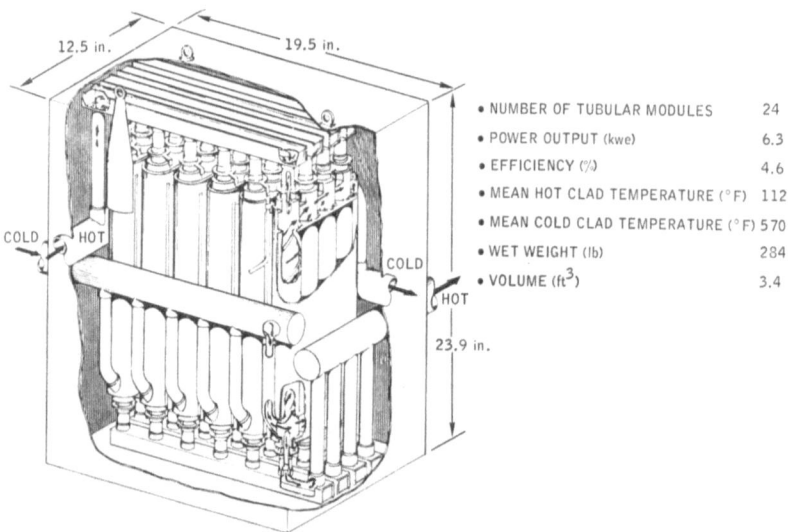

Fig. 10.   One quadrant of the thermoelectric power conversion system. The quadrant is comprised of 24 individual tubular modules.

TABLE I

25 kWe reactor thermoelectric system operating parameters

Design life, hours	20 000 min
Net power output, kWe	25
Voltage output, volts	56
Overall system efficiency, %	4.3
Radiator area, sq. ft	1280
System weight, unshielded, lb	7590
shield weight, lb	12 290
total weight, lb	19 880
Maximum primary coolant temperature, °F	1250
Average thermoelectric hot junction, temperature, °F	1115
Average radiator temperature, °F	563

governed by shielding considerations. The shield is sized taking into account direct radiation penetration; the scattering of radiation around the primary shield or off of satellite objects; and the radiation received when the astronauts rendezvous or depart from the space station. The shield is designed such that it provides a variable radiation field dependent upon the orientation from the space station. The expected radiation levels about the space station and within the crew quarters are indicated in Figure 11. The radiation levels which the astronauts will receive from the power system are less than the radiation levels normally received from the natural space environment.

Operationally the power plant is launched integrally with the unmanned space

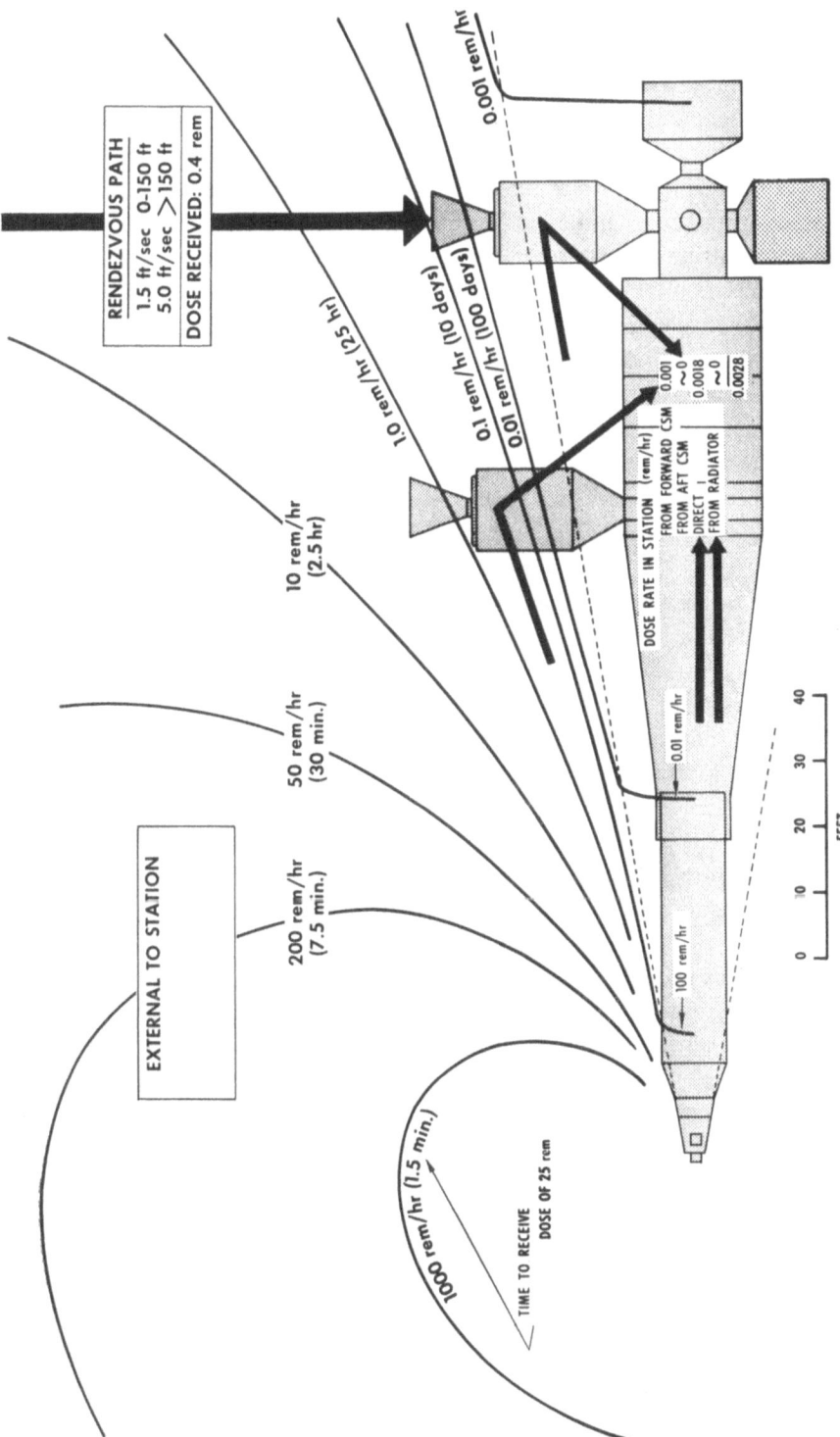

RENDEZVOUS PATH

1.5 ft/sec	0-150 ft
5.0 ft/sec	>150 ft

DOSE RECEIVED: 0.4 rem

0.001 rem/hr

1.0 rem/hr (25 hr)

0.1 rem/hr (10 days)

0.01 rem/hr (100 days)

10 rem/hr (2.5 hr)

50 rem/hr (30 min.)

200 rem/hr (7.5 min.)

1000 rem/hr (1.5 min.)

TIME TO RECEIVE DOSE OF 25 rem

EXTERNAL TO STATION

DOSE RATE IN STATION (rem/hr)	
FROM FORWARD CSM	0.001
FROM AFT CSM	~0
DIRECT	0.0018
FROM RADIATOR	~0
	0.0028

0.01 rem/hr

100 rem/hr

0   10   20   30   40
FEET

Fig. 11.   Radiation profile about and within the space station with the reactor at full power.

station and at the time of launch is totally nonradioactive. The power system is then started up in orbit either by a remote ground command or by the astronauts after the station is manned. The power system can be shut down and restarted without limitation, dependent only on the power requirements of the space station. The location of the power system on one end of the space station, with the heavy mass distribution of the reactor-shield at the very forward end, enhances the capability of the space station to operate in the gravity gradient stabilized mode.

The application of the reactor thermoelectric system to the space station concept obviously requires a total evaluation of the complete factory-to-flight sequence. Considerations involve how the system is assembled, what operations are conducted, what facilities are required, how the system is shipped, the kinds of pre-flight testing required, the mechanics of integration with the launch vehicle, and considerations of the launch facilities themselves. Reactor systems are particularly attractive and adaptive over this broad spectrum of pre-launch considerations. The basic reactor-shield-gallery is a small compact unit which can be readily tested. Physics tests of the system can be conducted at essentially zero power so that the unit remains non-radioactive, operational tests (i.e., PCS power output and temperature) can be conducted using auxiliary sources of heat. In this manner the entire spectrum of performance can be evaluated without inducing appreciable radioactivity prior to startup in orbit. In all operations; transportation to the launch site, vehicle integration, etc.; the power plant can be handled as any normal piece of space hardware. This concept was thoroughly proven and demonstrated on the earlier SNAP 10A flight. During the current studies it was found that normal transportation methods, and the currently available launch facilities at the U.S. Cape Kennedy launch site are adequate to handle reactor power plants of the size considered in these studies.

In summary, the total examination of the relative merits of different power systems requires the consideration of integration details and the effect of the power plant on the total system. It is interesting to note that the current joint AEC-NASA studies have not revealed any problem areas which would preclude the use of reactor-thermoelectric systems, and I have identified several positive integration benefits from the use of reactor space power systems. Included among these advantages are the following:

The mass distribution of the space station and the reactor power plant, in which there is a 100 to 150 ft moment arm, permits operation in the gravity gradient stabilized mode.

Stabilizing torques in the order of 0.7 ft lb/deg are available. This natural stability, in contrast to solar cell orientation requiring semi-continuous active control, can result in weight saving of several thousand pounds per year in propellants. In addition, of course, the safety is enhanced in that should failure of the control system occur, the space station would not go into uncontrolled motion as was experienced on one of the prior Gemini flights.

The radiator area of the reactor-thermoelectric system is about one-sixth that of an oriented solar cell array of equal average power capability. This smaller area permits a greater view factor from the space station, and less interference with docking and other extra-vehicular-activity. There are no requirements for orientation of the reactor system which substantially simplifies total space station control.

The greater structural rigidity of the reactor system, compared to solar cells, minimizes or eliminates control dynamics problems with the space station which are brought about by maneuvering a large nonrigid structure. The reactor system will perform as well on a space station with artificial gravity, which would be prohibitive with a large solar cell array.

The smaller area of the reactor-thermoelectric system results in substantially less drag penalty at the lower orbit altitudes, and therefore requires less station keeping propellant consumption.

The radiation levels induced by the reactor were found to be shieldable down to levels suitable for astronauts and were not found to impose any meaningful operating limitation on the space station.

The current technology status of the reactor-thermoelectric system will permit the overall mission planners to proceed to the higher power requirements with a high level of technical confidence.

Substantial growth capability exists.

The technology of the zirconium hydride-thermoelectric space power system is equally adaptable to a wide spectrum of power levels, space vehicles, and mission applications. The technology is in an advanced stage of development and programs using this electrical power source can do so with a high degree of confidence. A key to future space exploration and utilization will be the availability, in virtually unlimited quantities, of large amounts of electrical power. Man's past history has clearly demonstrated that power is his means of extending his ever-growing capabilities. In the hostile environment of space, the requirements for large, highly dependable power sources are unsurpassed. The future mission space planners have no alternative but to harness the power of the atom. The current studies of reactor systems are merely initial, but necessary, steps in the United States Space Program.

Acknowledgments

The authors wish to acknowledge the technical effort by members of the United States Atomic Energy Commission, the National Aeronautics and Space Administration's Marshall Space Flight Center, and Atomics International, a Division of North American Rockwell Corporation.

BATTERIES D'ACCUMULATEURS POUR ALIMENTATION EN ÉNERGIE DE SATELLITES

J. P. GOMIS

Société des Accumulateurs Fixes et de Traction, Romainville, France

Résumé. Diverses sources d'énergie sont utilisables à bord des satellites, mais les conditions propres à chaque mission restreignent considérablement le choix. L'ensemble cellules solaires – batterie d'accumulateurs – est la solution la plus fréquemment retenue.

Chaque type d'accumulateur pose des problèmes d'utilisation liés aux caractéristiques du couple électrochimique employé. Mais dans la grande majorité des cas les difficultés de fonctionnement proviennent de la nécessité de limiter ou d'éviter la surcharge. Plusieurs artifices de construction destinés généralement à augmenter l'aptitude à la surcharge des accumulateurs peuvent être utilisés dans la mesure où leur emploi ne conduit pas à une diminution de la durée de vie des batteries ou à poser des problèmes thermiques difficilement résolvables dans l'espace. D'autres solutions consistent à éviter la surcharge mais il faut pour cela disposer d'un signal de fin de charge facilement exploitable, qui permette d'interrompre la charge ou de diminuer l'intensité du courant de charge.

Aucun des systèmes existants n'est parfait, tous présentent des avantages et des inconvénients.

Les problèmes de charge et de surcharge des batteries d'accumulateurs n'ont reçu jusqu'à maintenant que des solutions partielles et tout progrès dans ce domaine améliorera la longévité et la fiabilité des batteries.

1. Sources d'énergie électrique spatiales

Les satellites, quelle que soit leur mission, consomment de l'électricité. Cette énergie électrique peut être obtenue à partir de l'énergie solaire (cellules solaires), de l'énergie nucléaire (générateurs isotopiques ou nucléaires), de l'énergie chimique (générateurs électrochimiques).

Les trois types de sources ont été utilisés à bord de satellites, démontrant ainsi leur faisabilité.

L'énergie solaire présente le grand avantage d'être 'gratuite' et de ne pas compter dans le devis poids du satellite. Par contre, son exploitation nécessite des structures de grande surface (il faut 50 cm² pour 1 W), d'orienter le satellite par rapport au soleil pour obtenir le rendement maximum des cellules et des sources d'énergie secondaires, pour alimenter le satellite de nuit et fournir les pointes de puissance pendant le jour.

L'énergie nucléaire a été peu utilisée (quelques satellites). Il est probable que les dispositifs permettant son utilisation ne seront pas pleinement compétitifs avant plusieurs années.

L'énergie chimique peut être stockée dans des générateurs électrochimiques dits 'primaires' où les réactions permettant la production d'énergie électrique sont irréversibles, ou 'secondaires' quand ces réactions sont réversibles et que le générateur peut être 'rechargé' en énergie électrique. Les générateurs primaires, auxquels on peut rattacher les piles à combustible, doivent comprendre une quantité de corps réactants proportionnelle à l'énergie à fournir pendant toute la durée de la mission, tandis que les générateurs secondaires ou accumulateurs peuvent avoir des dimensions

G. A. Partel (ed.), Proceedings of the Second International Conference on Space Engineering. All rights reserved.

et un poids en principe indépendants de la durée de la mission, puisqu'ils ne servent que de relais à l'énergie électrique. Par contre, ils nécessitent la présence d'une autre source d'énergie à bord du satellite.

L'association cellules solaires – accumulateurs électriques est très fréquemment employée dans les véhicules spatiaux.

Le choix d'une solution pour alimenter en énergie électrique un satellite, dépend essentiellement de la durée de la mission et de la puissance nécessaire. Le graphique suivant (figure 1) illustre les domaines respectifs des différents générateurs [1]. Pour des durées de quelques heures à quelques jours, les batteries primaires sont très

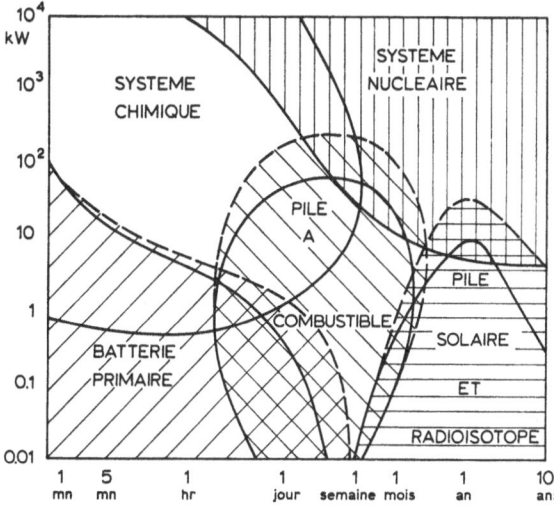

Fig. 1.   Domaine d'utilisation des sources de puissance pour l'espace [1]; actuel (———), projeté (----).

compétitives, et la mise au point récente de générateurs à anode de lithium à très haute énergie massique (>300 Wh/kg) [2] doit étendre ce domaine à des durées de l'ordre de 10 à 20 jours. Les piles à combustible apportent une solution pour les missions de l'ordre de 1 à 3 semaines, avec des puissances pouvant atteindre actuellement 10 à 20 kW. Pour des durées de fonctionnement allant de quelques mois à plusieurs années, seuls restent en concurrence les générateurs nucléaires et l'association cellules photovoltaïques – accumulateurs. Les générateurs nucléaires lourds s'imposeront certainement dans le cas des puissances élevées, supérieures à la dizaine de kW, mais dans l'état actuel de la technologie spatiale, ces puissances ne sont pas encore nécessaires, et les ensembles cellules solaires – accumulateurs demeurent la solution la plus intéressante.

2. Batteries d'accumulateurs

Les cellules solaires sont généralement calculées de façon à fournir au satellite, pendant les périodes de jour, l'énergie nécessaire à son fonctionnement durant une orbite complète pouvant comprendre une éclipse. La batterie d'accumulateurs assure l'alimentation en énergie électrique pendant l'éclipse, et fournit les pointes de courant durant les périodes d'ensoleillement, quand l'énergie demandée est supérieure à celle pouvant être fournie par les cellules solaires. Les cellules solaires vieillissent au cours de la vie du satellite, et leurs performances diminuent; pour tenir compte de ce phénomène, elles sont habituellement surdimensionnées, et pendant la plus grande partie de la mission fournissent à la batterie, en l'absence de systèmes de régulation, des quantités d'électricité supérieures à celles nécessaires à la recharge (surcharge).

Ces conditions générales d'utilisation permettent de définir les caractéristiques que doit posséder une batterie d'accumulateurs pour satellite:

- Aptitude au cyclage prolongé (autant de cycles que de périodes d'éclipse).
- Possibilité de délivrer des pointes de courant importantes.
- Aptitude à la recharge pendant la durée d'ensoleillement du satellite.
- Acceptation de la surcharge ou possibilité de l'éviter (principalement pour les orbites à ensoleillement variable).

Les conditions d'emploi sur satellite imposent par ailleurs:

- Aptitude au fonctionnement en étanche.
- Résistance aux contraintes mécaniques sévères qui apparaissent pendant le lancement.

Trois types d'accumulateurs satisfont à tout ou partie de ces impératifs. Ils utilisent les couples suivants:

$$NiOOH/KOH/Cd$$
$$AgO\text{-}Ag_2O/KOH/Cd$$
$$AgO\text{-}Ag_2O/KOH/Zn.$$

Les principales caractéristiques de ces accumulateurs sont rassemblées dans le tableau I.

L'accumulateur nickel–cadmium présente, par rapport aux accumulateurs à élec-

TABLEAU I

	NiOOH/KOH/Cd	AgO/Ag₂O/KOH/Cd	AgO/Ag₂O/KOH/Zn
Energie massique			
décharge en 5 h	30 Wh kg^{-1}	70 Wh kg^{-1}	120 Wh kg^{-1}
décharge en 15 mn	25 Wh kg^{-1}	55 Wh kg^{-1}	90 Wh kg^{-1}
Energie volumique			
décharge en 5 h	80 à 100 Wh dm^{-3}	100 à 150 Wh dm^{-3}	150 à 200 Wh dm^{-3}
Puissance massique	200 à 300 W kg^{-1}	200 à 400 W kg^{-1}	300 à 700 W kg^{-1}
Durée de vie: nombre cycles charge-décharge au régime de 5 h – Profondeur de décharge 80%	2000	500	100
Aptitude à la surcharge	bonne	faible	très faible

trode d'argent, l'avantage d'avoir une faible variation de la tension (figure 2) et d'être apte au fonctionnement à basse température.

L'accumulateur argent–cadmium peut être réalisé de telle sorte qu'il soit totalement amagnétique, son emploi s'impose donc dans tous les cas où l'amagnétisme des composants d'un satellite est prescrit. Cet accumulateur supporte mal le fonctionnement en étanche, et les dispositifs de charge doivent tenir compte de cette caractéristique.

L'accumulateur argent–zinc apparaît comme le plus énergétique, et par conséquent le plus intéressant pour les applications spatiales, d'autant plus qu'il peut être lui aussi amagnétique. En pratique, cet accumulateur a une durée de vie relativement faible et est mal adapté au fonctionnement en étanche. Il est donc surtout utilisé pour les missions de courte durée.

La comparaison des énergies massiques et volumiques de ces différents accumu-

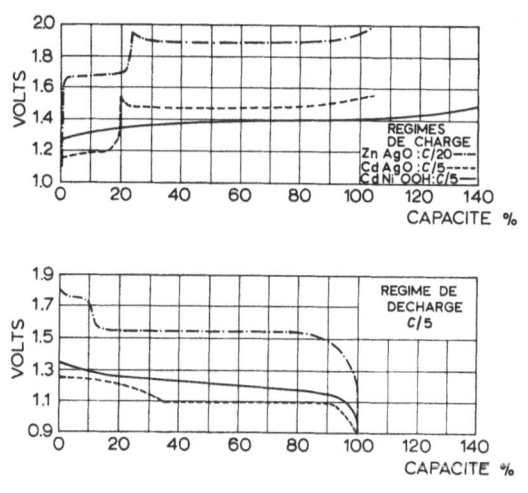

Fig. 2.

lateurs doit être pondérée par les paramètres durée de vie et profondeur de décharge. En effet, le nombre de cycles que peuvent supporter les accumulateurs dépend de la profondeur de cyclage: rapport de la quantité d'électricité déchargée à chaque cycle à la capacité nominale de l'accumulateur.

Les figures 3 et 4 donnent, d'après des essais SAFT, la durée de vie d'accumulateurs Ni–Cd et Ag–Cd en fonction de la profondeur de décharge pour des cycles de 100 minutes (charge 60 mn – décharge 40 mn) correspondant à des orbites basses. On constate par exemple, que pour une durée de vie de l'ordre de 2500 cycles, il est nécessaire pour le Ni–Cd de limiter la profondeur de décharge à 50%, et pour 10 à 15 000 cycles à 20%. L'énergie massique réelle dont il faudra tenir compte est donc diminuée de moitié par rapport à celle figurant dans le tableau, pour une mission de 6 mois, et divisée par 5 pour une durée de vie de 3 ans. Pour des missions de plus longue durée, la pénalisation est encore plus importante, et dans la conception d'une

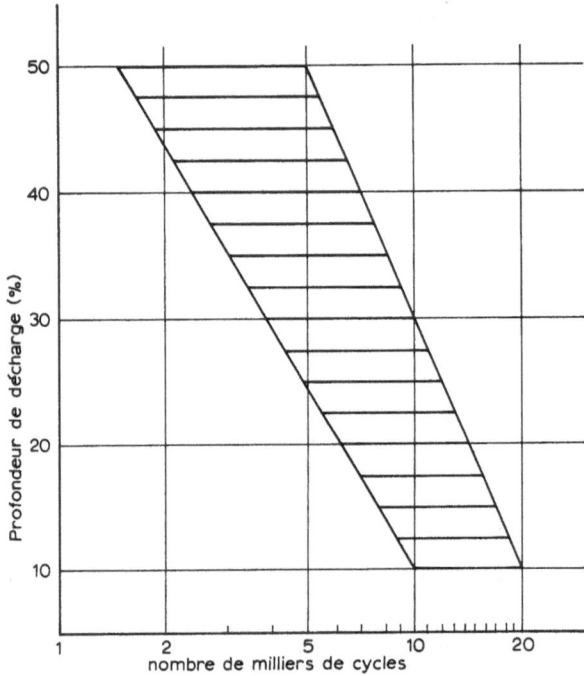

Fig. 3.   Accumulateurs nickel–cadmium pour l'espace. Durée de vie entre $+5°C$ et $+40°C$
(cycles de 1,5 h).

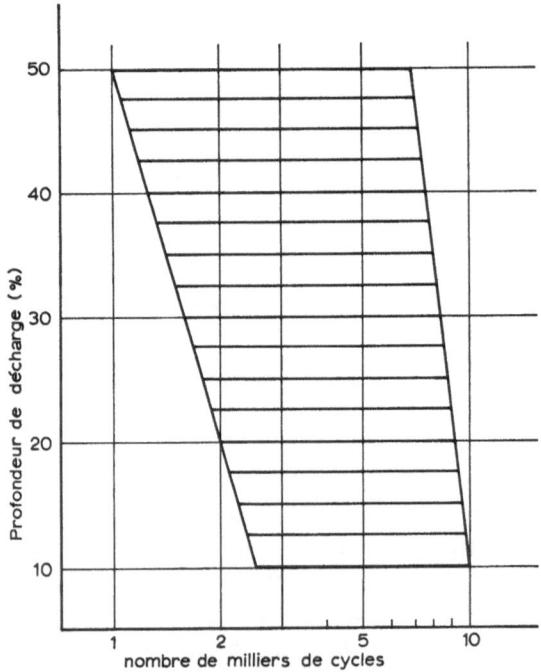

Fig. 4.   Accumulateurs argent–cadmium pour l'espace. Durée de vie entre $+5°C$ et $+40°C$
(cycles de 1,5 h).

batterie pour satellite, le choix doit se faire en fonction des performances réelles pour les conditions de cyclage retenues.

Les durées de vie indiquées sur les figures 3 et 4 concernent des cycles type orbite basse où la durée de charge, relativement courte, impose d'utiliser des courants forts. Dans ces conditions particulières, les accumulateurs sont systématiquement surchargés à fort régime. De très nombreux examens et analyses, effectués sur des accumulateurs ayant participé à ces essais de cyclage, ont permis de relier les dégradations des constituants (du séparateur principalement) à la surcharge.

3. Conditions de fonctionnement des accumulateurs en surcharge

Pour expliciter les causes du vieillissement des accumulateurs, il est nécessaire de rappeler les principes du fonctionnement en charge et surcharge.

La charge complète d'un accumulateur demande une quantité d'électricité plus importante que celle qui a été déchargée au cycle précédent. La quantité excédentaire fournie à l'accumulateur apparaît sous forme de gaz: oxygène si la capacité de l'électrode positive est plus faible que celle de l'électrode négative, hydrogène dans le cas contraire. Les accumulateurs sont généralement construits de telle sorte que le premier gaz apparaissant soit l'oxygène. Le fonctionnement en étanche est possible si ce gaz est consommé dans des réactions secondaires, au fur et à mesure de sa formation. Ceci se produit dans l'accumulateur Ni–Cd où l'oxygène apparaissant sur l'électrode positive est consommé sur l'électrode négative. Si la charge est poursuivie après transformation complète de l'électrode positive, il y a surcharge et l'énergie fournie à l'accumulateur est transformée en chaleur. Si la vitesse de consommation de l'oxygène est insuffisante, la pression monte dans l'accumulateur et, soit se stabilise si pour cette nouvelle pression la vitesse de consommation de l'oxygène augmentée devient suffisante, soit continue à croître jusqu'à destruction de l'accumulateur. C'est la vitesse de consommation de l'oxygène qui détermine le régime maximal admissible en surcharge. Ce régime est de l'ordre de $C/5$ (C = Capacité nominale en Ah de l'accumulateur) pour les accumulateurs Ni–Cd, mais est beaucoup plus faible pour les accumulateurs Ag–Cd, où la consommation de l'oxygène est rendue difficile par la présence de séparateurs freinant fortement le transport de l'oxygène entre les électrodes.

4. Amélioration de l'aptitude à la surcharge

La surcharge peut avoir une incidence mécanique sur la vie des accumulateurs. Des études ont été entreprises par différents laboratoires pour améliorer l'aptitude à la surcharge des accumulateurs, c'est à dire pour faciliter la consommation d'oxygène. La solution généralement retenue consiste à adjoindre à l'électrode négative des accumulateurs, une électrode à oxygène type pile à combustible, capable de consommer électrochimiquement ce gaz. Cette électrode est reliée électriquement à la négative. Ce dispositif permet d'augmenter les régimes admissibles en surcharge, donc de charger les accumulateurs en un temps plus court, mais il présente l'inconvénient de transformer l'énergie électrique fournie en chaleur difficilement évacuable dans un satellite.

5. Moyens utilisés pour éviter la surcharge

Divers procédés ayant pour but d'éviter la surcharge peuvent être employés, certains sont liés à l'apparition d'une pression d'oxygène dans l'accumulateur, d'autres à la variation de tension aux bornes pendant la charge.

Le système le plus simple consiste à détecter, par des moyens mécaniques, l'augmentation de pression interne dans l'accumulateur, mais ces dispositifs sont généralement peu fiables et supportent mal les contraintes élevées apparaissant au lancement du satellite. Dans un accumulateur Ni–Cd, l'apparition de l'oxygène est précoce, une interruption ou une limitation de la charge par la pression peut conduire à charger insuffisamment les accumulateurs. Ce système est donc peu utilisé.

La présence d'oxygène dans l'accumulateur peut aussi être détectée électrochimiquement, en mesurant le potentiel d'une électrode à oxygène placée à l'intérieur. Ce dispositif est simple, ne demande aucun assemblage mécanique, par contre il présente les inconvénients déjà signalés, liés à l'apparition précoce de l'oxygène. Le potentiel de l'électrode à oxygène dite 'signal' est mesuré par rapport à celui de l'électrode négative. Il varie linéairement avec la pression jusqu'à polarisation de l'électrode signal, il faut pour que cet indice soit facilement exploitable, que la variation de potentiel soit importante et reproductible de cycle en cycle. La figure 5 montre la variation de potentiel d'une électrode signal lors de cycles de 100 mn pour des accumulateurs Ni–Cd. Le système paraît maintenant au point, plus de 2000 cycles ont été effectués avec détection de la fin de charge par une électrode de ce type. Des études

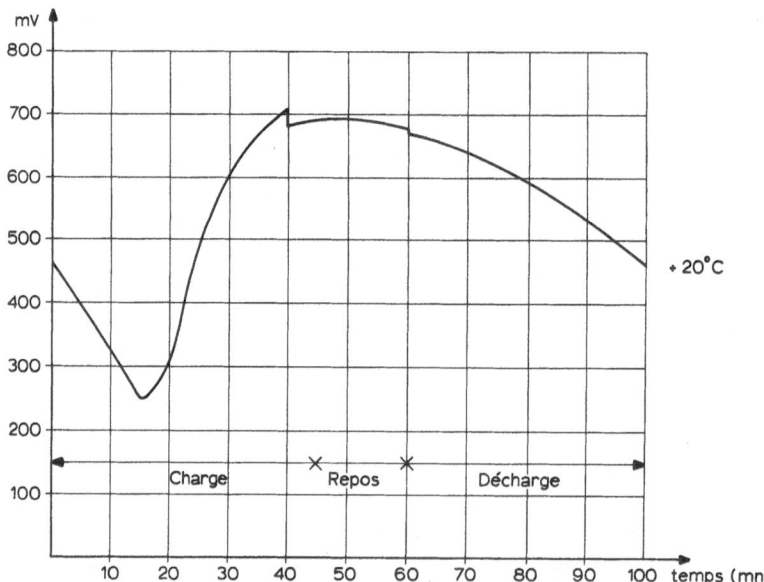

Fig. 5.   Potentiel de l'electrode signal. Cyclage de 100 mn d'accumulateurs VR 3.5 DS à ECS

d'optimisation de cette électrode, et de son comportement à différentes températures sont en cours. Ce dispositif est activement étudié aux Etats-Unis.

Les procédés basés sur les tensions des accumulateurs en fin de charge sont d'un emploi plus délicat. Un des plus simples consiste à associer à chaque accumulateur des diodes de caractéristiques telles qu'elles deviennent passantes quand la tension de l'accumulateur atteint la zone correspondant au dégagement gazeux [3]. Pour être pleinement efficace, il faudrait utiliser des diodes à caractéristiques rectangulaires ne dérivant pas de courant avant le seuil de tension fixé, et acceptant ensuite un fort débit. De plus, le coefficient de température des diodes est différent de celui des accumulateurs, et les dispersions normales des deux fabrications sont telles qu'un assortiment paraît impossible. Cette solution n'est pas utilisée, à notre connaissance, dans les applications spatiales.

Un autre procédé consiste à charger les accumulateurs à tension constante, la tension choisie étant légèrement inférieure à celle correspondant au dégagement gazeux. Cette solution est applicable surtout aux accumulateurs dont la tension varie de façon importante en charge (accumulateurs à électrode positive en argent), elle est d'un emploi plus délicat avec les accumulateurs Ni–Cd dont la variation de tension en charge est faible. Dans ces conditions, un accumulateur complètement déchargé reçoit un courant fort en début de charge, puis ce courant diminue quand le potentiel des électrodes augmente, pour devenir très faible quand l'accumulateur est complètement chargé (figure 6). Ce procédé a l'avantage de conduire à des courants de surcharge faibles, il est couramment utilisé pour les accumulateurs argent–cadmium. Néanmoins, les courants de surcharge étant encore trop élevés, une amélioration a dû être apportée, elle consiste à baisser brutalement la tension de charge, quand le

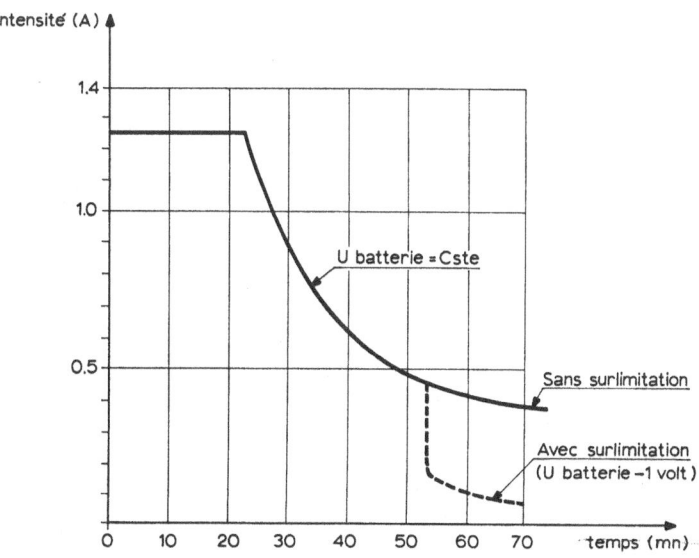

Fig. 6.   Charge d'un accumulateur Ag–Cd, avec et sans surlimitation.

courant devient inférieur à un seuil fixé. Les tensions de charge doivent être déterminées avec exactitude, et corrigées en fonction de la température.

6. Régulation de charge des batteries

Les divers procédés qui viennent d'être rapidement décrits pour la régulation de charge des accumulateurs, sont en principe applicables à des batteries constituées de *n* accumulateurs. En pratique, des problèmes liés aux dispersions des caractéristiques des accumulateurs composant une batterie se posent. On doit choisir très soigneusement les accumulateurs destinés à être assemblés en batterie, en tenant compte de nombreux paramètres tels que capacité, tensions en charge, rendement de charge, mais l'expérience prouve que ces paramètres évoluent de façon différente pendant la vie d'une batterie, et que l'assortiment ne se conserve pas.

Dans le cas d'accumulateurs Ni–Cd équipés d'électrodes signal, on peut soit surcharger, soit sous-charger systématiquement la batterie si le ou les accumulateurs pilotes ont un comportement différent des autres accumulateurs, mais ce défaut de fonctionnement n'est dangereux que pour des charges à des régimes élevés, en dehors du domaine permis pour les surcharges.

Dans le cas d'une batterie d'accumulateurs Ag–Cd chargée à tension constante, les divergences des caractéristiques des accumulateurs peuvent être rapidement catastrophiques. En effet, si nous appelons U la tension de charge optimum pour un accumulateur, la batterie de n accumulateurs sera chargée sous une tension nU. Si un accumulateur en cours de cyclage a un rendement de charge inférieur au rendement moyen de la batterie, sa tension sera plus faible et le courant dans l'ensemble de la batterie, plus élevé. Pour les accumulateurs ayant conservé le rendement de charge initial, il y aura surcharge alors que l'accumulateur responsable à tension faible sera sous-chargé. Les effet cumulatifs de plusieurs cycles dans ces conditions, peuvent devenir alors suffisamment importants pour provoquer l'explosion d'un ou plusieurs accumulateurs de la batterie. Ce problème n'est pas encore résolu, les études visent actuellement à maintenir la dispersion à l'intérieur des batteries dans des limites non dangereuses, et à augmenter l'aptitude à la surcharge des accumulateurs Ag–Cd.

7. Conclusions

L'association batterie d'accumulateurs – générateur solaire est actuellement la source d'énergie la plus utilisée dans les satellites, et le restera probablement encore pendant plusieurs années. De grands progrès ont été faits dans le domaine des batteries d'accumulateurs, mais leur fonctionnement dans les conditions spatiales pose encore de nombreux problèmes. En particulier, les problèmes de charge et de surcharge n'ont reçu jusqu'à maintenant que des solutions partielles et souvent peu satisfaisantes. A une époque où les satellites d'application météorologique ou de communication devant être économiquement rentables, se multiplient, il est primordial de pousser les recherches, tout progrès dans le domaine des sources d'énergie devant améliorer la longévité, donc la rentabilité du satellite.

References

[1] Daspet, H. et Goudot, D., 'Applications spatiales des générateurs électrochimiques', Paris, 4–7 décembre 1967.
[2] Gabano, J. P., Gerbier, G., et Laurent, J. F., 'High Energy Density Electrochemical Generators with Lithium Negative', à paraître, Power Sources Conference 1969.
[3] Mandel, H. J., 'Intersociety Energy Conversion Engineering Conference, Los Angeles, septembre 1966.

BATTERIES D'ACCUMULATEURS SOUPLES NICKEL–CADMIUM POUR ALIMENTATION DE BALLONS MÉTÉOROLOGIQUES

B. MONNERET

Société des Accumulateurs Fixes et de Traction, Romainville, France

Résumé. Le problème de l'alimentation électrique des équipements de mesure et de réponse des ballons du projet EOLE, élaboré par le CNES, a conduit la SAFT à mettre au point une batterie rechargeable d'accumulateurs entièrement nouveaux. En effet, cette batterie – comme d'ailleurs l'ensemble de chaque ballon – doit pouvoir subir une collision avec un avion sans l'endommager gravement.

Les impératifs massiques imposent une forme de couronne à axe vertical et l'emploi de matériaux de densité aussi faible que possible. La version retenue est un accumulateur souple, à enveloppe plastique, fretté sur un noyau cylindrique en polystyrène expansé, cet ensemble constituant un module EOLE.

Le fonctionnement électrique dans une gamme de températures étendue ($-60\,°C$ à $+40\,°C$) et à des régimes compris entre quelques mA et 1 A impose l'emploi du couple Ni-Cd.

La mise au point de l'accumulateur souple et celle du noyau, avec en particulier les problèmes posés par le poids des électrodes, la nécessité d'assurer l'étanchéité complète de l'enveloppe, le compromis entre la faible densité et la dureté superficielle du noyau, sont décrits. Les différents constituants du module, tels qu'ils résultent de ces mises au point, sont passés en revue.

Enfin les performances électriques, aux régimes prévus par le cahier des charges, sont exposées.

Tel qu'il est actuellement réalisé, l'accumulateur EOLE, est apte à assurer l'essentiel des missions pour lesquelles il a été conçu. Néanmoins, certains problèmes liés au fonctionnement à très basses températures et à forts régimes de décharge n'ont pas encore reçu une solution entièrement satisfaisante.

1. Le projet EOLE

Le déplacement des masses atmosphériques provoque des phénomènes météorologiques importants. Or, les mesures effectuées à partir de stations d'observation fixes ne permettent pas de décrire, avec précision, l'évolution des vents, en particulier dans les régions tropicales, où ces stations sont peu nombreuses.

Le projet EOLE, élaboré par le Centre National Français d'Etudes Spatiales (CNES) consiste à lancer dans l'hémisphère Sud plusieurs centaines de ballons à plafond constant. Ces ballons seront équipés d'instruments météorologiques et de répondeurs radioélectriques. Un satellite spécial assurera les liaisons avec ces ballons; à chaque passage il les localisera avec une précision supérieure à celle que l'on obtiendrait depuis le sol et il collectera les mesures effectuées par les instruments météorologiques embarqués sur les ballons.

La source d'énergie alimentant l'équipement de mesures et de réponse des ballons est constituée de cellules solaires et d'une batterie d'accumulateurs nickel–cadmium rechargeable.

La Société des Accumulateurs Fixes et de Traction (SAFT) dans le cadre d'un marché d'études avec le CNES a été chargée de mettre au point cette batterie dite 'batterie EOLE'.

G. A. Partel (ed.), Proceedings of the Second International Conference on Space Engineering. All rights reserved.

2. Impératifs techniques

L'expérience sera réalisée dans l'hémisphère sud où le trafic aérien est relativement faible. Néanmoins la probabilité de rencontre avec des avions est non négligeable. L'impératif technique primordial est donc celui de la sécurité aérienne.

Il impose une masse volumique aussi faible que possible et une géométrie de la batterie telle que la projection des masses sur tout plan vertical soit inférieure à $5\,g/cm^2$.

Ceci exclut l'emploi des matériaux métalliques de densité élevée dans toutes les parties de l'accumulateur où ceux-ci ne sont pas indispensables au fonctionnement, en particulier l'enveloppe. L'impératif géométrique a par ailleurs conduit à retenir, pour l'accumulateur, une forme de couronne circulaire à axe vertical.

L'accumulateur doit satisfaire aux exigences électriques suivantes:

fonctionnement en étanche entre $40°C$ et $-60°C$;

capacité de 1 Ah à $-40°C$ au régime de décharge de 10 mA avec arrêt de la décharge à 1 V;

possibilité de charge entre $+40°C$ et $-30°C$ au régime de 33 mA avec surcharge éventuelle à ce régime pendant 12 h;

décharge entre $+40°C$ et $-60°C$ au régime de 10 mA avec pointes de 1 A pendant 500 ms.

Le fonctionnement en étanche dans un tel domaine de température a imposé l'emploi du couple NiOOH–KOH–Cd dont la tension nominale est de 1,2 V.

Compte tenu de ces impératifs, la SAFT a étudié et mis au point un générateur conçu selon une technologie tout à fait nouvelle. Il s'agit d'un accumulateur souple NiCd, à enveloppe plastique, enroulé et fretté sur un noyau cylindrique de faible masse volumique, l'ensemble constituant un 'module EOLE'.

3. L'accumulateur souple

Cet accumulateur souple est constitué par une électrode positive, une électrode négative, un séparateur poreux et une solution d'électrolyte. L'ensemble est placé dans une enveloppe plastique, traversée par les sorties de courant qui sont assurées par des connexions soudées aux électrodes.

A. ELECTRODES

Elles sont fabriquées selon la technique habituelle SAFT de fabrication en continu d'électrodes minces. Un support constitué par un feuillard perforé et nickelé est enduit d'une pâte à base de poudre de nickel ex-carbonyle, qui est ensuite frittée à haute température.

Le poreux de nickel ainsi obtenu est imprégné d'hydroxyde de nickel ou d'hydroxyde de cadmium selon la polarité des électrodes désirées.

Après l'essai de différents supports, le feuillard finalement retenu est en acier nickelé, de 0,05 mm d'épaisseur. Son taux de perforation est de l'ordre de 40%, l'épaisseur des électrodes est de 0,5 mm.

De nombreux contrôles sont effectués aux divers stades de la fabrication des bandes dans lesquelles sont découpées les électrodes:

contrôle dimensionnel

contrôle d'aspect

contrôle électrochimique.

La capacité obtenue lors de ce dernier contrôle doit être supérieure ou égale à 1,8 Ah/dm² pour les électrodes positives et 2,0 Ah/dm² pour les électrodes négatives. Ce contrôle permet également d'apprécier la solidité de la bande après travail électrique.

B. SÉPARATEUR

Le séparateur est un feutre de polyamide de 25/100 mm d'épaisseur. Il isole électriquement les plaques entre elles et immobilise l'électrolyte qui n'est pas absorbé par les électrodes.

Pour chaque accumulateur, trois bandes de ce séparateur sont superposées et soudées selon leur plus grande dimension, pour constituer deux poches qui recoivent les électrodes positives et négatives.

C. ELECTROLYTE

C'est une solution aqueuse d'hydroxyde de potassium de densité 1.31 (7,5 N) dont le taux de K_2CO_3 doit être inférieur à 3 g/l.

D. ENVELOPPE

Initialement constituée par des feuilles de chlorure de polyvinyle (PVC), l'enveloppe est maintenant formée par deux feuilles rectangulaires de RILSAN de 20/100 mm d'épaisseur et soudées par haute fréquence.

E. CONNEXIONS

Leur étude a été liée à celle de l'enveloppe pour pouvoir assurer l'étanchéïté parfaite de l'accumulateur. Après l'essai infructueux de différents fils, feuillards et grillages, qui se sont avérés soit trop fragiles soit mauvais adhérents sur le matériau plastique, le métal déployé de nickel de 20/100 mm d'épaisseur a été adopté. Il assure une étanchéïté durable et il est relativement peu fragile.

Avant le montage de l'accumulateur, des bandes protectrices en téflon sont appliquées sur les bords des électrodes pour éviter les courts-circuits éventuels pouvant être provoqués par effritement des bords des électrodes.

Les caractéristiques de l'accumulateur souple sont les suivantes:

poids: 47g		
dimensions:	longueur hors tout	208 mm
	largeur (hors connexions)	64 mm
	épaisseur	1,7 mm.

4. Le module EOLE

Le module EOLE est constitué par l'accumulateur souple enroulé et fixé sur un noyau cylindrique. Ce noyau de très faible densité, doit présenter une dureté superficielle suffisante pour permettre, sans déformation le maintien de l'accumulateur. Il est fabriqué, par moulage à chaud, à partir de polystyrène expansé. Il a la forme d'un cylindre creux dont une extrémité comporte une collerette qui permet l'emboitage du noyau dans un autre.

Le frettage de l'accumulateur sur le noyau est assuré par un ruban adhésif à support polyester renforcé par des fils de verre, choisi pour son faible allongement, pour son bon pouvoir collant et sa bonne tenue à la température.

Avant l'assemblage du module EOLE, la formation électrique de l'accumulateur est réalisée. Elle comprend trois cycles charge – décharge – réalisés à température ambiante. Le premier consiste en une charge de 40 h à 0,095 A, un repos de 1 h et une décharge à 0,380 A jusqu'à 1 V. Les deux autres cycles consistent en une charge de 34 h à 0,057 A, un repos de 1 h et une décharge au régime de 0,190 A jusqu'à 1 V.

Les caractéristiques du module EOLE sont les suivantes:

poids	72 g
hauteur hors tout	75 mm
diamètre extérieur	66 mm.

5. La batterie EOLE

La tension moyenne de la batterie EOLE devant être de 12 V, celle-ci est donc composée de 10 modules, emboités les uns dans les autres, et de 2 supports. Ceux-ci sont des disques qui s'emboitent dans les modules terminaux. Ils sont obtenus comme les noyaux, par moulage de granulés de polystyrène et ils comportent, noyé dans le polystyrène, un insert destiné à collecter le courant.

La rigidité mécanique de la batterie est obtenue par collage des surfaces de contact des modules et des supports et par serrage dans un filet à mailles fines.

Les caractéristiques de la batterie sont les suivantes:

poids total	730 g
longueur	740 mm
diamètre	66 mm.

6. Performances électriques des accumulateurs EOLE

la capacité nominale à −40°C, au régime de décharge de 10 mA avec arrêt à 1 V, est supérieure à 1 Ah. D'ailleurs à des régimes de décharge plus élevés, la capacité est encore supérieure à 1 Ah. Par exemple, un groupe de 16 accumulateurs a subi le cycle suivant:

charge de 34 h à 20°C, au régime de 0,057 A ;

repos de 24 h à −40°C ;

décharge à −40°C, au régime de 0,030 A jusqu'à 1 V.

Les capacités restituées sont comprises entre 1.12 et 1.30 Ah.

Le fonctionnement en régime continu aux basses températures est possible. Une série de 16 accumulateurs, après les 3 cycles de formation à température ordinaire, a été essayée en décharge à −50°C et −60°C, après charge, chaque fois, à −30°C. Les régimes, en charge, étaient de 0,044 A pendant 33 h, et en décharge, de 0,044 A également, jusqu'à 1 V. Les résultats montré dans le tableau I ont été obtenus :

TABLEAU I

	Capacité en Ah	
	minimum	maximum
Formation 1° cycle	1,42	1,55
2° cycle	1,40	1,50
3° cycle	1,37	1,49
Charge à −30°C/décharge à −50°C	0,76	0,95
Charge à −30°C/décharge à −60°C	0,46	0,68

Enfin, le fonctionnement aux basses températures, avec pointes d'intensité de courant est également possible de façon satisfaisante jusqu'à −50°C. A températures plus basses, les résultats sont dispersés et la tension en fin de pointe devient vite inférieure au seuil fixé (0,800 V).

Le tableau II donne les résultats d'essais effectués à −50°C et −60°C sur un groupe de 9 accumulateurs. Ces 9 accumulateurs ont subi les cycles de formation puis ils ont été chargés à −30°C pendant 33 h au régime de 0,044 A.

Après repos de 24 h à −60°C, ils ont été soumis à l'essai de décharge avec pointes d'intensité, en 5 séquences.

Après décharge residuelle à +20°C, les accumulateurs ont été à nouveau chargés à −30°C, mis au repos 24 h à −50°C puis soumis au même essai de décharge à −50°C.

Il est à noter que, à −50°C, la dernière séquence a été prolongée de 7 h au delà des 30 premières minutes, et qu'au bout de ce temps, la tension moyenne était de 0,77 V.

7. Conclusion

Les accumulateurs, destinés à équiper les ballons du projet EOLE, devaient satisfaire à des impératifs mécaniques et électriques sévères. Le fonctionnement à −60°C d'un accumulateur dont l'électrolyte est une solution de potasse, la rigidité dont il doit être doté pour pouvoir être manipulé sans soins excessifs et la légèreté qu'il doit présenter, pour qu'en cas de chocs contre des avions la collision ne se transforme pas en catastrophe, ont été autant de problèmes auxquels la SAFT, en mettant au point un accumulateur EOLE, tout à fait original, a essayé de trouver une solution.

TABLEAU II

Séquence	Temps en mn	Tension moyenne en V	
		à $-60°C$	à $-50°C$
pointes de 1s à	0	0,85	0,98
1,2 A espacées de	4	0,84	0,98
30 sec à 0,012 A pendant	12	0,78	0,97
30 mn	20	0,70	0,97
	25	0,66	0,97
	30	0,60	0,96
Décharge de 10 mn	début	1,30	1,29
à 0,035 A	fin	1,29	1,27
pointes de 1 sec à	0	0,62	0,85
1. 2A espacées de	2	0,59	0,81
40 sec à 0,1 A pendant	5	0,64	0,78
10 mn	10	0,53	0,76
Décharge de 10 mn	début	1,29	1,26
à 0,035 A	fin	1,24	1,25
pointes de 1 sec à	0	0,72	0,89
1,2A espacées de	4	0,71	0,88
30 sec à 0,012 A	12	0,69	0,85
	20	0,66	0,83
	25	0,63	0,81
	30	0,61	0,79

Certes, des améliorations peuvent être apportées à l'accumulateur actuel. Des études en ce sens sont d'ailleurs en cours, en liaison avec le Centre National d'Etudes Spatiales. Elles concernent:

au point de vue fondamental: l'identification et l'explication des phénomènes électrochimiques régissant le comportement d'accumulateurs Ni–Cd à très basses températures;

au point de vue technologie: la recherche et l'expérimentation de nouveaux matériaux destinés à la fabrication des modules EOLE et permettant d'améliorer soit leurs performances électriques en diminuant leur résistance interne par exemple, soit leurs caractéristiques mécaniques en particulier la rigidité des noyaux et la légèreté de l'ensemble de la batterie;

Cependant, d'ores et déjà, l'accumulateur EOLE de la SAFT est apte à être utilisé pour diverses missions et, d'ailleurs, des ballons destinés à des essais préliminaires au projet EOLE ont été lancés en automne 1968, équipés de ces accumulateurs.

References

BREVETS CONCERNANT L'ACCUMULATEUR EOLE

Titres français et anglais du 1er brevet
– 'Générateur électrochimique a faible énergie d'impact'
– 'Improved électrochemical generator, with low impact energy'

Dates et numéros des dépôts de ce brevet

Brevet français n° 1.519.696 du 6/2/1967 et 1ère addition n° 134.357 du 29/12/1967

Dépôts a l'étranger:

Belgique	n° 54.059 du 31.1.1968
Canada	n° 10.936 du 27.1.1968
Italie	n° 12.219/A/68 du 31.1.1968
Japon	n° 6.940/68 du 6.2.1968
Luxembourg	dépôt du 25.1.1968
Pays-Bas	n° 6801675 du 6.2.1968.
U.R.S.S.	n° 1.215.897 du 6.2.1968
Allemagne Fédérale	n° 114.012 du 6.2.1968
Grande Bretagne	n° 5.953 du 6.2.1968
Etats Unis	SN 701.444 du 29.1.1968

LOW-POWER NUCLEAR ENERGY CONVERSION FOR LONG-DURATION SPACE MISSIONS

Comparison between Rankine and Brayton Cycles: Dynamic and MHD Power Conversion

JOHN W. BJERKLIE

Mechanical Technology, Inc., Latham, U.S.A.

Abstract. Several schemes of converting nuclear reactor energy are compared for overall suitability in long duration space power systems. System power levels under consideration are from 5 kWe to 400 kWe. Suitability is judged on the basis of approximate weight and size, and qualitative reliability as reflected in probability of success.

The four schemes differ in cycle and in power conversion machinery as follows:

(1) Dynamic Brayton cycle.
(2) Liquid metal magnetohydrodynamic (LMMHD) gas cycle.
(3) Dynamic Rankine cycle.
(4) LMMHD Rankine cycle.

Two basic methods of using liquid metal MHD energy conversion are compared – continuous acceleration by liquid metal injection into gas phase working fluid, and pulse acceleration of liquid metal slugs by gas phase working fluid.

Three different heat sources are used for the system comparisons: Pu-238 isotope (lightly shielded example), Co-60 isotope (heavily shielded example), and nuclear reactor.

Maximum probable temperature limits are used in the comparison so that all systems can be compared on their best footing.

Reliability is evaluated on the basis of probability of success in solving recognized development problems. The method assumes that all systems can be made reliable enough with sufficient development, but that the severity of technical problems is greater for some systems than others even for a conservative design. Such a method of evaluation is qualitative at present, but is sufficient for rating probable success and probable suitability of candidate space power systems.

It is shown that the dynamic Brayton machine and the two fluid Rankine continuous flow liquid metal MHD system deserve continuing development; that the low temperature dynamic Rankine system should be developed; and that equal emphasis should be placed on liquid metal dynamic Rankine and liquid metal Rankine intermittent flow LMMHD systems as desirable future power conversion devices.

1. Introduction

Over the last decade many difficulties have delayed the development of long duration multikilowatt space power systems of all kinds. At present rates of progress it may be another decade before any system can be said to be acceptable.

This is a good time to reconsider all space power systems potentially suitable in the next decade. Are we really looking to the right ones for long duration deep space manned missions? Are there others we should consider? Are there some that should be relegated to applied research programs because their promise is so low? While we cannot give direct answers of policy to these questions, we can certainly critique our present direction and suggest new ones. A comparison of development difficulties among the various systems, and of the weights of temperature sensitive components will show which systems should be considered for further development to meet real space vehicle requirements.

G. A. Partel (ed.), Proceedings of the Second International Conference on Space Engineering. All rights reserved.

Non-expendable heat sources – nuclear reactors and isotopes – are of major concern here. Power systems using these heat sources must be extremely reliable to be suitable for manned space vehicles. One indication of reliability is a measure of the probability of success. Both dynamic and static energy conversion systems can be used here.

Thermoelectric systems have achieved such low efficiency that their weight is too great for any power above a few watts. Thermionic systems, as yet unproven for large power and high efficiency, require such high temperatures for good efficiency that their development is still far from a certainty.

Plasma MHD power generation is as yet unproven, and requires either very high temperature or unsubstantiated physics for its operation.

These three systems, thermoelectric, thermionic and plasma MHD, are not considered further.

This paper indicates the expectations of performance, development problems, and approximate heat source and radiator weights of remaining types as follows:

Component weights considered in each case:

> Isotope heat source
> Nuclear reactor heat source
> Simplest armored heat rejection system

Power Systems:

Dynamic:	Closed Brayton Cycle	
	Low Temperature Rankine Cycle	
	Liquid Metal Rankine Cycle	
LMMHD:	Continuous flow system:	Single fluid Rankine (liquid metal)
	(Liquid Metal	Two fluid Rankine (liquid metal)
	Electromotive Fluid)	Gas cycle
	Intermittent flow system:	Low temperature Rankine
	(Liquid Metal	Single Fluid Rankine (liquid metal)
	Electromotive Fluid)	Gas Cycle.

The efficiency and weight comparisons are done in terms of achievable fraction of Carnot efficiency in all cases. The fraction of Carnot efficiency that has been achieved in practice or that logically can be expected is indicated for each type of system.

Then the systems are compared on the basis of the need for achieving success in component development and on the difficulty being encountered in achieving that success. By considering both the anticipated weight advantage of the various systems and the possibility of achieving development success a rational choice of systems for near term use can be made. The findings agree that the Brayton dynamic and the two fluid Rankine continuous flow LMMHD systems should continue to be developed. There is some question as to the immediate effort suitable for the dynamic liquid metal Rankine system. The low temperature dynamic Rankine system and the liquid metal Rankine intermittent flow LMMHD should be given a prominent role in future development, the latter device being aimed for future systems.

2. Factors Affecting System Weight

The major contributors to the weight of heat operated power systems for space systems are the heat source, the power conversion machinery, the condenser and/or radiator, adjunct heat exchangers such as regenerators and regenerative boiler feed heaters, power conditioning equipment, and structure. Only the heat source and radiator will be considered in this study. The weight of the rest of the system is relatively insensitive to temperature levels, unlike the radiator and isotope heat sources, so would tend to be the same for all power systems. Despite the nearly constant weight of the nuclear heat source it must be included to allow a one-to-one comparison with the isotope systems.

A. HEAT SOURCE

The heat source weights for high powered reactors and all isotope sources can be related to thermal power requirements. For any Carnot limited power systems (which the LMMHD and dynamic systems are) the thermal power can be related to output power by

$$P_{th} = \frac{P}{\eta_{system}} = \frac{P}{f \eta_{carnot}} = \frac{P}{f \left(1 - \dfrac{T_R}{T_H}\right)}$$

where P_{th} is the thermal power

P is the system output power (same units P_{th})
η_{system} is overall system efficiency
η_{carnot} is Carnot efficiency
f is the achievable fraction of Carnot efficiency
T_H is heat source temperature, absolute
T_R is heat rejection (minimum) temperature, absolute.

The factor, f, will be discussed later. Its value can be so low for some of the candidate systems that even the nuclear reactors must be larger than their minimum size even at power as low as 150 kW output. Normally the minimum size is limited by

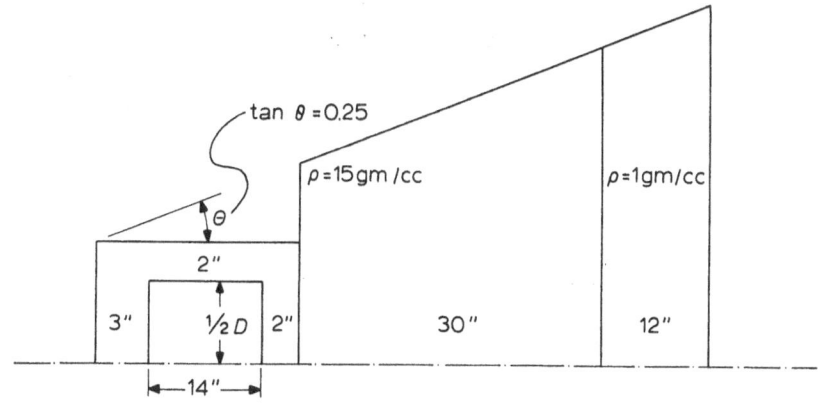

criticality considerations to about two thermal megawatts, which is big enough for all the more efficient systems even up to 400 kW output.

For shadow shielded nuclear reactors suitable for manned space vehicles shielding should be about the equivalent of 30 in. of lead and tungsten structure (having a specific gravity of 15) plus another foot of water outside of this ([1]). The reactor core diameter can be assumed to vary as shown in [1]. The shape and core weight will be assumed to be as in [2] and the half angle of shadow protection is 14° (tan $\theta = 0.25$) (see sketch on page 471). It has been assumed that the shielding requirement varies insignificantly with coolant outlet temperature. Thus reactor weight is:

$$W_R = 1000 \frac{D}{10} + \left\{ \left[\left(\frac{D}{10} \times 16 \right) + 15 \right]^3 - \left[\left(\frac{D}{10} \times 16 \right) + 11 \right]^3 \right\} \times \frac{62.4 \times 15}{1728}$$

$$+ \left\{ \left[\left(\frac{P}{10} \times 16 \right) + 21 \right]^3 - \left[\left(\frac{D}{10} \times 16 \right) + 19 \right]^3 \right\} \frac{62.4}{1728}$$

= Reactor weight + lead and tungsten shield weight

+ water shield weight

= $18.28D^2 + 407.4D + 1320$ lb

where 1000 = weight of 10″ diameter reactor without shielding

D = $7 + \frac{11}{7} \times \mathrm{MW}_{TH}$ in. for $D > 10″$

= 10 in. for $P_{th} < 2\ \mathrm{MW}_{th}$

= diameter of reactor core

P_{th} = thermal power.

It was similarly assumed that the isotope heat source must be shadow shielded sufficiently for use in a manned spacecraft. Two isotopes were chosen for investigation – Co-60 and Pu-238 – because their power density and half lives are good enough for long term space missions and their shielding needs vary from heavy for Co-60 to light for Pu-238. These two isotopes would represent the extremes of those which could be chosen. The shielding requirements were taken from [2, 3, 4, and 5], from which an estimate was made on a consistent basis of 5.5 in. of Uranium for Co-60 and 0.75 inches of Uranium for Pu-238. These shields provide man livable zones of 10 mrems/h at 1 m distance. The specific power output of the isotopes were derived from characteristics listed in [3]. Thus total weight of isotope plus shadow shielding weight is:

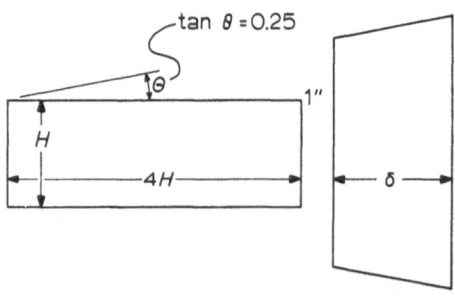

$$W_1 = \frac{2}{p}\left[\frac{P}{\eta_{\text{system}}}\right] + \varrho_u[\text{Vol}_u]$$

where p is specific power of the isotope

Vol_u = volume of shield, and is calculated from dimensions of shield

ϱ_u = density of shield material

or W_I = $4.4/p\ [P_{thi}] + 0.0571\ [\delta^3 + 3\delta^2 + 3\delta + 36H\delta + 18H\delta^2 + 108H^2\delta]$

δ = thickness of shield

H = small dimension of isotope heat source

and W_{co} = $3.71\ P_{th} + 15.6 + 42.4\ H_c + 33.9\ H_c^2$ lb

where H_c = $0.997\ [P_{th}/0.7]^{1/3}$ in.

P_{thi} = $P_{th}/0.7$ = initial thermal power. The factor $1/0.7$ allows for full required power at the end of three years with Co-60.

P = 1.7 W/g for Co-60

H_c = the height, width, and quarter length of the isotope volume

P = power output, kW.

P_{th} = thermal power, kW.

and for Pu-238 (3 years mission life)

$$W_{Pu} = 11.3\ P_{th} + 0.248 + 2.12\ H_p + 4.615\ H_p^2\ \text{lb}$$

where H_p = $1.575\ [P_{th}]^{1/3}$ in.

$P_{thi} \approx P_{th}$

In each case it is assumed that the unshielded heat source weighs twice as much as the isotope weight.

It has been tacitly assumed that these heat sources will be suitable at the high temperatures indicated later on for some systems.

B. RADIATOR

This study assumes that a total duration in deep space of about three years will be needed by the time any of the subject power plants are developed. Under this condition the radiator is the remaining most important weight contributor which is sensitive to the operating temperatures.

The simplest of meteoroid protection schemes will be used for this study; a thin stainless steel liner for containing the fluid, protected by a thick concentric beryllium armor, with solid beryllium fins, and with radiation from both sides of the radiator. The meteoroid flux is assumed to be isotropic and applicable to deep helio-centric space (as per [6]). For large low temperature radiators a model was chosen which included a full width fluid passage with no fins. The whole surface was armored on both sides with beryllium armor plate. The lighter of the two models was used in each case.

The internal heat transfer coefficient of condensing low temperature fluids (except steam) is about 500 BTU/h ft^2°F. For liquid metal and steam condensation this coefficient can be much higher, assumed to be 2000 BTU/h ft^2°F for this study. Under these conditions, ignoring those portions of the radiator having other than condensing heat transfer coefficients, ignoring solar heating, and for a fin effectiveness of 90% (as per [7]) and a constant thickness fin cross section the weight of the Rankine Cycle radiators is:

$$W_{RR} = (\pi\delta^2 L + \delta A_i)\,\varrho_{armor} + 2ltL\varrho_{fin} + A_i \times \frac{0.020}{12}\,\varrho_{tube} \left.\begin{array}{l} \text{whichever is} \\ \text{lighter} \end{array}\right\}$$

$$= A_o\delta'\varrho_{\text{rmor}}_a \times 2$$

where the steel inner tube has a 0.020 in. wall thickness

A_i is the internal area

$\quad = O_{rej}/(T_R - T_m)\,h_i$

$Q_{rej} = (1/\eta_{system} - 1)\,P \times 3413 = \sigma\varepsilon A_0\,(T_m^3 + T_m^2 T_s + T_m T_s^2 + T_s^3)(T_m + T_s)$
$\qquad\qquad\qquad\qquad\qquad\quad = h_0 A_0\,(T_m - T_s)$

$T_m\quad$ = metal temperature (assumed to be constant for simplicity)

$T_s\quad$ = sink temperature (assumed to be 0°R for simplicity)

$t\quad$ = 8 $(h_0 l^2/k_f)$, or $t = 0.015 \times 10^{-8} l^2\,T_m^3$ (feet) for beryllium fin and armor

$\delta\quad = a/2.54\,\gamma\,(6/\pi)^{-1/3}\,(62.4\varrho_p/\varrho_{armor})^\phi\,(V/c)^\theta\,([\alpha A_v\tau/-\ln(P_0)]^{1/(3\beta)}\varrho_p^{-1/3}$
$\qquad = 0.00415\,A_v^{1/4}\,\text{ft} = 0.00415\,(A_i + 2\pi\delta L)^{1/4}$

$h_i\quad$ = 500 BTU/h ft^2 °F for low temperature fluids
$\qquad$ = 2000 BTU/h ft^2 °F for liquid metals

$\delta'\quad = 0.00415\,(A_0)^{1/4}$

$A_0\quad$ = external area, ft^2

$A_v\quad$ = vulnerable area (external area of tube armor), ft^2

$h_i\quad$ = internal heat transfer coefficient, BTU/h ft^2 °F

$h_0\quad$ = effective outside heat transfer coefficient, BTU/h ft^2 °F

$L\quad$ = total length of tubes, ft

$Q\quad$ = heat rejection, BTU/h

$T_R\quad$ = rejection temperature

$T_m\quad$ = metal temperature (assumed constant) °R

$T_s\quad$ = sink temperature, °R (assumed 0°R for deep space)

$\sigma\quad$ = Stephan-Boltzmann constant, 0.173 × 10^{-8} BTU/h ft^2 (°F)^4

$\varepsilon\quad$ = surface emissivity (0.9 assumed)

$\delta\quad$ = armor thickness, ft

$l\quad$ = fin length, ft

$t\quad$ = fin thickness at root, ft

$\varrho_{fin}\quad$ = fin material density, lb/ft^3

$\varrho_{tube}\quad$ = tube material density, lb/ft^3

ϱ_{armor} = armor material density, lb/ft^3

$a\quad$ = empirical spalling factor

$\gamma\quad$ = empirical proportionality constant

$\tau\quad$ = time of exposure, seconds

$P(0)\;$ − probability of zero penetrations

α = experimental constant for meteoroid flux, g/m^2 sec
V = meteoroid velocity, m/sec
C = speed of sound in armor material, m/sec
ϱ_p = meteoroid particle density, g/cc.

It was found convenient to let the fluid tube diameters be $\frac{1}{8}$ in., since smaller tubes, as required for optimizing, become impractical.

The chosen model deviates from reality when large fins ensue from the calculations. They would be made hollow, and partially filled with a small amount of liquid metal so as to be used as heat pipes to maintain high fin efficiency. The results of the conservative model are, even so, usable for comparison purposes although they cannot be considered as real radiator weights. Considerable realism is retained for the lighter systems however.

The internal heat transfer coefficient for Brayton Cycles can be adjusted nearly at will limited only by allowed pressure drop and number of headers. Thus, for a large number of short tubes the pressure drop per unit length can be quite high, thus allowing very high heat transfer coefficients. Also, allowable pressure drop can be higher if the minimum cycle pressure is high, as it can be for some compact machines. Since there is no ready rationale for selecting a universally usable design criteria on pressure drop, number of headers, etc. an arbitrary choice was made to use an internal heat transfer coefficient of 150 BTU/h ft^2 °F. While this is much lower than for Rankine systems, it is considerably higher than for the usual terrestrial system. Higher heat transfer coefficients would begin to introduce compressibility effects in the flow, so for simplicity were not considered here.

The effective temperature of heat rejection can be related to the minimum cycle temperature easily after making the simplifying assumption that the entering temperature equals the compressor outlet temperature as it would be if a 100% effective regenerator were used. This assumption leads to a slightly lower effective temperature than will occur in reality. The real entering temperature will be somewhat higher since the regenerator's effectiveness is less than 1.0.

Thus

$$T_{eff}^4 = \frac{3(T_e - T_R)}{\dfrac{1}{T_e^3} - \dfrac{1}{T_R^3}}$$

for $T_s \approx 0$,

where

$$T_e = T_R \left[1 + \frac{(P_r^{(\gamma-1)/\gamma} - 1)}{\eta_c}\right]$$

and

T_R is the minimum cycle temperature
T_e is the temperature of the gas entering the radiator
P_r is the cycle pressure rated (assumed to be 3 for simplicity in this calculation)
η_c is the compressor efficiency (assumed to be 0.85).

Temperature T_{eff} can now be used in place of T_R found in the Rankine cycle equations. The ratio of specific heats, γ, is assumed to equal 1.67.

It should be mentioned that armor of the headers has been ignored in the weight equations for the sake of simplicity. For the Rankine systems the resulting error need not be large, but for the Brayton Cycle where specific volumes are higher the error is such that real radiators will be heavier than indicated.

The weights of the heat sources and the various radiators were calculated using these equations. Figure 1 shows the heat source weights as a function of thermal power to the cycle. Figures 2, 3, and 4 show the radiator weights of the liquid metal Rankine, low temperature Rankine, and Brayton cycles as a function of thermal power rejected and the minimum cycle temperature, T_R.

3. System Efficiency Considerations (Estimation of f Value)

A. DYNAMIC BRAYTON

The Brayton Cycle of interest for space power applications is closed and uses a fluid such as Argon, Krypton or mixtures of Xenon and Helium as the working fluid. The system description and performance has been well reported in the literature ([2, 4]). The basic system consists of a compressor, heater, expander and cooler.

The mechanism for accomplishing the compression and expansion can be turbo-machinery or positive displacement machinery. However, there is ample reason to believe that the very long-lived units should be turbo-machines. The most apparent evidence for this is the success of the turbo-jet airplane engine in subsonic aircraft resulting in part from the much better maintenance picture of the turbo-jets compared

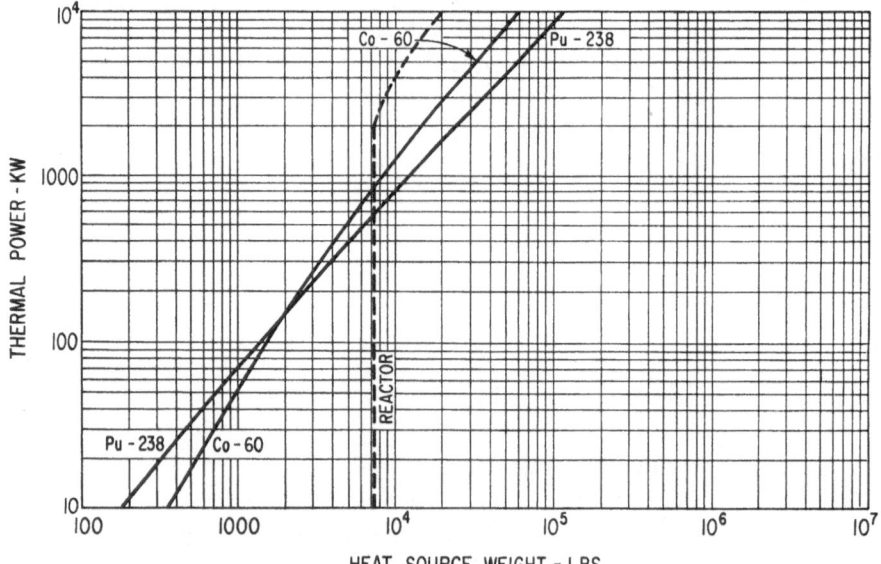

Fig. 1. Heat source weight.

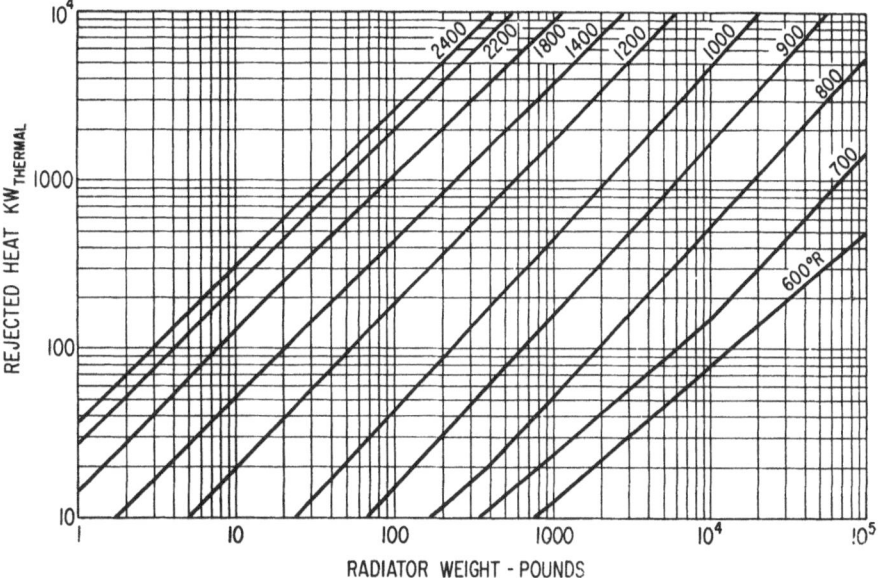

Fig. 2.   Radiator weight – Rankine system. Liquid metal cycles.

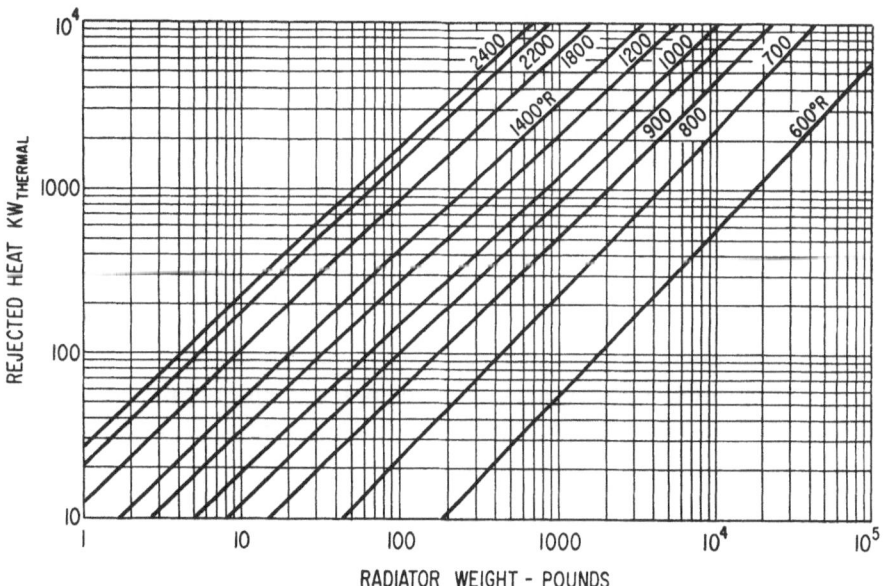

Fig. 3.   Radiator weight – Rankine system. Low temperature cycles.

to the reciprocating engine. For this paper, therefore, it will be assumed that the machinery is of the turbine type.

Turbo-machinery for low power closed Brayton cycles implies high rotating speeds, lightly loaded bearings, solid rotor alternators, high performance and low leakage in both the compressor and turbine components. Ability to withstand loads imposed by

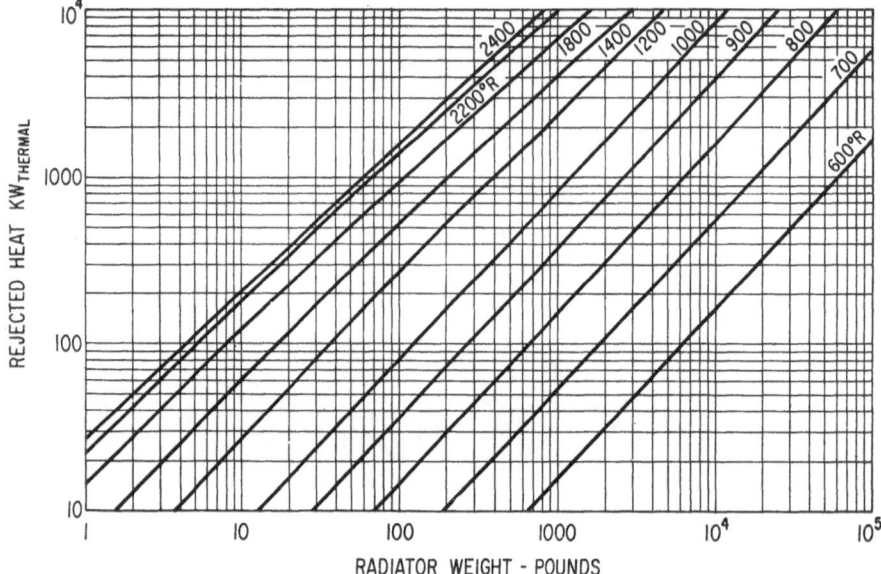

RADIATOR WEIGHT - POUNDS

Fig. 4.   Radiator weight – Brayton system.

launch and maneuvering must be built into the unit. The efficiency of the recuperated, closed Brayton system for ideal fluids is described by the following equation:

$$\frac{\eta_{\text{system}}}{\eta_{\text{elect}}} = \frac{\eta_t\left[1-\left(\dfrac{1+\zeta}{P_r}\right)^{(\gamma-1)/\gamma}\right] - \dfrac{T_R}{\eta_c T_H}\left(P_r^{(\gamma-1)/\gamma}-1\right)}{(1-\varepsilon)\left(1-\dfrac{T_R}{T_H}\right)+\varepsilon\eta_t\left[1-\left(\dfrac{1+\zeta}{P_r}\right)^{(\gamma-1)/\gamma}\right]-\dfrac{(1-\varepsilon)}{\eta_c}\dfrac{T_R}{T_H}\left[P_r^{(\gamma-1)/\gamma}-1\right]}$$

where

η = efficiency, subscripts; elect – electrical generator, t – turbine
         c – compressor
P_r = pressure ratio
ζ = pressure loss expressed as a fraction of the compressor pressure ratio
T = temperature, °R; subscripts; R – compressor inlet, H – turbine inlet
γ = ratio of specific heats, assumed = 1.67
ε = recuperator effectiveness.

Highest efficiency can be achieved for high γ values, such as 1.67 for the noble gases. For well designed machinery both the turbine and compressor efficiency can approach 85%. The recuperator effectiveness can logically be anywhere from 70% to 95% depending on the weight and cost penalty allowable. Usually, though, a value of 85% can be reached easily for space power systems.

Turbine and compressor efficiency have been widely discussed in the literature, the result being that semi-empirical design methods now can be used reliably. This is exemplified by the work in [9] and as extended in [10 and 11]. Considerable additional data is presented in [12]. Detailed loss factors from cascades at low Reynolds numbers (typical of low power machinery) have also been reported ([13 and 14]). These types of results can be used to modify the optimum design chosen by the normalizing methods of [9].

The power levels considered in this paper normally result in low specific speeds. That is the regime where the efficiency will increase as rotating speed and volume flow increase and as head change lowers. This means that high rotating speeds and/or multi-staging applies to the low specific speed components in order to obtain high efficiency. Since low power machinery should logically be simple machinery, also, the designer usually opts for high rotating speed to avoid the complexity of long shafts with many seals.

Figure 5 shows $f(=\eta_{\text{system}}/\eta_{\text{carnot}})$ vs. η_{carnot} at optimum P_r for the Brayton cycle.

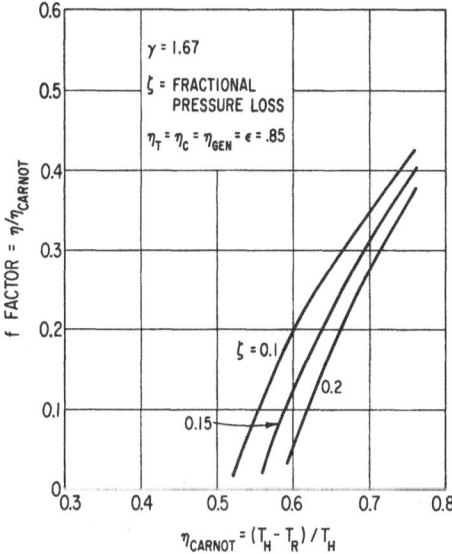

Fig. 5. Brayton cycle performance. Optimum pressure ratio.

The electrical efficiency was assumed to be 85%. The total pressure loss was assumed to be 10%, 15% and 20% of the pressure rise in the compressor as indicated on the curves. It will be noted that f is high for high Carnot efficiency (large temperature difference) but drops rapidly as Carnot efficiency decreases (T_R approaches T_H). This cycle is also quite sensitive to losses.

Other studies have indicated f factors that support the calculated values as in [2] (0.39 @ 2000°R = T_H, 0.27 @ 1760°R, 0.2 @ 1660°R), and in [4] (0.37 @ 2000°R).

CO_2 as a Brayton fluid holds considerable promise. It is possible for the f factor to approach 0.70 at overall Carnot efficiency of 70% and with high efficiency components and with minimum cycle temperatures below 100°F ([24]). This compares favorably with the best values indicated on Figure 5.

B. DYNAMIC RANKINE

The Rankine Cycle is best exemplified by the conventional central power plant using steam. The simplest systems consists of a pump, preheater-boiler, expander and condenser.

The literature on Rankine Cycle space power systems is voluminous. For instance, [15, 16, 17, 18, 19, and 20] are recent papers on power systems applicable to space.

Many variations of the Rankine Cycle exist which use reheat, regenerative feed water heating, economizers, and other heat transfer devices for conserving heat. Also, many fluids other than water can be used in the cycle. Some of these alternate fluids lead to simpler machinery and more efficient thermodynamic cycles than for steam or liquid metals.

Here again the pump and expander can take many forms. But for this paper it is assumed that turbo-machinery is to be considered on the grounds that ultimately this will lead to higher reliability than positive displacement equipment for very long duration missions.

Turbo-machinery for low power long duration space power systems implies high speed turbines, high speed pumps, lightly loaded bearings, solid rotor alternators and seals between the pump and turbine or alternator cavities.

Values of f for dynamic Rankine engines are quite readily obtained experimentally ([15, 16 and 17]). In the cases of the well designed systems where some effort is made to obtain excellent performance f factors range as high as 0.6. This obtains when the working fluid approaches being a Carnot fluid (Mercury, regenerated organic cycles) and when care is taken to optimize the turbine. The turbine and pump performance can be readily predicted and designs adjusted by use of the methods shown in [9] similarly to estimating performance of the Brayton cycle components. The effects on turbine efficiency of moisture in the expanded vapor when using some fluids is discussed in some detail in [12 and 21]. Moisture affects f factor because approximately 1% of turbine performance is paid for every percent of moisture in the expanded vapor. Losses due to windage, bearings, and leakage are readily estimated for hermetically sealed systems such as these would be.

The thermodynamic characteristics required for usable Rankine working fluids are indicated by [19]. Using the principles outlined there it is very easy to make an initial selection of fluids for particular purposes. Figure 6 is a reproduction of the fluids chart from [19].

Usable f values for regenerated organic fluid cycles can be considered to vary from 0.5 to 0.6. This value tends to be reasonably constant even if the operating temperatures change as long as there is no great amount of super-heating. For liquid metals the evidence points to usable values of 0.45 and less ([22 and 23]). This is partially brought on by superheating, thus causing the mean temperature of heat addition to be less than the peak temperature. As used here, f is evaluated on the basis of peak and minimum temperatures in the cycle. Moisture in the expanded vapor is a contributor to low f.

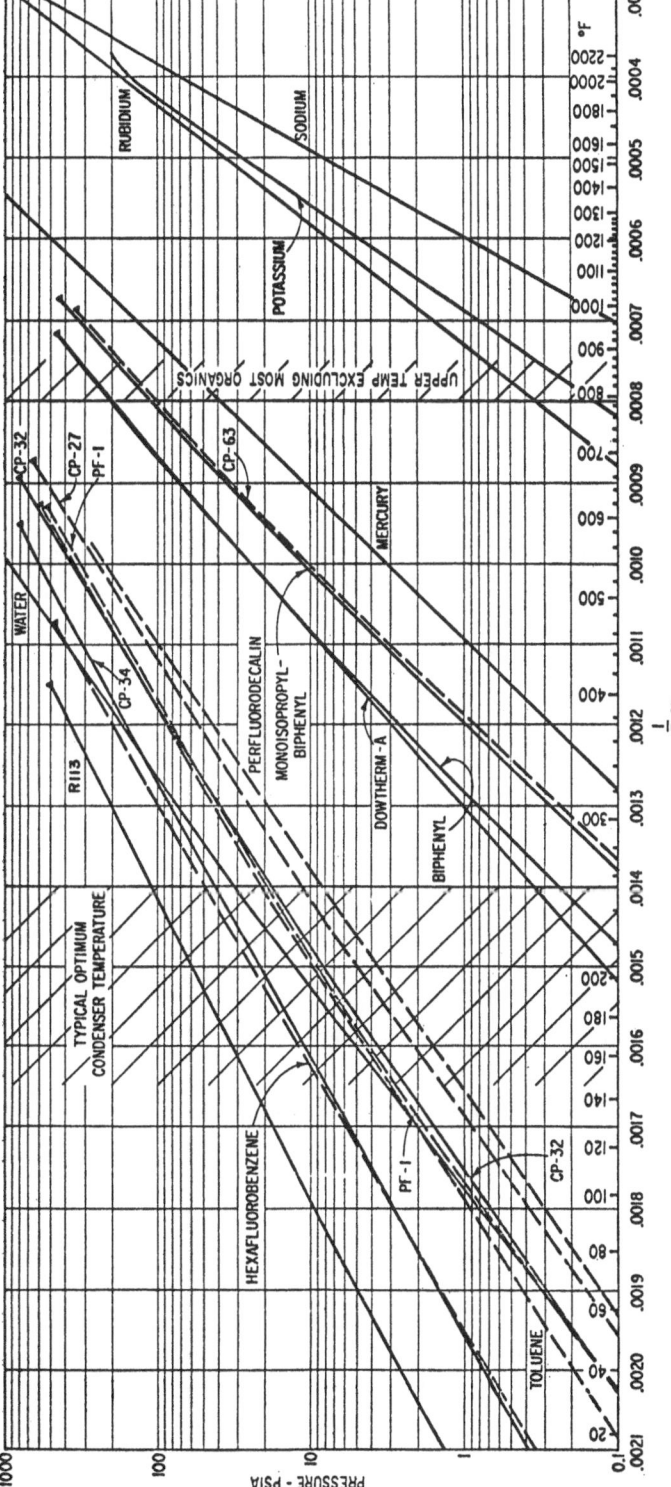

Fig. 6. Vapor pressure – Temperature relation for various Rankine cycle working fluids. Note: CP fluids – Monsanto Chemical Company designations. PF-1 – Mechanical Technology Inc. designation.

The supercritical Rankine cycle has been considered in some detail in [25 and 26] (the last one for CO_2 as the working fluid). Hexafluorobenzene and other highly fluorinated fluids show some promise here. In these cases the f factor depends greatly on the regenerator effectiveness that can be achieved. The use of a regenerator tends toward Carnot efficiency, so it's ineffectiveness destroys some of the potential benefits of rising f. But because the cycles are generally at quite high pressure the heat transfer rates are good in the regenerator. Thus, reasonable size can be held while obtaining regenerator effectiveness as high as 90%. With this value the f factors can be quite high. For hexafluorobenzene condensing at 300°F the f factor is about 0.55, and for 180°F condensing temperature it is between 0.60 and 0.65, for perfect electrical generators.

Use of CO_2 as a Rankine fluid for space is not practical because the condensing temperature has to be less than 88°F. However, if the low condensing temperature could be justified for some application the f value would be about 0.7.

C. LMMHD SYSTEMS

The thermodynamic cycle for LMMHD cycles can be similar to the corresponding dynamic system. In particular, Rankine LMMHD is analogous to Rankine dynamic. The LMMHD component merely replaces the turboalternator. The cycles can operate continuously or with intermittent expansion analogous to a reciprocating air cycle engine.

In the continuous flow systems the high pressure working fluid and the electromotive liquid can be mixed and expanded through a two phase nozzle. Heat is added to the mixture by virtue of either electromotive liquid or thermodynamic working fluid having been preheated. It is also possible to admit the working fluid in stages to improve the overall acceleration process efficiency.

After the mixture is accelerated it can be used in several ways:

(1) The foamy, frothy or bubbly mixture can be passed directly through the MHD generator if an electrical continuum can be maintained ([27]), or

(2) The working fluid and liquid can be separated while maintaining the high velocities attained in accelerating and let only the liquid pass through the MHD generator (represented by [28, 29 and 30]), or

(3) The working fluid can be completely condensed while accelerating the electromotive fluid (as in [31, 32, 33, 34, 35 and 36]), the complete flow then passing through the MHD generator.

In intermittent flow systems high pressure, high temperature working fluid is admitted into an expansion tube alternately with slugs of liquid metal. The expanding gas propels the liquid slug through the MHD generator, followed by separation of the two phases. The liquid metal can be admitted at either low or high pressure depending on whether or not the gas is allowed to expand before admitting the liquid.

A variation of this that shows promise is the internally heated intermittent flow LMMHD system. It consists of admitting a low pressure liquid or gas into a heating chamber at the upstream end of the expansion tube. A liquid metal slug closes the tube so that the injected liquid or gas is heated rapidly in a confined volume, thus achieving both high pressure and high temperature. The energy is removed from this trapped gas or vapor as the slug is accelerated down the tube similar to a bullet in a gun barrel. This system should achieve cycle efficiencies higher than any other LMMHD system, limited only by the volume ratio reached before the slug is penetrated by the expanding gas.

Yet another variation would be to trap a low pressure volume of gas in the tube and precompress it as in a two stroke (compression and expansion) cylinder closed at one end. In this case external heat would be added to the gas as it is compressed in the closed end of the tube ahead of a liquid metal slug accelerated upstream. After the gas is heated it begins to expand so that the liquid metal slug reverses direction and is propelled through the MHD generator to the condenser end of the tube. The cycle would approach being a Diesel cycle rather than a Brayton Cycle if it proved workable.

Another variation of all these intermittent flow cycles is to use a metal slug electromotive component lubricated by liquid metal. This would involve making a system in which the slug direction could be reversed, the lubricant could be circulated, and the working fluid injected either at high pressure with external heating, or at low pressure with internal heating.

Intermittent cycles should be usable to very low power levels. It should be emphasized that the large difference in efficiency between the internally heated intermittent flow systems and the continuous flow systems occur for two main reasons: (1) the intermittent systems in no way have to expend large power for circulating electromotive fluids, and (2) there are no large acceleration losses.

D. INTERMITTENT RANKINE CYCLE

The author has previously investigated an MHD system of this type using intermittent internal heating ([37]) and found it to be suitable for single module power to a few kilowatts. The efficiency was quite good – on the order of 15% – for an f factor of 0.33 corresponding to an MHD generator efficiency of 60%. The efficiency is represented by

$$\eta_{\text{system}} = \frac{\eta_{\text{gen}}\left[(U_{Hv} - U_{Hl}) - (H_{Rv} - H_{Rl})\right] - \text{losses}}{U_{Hv} - U_{Rl}}$$

where

   U_{Hv} is internal energy of vapor just before expansion
   U_{Rl} is internal energy of liquid just before injection
   H_{Rv} is enthalpy of vapor after expansion
   H_{Rl} is enthalpy of liquid after condensing and before injection

and the losses are:
   (1) Pumping the power fluid to high pressure for injection
   (2) Pumping the electromotive fluid to low pressure for circulation
   (3) Friction of the liquid metal slug moving down the power tube
   (4) Residual kinetic energy in the slug that is not recovered after leaving the MHD

generator

  (5) heat loss from the vapor to the liquid slug

  (6) power to hold the slug in place while the working vapor is being formed.

It was found that the most important losses are the residual kinetic energy and the power associated with the magnetic slug holding system. This system is expansion ratio limited because of the limited slug stability. The slug is eventually punctured as it travels down the tube, all the while sustaining a pressure differential.

The f factor should not vary greatly over a range of operating conditions since the losses tend to be proportional to power in the same way that flow rates are. For this study, then, the f factor can be considered fixed at 0.3 to 0.33.

A two fluid version of the intermittent Rankine cycle LMMHD system uses a low temperature fluid as the thermodynamic component, and liquid metal as the electromotive fluid. With regeneration as necessary, this system should achieve f factors nearly as high as the low temperature dynamic Rankine system. $f = 0.45$ will be used herein.

E. INTERMITTENT GAS CYCLE

The efficiency of this cycle cannot be expected to be as good as for the Rankine cycle with equal temperatures for two major reasons. First, the efficiency of the Brayton cycle itself is very loss dependent; second, the expansion ratio of the gas will not be as large as for the Rankine cycle unless very high pressures are used. The latter point arises because the specific volume of gas upon injection into the heating chamber usually will be much greater than for a vapor formed during constant volume vaporization of a liquid.

By using a high molecular weight gas, high pressure, very careful injection so as not to disturb the liquid metal slug, and assuming that extremely rapid methods of heating the gas can be found, the best (no loss) f factor at Carnot efficiencies of nearly 100% will approach 25% with an MHD generator efficiency of 60%. The efficiency of a cycle using external pressurization will be:

$$
\eta = \frac{\eta_{\text{gen}}\left[(H_1 - H_2)\right] - (H_5 - H_4) - \text{losses}}{U_1 - U_6}
$$

$$
= \frac{\eta_{\text{gen}}\gamma \dfrac{T_H}{T_R}\left[1 - \left(\dfrac{1}{P_r}\right)^{(\gamma-1)/\gamma}\right] - \dfrac{\gamma}{\eta_c}(P_r^{(\gamma-1)/\gamma} - 1) - \dfrac{\text{losses}}{C_v T_r}}{(1 - \varepsilon)\left[1 + \dfrac{1}{\eta_c}(P_r^{(\gamma-1)/\gamma} - 1)\right] + \varepsilon \dfrac{T_H}{T_R}\left(\dfrac{1}{P_R}\right)^{(\gamma-1)/\gamma}}
$$

where subscripts are 1 – temperature at beginning of expansion (end of heating)

  2 – power tube outlet (regenerator low pressure inlet)

  3 – regenerator low pressure outlet (radiator inlet)

  4 – radiator outlet (compressor inlet)

  5 – compressor outlet (regenerator high pressure inlet)

  6 – regenerator high pressure outlet (injection temperature)

and where

U is internal energy

H is enthalpy

η_{gen} is the MHD generator efficiency

η_c is the external compressor efficiency

ε is the regenerator effectiveness

P_r is the pressure ratio

and losses are similar to Rankine cycle losses.

The efficiency for an internally pressurized and heated system is

$$\eta = \frac{\left\{\left(1 - \dfrac{T_R}{T_H}\right) - (\gamma - 1)\left[1 - \left(\dfrac{T_R}{T_H}\right)^{(\gamma-1)/\gamma}\right]\right\}\eta_{gen} - \dfrac{\text{losses}}{C_v T_H}}{\left(1 - \dfrac{T_R}{T_H}\right)}.$$

For this version (assuming 10% losses and a PV diagram shape factor of 90%) f ranges from about 0.1 at low temperature differences to 0.15 at η_{carnot} approaching 1, with $f > 0.13$ to $\eta_{carnot} = 40\%$.

The condensation (heat loss to slug) loss found to be important in the Intermittent Rankine engine is missing in these cycles, although some heat transfer between the gas and liquid will be difficult to avoid.

Figure 7 shows a plot of f factor for these cycles. The externally heated version of this cycle was used in evaluating weights for comparing with other systems.

F. CONTINUOUS FLOW RANKINE SYSTEMS

Continuous flow systems all suffer from having to pump very large volumes of liquid metal to fairly high total pressures and/or by instituting large losses in the acceleration process due to large differences in gas and liquid velocities and in shock losses. Within these limits reasonable efficiencies can still be obtained with some systems.

The type is best exemplified by the two fluid system being developed at the Jet Propulsion Lab of the California Institute of Technology ([29 and 30]). This is a separation system, so inherent losses accrue in the separation mechanism. Other losses are similar to those that occur in other continuous flow systems. The reported efficiency figures – all calculated, but with some experimental verification of component efficiencies – go as high as 6.3% at 325 kW ([29]). This indicates an f factor of 0.25 with an MHD generator efficiency of about 63% ([30]). In this case some of the losses are not proportional to power – particularly the separator losses – so that f will vary somewhat. For this study, however, it will be assumed that it remains constant at lower power levels.

Because fluid losses are high the lower power limit of operation for this type of cycle is probably about 20 kW, although such studies have not been reported.

Single fluid systems for achieving continuous flow have been investigated rather widely [31, 32, 33, 34, 35 and 36]. The vapor is injected ahead of the MHD generator

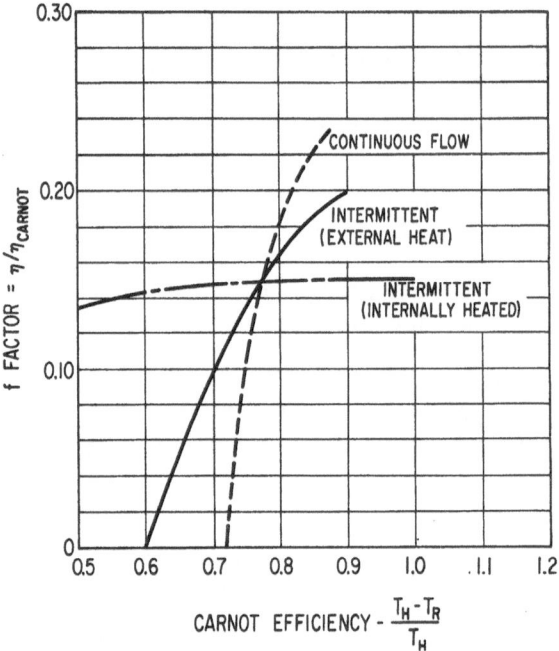

Fig. 7. Gas actuated LMMHD system performance. (Regenerated cycles.)

and condenses while expanding to accelerate the liquid. Calculation for multi-stage injection appear to yield high efficiencies, but actual measurement [33, 31, and 38] shows little promise of success. The basic problem seems to lie in the existence of high condensation shock losses and friction [33]. Achievable f factors now appear to be 0.06 [33] at best with MHD generator efficiency of 100%.

G. CONTINUOUS FLOW GAS CYCLE

The acceleration system of Compagnie Electro Mecanique [28] exemplifies this type of system. While not a true Brayton cycle, so that direct comparison with the dynamic Brayton cycle is not possible, it bears a resemblance, being an all gas system and in being sensitive to losses. The expansion process is isothermal, and the compression is isentropic with an external dynamic compressor. The following expression is an approximate, and optimistic, expression of the cycle efficiency:

$$\eta_{\text{cycle}} = \frac{[\eta_{\text{gen}} + \eta_D - 1]\left(\dfrac{\gamma - 1}{\gamma}\right)\ln\left(\dfrac{P_H}{P_R}\right) - \dfrac{T_R}{\eta_c T_H}\left[\left(\dfrac{P_H}{P_R}\right)^{(\gamma-1)/\gamma} - 1\right]}{(1 - \varepsilon)\left\{\dfrac{T_H - T_R}{T_H} - \dfrac{T_R}{\eta_c T_H}\left[\left(\dfrac{P_H}{P_R}\right)^{(\gamma-1)/\gamma} - 1\right]\right\} + \dfrac{\gamma - 1}{\gamma}\ln\left(\dfrac{P_H}{P_R}\right)}$$

where η_D is the diffuser efficiency, P_H is heat addition pressure, P_R is heat rejection pressure and the other terms are as previously noted.

There are losses in the process due to skin friction, diffuser losses, and ineffectiveness of the regenerator but the overall f factor with a 60% MHD generator can be about 0.2. Calculated (optimistic) f values with high component performance are shown in Figure 7. A characteristic of this cycle is that the Carnot efficiency with the component performance chosen must be over approximately 75% before it will work.

4. Weight Comparison

Using the f values indicated for each type of system, the weights of the heat sources and radiators can be estimated as a function of peak temperature and minimum cycle temperature. Probable peak operating temperatures were chosen for each system. The optimum minimum temperature was then found. Tables I, II, and III show T_H, T_R, f, and combined heat source and radiator weights for 25, 50, 100, and 400 kW using the three heat sources. It was found that the radiator weight varied little whether one or more units were used to achieve a given power.

These results (except with the 400 kW weights) are repeated in Figures 8, 9, and 10. The curves in Figures 8, 9, and 10 pictorially present that the weights fall into fairly distinct bands. In particular:

for Pu-238 systems:

(1) the liquid metal Rankine and Brayton Dynamic systems are lightest

(2) the liquid metal intermittent LMMHD system is next lightest

for Co-60 systems:

(1) the liquid metal and Brayton dynamic systems and the liquid metal intermittent LMMHD systems are all lightest.

for Nuclear systems:

(1) the liquid metal intermittent LMMHD and the two fluid Rankine continuous flow LMMHD systems are lightest.

5. Probability of Success

In Table IV are listed the types of units under consideration and the development and operating problems associated with each. It is of value to attempt evaluating the problems in developing these systems to get an idea of their probability of success. The method used here is meant to be a screening method, only, but one which is realistic. Each problem is given a rating factor as follows:

(1) Engineering development required, will take some time and money.

(2) Applied research required.

(3) Fundamental problem: basic research required.

Furthermore, the desirability of finding solutions to each problem can be evaluated as follows:

(1) Satisfactory solutions or alternates available.

(2) Poor solution available which will decrease desirability.

(3) New solution required for success.

If both rating numbers are accorded each known problem and multiplied together, the sum of all the products for a given type gives a hint of the probability of success.

TABLE I

Optimum weights (Σ heat source + radiator) Pu-238 isotope heat source

	Dynamic			LMMHD Intermittent				Continuous		
	Low Temp R	Liquid Met R	Brayton	Low Temp R	Liquid Met R		Gas	2 Fluid R	1 Fluid R	Gas
T_H (max. usable)	1200°R	2500°R	2500°R	1200°R	2700°R		3000°R	2700°R	2700°R	2700°R
T_R (optimum)	700°R	1000°R	800°R	700°R	1000°R (optimum)	1400°R (usable)	800°R	1000°R	1000°R	600°R
f	0.5	0.45	0.555	0.45	0.33		0.125	0.25	0.06	0.15
Power										
5 kW	453	332	280	505	420	485	1120			
25 kW	2020	1445	1205	2270	1815	2128	4800	2405	9200	14 400
50 kW	3970	2710	2350	4380	3540	4060	10 350	4700	19 150	30 000
100 kW	7750	5310	4530	8570	6890	8023	20 050	9200	38 000	60 700
400 kW	30 100	15 650	12 650	33200	26 800	31 030	76 500	35 600	143 000	243 200

TABLE II

Optimum weights (Σ heat source + radiator) Co-60 isotope heat source

	Dynamic			LMMHD Intermittent			Continuous			
	Low Temp. R	Liquid Met. R	Brayton	Low Temp. R	Liquid Met. R		Gas	2 Fluid R	1 Fluid R	Gas
T_H (max. usable)	1200°R	2500°R	2500°R	1200°R	2700°R		3000°R	2700°R	2700°R	2700°R
T_R (optimum)	700°R	1000°R	800°R	700°R	1000°R (optimum)	1400°R (most usable)	800°R	1200°R	1200°R	600°R
f	0.5	0.45	0.555	0.45	0.33		0.125	0.25	0.06	0.15
Power										
5 kW	683	542	470	735	645	725	1340			
25 kW	2120	1605	1435	2300	1914	2128	4300	2344	6905	14 200
50 kW	3620	2660	2370	3930	3220	3460	8550	3870	12 430	28 900
100 kW	6350	4410	4080	6970	5540	5823	15 350	6745	22 210	57 200
400 kW	20 600	11 150	17 150	22 600	17 400	18 330	53 300	21 430	75 600	224 300

TABLE III

Optimum weights (Σ Heat source + radiator) Nuclear heat source

	Dynamic			LMMHD Intermittent			Continuous		
	Low Temp. R	Liquid Met. R	Brayton	Low Temp. R	Liquid Met. R	Gas	2 Fluid R	1 Fluid R	Gas
T_H (max. usable)	1200°R	2500°R	2500°R	1200°R	2700°R	3000°R	2700°R	2700°R	2700°R
T_R (optimum)	900°R [a]	1000°R	800°R [a]	1000°R [a]	1400°R [a]	1000°R [a]	2000°R [a]	1400°R	600°R
f	0.5	0.45	0.555 [a]	0.45	0.33	0.07	0.25	0.06	0.15
Power									
5 kW	7231	7232	7260	7241	7205	7 330			
25 kW	7375	7355	7475	7420	7228	7 810	7223	7400 [a]	18 800
50 kW	7580	7510	7750	7680	7260	8 460	7247	7610	31 800
100 kW	8010	7810	8280	8240	7323	9 850	7300	9960	57 200
400 kW	14 800	9550	11 350	16 900	8530	26 800	9760	29 700	209 400
							(1400°R = T_B)		

[a] Σ weight is nearly constant over wide T_R range.

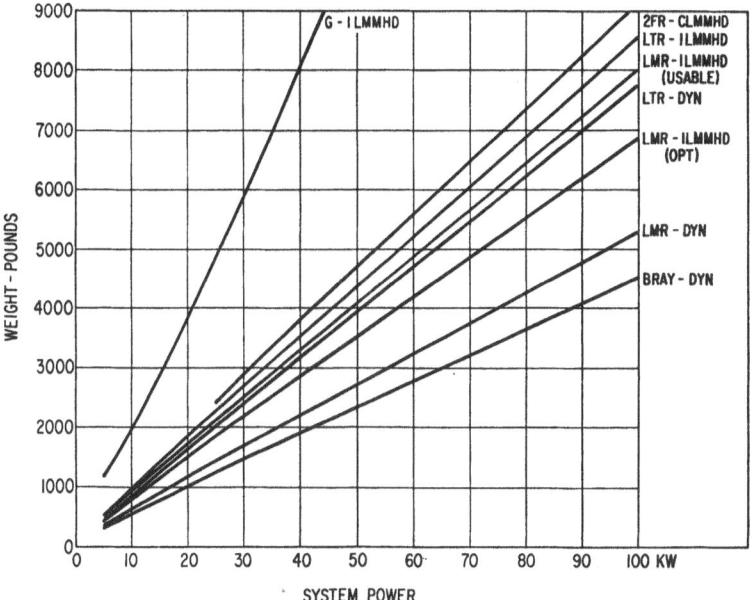

Fig. 8.   Weight of Pu-238 heat source + radiator.

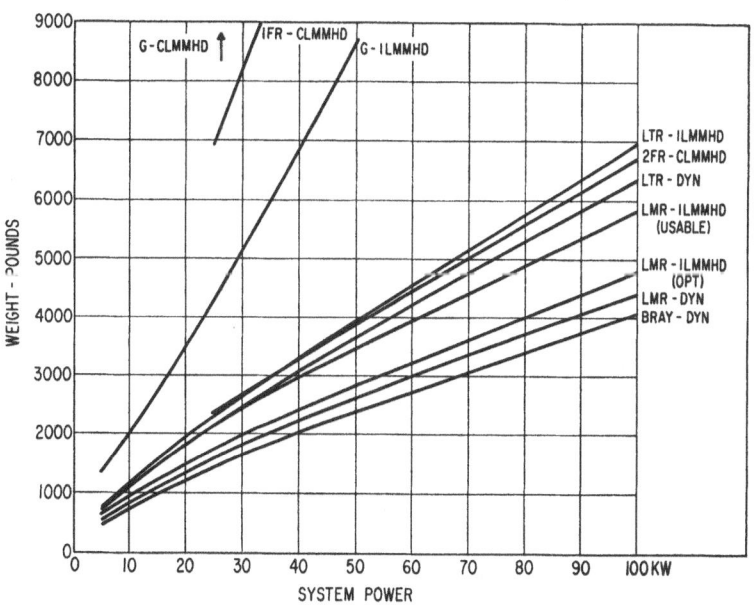

Fig. 9.   Weight of Co-60 heat source + radiator.

Three points stand out when this is completed for all the systems as in Table IV:
(1) the low temperature Rankine and the Brayton cycle appear outstanding for most
probability of success; (2) the single fluid continuous flow and the intermittent gas
LMMHD systems appear least probable of success; (3) the liquid metal Rankine
and two fluid continuous flow LMMHD systems presently being developed as well

TABLE
Probability

Dynamic									LMMHD–intermittent					
Rankine						Brayton			Rankine					
Low temp. fluid			Liquid metal						Low temp. fluid			Liquid metal		
	A	B		A	B		A	B		A	B		A	B
Zero-g boil	2	1	Zero-g boil	2	1	Turbine performance	1	1	Heat loss in expansion	2	2	Heat loss in expansion	2	2
Zero-g cond.	2	1	Zero-g cond.	2	1				Slug stability	2	2	Slug stability	2	2
Bearings in shock and vibration	1	1	High temperature bearings	1	2	Comp. performance	1	1	MHD shorting	2	2	Erosion and corrosion	2	2
						Bearings in shock and vibration	1	1						
Turbine performance	1	1	Bearings in shock and vibration	1	1	High temperature bearings	1	2	Ht. trans to spray	2	3	MHD shorting	2	2
Pump cavitation	1	1	Syst. erosion and corrosion	3	3	Fluid loss	3	2	Mech. valving	1	1	Heat transfer to gas spray	2	3
Fluid stability	3	1	Turbine performance	1	1	Start up	1	2	Mag. holding devices	1	2	Valving	1	
Bearing erosion	1	2	Pump cavitation	1	1	Turbine creep	1	3	Fluid compatibility	1	3	Mag. holding devices	1	2
Start up	1	2	Erosion in turbine	2	2	Meteor. prot.	2	1	Σ = 24			Σ = 25		
Meter hours	2	1				Σ = 18								
Σ = 16			Start up	1	2									
			Turbine creep	1	2									
			Meteor. protection	2	1				(With use of LM lubed solid metal slug) Eliminate slug stability, add lube system at $1 \times 1 = 1$					
			Alt. protection	2	2				Σ = 21			Σ = 22		
			Σ = 32											

IV
of Success

Gas	LMMHD – Continuous			
	1 Fluid	Rankine 2 Fluid	Gas	
A B	**A B**	**A B**	**A B**	
Heat loss in expansion 2 2	Heat loss in expansion 2 2	Fl dyn loss in separation 1 3	Fl dyn loss in separation 1 3	A
Slug stability 2 2	Shock loss in condensing jet 3 3	Heat loss in expansion 2 2	Heat loss in expansion 2 2	1 – Engineering development
Erosion and corrosion 2 2	Erosion and corrosion 3 3	Erosion and corrosion 3 3	Shock losses 2 2	2 – Applied research
MHD shorting 2 2	Diff. losses 1 1	Diff. losses 1 1	Erosion and corrosion 3 3	3 – Fundamental problem
Fluid loss 3 2	MHD end Losses 1 1	MHD end losses 1 1	Diffuser loss 1 1	
Ht. trans to gas 3 3	MHD friction 1 1	MHD friction 1 1	MHD end losses — 1 1	B
Gas compression 1 2	Σ = 25	Σ = 19	MHD friction 1 1	1 – Satisfactory solutions or alternates available
Mech. valving 1 1			Gas loss 2 2	2 – Poor solution available which will decrease desirability
Mag. holding device 1 2			Σ = 27	3 – New solution required for success
Σ = 36				
Σ = 33				

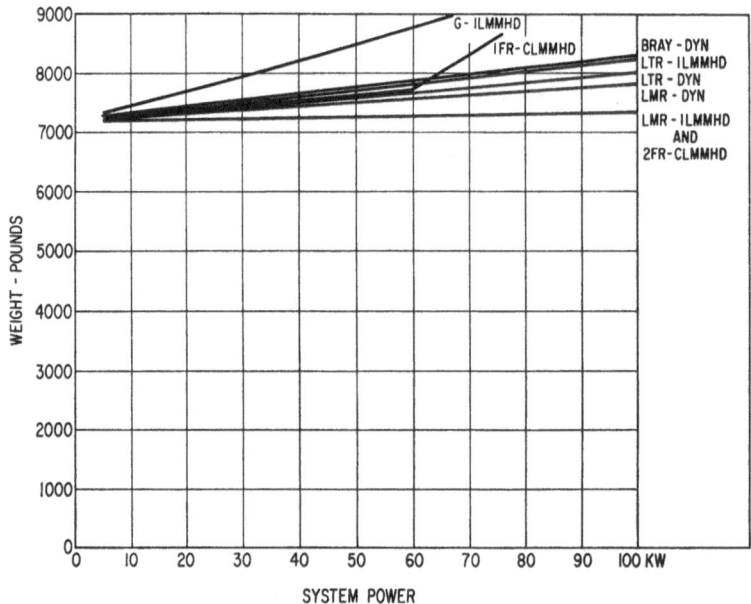

Fig. 10.   Weight of reactor heat source + radiator.

as the two intermittent Rankine LMMHD and gas type continuous flow LMMHD systems are all about equally easy to develop.

Some comments are in order for each problem area noted in Table IV. These are given in Appendix A. Only those problems having to do with energy conversion are considered there. Problems with the development of reactor and isotope heat sources, their treatment before and after the launch and the mission, and policy problems are beyond the scope of this paper.

The list of problems and their status is meant only to indicate the major ones. Development problems always occur when actual work begins such that each type of power system will have its own specific problems that are not typical of a class of systems, but only of the particular design.

It is obvious that some technological achievements can change the status of these technological considerations. Principally materials improvements could change some fundamental problems, or problems with poor alternatives, to purely engineering problems. The competition between systems would then be nearly reduced to purely thermodynamic and fluid dynamic differences. Although drastic improvements in heat transfer limits, meteoroid armor requirements and injector performance might cause large changes in policy, these are not to be expected immediately.

6. Overall Rating

It is abundantly clear that, despite the many simplifying assumptions made in reaching conclusions, some of the new underdeveloped systems should be pursued further.

From a weight point of view the following systems are meritorious with one or more of the heat sources

Brayton dynamic   Pu-238, Co-60
Liquid metal Rankine dynamic   Pu-238, Co-60, nuclear
Liquid metal Rankine intermittent
    LMMHD (new system)   Pu-238, Co-60, nuclear
Two fluid Rankine continuous flow
    LMMHD   nuclear

From the ease of development point of view the following systems are particularly meritorious

Low temperature Rankine dynamic
Brayton dynamic
Two fluid Rankine continuous flow LMMHD

It is at least instructive to attempt combining the weight and probability of success ratings. To do this the success ratings on Table IV were first normalized by dividing all the summed product values by the lowest one. Using the 400 kW combined weights shown in Tables I, II, and III, the weights were similarly normalized for each kind of heat source. The product of these normalized ratings should be a first approximation of overall desirability, it being necessary to assume only that weight and probability of success have equal value to a space program. These products are shown in Table V.

In all cases the Dynamic Brayton system appears superlative. In addition, for the nuclear heat source the following systems are superior:

two fluid Rankine continuous flow LMMHD
liquid metal Rankine intermittent flow LMMHD
low temperature Rankine dynamic

The low temperature Rankine dynamic system also appears very satisfactory for the Co-60 isotope heat source.

7. Conclusions

After comparing performance, normalized weight of heat source plus radiator, and the probability of success of various power systems suitable for long duration manned space vehicles it is clear that a shift in development emphasis is desirable. The following table summarizes the situation:

(1) Light, easy to develop (high probability of success) sytems should be selected where isotopes are the heat source. That is, the isotope heated system with the highest overall rating should be most desirable:

Co-60 isotope heat source (heavily shielded example)
    Low temperature Rankine dynamic
    Brayton dynamic

TABLE V

Overall rating

	Dynamic			LMMHD				Continuous		
	Low Temp. R / Liq. Met. R		Brayton	Intermittent Low Temp. R / Liq. Met. R			Gas	2FR	1FR	Gas
Normalized success rating S	1	2	1.125	1.5		1.56	2.37	1.18	1.56	1.69
Pu-238 Weight rating (W)	2.38	1.24	1.0	2.62	(opt.) 2.12	(most) (usable) 2.46	6.05	2.82	11.3	19.2
Overall rating (S × W)	2.38	2.48	1.125	3.93	3.3	3.84	14.32	3.33	17.6	32.4
Co-60 W	1.85	1	1.54	2.03	1.56	1.65	4.78	1.92	6.79	20.1
Co-60 S × W	1.85	2	1.74	3.05	2.43	2.57	11.33	2.26	10.6	34
Nuclear W	1.74	1.12	1.33	1.98		1	3.14	1.14	3.48	24.6
Nuclear S × W	1.74	2.24	1.5	2.97		1.56	7.44	1.35	5.43	41.6

Pu-238 isotope heat source (lightly shielded example)
Brayton dynamic

(2) Highest probability of success is most important with nuclear heat sources since lowest weight is a minor virtue when all weights are roughly the same (see Figure 10) as for the nuclear system.

	Peak Temp.
Low temperature Rankine dynamic	1200°R
Brayton dynamic	2500°R
Two fluid continuous flow LMMHD	2700°R

(3) Systems showing up well in general also include:

liquid metal Rankine intermittent flow LMMHD	2700°F
liquid metal Rankine dynamic	2500°F

The results show that the liquid metal Rankine intermittent LMMHD and the low temperature Rankine dynamic systems should be taken up as viable space power plant projects to supplement the active work already being done with Brayton dynamic, two fluid Rankine continuous flow LMMHD and the liquid metal Rankine dynamic systems.

Especially the low temperature systems are suitable back ups for the higher temperature ones. The weights of the low temperature Rankine dynamic system is sufficiently close to the lightest systems that there is little penalty for their use.

For some applications, the low temperature Rankine intermittent system LMMHD answers the same purpose, and from this consideration alone, should be considered in further analyses.

Acknowledgement

The author wishes to thank the Management of Mechanical Technology for making time and facilities available for the preparation of this paper. In particular the services of Peter Groot are appreciated for making the necessary computations.

Appendix A: Technological Status of Problem Areas

The following brief statements are offered in support of development status ratings for the various problem areas cited in Table IV:

1. *Zero-g boiling and condensing:*
   Work at NASA [39] and some definitive work in the U.S.S.R. ([40]) on the use of surface tension and surface area for controlling the position of liquids in zero-g has logically led to the development of the heat pipe for rapid heat transfer from end to end of a pipe heated at one end. Virtually the same technology can be used to control the liquid position at the walls of tubes within which boiling and

condensing are occurring. The additional restraint of through-flow requires considerations of allowed pressure loss, however.

2. *Bearings in shock and vibration:*
   The most sensitive bearing for this type of environment is the gas bearing, typical of what is needed for the dynamic Brayton cycle. Tests reported in [41, 42, and 43] very ably illustrate the potential of gas bearings to operate under shock and vibration conditions. In many cases damage has occurred to shafts and bearing supports during rubbing contact in shock and vibration while the bearing surfaces are left undamaged.
   The ability to operate under sustained high-g loads typical of launch has not yet been proven.

3. *Bearing erosion in liquid metals:*
   So far this has been a very difficult problem at liquid metal temperatures over 1000°F. Investigation is continuing ([44]).

4. *High temperature fluid film bearing operation:*
   Distortions introduced in close tolerance fluid film bearings by coming up to temperature and by the bearings' own losses make a gas bearing, or even a liquid film bearing inoperative. But tests to 1900°F with controlled deviations in film thickness indicate generally that design for mean clearance, as long as actual touching does not occur, adequately predicts performance ([45]).

5. *Turbine and compressor performance:*
   This art is well on the way to reduction to routine work, largely by the efforts reported in [9, 10, and 11], but also by books as [12]. Partial and full admission impulse and partial reaction axial flow turbines answer most requirements for low power systems of both the dynamic Rankine and Brayton types. Methods of normalizing now allow evaluating effects of shock loss and losses due to low Reynolds number as well as the losses due to wheel friction, blade losses, carryover losses and the hydraulic losses.

6. *Pump cavitation:*
   This problem has been attacked on an engineering-design basis in [9] However, use of the jet condenser [17] also relieves this problem, but at the expense of some power consumption.

7. *Thermal stability of low temperature working fluids:*
   This problem has been explored extensively for biphenyl and Dowtherm A ([46]) and somewhat less extensively (and with less promise) for mono-iso-propylbiphenyl ([16]). Other work ([47, 48, and 49]) has not been as extensive, but shows good promise that fluids other than those used in the past few years should be

satisfactory as working fluids for low temperature Rankine systems for space power. The freezing points and thermodynamic properties of some of the newer fluids is superior to biphenyl and Dowtherm A. Thus chlorobenzene, toluene, perfluorodecalin, hexafluorbenzene, thiophene and pyridine can all be considered for future applications.

8. *Start up of dynamic systems and continuous flow LMMHD systems:*
Starting methods such as full charge injection have proven successful in some dynamic unit laboratory tests [17], but have yet to be proven in unattended operation. Alternates exists, such as using a battery driven pump (or battery driven compressor for gas systems), but these can be cumbersome.

9. *Meteoroid protection:*
The meteoroid protection scheme used for evaluation in this report is the old, cumbersome method. However, there is insufficient knowledge at this time to design effective bumpers, or even to thoroughly evaluate the true hazard. For small power systems the methods used herein are probably adequate even far in the future. However, at larger unit power levels it becomes practical to think of methods such as the moving belt or rotating disk. It may also become practical to speculate on new, less cumbersome systems, such as 'bag bumpers' which would be plastic or metal bags having sufficient thickness to act as a bumper, and also acting as a large, light weight intermediate heat transfer surface between the heat rejection system and the space sink. Whether this is lighter and easier to handle than other alternate schemes remains to be seen.

10. *Turbine creep:*
Use of new metals suitable for even higher temperatures has been the only solution to this problem. For the lowest power systems this will remain the case. At sufficiently high power it is conceivable to build in wheel and blade cooling. The weight trade-off on using cooling depends almost entirely on the willingness to increase total radiator area to increase turbine life. It is beyond the scope of this paper to evaluate cooling methods, but [50 and 51] discuss schemes that should be suitable for space power plants.

11. *Alternator protection:*
Windings and some magnetic materials must be protected from liquid metals and their vapors. This can be done with solid core rotors if an impervious material is placed in the gap between rotor and stator. This nonmagnetic 'canning' material can be metallic or ceramic, but each poses construction problems because of the necessity of being thin. The metallic cans tend to warp, the ceramic cans are fragile and also require very special joining methods with the metallic parts. Thus, while solutions are available, so far they are either somewhat cumbersome and decrease efficiency, or they are risky.

12. *Light gas loss in hot gas systems:*
The noble gases have been used in analyses of gas cycles because they result in high cycle efficiency. However, mixtures of helium and higher molecular weight noble gases have sometimes been proposed since the molecular weight can be set at any value, and the transport properties of the mixture are superior to that of a pure gas of the same molecular weight. In a real system, however, there is a tendency to lose the helium by diffusion through the hot parts. Because of this it may prove more satisfactory to use pure gases of the desired molecular weight – one which is chosen for its ease of use in turbines and compressors.

13. *Two-fluid continuous flow LMMHD system separator losses:*
Work is continuing at the Jet Propulsion Laboratory of the California Institute of Technology to reduce these losses.

14. *Heat and shock loss in expansion of gas-liquid mixture in continuous flow systems:*
While there are unavoidable heat losses in the two fluid systems during expansion, shock losses do not appear to be severe. However, in the jet condenser accelerators being tried for single fluid systems the initial mixing zone shock losses, fluid mixing losses and condensation shock losses are severe ([33 and 38]). Methods of achieving marginal improvement are discussed in [32 and 33], but the chances of great improvement appear remote.

15. *Erosion and corrosion in high speed liquid metal flow:*
The literature is filled with many applicable reports which generally indicate that high temperatures and high velocities pose very severe erosion and corrosion problems. There are some specific solutions, however, for instance [52]. Specific materials must be found to solve specific problems. This means that at best the use of high velocity and high temperature liquid metals is difficult.
Successful initial tests of refractory materials to 2200°F–2400°F are reported by [53], which begins to appear satisfactory for all the high temperature liquid metal systems. This class of problems then applies directly to all the continuous flow LMMHD systems and to the liquid metal Rankine system. The solutions are more difficult to find as the peak temperature climbs.
In the Rankine system it is always necessary to encounter this problem in fluid film bearings and in the boiler feed pump. The boiler can be designed to avoid the problem to some degree, but not entirely. In the continuous flow LMMHD systems it can only be avoided with low temperature liquid – which interacts to produce large heat transfer losses in the expansion process.
The intermittent flow systems must also encounter the problem but there will be an order of magnitude reduction in total exposure time compared to the continuous flow system.

16. *Fluid dynamic losses in diffusers and channels in continuous flow LMMHD systems:*
This is basically an engineering problem which should be resolved by design.

17. *MHD end losses in continuous flow LMMHD:*
Considerable work is going on, and solutions appear to be at hand ([30, 54]).

18. *Liquid metal slug stability in intermittent LMMHD:*
Most of the reported work has been carried out at Brookhaven National Laboratory ([55]). High surface tension liquids, such as lithium, would tend to alleviate the problems of slug stability.

19. *MHD electrode shorting in intermittent LMMHD:*
A layer of liquid metal is deposited on the wall as the slug travels down the channel. This layer is thin, but it shorts out one segment of the generator with the others in the segmented Faraday generators used to date. This can be cured by using a small amount of gas or vapor to break the film sheet at each segment, or merely by adjusting the liquid temperature, etc. to keep the film so thin as to have very high resistance.

20. *Heat transfer to spray or gas in intermittent Rankine LMMHD:*
Very high heat transfer rates can be obtained under some conditions of a spray played upon a hot surface [56]. This should be studied more for conditions more directly applicable to the case at hand.
As for intermittent gas cycles very rapid heating of gas under scrubbing conditions has not been reported, and may never be shown to be adequate for an intermittent system.

21. *Mechanical valving in intermittent LMMHD:*
Most of the mechanical valves in these systems are check valves. Engineering development should be able to produce satisfactory ones.

22. *Magnetic holding devices in intermittent LMMHD:*
The most sophisticated, and probably the best, solution for this problem would be a superconducting electromagnet, although a permanent magnet device could be used ([37]).

23. *Fluid compatibility in low temperature intermittent Rankine LMMHD system:*
In this system the thermodynamic working fluid comes in contact with the liquid metal electromotive fluid. Therefore they must be compatible. Since this system is a low temperature cycle many thermodynamic fluids become usable (Figure 6). The electromotive fluid can be any of a variety of conducting materials that melt at low enough temperatures. A number of these exist – some being quinary eutectics that melt at low temperature and are electrical conductors.

References

[1] Luchter, Stephen, 'A Weight and Volume Comparison Between Internally and Externally Mounted Nuclear Power Plants for Deep Submersibles', AIAA Paper No. 65-233, AIAA/USN Marine Systems and ASW Conference, San Diego, California, March 8–10, 1965.

[2] Coombs, M. G. and Norman, L. W., 'Application of the Brayton Cycle to Nuclear Electric Space Power Systems', AIAA Paper 64–757, Third Biennial Aerospace Power Systems Conference, Philadelphia, Pennsylvania, September 1–4, 1964.

[3] Rohrmann, C. A. and Sayre, E. D., 'Radioisotopic Space Power – Prospects and Limitations', AIAA Paper No. 64–453, First AIAA Annual Meeting, Washington, D.C., June 29 – July 2, 1964.

[4] Tonelli, A. Duane and Regnier, Edward P., 'Radioisotope Power-Generating System for a Manned Space Laboratory', ASME Paper 65-AV-18, Aviation and Space Conference, Los Angeles, California, March 14–18, 1965.

[5] Sternlicht, B. and Bjerklie, J. W., 'Comparison of Dynamic and Static Power Conversion Systems for Undersea Missions', ASME Paper 66-GT/CLC-11, *Trans. ASME, J. Eng. Power* **88**, Series A, No. 4 (October 1966).

[6] Loeffler, I. J., Clough, Nestor, and Lieblein, Seymour, 'Recent Developments in Meteoroid Protection for Space Power Systems', AIAA Paper 64–759 presented at the Third Biennial Aerospace Power Systems Conference, Philadelphia, Pennsylvania, September 1–4, 1964.

[7] McAdams, W. H., 'Heat Transmission', third edition, McGraw-Hill, New York, 1954, pp. 268–271.

[8] *Progress in Astronautics and Aeronautics*, Vol. 11: *Power Systems for Space Flight* (ed. by Morris A. Zipkin and Russell N. Edwards), Academic Press, New York, 1963.

[9] Balje, O. E., 'A Study of Design Criteria and Matching of Turbomachines: Part A – Similarity Relations and Design Criteria of Turbines; Part B – Compressor and Pump Performance and Matching of Turbocomponents', *Eng. for Power* (January 1961), 83–114.

[10] Balje, O. E. and Binsley, R. L., 'Axial Turbine Performance Evaluation, Part A – Loss Geometry Relationships and Part B – Optimization with and without Constraints', ASME papers 68-GT-13 and 68-GT-14.

[11] Balje, O. E., 'Axial Cascade Technology and Application to Flow Path Designs. Part I – Axial Cascade Technology' and 'Part II – Application of Data to Flow Path Designs', ASME Papers 68-GT-5 and 68-GT-6.

[12] Deych, M. Y. and Troyanovsky, B. M., *Investigation and Calculation of Axial Turbine Stages*, Izdatel'stvo 'Mashmostroyeniye', Moscow, 1964.

[13] Mukhtarov, M. Kh., 'Experimental Study of the Boundary Layer in Turbine Cascades at Low Reynolds Numbers', *Teploenergetiko* **13**, No. 10 (1966) 44–49.

[14] Mukhtarov, M. Kh. and Baranov, P. I., 'Experimental Investigations of Losses in Turbine Cascades at Low Reynolds Numbers', *Teploenergetiko* No. 9 (1968) 56–58.

[15] Degner, V. R. and Velie, W. W., 'Demonstration of a Self-Contained Organic Rankine Silent Engine', SAE Paper No. 690062.

[16] Linhardt, Hans D. and Carver, G. P., 'Development Progress of Organic Rankine Cycle Power Systems' in *Advances in Energy Conversion Engineering*, The American Society of Mechanical Engineers, 1967, pp. 103–115.

[17] Macauley, B. T. and Marick, J. J., 'ORACLE – Technical Assessment of an Organic Rankine Power Conversion System Operated as a Breadboard Engine', 1968 Intersociety Energy Conversion Engineering Conference, Boulder, Colorado. IEEE.

[18] Bjerklie, J. W., 'Working Fluid as a Design Variable for a Family of Small Rankine Power Systems', ASME Paper 67-GT-6.

[19] Bjerklie, J. and Luchter, S., 'Rankine Cycle Working Fluid Selection and Specification Rationale', SAE Paper 690063.

[20] NASA SP-5057. 'Selected Technology for the Electric Power Industry', Chapter 2, Lewis Research Center, September 11–12, 1968, Cleveland, Ohio.

[21] Dorogov, B. S., *Erosion of Blades in Steam Turbines*, Izdatel'stvo "Energiya", Moscow, 1965. Translated as FTD-MT-67-37, Foreign Technology Division, WPAFB, Ohio.

[22] Nichols, K. E., '15 kW Advanced Solar Turbo-Electric Concept', in *Progress in Astronautics and Aeronautics*, Volume 11: *Power Systems for Space Flight*, Academic Press, New York, 1963.

[23] NASA SP-131, 'Space Power Systems Advanced Technology Conference', Lewis Research Center, Cleveland, Ohio, August 23–24, 1966, Chapter VI.

[24] Feher, E. G., 'The Supercritical Thermodynamic Power Cycle', in *Advances in Energy Conversion Engineering*, ASME, 1967, pp. 37–44.

[25] Angelino, G., 'Liquid-Phase Compression Gas Turbine for Space Power Applications', *J. Spacecraft* **4**, No. 2, AIAA.

[26] Angelino, G., 'Carbon Dioxide Condensation Cycles for Power Production', ASME Paper 68-GT-23.

[27] Petrick, M. and Roberts, J. J., 'Analytical and Experimental Studies of Liquid Metal Faraday Generators', Paper SM-107/20, Symposium on Magnetohydrodynamic Electrical Power Generation, Warsaw, 24–30 July, 1968.

[28] Bidard, R. and Sterlini, J. 'Etude des milieux biphases en vue de leur utilization dans des dispositifs magnetohydrodynamiques', Paper SM-107/75, *Ibid.*

[29] Cerini, D. J., 'Circulation of Liquids for MHD Power Generation', SM-107/40, *ibid.*

[30] Elliott, D. G. 'Performance Capabilities of Liquid-Metal MHD Induction Generators', SM-107/41, *ibid.*

[31] Radebold, R., *et al.*, 'Energy Conversion with Liquid Metal Working Fluids in the MHD-Staustrahlrohr', SM-107/5, *ibid.*

[32] Freund, J., 'Investigation of the Liquid Metal Multi-Stage Injection Process', SM-107/10, *ibid.*

[33] Grolmes, M. A., Petrick, M., and Jerger, E. W., 'Condensing Injector Experiments and Analysis of Performance with Supersonic Inlet Vapor', SM-107/21, *ibid.*

[34] Bayer, Z., 'The Thermal Efficiency of Liquid Metal MHD Generator Cycles', SM-107/14, *ibid.*

[35] Shpil'rayn, E. E., *et al.*, 'The Thermodynamics of Multistage Cycles of MHD Installations with Heat Regeneration', from *The Magneto-Hydrodynamic Method of Generating Electric Power*, Moscow, 1968, pp. 421–433.

[36] Stepanchuk, V. F. and Tsiklauri, G. V., 'Evaluating the Effectiveness of Liquid–Metal MHD Installations in the Presence of Discontinuities of Injector Operation in the Mixing Chamber of the Accelerator', from *The Magneto-Hydrodynamic Method of Generating Electric Power*, Moscow, 1968, pp. 468–476.

[37] Bjerklie, J. and Powell, J. R., Jr., 'A Liquid Metal MHD Power Generating Scheme Using Intermittent Vaporization', SM-107/212, Symposium on Magnetohydrodynamic Electrical Power Generation, Warsaw, 24–30 July, 1968.

[38] Deych, M. E., *et al.*, 'Experimental Study of an Injector Model to be Used as Accelerating Device in MHD Installations', from *The Magneto-Hydrodynamic Method of Generating Electric Power*, Moscow, 1968, pp. 433–444.

[39] Macosko, R. P., *et al.*, NASA Technical Notes TN-D2830 and TN-D4023.

[40] Serebryakov, V. N., 'On the Control of the Dynamics of a Liquid–Gas System Under Weightlessness Conditions by Surface Effects', *Kosmicheskiye issledovaniyo* **4**, No. 5 (1966), 713–721.

[41] Curwen, P. W., 'Research and Development of High-Performance Axial-Flow Turbomachinery, Vol. 2: Design of Gas Bearings', NASA Contractor Report, NASA CR-801, May, 1968.

[42] Frost, A., Lund, J. W., and Curwen, P. W., 'Research and Development of High Performance Turboalternator and Associated Hardware, Vol. 2: Design of Gas Bearings', Mechanical Technology Incorporated Report MTI-67TR29 (NASA Contractor Report), July, 1968.

[43] Curwen, P. W. and Frost, A., 'An Investigation of the Performance of Gas-Bearing Machinery Subjected to Low-Frequency Vibration and Shock', *The Shock and Vibration Bulletin* (August 1968) Part 3.

[44] *Bearing and Seal Design in Nuclear Power Machinery* (ed. by R. A. Burton), ASME Symposium on Lubrication in Nuclear Applications, June 5–7, 1967, Miami Beach, Florida. Especially pages 93, 110 and 462.

[45] Eusepi, M., Wilson, D., and Murray, F., 'Gas Lubrication Research for 1900°F Non-Isothermal Operation: Bearing Distortion Effects on Performance and High Temperature Material Investigations', Air Force Report AFAPL-TR-67-57, July, 1968, Air Force Aero Propulsion Laboratory, Air Force Systems Command, WP-ARB, Ohio.

[46] Adam, A. W., Niggemann, R. E., and Sibert, L. W., 'Thermal Stability Determination of Biphenyl and the Eutectic of Biphenyl and Phenyl Ether in a Rankine Cycle System', 1968 Intersociety Energy Conversion Engineering Conference, Boulder, Colorado, IEEE.

[47] Blake, S., *et al.*, 'Thermal Stability as a Function of Chemical Structure', *J. Chem. Eng. Data* **6**, No. 1 (1961), 87–98.

[48] Johns, I. B., McElhill, E. A., and Smith, J. O., 'Thermal Stability of Some Organic Compounds', *J. Chem. Eng. Data*, **7**, No. 2 (1962), 277–281.

[49] Cullis, C. F. and Manton, J. E., 'The Pyrolysis of Chlorobenzene', *Trans. Faraday Soc.* **54** (1958) 391–389.

[50] Lyubchenko, I. S., 'Temperature Field of a Turbine Blade with Transverse Air Cooling', *Kazan. Aviatsionnyy Institut. Trudy* No. 89 (1965), Matematika i Mekhanikha.

[51] Lokay, V. I., Bodunow, M. N., and Devyatov, V. I., 'The Problem of Cooling Individual Components of Gas Turbine Units', *ANSSSR Izvestiya Energetika i Transport*, No. 3 (1968), 87–96.

[52] Hays, L. G., Effect of High Velocity Lithum on Structural Materials', SM-107/42, Symposium on Magnetohydrodynamic Electrical Power Generation, Warsaw, 24–30 July, 1968.

[53] NASA SP-131, 'Space Power Systems Advanced Technology Conference', Lewis Research Center, Cleveland, Ohio, August 23–24, 1966, Chapters VII, VIII.

[54] Moszynski, J. R. and Agrawal, J. E., 'Electrical End Losses in Liquid–Metal MHD Generators With Variable Conductivity', SM-107/38, Symposium on Magnetohydrodynamic Electrical Power Generation, Warsaw, 24–30 July, 1968.

[55] Powell, J. R., Jr., Zucker, M., Palmer, J., and Becker, W., 'Studies of a Repetitive Liquid Metal Slug MHD Generator', Engineering Developments in Energy Conversion (First International Conference on Energetics), ASME, University of Rochester, August 18–20, 1965.

[56] Druge, H. L., 'High Heat Flux Removal by Liquid Metal Spray Cooling of Surfaces', *Chemical Engineering Progress Symposium Series*, **61**, No. 59, (1965) 115.

PART VIII

TESTING, SUPPORT AND ACCESSORIES, I

Chairman: B. N. Petrov, U.S.S.R.

INVESTIGATION OF THE VIBRATION MODES OF A THREE DEGREES OF FREEDOM GROUND TESTING APPARATUS FOR THE ATTITUDE-CONTROL SYSTEM OF SATELLITES

D. DINI and E. GIMELLI

Nuova San Giorgio S.p.A., Divisione Servosistemi ed Elettronica

and

R. C. MICHELINI and R. GHIGLIAZZA

Istituto di Meccanica Applicata alle Macchine, Università di Genova

Abstract. A ground testing apparatus for the attitude-control system of satellites is briefly presented and discussed.

The system has three servo-controlled possibilities of motion and therefore it is able to follow the satellite in pitch, yaw and roll.

In order to properly design the high performances servo-followers in pitch and roll, it has been necessary to establish a full mathematical model of the testing apparatus.

The note presents the results both of the computation of the principal modes via digital computer, and of the simulation of the dynamic performances via analog computer.

It has to be pointed out that the results obtained by the two ways accord completely.

1. Introduction

The artificial satellite ground testing apparatus presented here aims to perform the overall tests on the attitude control system, the relative lining-up of the different subsystems and the final checks on the dynamic behaviour of the satellite driven by the motors for orbital and attitude corrections.

The apparatus was primarily designed under the Italian National Program based on the satellite SIRIO, that is for a spin-stabilized telecommunications satellite. However a small change in a feedback transducer closing a third error loop will modify the apparatus and make it suitable for three-axes satellites.

Figure 1 presents the general mechanization of the equipment where the following parts may be distinguished:

- *A* substructure for groundwork anchorage
- *B* 1st slide with cylindrical follow-up motion
- *C* follow-up system of the 1st slide
- *D* 2nd slide with cylindrical follow-up motion
- *E* follow-up system of the 2nd slide
- *F* mounting of the satellite
- *G* spin motion
- *H* spherical hydrostatic bearing
- *I* two-axis feedback transducer of the angular motion
- *J* adapter to fit the satellite to the spherical bearing.

G. A. Partel (ed.), Proceedings of the Second International Conference on Space Engineering. All rights reserved.

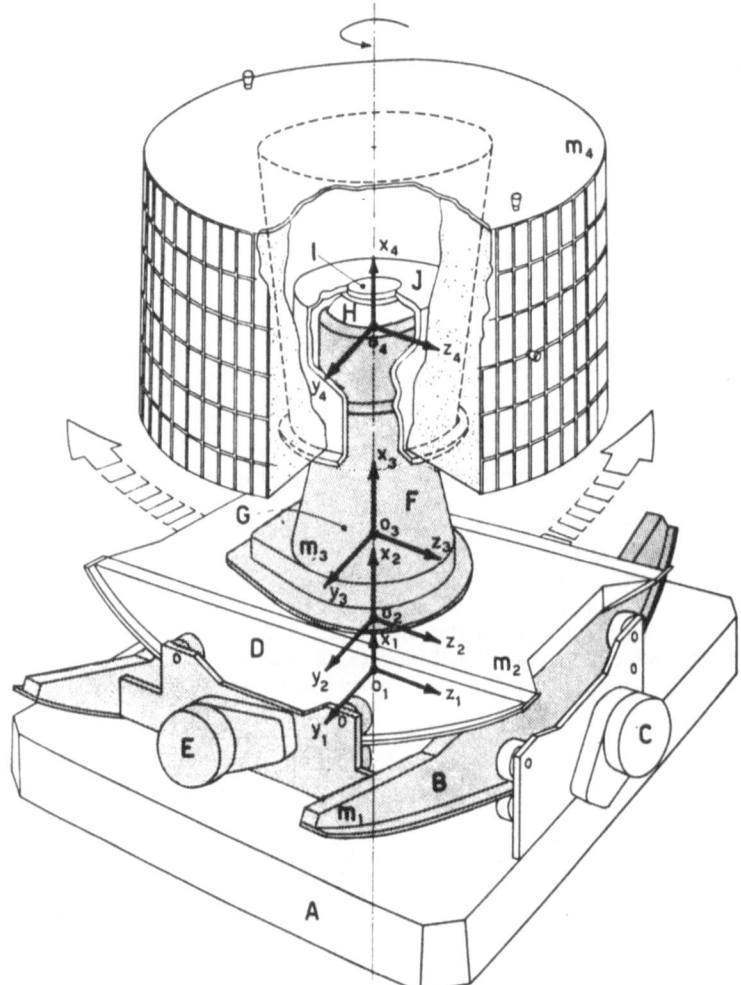

Fig. 1.   Overall mechanization of the equipment.

2. Design Specifications

During the designing of the structure one of the problems was to establish the optimum geometrical form of the two servo-slides B and D to maximize its first vibration mode and to minimize the moments of inertia of the parts moving around point O, center of the spherical motion.

Point O is coincident with the satellite's center of mass, and for that purpose the apogee motor has been replaced by a specially designed adapter in order to realize exactly the spherical motion supported by the hydrostatic bearing.

As a design specification, the slide lowest vibration modes ought to be kept as high as 3 to 3.5 times the cross-over frequency of the servo-followers which have to be about 5 to 6 Hz.

The optimal design will, accordingly, emerge from a compromise between the two specifications. The elements are light alloy box-structures stiffened by ribs and girders. The overall apparatus is quite compact even if it allows the degrees of freedom with angular movement of $\pm 20°$ in pitch and in roll while complete freedom is permitted around the yaw axis.

All the elastic and the inertial characteristics for each component were evaluated during the design phase: and this has been done both with reference to distributed and to lumped parameter subsystems.

Further the distributed parameter subsystems were reduced to dynamically equivalent lumped parameter ones, since only a few fundamental vibration modes are of interest.

Obviously the provisions that can modify the elastic and inertial characteristics of subassemblies are limited by the geometrical dimensions, the moving masses, the degrees of freedom, and the compatibility with the structures to be connected and with which they have to be integrated.

Clearance and backlashes are not explicitly considered, since these were avoided by employing the usual methods for their recapture.

Before deciding on the final design, a parametric investigation on the inertial and elastic coefficients was performed for different layouts.

The investigation led to the configuration sketched in Figure 2 where the inertial and elastic coefficients are localized and evidenced according to a lumped parameter model.

To characterize the system dynamically the above model has been considered both as dissipationless (in order to compute the vibration modes individually) and with reference to a proper amount of viscous damping of structural and of environmental nature (in order to investigate the harmonic response of the system).

3. Definition of the Characteristic Parameters

With reference always to Figure 2, the values of the parameters emerging from the principles just stated are given in Tables I and II. These were used both for programming the simulation of the system by analog computer and for reckoning the vibration modes by digital computer.

We will now briefly recall the general criteria for establishing the values given in the two tables with reference to components, subassemblies, and to structures.

3.1. BALL BEARINGS

The elastic constants are obtained from the stiffness verifying the results with the Föppl theory. The bearings were further tested singularly in the laboratory with a universal testing machine.

3.2. HYDROSTATIC SPHERICAL BEARING

The elastic constants were measured on a prototype for different positions of the

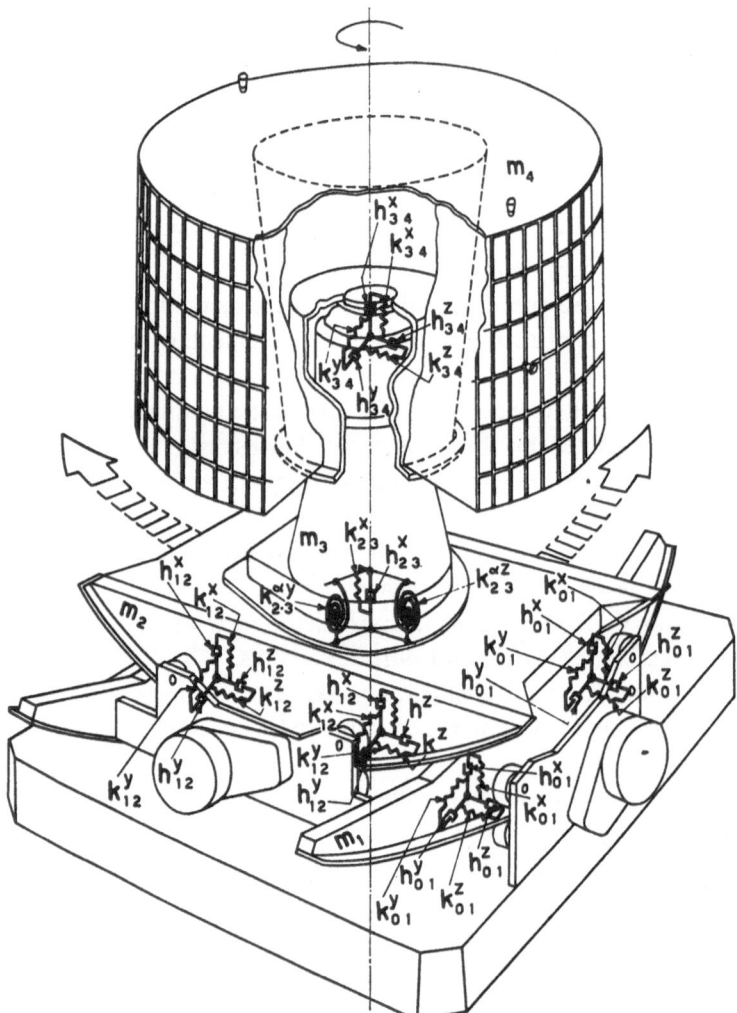

Fig. 2.   Inertial, elastic and damping characteristics localization.

TABLE I

$m_1 = 552$	kg	$J_1^z = J_1^y = 80$	kg m^2
$m_2 = 350$	,,	$J_2^z = J_2^y = 35$	,,
$m_3 = 220$	,,	$J_3^z = J_3^y = 10$	,,
$m_4 = 400$	.,	$J_4^z = J_4^y = 38$	,,

TABLE II

$K_{34}^z = 1.2 \cdot 10^7$	N/m	$K_{23}^y = 2 \cdot 15.18^8$	N/m
$K_{34}^x = 1.2 \cdot 10^7$	,,	$K_{23}^y = 5 \cdot 10^7$	,,
$K_{34}^x = 1.2 \cdot 10^7$	,,	$K_{23}^y = 5 \cdot 10^7$	,,
$K_{34}^y = 1.2 \cdot 10^7$	,,	$K_{23}^z = 5 \cdot 10^7$	,,
$K_{12}^z = 4 \cdot 10^8$	,,	$K_{01}^x = 4 \cdot 10^8$	,,
$K_{12}^x = 8.7 \cdot 10^8$	,,	$K_{01}^y = 4 \cdot 10^8$	,,
$K_{12}^z = 4 \cdot 10^8$	,,	$K_{01}^z = 4.2 \cdot 10^8$	,,

regulating feeding devices. The measures have been performed both statically, imposing loads of known intensity; and during transients, applying position-steps and plotting the output waveforms with high precision linear transducers.

During the dynamic tests the damping constants of the bearing were also measured and it was found that, in a first approximation, these can be supposed proportional to the first derivatives of the mutual displacements.

3.3. GEARING

The elastic constants were determined as static deflections of teeth in keeping with the mesh ratio.

3.4. SHAFTINGS

Both torsion and bending compliances were evaluated for shaftings in conditions of non-motion of the prime mover (locked rotor of the low inertia motors).

The results are composed with gearing elasticity.

3.5. BEARING HOUSINGS

The elastic constants were determined as static characteristics in the directions of the preimposed degrees of freedom.

3.6. SLIDES

Here too, the elastic constants were evaluated taking into consideration the load direction and their distribution. To decide which structural model would best represent the real structure, the natural frequencies of the slides were first determined; second, the equivalent lumped parameter models (not disregarding the cross-coupling effects); and third, the natural frequencies of the slides as components connected with the other subassemblies.

3.7. SATELLITE MOUNTING

The static characteristics were determined taking into consideration the particular geometrical form adopted (see in Figure 1 the part J). The computation of the masses, of the bulk and surface inertia was done analytically.

4. Dynamic Equations

The basic philosophy adopted in the previous analysis is based on the idea of breaking down a complex system into component parts with simple dynamic properties of mass, stiffness and damping.

Within the above assumptions general dynamic equations are deduced for a lumped parameter system represented by rigid bodies elastically constrained and viscously damped, thus:

$$F_i - f_i - A_{ik}x_k - B_{ij}\vartheta_j = \dot{q}_i$$
$$M_j - \mu_j - B_{ij}x_i - C_{jk}\vartheta_k = \dot{p}_j$$

where:

F_i, M_j	are the components of the resulting force and moment applied with reference to a fixed frame x_i. The pole of M is the center of mass of each rigid subsystem;
f_i, u_j	are the viscous forces and moments due to structural damping (i.e. proportional to the relative velocities of the structure components) and to environment damping;
A_{ik}, B_{ik}, C_{ik}	are the components of the stiffness tensors of order 0, 1st and 2nd respectively, for linear elastic constraints.
q_i, p_i	are the components of the momentum and of the angular momentum of each rigid subsystem with the center of mass.

As a general rule, the angular momentum derivatives depend both on angular accelerations and velocities. However, we will accept the usual assumptions of the dynamics of vibrations; thus the mobile frames through the centers of masses can be accepted as static (reference) frames, and the angular velocities can be neglected with respect to angular accelerations. The crosscoupling as regards inertia terms referring to rotational and translatory movements is accordingly broken.

The crosscouplings due to the constraints still stand. As a general trend the influence of the viscous terms is small and can be disregarded. In keeping with the elastic terms, in order to have dynamically uncoupled systems, the components B_{ik} of the first order stiffness tensor ought to be all zero with reference to a frame through the centers of mass and principal of inertia (the condition leads to the coincidence of the center and of the axes directions of both to inertia and the stiffness ellipsoid).

In order to establish the system's dynamic equations let us express the components of the stiffness tensors in terms of the elastic constants previously discussed and geometrical dimensions.

We would mention some other points. First, only the motion in the plane (x, y) is considered, and not the motion in the plane (x, z) since the structure is openly symmetric. Further the elastic constants k^y appear to be slightly smaller than some of the k^z and thus the vibrating modes of the system are expected to be lower with reference to the (x, y)-plan.

Second, on the previous assumptions and given the symmetry of the constraints, the vertical motion is uncoupled from the other movements and thus this system of equation can be resolved separately.

Third, some uncertainty arises when masses m_2 and m_3 are considered, since the pattern of their mutual constraints is not straightforward. Thus two different models were considered, sketched in Figures 3 and 4 respectively. In effect mass m_3 of the turret has an elastically-constrained vertical motion due to the structure compliance in this direction but the section of connection with the upper slide m_2 does not allow transverse motion; an angular relative compliance may be superimposed since the turret may be regarded as a flexible cantilever connected to slide m_2. As a limit case, the two masses m_3 and m_2 may be considered as rigidly connected and so when

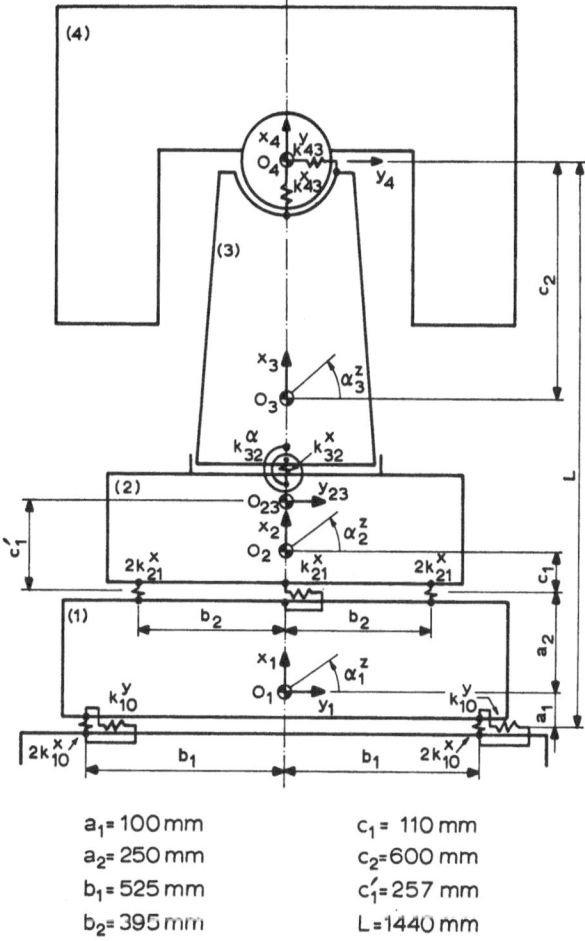

$a_1 = 100$ mm $c_1 = 110$ mm
$a_2 = 250$ mm $c_2 = 600$ mm
$b_1 = 525$ mm $c_1' = 257$ mm
$b_2 = 395$ mm $L = 1440$ mm

Fig. 3.   Four-masses model schematic.

transverse and angular motion are considered, the dynamic behaviour may be depicted by a three-mass system. The two Figures 3 and 4 give the two schematics with three- and four-mass systems. Lastly it must be pointed out that the rotational motion of mass m_4 of the satellite can be disregarded since this is supported by a spherical bearing whose center coincides with the center of the mass of the satellite.

Thus the dynamic equations for the vertical movement are:

$$
\begin{aligned}
&m_1\ddot{x}_1 + h_{21}^x(\dot{x}_1 - \dot{x}_2) + 4k_{10}^x x_1 + 4k_{21}^x(x_1 - x_2) = 0 \\
&m_2\ddot{x}_2 + h_{21}^x(\dot{x}_2 - \dot{x}_1) + h_{32}^x(\dot{x}_2 - \dot{x}_3) \\
&\quad + 4k_{21}^x(x_2 - x_1) + k_{32}^x(x_2 - x_3) = 0 \\
&m_3\ddot{x}_3 + h_{32}^x(\dot{x}_3 - \dot{x}_2) + h_{34}^x(\dot{x}_3 - \dot{x}_4) \\
&\quad + k_{32}^x(x_3 - x_2) + k_{34}^x(x_3 - x_4) = 0 \\
&m_4 x + h_4^x\ddot{x}_4 + h_{34}^x(\dot{x}_4 - \dot{x}_3) + k_{34}^x(x_4 - x_3) = 0 .
\end{aligned}
\tag{1}
$$

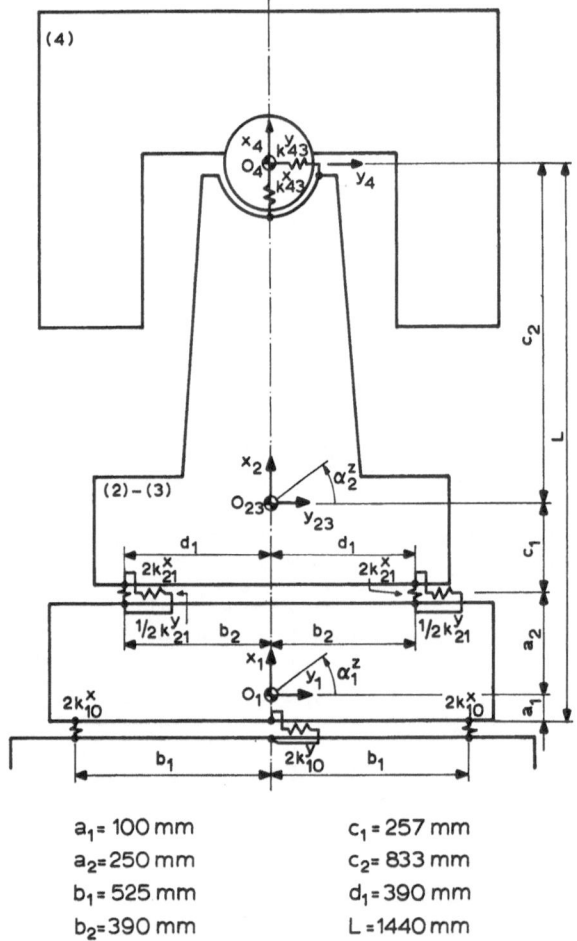

$a_1 = 100\ \text{mm}$ $c_1 = 257\ \text{mm}$

$a_2 = 250\ \text{mm}$ $c_2 = 833\ \text{mm}$

$b_1 = 525\ \text{mm}$ $d_1 = 390\ \text{mm}$

$b_2 = 390\ \text{mm}$ $L = 1440\ \text{mm}$

Fig. 4.   Three-masses model schematic.

For the transverse motion the dynamic equations should be written simultaneously with the related angular motion. For dimension likeness let us define:

$$y_5 = a_1 \alpha_1$$
$$y_6 = c_1 \alpha_2$$
$$y_7 = c_2 \alpha_3$$

then we have:

$$m_1 \ddot{y}_1 + h_{21}^y(\dot{y}_1 - \dot{y}_2) + 2k_{10}^y(y_1 + y_5) + k_{21}^y\left(y_1 - y_2 - \frac{a_2}{a_1}y_5 - y_6\right) = 0$$

$$(m_2 + m_3)\,\ddot{y}_2 + h_{21}^y(\dot{y}_2 - \dot{y}_1) + h_{43}^y(\dot{y}_2 - \dot{y}_4)$$
$$+ k_{21}^y\left(y_2 - y_1 + \frac{a_2}{a_1}y_5 + y_6\right) + k_{43}^y(y_2 - y_4 - y_7) = 0$$

$$m_4 \ddot{y}_4 + h_4^y \dot{y}_4 + h_{43}^y(\dot{y}_4 - \dot{y}_2) + k_{43}^y(y_4 - y_2 - y_7) = 0$$

$$J_1^z \ddot{y}_5 + h_{21}^\alpha (\dot{y}_5 - \dot{y}_6) + 4k_{10}^x b_1^2 y_5 + 2k_{10}^y a_1^2 (y_1 + y_5) + 4k_{21}^x b_2$$

$$\times \left(b_2 y_5 - d_1 \frac{a_1}{c_1} y_6 \right) - k_{21}^y a_2 (y_1 - y_3 - a_2 y_5 - a_1 y_5) = 0 \tag{2}$$

$$J_2^z \ddot{y}_6 + h_{21}^\alpha (\dot{y}_6 - \dot{y}_5) + 4k_{21}^x d_1 \left(d_1 y_6 - b_2 \frac{c_1}{a_1} y_5 \right) + k_{21}^y c_1^2$$

$$\times \left(y_2 - y_1 + y_6 - \frac{a_2}{a_1} y_6 \right) + k_{32}^\alpha c_1 c_2 (c_2 y_6 - c_1 y_7) = 0$$

$$J_3^z \ddot{y}_7 + k_{32}^y c_1 c_2 (c_1 y_7 - c_2 y_6) - k_{43}^x c_2 (y_2 - y_4 - y_7) = 0.$$

To characterize the system dynamically in the next section of the paper, the poles are reckoned for the different configurations given to the apparatus schematics. Only the lower poles are of interest for the actual modes of the vibrating structure, nevertheless a general computation has been carried out for completeness.

In fact, the presence of the different natural frequencies can be shown in transients and when very small (or zero) damping is provided. For that purpose some analog computer records of responses to Dirac-pulses are presented.

Higher harmonics rapidly fade away with little damping so that they can be neglected when designing the servodirections in roll and in yaw. In the last section, amplitude vs. frequency diagrams obtained with analog computations are plotted.

Along with the six-equations system (2) a reduced system can be considered when masses m_2 and m_3 are considered rigidly connected; in this case the last two equations reduce to:

$$J_{2/3}^z \ddot{y}_6 + h_{21}^\alpha (\dot{y}_6 - \dot{y}_5) + 4k_{21}^x d_1 \left(d_1 y_6 - b_2 \frac{e_1}{a_1} y_5 \right)$$

$$+ k_{21}^y c_1 c_2 \left(y_2 - y_1 + y_6 + \frac{a_2}{a_1} y_5 \right) + k_{32}^\alpha c_1 c_2 \left(y_2 - y_3 - \frac{e_2}{e_1} y_6 \right) = 0.$$

When the reduced three-mass configuration is considered two subcases are discussed.

subcase (a): mass 2 and mass 3 are rigidly connected and no further modification in the dynamic model are introduced;

subcase (b): mass 2 and mass 3 are connected but the vertical compliance among the two masses is added to the one existing between mass 1 and mass 2, thus:

$$(4k_{21}^x)_b = \frac{k_{32}^x \cdot 4k_{21}^x}{4k_{21}^x + k_{32}^x} = 1.895 \times 10^8 N/m.$$

Of course, the new value of this elastic constant will also affect the transverse motion of the structure.

5. Vibration Modes Computations

In order to compute the poles of the system we will take the Laplace transform of the dynamic equations allowing zero damping.

Both the vertical motion system and the transverse system are linear, and with some cumbersome algebraic manipulations can be put in the transfer form:

$$x_j = \frac{A_j^x}{A^x} \quad \text{and} \quad y_k = \frac{B_k^y}{B^y}$$

that explicitly give each movement in terms of the structure parameter.

The terms of the above matrices have been deduced and are given in the Appendix I for the most relevant configuration of the structure.

The characteristic matrices A^x and B^y are respectively polynomials in (s^2) – the Laplace operator, whose coefficients depend on the structure parameters. The poles computation has been performed putting both the denominator into the canonical form: $(s^2 + \omega_1^2)(s^2 + \omega_2^2)...(s^2 + \omega_k^2)$ developing the brackets and identifying equal power coefficients. We are then reduced to sets of (nonlinear) algebraic equations – given in Appendix II – that have been numerically resolved iteratively with the Newton-Raphson method.

It should be pointed out that a routine-program can be established for digital computers if the poles are all distinct, since the partial derivatives of the Newton-Raphson method have a recurrent structure. However, the procedure here summarized is tedious due to the fact that the dynamic equations ought to be put in the transfer-form.

The results of the computation performed with the parameter of our structure are, for the vertical motion:

ω^x	167.2	844.2	1458	3030	rad/sec
f^x	26.6	134	232	482	Hz

and for the transverse motion:

ω^y	128.4	404.9	1268	1603	2539	3684	rad/sec
f^y	20.4	64.5	202	255.5	405	587	Hz

The reduced three-mass configuration has been further investigated. As a matter of fact, the shifting to the lower frequencies is not very great. Both the subcases (a) and (b) were considered, and full analytical expressions were worked out; the results are, for subcase (a):

ω^x	163.4	1043	2750	rad/sec
f^x	26.2	166	437	Hz

and for subcase (b):

ω^x	164.3	565	1812	rad/sec
f^x	26.2	90	288	Hz

Insofar as angular and transverse motions are concerned subcase (a) yields the natural frequencies:

ω^y	150.8	396.1	1270	1515	3230	rad/sec
f^y	24	63	202	241	514	Hz

and subcase (b):

ω^y	138.4	394	803.7	1274	2466	rad/sec
f^y	22.1	62.8	128	203	392	Hz

It is interesting to note that, as far as the vertical motion is concerned, the first harmonic is practically unaffected, whereas the second harmonics of the two reduced systems diverge in opposite directions from the corresponding of the basic system. This is less true for the transverse motion, but even in this case the first harmonic differs very little from the corresponding of the basic system.

All the results are presented, since it is well known that the most important problem during dynamic analysis of mechanical structures is certainly its identification. Some analytical assumptions may sometimes be misleading and the results are accordingly not always acceptable.

In our investigation the basic five-mass case and the two four-mass cases yield practically the same results, and these should be accepted for design purposes.

6. Analog Simulation Results

In order to have a straightforward picture of the dynamic behaviour of the structure, the general Equations (1) and (2) have been patched on a high precision analog computer. The purpose of the simulation was twofold. On the one hand, it was decided to investigate the perturbation on the satellite mounted on the testing apparatus due to spurious inputs from the environment that is applied on the two slides or on the satellite itself.

On the other hand, it was necessary to decide up to what frequency it would be safe to push the crossover frequency of the servos of the follow-up systems without interfering with the band of the vibrating modes of the structure.

To carry out the first investigation reference was made to the dynamic model with zero damping. Dirac-like excitations were applied to the lower slide, to the upper slide and to the satellite, and the resulting perturbations on the satellite itself were plotted against time. The results of the investigation are quite general since the systems are linear and the principle of superposition of effects holds. A Dirac-like disturbance can derive both from environmental agent or from particular excitations on the servodirections.

The records for the basic five-mass cases and for the two four-mass cases are

qualitatively quite different, obviously; however, quantitatively, when only amplification ratios – i.e. ratios of the satellite amplitude of perturbation to the applied
Dirac, are considered – the results show a remarkably close affinity.

The three diagrams of Figure 5 present the satellite disturbances due to Dirac-like
vertical displacements on the slides and on the satellite itself. In the last case the
excitation of the first harmonic is remarkably pure.

The five diagrams of Figure 6 show the satellite disturbances due to Dirac-like
transverse displacements (6.1, 6.3) and normalized angular displacements (6.2, 6.4) on
the slides. A transverse displacement (6.5) was also applied to the satellite, and here again
the motion excited practically coincides with that described by the first mode
alone.

The ratios of the perturbation amplitude on the applied Dirac are, for the vertical

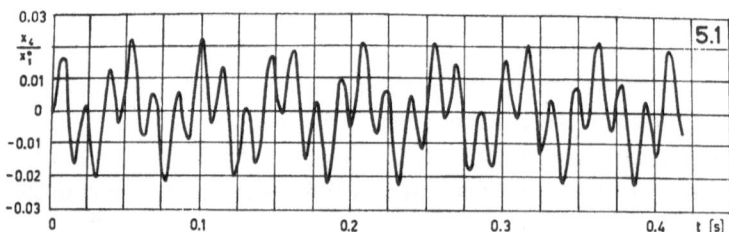

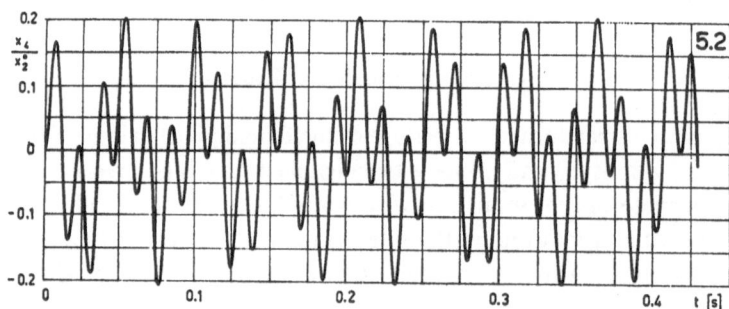

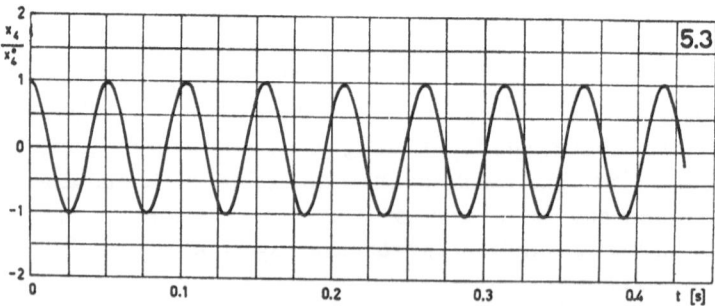

Fig. 5.   Vertical motion responses to Dirac-like disturbances.

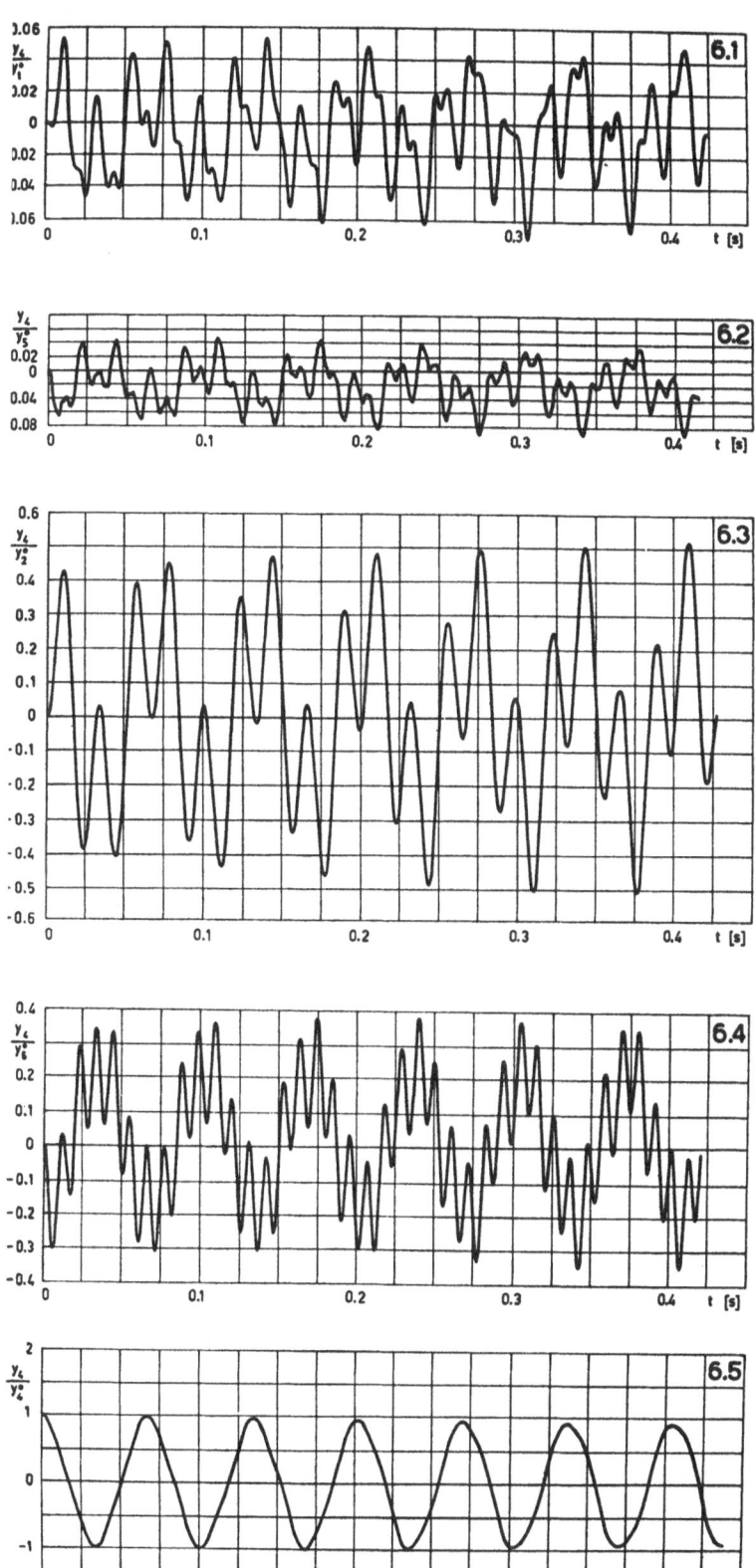

Fig. 6. Transverse motion responses to Dirac-like disturbances.

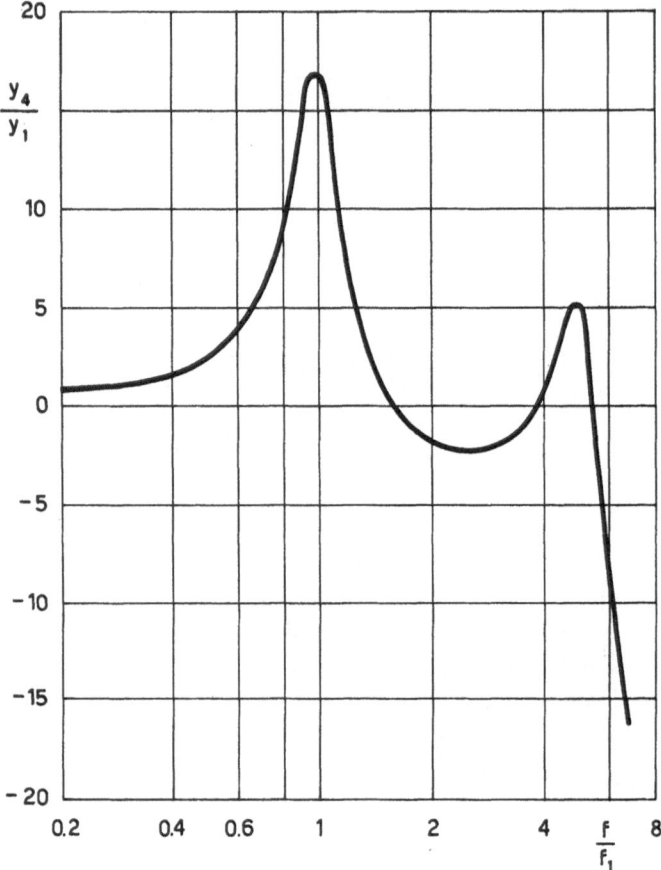

Fig. 7. Amplitude vs. frequency diagrams for vertical motion.

motion:

$$\frac{x_4}{x_1^0} = 0.023 \qquad \frac{x_4}{x_2^0} = 0.208 \qquad \frac{x_4}{x_4^0} = 1$$

and for the transverse motion:

$$\frac{y_4}{y_1^0} = 0.067 \qquad \frac{y_4}{y_5^0} = 0.092 \qquad \frac{y_4}{y_2^0} = 0.536 \qquad \frac{y_4}{y_6^0} = 0.384 \qquad \frac{y_4}{y_4^0} = 1$$

In order to decide the band amplitude acceptable for the servodirections, the frequency-response of the structure was investigated. To that end suitable amounts of damping were introduced that the mathematical model would fit real conditions.

The evaluation of the exact damping characteristics of each element of the testing

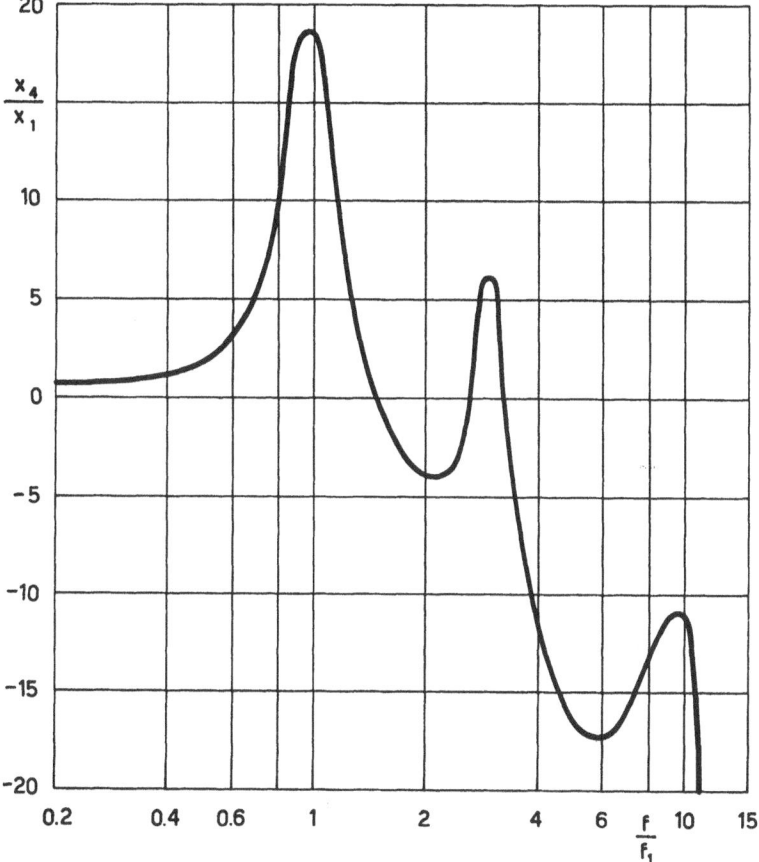

Fig. 8. Amplitude vs. frequency diagrams for transverse motion.

apparatus is arduous and practically meaningless since they are subject to change according to parameters that are not generally controllable.

Besides, the determination of the amounts of damping are non-critical in regions sufficiently far from the resonances. Thus equal-value damping characteristics are introduced in to the analog model and are equal to five percent of the local relative velocities.

As a result of the simulation the two amplitude vs. frequencies diagrams of Figures 7 and 8 have been plotted, the first with reference to the vertical motion, the second with reference to the transverse motion.

It should be emphasized that the vibrating modes previously reckoned by digital computer exactly matched the ones found by analog simulation, even if the higher modes can hardly be discovered since the output amplitudes practically do not exist.

Further it is readily seen that in order to have moderate interferences between the follow-up and the excited oscillations in the structure, the crossover frequencies of the servosystems must be of the order of 6 Hz.

Appendix 1: System Transfer Equations

1.1. VERTICAL MOTION

$$x_j^{\rightharpoonup}A^x = A_j^x \quad (j = 1, 2, 3, 4)$$

$$A^x = \begin{vmatrix} s^2 + a_{11} & a_{12} & 0 & 0 \\ a_{21} & s^2 + a_{22} & a_{23} & 0 \\ 0 & a_{32} & s^2 + a_{33} & a_{34} \\ 0 & 0 & a_{43} & s^2 + a_{44} \end{vmatrix}$$

$$a_{11} = \frac{4k_{10}^x}{m_1} + \frac{4k_{21}^x}{m_1} \qquad a_{32} = -\frac{k_{32}^x}{m_3}$$

$$a_{12} = -\frac{4k_{21}^x}{m_1} \qquad a_{33} = \frac{k_{32}^x}{m_3} + \frac{k_{43}^x}{m_3}$$

$$a_{21} = -\frac{4k_{21}^x}{m_2} \qquad a_{34} = -\frac{k_{43}^x}{m_3}$$

$$a_{22} = \frac{4k_{21}^x}{m_2} + \frac{k_{32}^x}{m_2} \qquad a_{43} = -\frac{k_{43}^x}{m_4}$$

$$a_{23}^{\blacksquare} = -\frac{k_{32}^x}{m_2} \qquad a_{44} = \frac{k_{43}^x}{m_4}$$

$$A_1^x = \begin{vmatrix} sx_1^0 & a_{12} & 0 & 0 \\ sx_2^0 & s^2 + a_{22} & a_{23} & 0 \\ sx_3^0 & a_{32} & s^2 + a_{33} & a_{34} \\ sx_4^0 & 0 & a_{43} & s^2 + a_{44} \end{vmatrix}$$

$$A_2^x = \begin{vmatrix} s^2 + a_{11} & sx_1^0 & 0 & 0 \\ a_{21} & sx_2^0 & a_{23} & 0 \\ 0 & sx_3^0 & s^2 + a_{33} & a_{34} \\ 0 & sx_4^0 & a_{43} & s^2 + a_{44} \end{vmatrix}$$

$$A_3^x = \begin{vmatrix} s^2 + a_{11} & a_{12} & sx_1^0 & 0 \\ a_{21} & s^2 + a_{22} & sx_2^0 & 0 \\ 0 & a_{32} & sx_3^0 & a_{34} \\ 0 & 0 & sx_4^0 & s^2 + a_{44} \end{vmatrix}$$

$$A_4^x = \begin{vmatrix} s^2 + a_{11} & a_{12} & 0 & 0 \\ a_{21} & s^2 + a_{22} & a_{23} & 0 \\ 0 & a_{32} & s^2 + a_{33} & a_{34} \\ 0 & 0 & a_{43} & s^2 + a_{44} \end{vmatrix}$$

1.2. TRANSVERSE AND ANGULAR MOTION

$$y_j B^y = B_j^y \quad (j = 1, 2, 3, 4, 5, 6, 7)$$

$$B^y = \begin{vmatrix} s^2 + b_{11} & b_{12} & b_{13} & b_{14} & 0 & 0 \\ b_{21} & s^2 + b_{22} & b_{23} & b_{24} & 0 & 0 \\ b_{31} & b_{32} & s^2 + b_{33} & b_{34} & b_{35} & b_{36} \\ b_{41} & b_{42} & b_{43} & s^2 + b_{44} & b_{45} & 0 \\ 0 & 0 & b_{53} & b_{54} & s^2 + b_{55} & b_{56} \\ 0 & 0 & b_{63} & 0 & b_{65} & s^2 + b_{66} \end{vmatrix}$$

$$b_{11} = \frac{2k_{10}^y}{m_1} + \frac{k_{21}^y}{m_1}; \qquad b_{42} = -\frac{4k_{21}^x}{J_2^z} c_1 d_1 \frac{b_2}{a_2} - \frac{k_{21}^y}{J_2^z} c_1^2;$$

$$b_{12} = \frac{2k_{10}^y}{m_1} \frac{a_1}{a_2} - \frac{k_{21}^y}{m_1}; \qquad b_{43} = \frac{k_{21}^y}{J_2^z} c_1^2;$$

$$b_{13} = -\frac{k_{21}^y}{m_1}; \qquad b_{44} = \frac{4k_{21}^x}{J_2^z} d_1^2 + \frac{k_{21}^y}{J_2^z} c_1^2;$$

$$b_{14} = -\frac{k_{21}^y}{m_1}; \qquad b_{45} = -\frac{k_{32}^\alpha}{J_2^z} \frac{c_1}{c_2};$$

$$b_{21} = \frac{2k_{10}^y}{J_1^z} a_1 a_2; \qquad b_{53} = -\frac{k_{43}^y}{J_3^z c_2^2};$$

$$b_{22} = \frac{4k_{10}^x}{J_1^z} b_1^2 + \frac{2k_{10}^y}{J_1^z} a_1^2 + \frac{4k_{21}^x}{J_1^z} b_2^2 + \frac{k_{21}^y}{J_1^z} a_2^2;$$

$$b_{23} = \frac{k_{21}^y}{J_1^z} a_2^2; \qquad b_{54} = -\frac{k_{32}^\alpha}{J_3^z} \frac{c_2}{c_1};$$

$$b_{24} = \frac{k_{21}^y}{J_1^z} a_1^2 - \frac{4k_{21}^x}{J_1^z} a_2 b_2 \frac{d_1}{c_1};$$

$$b_{31} = -\frac{k_{21}^y}{m_2 + m_3}; \qquad b_{55} = \frac{k_{32}^\alpha}{J_3^z} + \frac{k_{43}^y}{J_3^z} c_2^2;$$

$$b_{32} = \frac{k_{21}^y}{m_2 + m_3}; \qquad b_{56} = \frac{k_{43}^y}{J_3^y} c_2^2;$$

$$b_{33} = \frac{k_{21}^y}{m_2 + m_3} + \frac{k_{43}^y}{m_2 + m_3}; \qquad b_{63} = -\frac{k_{43}^y}{m_4};$$

$$b_{34} = \frac{k_{21}^y}{m_2 + m_3}; \qquad b_{65} = \frac{k_{43}^y}{m_4};$$

$$b_{35} = \frac{k_{43}^y}{m_2 + m_3}; \qquad b_{66} = \frac{k_{43}^y}{m_4};$$

$$b_{36} = \frac{k_{43}^y}{m_2 + m_3};$$

$$b_{41} = -\frac{k_{21}^y}{J_2^z} c_1^2;$$

$$B_1^y = \begin{vmatrix} sy_1^0 & b_{12} & b_{13} & b_{14} & 0 & 0 \\ 0 & s^2+b_{22} & b_{23} & b_{24} & 0 & 0 \\ sy_{23}^0 & b_{32} & s^2+b_{33} & b_{34} & b_{35} & b_{36} \\ 0 & b_{42} & b_{43} & s^2+b_{44} & b_{45} & 0 \\ 0 & 0 & b_{53} & b_{54} & s^2+b_{55} & b_{56} \\ sy_4^0 & 0 & b_{63} & 0 & b_{65} & s^2+b_{66} \end{vmatrix}$$

$$B_2^y = \begin{vmatrix} s^2+b_{11} & sy_1^0 & b_{13} & b_{14} & 0 & 0 \\ b_{21} & 0 & b_{23} & b_{24} & 0 & 0 \\ b_{31} & sy_{23}^0 & s^2+b_{33} & b_{34} & b_{35} & b_{36} \\ b_{41} & 0 & b_{43} & s^2+b_{44} & b_{45} & 0 \\ 0 & 0 & b_{53} & b_{54} & s^2+b_{55} & b_{56} \\ 0 & sy_4^0 & b_{63} & 0 & b_{65} & s^2+b_{66} \end{vmatrix}$$

$$B_3^y = \begin{vmatrix} s^2+b_{11} & b_{12} & sy_1^0 & b_{14} & 0 & 0 \\ b_{21} & s^2+b_{22} & 0 & b_{24} & 0 & 0 \\ b_{31} & b_{32} & sy_{23}^0 & b_{34} & b_{35} & b_{36} \\ b_{41} & b_{42} & 0 & s^2+b_{44} & b_{45} & 0 \\ 0 & 0 & 0 & b_{54} & s^2+b_{55} & b_{56} \\ 0 & 0 & sy_4^0 & 0 & b_{65} & s^2+b_{66} \end{vmatrix}$$

$$B_4^y = \begin{vmatrix} s^2+b_{11} & b_{12} & b_{13} & sy_1^0 & 0 & 0 \\ b_{21} & s^2+b_{22} & b_{23} & 0 & 0 & 0 \\ b_{31} & b_{32} & s^2+b_{33} & sy_{23}^0 & b_{35} & b_{36} \\ b_{41} & b_{42} & b_{43} & 0 & b_{45} & 0 \\ 0 & 0 & b_{53} & 0 & s^2+b_{55} & b_{56} \\ 0 & 0 & b_{63} & sy_4^0 & b_{65} & s^2+b_{66} \end{vmatrix}$$

$$B_5^y = \begin{vmatrix} s^2+b_{11} & b_{12} & b_{13} & b_{14} & sy_1^0 & 0 \\ b_{21} & s^2+b_{22} & b_{23} & b_{24} & 0 & 0 \\ b_{31} & b_{32} & s^2+b_{33} & b_{34} & sy_{23}^0 & b_{36} \\ b_{41} & b_{42} & b_{43} & s^2+b_{44} & 0 & 0 \\ 0 & 0 & b_{53} & b_{54} & 0 & b_{56} \\ 0 & 0 & b_{63} & 0 & sy_4^0 & s^2+b_{66} \end{vmatrix}$$

$$B_6^y = \begin{vmatrix} s^2+b_{11} & b_{12} & b_{13} & b_{14} & 0 & sy_1^0 \\ b_{21} & s^2+b_{22} & b_{23} & b_{24} & 0 & 0 \\ b_{31} & b_{32} & s^2+b_{33} & b_{34} & b_{35} & sy_{33}^0 \\ b_{41} & b_{42} & b_{43} & s^2+b_{44} & b_{45} & 0 \\ 0 & 0 & b_{53} & b_{54} & s^2+b_{55} & 0 \\ 0 & 0 & b_{63} & 0 & b_{65} & sy_4^0 \end{vmatrix}$$

Appendix 2: System Characteristic Equations

2.1. VERTICAL MOTION

$$x_1 + x_2 + x_3 + x_4 = B_1$$
$$x_2x_2 + x_1x_3 + x_1x_4 + x_2x_3 + x_2x_4 + x_3x_4 = B_2$$

$$x_1 x_2 x_3 + x_1 x_2 x_4 + x_1 x_3 x_4 + x_2 x_3 x_4 = B_3$$

$$x_1 x_2 x_3 x_4 = B_4$$

$$B_1 = \frac{4 k_{10}^x}{m_1} + \frac{4 k_{21}^x}{m_1} + \frac{4 k_{21}^x}{m_2} + \frac{k_{32}^x}{m_2} + \frac{k_{32}^x}{m_3} + \frac{k_{43}^x}{m_3} + \frac{k_{43}^x}{m_4}$$

$$B_2 = \left(\frac{4 k_{21}^x}{m_2} + \frac{k_{32}^x}{m_2} \right) \left(\frac{k_{32}^x}{m_3} + \frac{k_{43}^x}{m_3} + \frac{k_{43}^x}{m_4} \right) + \left(\frac{k_{32}^x}{m_3} + \frac{k_{43}^x}{m_3} \right)$$

$$\times \left(\frac{4 k_{10}^x}{m_1} + \frac{4 k_{21}^x}{m_1} + \frac{k_{43}^x}{m_4} \right) + \left(\frac{4 k_{10}^x}{m_1} + \frac{4 k_{21}^x}{m_1} \right) \left(\frac{4 k_{21}^x}{m_2} + \frac{k_{32}^x}{m_2} + \frac{k_{43}^x}{m_4} \right)$$

$$- \frac{k_{43}^x k_{43}^x}{m_3 m_4} - \frac{k_{32}^x k_{32}^x}{m_2 m_3} + \frac{16 (k_{21}^x)^2}{m_1 m_2}$$

$$B_3 = \left(\frac{4 k_{21}^x}{m_2} + \frac{k_{32}^x}{m_2} \right) \left[\left(\frac{k_{32}^x}{m_3} + \frac{k_{43}^x}{m_3} \right) \frac{k_{43}^x}{m_4} - \frac{k_{43}^x k_{43}^x}{m_3 m_4} \right]$$

$$- \frac{k_{32}^x k_{32}^x k_{43}^x}{m_2 m_3 m_4} + \left[\left(\frac{4 k_{10}^x}{m_1} + \frac{4 k_{21}^x}{m_1} \right) \left(\frac{4 k_{21}^x}{m_2} + \frac{k_{32}^x}{m_2} \right) \right]$$

$$\times \left[\left(\frac{k_{32}^x}{m_3} + \frac{k_{43}^x}{m_3} \right) + \frac{k_{43}^x}{m_4} \right] - \frac{16 (k_{21}^x)^2 k_{43}^x}{m_1 m_2 m_4} + \left(\frac{4 k_{10}^x}{m_1} + \frac{4 k_{21}^x}{m_1} \right)$$

$$\times \left[\left(\frac{k_{32}^x}{m_3} + \frac{k_{43}^x}{m_3} \right) \frac{k_{43}^x}{m_4} - \frac{k_{43}^x k_{43}^x}{m_3 m_4} - \frac{k_{32}^x k_{32}^x}{m_2 m_3} \right] - \frac{16 (k_{21}^x)^2}{m_1 m_2} \left(\frac{k_{32}^x}{m_3} + \frac{k_{43}^x}{m_4} \right)$$

$$B_4 = \left[\left(\frac{4 k_{10}^x}{m_1} + \frac{4 k_{21}^x}{m_1} \right) \left(\frac{4 k_{21}^x}{m_2} + \frac{k_{32}^x}{m_2} \right) \right] \left[\left(\frac{k_{32}^x}{m_3} + \frac{k_{43}^x}{m_3} \right) \frac{k_{43}^x}{m_4} \right.$$

$$\left. - \frac{(k_{43}^x)^2}{m_3 m_4} \right] - \left(\frac{4 k_{10}^x}{m_1} + \frac{4 k_{21}^x}{m_1} \right) \frac{(k_{32}^x)^2 k_{34}^x}{m_2 m_3 m_4}$$

$$+ \frac{16 (k_{21}^x)^2 (k_{43}^x)^2}{m_1 m_2 m_3 m_4} - \left(\frac{k_{32}^x}{m_3} + \frac{k_{43}^x}{m_3} \right) \frac{16 (k_{21}^x)^2 k_{43}^x}{m_1 m_2 m_4}$$

2.2. Transverse and angular motion

$$x_1 + x_2 + x_3 + x_4 + x_5 + x_6 = C_1$$

$$x_1 x_2 + x_1 x_3 + x_1 x_4 + x_1 x_5 + x_1 x_6 + x_2 x_3 + x_2 x_4 + x_2 x_5 + x_2 x_6$$

$$+ x_3 x_4 + x_3 x_5 + x_3 x_6 + x_4 x_5 + x_4 x_6 + x_5 x_6 = C_2$$

$$x_1 x_2 x_3 + x_1 x_2 x_4 + x_1 x_2 x_5 + x_1 x_2 x_6 + x_1 x_3 x_4 + x_1 x_3 x_5$$

$$+ x_1 x_3 x_6 + x_1 x_4 x_5 + x_1 x_4 x_6 + x_1 x_5 x_6 + x_2 x_3 x_4$$

$$+ x_2 x_3 x_5 + x_2 x_3 x_6 + x_2 x_4 x_5 + x_2 x_4 x_6 + x_2 x_5 x_6$$

$$+ x_3 x_4 x_5 + x_3 x_4 x_6 + x_3 x_5 x_6 + x_4 x_5 x_6 = C_3$$

$$x_1 x_2 x_3 x_4 + x_1 x_2 x_3 x_5 + x_1 x_2 x_3 x_6 + x_1 x_2 x_4 x_5 + x_1 x_2 x_4 x_6$$

$$+ x_1 x_2 x_5 x_6 + x_1 x_3 x_4 x_5 + x_1 x_3 x_4 x_6 + x_1 x_4 x_5 x_6$$

$$+ x_2 x_3 x_4 x_5 + x_2 x_3 x_4 x_6 + x_2 x_3 x_5 x_6 + x_2 x_4 x_5 x_6$$

$$+ x_3 x_4 x_5 x_6 = C_4$$

$$x_1x_2x_3x_4x_5 + x_1x_2x_3x_4x_6 + x_1x_2x_3x_5x_6 + x_1x_2x_4x_5x_6$$
$$+ x_1x_3x_4x_5x_6 + x_2x_3x_4x_5x_6 = C_5$$

$$x_1x_2x_3x_4x_5x_6 = C_6$$

$$C_1 = \frac{2k_{10}^y}{m_1} + \frac{k_{21}^y}{m_1} + \frac{4k_{10}^x}{J_1^z}b_1^2 + \frac{2k_{10}^y}{J_1^z}a_1^2 + \frac{4k_{21}^x}{J_1^z}b_2^2$$

$$+ \frac{k_{21}^y}{J_1^z}a_2^2 + \frac{k_{21}^y}{m_2+m_3} + \frac{k_{43}^y}{m_2+m_3}$$

$$C_2 = \left(\frac{2k_{10}^y}{m_1} + \frac{k_{21}^y}{m_1}\right)\left(\frac{4k_{10}^x}{J_1^z}b_1^2 + \frac{2k_{10}^y}{J_1^z}a_1^2 + \frac{4k_{21}^x}{J_1^z}b_2^2 + \frac{k_{21}^y}{J_1^z}a_2^2\right.$$

$$\left. + \frac{k_{21}^y}{m_2+m_3} + \frac{k_{43}^y}{m_2+m_3}\right) + \left(\frac{4k_{10}^x}{J_1^z}b_1^2 + \frac{2k_{10}^y}{J_1^z}a_1^2 + \frac{4k_{21}^x}{J_1^z}b_2^2 + \frac{k_{21}^y}{J_1^z}a_2^2\right)$$

$$\times\left(\frac{k_{21}^y}{m_2+m_3} + \frac{k_{43}^y}{m_2+m_3}\right) - \left(\frac{2k_{10}^y}{m_1}\frac{a_1}{a_2} - \frac{k_{21}^y}{m_1}\right)\left(\frac{2k_{10}^y}{J_1^z}a_1a_2 - \frac{k_{21}^y}{J_1^z}a_2^2\right)$$

$$- \frac{k_{21}^y}{m_2+m_3}\left(\frac{k_{21}^y}{J_1^z}a_2^2 + \frac{k_{21}^y}{m_1}\right) + \left(\frac{2k_{10}^y}{m_1} + \frac{k_{21}^y}{m_1} + \frac{4k_{10}^x}{J_1^z}b_1^2\right.$$

$$\left. + \frac{2k_{10}^y}{J_1^z}a_1^2 + \frac{4k_{21}^x}{J_1^z}b_2^2 + \frac{k_{21}^y}{J_1^z}a_2^2 + \frac{k_{21}^y}{m_2+m_3} + \frac{k_{43}^y}{m_2+m_3}\right)$$

$$\times\left(\frac{4k_{21}^x}{J_2^z}d_1^2 + \frac{k_{21}^y}{J_2^z}c_1^2 + \frac{k_{32}^\alpha}{J_2^z} + \frac{k_{32}^\alpha}{J_3^z} + \frac{k_{43}^y}{J_3^z}c_2^2 + \frac{k_{43}^y}{m_4}\right)$$

$$+ \left(\frac{4k_{21}^x}{J_2^z}d_1^2 + \frac{k_{21}^y}{J_2^z}c_1^2 + \frac{k_{32}^\alpha}{J_2^z}\right)\left(\frac{k_{32}^\alpha}{J_3^z} + \frac{k_{43}^y}{J_3^z}c_2^2 + \frac{k_{43}^y}{m_4}\right)$$

$$+ \frac{k_{43}^y}{m_4}\frac{k_{32}^\alpha}{J_3^z} - \frac{k_{32}^\alpha}{J_2^z}\frac{k_{32}^\alpha}{J_3^z} - \frac{k_{21}^y}{J_2^z}c_1^2\frac{k_{21}^y}{m_2+m_3}$$

$$- \left(\frac{4k_{21}^x}{J_1^z}\frac{a_2b_2d_1}{c_1} - \frac{k_{21}^y}{J_1^z}a_2^2\right)\left(\frac{4k_{21}^x}{J_2^z}\frac{c_1d_1b_2}{a_2} + \frac{k_{21}^y}{J_2^z}c_1^2\right)$$

$$- \frac{k_{21}^y}{m_1}\frac{k_{21}^y}{J_2^z}c_1^2 - \frac{k_{43}^y}{m_2+m_3}\frac{k_{43}^y}{J_3^z}c_2^2 - \frac{k_{43}^y}{m_2+m_3}\frac{k_{43}^y}{m_4}$$

$$C_3 = \left(\frac{4k_{10}^x}{J_1^z}b_1^2 + \frac{2k_{10}^y}{J_1^z}a_1^2 + \frac{4k_{21}^x}{J_1^z}b_2^2 + \frac{k_{21}^y}{J_1^z}a_2^2\right)\left[\left(\frac{2k_{10}^y}{m_1} + \frac{k_{21}^y}{m_1}\right)\right.$$

$$\times\left(\frac{k_{21}^y}{m_2+m_3} + \frac{k_{43}^y}{m_2+m_3}\right) - \frac{k_{21}^y}{m_1}\frac{k_{21}^y}{m_2+m_3} + \frac{k_{43}^y}{m_2+m_3}\frac{k_{43}^y}{J_3^z}c_2^2$$

$$+ \frac{k_{21}^y}{m_2+m_3}\frac{k_{21}^y}{J_2^z}c_1^2 + \frac{k_{32}^\alpha}{J_3^z}\frac{k_{43}^y}{m_4} + \frac{k_{32}^\alpha}{J_2^z}\frac{c_1}{c_2}\frac{k_{32}^\alpha}{J_2^z}\frac{c_1}{c_2}$$

$$- \frac{k_{21}^y}{m_1}\frac{k_{21}^y}{J_2^z}c_1^2 - \frac{k_{43}^y}{m_2+m_3}\frac{k_{43}^y}{m_4} + \left(\frac{4k_{21}^x}{J_2^z}d_1^2 + \frac{k_{21}^y}{J_2^z}c_1^2 + \frac{k_{32}^\alpha}{J_2^z}\right)$$

$$\left.\times\left(\frac{k_{32}^\alpha}{J_3^z} + \frac{k_{43}^y}{J_3^z}c_2^2 + \frac{k_{43}^y}{m_4}\right)\right] + \left(\frac{4k_{21}^x}{J_2^z}d_1^2 + \frac{k_{21}^y}{J_2^z}c_1^2 + \frac{k_{32}^\alpha}{J_2^z} + \frac{k_{32}^\alpha}{J_3^z}\right)$$

$$+ \frac{k_{43}^y}{J_3^z} c_2^2 + \frac{k_{43}^y}{m_4} \Bigg) \Bigg[\left(\frac{4k_{10}^x}{J_1^z} b_1^2 + \frac{2k_{10}^y}{J_1^z} a_1^2 + \frac{4k_{21}^x}{J_1^z} b_2^2 + \frac{k_{21}^y}{J_1^z} a_2^2 \right)$$

$$\times \left(\frac{2k_{10}^y}{m_1} + \frac{k_{21}^y}{m_1} + \frac{k_{21}^y}{m_2 + m_3} + \frac{k_{43}^y}{m_2 + m_3} \right) + \left(\frac{2k_{10}^y}{m_1} + \frac{k_{21}^y}{m_1} \right)$$

$$\times \left(\frac{k_{21}^y}{m_2 + m_3} + \frac{k_{43}^y}{m_2 + m_3} \right) - \left(\frac{2k_{10}^y}{m_1} \frac{a_1}{a_2} - \frac{k_{21}^y}{m_1} \right) \left(\frac{2k_{10}^y}{J_1^z} a_1 a_2 \right.$$

$$\left. - \frac{2k_{10}^y}{J_1^z} a_1^2 \right) - \frac{k_{21}^y}{m_2 + m_3} \left(\frac{k_{21}^y}{J_1^z} a_2^2 - \frac{k_{21}^y}{m_1} \right) \Bigg] + \left(\frac{2k_{10}^y}{m_1} + \frac{k_{21}^y}{m_1} + \frac{k_{21}^y}{m_2 + m_3} \right.$$

$$\left. + \frac{k_{43}^y}{m_2 + m_3} \right) \Bigg[\left(\frac{4k_{21}^x}{J_2^z} d_1^2 + \frac{k_{21}^y}{J_2^z} c_1^2 + \frac{k_{32}^\alpha}{J_2^z} \right) \left(\frac{k_{32}^\alpha}{J_2^z} + \frac{k_{43}^y}{J_3^z} c_2^2 + \frac{k_{43}^y}{m_4} \right)$$

$$+ \left(\frac{k_{32}^\alpha}{J_3^z} \frac{k_{43}^y}{m_4} + \frac{k_{32}^\alpha}{J_2^z} \frac{c_1}{c_2} \frac{k_{32}^\alpha}{J_3^z} \frac{c_2}{c_1} \right) \Bigg] + \left(\frac{4k_{21}^x}{J_2^z} d_1^2 + \frac{k_{21}^y}{J_2^z} c_1^2 + \frac{k_{32}^\alpha}{J_2^z} \right)$$

$$\times \left(\frac{k_{32}^\alpha}{J_3^z} \frac{k_{43}^y}{m_4} - \frac{k_{43}^y}{m_2 + m_3} \frac{k_{43}^y}{J_3^z} c_2^2 - \frac{k_{43}^y}{m_2 + m_3} \frac{k_{43}^y}{m_4} \right) - \left(\frac{2k_{10}^y}{J_1^z} a_1 a_2 \right.$$

$$\left. - \frac{k_{21}^y}{J_1^z} a_2^2 \right) \Bigg[\left(\frac{2k_{10}^y}{m_1} \frac{a_1}{a_2} - \frac{k_{21}^y}{m_1} \right) \left(\frac{k_{21}^y}{m_2 + m_3} + \frac{k_{43}^y}{m_2 + m_3} \right)$$

$$+ \frac{k_{21}^y}{m_1} \frac{k_{21}^y}{m_2 + m_3} \Bigg] - \frac{k_{21}^y}{m_2 + m_3} \frac{k_{21}^y}{J_2^z} c_1^2$$

$$\times \left(\frac{2k_{10}^y}{m_1} + \frac{k_{21}^y}{J_1^z} a_2^2 + \frac{k_{32}^\alpha}{J_3^z} + \frac{k_{32}^\alpha}{J_3^z} \frac{c_2}{c_1} + \frac{k_{43}^y}{m_4} \right)$$

$$+ \frac{k_{21}^y}{m_1} - \frac{4k_{21}^x}{J_1^z} a_2 b_2 \frac{d_1}{c_1} + \frac{k_{21}^y}{J_1^z} a_2^2 \right) - \frac{k_{21}^y}{m_1 + m_3} \left(\frac{k_{21}^y}{J_1^z} a_2^2 \frac{4k_{21}^x}{J_2^z} \right.$$

$$\times c_1 d_1 \frac{b_2}{a_2} - \frac{k_{32}^\alpha}{J_2^z} \frac{c_1}{c_2} \frac{k_{43}^y}{J_3^z} c_2^2 \right) - \frac{k_{32}^\alpha}{J_3^z} \frac{c_2}{c_1}$$

$$\times \left(\frac{k_{32}^\alpha}{J_2^z} \frac{c_1}{c_2} \frac{k_{43}^y}{m_4} - \frac{k_{43}^y}{m_2 + m_3} \frac{k_{21}^y}{J_2^z} c_1^2 \right) + \left(\frac{k_{21}^y}{m_2 + m_3} + \frac{k_{43}^y}{m_2 + m_3} \right)$$

$$+ \frac{k_{32}^\alpha}{J_3^z} + \frac{k_{43}^y}{J_3^z} c_2^2 + \frac{k_{43}^y}{m_4} \Bigg) \Bigg[\left(\frac{4k_{21}^x}{J_1^z} a_2 b_2 \frac{d_1}{c_1} - \frac{k_{21}^y}{J_1^z} a_2^2 \right)$$

$$\times \left(\frac{4k_{21}^x}{J_2^z} c_1 d_1 \frac{b_2}{a_2} + \frac{k_{21}^y}{J_2^z} c_1^2 \right) - \frac{k_{21}^y}{m_1} \frac{k_{21}^y}{J_2^z} c_1^2 \Bigg]$$

$$+ \left(\frac{4k_{21}^x}{J_2^z} c_1 d_1 \frac{b_2}{a_2} + \frac{k_{21}^y}{J_2^z} c_1^2 \right) \Bigg[\frac{k_{21}^y}{m_1} \left(\frac{2k_{10}^y}{J_1^z} a_1 a_2 - \frac{k_{21}^y}{J_1^z} a_2^2 \right) \Bigg.$$

$$- \left(\frac{4k_{21}^x}{J_1^z} a_2 b_2 \frac{d_1}{c_1} - \frac{k_{21}^y}{J_1^z} a_2^2 \right) \left(\frac{2k_{10}^y}{m_1} + \frac{k_{21}^y}{m_1} \right) \Bigg] + \frac{k_{21}^y}{J_1^z}$$

$$\times a_2^2 \left(\frac{2k_{10}^y}{m_1} \frac{a_1}{a_2} - \frac{k_{21}^y}{m_1} \right) \left(\frac{4k_{21}^x}{J_1^z} a_2 b_2 \frac{d_1}{c_1} + \frac{k_{21}^y}{J_1^z} a_2^2 \right)$$

$$-\frac{k_{43}^y}{m_2+m_3}\frac{k_{43}^y}{m_4}\frac{k_{32}^\alpha}{J_3^z}-\left(\frac{2k_{10}^y}{m_1}+\frac{k_{21}^y}{m_1}\right)$$

$$\times\left(\frac{k_{43}^y}{m_2+m_3}\frac{k_{43}^y}{m_4}-\frac{k_{43}^y}{m_2+m_3}\frac{k_{43}^y}{J_3^z}c_2^2\right).$$

$$C_4=\left(\frac{4k_{21}^x}{J_2^z}d_1^2+\frac{k_{21}^y}{J_2^z}c_1^2+\frac{k_{32}^\alpha}{J_2^z}+\frac{k_{32}^\alpha}{J_3^z}+\frac{k_{43}^y}{J_3^z}c_2^2+\frac{k_{43}^y}{m_4}\right)$$

$$\times\left[\left(\frac{2k_{10}^y}{m_1}+\frac{k_{21}^y}{m_1}\right)\left(\frac{4k_{10}^x}{J_1^z}b_1^2+\frac{2k_{10}^y}{J_1^z}a_1^2+\frac{4k_{21}^x}{J_1^z}b_2^2+\frac{k_{21}^y}{J_1^z}a_2^2\right)\right.$$

$$\times\left(\frac{k_{21}^y}{m_2+m_3}+\frac{k_{43}^y}{m_2+m_3}\right)-\left(\frac{2k_{10}^y}{m_1}\frac{a_1}{a_2}-\frac{k_{21}^y}{m_1}\right)\left(\frac{2k_{10}^y}{m_1}\frac{a_1}{a_2}-\frac{k_{21}^y}{J_1^z}a_2^2\right)$$

$$\times\left(\frac{k_{21}^y}{m_2+m_3}+\frac{k_{43}^y}{m_2+m_3}\right)-\frac{k_{21}^y}{J_1^z}a_2^2\frac{k_{21}^y}{m_2+m_3}\left(\frac{2k_{10}^y}{m_1}+\frac{2k_{10}^y}{m_1}\frac{a_1}{a_2}\right)$$

$$-\frac{k_{21}^y}{m_1}\frac{k_{21}^y}{m_2+m_3}\left(\frac{4k_{10}^x}{J_1^z}b_1^2+\frac{2k_{10}^y}{J_1^z}a_1a_2+\frac{2k_{10}^y}{J_1^z}a_1^2+\frac{4k_{21}^x}{J_1^z}b_2^2\right)\Bigg]$$

$$+\left[\left(\frac{2k_{10}^y}{m_1}+\frac{k_{21}^y}{m_1}\right)\left(\frac{4k_{10}^x}{J_1^z}b_1^2+\frac{2k_{10}^y}{J_1^z}a_1^2+\frac{4k_{21}^x}{J_1^z}b_2^2+\frac{k_{21}^y}{J_1^z}a_2^2\right)\right.$$

$$+\left(\frac{2k_{10}^y}{m_1}+\frac{k_{21}^y}{m_1}\right)\left(\frac{k_{21}^y}{m_2+m_3}+\frac{k_{43}^y}{m_2+m_3}\right)+\left(\frac{k_{21}^y}{m_2+m_3}+\frac{k_{43}^y}{m_2+m_3}\right)$$

$$\times\left(\frac{4k_{10}^x}{J_1^z}b_1^2+\frac{2k_{10}^y}{J_1^z}a_1^2+\frac{4k_{21}^x}{J_1^z}b_2^2+\frac{k_{21}^y}{J_1^z}a_2^2\right)-\left(\frac{2k_{10}^y}{m_1}\frac{a_1}{a_2}-\frac{k_{21}^y}{m_1}\right)$$

$$\times\left(\frac{2k_{10}^y}{J_1^z}a_1a_2-\frac{k_{21}^y}{J_1^z}a_2^2\right)-\frac{k_{21}^y}{J_1^z}a_2^2\frac{k_{21}^y}{m_2+m_3}-\frac{k_{21}^y}{m_1}\frac{k_{21}^y}{m_2+m_3}\Bigg]$$

$$\times\left[\left(\frac{4k_{21}^x}{J_2^z}d_1^2+\frac{k_{21}^y}{J_2^z}c_1^2+\frac{k_{32}^\alpha}{J_2^z}\right)\left(\frac{k_{32}^\alpha}{J_3^z}+\frac{k_{43}^y}{J_3^z}c_2^2\right)+\frac{k_{43}^y}{m_4}\right.$$

$$\times\left(\frac{4k_{21}^x}{J_2^z}d_1^2+\frac{k_{21}^y}{J_2^z}c_1^2+\frac{k_{32}^\alpha}{J_2^z}\right)+\frac{k_{43}^y}{m_4}\frac{k_{32}^y}{J_3^z}-\frac{k_{32}^\alpha}{J_2^z}$$

$$\times\frac{c_1}{c_2}\frac{k_{32}^\alpha}{J_3^z}\frac{c_2}{c_1}\Bigg]+\left(\frac{2k_{10}^y}{m_1}+\frac{k_{21}^y}{m_1}+\frac{4k_{10}^x}{J_1^z}b_1^2+\frac{2k_{10}^y}{J_1^z}\right)$$

$$\times a_1^2+\frac{4k_{21}^x}{J_1^z}b_2^2+\frac{k_{21}^y}{J_1^z}a_2^2+\frac{k_{21}^y}{m_2+m_3}+\frac{k_{43}^y}{m_2+m_3}\right)$$

$$\times\left[\left(\frac{4k_{21}^x}{J_2^z}d_1^2+\frac{k_{21}^y}{J_2^z}c_1d_1\frac{b_2}{a_2}+\frac{k_{21}^y}{J_2^z}c_1^2\right)\frac{k_{32}^\alpha}{J_3^z}\frac{k_{43}^y}{m_4}\right.$$

$$-\frac{k_{32}^\alpha}{J_2^z}\frac{c_1}{c_2}\frac{k_{32}^\alpha}{J_3^z}\frac{c_2}{c_1}\frac{k_{43}^y}{m_4}\Bigg]+\frac{k_{21}^y}{m_2+m_3}\left[\frac{k_{21}^y}{J_2^z}c_1^2\right.$$

$$\times\left(\frac{2k_{10}^y}{m_1}\frac{a_1}{a_2}-\frac{k_{21}^y}{m_1}\right)\left(\frac{2k_{10}^y}{J_1^z}a_1a_2-\frac{k_{21}^y}{J_1^z}a_2^2\right)-\frac{k_{21}^y}{J_2^z}c_1^2$$

$$\times \left(\frac{2k_{10}^y}{m_1} + \frac{k_{21}^y}{m_1}\right)\left(\frac{4k_{10}^x}{J_1^z} b_1^2 + \frac{2k_{10}^y}{J_1^z} a_1^2 + \frac{4k_{21}^x}{J_1^z} b_2^2 + \frac{k_{21}^y}{J_1^z} a_2^2\right)$$

$$- \frac{k_{21}^y}{J_1^z} a_2^2 \left(\frac{4k_{21}^x}{J_2^z} c_1 d_1 \frac{b_2}{a_2} + \frac{k_{21}^y}{J_2^z} c_1^2\right)\left(\frac{2k_{10}^y}{m_1} + \frac{k_{21}^y}{m_1}\right) - \frac{k_{21}^y}{m_1}$$

$$\times \left(\frac{2k_{10}^y}{J_1^z} a_1 a_2 - \frac{k_{21}^y}{J_1^z} a_2^2\right)\left(\frac{4k_{21}^x}{J_2^z} c_1 d_1 \frac{b_2}{a_2} + \frac{k_{21}^y}{J_2^z} c_1^2\right) + \frac{k_{21}^y}{J_1^z} a_2^2 \frac{k_{21}^y}{J_2^z} c_1^2$$

$$\times \left(\frac{2k_{10}^y a_1}{m_1 a_2} - \frac{k_{21}^y}{m_1}\right) + \frac{k_{21}^y}{m_1} \frac{k_{21}^y}{J_2^z} c_1^2 \left(\frac{4k_{10}^x}{J_1^z} b_1^2 + \frac{2k_{10}^y}{J_1^z} a_1^2 + \frac{4k_{21}^x}{J_1^z} b_2^2\right)$$

$$+ \frac{k_{21}^y}{J_1^z} a_2^2\Bigg)\Bigg] + \Bigg[- \frac{k_{21}^y}{J_2^z} c_1^2 \frac{2k_{10}^y}{m_1} - \frac{k_{21}^y}{J_2^z}\left(\frac{4k_{10}^x}{J_1^z} b_1^2 + \frac{2k_{10}^y}{J_1^z} a_1^2\right)$$

$$+ \frac{4k_{21}^x}{J_1^z} b_2^2 + \frac{k_{21}^y}{J_1^z} a_2^2\right) - \frac{k_{21}^y}{J_1^z} a_2^2 \left(\frac{4k_{21}^x}{J_2^z} c_1 d_1 \frac{b_2}{a_2} + \frac{k_{21}^y}{J_2^z} c_1^2\right)\Bigg]$$

$$\times \Bigg[\frac{k_{21}^y}{m_2 + m_3}\left(\frac{k_{32}^\alpha}{J_3^z} + \frac{k_{43}^y}{J_3^z} c_2^2 + \frac{k_{43}^y}{m_4}\right) - \frac{k_{43}^y}{m_2 + m_3} \frac{k_{32}^\alpha}{J_3^z} \frac{c_2}{c_1}\Bigg]$$

$$+ \frac{k_{21}^y}{m_2 + m_3}\left(\frac{k_{21}^y}{m_1} - \frac{4k_{21}^x}{J_1^z} a_2 b_2 \frac{d_1}{c_1} + \frac{k_{21}^y}{J_1^z} a_2^2\right)$$

$$\times \Bigg[\frac{k_{21}^y}{J_2^z} c_1^2 \left(\frac{k_{32}^\alpha}{J_3^z} + \frac{k_{43}^y}{J_3^z} c_2^2\right) + \frac{k_{21}^y}{J_2^z} c_1^2 \frac{k_{43}^y}{m_4} - \frac{k_{32}^\alpha}{J_2^z} \frac{c_1}{c_2} \frac{k_{43}^y}{J_3^z} c_2^2\Bigg]$$

$$+ \frac{k_{21}^y}{J_2^z} c_1^2 \Bigg[\frac{k_{21}^y}{m_1} \frac{k_{21}^y}{m_2 + m_3}\left(\frac{4k_{10}^x}{J_1^z} b_1^2 + \frac{2k_{10}^y}{J_1^z} a_1 a_2 + \frac{2k_{10}^y}{J_1^z} a_1^2\right)$$

$$- \frac{k_{21}^y}{m_2 + m_3}\left(\frac{4k_{21}^x}{J_1^z} a_2 b_2 \frac{d_1}{c_1} - \frac{k_{21}^y}{J_1^z} a_2^2\right)\left(\frac{2k_{10}^y}{m_1} + \frac{2k_{10}^y a_1}{m_1 a_2}\right)\Bigg]$$

$$- \Bigg[\frac{k_{21}^y}{m_1} \frac{k_{21}^y}{J_2^z} c_1^2 + \left(\frac{4k_{21}^x}{J_2^z} c_1 d_1 \frac{b_2}{a_2} + \frac{k_{21}^y}{J_2^z} c_1^2\right)$$

$$\times \left(\frac{4k_{21}^x}{J_1^z} a_2 b_2 \frac{d_1}{c_1} - \frac{k_{21}^y}{J_1^z} a_2^2\right)\Bigg]\Bigg[\frac{k_{21}^y}{m_2 + m_3}\left(\frac{k_{32}^\alpha}{J_3^z} + \frac{k_{43}^y}{J_3^z} c_2^2\right) + \frac{k_{32}^\alpha}{J_3^z}$$

$$\times \frac{k_{43}^y}{m_2 + m_3} + \frac{k_{43}^y}{m_4}\left(\frac{k_{32}^\alpha}{J_3^z} + \frac{k_{21}^y}{m_2 + m_3}\right)\Bigg] + \Bigg[\frac{k_{21}^y}{m_1}\left(\frac{2k_{10}^y}{J_1^z} a_1 a_2 - \frac{k_{21}^y}{J_1^z} a_2^2\right)$$

$$\times \left(\frac{4k_{21}^x}{J_2^z} c_1 d_1 \frac{b_2}{a_2} + \frac{k_{21}^y}{J_2^z} c_1^2\right) - \left(\frac{4k_{21}^x}{J_1^z} a_2 b_2 \frac{d_1}{c_1} - \frac{k_{21}^y}{J_1^z} a_2^2\right)$$

$$\times \left(\frac{4k_{21}^x}{J_2^z} c_1 d_1 \frac{b_2}{a_2} + \frac{k_{21}^y}{J_2^z} c_1^2\right)\left(\frac{2k_{10}^y}{m_1} + \frac{k_{21}^y}{m_1}\right) - \frac{k_{21}^y}{m_1} \frac{k_{21}^y}{J_2^z} c_1^2$$

$$\times \left(\frac{4k_{10}^x}{J_1^z} b_1^2 + \frac{2k_{10}^y}{J_1^z} a_1^2 + \frac{4k_{21}^x}{J_1^z} b_2^2 + \frac{k_{21}^y}{J_1^z} a_2^2\right) + \frac{k_{21}^y}{J_2^z} c_1^2$$

$$\times \left(\frac{2k_{10}^y a_1}{m_1 a_2} - \frac{k_{21}^y}{m_1}\right)\left(\frac{4k_{21}^x}{J_1^z} a_2 b_2 \frac{d_1}{c_1} - \frac{k_{21}^y}{J_1^z} a_2^2\right)\Bigg]$$

$$
\times \left(\frac{k_{21}^y}{m_2+m_3} + \frac{k_{43}^y}{m_2+m_3} + \frac{k_{32}^\alpha}{J_3^z} + \frac{k_{43}^y}{J_3^z} c_2^2 + \frac{k_{43}^y}{m_4} \right) + \frac{k_{43}^y}{m_2+m_3}
$$

$$
\times \left[\frac{k_{43}^y}{J_3^z} c_2^2 \left(\frac{2k_{10}^y a_1}{m_1 a_2} - \frac{k_{21}^y}{m_1} \right) \left(\frac{2k_{10}^y}{J_1^z} a_1 a_2 - \frac{k_{21}^y}{J_1^z} a_2^2 \right) - \frac{k_{43}^y}{J_3^z} c_2^2 \right.
$$

$$
\left. \times \left(\frac{2k_{10}^y}{m_1} + \frac{k_{21}^y}{m_1} \right) \left(\frac{4k_{10}^x}{J_1^z} b_1^2 + \frac{2k_{10}^y}{J_1^z} a_1^2 + \frac{4k_{21}^x}{J_1^z} b_2^2 + \frac{k_{21}^y}{J_1^z} a_2^2 \right) \right]
$$

$$
+ \left[-\frac{k_{43}^y}{J_3^z} c_2^2 \left(\frac{2k_{10}^y}{m_1} + \frac{k_{21}^y}{m_1} \right) - \frac{k_{43}^y}{J_3^z} c_2^2 \left(\frac{4k_{10}^x}{J_1^z} b_1^2 + \frac{2k_{10}^y}{J_1^z} a_1^2 + \frac{4k_{21}^x}{J_1^z} \right.\right.
$$

$$
\left.\left. \times b_2^2 + \frac{k_{21}^y}{J_1^z} a_2^2 \right) \right] \left[\frac{k_{21}^y}{m_2+m_3} \left(\frac{4k_{21}^x}{J_2^z} d_1^2 + \frac{k_{21}^y}{J_2^z} c_1^2 + \frac{k_{32}^\alpha}{J_2^z} \right) \right.
$$

$$
\left. - \frac{k_{21}^y}{m_2+m_3} \frac{k_{32}^\alpha}{J_2^z} \frac{c_1}{c_2} \right] + \frac{k_{21}^y}{m_2+m_3} \frac{k_{43}^y}{J_3^z} c_2^2 \frac{k_{32}^\alpha}{J_2^z} \frac{c_1}{c_2} \frac{k_{43}^y}{m_4} - \frac{k_{43}^y}{m_2+m_3}
$$

$$
\times \frac{k_{43}^y}{m_4} \left[\left(\frac{2k_{10}^y}{m_1} + \frac{k_{21}^y}{m_1} \right) \left(\frac{4k_{10}^x}{J_1^z} b_1^2 + \frac{2k_{10}^y}{J_1^z} a_1^2 + \frac{4k_{21}^x}{J_1^z} b_2^2 + \frac{k_{21}^y}{J_1^z} a_2^2 \right) \right.
$$

$$
\left. - \left(\frac{2k_{10}^y}{m_1} \frac{a_1}{a_2} - \frac{k_{21}^y}{m_1} \right) \left(\frac{2k_{10}^y}{J_1^z} a_1 a_2 - \frac{k_{21}^y}{J_1^z} a_2^2 \right) \right] - \frac{k_{43}^y}{m_2+m_3} \frac{k_{43}^y}{m_4}
$$

$$
\times \left[\left(\frac{2k_{10}^y}{m_1} + \frac{k_{21}^y}{m_1} \right) + \left(\frac{4k_{10}^x}{J_1^z} b_1^2 + \frac{2k_{10}^y}{J_1^z} a_1^2 + \frac{4k_{21}^x}{J_1^z} b_2^2 + \frac{k_{21}^y}{J_1^z} a_2^2 \right) \right]
$$

$$
\times \left(\frac{4k_{21}^x}{J_2^z} d_1^2 + \frac{k_{21}^y}{J_2^z} c_1^2 + \frac{k_{32}^\alpha}{J_2^z} + \frac{k_{32}^\alpha}{J_3^z} \right) + \frac{k_{43}^y}{m_4} \left[\frac{k_{43}^y}{m_2+m_3} \frac{k_{32}^\alpha}{J_2^z} \frac{c_1}{c_2} \right.
$$

$$
\times \frac{k_{32}^\alpha}{J_3^z} \frac{c_2}{c_1} - \frac{k_{21}^y}{m_2+m_3} \frac{k_{32}^\alpha}{J_2^z} \frac{c_1}{c_2} \frac{k_{43}^y}{J_3^z} c_2^2 - \frac{k_{43}^y}{m_2+m_3} \frac{k_{32}^\alpha}{J_3^z}
$$

$$
\times \left(\frac{4k_{21}^x}{J_2^z} d_1^2 + \frac{k_{21}^y}{J_2^z} c_1^2 + \frac{k_{32}^\alpha}{J_2^z} \right) \right] + \frac{4k_{21}^x}{J_1^z} a_2 b_2 \frac{d_1}{c_1} \frac{k_{21}^y}{m_1} \frac{k_{21}^y}{m_2+m_3}
$$

$$
\times \left(\frac{4k_{21}^x}{J_2^z} c_1 d_1 \frac{b_2}{a_2} + 2 \frac{k_{21}^y}{J_2^z} c_1^2 \right)
$$

$$
C_5 = \left\{ \left(\frac{k_{21}^y}{m_2+m_3} + \frac{k_{43}^y}{m_2+m_3} \right) \left[\left(\frac{2k_{10}^y}{m_1} + \frac{k_{21}^y}{m_1} \right) \right.\right.
$$

$$
\times \left(\frac{4k_{10}^x}{J_1^z} b_1^2 + \frac{2k_{10}^y}{J_1^z} a_1^2 + \frac{4k_{21}^x}{J_1^z} b_2^2 + \frac{k_{21}^y}{J_1^z} a_2^2 \right) - \left(\frac{2k_{10}^y}{m_1} \frac{a_1}{a_2} - \frac{k_{21}^y}{m_1} \right)
$$

$$
\times \left. \left(\frac{2k_{10}^y}{J_1^z} a_1 a_2 - \frac{k_{21}^y}{J_1^z} a_2^2 \right) \right] - \frac{k_{21}^y}{J_1^z} a_2^2 \frac{k_{21}^y}{m_2+m_3} \left(\frac{2k_{10}^y}{m_1} + \frac{2k_{10}^y}{m_1} \frac{a_1}{a_2} \right)
$$

$$
\left. - \frac{k_{21}^y}{m_1} \frac{k_{21}^y}{m_2+m_3} \left(\frac{4k_{10}^x}{J_1^z} b_1^2 + \frac{2k_{10}^y}{J_1^z} a_1 a_2 + \frac{2k_{10}^y}{J_1^z} a_1^2 + \frac{4k_{21}^x}{J_1^z} b_2^2 \right) \right\}
$$

$$
\times \left[\left(\frac{4k_{21}^x}{J_2^z} d_1^2 + \frac{k_{21}^y}{J_2^z} c_1^2 + \frac{k_{32}^\alpha}{J_2^z} \right) \left(\frac{k_{32}^\alpha}{J_3^z} + \frac{k_{43}^y}{J_3^z} c_2^2 \right) + \frac{k_{43}^y}{m_4} \right.
$$

$$\times \left(\frac{4k_{21}^x}{J_2^z} + \frac{k_{21}^y}{J_2^z} c_1^2 + \frac{k_{32}^\alpha}{J_2^z} + \frac{k_{32}^\alpha}{J_3^z} \right) - \frac{k_{32}^\alpha}{J_2^z} \frac{c_1}{c_2} \frac{k_{32}^\alpha}{J_3^z} \frac{c_2}{c_1} \Bigg]$$

$$+ \left[\left(\frac{4k_{10}^x}{J_1^z} b_1^2 + \frac{2k_{10}^y}{J_1^z} a_1^2 + \frac{4k_{21}^x}{J_1^z} b_2^2 + \frac{k_{21}^y}{J_1^z} a_2^2 \right) \right.$$

$$\times \left(\frac{2k_{10}^y}{m_1} + \frac{k_{21}^y}{m_1} + \frac{k_{21}^y}{m_2 + m_3} + \frac{k_{43}^y}{m_2 + m_3} \right) + \frac{2k_{10}^y}{m_1} \left(\frac{k_{21}^y}{m_2 + m_3} \right.$$

$$\left. + \frac{k_{43}^y}{m_2 + m_3} \right) + \frac{k_{21}^y}{m_1} \frac{k_{43}^y}{m_2 + m_3} - \left(\frac{2k_{10}^y}{m_1} \frac{a_1}{a_2} - \frac{k_{21}^y}{m_1} \right) \left(\frac{2k_{10}^y}{J_1^z} a_1 a_2 \right.$$

$$\left. - \frac{k_{21}^y}{J_1^z} a_2^2 \right) - \frac{k_{21}^y}{J_1^z} a_2^2 \frac{k_{21}^y}{m_2 + m_3} \Bigg] \left[\frac{k_{43}^y}{m_4} \frac{k_{32}^\alpha}{J_3^z} \left(\frac{4k_{21}^x}{J_2^z} d_1^2 + \frac{k_{21}^y}{J_2^z} c_1^2 + \frac{k_{32}^\alpha}{J_2^z} \right) \right.$$

$$\left. - \frac{k_{32}^\alpha}{J_2^z} \frac{c_1}{c_2} \frac{k_{32}^\alpha}{J_3^z} \frac{c_2}{c_1} \frac{k_{43}^y}{m_4} \right] + \left[\frac{k_{21}^y}{J_2^z} c_1^2 \frac{2k_{10}^y}{J_1^z} a_1 a_2 \left(\frac{2k_{10}^y}{m_1} \frac{a_1}{a_3} - \frac{k_{21}^y}{m_1} \right) \right.$$

$$- \frac{2k_{10}^y}{m_1} \frac{k_{21}^y}{J_2^z} c_1^2 \left(\frac{4k_{10}^x}{J_1^z} b_1^2 + \frac{2k_{10}^y}{J_1^z} a_1^2 + \frac{4k_{21}^x}{J_1^z} b_2^2 + \frac{k_{21}^y}{J_1^z} a_2^2 \right)$$

$$\left. - \left(\frac{4k_{21}^x}{J_2^z} c_1 d_1 \frac{b_2}{a_2} + \frac{k_{21}^y}{J_2^z} c_1^2 \right) \left(\frac{2k_{10}^y}{m_1} \frac{k_{21}^y}{J_1^z} a_2^2 - \frac{k_{21}^y}{m_1} \frac{2k_{10}^y}{J_1^z} a_1 a_2 \right) \right]$$

$$\times \left[\frac{k_{21}^y}{m_2 + m_3} \left(\frac{k_{32}^\alpha}{J_3^z} + \frac{k_{43}^y}{J_3^z} c_2^2 + \frac{k_{43}^y}{m_4} \right) - \frac{k_{43}^y}{m_2 + m_3} \frac{k_{32}^\alpha}{J_3^z} \frac{c_2}{c_1} \right]$$

$$+ \left[-\frac{k_{21}^y}{J_2^z} c_1^2 \left(\frac{2k_{10}^y}{m_1} + \frac{4k_{10}^x}{J_1^z} b_1^2 + \frac{2k_{10}^y}{J_1^z} a_1^2 + \frac{4k_{21}^x}{J_1^z} b_2^2 + 2 \frac{k_{21}^y}{J_1^z} a_2^2 \right) \right.$$

$$\left. - \frac{k_{21}^y}{J_1^z} a_2^2 \frac{4k_{21}^x}{J_2^z} c_1 d_1 \frac{b_2}{a_2} \right] \frac{k_{21}^y}{m_2 + m_3} \frac{k_{43}^y}{m_4} \frac{k_{32}^\alpha}{J_3^z} + \frac{k_{32}^\alpha}{J_3^z} \frac{k_{21}^y}{m_2 + m_3}$$

$$\times \frac{k_{21}^y}{J_2^z} c_1^2 \frac{k_{43}^y}{m_4} \left(\frac{k_{21}^y}{m_1} - \frac{4k_{21}^x}{J_1^z} a_2 b_2 \frac{d_1}{c_1} + \frac{k_{21}^y}{J_1^z} a_2^2 \right) + \frac{k_{21}^y}{m_2 + m_3} \frac{k_{21}^y}{J_2^z}$$

$$\times c_1^2 \left[\frac{k_{21}^y}{m_1} \left(\frac{4k_{10}^x}{J_1^z} b_1^2 + \frac{2k_{10}^y}{J_1^z} a_1 a_2 + \frac{2k_{10}^y}{J_1^z} a_1^2 + \frac{4k_{21}^x}{J_1^z} b_2^2 \right) \right.$$

$$\left. - \left(\frac{4k_{21}^x}{J_1^z} a_2 b_2 \frac{d_1}{c_1} - \frac{k_{21}^y}{J_1^z} a_2^2 \right) \left(\frac{2k_{10}^y}{m_1} + \frac{2k_{10}^y}{m_1} \frac{a_1}{a_2} \right) \right]$$

$$\times \left(\frac{k_{32}^\alpha}{J_3^z} + \frac{k_{43}^y}{m_4} + \frac{k_{43}^y}{J_3^z} c_2^2 - \frac{k_{32}^\alpha}{J_2^z} \frac{c_1}{c_2} \frac{k_{43}^y}{J_3^z} c_2^2 \right) + \frac{k_{43}^y}{m_4}$$

$$\times \left[\left(\frac{4k_{21}^x}{J_2^z} c_1 d_1 \frac{b_2}{a_2} + \frac{k_{21}^y}{J_2^z} c_1^2 \right) \left(\frac{4k_{21}^x}{J_1^z} a_2 b_2 \frac{d_1}{c_1} - \frac{k_{21}^y}{J_1^z} a_2^2 \right) \right.$$

$$- \frac{k_{21}^y}{m_1} \frac{k_{21}^y}{J_2^z} c_1^2 \Bigg] \frac{k_{32}^\alpha}{J_3^z} \frac{k_{21}^y}{m_2 + m_3} + \left\{ \left(\frac{4k_{21}^x}{J_2^z} c_1 d_1 \frac{b_2}{a_2} + \frac{k_{21}^y}{J_2^z} c_1^2 \right) \right.$$

$$\times \left[\frac{k_{21}^y}{m_1} \left(\frac{2k_{10}^y}{J_1^z} a_1 a_2 - \frac{k_{21}^y}{J_1^z} a_2^2 \right) - \left(\frac{4k_{21}^x}{J_1^z} a_2 b_2 \frac{d_1}{c_1} - \frac{k_{21}^y}{J_1^z} a_2^2 \right) \right.$$

$$\times \left(\frac{2k_{10}^y}{m_1} + \frac{k_{21}^y}{m_1} \right) \Big] - \frac{k_{21}^y}{m_1} \frac{k_{21}^y}{J_2^z} c_1^2 \left(\frac{4k_{10}^x}{J_1^z} b_1^2 + \frac{2k_{10}^y}{J_1^z} a_1^2 + \frac{4k_{21}^x}{J_1^z} \right)$$

$$\times b_2^2 + \frac{k_{21}^y}{J_1^z} a_2^2 \right) + \frac{k_{21}^y}{J_2^z} c_1^2 \left(\frac{2k_{10}^y}{m_1} \frac{a_1}{a_2} - \frac{k_{21}^y}{m_1} \right)$$

$$\times \left(\frac{4k_{21}^x}{J_1^z} a_2 b_2 \frac{d_1}{c_1} - \frac{k_{21}^y}{J_1^z} a_2^2 \right) \Big\} \left[\left(\frac{k_{21}^y}{m_2 + m_3} + \frac{k_{43}^y}{m_2 + m_3} \right) \right.$$

$$\times \left(\frac{k_{32}^\alpha}{J_3^z} + \frac{k_{43}^y}{J_3^z} c_2^2 \right) + \frac{k_{21}^y}{m_2 + m_3} \frac{k_{43}^y}{m_4} + \frac{k_{32}^\alpha}{J_3^z}$$

$$\times \frac{k_{43}^y}{m_4} - \frac{k_{43}^y}{m_2 + m_3} \frac{k_{43}^y}{J_3^z} c_2^2 \Big] + \left[\frac{k_{43}^y}{J_3^z} c_2^2 \left(\frac{2k_{10}^y}{m_1} \frac{a_1}{a_2} - \frac{k_{21}^y}{m_1} \right) \right.$$

$$\times \left(\frac{2k_{10}^y}{J_1^z} a_1 a_2 - \frac{k_{21}^y}{J_1^z} a_2^2 \right) - \frac{k_{43}^y}{J_3^z} c_2^2 \left(\frac{2k_{10}^y}{m_1} + \frac{k_{21}^y}{m_1} \right)$$

$$\times \left. \left(\frac{4k_{10}^x}{J_1^z} b_1^2 + \frac{2k_{10}^y}{J_1^z} a_1^2 + \frac{4k_{21}^x}{J_1^z} b_2^2 + \frac{k_{21}^y}{J_1^z} a_2^2 \right) \right]$$

$$\times \left[\frac{k_{43}^y}{m_2 + m_3} \left(\frac{4k_{21}^x}{J_2^z} d_1^2 + \frac{k_{21}^y}{J_2^z} c_1^2 + \frac{k_{32}^\alpha}{J_2^z} \right) - \frac{k_{21}^y}{m_2 + m_3} \frac{k_{32}^\alpha}{J_2^z} \frac{c_1}{c_2} \right]$$

$$+ \frac{k_{21}^y}{m_2 + m_3} \frac{k_{32}^\alpha}{J_2^z} \frac{c_1}{c_2} \frac{k_{43}^y}{m_4} \frac{k_{43}^y}{J_3^z} c_2^2 \left(\frac{2k_{10}^y}{m_1} + \frac{k_{21}^y}{m_1} + \frac{4k_{10}^x}{J_1^z} \right)$$

$$\times b_1^2 + \frac{2k_{10}^y}{J_1^z} a_1^2 + \frac{4k_{21}^x}{J_1^z} b_2^2 + \frac{k_{21}^y}{J_1^z} a_2^2 \right) + \frac{k_{43}^y}{m_4} \frac{k_{43}^y}{m_2 + m_3} \left[\left(\frac{2k_{10}^y}{m_1} + \frac{k_{21}^y}{m_1} \right) \right.$$

$$\times \left(\frac{4k_{10}^x}{J_1^z} b_1^2 + \frac{2k_{10}^y}{J_1^z} a_1^2 + \frac{4k_{21}^x}{J_1^z} b_2^2 + \frac{k_{21}^y}{J_1^z} a_2^2 \right) - \left(\frac{2k_{10}^y}{m_1} \frac{a_1}{a_2} - \frac{k_{21}^y}{m_1} \right)$$

$$\times \left. \left(\frac{2k_{10}^y}{J_1^z} a_1 a_2 - \frac{k_{21}^y}{J_1^z} a_2^2 \right) \right] \left[- \frac{4k_{21}^x}{J_2^z} d_1^2 - \frac{k_{21}^y}{J_2^z} c_1^2 - \frac{k_{32}^\alpha}{J_2^z} - \frac{k_{32}^\alpha}{J_3^z} \right]$$

$$+ \frac{k_{43}^y}{m_4} \left(\frac{2k_{10}^y}{m_1} + \frac{k_{21}^y}{m_1} + \frac{4k_{10}^x}{J_1^z} b_1^2 + \frac{2k_{10}^y}{J_1^z} a_1^2 + \frac{4k_{21}^x}{J_1^z} b_2^2 + \frac{k_{21}^y}{J_1^z} a_2^2 \right)$$

$$\times \left[\frac{k_{43}^y}{m_2 + m_3} \frac{k_{32}^\alpha}{J_2^z} \frac{c_1}{c_2} \frac{k_{32}^\alpha}{J_3^z} \frac{c_2}{c_1} - \frac{k_{21}^y}{m_2 + m_3} \frac{k_{32}^\alpha}{J_2^z} \frac{c_1}{c_2} \frac{k_{43}^y}{J_3^z} c_2^2 - \frac{k_{32}^\alpha}{J_3^z} \frac{k_{43}^y}{m_2 + m_3} \right.$$

$$\times \left. \left(\frac{4k_{21}^x}{J_2^z} d_1^2 + \frac{k_{21}^y}{J_2^z} c_1^2 + \frac{k_{32}^\alpha}{J_2^z} \right) \right] + \left(\frac{k_{32}^\alpha}{J_3^z} + \frac{k_{43}^y}{J_3^z} c_2^2 + \frac{k_{43}^y}{m_4} \right)$$

$$\times \frac{4k_{21}^x}{J_1^z} a_2 b_2 \frac{d_1}{c_1} \frac{k_{21}^y}{m_1} \frac{k_{21}^y}{m_2 + m_3} \left(\frac{4k_{21}^x}{J_2^z} c_1 d_1 \frac{b_2}{a_2} + 2 \frac{k_{21}^y}{J_2^z} c_1^2 \right).$$

$$C_6 = \left\{ \left(\frac{k_{21}^y}{m_2 + m_3} + \frac{k_{43}^y}{m_2 + m_3} \right) \left[\left(\frac{2k_{10}^y}{m_1} + \frac{k_{21}^y}{m_1} \right) \right. \right.$$

$$\times \left(\frac{4k_{10}^x}{J_1^z} b_1^2 + \frac{2k_{10}^y}{J_1^z} a_1^2 + \frac{4k_{21}^x}{J_1^z} b_2^2 + \frac{k_{21}^y}{J_1^z} a_2^2 \right) - \left(\frac{2k_{10}^y}{m_1} \frac{a_1}{a_2} - \frac{k_{21}^y}{m_1} \right)$$

$$\times \left(\frac{2k_{10}^y}{J_1^z} a_1 a_2 - \frac{k_{21}^y}{J_1^z} a_2^2 \right) \bigg] - \frac{k_{21}^y}{J_1^z} a_2^2 \frac{k_{21}^y}{m_2 + m_3} \left(\frac{2k_{10}^y}{m_1} + \frac{2k_{10}^y a_1}{m_1 a_2} \right)$$

$$- \frac{k_{21}^y}{m_1} \frac{k_{21}^y}{m_2 + m_3} \left(\frac{4k_{10}^x}{J_1^z} b_1^2 + \frac{2k_{10}^y}{J_1^z} a_1 a_2 + \frac{2k_{10}^y}{J_1^z} a_1^2 + \frac{4k_{21}^x}{J_1^z} b_2^2 \right) \Bigg\}$$

$$\times \left[\frac{k_{43}^y}{m_4} \frac{k_{32}^\alpha}{J_3^z} \left(\frac{4k_{21}^x}{J_2^z} d_1^2 + \frac{k_{21}^y}{J_2^z} c_1^2 + \frac{k_{32}^\alpha}{J_2^z} \right) - \frac{k_{32}^\alpha}{J_2^z} \frac{c_1}{c_2} \right.$$

$$\times \frac{k_{32}^\alpha}{J_3^z} \frac{c_2}{c_1} \frac{k_{43}^y}{m_4} \Bigg] + \left[\frac{k_{21}^y}{J_2^z} c_1^2 \frac{2k_{10}^y}{J_1^z} a_1 a_2 \left(\frac{2k_{10}^y}{m_1} \frac{a_1}{a_2} - \frac{k_{21}^y}{m_1} \right) \right.$$

$$- \frac{2k_{10}^y}{m_1} \frac{k_{21}^y}{J_2^z} c_1^2 \left(\frac{4k_{10}^x}{J_1^z} b_1^2 + \frac{2k_{10}^y}{J_1^z} a_1^2 + \frac{4k_{21}^x}{J_1^z} b_2^2 + \frac{k_{21}^y}{J_1^z} a_2^2 \right)$$

$$- \left(\frac{4k_{21}^x}{J_2^z} c_1 d_1 \frac{b_2}{a_2} + \frac{k_{21}^y}{J_2^z} c_1^2 \right) \left(\frac{2k_{10}^y}{m_1} \frac{k_{21}^y}{J_1^z} a_2^2 - \frac{k_{21}^y}{m_1} \frac{2k_{10}^y}{J_1^z} a_1 a_2 \right) \Bigg]$$

$$\times \frac{k_{21}^y}{m_2 + m_3} \frac{k_{43}^y}{m_4} \frac{k_{32}^\alpha}{J_3^z} + \frac{k_{21}^y}{m_2 + m_3} \frac{k_{21}^y}{J_2^z} c_1^2 \frac{k_{43}^y}{m_4} \frac{k_{32}^\alpha}{J_3^z}$$

$$\times \left[\frac{k_{21}^y}{m_1} \left(\frac{4k_{10}^x}{J_1^z} b_1^2 + \frac{2k_{10}^y}{J_1^z} a_1 a_2 + \frac{2k_{10}^y}{J_1^z} a_1^2 + \frac{4k_{21}^x}{J_1^z} b_2^2 \right) \right.$$

$$- \left(\frac{4k_{21}^x}{J_1^z} a_2 b_2 \frac{d_1}{c_1} - \frac{k_{21}^y}{J_1^z} a_2^2 \right) \left(\frac{2k_{10}^y}{m_1} + \frac{2k_{10}^y}{m_1} \frac{a_1}{a_2} \right) \Bigg]$$

$$+ \left\{ \left(\frac{4k_{21}^x}{J_2^z} c_1 d_1 \frac{b_2}{a_2} + \frac{k_{21}^y}{J_2^z} c_1^2 \right) \left[\frac{k_{21}^y}{m_1} \left(\frac{2k_{10}^y}{J_1^z} a_1 a_2 - \frac{4k_{21}^x}{J_1^z} a_2 b_2 \frac{d_1}{c_1} \right) \right. \right.$$

$$- \frac{2k_{10}^y}{m_1} \left(\frac{4k_{21}^x}{J_1^z} a_2 b_2 \frac{d_1}{c_1} - \frac{k_{21}^y}{J_1^z} a_2^2 \right) \Bigg] - \frac{k_{21}^y}{m_1} \frac{k_{21}^y}{J_2^z} c_1^2 \left(\frac{4k_{10}^x}{J_1^z} b_1^2 + \frac{2k_{10}^y}{J_1^z} a_1^2 \right.$$

$$+ \frac{4k_{21}^x}{J_1^z} b_2^2 + \frac{k_{21}^y}{J_1^z} a_2^2 \Bigg) + \frac{k_{21}^y}{J_2^z} c_1^2 \left(\frac{2k_{10}^y}{m_1} \frac{a_1}{a_2} - \frac{k_{21}^y}{m_1} \right)$$

$$\times \left(\frac{4k_{21}^x}{J_1^z} a_2 b_2 \frac{d_1}{c_1} - \frac{k_{21}^y}{J_1^z} a_2^2 \right) \Bigg\} \frac{k_{43}^y}{m_4} \frac{k_{21}^y}{m_2 + m_3} \frac{k_{32}^\alpha}{J_3^z} - \frac{k_{43}^y}{J_3^z} c_2^2$$

$$\times \left[\left(\frac{2k_{10}^y}{m_1} \frac{a_1}{a_2} - \frac{k_{21}^y}{m_1} \right) \left(\frac{2k_{10}^y}{J_1^z} a_1 a_2 - \frac{k_{21}^y}{J_1^z} a_2^2 \right) - \left(\frac{2k_{10}^y}{m_1} + \frac{k_{21}^y}{m_1} \right) \right.$$

$$\times \left(\frac{4k_{10}^x}{J_1^z} b_1^2 + \frac{2k_{10}^y}{J_1^z} a_1^2 + \frac{4k_{21}^x}{J_1^z} b_2^2 + \frac{k_{21}^y}{J_1^z} a_2^2 \right) \Bigg] \frac{k_{21}^y}{m_2 + m_3} \frac{k_{32}^\alpha}{J_2^z} \frac{c_1}{c_2} \frac{k_{43}^y}{m_4}$$

$$+ \frac{k_{43}^y}{m_4} \left[\left(\frac{2k_{10}^y}{m_1} + \frac{k_{21}^y}{m_1} \right) \left(\frac{4k_{10}^x}{J_1^z} b_1^2 + \frac{2k_{10}^y}{J_1^z} a_1^2 + \frac{4k_{21}^x}{J_1^z} b_2^2 + \frac{k_{21}^y}{J_1^z} a_2^2 \right) \right.$$

$$- \left(\frac{2k_{10}^y}{m_1} \frac{a_1}{a_2} - \frac{k_{21}^y}{m_1} \right) \left(\frac{2k_{10}^y}{J_1^z} a_1 a_2 - \frac{k_{21}^y}{J_1^z} a_2^2 \right) \Bigg] \left[\frac{k_{32}^\alpha}{J_2^z} \frac{c_1}{c_2} \right.$$

$$\times \left(\frac{k_{43}^y}{m_2 + m_3} \frac{k_{32}^\alpha}{J_3^z} \frac{c_2}{c_1} - \frac{k_{21}^y}{m_2 + m_3} \frac{k_{43}^y}{J_3^z} c_2^2 \right) - \frac{k_{43}^y}{m_2 + m_3} \frac{k_{32}^\alpha}{J_3^z}$$

$$\times \left(\frac{4k_{21}^x}{J_2^z} d_1^2 + \frac{k_{21}^y}{J_2^z} c_1^2 + \frac{k_{32}^\alpha}{J_2^z} \right) \Bigg] + \frac{k_{43}^y}{m_4} \frac{k_{32}^\alpha}{J_3^z} \frac{4k_{21}^x}{J_1^z} a_2 b_2$$

$$\times \frac{d_1}{c_1} \frac{k_{21}^y}{m_1} \frac{k_{21}^y}{m_2 + m_3} \left(\frac{4k_{21}^x}{J_2^z} c_1 d_1 \frac{b_2}{a_2} + 2 \frac{k_{21}^y}{J_2^z} c_1^2 \right).$$

References

[1] On, F. J., 'Mechanical Impedance Analysis for Lumped Parameter Multi-Degree of Freedom/ Multi-Dimensional Systems', NASA-Tech. Note TN D-3856 – May 1967.
[2] Jacobsen, L. S. and Ayre. R. S., *Engineering Vibrations*, McGraw-Hill, New York, 1958.
[3] Michelini, R. C., 'I mezzi e i metodi di calcolo automatico nella progettazione dei sistemi dinamici', EDIME, Milano, 1968.
[4] Cannon, Jr., R. H., *Dynamics of Physical Systems*, McGraw-Hill, New York, 1967.
[5] Sherby T. A. and Chmielewski, J. F., 'Generalized Vector Derivatives for Systems with Multiple Relative Motion' *Trans. ASME*, Series E (March 1968).

SATELLITES AND HEAT-SHIELDS TEST FACILITIES
AT FIAT AEROSPACE LABORATORY

I. CAPRIOLO

Sezione Velivoli, FIAT, Italy

Abstract. FIAT aerospace test facilities to qualify whole satellite and heat-shield structures are illustrated.

The spacecraft structures may meet intricate requirements as load stability, stiffness, vibrations, medium- and high-temperature operation, thermal and electrical conductivity, etc.

Many other qualification and reliability tests at simulated space and at ambient conditions are performed on space assemblies and sub-assemblies.

1. Introduction

The main FIAT activity in the space field was the manufacturing of the top portion of the Eldo Europa 1 vector, for which the satellite structure and heat shields were designed, built and tested.

The heat shields are made by two half shells, of sandwich structure, with phenolic-glass fabric skins bonded to a honeycomb core. All materials are resistant to high temperature and have a complete radar transparency. The two half shells are joined to the lower vector stage by a coupling formed by two half rings locked by two explosive bolts; four explosive actuators hold the two half shells together. When the vector reaches a certain altitude (about 100 km) an electronic device ignites first the explosive bolts, and then the explosive actuators. Thus in the extremely thin air, the fairings are jettisoned and the satellite is fully uncovered. The satellite body is also sandwich structure made.

Structures of FIAT satellites are mainly made by a cylinder or by a conic central part built of light alloy sheet or of a light alloy sandwich; to them are fixed a certain number of shelves on which the satellite instrumentation is fastened. The perimetral struts complete the satellite structures. The solar cell panels are fixed to these struts.

2. The Aerospace Structure Laboratory

The qualification of space structures was carried out completely at the Structure Laboratory of FIAT Aviation Division. In this kind of tests, the Lab. can be considered as the only one in Italy and one of the most important in Europe. The main building of this laboratory has a working area of about 3000 m² and a wide part of it has a vertical clearance higher than 15 m.

The Laboratory is provided with new and sophisticated equipment for the simulation of environmental conditions, such as loading, cooling, heating, vacuum etc., and for the acquisition, record and elaboration on real time bases of the data coming from the structures being tested.

G. A. Partel (ed.), Proceedings of the Second International Conference on Space Engineering. All rights reserved.

The tests performed in the Lab., whose results were fully confirmed by the launches from the Australian base of Woomera, can be divided into static, transient and in-space environmental tests (heat and vacuum). It has to be pointed out that every test has always required a high technical knowledge to develop the test installation, to carry out the test and to interpret the results. Sometimes, the encountered difficulties were subjected to collateral studies more important and binding than the whole test.

Fig. 1. Advanced material evaluation devices at the Technological Laboratory.

Therefore, the Structure Laboratory is assisted by other facilities, which can evaluate the parameters that affect the design performances both from materials and components points of view.

Among these collateral Labs, one worthy of mention being the Technological Laboratory, which by means of up-to-date investigation equipment (Fig. 1), carried out long and detailed studies on structural specimens for the purpose of:

establishing the most favorable curing conditions of advanced plastic materials used on structures (laminates, adhesives, thermoprotective, varnishes, etc.)

ascertaining the physical and mechanical properties of structural elements (peel strength, flatwise, interlaminar shear, specific heat, thermoconductivity, electrical transparency, etc.)

ascertaining the mechanical, thermal and dynamic properties of structural parts (compression load, thermal balance, damping properties, etc.).

Thus, within FIAT Aviation Division, a team of highly skilled engineers have been trained to work in the design, fabrication, experimentation of the structures which will go into Space.

3. Facilities for Satellite Structure Tests

3.1. GENERAL

It is known that a satellite is subjected to very high stresses on launching, due to dynamic and vibration environments during its initial firing. In space, it is submitted to other environments such as solar and aerodynamic heating, as well as heat produced by equipment carried by the satellite.

These working conditions of the satellite structure, make the development and the accomplishment of a suitable qualification test program very difficult.

As it is not possible to simulate simultaneously the different conditions displaced during the launch and flight phases, separate tests must be performed using safety limits, which take into account a possible overlapping of the maximum values of the various stresses.

3.2. STATIC TESTS

The satellite structures were fitted to a mock-up of the vector's third stage, using the interconnection rings (Fig. 2).

Fig. 2.   Static test of STV-SL satellite.

The satellite equipment was simulated by dummy loads so as to have the applied forces passing exactly through each instrument's center of g. The different forces, simulating the mass at a certain vector acceleration, were grouped through a complicated linkage, which in turn led to a number of hydraulic cylinders, each of them controlled by an electric force gage. By varying the supply pressures of these cylinders, the changing of the test conditions was made feasible.

During the loading tests, several deformation surveys were carried out by means of linear transducers and strains were surveyed by means of photoelastic-coating or with strain-gages.

Besides testing the whole satellite structure, additional tests were performed on structural parts in order to verify the design assumptions and the accuracy of the bonding procedure, e.g., the optimization tests of the tapped shur-locks in the loaded shelves to which the delicate satellite instruments are fastened.

Finally, let us mention the tests for determining the satellite structure barycenter, whose extreme location survey accuracy ($\pm\frac{1}{10}$ mm) required the use of very sensitive piezoelectric weight cells, having followed the multiple weight method with different structure orientations.

3.3. DYNAMIC TESTS

The vibration tests of the satellite structure equipped with dummy loads simulating the true instrumentation, were conducted by CRA Organization in Rome in accordance with U.S.A. policy.

The FIAT Aerospace Laboratory has instead carried out vibration tests for the optimization of the design philosophy of the shelves sandwich structure and other very interesting tests for the purpose of determining the dynamic damping (Fig. 3) in still air and at high vacuum. In addition, it has carried out tests on some equipment and antennas developed by Centro Elettronico Avio.

3.4. THERMAL CONDUCTIVITY TESTS

The Laboratory has a remarkable experience in tests associated with the study of the thermal phenomena in as much as it dealt with thermal problems, in particular to radiation and conductivity ones, which are of most importance for space structures.

Among them, it is worth mentioning the tests on the structure of the shelves which bear the instruments using a device for the thermal conductivity measurement placed inside the normal high vacuum cell (5×10^{-11}). This equipment can be used up to a temperature ($500\,°C$) which, while being by far higher than the operating range of the satellites, is just sufficient for identical tests on the structure of the heat shields.

4. Facilities for Heat Shield Tests

4.1. GENERAL

During the first launching phases, a space vector and, in particular, the nose of it are subjected to the aerodynamic and inertial stresses as well as the heating encountered while passing through the atmospheric dense layers.

Fig. 3.   Sandwich structure's dampening test in vacuum.

The most stressed part of the vector during this phase is the nose (heat shields), which is made of special material and is designed to protect the pay load (satellite). Above a certain altitude the heat shields must be jettisoned otherwise they would transfer to the satellite their stored heat in an amount higher than that the uncovered satellite would receive, through kinetic effect, from the extremely rarefied air.

A facility for ground simulation of the heat shield environment should therefore evaluate:

stiffness and static strength requirements at ultimate loads

strength requirement at limit load and temperatures

dynamic requirements

reliability requirements of the jettisoning system.

All tests should be performed on full scale structures and as far as possible, fully assembled. Some of these tests, however, should be conducted in a sufficiently rarefied air, i.e., under the same environmental conditions encountered in space. The accomplishment of these tests on very large structures implies the availability of very expensive facilities which would rarely be used.

Therefore, when the environmental effect can not be ignored, then the test on the full structure will be superseded by tests on structural components only, the size of which fits the available chamber facility.

4.2. STRENGTH AND STIFFNESS STRESSES

A structure which during service is subjected to dynamic and static loads which must show its capabilities during tests simulating the load distribution and intensity that it must withstand.

For this purpose the two half-shells of the heat shield were enclosed in a special cage-shaped fixture, whose function was:

to support the shells

to permit the load application

to act as a supporting element for reacting the various and asymmetric loads applied

to support, inside the heat shields, a frame designed to sustain the displacement measurement devices.

The test was performed by applying the loads with rubber pads bonded to the external surfaces of the areas where the resultants of the aerodynamic loads and of the inertial forces were directed outward; with compressed air rubber sacks in those areas where loads and forces were directed inward (Fig. 4).

Fig. 4.   Heat-shield static test loading frame.

The supply pressure of the hydraulic cylinders utilized to apply the loads on the rubber pads and the supply pressure of the rubber sacks, were synchronized by means of a sophisticated servo-valves loop.

During the test the strains and stresses in several areas of the structure were measured.

Previously, on the same structure some tests were conducted in order to investigate possible local phenomena such as the opening of the top of the heat shields thus allowing the air to enter inside with catastrophic effect on the satellite.

The following tests were carried out:

simulation of the dynamic stresses on the heat shields surface,

determination of the best preload conditions of the ballistic actuators,

evaluation of the structure stiffness of the section between the top of the heat shields and the plane where the first two actuators are located.

Also the actuator fittings on the structures, though thoroughly studied during designing, required a long experimentation in order to avoid that the thrust developed by the internal explosion could locally damage the structure and affect the shield jettisoning.

These very expensive tests were justified by the special materials used for the realization of the structure. These materials had to satisfy definite design requirements and, therefore, created problems unusual to the designer and the experimenter as well. It is worth mentioning as an example the simple test of determining the center of gravity location by weighing. Because of the structure deformability this test required a complex supporting and reference base whose purpose was to keep the half-shell shape unchanged while varying their orientation.

4.3. HEATING TESTS

The FIAT test facilities essentially consists of the following three systems:

The heating system which is fitted to the test structure and includes the unit for the application of the stresses originated by the aerodynamic and inertial (loads) effects.

The automatic and programmed system for the supply and simultaneous control of the heating and loads (Fig. 5).

The synchronized system for the measurement of the thermal flows, temperatures and structural stresses and strains.

In general this facility allows the accomplishment either of real time tests to reproduce the thermal stresses, or the heat flow on structures with the possible simultaneous application of loads, or of tests with cyclic heating, to reproduce the thermal fatigue effects at stabilized constant temperatures for functional tests, or for evaluating the mechanical properties of materials and structures.

During the heating tests the heat shields were resting on an approximately 1 m high metal base by means of a ring which connect to the vector third stage.

Besides supporting the heat shields, this metal base carried, also, the three hinged arms designed to hold the heating reflectors and the three relevant controlling cylinders. In addition, it acted as an element of reaction for the forces applied on the

Fig. 5.   Heat-shield heating-loading installation.

structure and it was used as a support for the several wiring harnesses, for the electric power distribution and all the safety and warning devices (Fig. 6).

A sufficient number of pressure and return ports were provided on the top of the base for supplying nitrogen and fire extinguishing foam between shield structure and reflectors in case of an emergency.

The infrared radiation on the external shield surface was accomplished by means of about 3500 infrared lamps of three different ratings (0.5 kW, 1 kW and 2 kW) installed on suitable holders fitted on three reflectors. These reflectors were made of heat resistant resin impregnated fiberglass, and protected by a special ceramic compound (silica and alumina) which besides being a good insulating material, has a very high reflecting capability.

The radiating system dome was made by a silica–alumina single piece of a conical shape because of the very high heat density (about 400 kW/m^2). A metal hood fitted with an exhaust fan was placed above the dome.

The infrared lamps were divided into thirty sections each connected to a channel of the supply system. Each group of three sections was connected to one of the ten load programmers mutually synchronized (Fig. 7).

This facility, that can be considered among the best ones, has a 3500 KVA continuous rating and a 7000 KVA short time (5 min) capability.

During the test, two hundred measurements (temperatures, stresses, strains etc.)

Fig. 6. Top and bottom portions of heating-loading frame.

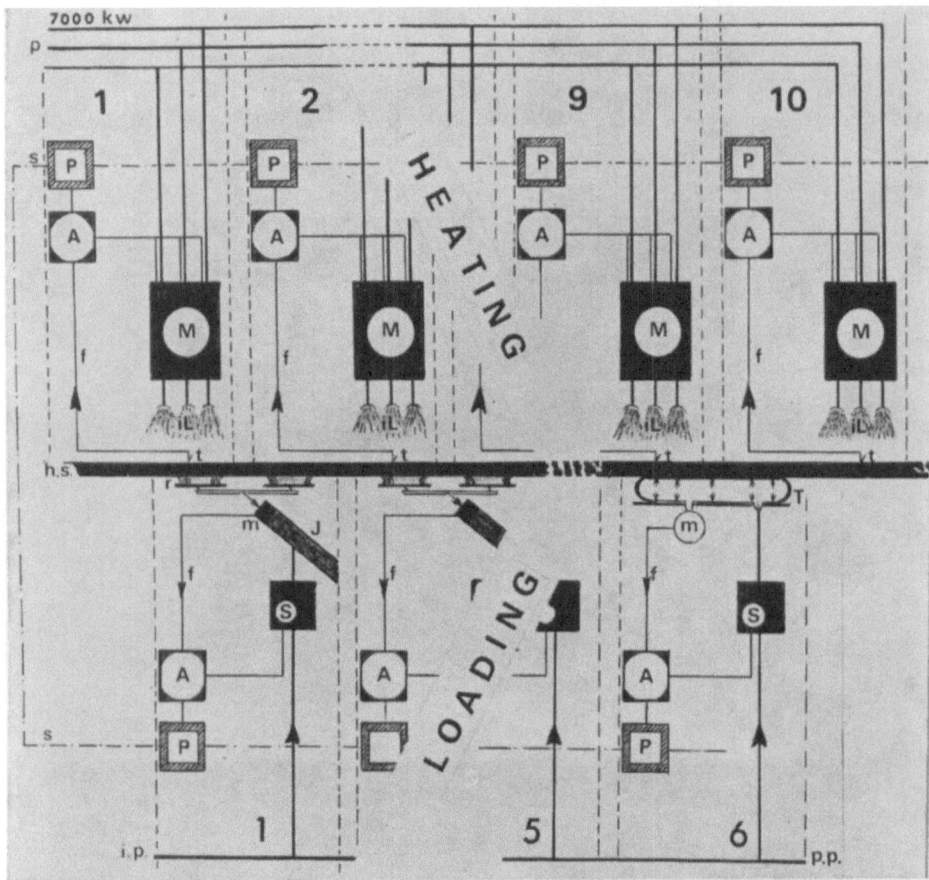

Fig. 7. Block diagram of 10 heating + 6 loading channels facility. A – Differential amplifier. J –
Hydraulic jack. M – Electric Modulator. P – Heating or loading programmer. S – Hydraulic or
pneumatic servovalve. T – Flexible tank. f – Feed back. iL – Infrared lamp. i.p. – Hydraulic supply.
h.s. – Heat shield structure. m – Load or pressure transducer. p – Electric supply. p.p. – Pneumatic
supply. r – Rubber pad. s – Program synchronism. t – Temperature transducer.

were taken in different areas of the structure. All data were conditioned and sent to a
data acquisition digital system while, in parallel, the basic ones were kept under
immediate and direct verification by means of analogue recorders.

In order to verify the reciprocal movement of the two half shells caused by the
thermal expansion, closed circuit television cameras and high speed movie cameras
were provided inside the test facility.

Since the test was particularly important and allowed no repetition, it was necessary
to train the personnel on the correct accomplishment of the test. This was done by a
series of pre-tests and, at the end, a program was recorded on a magnetic tape having
the purpose of coordinating the complete test realization.

Simultaneously to the infrared radiation, the heat shields were subjected to the

Fig. 8. Heat shield vibration test.

operational loads distributed on the conical and spherical section (inward loads) by means of rubber pads bonded to the shield's internal surface and of rubber pockets on the cylindrical section (outward loads). The loads on the rubber pads were so grouped as to have three resultants individually connected to a hydraulic cylinder. Of the rubber sacks, three groups were made, each assigned to three different pneumatic supply lines. These latter ones together with the three hydraulic lines, were controlled by a programming unit having six channels all synchronized with the heating program.

For the peculiar type of facility used, the pneumatic and hydraulic systems, as well as the heating ones, were automatically verified by the numerous sensors located on the structure, supply lines and reflectors. Some of these sensors, such as the thermal flow meters, required a proper development study.

4.4. DYNAMIC TESTS

The impossibility of carrying out complete vibration tests on the heat shields, due to their own sizes and weight, suggested a special program whose purpose was to ascertain the capability of the structure and the installations, to withstand properly the vibrations occurring during launch and flight.

The program was divided into the following two phases:

vibration tests to establish the structure own modes and correspondent amplification factors (Fig. 8).

Fig. 9.   Detail of performance test-bench for explosive actuator.

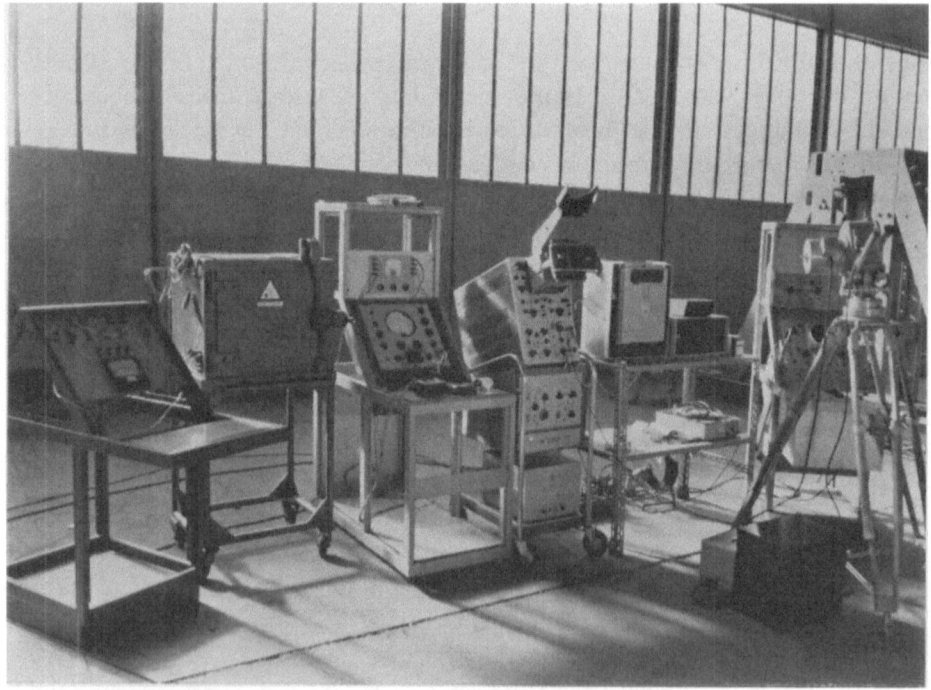

Fig. 10.   Electronic equipment for explosive bolt qualification test.

Fig. 11. Data acquisition system at Aerospace Laboratory.

vibration endurances tests of structural samples with transducers fittings, wiring harness clamps, etc.

The technique used in the first phase was the traditional one derived from vibration testing, currently carried out on aircraft. In this particular case, the heat shields were secured to a rigid fixture by means of the rings to connect to the vector third stage; the two half-shells were facing each other with a pressure equal to that developed by the jettisoning actuators on the structure.

The structure was shaken by means of electrodynamic shakers working either in phase or not (according to the studied modes) which vibrated the structure at different locations.

A mobile unit acquired and recorded the data which came from several accelerometers.

In the second phase, the structural elements, completely similar to those of the heat shields, except for having flat surfaces in lieu of curved ones, were instead, subjected to vibrations at high load factor and at suitable endurance by means of a normal 750 lb vibrating system.

By the above test the shur-locks arrangement for the transducer mounting, as well as the installation of wiring and plumbing, were optimized.

4.5. Jettisoning tests

4.5.1. *Preliminary Tests*

The full scale jettisoning tests of the heat shields were proceded by several severe tests on the ballistic devices which actuate the jettisoning.

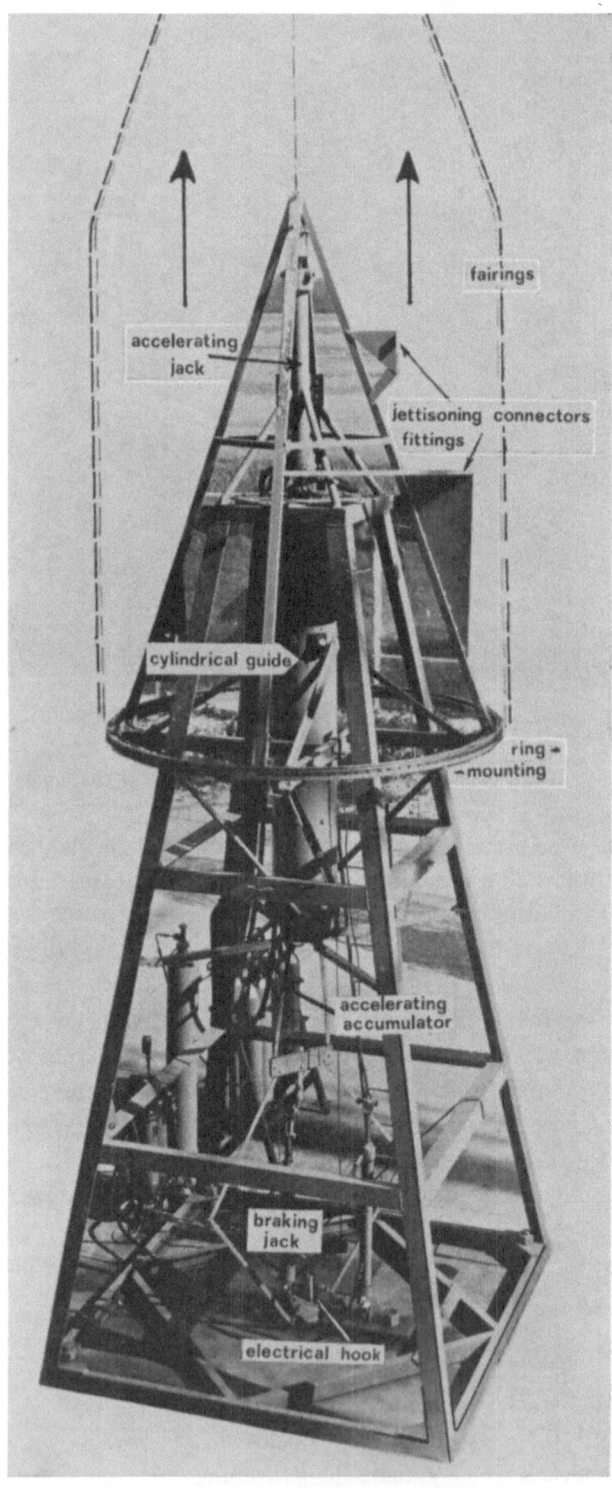

Fig. 12.   Accelerating tower for jettisoning test of heat-shield.

The ballistic devices used on the ELDO heat shields are explosive actuators and bolts.

A number of tests were carried out on the ballistic actuators, which were designed and manufactured by an European firm with FIAT technical assistance, in order to become aware of the relationship between charge explosiveness, charge density and shear ring material.

Finally, other tests were performed to optimize the fitting shape installed on the shield structure. The purpose of all these tests was to establish the actuator reliability, with no damage to the surrounding structures.

To determine the actuator dynamic characteristics (force, time, acceleration) a slide fitted test bench was built. Two wheeled carts of a known mass (equal to that of a section of the structure) could move freely on the slide under the action of a single actuator when fired (Fig. 9).

Both carts consisted of a metal frame, a ballast and a piece of structure equal to the heat shield structure to which the true fitting for attaching the actuator to the shield, was secured. During the test the two carts were tight to each other with a force equal to that which holds the two half-shells tight, by suitably preloading the actuator.

Near the fittings, accelerometers and break-through event markers with progressively variable length, were provided.

At both ends of the bench, two thick meter-scaled expanded polyuretane carpets were provided. The purpose of these carpets was to dampen the cart falling and take the impact impression, which was a reference for a ready evaluation of the trajectory, even though the test was filmed by a number of high speed movie cameras arranged for different angle shots.

The tests carried out in air were followed, though on a much less sophisticated basis, by ballistic tests in the vacuum chamber (after 24 hours at 10^{-3} torr) in order to verify the actual tightness for a proper operation of the system.

The explosive bolts, instead, already had the U.S.A. qualification and, therefore, required only some particular tests in order to determine the pretorque optimum value and the axial thrust diagram from the time of ignition to the time the effect stops (Fig. 10).

These tests were performed with the section of structure involved, so as to maintain the actual elasticity of the system inside a special armored chamber using an adequate test instrumentation (Fig. 11). More binding were the tests conducted to optimize and to qualify the explosive electrical connectors used in the heat shield instrumentation. In addition to the normal vibrations, shocks and electrical tests, the connectors were subjected to humidity tests, ultra high vacuum exposure (10^{-7} torr) and disconnecting force surveys, in order to verify the design requirements.

4.5.2. Full Scale Tests

Owing to the complexity and sensitivity of the jettisoning devices and because of the importance of the correct firing sequence of the explosive bolts and actuators, etc., numerous full scale jettisoning tests were made on the Europe 1 vector heat shields.

Fig. 13.   Jettisoning test at 0.22 sec after ignition.

These tests were essential for determining the effects and the reliability of the various devices and of the shields move-away trajectory to verify that no possibility of collision with the satellite exists.

To carry out the tests, a tower was built with a mobile top portion, which supported the heat shields therefore enabling them to accelerate at the same rate they are subjected to at the time of jettison. The mobile tower portion could slide freely on a support to which the acceleration devices together with other inspection and control systems were attached. Inside this mobile portion a dummy satellite made of friable material was placed, whose overall dimensions were equal to the maximum expected dimensions of the satellite (Fig. 12).

The whole was placed above a wide nylon net whose purpose was to dampen the impact and prevent the heat shields from falling on the ground after the jettisoning.

The accomplishment of these tests made it necessary to develop many electronic instruments to coordinate and automate all preparation control and inspection operations of the tests.

Numerous tests were performed, some under perfectly identical conditions, for instance, the failure of one of the jettisoning devices. For each test the accelerations on parts of the structure were verified and recorded. The heat shields trajectories were verified and recorded accurately during the first move-away phase and coarsely, immediately after (Fig. 13).

JETTISONING TEST EQUIPMENT AT
FIAT AEROSPACE LABORATORY

V. SACCHI

Sezione Velivoli, FIAT, Italy

Abstract. To qualify the jettisoning system for heat shields, a test facility has been developed.
So far several tests were performed to cover every heat shields configuration and system reliability.
The test facility consists of an accelerating tower, an automatic sequence control and an optical
and/or electronic data recording system. The system is outlined in details with some test results.

1. Introduction

The purpose of this paper is to illustrate the equipment required for the qualification
tests of the satellite fairings jettisoning system, developed and experimented with
success at FIAT Aerospace Laboratory.

As is well known, once the launcher has carried the satellite out of the earth's
atmosphere the fairings are no longer required and are therefore jettisoned.

The integrity of the satellite during the jettisoning phase is entirely dependent on
the correct operation of the flight sequence system. In fact, even if the sequence
results incorrect, in most cases the fairings are still jettisoned but the satellite may be
damaged by a collision with the fairings if the motion of these interferes with the
flight path of the launcher after jettisoning. For example in the case of a strong
initial rotation which would tend to make the tops of the fairings reapproach one
another.

In order to test the jettisoning system, ground simulation equipments have been
developed. These equipments are capable of supplying all the necessary elements for
the evaluation of the exact time sequences and plan configurations which occur
during jettisoning.

2. Accelerating Tower

In order to supply the acceleration required, the fairings are assembled on a tower
which is a frame constructed with extruded and rolled steel shapes and is divided
into two parts; one part is fixed to the ground and the other part, which is vertically
movable, contains the element that connects the fairings to the launcher. Therefore
the base of the fairings is situated at about 3 m from the ground. Before the test, the
movable part is anchored to the ground by a steel cable and an electrical hook.

As shown in the sketch of Figure 1, a pneumatic jack with a useful area of 50.3 cm^2
connects the two parts of the tower. This jack is fed by a tank the pressure of which
is experimentally determined in preliminary acceleration tests without jettisoning
during which the time delay between the opening of the electrical hook and the
attainment of the vertical acceleration required for the test is obtained by means of
accelerometers and a set of event-markers. The useful maximum travel is 600 mm;

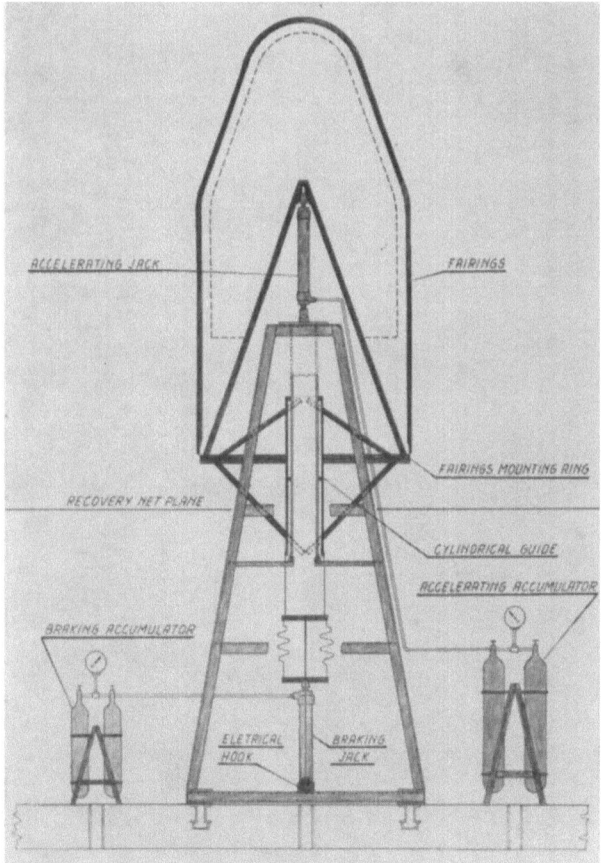

Fig. 1.   Accelerating tower sketch.

at the end of this travel the braking action of another pneumatic jack occurs and finally a twofold articulated beam (not illustrated in the sketch of Figure 1) stops the movable part.

On the top of the tower an expanded polystyrene profile represents the overall size of the useful load in the three main directions. It is therefore possible, from any evidence of damage of the load profile, to evaluate the importance of the interferences in the jettisoning trajectories.

Around the tower, a nylon net of about 130 m², supported by the tower itself and by four posts, provides for the recovery of the fairings. In Figure 2, a picture of the tower during the construction phase is shown and the twofold articulated beam which works as a mechanical stop, can be seen. The long hydraulic jack within the base and on the right hand side, serves to quickly tie down the movable part during the preliminary acceleration tests without completely loosing the energy stored in the accumulators. In this way the preliminary acceleration tests can be repeated approximately every 20 min.

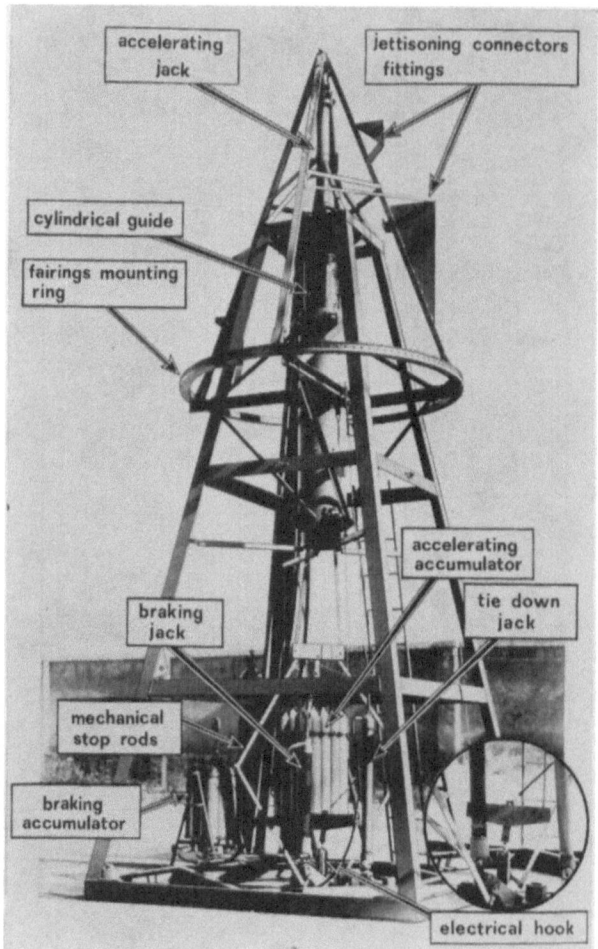

Fig. 2. Accelerating tower during construction.

In Figures 3 and 4 the fairings are shown before and during jettisoning. In Figure 4 the aforementioned profile and the attachments for the event-markers can also be seen. The electromechanical numerator, which appears on the left hand side of the picture, scans the tenths of a second as time reference for the cine-cameras which are located on the separation plane.

3. Recording and Control Instrumentation

The control systems for about 200 circuits which form the installation, the test sequencer, the signals calibration panels and the electronic recording systems are in the control compartment which is located at one end of the test area.

The test sequencer, developed by FIAT, supplies the start and stop impulses for

Fig. 3.   Fairings before jettisoning.

both the electronic and optical recording systems; it opens the electrical hook which releases the movable part of the tower, gives the starting impulse to the flight sequencer unit and can also be used to replace the latter for reliability tests on the remaining part of the jettisoning system.

The recording instrumentation can be divided into three groups according to the function accomplished:

3.1. JETTISONING INITIAL KINEMATICS RECORDING

Seventy-two event-markers (wires having a breakage length variable with constant increments from 0 to 425 mm) are mounted in correspondence with the separation plane. They are divided into 4 groups in such a way that their signals give a measurement of the fairings reciprocal rotation and move away during jettisoning.

The recording of the event-markers' signals is obtained by an event recorder with an 80 channel capacity, in which the tape velocity is such as to permit the evaluation of 0.002 sec time delays.

Figure 5 shows the processings of signals obtained during a standard test. The upper left-hand side shows how the signals are recorded in such a way that the accuracy of the jettisoning initial trajectories can be immediately evaluated. In fact,

Fig. 4.   Fairings during jettisoning.

in such a case the signals show an envelope symmetrical with respect to an axis which passes through point B and moreover, the points A, B, C, D and E are aligned on two straight lines perpendicular to this axis. In this case, as can be seen in processings of Figure 5, the tops of the fairings present a spring back of about 0.01 sec. In Figure 6, the detailed picture of the attachments relative to a unit of 18 event-markers is shown.

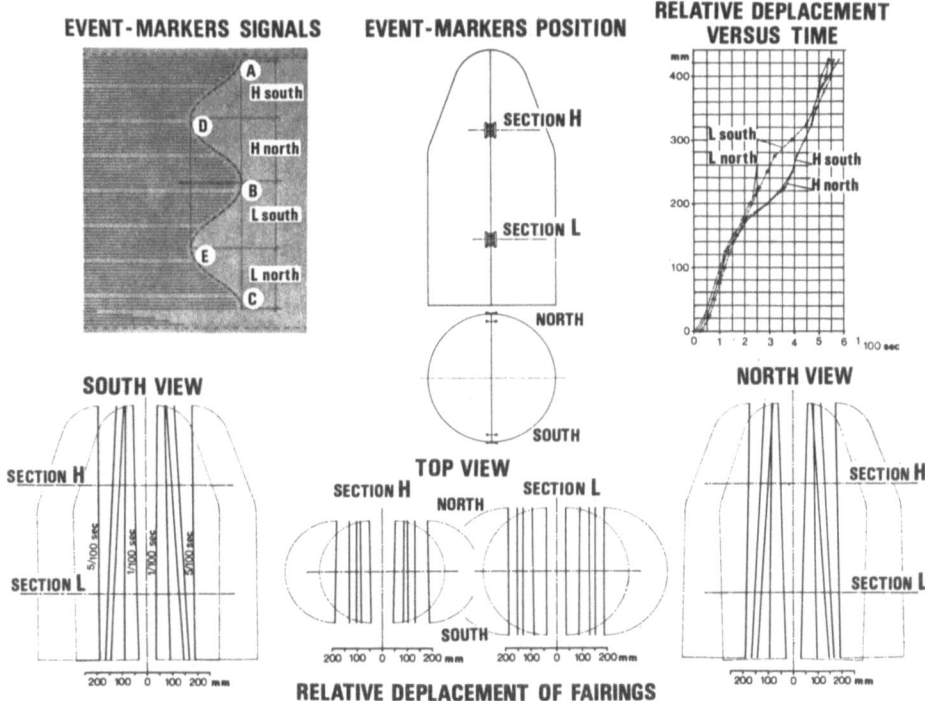

Fig. 5.   Event-markers signals and processing.

3.2. OPERATIVE SEQUENCE RECORDING

U.V. recorders, whose galvanometers have a sensitivity that is suitable for the type and level of the signal to be recorded, are available for recording of the operative sequence and a 14 channel magnetic recorder is available for recording the internal working characteristics of the flight sequencer.

Figure 7 illustrates some signals obtained during the qualification tests of the well-known ELDO program fairings jettisoning system for the EUROPA 1 satellite. In this case the system consisted of 2 explosive bolts and 4 ballistic actuators.

It was therefore necessary to obtain following recordings:

(a) the initiating pulse at which the electrical hook releases the movable part of tower,

(b) the initiating signal that starts the operation of the flight sequencer,

(c) the time firing sequences of the explosive bolts,

(d) the time firing sequences of the ballistic actuators,

(e) the working signals of the microswitches, the later being connected in series in the firing circuits of the squibs installed on the electrical connector which joins the fairings with the launcher,

(f) the time at which the disengagement of the electrical connector occurs (this is the final separation of the fairings from the launcher).

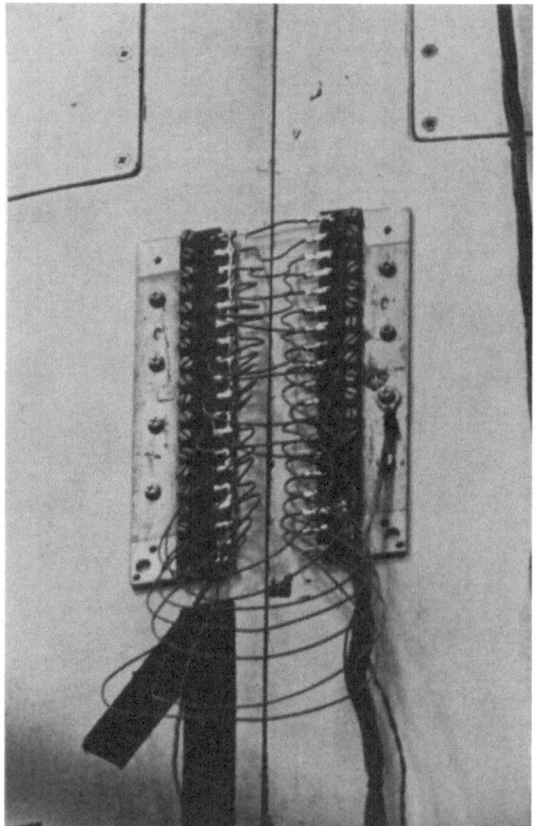

Fig. 6.   Event-markers mounting detail.

The first of these signals which can be assumed as the zero time for the entire sequence, is obtained by means of a shunt on the feeding circuit of the electrical hook and is simultaneously sent to all the recording systems as a reference.

Likewise, by means of galvanometers placed in parallel on resistors of about a few milliohms, which are inserted in the feeding lines of the explosive charges, the other signals required for the reproduction of the entire operative sequence are found.

The recording of Figure 7 was carried out on a U.V. recorder with a 2 m/sec velocity of the photo sensitive tape, so as to permit the measurement of the 0.0005 sec time delays.

Naturally, this makes it necessary for the test sequencer to supply the start impulse to the recorder a few seconds before the recording starts in order to stabilize the tape velocity.

By examining Figure 7 from the top towards the bottom, the following signals can be seen: the signals (a) and (b), the four (c) signals (the explosive charge of each bolt contains two bridgewires which are separately fed for better reliability) and the eight (d) signals (in this case there are also 2 signals for each actuator).

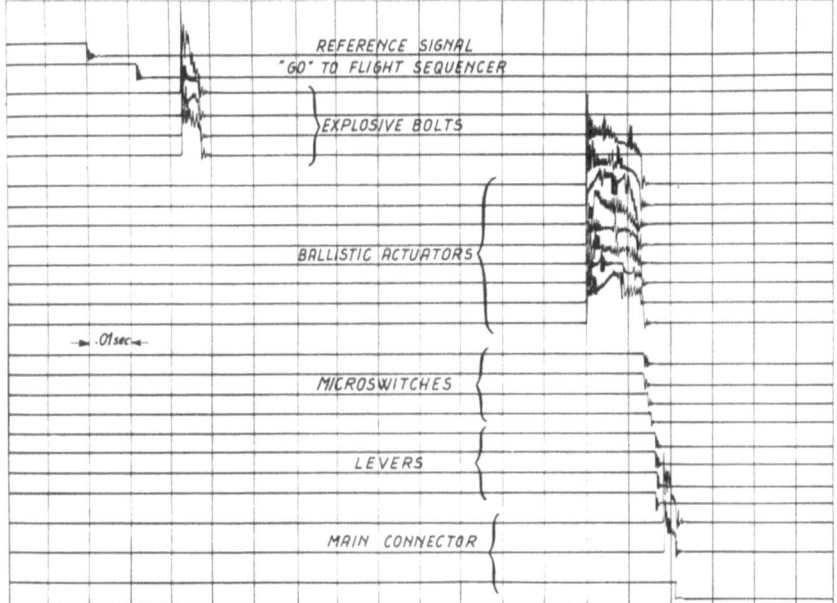

Fig. 7.   UV record.

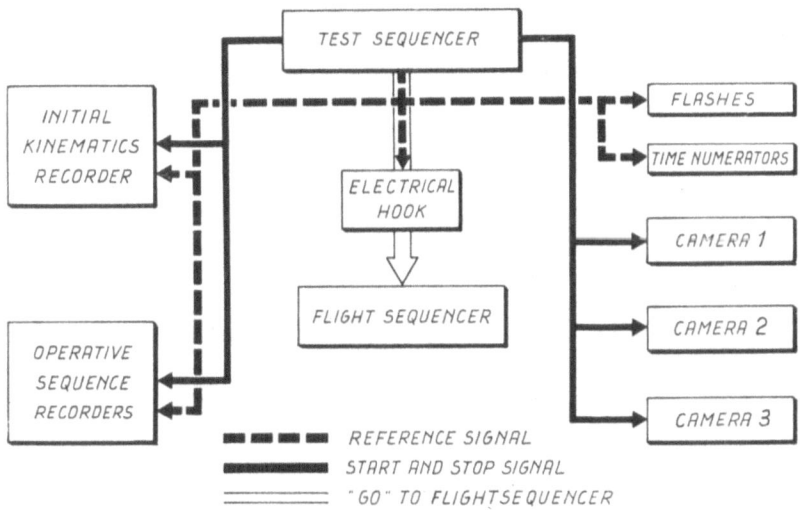

Fig. 8.   Control and record systems.

At this point, 8 signals to control the operation of the micro-switches can be seen; in fact, for each microswitch, besides the time at which the contact close, the mechanical opening instant of the relative lever has also been recorded.

Recently, the recording of these signals was used to replace a stock of microswitches which were found to be faulty and consequently caused a delay in the disengagement of the main connector. The last three signals of Figure 7 concern the

Fig. 9. Foreshortening of instrumentation.

explosion of the charges of the connector and its final mechanical separation which terminates the operative sequence.

Figures 8 and 9 show a simple block diagram of the various parts of the test installation and a foreshortening of the instrumentation located in the control compartment.

3.3. Optical records

Two cine-cameras directly actuated by the test sequencer are situated on the fairings separation plane in opposite positions with respect to the tower. The maximum shooting speed obtainable is 64 photograms/sec. At the time that the electrical hook is opened, the film is exposed to a flash while a numerator controlled by an electronic timer scans the tenths of a second in the cine-cameras visible field. Moreover, a cine-

→

Fig. 10. Jettisoning movie record.

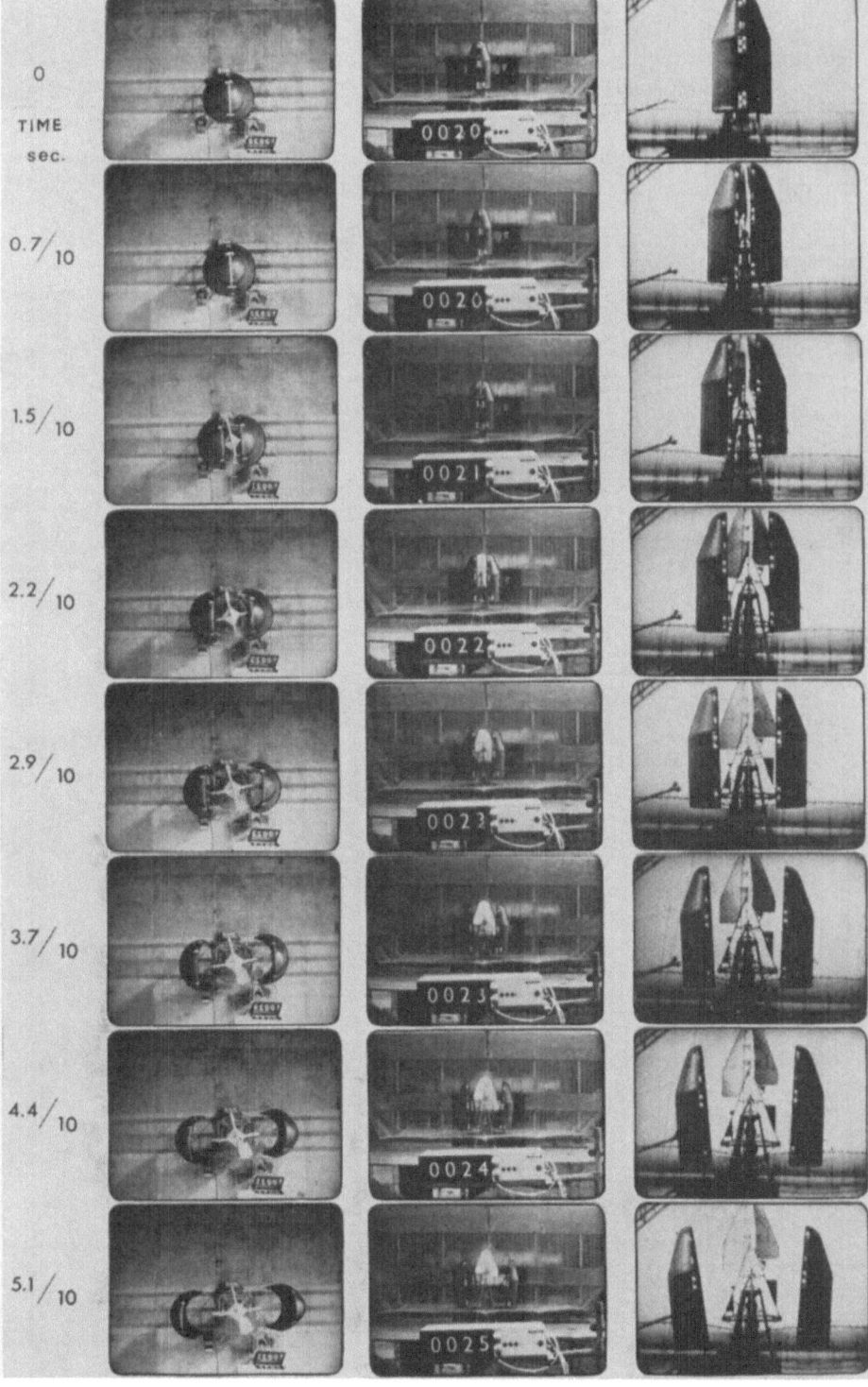

camera with a maximum speed of 1500 photograms/sec is available to shoot interesting details, such as the disengagement of the main connector, the movements of the actuators etc.

The optical unit is completed by one or more close circuit tele-cameras with simultaneous audio-video recording.

In Figure 10, a set of photograms obtained with all the cine-cameras pointed on the separation plane is shown.

A COMPARISON OF SATELLITE DRAG
MEASUREMENT TECHNIQUES

DANIEL B. DeBRA

Dept. of Aeronautics and Astronautics, Stanford University, Stanford, Calif., U.S.A.

Abstract. The effect of atmospheric drag on a satellite makes it possible to measure atmospheric density. Two approaches have been proposed: (1) Accelerometers may be mounted in the satellite to measure the vehicle acceleration which, together with the satellite mass, determines the drag force; (2) Alternatively, a drag-free satellite control system may be used. In this approach, an unsupported, shielded proof mass is used as a reference. The satellite is forced to follow the proof mass by thrustors which are actuated whenever the satellite approaches the proof mass. The proof mass is shielded by the satellite and is therefore free of drag, and the satellite follows the same orbit. Therefore, the control thrust must equal the drag force on the average. By measuring the control, the drag can be determined. Each approach has advantages and disadvantages. The relative measurement accuracy and bandwidth, calibration techniques, operating convenience and the effect of orbit longevity as a function of payload are discussed.

1. Introduction

The most important region of the upper atmosphere lies between 100 and 250 km altitude. It is in this region that the energy from the sun that affects the upper atmosphere is absorbed and distributed. Ultraviolet and corpuscular radiation from the sun and possibly gravity waves, deposit energy which is then redistributed through the diffusion of atomic and molecular oxygen and by radiation. Diffusion times are large and it is not clear what role convective transfer plays. As a result, it is difficult to model particularly, the time-dependent behavior in this region – a situation which is compounded by the lack of experimental measurements of density and composition.

Until very recently, a gap existed between sounding rocket data which gave measurements below 120 km with a very restricted geographical distribution, and satellite data which was primarily above 200 km. However, two recent flights – The Atmospheric Research Satellites (ARS) OV 1-15 and OV 1-16 – were devoted to measurements including part of this region (Elliott *et al.* [9]; Champion and Marcos [4]), and King-Hele and Hingston [12] have discussed the satellite 1966-101G which had a perigee height of 140 km. These measurements have not been consistent with the theoretical models. For example, the data presented by King-Hele indicates daytime density is 1.71 larger than nighttime density at 155 km altitude, while the CIRA 1965 model predicts only small day-night variations. The need for continued data gathering is clear and should include temperature and composition, as well as density to provide the basis for improved models (Moe [16]), particularly low altitude polar orbits (Thomas and Ching [24]).

Significant latitude and seasonal variations have been predicted by Newton *et al.* [17] and Champion *et al.* [3]. Experimental data, therefore, should include orbits with large latitude coverage and flights at different times of the year. Hence, an orbit which would give the required information should be polar to cover all latitudes, and have as low a perigee as possible consistent with a minimum of one-week life to insure that at least one solar storm will occur during the satellite's life so that correlation of the atmospheric response to solar activity can be made. Perigee altitudes in the range of 120 to 150 km appear practical but the orbit must be eccentric to obtain the required life. Apogees of at least 400 to 500 km are required. A great deal more information is obtained about the structure of the atmosphere if the spatial resolution of atmospheric density variations is fine, say, on the order of a scale height along the orbit. By contrast, density inferred from orbit decay data is averaged over one or more orbits and spatial resolution is only possible if the orbit is eccentric – thereby concentrating the drag at perigee. The data of principal interest will be at perigee but to obtain data around the whole orbit, the dynamic range of measurement must be 3 or 4 orders of magnitude. A noon orbit makes possible the observation of sunlight and darkness on the densities in these regions.

Drag measurements at these low altitudes can be made with conventional satellites using accelerometers, but it must be very dense to obtain adequate drag life (Champion and Marcos [4]), or fly at a high altitude (Broglio [1]). A second approach is to use a Drag-Free Satellite (Lange [13]). The reference for Drag-Free Satellite control is an unsupported proof mass shielded by the satellite from all nongravitational forces (air drag, solar pressure, etc.). Since only gravitational forces act on the proof mass, it follows a purely gravitational orbit. A control system in the Satellite senses the relative position of the Satellite with respect to the proof mass and actuates reaction jets forcing the Satellite to follow the proof mass without touching it; hence, the Satellite also follows a purely gravitational orbit unaffected by the nongravitational forces (air drag, solar pressure, etc.). By measuring the control forces, drag is inferred because the control has to equal the drag.

First, the requirements for measuring drag are discussed and a comparison of the two approaches is given.

2. Drag Determination of Atmospheric Density

In using the drag on a satellite to determine atmospheric density, two requirements must be met: First, that the drag force can be separated from the other nongravitational forces, and second, that the drag force (or acceleration) can be related to atmospheric density.

Below approximately 400 km altitude, atmospheric drag dominates as the principal force by two orders of magnitude and hence, to an accuracy of 1%, the other forces can be neglected. Above that altitude, a calculation of the other forces is needed to correct the data to 1%. For example, if radiation pressure forces are known to 10%, the identification of drag to 1% can be accomplished below about 550 km altitude (see Figure 1, second quadrant).

The atmospheric drag force, F, is related to the atmospheric density, ϱ, by the standard expression

$$F = \tfrac{1}{2} C_D \varrho V^2 A$$

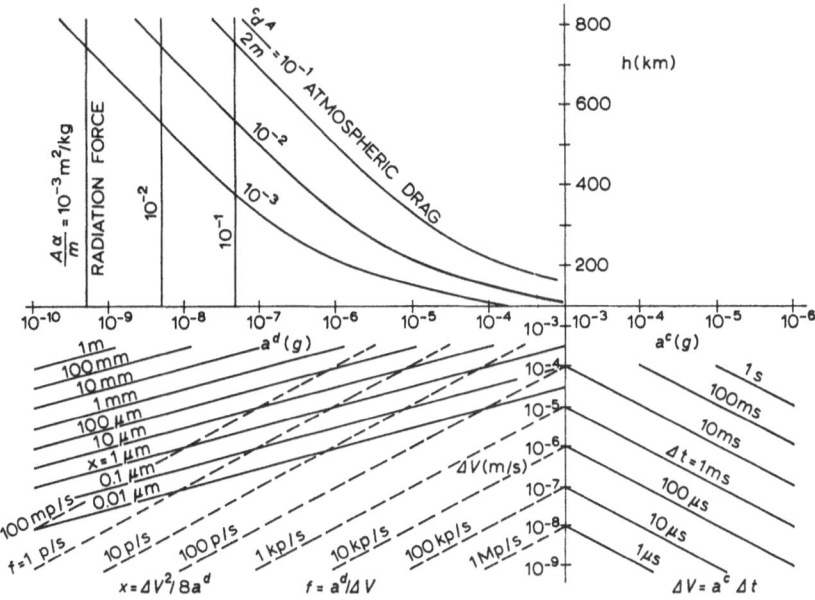

Fig. 1. Disturbance and control parameters.

where C_D = drag coefficient
V = orbital speed
A = satellite frontal area.

The force may be measured indirectly by accelerometers which sense (F/m)

$$F = (F/m)\, m$$

where m is the satellite mass, or directly in a Drag-Free Satellite. The measurement of F or F/m is the principal subject of this paper and is discussed later.

The frontal area of a satellite, A, is known to at least 1% if the satellite is spherical.

The speed of the satellite with respect to the atmosphere V, depends on knowledge of the motion of both the satellite and the air. In a polar orbit, the rotation of the earth causes a relative velocity component normal to the orbit of less than 0.5 km/sec at the equator compared with the orbital speed of approximately 8 km/sec. Even without correction (which can be done quite accurately), the change in speed due to this crosswind is only 0.2%. The principal uncertainty is due to winds relative to the earth which may be of the order of 100 m/sec or greater near the 120 km altitude region. Hence, the uncertainty in V is approximately 1% and the uncertainty in V^2 about 2%.

Satellite mass, which is needed for accelerometer measurements, is known to fractions of a percent.

Then, if F or F/m is known with 1% accuracy, the relative error in $C_{D\varrho}$

$$\frac{dC_{D\varrho}}{C_{D\varrho}} = \frac{dF}{F} - \frac{dA}{A} - \frac{2dV}{V} = \frac{d(F/m)}{(F/m)} + \frac{dm}{m} - \frac{dA}{A} - \frac{2dV}{V}$$

is less than 3% if the error sources are independent.

In some instances, the interest in drag may only require models involving $C_D \varrho$ but generally the density ϱ is desired and C_D must be calculated or measured to determine ϱ.

The value of the drag coefficient C_D, deserves a more detailed discussion. The drag coefficient of a spherical satellite, say, 70 cm in diameter is known with an accuracy conservatively estimated as 15% in absolute value (with an extreme bound of 33%) over an altitude range from 120 km (Knudsen number = 5) to 500 km (Knudsen number = 10^4) or higher. The relative accuracy should not exceed 5% in this altitude range. The uncertainties are principally due to our ignorance of the free molecular flow value. The experimental data show that the change in approaching transition flow (at low Knudsen numbers) are relatively smaller.

The momentum exchange between the atmosphere and a satellite can be considered in two parts in free molecular flow. The incident flux of particles with respect to the satellite is not collimated due to the thermal speed of the freestream molecules. They therefore 'wet' a somewhat larger surface than the projected area. This correction which depends on the molecular speed ratio amounts to from 1% to 5% around the orbit. The uncertainty in this correction is negligible.

The second contribution to the drag is the momentum flux leaving the satellite. Tests in space to measure the sputtering that might add mass to the momentum flux indicate this is an entirely negligible consideration. It is therefore necessary to know the departing velocity distribution of the atmospheric molecules which have collided with the satellite. Models by Schamberg [23] and Nocilla [18 and 19], theoretical guides based on particle-surface interaction by Oman [20, 21, and 22], and Hurlbut [10], partially applicable laboratory data reviewed by Hurlbut and Sherman [11] and orbital data on the despin of paddle-wheel satellites due to re-emission correlated with re-emission models by Moe [15], show a significant spread in results which prompt the earlier estimate on inaccuracy. Cook [5 and 6] has predicted values of drag coefficient with much smaller estimated uncertainty and the data on density obtained from a satellite now may be improved in the future as the role of geometry, surface finish, material, and other factors (perhaps yet to be identified) become better understood and allow us to predict the drag coefficient with greater certainty. Therefore, although drag coefficient uncertainty is now the limiting factor, it need not be in the future.

3. Requirements for Measuring F or F/m

The specifications of a useful orbit for an aeronomy satellite were given. They effect the magnitude, range, and bandwidth of the measurements to be made. The polar inclination has no significant effect, but the combined requirements for a low perigee and long life do: they result in an eccentric orbit for either a Drag-Free Satellite or a drag life (accelerometer) satellite. The apogee must be kept as low as possible to maintain the maximum latitude coverage in the important altitude range near perigee. Therefore, a dense satellite must give up sensitivity in order to minimize its ballistic coefficient, $C_D A/2m$, resulting in maximum accelerations of say, 10^{-5} to 10^{-4} g. Typically for a perigee of 130 km, an apogee height of nearly 500 km is required for either a drag-free or a drag-life satellite which corresponds to a range in atmospheric density of four orders of magnitude. For density measurements over the entire orbit to 1% resolution requires a dynamic range on the order of 10^6. This would be very difficult in a single-range analog instrument. At ONERA, a multirange analog accelerometer is under development (Delattre [8]), but this discussion will be limited to pulse rebalance techniques where the wide dynamic range is obtained by time averaging.

4. Information Bandwidth

The spatial resolution of one scale height requires bandwidths of approximately 1 measurement per second at perigee and 0.1 measurement per second at apogee. To allow for data smoothing, 10 and 1 pulses per second respectively should be adequate. This is a more stringent requirement on bandwidth than the temporal variations of interest.

Figure 1 shows three plots. In the second quadrant, the atmospheric drag and radiation pressure are shown as a function of altitude. They are drawn parametrically in $C_D A/2m$ for drag, and for radiation pressure the parameter is $\alpha A/m$ where α is the coefficient that determines the fraction of the incident electromagnetic momentum flux transferred to the vehicle. For example $\alpha=0$ for a transparent vehicle, $\alpha=1$ for total absorption, and $\alpha=2$ for total specular reflection from a flat plate, etc. The fourth quadrant shows the control acceleration (thrust/mass for a Drag-Free Satellite or saturation level for an accelerometer) versus the velocity change, ΔV, produced as a function of on-time. As this on-time is short compared with the off-time of the control, the velocity change can be considered impulsive in what follows. The third quadrant shows parabola parameters for the relative motion of a satellite and its proof mass under uniform relative acceleration due to drag. The amplitude and reciprocal time of a parabola are shown as parameters on a plot of relative acceleration versus the velocity change produced by a control impulse. It is assumed the parabolas are uniform and repeating so that the initial velocity on the parabola is half the impulsive velocity. The motion represents single-axis behavior of the satellite with respect to the proof mass for a Drag-Free Satellite or the proof mass with respect to the satellite for an accelerometer. In both cases, the proof mass is shielded and hence is in free fall. For the Drag-Free Satellite, pulses are applied to the satellite to make it follow the proof mass whereas in the accelerometer, the pulses force the proof mass to follow the satellite. In either case, a parabolic relative trajectory results due to the drag acceleration of the satellite. But, typical values for the two approaches are quite different.

Figure 1 may be used to size a Drag-Free Satellite control system or an accelerometer for an aeronomy mission. Consider an orbit with perigee at 130 km and a factor of safety of say, 5, in selecting the maximum control relative to the nominal drag at nominal perigee.

Assume that to extend the life of the drag-life satellite, it has an area-to-mass ratio of 10^{-3} m^2/kg which might be 0.2 of the value for a Drag-Free Satellite. Then, from Figure 1, the drag acceleration at a 500 km apogee is slightly greater than 10^{-8} g for a drag-life satellite, and for pulse rates of once per second, a ΔV is established which is consistent with an on-time of 30 μsec and saturation at 2.5×10^{-4} g. Perigee at 150 km decreases the saturation by a factor of 3 and increases the on-time to 100 μsec. These values are comfortably within the state of the art for pulsing circuits and for an electrically supported proof mass should offer no difficulty in an accelerometer. It is therefore possible to cover the entire range of altitudes with the required bandwidth with a single range pulse-rebalanced accelerometer.

A Drag-Free Satellite on the other hand cannot achieve 30 μsec pulses with cold gas or any valved fluids (pulsed plasma jets have this kind of response but do not appear promising for thrust levels above about 5×10^{-3} N). Using the data on Figure 1 with a minimum on-time for thrustors of say, 10 msec, the 10 p/sec pulsing rate is met at perigee but not at apogee. To provide the additional bandwidth, it is necessary to observe the relative trajectory of the satellite with respect to the proof mass and to infer the drag acceleration from its curvature. The pulses then become more a calibration for the pick-off scale factor being frequent enough to overcome slowly varying drift. Analyzing the pick-off signal requires good signal-to-noise ratio. The 10 msec pulses result in parabola amplitudes of slightly greater than 1 mm and 0.1 mm for the 130 km and 150 km perigees respectively. These are small enough that two-sided limit cycles can be avoided (i.e., deadbands can be this large) and large enough to give a clean signal. Nearer perigee, the amplitude of motion gets smaller until it is approximately 1 μm at the point where the pulse frequently has increased to 10 p/sec. This requires a pick-off with a wide dynamic range and quiet output for the smoothing process.

Pick-off noise is also a concern in an accelerometer. The limit cycle amplitude would be 0.01 μm at apogee and would be about 3×10^{-12} m at 130 km. Pulse rates would certainly vary with modest noise for these extremely small displacements. A more representative bandwidth at perigee would be estimated by following the parabola amplitude, 0.01 μm, at apogee to the perigee acceleration giving an effective pulsing rate of slightly less than 100 p/sec. The actual pulsing rate is still about 4 kp/sec so some form of data handling is desired to prevent excessive bandwidth in the telemetry. For example, a counter can be read 10 times/sec which would automatically provide the desired smoothing and keep the telemetry bandwidth a constant and adequate for the job. Similarly, the analog information on the Drag-Free Satellite can have a comparable fixed bandwidth (or sampling rate).

In the discussion of the accelerometer so far, it has been tacitly assumed that the pulse frequency changes in step with the disturbing acceleration. A typical integral pulse frequency mechanization has a region of modulation proportional to the commanding error signal. The slope, K/m, of this region and the displacement for saturation, x_s, determine the realizable bandwidth. The 'natural frequency' ω is $(k/m)^{1/2}$ or

$$\omega^2 = (Kx_s/m)/x_s$$

where Kx_s/m is the full scale capability of the instrument. Noise from the pickoff and mechanical vibrations limit how small x_s can be and still have a region of pulse frequency modulation as opposed to a bang-bang control. (The deadband will be typically of the same order of magnitude.) This means, for example, that if the bandwidth for a 10 g full-scale accelerometer was 1000 rad/sec, for a constant x_s the bandwidth at 10^{-5} g full scale would be 1 rad/sec.

5. Additional Considerations

A. CALIBRATION

In a single-axis accelerometer, the support forces may couple into the sensitive axis. This is one source of bias which has no counterpart in a Drag-Free Satellite. There is little information available on the bias levels of accelerometers designed for space application. There does not appear to be any fundamental reason, however, why an instrument cannot be developed that will have acceptable performance for an aeronomy experiment which requires a relatively modest change in range from earth applications compared with other space applications.

In a Drag-Free Satellite, the calibration of thrustors performed in the laboratory may be carried out to a high degree of accuracy, and there is reasonable confidence about the repeatability and stability of these calibrations on orbit. However, a device which directly measures the thrust may be more complicated than is necessary. It may become possible to adequately correlate thrust-time histories with chamber-pressure-time histories to obtain the desired impulse. This would result in simpler satellite mechanization.

The calibration problem for an accelerometer is somewhat different. The thrustors are calibrated in an environment where $1-g$ has little effect on the measurements. In an accelerometer, however, it is necessary to support the proof mass in the $1-g$ environment and at the same time measure the extremely small rebalance forces required for orbital operation. This may be done by calibrating an instrument operating at $1-g$ full scale and scaling its performance for the orbital environment. This technique will always leave some question marks on the validity of the calibration. A free fall calibration has been developed at ONERA (Delattre [7 and 8]) and a technique of supporting most of the weight of the proof mass by flotation in a dense gas has been developed by Likeness and Cannon [14]. At the present time, however, some method of orbital calibration is desirable as a cross check on these other methods.

B. OPERATIONAL CONSIDERATIONS

In a Drag-Free Satellite, the vehicle must be built around the proof mass at the mass center. The cost of supplying propellant to move the vehicle's mass center so that the pickoff (proof mass location) is in free fall is normally prohibitive if the vehicle has significant attitude motion. Therefore, there is greater freedom in installing accelerometers in an existing satellite to make atmospheric drag measurements than incorporating a drag-free control system.

The reduced dynamic range in a Drag-Free Satellite is caused by the minimum on-time possible with thrustors. This results in a larger deadband and greater relative motion of the proof mass with respect to the satellite. These dimensions are comparable to the uncertainties in location of the mass center of the satellite and hence, the coupling of vehicle dynamics with the output of the Drag-Free Satellite control system is relatively small. In an accelerometer, it is necessary to correct instrument readings for the vehicle dynamics. Furthermore, when three single-axis accelerometers are used, it is not possible to have all the proof masses located at the vehicle mass center which increases the number of parameter uncertainties.

A major difference which could be significant for some missions is the fact that a Drag-Free Satellite retains its same orbit throughout its active life. The decaying orbit of a drag-life satellite does not permit comparison of apogee and perigee data throughout the life of the satellite, but near the end of its life, has small eccentricity which could result in interesting data.

On an aeronomy experiment, it is frequently desirable to make other measurements simultaneously with the determination of drag. A drag-free control system expels propellant which could contaminate, for example, the atmospheric composition measurements of a spectrometer on board the satellite. This can be acceptable if a noble gas is used as the propellant although there is an attendant loss in the overall propulsion system's specific impulse. Krypton is desirable but may be prohibitive in cost. Argon or neon are acceptable although a greater weight penalty or equivalently, a penalty in life-time must be accepted with their use.

C. LAUNCH WEIGHT CONSIDERATION

It would be convenient to be able to compare the relative lifetimes that are possible for drag life and Drag-Free Satellites, but the lifetime is quite dependent on the orbit selected. The two approaches achieve their lifetimes in different ways. A Drag-Free Satellite carries propellant with which to cancel drag whereas the energy to be dissipated by a dense, drag-life satellite is stored in the orbital energy of the satellite. In an eccentric orbit, the energy is predominantly dissipated at perigee and the apogee height decays. The height of apogee over perigee is therefore a measure of the usable kinetic energy stored in a satellite and hence, its life. This energy is proportional to its mass just as the propellant mass in the Drag-Free Satellite is a measure of its life. To estimate the launch requirements to achieve the same life (albeit a decaying orbit for the dense satellite), the total mass of a dense satellite can be compared with the propellant required for a Drag-Free Satellite. The propellant requirements for a Drag-Free Satellite is

$$m^p = \frac{C_D A^{DF}}{c} \int_0^T q \, dt$$

where m^p = the propulsion system mass,

c = the specific impulse (effective exhaust velocity) reduced to include the
cost of tankage and thrustors,

C_D = drag coefficient,

q = dynamic pressure = $\varrho V^2/2$

T = useful lifetime for measurement

A^{DF} = frontal area of the Drag-Free Satellite.

The total satellite mass including dead weight to increase density for a drag-life
satellite using accelerometers is

$$m^s + m^{dw} \cong \frac{2C_D A^{DL}}{V} \frac{a}{\varDelta a} \int_0^T q\, dt$$

where $m^s + m^{dw}$ = mass of satellite and deadweight

A^{DL} = frontal area of the drag-life satellite

V = orbital speed at perigee

$\varDelta a/a = 2e$ = relative increase in semi-major axis to provide energy for orbit
decay and e is the orbit eccentricity (fixed perigee height),

and the approximation is made that $\varDelta a/a$ is small and the variation in V is neglected.
Initially, the apogee of the decaying orbit changes slowly so an approximate comparison of the launch requirements can be made by comparing the satellite masses
assuming the integral

$$\int_0^T q\, dt,$$

is the same for both satellites for equal useful lifetimes, T. Then

$$\frac{m^s + m^{dw}}{m^p} = \frac{2c}{V} \frac{a}{\varDelta a} \frac{A^{DL}}{A^{DF}}.$$

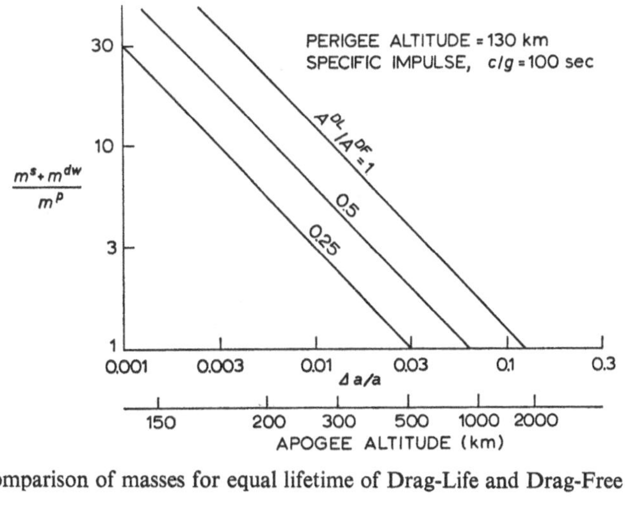

Fig. 2.   Comparison of masses for equal lifetime of Drag-Life and Drag-Free Satellites.

The results are plotted in Figure 2. Since the propellant-plus-tank would be approximately one-half of a Drag-Free Satellite, the launch weight requirements for the two satellites are comparable when the ratio $(m^s + m^{dw})/m^p$ is two, which corresponds to the nominal orbit with perigee at 130 km and apogee at 500 km. At lower altitudes, the Drag-Free Satellite gives longer life and at higher altitudes, the drag-life satellite would last longer for equal launch weight.

6. Conclusions

The instrumentation of a satellite to make atmospheric density measurements by measuring drag can be accomplished with a drag-free control system or with an accelerometer. Each approach has relative advantages over the other and selection will be mission dependent.

Acknowledgments

The research described in this paper was partially supported by National Aeronautics and Space Administration, Grant No. NsG 582, and the United States Air Force Contract No. AF 33(615)-1411.

References

[1] Broglio, L., 'Air Density Between 200 and 300 km Obtained by San Marco 1 Satellite', *Space Research*, Vol. VII (ed. by R. L. Smith-Rose, S. A. Bowhill, and J. W. King), North-Holland Publ. Co., Amsterdam, 1967, pp. 1135–1147.

[2] Champion, K. S. W., 'Variations With Season and Latitude of Density, Temperature and Composition in the Lower Thermosphere', Air Force Cambridge Research Laboratories, Bedford, Mass., 1966.

[3] Champion, K. S. W., Marcos, F. A., and Slowey, J. 'New Model Atmospheres Giving Latitudinal and Seasonal Variations in the Thermosphere', Air Force Cambridge Research Laboratories Bedford, Mass., 1966.

[4] Champion, K. S. W. and Marcos, F. A., 'The Air Force (OAR) Satellite OV1-16 (Cannonball)', *Trans. Am. Geophys. Union* (Seventh National Fall Meeting), **49**, no. 4, (Dec. 2–5, 1968) 637 and 724.

[5] Cook, G. E., 'Satellite Drag Coefficients', *Planet. Space Sci.* **13** (1965) 929–946.

[6] Cook, G. E., 'Drag Coefficients of Spherical Satellites', *Ann. Geophys.* **22**, No. 1 (Janvier–Mars, 1966).

[7] Delattre, Michel, 'L'accéléromètre O.N.E.R.A. a grande sensibilité', Paper presented at the Mesucora Congress, Paris, 17–21 April, 1967.

[8] Delattre, Michel, 'L'accéléromètre O.N.E.R.A. a grande sensibilité', Communication présentée a la réunion du Groupe Pilotage et Guidage de l'AGARD, Braunschweig, 7–9 Mai, 1968.

[9] Elliott, D. D., Becker, R. A., Champion, K. S. W., and Marcos, F., 'The Atmospheric Research Satellite OV 1-15 (1968-059A)', *Trans. Am. Geophys. Union* (Seventh National Fall Meeting), **49**, no. 4, (Dec. 2–5, 1968) 637 and 721.

[10] Hurlbut, F. C., 'Particle Surface Interactions in Hypervelocity Flight, An Annotated Bibliography', RAND Report, 1966.

[11] Hurlbut, F. C. and Sherman, F. S., 'Application of the Nocilla Wall Reflection Model to Free Molecule Kinetic Theory', *Phys. Fluids* **11** (1968) 486–496.

[12] King-Hele, D. G. and Hingston, Janice, 'Variations in Air Density at Heights Near 150 km, from the Orbit of the Satellite 1966–101G', *8th COSPAR International Space Science Sym.*, London, 24–28 July, 1967.

[13] Lange, Benjamin O., 'The Drag-Free Satellite', *AIAA J.* **2**, no. 9 (Sept. 1964) 1590–1606.

[14] Likeness, B. K. and Cannon, Jr., Robert H., 'Flotation Technique for Laboratory Calibration of Low-Level Accelerometers', paper presented at the AIAA Guidance, Control, and Flight Dynamics Conf., Pasadena, Calif., Aug. 12–14, 1968.

[15] Moe, Kenneth, 'Absolute Atmospheric Densities Determined from the Spin and Orbital Decays of Explorer VI', paper presented at the COSPAR Meeting in Vienna, May 1966. Published in *Planet. Space Sci.*

[16] Moe, Kenneth, 'A Review of Atmospheric Models in the Altitude Range 100 to 1000 km', paper presented at the AIAA 7th Aerospace Sciences Meeting, New York, New York, Jan. 20, 1969.

[17] Newton, George P., Horowitz, Richard, and Priester, Wolfgang, 'Atmospheric Density and Temperature Variations from the Explorer XVII Satellite and a Further Comparison with Satellite Drag', *Planet. Space Sci.* **13** (1965) 599–616.

[18] Nocilla, Silvio, 'On the Interaction Between Stream and Body in Free-Molecule Flow', *Rarefied Gas Dynamics*, Supplement 1, Academic Press, New York, 1961.

[19] Nocilla, Silvio, 'The Surface Re-Emission Law in Free Molecule Flow', in *Rarefied Gas Dynamics, Third Symposium*, Supplement 2, Vol. 1, Academic Press, New York, 1963.

[20] Oman, Richard A., Bogan, Alexander, Weiser, Calvin H., and Li, Chou H., 'Interactions of Gas Molecules with an Ideal Crystal Surface', *AIAA J.* **2** (Oct. 1964) no. 10, 1722–1730.

[21] Oman, Richard A. and Clia, Vincent S., 'Research in Gas-Surface Interaction, Part II: A Shock Tube Driven Molecular Beam for Gas-Surface Interaction Experiments', Grumman Aircraft Engineering Corp., Bethpage, New York, Final Report submitted to NASA 1964–1965, NASA-CR-67541, 1965.

[22] Oman, Richard A., Bogan, Alexander, and Li, Chou H., 'Theoretical Prediction of Momentum and Energy Accommodation for Hypervelocity Gas Particles on an Ideal Crystal Surface', *Rarefied Gas Dynamics, Fourth Symposium*, Supplement 3, Vol. 2, Academic Press, New York, 1966, pp. 397–416.

[23] Schamberg, R., 'A New Analytic Representation of Surface Interaction for Hyperthermal Free Molecule Flow with Application to Neutral-Particle Drag Estimates of Satellites', Rand Res. Mem., RM-2313, 1959.

[24] Thomas, G. E. and Ching, B. K., 'Upper Atmospheric Response to Transient Heating', *J. Geophys. Res., Space Phys.* **74**, no. 7 (Apr. 1969) 1796–1811.

SATELLITE GRAVITATIONAL STABILIZATION SYSTEMS
WITH MAXIMUM DAMPING RATE

V. A. SARYCHEV

Academy of Sciences, U.S.S.R.

Abstract. The dynamics of a satellite with a gravitational stabilization system is investigated. The parameters providing minimum duration of the natural oscillation damping of the system are determined. Main attention is given to the analysis of oscillation of the system in the orbit plane.

The dynamics of a satellite with a gravitational stabilization system is investigated. Optimal parameters providing minimum duration of the damping of the system are determined using the concept of 'the degree of stability'. Special attention is paid to the analysis of oscillation of the system in the orbital plane.

A typical scheme of the satellite's gravitational stabilization system was suggested in papers [1, 2]. In this scheme (Figure 1) a stabilizer represents two identical bars rigidly connected with each other with equal weights at the ends. This stabilizer joint the satellite body by means of spherical gimbal P. O_1 and O_2 are the centers of mass of the satellite and stabilizer. The axes of the coordinate system $O_1x_1y_1z_1$ and $O_2x_2y_2z_2$

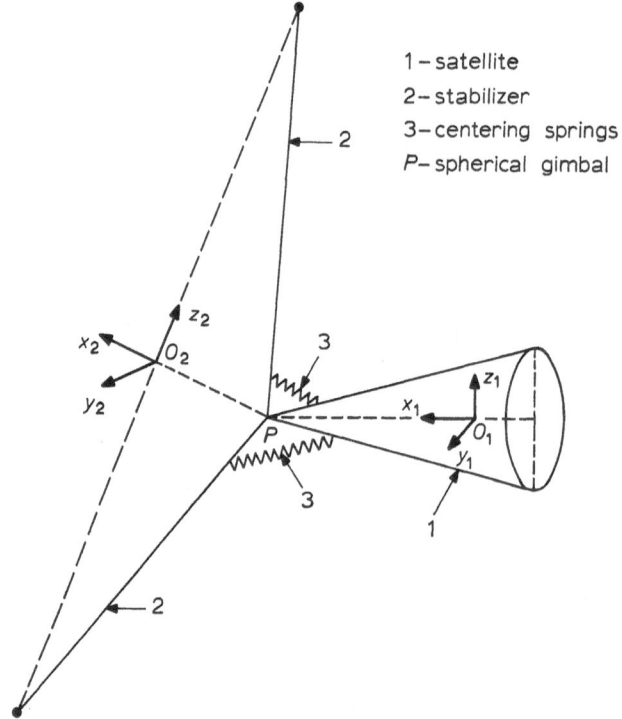

1 – satellite
2 – stabilizer
3 – centering springs
P – spherical gimbal

Fig. 1.   The scheme of the satellite-stabilizer system.

G. A. Partel (ed.), Proceedings of the Second International Conference on Space Engineering. All rights reserved.

direct along the major axes of inertia of the satellite and stabilizer correspondently. The position of the stabilizer relative to the satellite body is fixed by centering springs.

Linearized equations determining oscillations of the satellite-stabilizer system in a circular orbit plane are in the form [2]:

$$
\begin{aligned}
(B_1 + Ma_1^2)\,\ddot{\alpha}_1 - Ma_1a_2\ddot{\alpha}_2 + 3\,[(A_1 - C_1) - Ma_1^2]\,\omega_0^2\alpha_1 \\
+ 3Ma_1a_2\omega_0^2\alpha_2 + \bar{k}_1(\dot{\alpha}_1 - \dot{\alpha}_2) + \bar{k}_2(\alpha_1 - \alpha_2) = 0, \\
- Ma_1a_2\ddot{\alpha}_1 + (B_2 + Ma_2^2)\,\ddot{\alpha}_2 + 3Ma_1a_2\omega_0^2\alpha_1 + 3\,[(A_2 - C_2) \\
- Ma_2^2]\,\omega_0^2\alpha_2 - \bar{k}_1(\dot{\alpha}_1 - \dot{\alpha}_2) - \bar{k}_2(\dot{\alpha}_1 - \dot{\alpha}_2) = 0.
\end{aligned}
\tag{1}
$$

Here

$$
M = \frac{M_1 M_2}{M_1 + M_2},
$$

M_1, A_1, B_1, C_1 and M_2, A_2, B_2, C_2 – are masses and the main central moments of inertia of the system body, a_1, a_2 – is a gimbal abscissa P in the coordinate systems connected with both the bodies, $\bar{k}_1$ – is a friction coefficient, $\bar{k}_2$ – is an elasticity coefficient, ω_0 – is an angular velocity of motion of the system's center of mass in a circular orbit, α_1, α_2 – are axis angles O_1x_1 and O_2x_2 with tangent to the orbit. Differentiation with respect to time t is designated by point.

It is more convenient to introduce dimensionless parameters instead of dimension ones:

$$
s_1 = a_1\sqrt{\frac{M}{B_1}}, \qquad s_2 = a_2\sqrt{\frac{M}{B_2}}, \qquad \mu^2 = \frac{B_2}{B_1}, \qquad p_1 = \frac{A_1 - C_1}{B_1}
$$

$$
p_2 = \frac{A_2 - C_2}{B_2}, \qquad k_1 = \frac{\bar{k}_1}{\omega_0 B_1}, \qquad k_2 = \frac{\bar{k}_2}{\omega_0^2 B_1}.
\tag{2}
$$

Then, at $\omega_0 t = \tau$, the system (1) is in the form:

$$
\begin{aligned}
(1 + s_1^2)\,\ddot{\alpha}_1 - \mu s_1 s_2\ddot{\alpha}_2 + 3\,(p_1 - s_1^2)\alpha_1 + 3\mu s_1 s_2\alpha_2 \\
+ k_1(\dot{\alpha}_1 - \dot{\alpha}_2) + k_2(\alpha_1 - \alpha_2) = 0, \\
- \mu s_1 s_2\ddot{\alpha}_1 + \mu^2(1 + s_2^2)\,\ddot{\alpha}_2 + 3\mu s_1 s_2\alpha_1 + 3\mu^2(p_2 - s_2^2)\,\alpha_2 \\
- k_1(\dot{\alpha}_1 - \dot{\alpha}_2) - k_2(\alpha_1 - \alpha_2) = 0.
\end{aligned}
\tag{3}
$$

Zero solution of system (3) is stable asymptotically if [2]

$$
\begin{aligned}
&k_1 > 0, \\
&p_1 + \mu^2 p_2 - (s_1 - \mu s_2)^2 > 0, \\
&k_2\,[p_1 + \mu^2 p_2 - (s_1 - \mu s_2)^2] + 3\mu^2\,(p_1 p_2 - p_1 s_2^2 - p_2 s_1^2) > 0, \\
&[\mu - (s_1 - \mu s_2)\,s_2]\,p_1 - \mu\,[1 + s_1\,(s_1 - \mu s_2)]\,p_2 \\
&\qquad\qquad\qquad\qquad - (s_1 - \mu s_2)\,(\mu s_1 + s_2) \neq 0.
\end{aligned}
\tag{4}
$$

To estimate damping rate of system (3) it is convenient to use the concept of the degree of stability [3] which is represented as an absolute value of minimum real part of the roots of the characteristic equation

$$
\begin{aligned}
&\mu^2 (1 + s_1^2 + s_2^2)\, \lambda^4 + k_1 \left[(1 + \mu^2) + (s_1 - \mu s_2)^2 \right] \lambda^3 \\
&+ \{ k_2 \left[(1 + \mu^2) + (s_1 - \mu s_2)^2 \right] + 3\mu^2 \left[p_1 (1 + s_2^2) + p_2 (1 + s_1^2) \right] \\
&- 3\mu^2 (s_1^2 + s_2^2) \} \lambda^2 + 3 k_1 \left[p_1 + \mu^2 p_2 - (s_1 - \mu s_2)^2 \right] \lambda \\
&+ 3 k_2 \left[p_1 + \mu^2 p_2 - (s_1 - \mu s_2)^2 \right] + 9\mu^2 (p_1 p_2 - p_1 s_2^2 - p_2 s_1^2) = 0.
\end{aligned}
\tag{5}
$$

Maximum degree of stability of the system is shown [3] to be achieved at such choice of free parameters when all the roots of the characteristic equation are real and equal to one another. It becomes possible if the equations

$$
\begin{aligned}
&\xi_m^2 = 3\, \frac{p_1 + \mu^2 p_2 - (s_1 - \mu s_2)^2}{(1 + \mu^2) + (s_1 - \mu s_2)^2}, \\
&5\xi_m^4 - 3\xi_m^2\, \frac{p_1 (1 + s_2^2) + p_2 (1 + s_1^2) - (s_1^2 + s_2^2)}{1 + s_1^2 + s_2^2} \\
&\quad + 9\, \frac{p_1 p_2 - p_1 s_2^2 - p_2 s_1^2}{1 + s_1^2 + s_2^2} = 0, \\
&k_1 = 4\xi_m\, \frac{\mu^2 (1 + s_1^2 + s_2^2)}{(1 + \mu^2) + (s_1 - \mu s_2)^2}, \\
&k_2 = 3\, \frac{\mu^2 (1 + s_1^2 + s_2^2)}{(1 + \mu^2) + (s_1 - \mu s_2)^2} \left[2\xi_m^2 - \frac{p_1 (1 + s_2^2) + p_2 (1 + s_1^2) - (s_1^2 + s_2^2)}{1 + s_1^2 + s_2^2} \right].
\end{aligned}
\tag{6}
$$

are fulfilled. Here ξ_m – is the system's degree of stability.

To determine maximum value of the degree of stability ξ_m let us allow the first two equations in (6) relative to p_1 and p_2:

$$
\begin{aligned}
&p_1 = \tfrac{1}{3}\xi_m^2 + (1 + \tfrac{1}{3}\xi_m^2) s_1 (s_1 - \mu s_2) \pm \frac{2\mu}{3} \xi_m^2 \sqrt{1 + s_1^2 + s_2^2}, \\
&p_2 = \frac{\tfrac{1}{3}\mu \xi_m^2 - (1 + \tfrac{1}{3}\xi_m^2)(s_1 - \mu s_2) s_2 \mp \tfrac{2}{3}\xi_m^2 \sqrt{1 + s_1^2 + s_2^2}}{\mu}.
\end{aligned}
\tag{7}
$$

For the physically real systems the parameters p_1 and p_2 must satisfy the inequalities

$$
-1 \leqslant p_1 \leqslant 1, \qquad -1 \leqslant p_2 \leqslant 1.
\tag{8}
$$

Substituting into (8) the expressions for p_1 and p_2 from (7) at upper sign before the root we obtain the system of inequalities

$$
\begin{aligned}
&(1 + \mu \sqrt{1 + s_1^2 + s_2^2})\, \frac{2\xi_m^2}{3 + \xi_m^2} \leqslant 1 - s_1 (s_1 - \mu s_2), \\
&\frac{2\xi_m^2}{3 + \xi_m^2} \mu \sqrt{1 + s_1^2 + s_2^2} + 1 + s_1 (s_1 - \mu s_2) \geqslant 0, \\
&(\mu - \sqrt{1 + s_1^2 + s_2^2})\, \frac{2\xi_m^2}{3 + \xi_m^2} \leqslant \mu + (s_1 - \mu s_2) s_2, \\
&\frac{2\xi_m^2}{3 + \xi_m^2} \sqrt{1 + s_1^2 + s_2^2} \leqslant \mu - (s_1 - \mu s_2) s_2,
\end{aligned}
\tag{9}
$$

which determines solution 1 and at lower sign before the root we obtain the one

$$(1 - \mu\sqrt{1 + s_1^2 + s_2^2})\frac{2\xi_m^2}{3 + \xi_m^2} \leqslant 1 - s_1(s_1 - \mu s_2),$$

$$\frac{2\xi_m^2}{3 + \xi_m^2}\mu\sqrt{1 + s_1^2 + s_2^2} \leqslant 1 + s_1(s_1 - \mu s_2),$$

$$(\mu + \sqrt{1 + s_1^2 + s_2^2})\frac{2\xi_m^2}{3 + \xi_m^2} \leqslant \mu + (s_1 - \mu s_2)s_2,$$

$$\frac{2\xi_m^2}{3 + \xi_m^2}\sqrt{1 + s_1^2 + s_2^2} + \mu - (s_1 - \mu s_2)s_2 \geqslant 0,$$

$$(10)$$

which determines solution 2. Solution 1 turns into 2 and vice versa, when $B_1 \rightleftarrows B_2$, $s_1 \rightleftarrows s_2$.

The analysis of the inequalities (9) and (10) is very complicated at the arbitrary values of the parameters μ, s_1, s_2. That is why we consider only the most interesting spatial cases:

$$s_1 = 0, \qquad s_2 = 0;$$
$$s_1 \neq 0, \qquad s_2 = 0;$$
$$s_1 = 0, \qquad s_2 \neq 0.$$

Let $s_1 = s_2 = 0$. This variant of construction corresponds to the coincidence of the center of mass O_1 of the satellite and O_2 of the stabilizer with the position of the gimbal P. For the considered case solution 1 is in the form of:

$$\xi^2 = \frac{3\mu}{2 - \mu}, \qquad p_1 = \frac{\mu(1 + 2\mu)}{2 - \mu}, \qquad p_2 = -1 \qquad (11)$$

in the interval

$$0 \leqslant \frac{B_2}{B_1} \leqslant \frac{3 - \sqrt{5}}{2};$$

$$\xi^2 = \frac{3}{1 + 2\mu}, \qquad p_1 = 1, \qquad p_2 = \frac{\mu - 2}{\mu(1 + 2\mu)} \qquad (12)$$

when

$$\frac{B_2}{B_1} \geqslant \frac{3 - \sqrt{5}}{2}.$$

Here maximum value of the system's degree of stability ξ_m at the fixed value μ is designated as ξ. Optimization ξ with respect to μ lead to a maximum value of the degree of stability

$$\bar{\xi} = \max_{\mu} \xi = \frac{\sqrt{3}}{\sqrt[4]{5}}, \qquad (13)$$

which is obtained at $B_2/B_1 = (3 - \sqrt{5})/2$. In this case $p_1 = 1$, $p_2 = -1$ i.e. the satellite represents gravitational stable bars and stabilizer – gravitational unstable ones.

Solution 2 is written in the form of:

$$\xi^2 = \frac{3\mu}{2+\mu}, \qquad p_1 = \frac{\mu(1-2\mu)}{2+\mu}, \qquad p_2 = 1 \qquad (14)$$

in the interval

$$0 \leqslant \frac{B_2}{B_1} \leqslant \frac{3+\sqrt{5}}{2};$$

$$\xi^2 = \frac{1}{\mu}, \qquad p_1 = -1, \qquad p_2 = \frac{2+\mu}{\mu(2\mu-1)} \qquad (15)$$

when

$$\frac{B_2}{B_1} \geqslant \frac{3+\sqrt{5}}{2}.$$

Maximum value of the degree of stability

$$\bar{\xi} = \max_{\mu} \xi = \frac{\sqrt{3}}{\sqrt[4]{5}}$$

is obtained when $B_2/B_1 = (3+\sqrt{5})/2$. In this case $p_1 = -1$, $p_2 = 1$ i.e. the satellite represents gravitational unstable bars and stabilizer – gravitational stable ones. It should be noticed that solution 1 turn into 2 and vice versa, when $B_1 \rightleftarrows B_2$.

The case $s_1 = s_2 = 0$ is considered in [4] but in different designations and by different way.

It is well known that the obtained optimal values of the parameters satisfy the conditions of stability (4). It is necessary to study only the sign of the coefficient k_2 which cannot be negative according to its physical sense. It is found that the inequality $k_2 \geqslant 0$ for solution 1 is fulfilled in the interval $0 \leqslant B_2/B_1 \leqslant 3+2\sqrt{2}$ and for solution 2 – $(B_2/B_1) \geqslant 3-2\sqrt{2}$.

Figures 2–6 illustrate the results of the analysis carried out. For solution 1 (Figure 2) in the interval $0 \leqslant B_2/B_1 \leqslant \frac{1}{2}(3-\sqrt{5})$ the value ξ increases rapidly and at the end of the interval it achieves maximum which is equal to $\sqrt[4]{1.8}$. Then maximum point corresponds to the satellite in the form of the gravitational stable bar ($p_1 = 1$), and stabilizer in the form of the gravitational unstable one ($p_2 = -1$). In the interval $\frac{1}{2}(3-\sqrt{5}) \leqslant B_2/B_1 \leqslant 3+2\sqrt{2}$, ξ decreases slowly. When $B_2/B_1 > 3+2\sqrt{2}$, $k_2 < 0$ and an optimal solution does not exist.

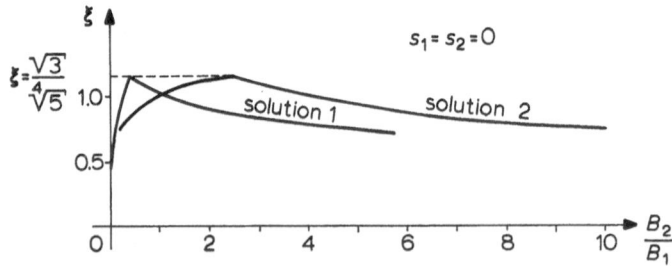

Fig. 2.   The maximum degree of stability of the system.

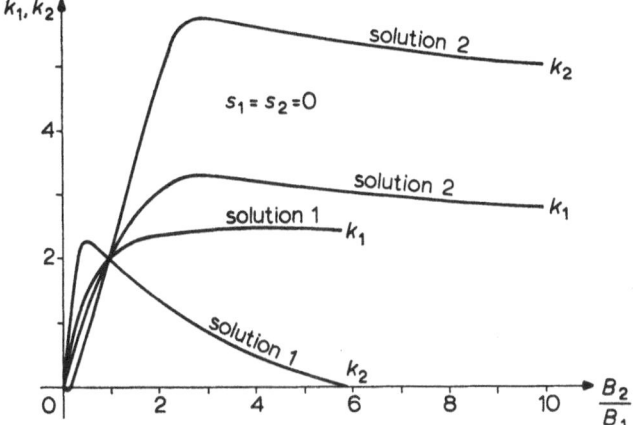

Fig. 3.   The optimal values of the parameters k_1, k_2.

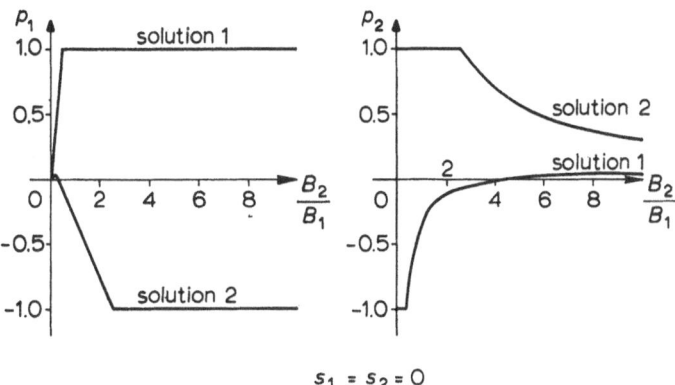

$s_1 = s_2 = 0$

Fig. 4.   The optimal values of the parameters p_1, p_2.

Solution 2 (Figure 2) exists when $B_2/B_1 \geqslant 3 - 2\sqrt{2}$. For this solution the value ξ increases slowly to maximum which is equal to $\sqrt[4]{1.8}$ at $(B_2/B_1 = \tfrac{1}{2}(3 + \sqrt{5})$ and then slowly tending to zero at $B_2/B_1 \to \infty$. In the point of maximum $p_1 = -1$, $p_2 = 1$.

Figures 3–4 illustrate the optimal values of the parameters k_1, k_2, p_1, p_2 for solution 1 and 2. For the different values of the parameters (Figures 5–6) the optimal transitional processes of the satellite-stabilizer system were obtained by the numerical integration. There are following initial conditions in every variant: $\alpha_1(0) = 20°$, $\alpha_2(0) = 10°$, $\dot{\alpha}_1(0) = \dot{\alpha}_2(0) = 0$.

Let $s_1 \neq 0$, $s_2 = 0$. This variant of construction corresponds to the coincidence of the center of mass of the stabilizer with the gimbal position P. For the considered case solution 1 exists only when $s_1^2 \leqslant 1$ and then it is in the form:

$$\xi^2 = \frac{3\mu}{2\sqrt{1 + s_1^2} - \mu}, \qquad p_1 = \frac{\mu + 2(\mu^2 + s_1^2)\sqrt{1 + s_1^2}}{2\sqrt{1 + s_1^2} - \mu}, \qquad p_2 = -1 \quad (17)$$

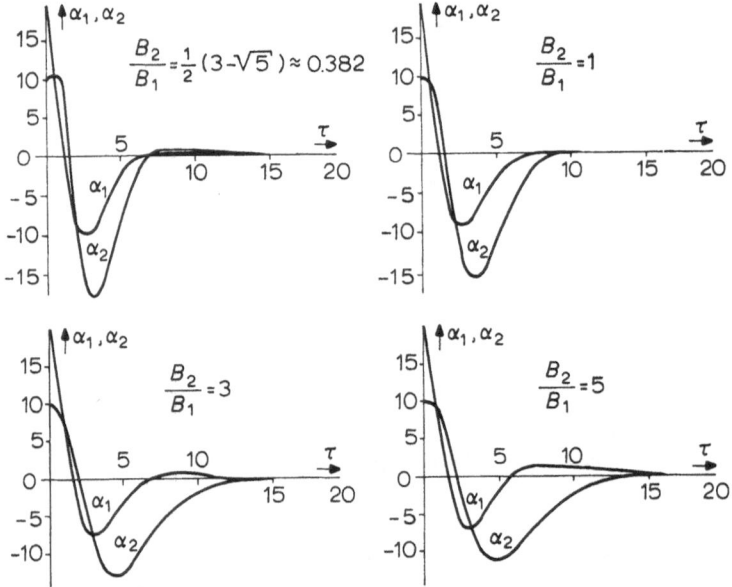

Fig. 5.   The optimal transitional processes of the satellite-stabilizer system for solution 1.

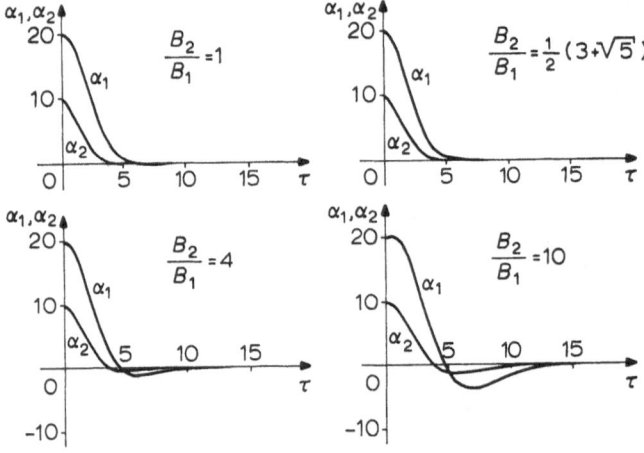

Fig. 6.   The optimal transitional processes of the satellite-stabilizer system for solution 2.

in the interval

$$0 \leqslant \frac{B_2}{B_1} \leqslant \frac{(3 - 2s_1^4) - \sqrt{5 - 4s_1^4}}{2(1 + s_1^2)};$$

$$\xi^2 = \frac{3(1 - s_1^2)}{\sqrt{1 + s_1^2}(\sqrt{1 + s_1^2} + 2\mu)}, \quad p_1 = 1, \quad p_2 = \frac{(\mu - 2\sqrt{1 + s_1^2})(1 - s_1^2)}{\mu\sqrt{1 + s_1^2}(\sqrt{1 + s_1^2} + 2\mu)}$$

$$(18)$$

when

$$\frac{B_2}{B_1} \geqslant \frac{(3 - 2s_1^4) - \sqrt{5 - 4s_1^4}}{2(1 + s_1^2)}.$$

Maximum degree of stability

$$\bar{\xi} = \max_{\mu} \xi = \left[\frac{3}{5} \frac{\sqrt{5 - 4s_1^4} - s_1^2}{1 + s_1^2}\right]^{1/2} \tag{18'}$$

is achieved at

$$\frac{B_2}{B_1} = \frac{(3 - 2s_1^4) - \sqrt{5 - 4s_1^4}}{2(1 + s_1^2)}.$$

In this case $p_1 = 1$, $p_2 = -1$. When $s_1 \neq 0$, $\bar{\xi} < \sqrt{3/\sqrt[4]{5}}$

Solution 2 has more complicated form

$$\xi^2 = \frac{3\mu}{2\sqrt{1 + s_1^2} + \mu}, \quad p_1 = \frac{\mu(1 + 2s_1^2) - 2(\mu^2 - s_1^2)\sqrt{1 + s_1^2}}{2\sqrt{1 + s_1^2} + \mu}, \quad p_2 = 1 \tag{19}$$

in the following regions:

(1) $0 \leqslant s_1^2 < \dfrac{2}{\sqrt{5}}, \qquad 0 \leqslant \dfrac{B_2}{B_1} \leqslant \dfrac{3 + \sqrt{5}}{2}(1 + s_1^2)$

(2) $\dfrac{2}{\sqrt{5}} \leqslant s_1^2 \leqslant 1, \qquad 0 \leqslant \dfrac{B_2}{B_1} \leqslant \dfrac{(3s_1^4 - 2) - \sqrt{5s_1^4 - 4}}{2(1 + s_1^2)},$

$$\frac{(3s_1^4 - 2) + \sqrt{5s_1^4 - 4}}{2(1 + s_1^2)} \leqslant \frac{B_2}{B_1} \leqslant \frac{3 + \sqrt{5}}{2}(1 + s_1^2)$$

(3) $s_1^2 > 1, \qquad \dfrac{(3s_1^4 - 2) + \sqrt{5s_1^4 - 4}}{2(1 + s_1^2)} \leqslant \dfrac{B_2}{B_1} \leqslant \dfrac{3 + \sqrt{5}}{2}(1 + s_1^2).$

Maximum value of the degree of stability

$$\bar{\xi} = \max_{\mu} \xi = \frac{\sqrt{3}}{\sqrt[4]{5}} \tag{20}$$

does not depend on s_1 and is achieved when

$$\frac{B_2}{B_1} = \tfrac{1}{2}(3 + \sqrt{5})(1 + s_1^2).$$

Then $p_1 = -1, p_2 = 1$.

$$\xi^2 = \frac{3\sqrt{1 + s_1^2}}{2\mu - \sqrt{1 + s_1}}, \quad p_1 = -1, \quad p_2 = \frac{(2\sqrt{1 + s_1^2} + \mu)\sqrt{1 + s_1^2}}{\mu(2\mu - \sqrt{1 + s_1^2})} \tag{21}$$

when

$$\frac{B_2}{B_1} \geqslant \frac{3+\sqrt{5}}{2}(1+s_1^2).$$

Maximum value of the degree of stability is determined by formula (20).

$$\xi^2 = \frac{3(1-s_1^2)}{\sqrt{1+s_1^2}(\sqrt{1+s_1^2}-2\mu)}, \quad p_1 = 1, \quad p_2 = \frac{(2\sqrt{1+s_1^2}+\mu)(1-s_1^2)}{\mu\sqrt{1+s_1^2}(\sqrt{1+s_1^2}-2\mu)}$$

$$(22)$$

when

$$\frac{2}{\sqrt{5}} \leqslant s_1^2 \leqslant 1, \quad \frac{(3s_1^4-2)-\sqrt{5s_1^4-4}}{2(1+s_1^2)} \leqslant \frac{B_2}{B_1} \leqslant \frac{(3s_1^4-2)+\sqrt{5s_1^4-4}}{2(1+s_1^2)}.$$

Maximum value of the degree of stability

$$\bar{\xi} = \max_{\mu} \xi = \left[\frac{3}{5}\frac{\sqrt{5s_1^4-4}+1}{1+s_1^2}\right]^{1/2}$$

$$(23)$$

is achieved when

$$\frac{B_2}{B_1} = \frac{(3s_1^4-2)+\sqrt{5s_1^4-4}}{2(1+s_1^2)}.$$

In this case $p_1 = p_2 = 1$. It is easy to show that $\bar{\xi} \leqslant \sqrt{\frac{3}{5}}$ and maximum value $\bar{\xi} = \sqrt{\frac{3}{5}}$ is achieved at $s_1^2 = 1$.

The condition of positiveness of the elasticity coefficient k_2 lead to the inequality

$$0 \leqslant \frac{B_2}{B_1} \leqslant (3+2\sqrt{2})(1+s_1^2)$$

$$(24)$$

in solution 1 and to the inequality

$$\frac{B_2}{B_1} \geqslant (3-2\sqrt{2})(1+s_1^2)$$

$$(25)$$

in solution 2.

Let $s_1 = 0$, $s_2 \neq 0$. This variant of construction corresponds to the coincidence of the center of mass O_1 of the satellite with the gimbal position P. For the considered case solution 1 and 2 coincide correspondingly with solution 2 and 1 of the previous case if we carry out the substitution $s_1^2 \rightarrow s_2^2$, $B_1 \rightleftarrows B_2$ in formulae (17–25).

In conclusion let us consider the general variant of the construction when $s_1 \neq 0$, $s_2 \neq 0$. In this case the general solution of the inequality systems (9) and (10) is very complicated and that is why it will not represent here. But it is not difficult to show that in general case the inequality $\xi_m \leqslant \sqrt{3}/\sqrt[4]{5}$ is also fulfilled.

It means that maximum degree of stability which is achieved when $a_1 = a_2 = 0$, can not be increased displacing the gimbal from the general center of mass of the satellite and stabilizer.

References

[1] Okhotsimsky, D. E. and Sarychev, V. A., 'A Gravitational Stabilization System of the Artificial Satellites', in Col. *Artificial Earth Satellite*, Vol. 16, U.S.S.R. Academy of Sciences, Moscow, 1963, p. 5–9.
[2] Sarychev, V. A., 'Investigation of the Dynamics of a Gravitational Stabilization System', in Col. *Artificial Earth Satellite*, Vol. 16, U.S.S.R. Academy of Sciences, Moscow, 1963, p. 10–33.
[3] Tsypkin, Ja. Z. and Bromberg, P. V., 'The Degree of Stability of Linear System', NISO, N9, 1946.
[4] Zajac, E. E., 'Damping of a Gravitationally Oriented Two-Body Satellite', *ARS J* **32** (1962) 1871–1875.

AN ANALYSIS AND SIMULATION OF MARS LANDING

E. KANE CASANI

Jet Propulsion Laboratory, California Institute of Technology, U.S.A.

Abstract. A novel approach to simulate a landing on Mars has been developed, and a series of tests conducted on a prototype of an unmanned Mars lander. Recently, an advanced development program, in which a Mars entry and landing system was fabricated and tested, was successfully completed. As part of this program, the landing event was simulated by dropping the lander from a helicopter hovering at 250 feet, creating the design impact velocity of 120 feet/sec. The lander was dropped first on a hardpan, dry lake bed, and in a second test, on an unyielding asphalt surface. Unique techniques were successfully developed to release the lander from the helicopter and accurately impact the target. In addition to the actual testing, a digital simulation of the landing event was developed using Monte Carlo techniques. This landing simulation program includes such random variables as the atmospheric temperature, pressure and density, the surface slopes, rocks, and bearing strength, and the winds.

1. Introduction

In the spring of 1967, the CSAD project (Capsule System Advanced Development) was initiated at the Jet Propulsion Laboratory (JPL) to design, fabricate, and test a partially functional Mars entry and landing capsule. A theoretical analysis of the impact event was carried out under this program. In the course of the design tradeoff studies, the spherical lander, envisioned in previous studies, evolved into a disc-shaped lander surrounded by a balsa impact limiter. A hardwood mockup of this disc-shaped lander was constructed to (1) test two different impact limiter materials; (2) determine acceleration levels in the lander at impact; and (3) test impact conditions on several different surfaces. Then, a functional Feasibility Model of this lander was constructed and functionally tested under impact conditions by dropping it from a hovering helicopter onto a hardpan, dry lake bed and, later, onto an asphalt runway. This paper discusses the analysis and test associated with the development of the landing system used on the Feasibility Model.

2. Conclusions

The analysis, design, fabrication, and test of the JPL (CSAD) Mars rough lander demonstrated that

(1) A Mars rough lander impact limiter designed to absorb the energy of a 112-ft/sec-impact against an unyielding surface will be more than adequate to decelerate an instrument payload under 95% of the landing conditions encountered.

(2) The analysis of impact limiter behavior is quite conservative, and limiter energy absorption capabilities are much higher than predicted.

(3) Survivable impacts at the highest expected velocities can be made on surfaces thought to correspond to the light-colored areas of Mars.

(4) Surface rocks of the critical size range do not present a landing hazard.

(5) Horizontal velocities do not introduce unique or hazardous conditions.

(6) No deep cratering or burying occurred in impacts against soft surfaces.

(7) Care must be taken at the interface of the lander and impact limiter to ensure the cleanliness of the contacting surfaces, the quality of the adhesive, and the integrity of the bond.

3. Design Philosophy

A. DESIGN GOALS

This lander was designed to a specific set of design requirements and criteria. The most significant requirements were

(1) To provide access to the ambient environment.

(2) To allow the erection of a six-foot boom with a wind instrument and a water vapor detector; the boom is to be erected normal to the local slope.

(3) To minimize the operations necessary immediately after landing to achieve partial mission success.

(4) To provide locations for the six antennas of the omni-directional antenna system.

The impact attenuator was required to limit the acceleration transmitted to the lander package to 2500 g. To accomplish this, the design criteria called for the impact limiter to withstand a vertical impact of 112 ft/sec on a flat, unyielding surface. This criterion was derived by carrying out a weight-optimization study on a combined parachute-impact-limiter system in the most tenuous (5 mbar surface pressure) Mars atmosphere. It was argued and subsequently proved that if the limiter were designed to this criterion, any realistic combination of wind, slopes, rocks, surface density, and bearing strength would combine to give a less severe impact condition. Clearly, it would be possible to construct a landing environment for which this design would be inadequate by combining all the 'worst-case' conditions; but to design such an unrealistic model would be so costly in system performance that the mission would lose significance.

B. DESIGN RATIONALE

Several nonspherical lander configurations and several techniques for satisfying the orientation requirements were studied. The disc-shape (Figures 1 and 2) was chosen as the configuration that best satisfies the design requirements.

The required impact limiter weight for a spherical lander and the parachute weight can both be expressed as a function of impact velocity. The sum of these two functions will then optimize at a velocity which minimizes the overall lander weight. For this particular design the optimum velocity was 112 ft/sec, although the weight is not particularly sensitive at this velocity.

After the required amount of balsa is established, the actual crushing geometry must be determined. Since the lander is really disc-shaped, not spherical, different impact orientations must be considered. This was done for three different orientations of 0°, 45° and 90° from the roll axis.

Several iterations on the geometry were made to establish the required footprint and desirable stroke. If the footprint is too large, the crushing stress at impact will

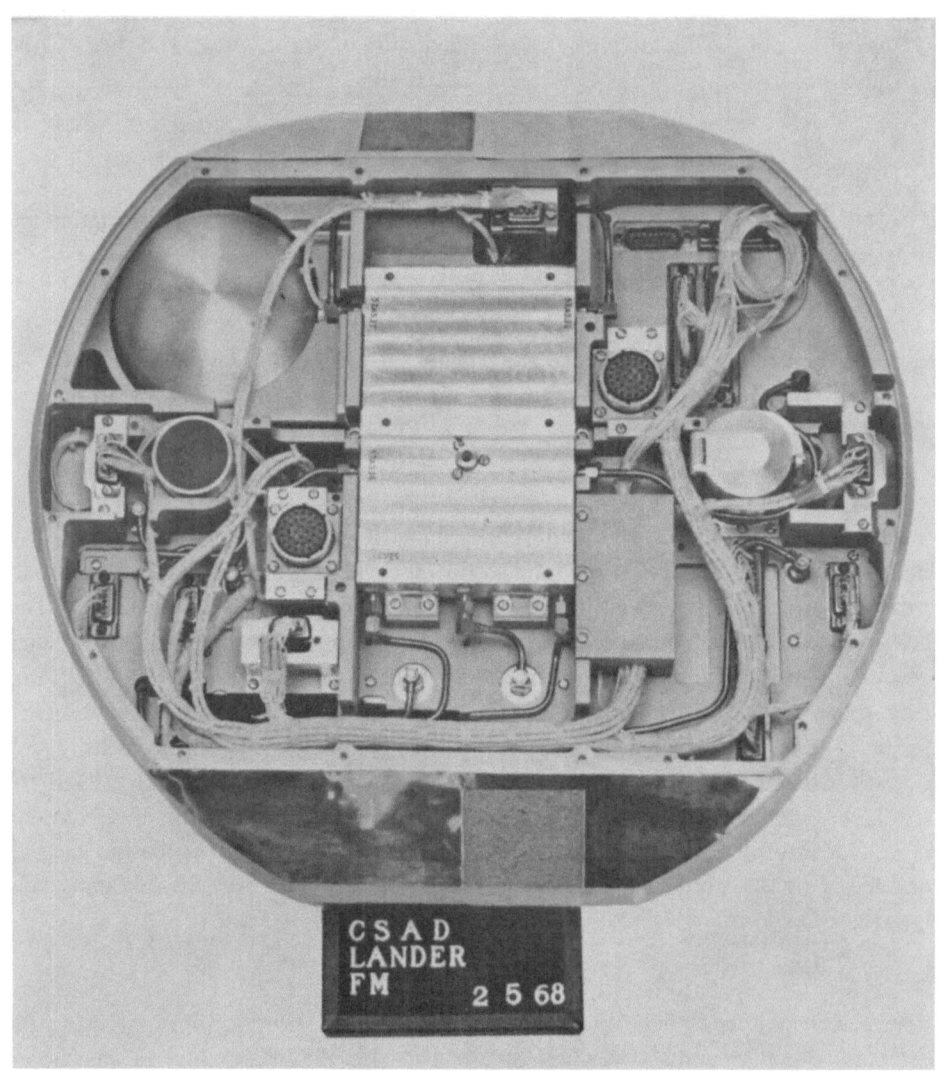

Fig. 1. Lander, side 1.

not be high enough to crush the balsa, and it will not absorb the impact energy. In determining the geometry, it was also assumed that for the 0° impact orientation, only the balsa directly under the lander would be effective in the energy dissipation. The grain orientation and the specific energy of the balsa as a function of the impact angle with respect to the grain are also considered in determining the geometry. This process establishes the geometry and the stroke for different impact orientations.

The disc-shaped configuration was refined and modified as a result of initial drop-testing and is shown in Figures 1 and 2. Some of the major advantages of this configuration are:

Fig. 2. Lander, side 2.

(1) Better packaging and cabling configuration. (2) Improved accessibility to equipment. (3) Simpler electrical interface. (4) More reliable mechanical interface. (5) Sample acquisition capability. (6) No post-landing orientation. (7) Lower weight. (8) Easier subsystem and chassis fabrication. (9) Stronger and simpler limiter installation.

4. Rough Impact Model and Simulation

Many atmospheric and planetary surface parameters are involved in the design of a hard lander for a Mars mission. When all worst-case values of these parameters are combined and used as the design criteria for the lander, an extremely conservative design is obtained for the impact attenuator and a high probability of successful landing is achieved; but this conservatism entails a performance penalty and requires complicated post-landing operations, such as removing the impact attenuator. In effect, this design approach maximizes the probability of a successful landing but decreases the probability of successfully performing any post-landing operations by complicating these operations and minimizing the useful landed payload.

To arrive at a better overall design, a statistical approach was taken which randomly generated the significant parameters in order to predict the expected impact conditions. This technique allowed the performance of a specific impact limiter design to be evaluated for the expected landing environments and permitted an assessment of the sensitivity of the design to specific distribution functions of the environmental parameters. This analysis confirmed the original limiter design, which was derived from a parachute-limiter weight optimization and which used fixed environmental parameters – e.g., a 5 mbar surface pressure.

The velocity of the lander at impact was dependent upon the combined values of several environmental parameters. With the use of Monte Carlo techniques, the landing was simulated repeatedly with each parameter being randomly generated in accordance with its distribution function. After a very large number of samples, the effect of all the parameters on the velocity at landing was obtained. A brief analysis of the impact event and a description of the program follow.

In this analysis, it will be assumed that the lander descends on a parachute through a Martian atmosphere with a surface density probability given by Figure 3. Hori-

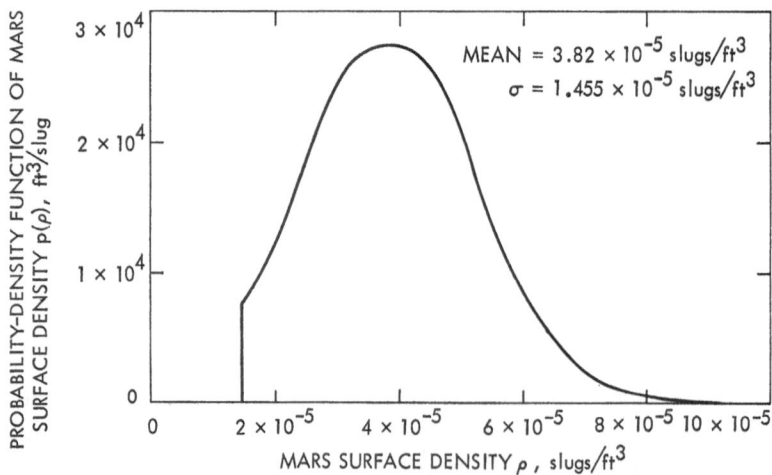

Fig. 3.   Mars atmosphere surface-density probability-density function.

zontal surface winds (assumed uniform between 0 and 100 m altitude) are given by the probability curve of Figure 4. The direction of the wind is assumed uniformly distributed between 0° and 360°. The bearing strength of the soil is distributed according to the curve of Figure 5.

A stress-strain curve for the assumed Martian surface is shown in Figure 6. The elastic portion of the soil's deformation characteristic only applies at low stress levels and at small strains. The plastic portion of the deformation characteristic occurs at a stress level defined by the bearing strength b_s. This is the most likely condition under impact situations. At high deformation levels, the soil becomes compacted or the penetration extends to a rock layer so that an almost impenetrable surface is reached. These three conditions are indicated in Figure 6. This deformation characteristic will be idealized by assuming that the elastic portion is negligible, and that plastic deformation continues from the first contact on the surface until a depth Δ_s is reached.

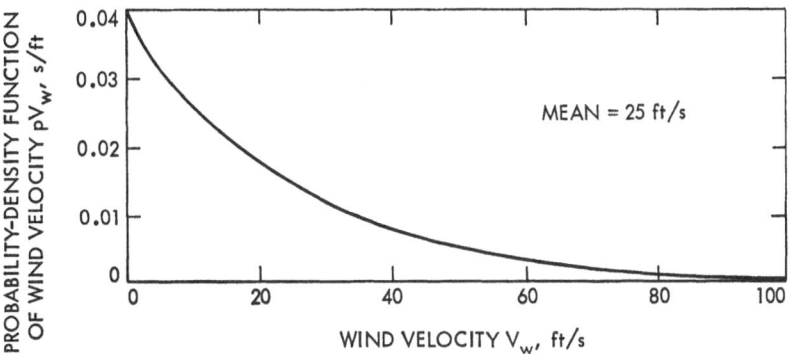

Fig. 4.   Mars surface winds probability-density function.

No deformation beyond Δ_s is allowed. Any remaining energy must be absorbed by the impact limiter. The impact limiter is assumed to have a similar idealized deformation characteristic with its own appropriate values for b_l and Δ_l.

The lander impacts with a kinetic energy of $(mv_t^2)/2$. The impact energy absorbed by the planet is $b_s A_c \delta_s$ where

b_s = soil bearing strength

A_c = effective contact area between lander and surface

δ_s = deformation depth.

If $[(mv_t^2)/2] < b_s A_c \Delta_s$, no energy is required to be absorbed by the lander because the soil layer has not 'bottomed out' yet. If $[(mv_t^2)/2] > b_s A_c \Delta_s$ the energy absorbing capacity of the surface has been exhausted and the remaining kinetic energy must be absorbed by the lander. The effective velocity required to be attenuated by the lander is then merely the velocity associated with this excess kinetic energy. The excess kinetic energy is

$$\frac{mv_e^2}{2} = \frac{mv_t^2}{2} - b_s A_c \Delta_s$$

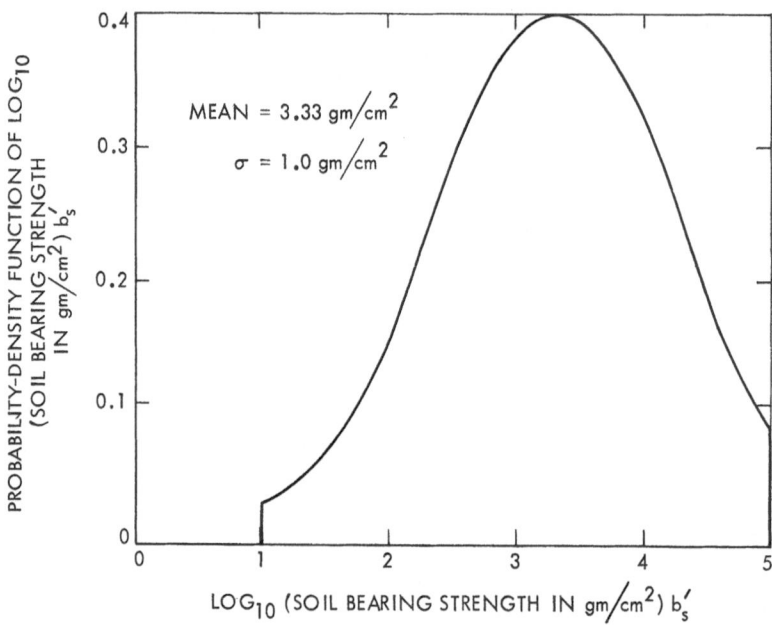

Fig. 5.   Mars surface-bearing-strength probability-density function.

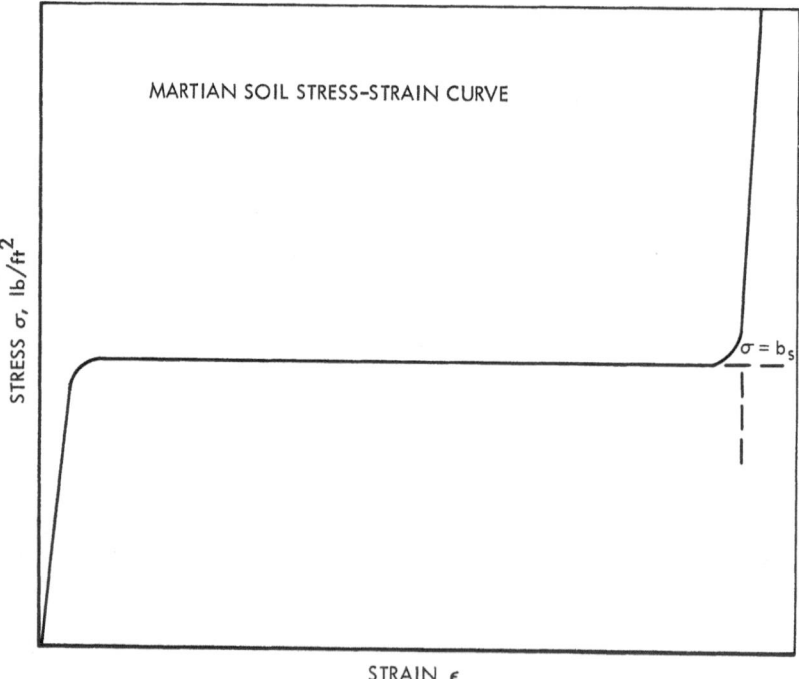

Fig. 6.   Mars surface stress-strain curve.

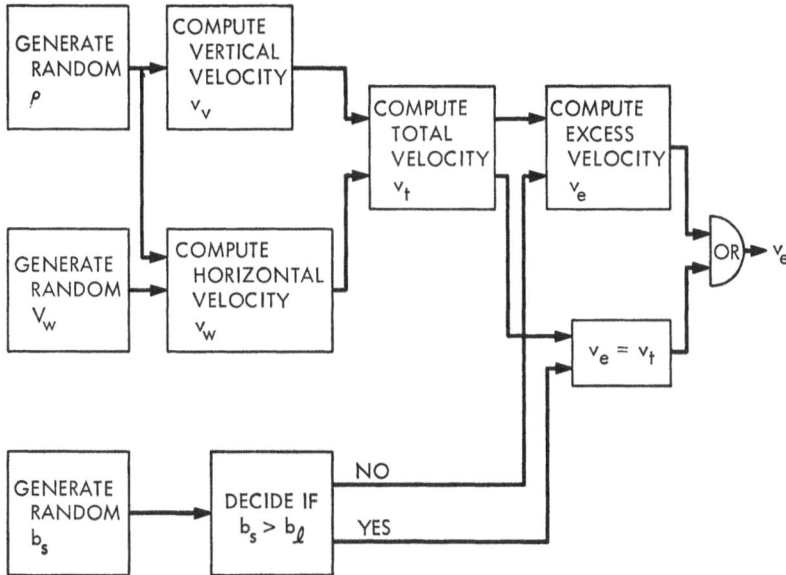

Fig. 7. Flow chart for Monte Carlo computer simulation of rough lander.

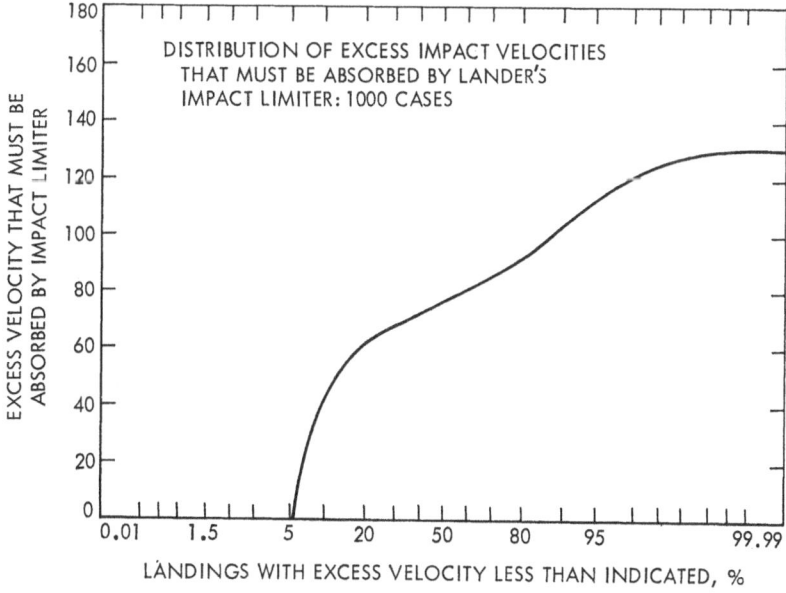

Fig. 8. Distribution of impact velocities that must be absorbed by impact limiter.

Fig. 9.   CSAD lander after drop-test showing instrument boom and anemometer deployed.

so that the excess velocity that must be attenuated is

$$v_e = \left[v_t^2 - \frac{2 b_s A_c A_s}{m} \right]^{1/2}.$$

It is possible that the bearing strength of the soil is greater than the bearing strength of the impact limiter, i.e., that $b_s > b_l$. In this situation, it is assumed that all the kinetic energy must be attenuated by the impact limiter, i.e., that $v_e = v_t$. If a maximum g-level has been established for the lander payload, the characteristics of the impact attenuator can be determined. The acceleration produced by the crushing of the impact limiter

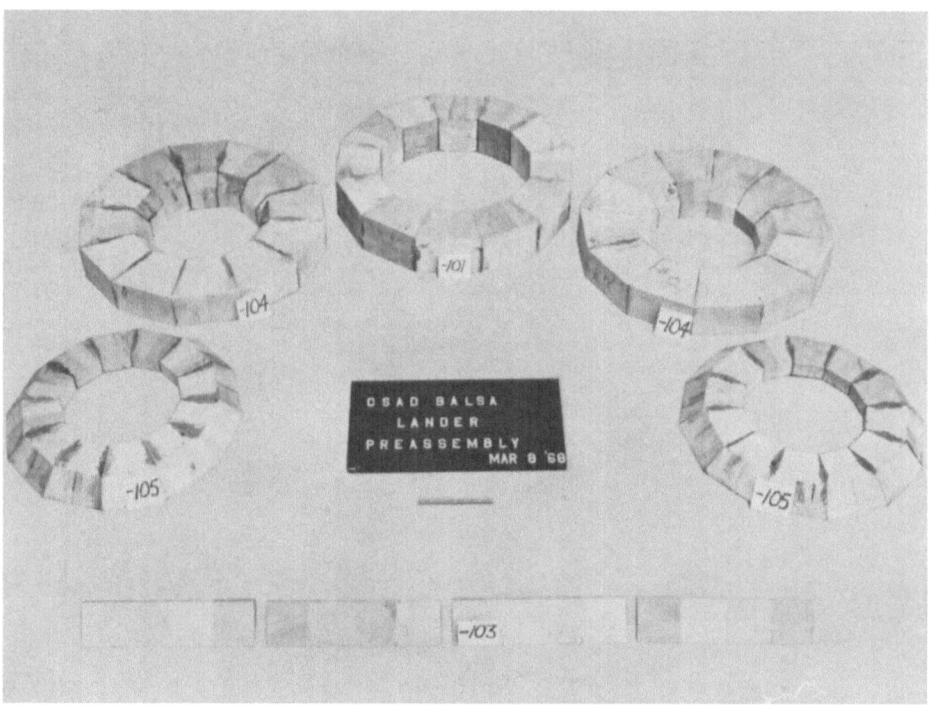

Fig. 10. Impact limiter balsa rings.

will be assumed to be the crushing force (bearing strength of limiter times contact area) divided by the lander's mass; therefore the required limiter bearing strength is

$$b_l \leqslant \frac{g_e g_{\max} m}{A_c}$$

where

g_e = acceleration of gravity at surface of earth

$g_{\max}$ = maximum allowable lander payload g-level.

The limiter thickness is determined by the amount of energy required to be absorbed by the limiter, i.e., by the excess velocity that has not been absorbed by the ground. This energy relationship can be written

$$\frac{m v_e^2}{2} = b_l A_c f_c \Delta_l$$

where f_c is the fraction of the limiter that is allowed for crushing. This fraction is really a sort of safety factor. The required impact limiter thickness is then

$$\Delta_l \geqslant \frac{m v_e^2}{2 b_l A_c f_c}.$$

Fig. 11.   Impact limiter rings installed on lander.

A Monte Carlo program for statistically evaluating the likely values of excess velocity that must be absorbed by the impact limiter has been written. It follows the flow chart of Figure 7. The results of this computer simulation are shown in Figure 8. The impact limiter can be designed to accommodate any value of excess velocity that a confidence limit sets. The limiter can be designed to accommodate the mean expected v_e, the 95%-worst-case v_e, the 3σ-worst-case v_e, or any desired confidence limit.

The constants that were used in this computation were

$$m = 3.6 \text{ slugs}$$
$$g_m = 12.2 \text{ ft/s}^2$$
$$g_e = 32.2 \text{ ft/s}^2$$
$$C_{D_v} = 0.5$$

$Av = 700 \text{ ft}^2$

$A_s = 350 \text{ ft}^2$

$h = 100 \text{ m}$

$\Delta_s = 6 \text{ in.}$

$b_l = 10^3 \text{ lb/in.}^2$

$A_c = 50 \text{ in.}^2$

For these assumed values of the constants, and for the assumed random variable distributions in Figures 3 through 5, the excess velocity distribution of Figure 8 indicates that the upper 95% excess velocity limit is 112 ft/sec. If the impact limiter is designed to absorb the energy of a 112-ft/sec impact against an unyielding surface, it will be more than adequate to decelerate the assumed payload under 95% of the landing conditions encountered.

Following the design, construction, test, and analysis of a hardwood mockup lander, the CSAD effort undertook the development of a functional Feasibility Model lander. This Feasibility Model was the same shape as the hardwood mockup, but internally was a functional machined aluminum chassis containing numerous functional flight-type electronic and mechanical subsystems.

Figure 9 shows the Feasibility Model lander. The impact limiter on the Feasibility Model lander is similar to the hardwood mockup limiter. It is made of laminated sections of balsa. The wood grain is oriented so that most of the impact is taken end-on. The specific energy absorption capability of the balsa is highest when crushed end-on.

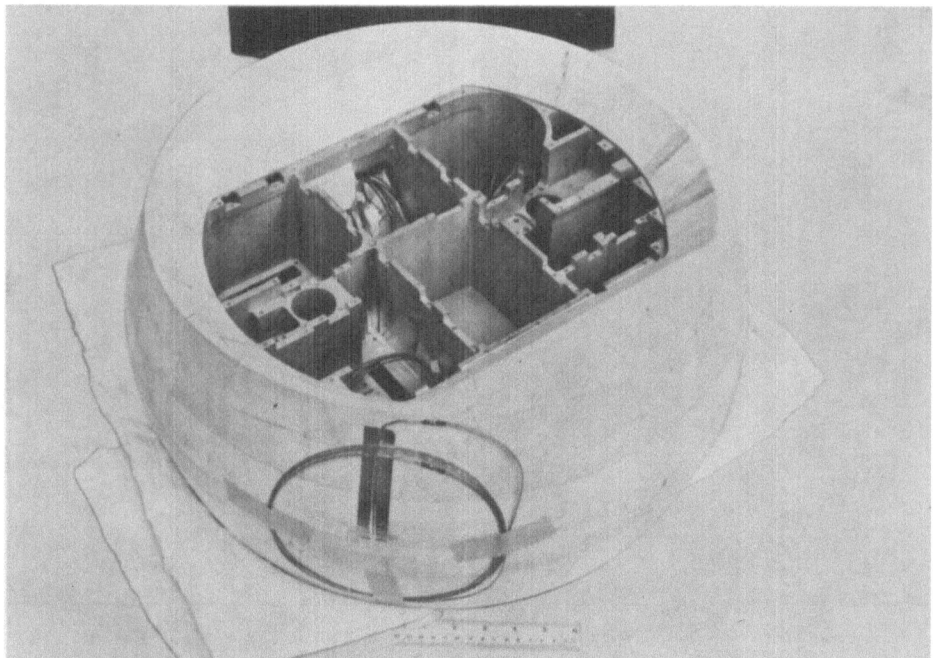

Fig. 12. Impact limiter installed and machined to desired contour.

The balsa is first dried, then cut into $3 \times 3 \times 6$-in. rectangular parallelepipeds. These are then weighed and segregated according to density. Only the blocks with a density between 6.5 and 7.5 lb/ft^3 are used in the limiter construction.

The blocks are cut into trapezoids, which are arranged into rings (Figure 10). The main outer limiter is constructed from these rings, the center ring being machined to the inside diameter of the lander. The upper and lower rings are stepped to the lander inside diameter and provide the necessary overlap. After the rings are installed (Figure 11), the entire limiter is machined to the desired contour (Figure 12).

The upper and lower covers have balsa on them which is installed following a similar procedure. The covers are attached to the lander chassis with screws and can be removed without disturbing the limiter.

The impact tests of the Feasibility Model lander were conducted at Goldstone: the first was on the dry lake bed; the second was on the asphalt runway. The first test was representative of a nominal Mars landing surface although the impact velocity was about 120 ft/sec. The impact limiter was hardly crushed, and most of the impact energy was absorbed by the ground. The limiter crushed about 0.1 in. and developed several longitudinal cracks.

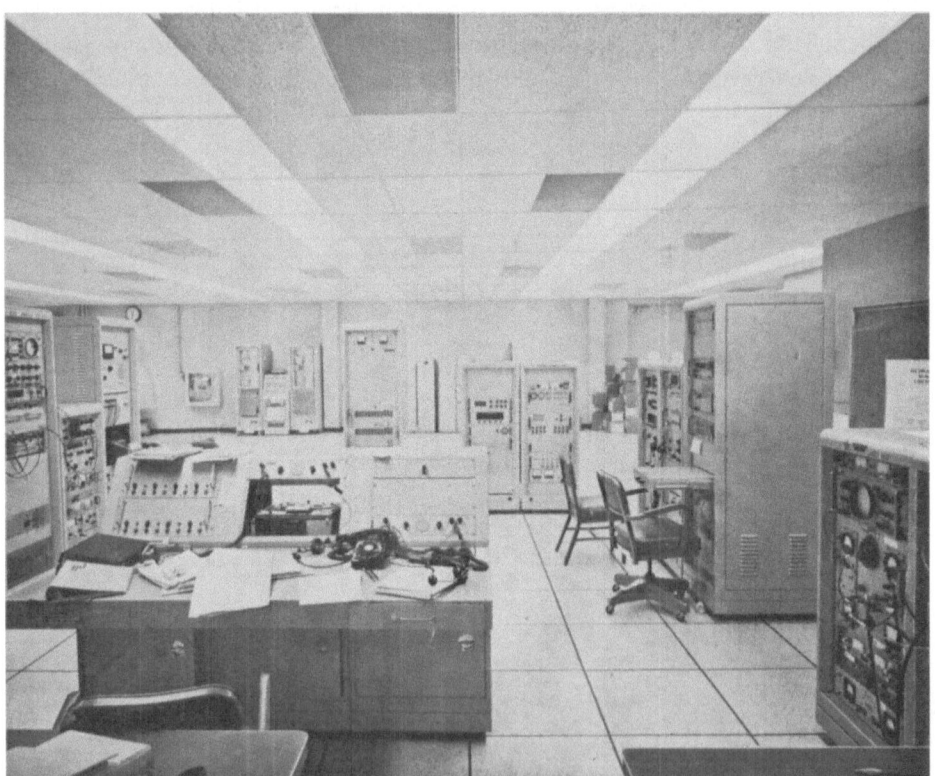

Fig. 13.   CSAD system test complex at the Jet Propulsion Laboratory.

Immediately after landing, the lander sequenced itself through a landed mission profile for 24 h. All subsystems performed as expected after the impact.

The lander was returned to JPL for more in-depth system testing which can only be done with the CSAD system text complex (Figure 13). In addition to functional testing, standard inspections were conducted on the equipment at the system level.

Inspection of the impact limiter following the drop on the dry lake bed revealed (1) some crushing and splintering of the balsa which was bonded to the cover plate that sustained contact during impact, and (2) a crack approximately $\frac{1}{16}$ in. wide transverse to and running the full depth of the peripheral section of the limiter. This crack widened to approximately $\frac{1}{8}$ in. during the week of the system tests following the drop. There was a 15° sector in the vicinity of the crack where the bond joint between the chassis and limiter had separated.

Before the second sterilization cycle and drop-test on the asphalt road, the following repairs were made: (1) all the balsa on the cover was replaced, and (2) the crack and bond joints were repaired by first filling the voids with Epoxic Shell 112 adhesive, and then putting the entire portion of the limiter in compression by means of an external band while the adhesive hardened.

Fig. 14. Dry lake drop-test impact site.

At this time, there is no way of knowing the condition of the bond joints before or after sterilization other than a visual inspection of the exposed edges.

Improved and controlled techniques for applying the adhesives are required. In addition, improvement of the adhesives and a better understanding of what is optimum in the preparation of the chassis surface are required. Finally, some inspection techniques such as the use of X-rays or, perhaps, ultrasonics should be developed.

The repaired lander was then taken back to the desert for the second impact test, which was conducted on a hard, flat, unyielding asphalt surface; again the impact velocity was 120 ft/sec.

After the impact, the lander was successfully taken through the steps of its mission simulation and again returned to JPL for extensive testing and inspection.

Figure 14 shows the impact crater on the dry lake which absorbed the energy on landing. Figure 15 shows no evidence of cratering; here the energy was absorbed by the balsa.

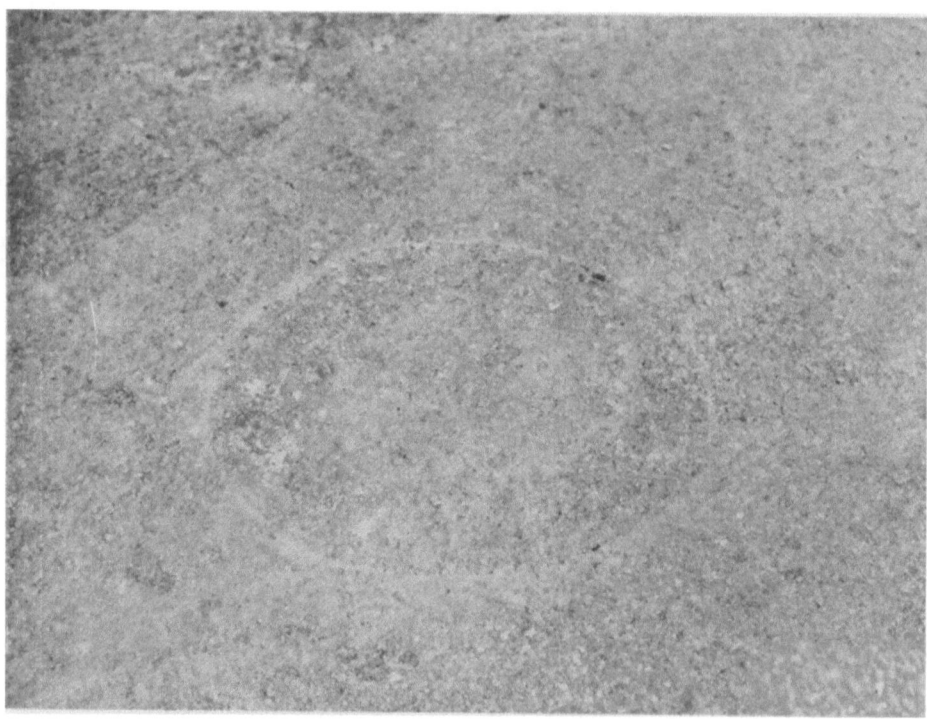

Fig. 15.   Asphalt road drop-test impact site.

THE DEEP SPACE NETWORK*

An Instrument for Radio Navigation for the Mariner Mission to Mars, 1969

J. P. FEAREY and N. A. RENZETTI

Jet Propulsion Laboratory, California Institute of Technology, U.S.A.

Abstract. The Deep Space Network, as it is configured for the mission to Mars, is described in general terms with particular attention to the metric data capability. The science objectives are described in relation to the requirements for navigational accuracy and scan platform pointing. The methods for validating the raw metric data as supplied to the large-scale, general purpose digital computers which are used for computing orbits are delineated. A brief discussion of the evolution of the astrodynamical constants used in the orbital computation process is presented. Finally, current in-flight status of the mission is reviewed and it will include the latest prediction based on flight data of the navigation accuracy which will be achieved upon encounter with the Planet Mars during the period 31 July – 5 August 1969.

I. The MM69 Mission

The Jet Propulsion Laboratory of the California Institute of Technology, as a prime contractor to the National Aeronautics and Space Administration, exercises the project responsibility for the 1969 Mariner Mission to Mars. The Jet Propulsion Laboratory is also responsible for the development and operation of the Deep Space Network, which is a world-wide precision data acquisition system, the major functions of which are

(1) To provide engineering and scientific data acquisition capability from spacecraft in interplanetary and planetary missions,

(2) To permit control of the spacecraft,

(3) To provide precision navigational information required to achieve the scientific and engineering objectives of the mission.

The MM69 mission primary objective is to make exploratory investigations of Mars which will set the basis for future experiments, particularly those relevant to the search for extra terrestrial life. The secondary objective is to develop technology needed for the succeeding Mars missions.

The mission design is based on an integrated package of five complementary planetary experiments whose purpose is to extend the information acquired by the Mariner flyby mission to Mars in 1965 and to shed light on some of the questions raised by that mission. (There is a sixth experiment, in Celestial Mechanics, which is concerned with the investigations of the astrodynamical constants of the solar system.)

The Mariner flyby mission to Mars 1964 TV instrument indicated that the planet had a cratered surface somewhat like that of the moon. The S-band Occultation

* This paper presents the results of work carried out at the Jet Propulsion Laboratory, California Institute of Technology, under Contract No. NAS7-100 sponsored by the National Aeronautics and Space Administration.

Experiment indicated that the surface atmospheric pressure was at the low end of the previously postulated bounds (about 6 mb).

As a consequence, the Mariner flyby mission to Mars 1969 scientific package is concentrating on the physical, chemical, and thermal properties of the planet's surface and atmosphere, with the goal of determining the environment on and near the surface of the planet in which biological activity might have, may, or might eventually exist.

Table I-1a through f presents a summary of the MM69 experiments. It is taken from a paper delivered to the AIAA by H. M. Schurmeier, the Mariner Mars 1969 Project Manager for JPL [1].

TABLE I-1a

Experiment	Visual imaging
Principal investigator	R. B. Leighton
Condensed objective	Physiographic survey of most of surface at better than Earth-based resolution, categorize light and dark areas; obtain data at higher resolution for study of origin, history.
Instrumentation	Wide angle (11×14 deg) TV camera red/green/blue filters 704×945 elements; narrow-angle (1.1×1.4 deg) TV camera blue band filter.
Desired mission characteristics	(1) Camera pointing at planet during approach to give complete planet coverage at better resolution than available from Earth.
	(2) Coverage of maximum number of light and dark surface features as well as the southern polar cap.
	(3) Capability for late selection of Mars longitude of closest approach.
	(4) Capability to update camera pointing direction just prior to encounter.
	(5) Low approach speed and low altitude of closest approach for hi-resolution pictures.
	(6) Sufficient overlap between subsequent low resolution pictures to include hi-resolution frame, and provide color overlap.
Navigational accuracy implications	$3\sigma_B = 500$ km (to control picture sequence)
	$3\sigma_{PB} \approx 300$ to 400 km (to achieve a definite control of swath latitude)

TABLE I-1b

Experiment	Surface temperature
Principal investigator	G. Neugebauer
Condensed objective	Temperature map along TV swath; surface cooling curve; dark-side and polar cap temperatures.
Instrumentation	Bimetallic thermopile detector 8–12 and 18–25 μ, boresighted with TV. 0.7×0.7 deg field of view.
Desired mission	(1) Coverage of maximum number of light and dark surface features as well as the southern polar cap.
	(2) Boresighted with TV cameras to provide maximum correlation capability.
	(3) Scan trace must be nearly perpendicular to the terminator at terminator crossing and not on polar cap.
	(4) Significant amount of dark side viewing.
Navigational accuracy implications	$3\sigma_B \approx 500$ km
	$3\sigma_{PB} \approx 300$–400 km (see TV)

TABLE I-1c

Experiment	Infrared spectroscopy
Principal investigator	G. C. Pimentel
Condensed objective	Detection and measurement of polyatomic atmospheric components, including life-related molecules; determine variation with locale; take surface composition and other data.
Instrumentation	IR Spectrometer, two detectors, 1.9–14.3 μ. One radiation cooled the other Joule Thompson cooled 2.1×0.06 deg field of view.
Desired mission	(1) Equatorial scan trace with maximum number of light and dark surface features.
	(2) Slant range should be small.
	(3) Approach speed should be low to improve spacial resolution.
	(4) Scan trace should follow closely the TV scan trace to obtain data on the upper atmosphere and still provide correlation capability with TV pictures.
	(5) Significant amount of dark side viewing at not too oblique angles.
	(6) Scan trace should cross terminator close to equator.
	(7) Data recording interval should extend from above atmosphere on lighted limb to above atmosphere on dark limb.
Navigational accuracy implications	$3\sigma_B \approx 500$ km
	$3\sigma_{PB} = 300\text{–}400$ km (see TV)

TABLE I-1d

Experiment	Ultraviolet spectroscopy
Principal investigator	C. A. Barth
Condensed objective	Detection and scale height of atoms, ions, and molecules in upper atmosphere; Rayleigh scattering and surface reflectivity.
Instrumentation	UV Spectrometer, two detectors, 1100–4350 Å. 0.23×2.29 deg field of view.
Desired mission	(1) Range must be as small as possible, particularly at the lighted limb crossing.
	(2) Slit should be aligned to within $3°$ of local horizontal plane at 100 km altitude above lighted limb with 90% confidence.
	(3) Approach speed should be low to improve spacial resolution.
	(4) Scan trace should be as near to the subsolar point at near limb crossing.
	(5) Scan trance should follow closely the scan traces of the other instruments to provide correlation capability.
	(6) Data recording interval should extend from above the atmosphere on the lighted limb to above the atmosphere on the dark limb.
Navigational accuracy implications	$3\sigma_{PB} = 300$ km.
	$3\sigma_B \simeq 500$ km (timing of recording dark side data)

II. The Spacecraft for the Mariner Mission 1969

The Mariner Spacecraft-1969 Model is the latest member of a family of spacecraft which go back to Mariner II mission to Venus 1962 [2]. Like its predecessors, this Mariner is an almost fully automatic spacecraft capable of operating from launch

TABLE I-1e

Experiment	S-band occultation
Principal investigator	A. J. Kliore
Condensed objective	Pressure and density profiles of atmosphere, electron density profile of ionosphere, radius of solid planet, at four locations.
Instrumentation	Existing S-band (13-cm) radio link, specialized ground equipment.
Desired mission	(1) Radio beam tangent points at entrance and exit of occultation should be widely separated in absolute value of latitude.
	(2) Absolute value of difference between solar zenith angle at tangency points at entrance and exit should be large.
	(3) Range to tangent points should not be large.
	(4) Elevation angles at Deep Space Stations should be high.
Navigational accuracy implications	$3\sigma_B = 3\sigma_{PB} = 1$ km (Design goal or better at $E + 15$ days).

TABLE I-1f

Experiment	Celestial mechanics
Principal investigator	J. D. Anderson
Condensed objective	Improvement in accuracy of planetary and solar system parameters and ephemerides. Primarily the AU and Mars/Sun mass ratio.
Instrumentation	Existing S-band tracking and ranging capabilities.
Desired mission characteristics	(1) Periapsis of spacecraft trajectory should be close to Mars.
	(2) Low approach speed to allow maximum trajectory bending.
	(3) Minimum time of occultation.
	(4) Minimum spacecraft produced orbit perturbations during encounter.
	(5) Tracking line-of-sight (Earth-to-spacecraft) to lie within aerocentric trajectory plane of motion.
Navigational accuracy implications	Places requirements on quantity and quality of radio tracking data samples. No navigation accuracy requirements.

through cruise, encounter, encounter play back, and post flyby cruise without direction from earth with the exception of the trajectory correction maneuvers whose parameters must be based on knowledge of the orbit which, in turn, is derived from DSN metric tracking data.

This Mariner exhibits the results of technological improvement over its predecessors in weight, power, quantity and degree of sophistication of scientific instrumentation and functional capability.

This spacecraft is composed of 15 major subsystems. The most interesting and significant new features are those which enhance the information acquisition and processing capability of the spacecraft. (See Figures II-1 and 2.)

1. *Scan Platform*

All five of the planetary scientific experiments are mounted on a 2 deg of freedom platform. The platform has a planet tracker which keeps the platform aimed at the center of the lighted portion of the planet during the last three days before planetary

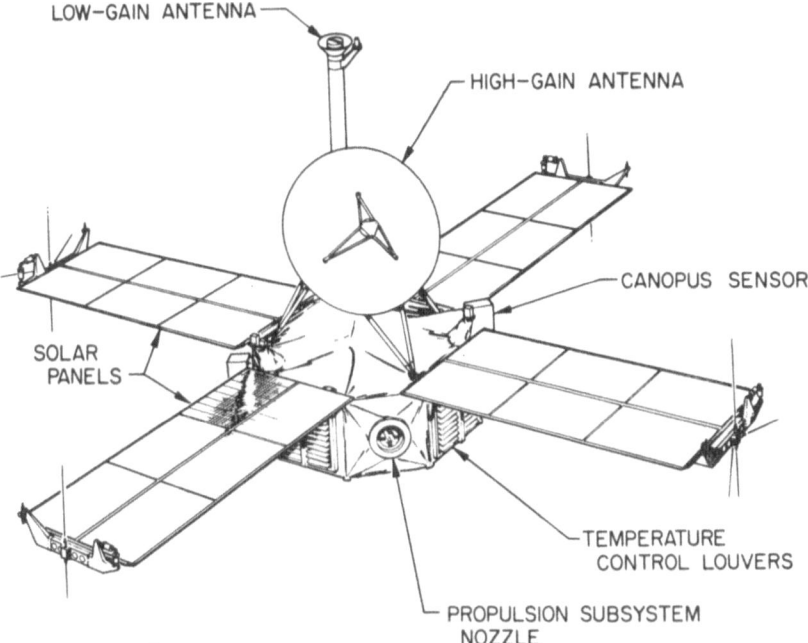

Fig. II-1.

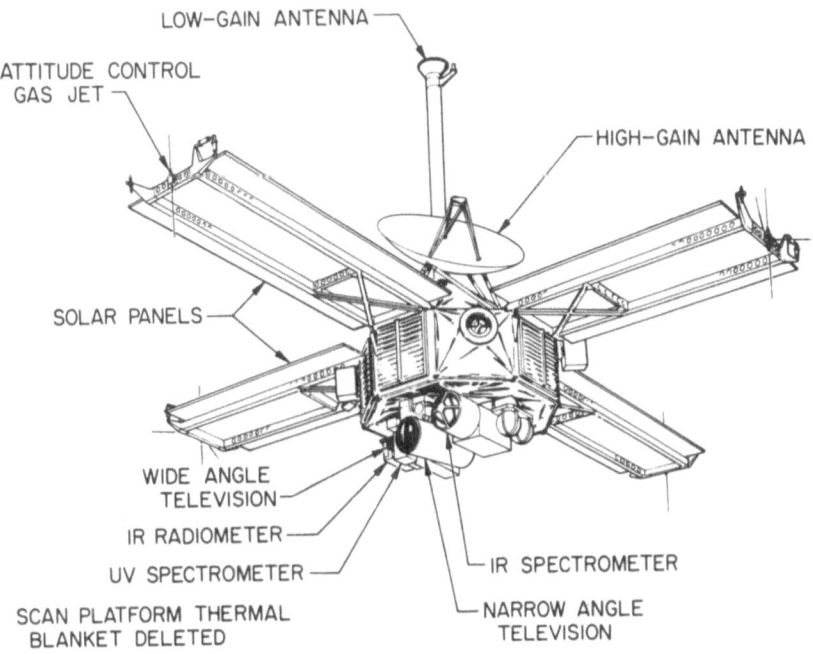

Fig. II-2.

encounter. In addition, it has a stored program capability for several scan patterns for use during the near encounter data gathering.

2. *Analog-Digital Recorders*

TV data is recorded directly on magnetic tape where it is stored for later analog to digital conversion and playback. The TV data is also digitized in real time and recorded on digital tape for later playback. The contents of the analog tape may be digitized and recorded on digital tape.

3. *High Data Rate Communication Capability*

Increased spacecraft power and an improved high gain antenna and antenna pointing, plus block encoding of the data will allow a 16.2K bit data rate at encounter. This will actually increase the reliability of information transfer as it will allow several replays of the recorded encounter data after flyby.

4. *Programmable Computer Controller and Sequencer*

The Mariner 69 Central Computer and Sequencer (CC&S) is a programmable special purpose computer processor which can be reprogrammed in flight. This increases spacecraft flexibility and affords flight controllers additional options in the event of a non-standard mission.

The spacecraft is three-axis-stabilized during its journey. The attitude control subsystem points the spacecraft roll axis toward the sun, and control about this axis is achieved by acquiring and locking onto the star Canopus. Power is supplied by photovoltaic cells on the solar panels and also by a battery that is used from liftoff until sun acquisition, during trajectory correction maneuvers, and to provide shared power during peak loads. The radio frequency and flight telemetry subsystems provides telemetry to earth. Earth-based space communication stations provide angle, doppler, and ranging measurements for orbit determination.

Trajectory correction maneuvers are provided by the propulsion subsystem and controlled by the CC&S acting on stored data inserted by ground command through the flight command subsystem. The power subsystem provides master timing which the CC&S uses to clock out events in the sequence.

The spacecraft has the capability to perform a maximum of two in-flight trajectory correction maneuvers should these be required to deliver the spacecraft to the proper aiming zone near Mars.

During the period when science data are being collected at Mars, the data automation subsystem will govern the data gathering activities of the scientific instruments, and the data storage subsystem will place the data on the appropriate tape recorders.

III. Mission Design

The mission was conceived as using two Mariner-class spacecrafts which would conduct their scientific operations during the brief period in which they are close to the planet. Requirements for improved operational flexibility for the encounter, and the storage and return to earth of vastly increased quantities of planetary data, were the major factors dictating the spacecraft design.

Some of the desirable characteristics for trajectory selection were – short flight time; short communication distance; slow pass by Mars; favorable encounter geometry for the various experiments; a late launch to better match the spacecraft development schedule to the availability of funds; a long launch period to provide time to correct problems prior to the second launch should they occur with the first launch; near optimum near-earth coverage by the Tracking and Data System; and a direct ascent mode for the launch to enhance the launch vehicle reliability. Many of these are competing characteristics. The design process that led to the selection of the specific trajectories is described elsewhere. [12]

The mission design task centered around evolving a design that satisfies to the maximum degree practical the requirements of the selected experiments. This had to be done within the constraints imposed by the solar system characteristics, the launch vehicle capability, spacecraft engineering considerations, tracking and data system limitations, operational considerations, and resource limitations. The specific characteristics to satisfy each of the six experiments were numerous, varied, and often competing. Several of the key ones for each experiment are shown in Table I-1.

The resulting flight path design consisted of Type I trajectories (with an arc of travel about the sun less than 180°) with a flight time of 5 to 6 months; arrival at Mars between 29 July and 15 August 1969; and launch from Cape Kennedy between 24 February and 8 April, GMT.

During the early portion of the launch period, the trajectories had negative geocentric declinations, requiring injection azimuths far more southerly than can be safely flown from Cape Kennedy. The use of powered flight yaw maneuvers (or 'dog leg' trajectories) permitted the vehicle to fly around the corner in the safe launch corridor and meet these injections requirements using a single burn, direct ascent powered flight trajectory.

Planetary quarantine policy prescribed that the probability of contaminating Mars be less than 6×10^{-5} for both flights. The initial aiming point was biased away from the desired aiming point to ensure maintaining biological integrity of the planet. The aiming points were brought close to the planet by the midcourse maneuver. The aiming zones and their relations to Mars and several of the constraints are shown in Figures III-1, 2, 3. The preferred zones for the two spacecraft shown as windows in the shaded mission zone, indicate that one mission is to make an equatorial pass, while the other permits the instruments to view the southern latitude and the edge of the south polar cap. Figures III-1, 2, 3 show the relative size of the aiming zones for previous U.S. planetary missions compared with that for MM69.

IV. The Tracking and Data System (T&DS) for MM69

A. THE MM69 T&DS

The T&DS for a mission is a subset of the facilities of the Air Force Eastern Test

TABLE III-1

	Atlas/Centaur		Midcourse	
	Aiming point	Determined from actual data[a]	Aiming point	Determined from actual data[a]
Mariner VI				
$B \cdot T$ (km)	17 000	3 619.95	7452	7 679.14
$B \cdot R$ (km)	− 13 000	− 12 909.08	− 643	− 370.30
Time of closest approach (GMT)	05:01:32	04:40:14.379	05:17:50	05:17:37.123
Mariner VII				
$B \cdot T$ (km)	2402	− 6777.44	6528	6421.24
$B \cdot R$ (km)	20 994	29 309.03	3440	3565.97
Time of closest approach (GMT)	04:31:32	04:48:22.005	05:01:09	04:59:14.300

[a] Best determination as of 22 April 1969

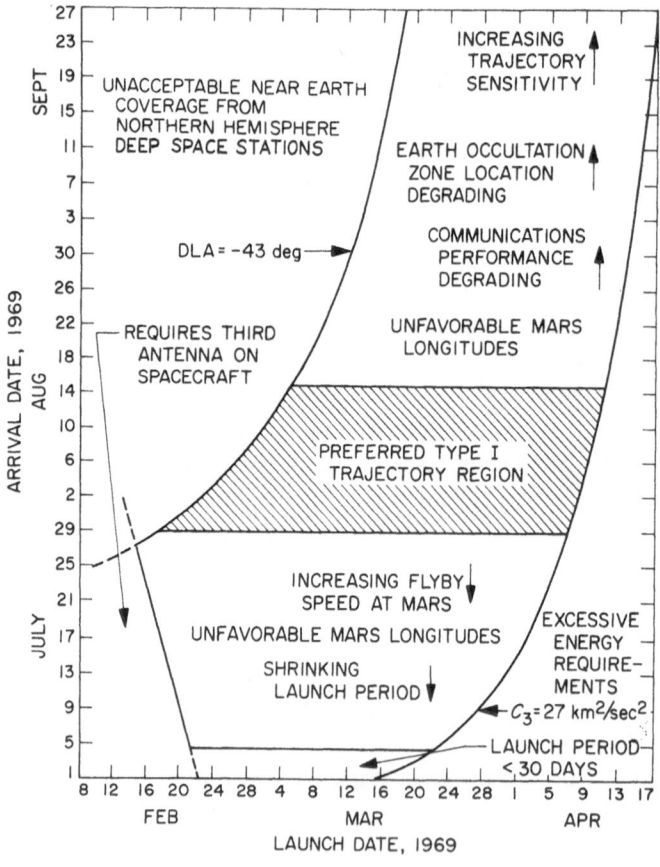

Fig. III-1.

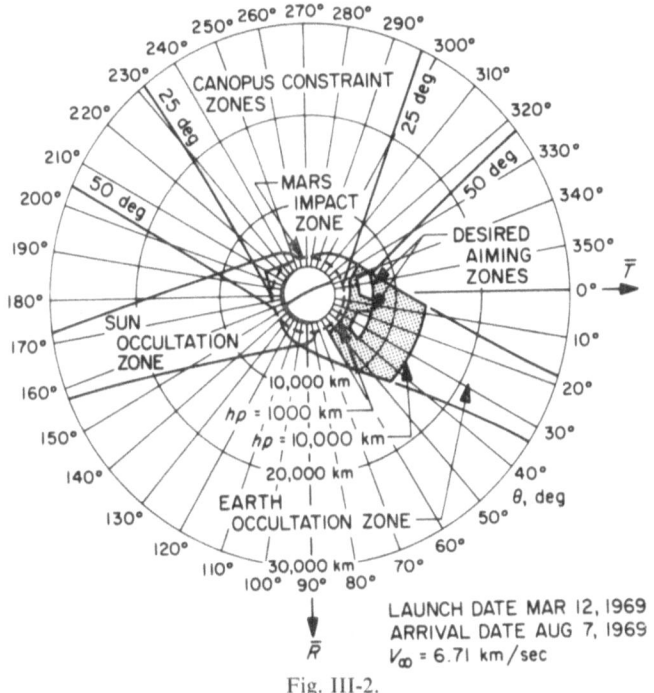

LAUNCH DATE MAR 12, 1969
ARRIVAL DATE AUG 7, 1969
V_∞ = 6.71 km/sec

Fig. III-2.

RELATIVE SIZES OF DIFFERENT MISSION AIMING ZONES

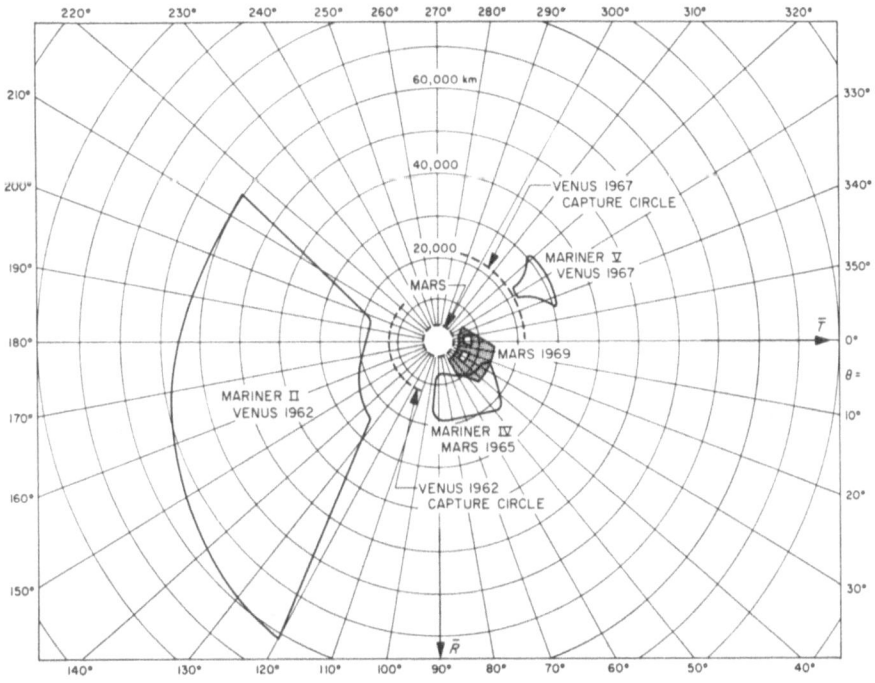

Fig. III-3.

Range (AFETR), Manned Space Flight Network (MSFN), NASA Communications (NASCOM), and Deep Space Network (DSN) which are brought together to meet the requirements of the mission. The T&DS deep space phase for Mariner 69 is illustrated in Figure IV-1. It comprises five Deep Space Stations (DSS), the Ground Communications Facility (GCF), the Space Flight Operations Facility (SFOF), and some other functions. Figure IV-2 shows the key characteristics of the DSN as configured for a mission.

DEEP SPACE T&DS CONFIGURATION FOR MISSION TO MARS IN 1969

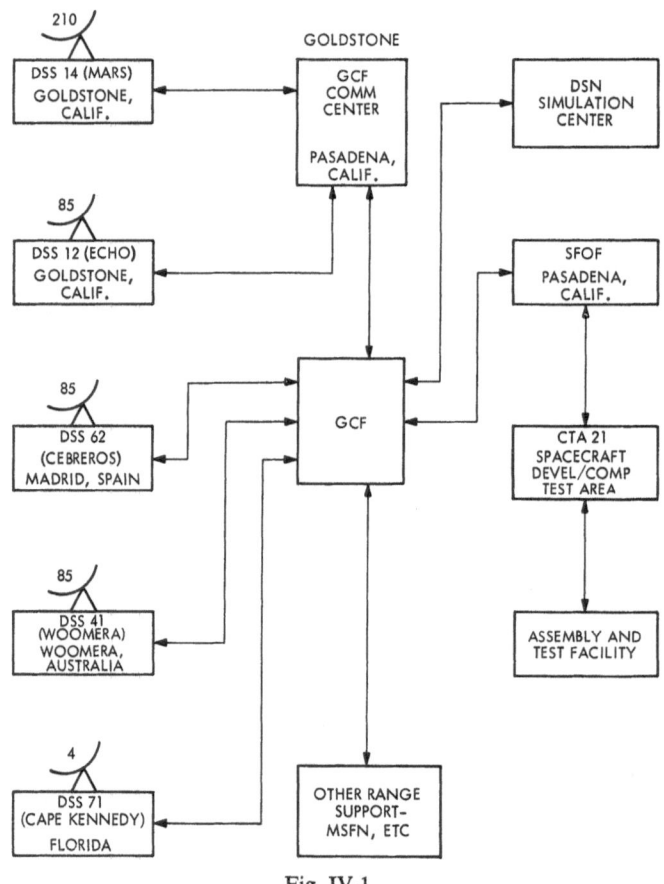

Fig. IV-1.

The capabilities of a typical DSS are illustrated in Figure IV-3. Included are metric data flow paths. The DSS radio system is a CW phase coherent mono pulse tracking system. The system has a liquid helium cooled ruby Maser preamplifier and a double superhetrodyned phase locked receiver. Table IV-1 gives some of the pertinent performance characteristics.

The GCF provides all of the communications throughout the DSN. Figure IV-4

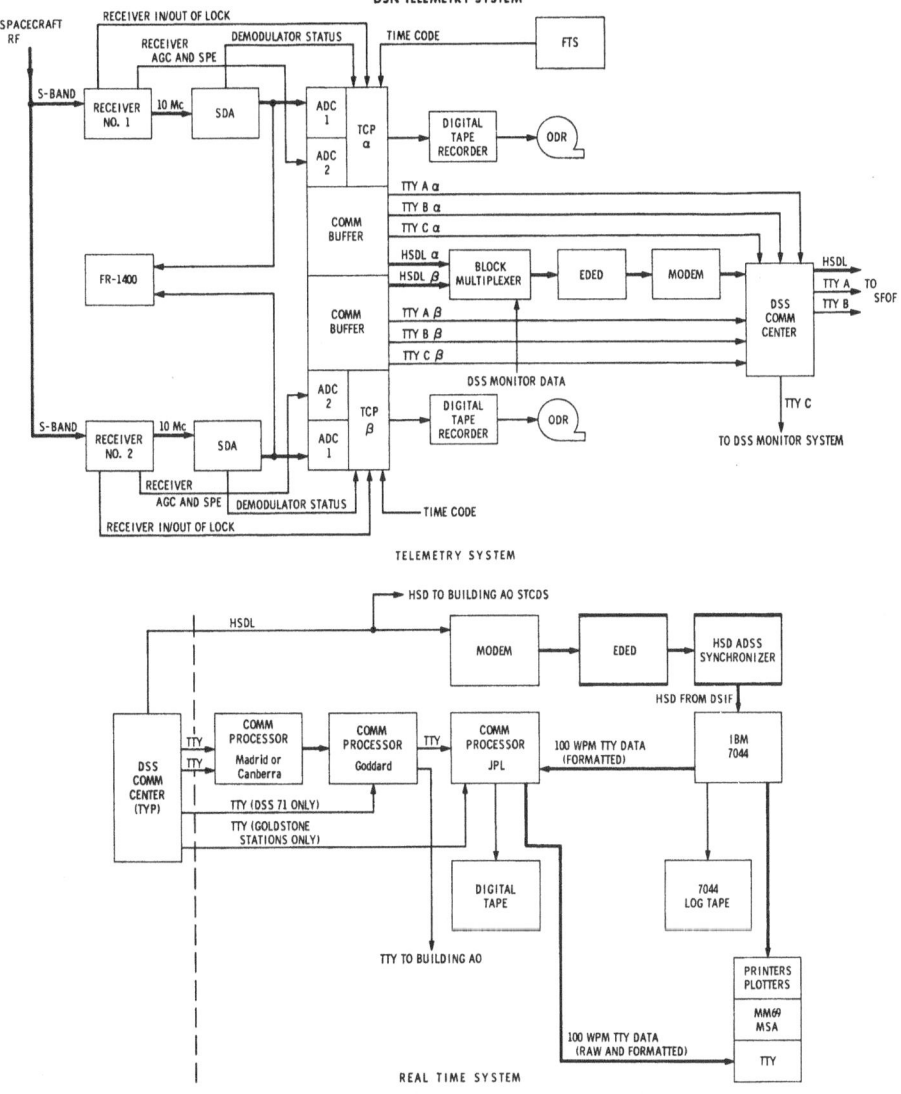

Fig. IV-2a.

shows the GCF configured for MM69. All teletype circuits are 100 words/minute full duplexed circuits. The High Speed Data Lines (HSDL) are 2400 bit capacity. NASCOM (Goddard Space Flight Center) provides all facilities of the GCF except the Goldstone to JPL (SFOF) circuits leased from a common carrier.

The SFOF provides communication and display equipment for network and mission control. The network also provides in this facility extensive computing equipment necessary for the flight path analysis and engineering analysis of the spacecraft data in real time. The support of MM69 is being provided by IBM 7044/7094 computer systems.

DSN TRACKING SYSTEM

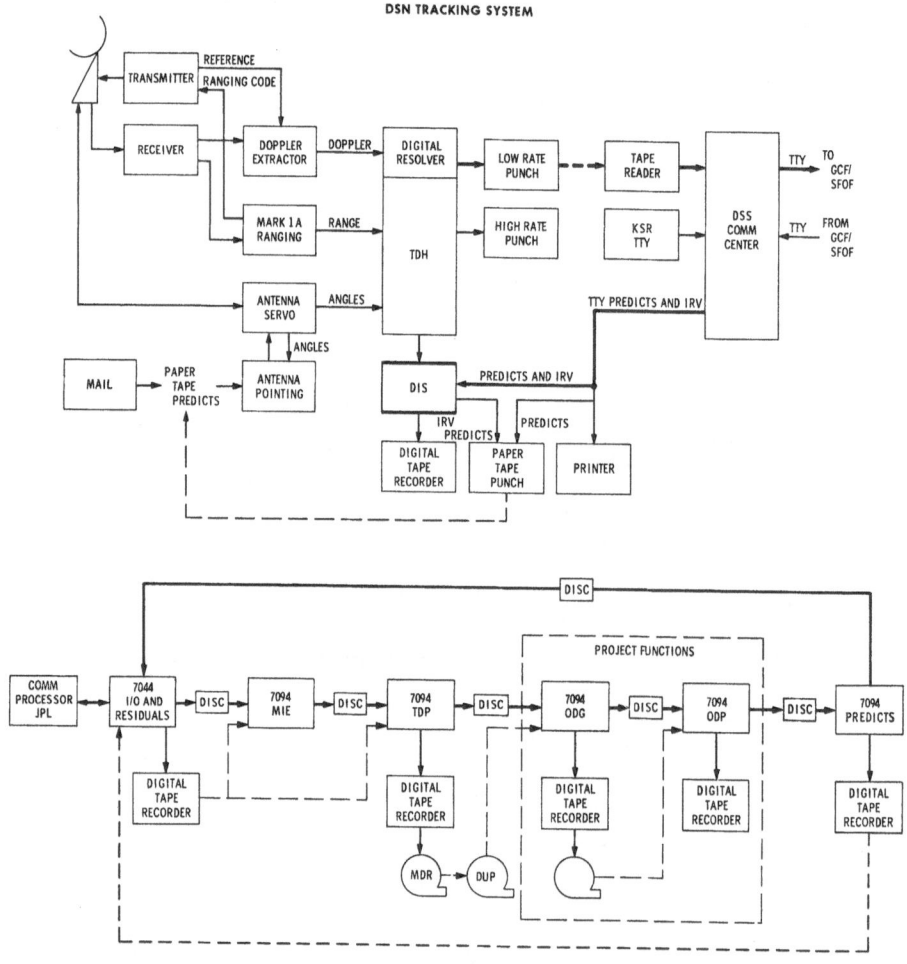

Fig. IV-2b.

B. DSN METRIC DATA

In discussing DSN metric data, it is convenient to distinguish two basic categories of data.

Raw Metric Data (RMD) is metric data as it comes from the measuring equipment, Figure IV-5. Three basic types of RMD will be discussed in this paper. They are Time, Doppler Frequency, and Ranging Time Delay.

Processed Metric Data (PMD) includes the final navigational parameters derived from processing the RMD. PMD also refers to data accuracies of RMD data types after they have been processed. This distinction is made since the processed metric data (PMD) accuracies are generally different from the RMD accuracies due to the nature of the data processing system. It is generally true that the PMD accuracies are the more important of the two in determining whether or not mission accuracy requirements can be met.

TYPICAL DSS BLOCK DIAGRAM

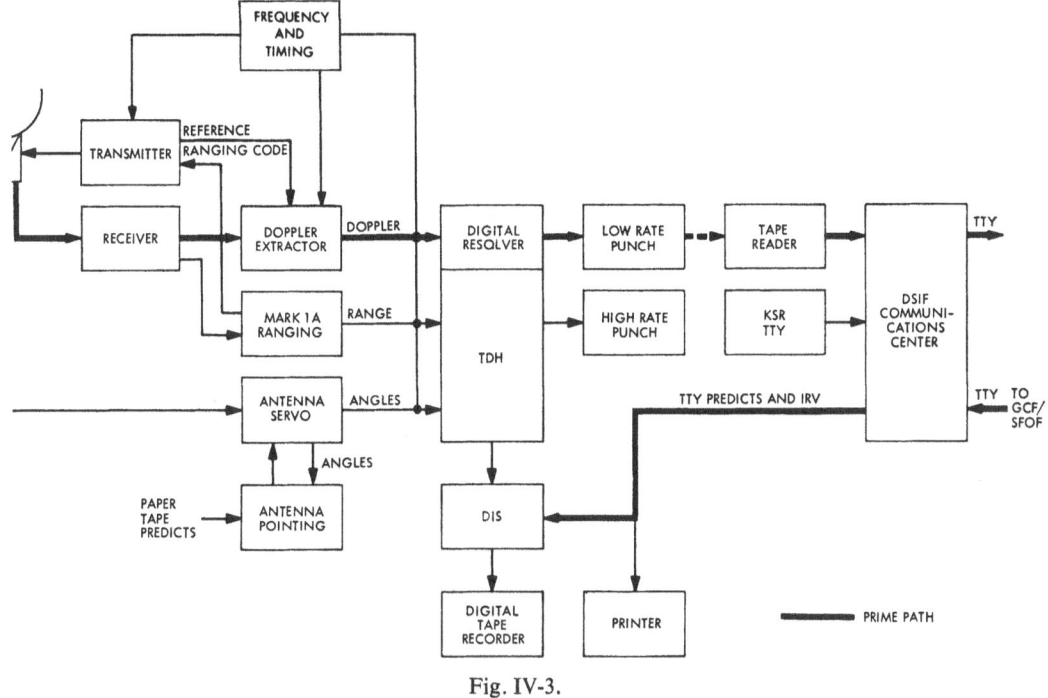

Fig. IV-3.

V. Raw Metric Data

A. FREQUENCY AND TIME

The three raw metric data types discussed herein are either direct time measurements or time derived phase measurements. Highly accurate, stable frequency and absolute time measurements are of fundamental importance to the tracking system and to the data reduction.

Figure V-1 shows the frequency and timing system of a typical DSN tracking station. The primary frequency sources for the station are two rubidium vapor frequency standards. These standards are continuously calibrated against each other, primarily for the purpose of detecting malfunctions or deteriorations in the frequency standards themselves. The remainder of the system as shown in Table V-1 is used for linking the local rubidium standards to the outside world.

1. WWV

The path labeled WWV is the standard worldwide time source for UTC (Table V-1) broadcast from Fort Collins, Colorado. The station clocks can be set to the WWV time pulses by means of an oscilloscope to an accuracy of approximately 5 msec. The chief source of error in this technique is the uncertainty in time delay between the station clock and the WWV transmitter due to variations in the height of the signal reflective layer of the ionosphere.

TABLE IV-1

S-Band microwave and antenna performance

	85-f antenna	210-f antenna
Antenna gain (db relative to isotropic, matched polarization)		
Transmit (2110–2120 MHz)	52.1 ± 0.9	59.8 ± 0.5
Receive (2290–2300 MHz)	53.5 ± 0.6	61.4 ± 0.3 diplexed 61.6 ± 0.3 listen only
Polarization	RCP or LCP or rotatable LIN (selectable)	RCP or LCP or rotatable LIN (selectable)
Beam width (half power)		
Transmit	0.36 ± 0.03 deg	0.145 deg
Receive	0.33 ± 0.03 deg	0.135 deg
Transmit power	10 kW	Single carrier 400 kW or two carriers 100 kW each
System temperature:		
Diplexed (10 kW for 85-ft; 20 kW for 210-ft)		
At 10° elevation	$\leqslant 65°C$	$\leqslant 55K$
At 25° elevation	—	$\leqslant 41K$
At zenith	$\leqslant 45K$	$\leqslant 35K$
Listen Only		
At 10° elevation		$\leqslant 38K$
At 25° elevation		$\leqslant 26 K$
At zenith		$\leqslant 20K$

2. *WWVL*

The WWVL Absolute Time reference system transmits a very low frequency (20 KHz) signal which travels over the surface of the earth, thus eliminating the major error source of WWV time. It also provides UTC time. The absolute time accuracy of this system is limited by the data reduction technique and uncalibrated phase errors in the station receiver to about 50 μsec. A further difficulty in the use of this system is that it is not always possible to acquire a WWVL signal at the station to lock up the VLF phase receiver.

3. *X-Band Lunar & S-Band Ranging Time Correlation*

The X-band Lunar Radar Absolute Time Synchronization is perhaps the most interesting, and potentially the most effective, time calibration system available to the station.

The system (Figure V-2) consists of a master clock transmitter at Goldstone, California, which transmits timing signals to the DSN tracking stations via the moon during the mutual moon viewing period between the master and the stations. The stations are able to detect their clock error with respect to the master. The master, in turn, is synched to the cesium standard at the NBS Laboratory in Boulder, Colorado by the same system.

NASCOM DSN/GCF CONFIGURATION FOR MM 69

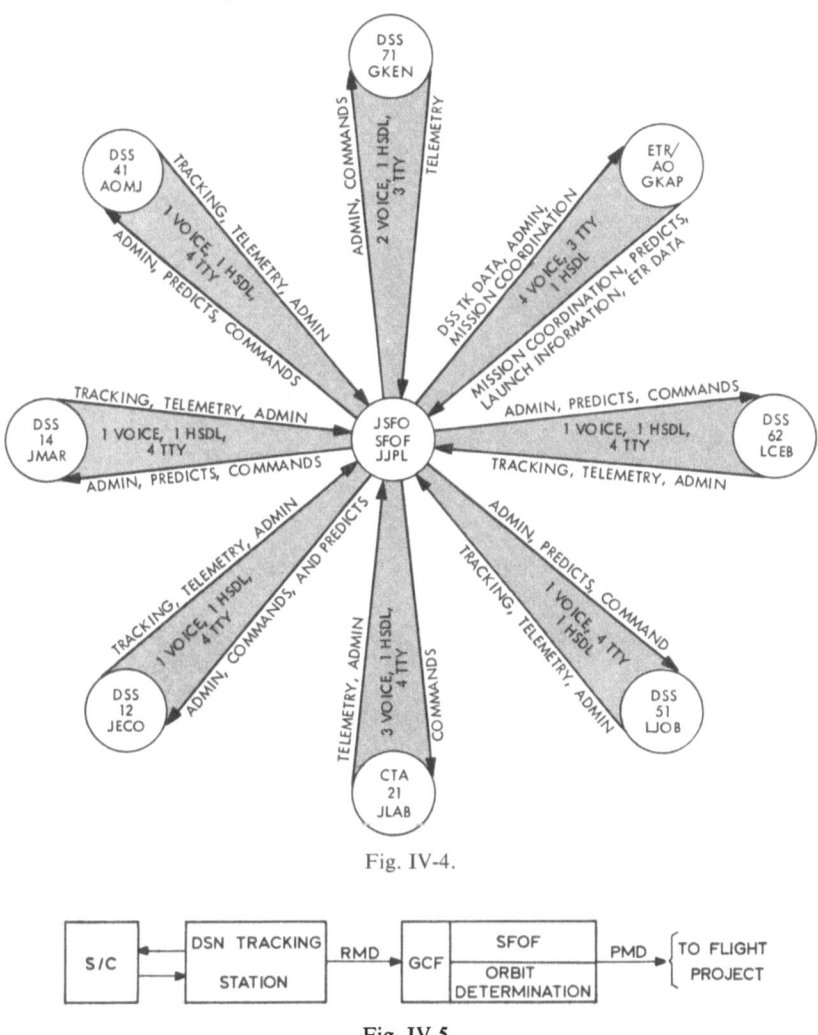

Fig. IV-4.

Fig. IV-5.

A radio signal, biphase modulated by a pseudo-random (PN) code which is controlled in time by the master clock at the transmitter, is sent via the front cap of the moon to the receiver. The receiver signal is compared by cross-correlation with a similar code generated locally and controlled in time by the receiver clock. Energy is detected at the receiver only when the two codes are in step or correlated. This condition can exist if the transmitted code is advanced in time by the transit time via the moon to the receiver and the clocks are perfectly synchronized. The necessary transmitter time advance is accomplished by a computer with time and lunar ephemeris inputs. Since the range is continuously changing, the code timing must be continuously

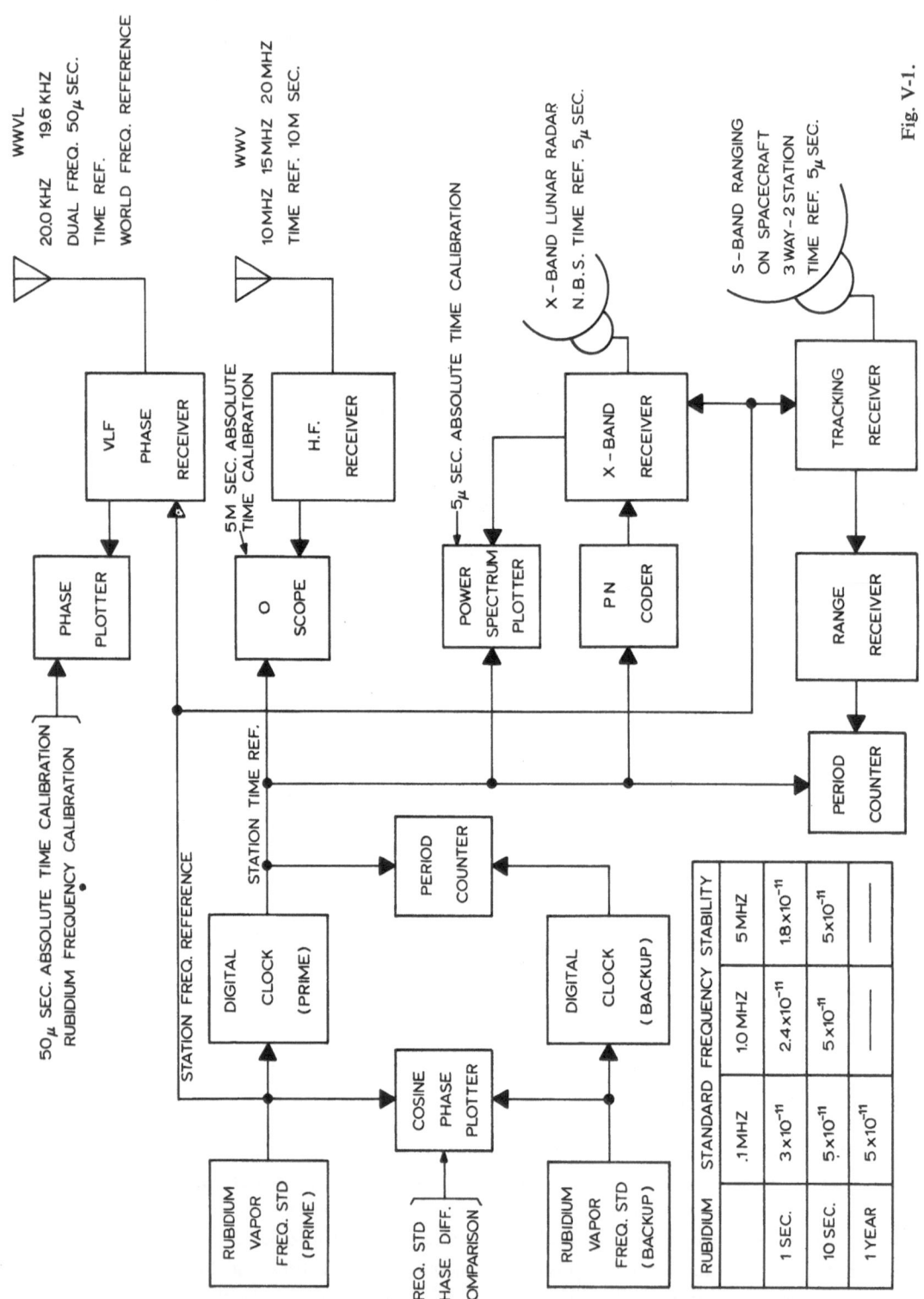

Fig. V-1.

TABLE V-1

Time	Description	Comment
UT 1	Rotational position of earth derived from star transits.	Uncertainties in $UT1 - t_s$ result in degradation of the apparent quality of the RMD, incorrect solution for station location, and
t_s	Station Time, hopefully equal to UTC or A1.	erroneous prediction of probe coordinates near planetary encounter.
UT 2	UT1 − predicted seasonal variations. UT2 $\underline{\triangle}$ Greenwich Civil Time	
UTC	A1 +fixed frequency ratio offset from A1. UTC is adjusted frequently so that the offset from A1 remains less than about 0.1 sec. Radio timing signals are UTC.	

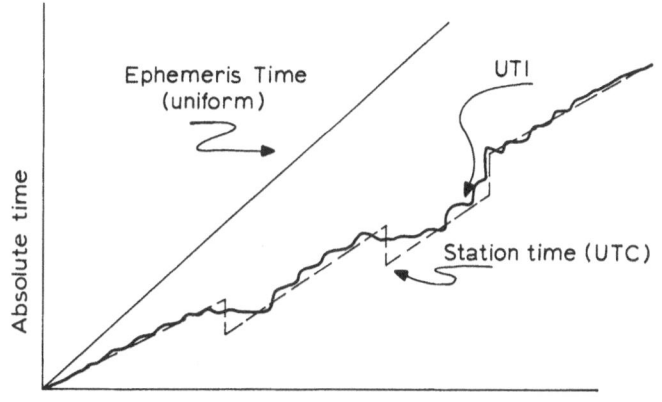

Relationship between time systems

A1 Time (uniform)

A-1	NBS, Boulder, Colorado, Cesium Frequency STD Time. A1 = UT 2 (1958.0) The A1 Second is invariant in length, and was chosen equal to the best available estimate of the Ephemeris second in 1958. (i.e. A1 is a uniform time)	An error in t_s − A1 causes a 'shift' in the coordinates of the ephemeris bodies at the time the computer program believes the data was taken. With X-Band Synch, UTI–A1 errors become dominant sources of timing errors.
ET	Ephemeris Time is a uniform time which determines the positions of the celestial bodies.	Independent variable of probe equations of motion. Accumulated errors in ET affect probe orbit in complex ways. ET is assumed identical to A1 in orbit computation.

Reference 3

updated. In practice, the transmitter code is started 30 μsec late and advanced 1 μsec/sec during the 1-min transmission period by the computer until, on the 30th sec, the code is arriving at the receiver in the correct time and, at the end of the minute, it is arriving 30 μsec early. Thus a range of 60 μsec is searched. If the receiving clock is

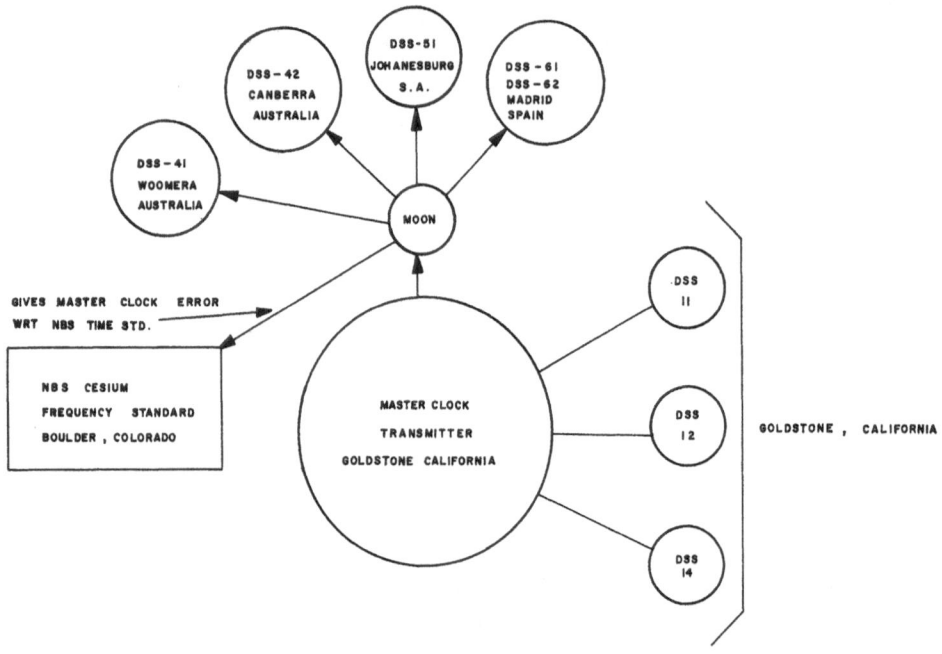

Fig. V-2.

10 μsec behind the transmitting station, clock correlation will occur 10 sec ahead of the 30th sec. It is, then, only necessary to measure resolution of clock synchronization.

The output of the receiver is recorded on a strip chart (Figure V-3) along with a timing track containing 1-min and 1-sec marks. In the example of Figure V-3, the receiving clock was 8 μsec behind the transmitting clock resulting in code correlation 8 sec earlier than true synchronization. A 10-sec time constant filter ahead of the recorder effectively removes all receiver noise as indicated by the smooth line before the range gate reaches the moon. The roughness of the curve after this point is caused by multiple-path lunar reflections of such magnitude that the filter does not entirely remove them. The large pulse of energy appearing at the start of each minute is a 1-sec transmission of continuous wave signal (no biphase modulation) transmitted to verify antenna pointing, correct frequency, and system operation. To determine the time of day from the 1-min recording of Figure V-3, a straight line is drawn along the rise of the receiver output curve of its intersection with the base line. The number of 1-sec marks from the center, or 30-sec, mark are counted. This is the receiver clock error, in microseconds.

The principal error sources in this system are in the lunar ephemeris and roughness in the recording due to lunar multipath and reflectivity changes. The 1σ system errors due to all sources are on the order of 2 to 5 μsec. S-Band Ranging Time correlation will be discussed in the section on Ranging.

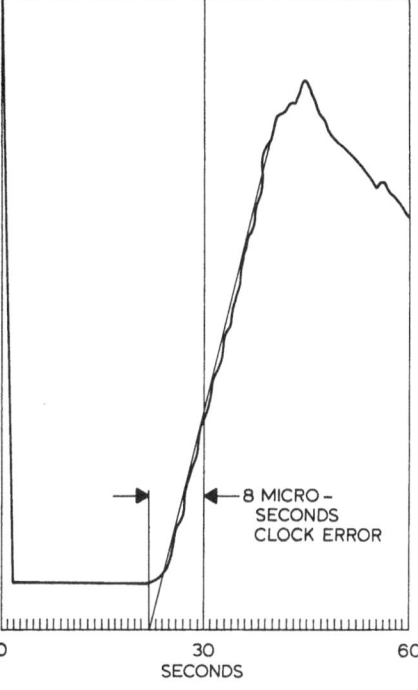

Fig. V-3.

B. DOPPLER

1. Functional Block Diagram

Historically, the most important raw metric data type is radial velocity. The tracking system measures a phase difference between the transmitted and received signal which is proportional to spacecraft radial velocity. There are many ways of treating the doppler measurement function. The method illustrated in Figure V-4 is chosen to emphasize that the measurement is basically one of *phase* comparison.

The Frequency multipliers K, and K_2 are required so that the transmitted frequency will be sufficiently different from the received frequency to avoid interference. A 1 MHz bias frequency is added to the output of the coherent phase differencer before the counter so that the counted doppler will always be positive. It is not shown in the figure.

$$C(t_{i+1}) - C(t_i) = 2\pi K_1 K_2 f_r \left[\tau(t_{i+1}) - \tau(t_i) + \phi(t_{i+1}) - \phi(t_i) \right] \qquad \text{(V-1)}$$

where $\phi(t)$ is Phase Noise on the Doppler Count.

Let

$$\tau(t_{i+1}) - \tau(t_i) = \frac{2}{c} \left[r(t_i) + \dot{r}(t_i)(t_{i+1} - t_i) + \cdots - r(t_i) \right]$$

where

c = Speed of Light and r is Probe Station Slant Range

$$\left. \begin{array}{l} \\ \doteq \dfrac{2}{c} \dot{r}(t_i)(t_{i+1} - t_i) \end{array} \right\} \qquad \text{(V-2)}$$

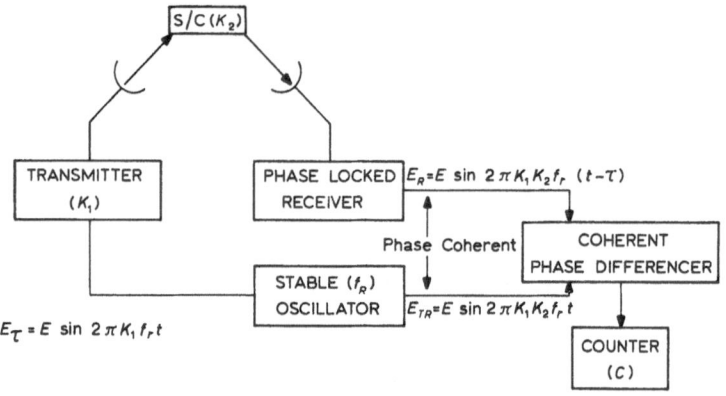

$$C(t) \approx 2\pi K_1 K_2 fr\, T\,(t) + \phi\,(t)$$

Fig. V-4.

$$\dot{r}(t_i) \doteq \frac{1}{2\pi K_1 K_2 f_r} \left(\frac{c}{2}\right) \underbrace{\frac{C(t_{i+1}) - C(t_i)}{T_s}}_{\text{Average range rate}} + \underbrace{(\text{Const.}) \left[\phi(t_{i+1}) - \phi(t_i)\right]}_{\text{Range rate error due to phase error}}$$

$$(\text{V-3})$$

where $T_s = t_{i+1} - t_i$, the data system sample interval (Doppler Averaging Time).

4. *Doppler Resolver*

The doppler counter increments its tally at each positive going crossing of the coherent phase differencer output. Transfer of the tally from the counter to the output register takes place at the positive going crossing following a sample pulse. The time tag on the tally in the output register is that of the sample pulse, an integral number of hours, minutes, and seconds of UTC. Thus the doppler tally is in error with respect to the time tag by plus or minus some fraction of a cycle [4].

The Doppler Resolver is mechanized to minimize this quantization error. It performs a very accurate measurement of the *time* between the sample pulse and the transfer pulse, thus enabling the data user to determine the actual time interval which corresponds to a given doppler counter difference (i.e., T_s in Eq. V-3).

Figure V-5 illustrates the process. The actual resolution of the doppler becomes 0.01 Hz or 0.65 mm/sec.

2. *Error Sources*

The errors in the doppler measurement are all, of course, phase errors. Three major sources of phase errors are considered here.

a. *Phase Locked Loop RMS Phase Jitter*. Thermal noise produces 'phase jitter' in the receiver coherent phase tracking loop. It is dependent upon tracking system noise

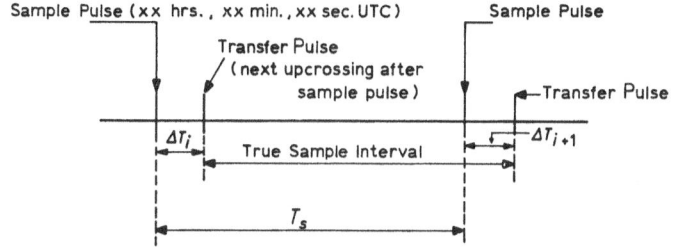

Fig. V-5. The sample interval $= T_s + (\Delta T_{i+1} - \Delta T_i) \times 10^{-8}$ sec. ΔT_i is measured by a 100 MHz counter to a resolution of 10^{-8} sec. This is equivalent to 0.01 Hz at the doppler bias frequency of 1 MHz.

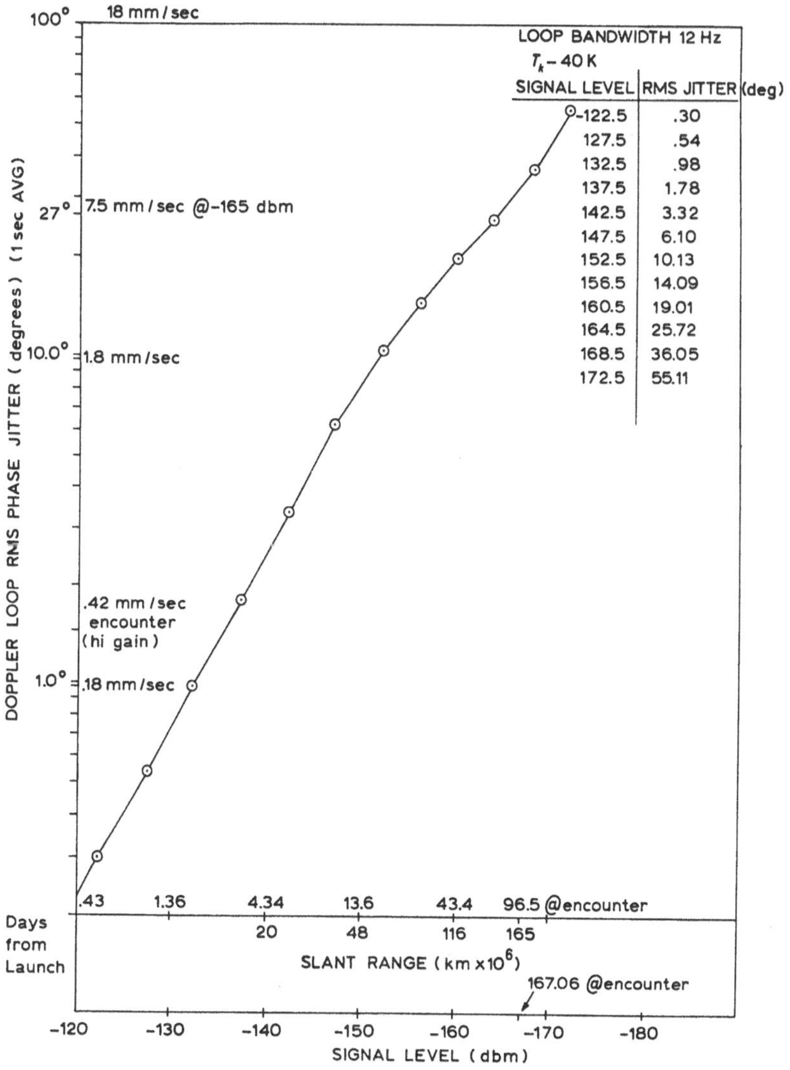

LOOP BANDWIDTH 12 Hz

$T_k - 40$ K

SIGNAL LEVEL	RMS JITTER (deg)
-122.5	.30
127.5	.54
132.5	.98
137.5	1.78
142.5	3.32
147.5	6.10
152.5	10.13
156.5	14.09
160.5	19.01
164.5	25.72
168.5	36.05
172.5	55.11

100° — 18 mm/sec

27° — 7.5 mm/sec @ -165 dbm

10.0° — 1.8 mm/sec

.42 mm/sec encounter (hi gain)

1.0° — .18 mm/sec

DOPPLER LOOP RMS PHASE JITTER (degrees) (1 sec AVG)

Days from Launch

.43   1.36   4.34   13.6   43.4   96.5 @encounter
        20   48   116   165

SLANT RANGE (km x10⁶)

167.06 @encounter

-120   -130   -140   -150   -160   -170   -180

SIGNAL LEVEL (dbm)

Fig. V-6.

temperature, T_k, tracking loop noise bandwidth, and signal strength. Since the phase tracking loop is highly non linear in the signal strength regions of interest, no attempt will be made to provide a mathematical description of Loop RMS Phase Jitter. Figure V-6 is a plot of Loop RMS Jitter versus received signal level for a typical MM69 trajectory. The loop bandwidth used was 12 Hz. T_k was 40 K. Slant range and days from launch also appear on the abscissa of the plot. The ordinate shows Doppler Loop RMS Phase Jitter in degrees of phase of a single hertz of doppler. Also plotted on the ordinate is the equivalent range rate error in millimeters/second for a 1 sec averaging time (1 Hz of doppler = 0.065 m/sec of range rate). It will be shown in subsequent discussion that the shape of the auto correlation function of the noise is very important to the user of the RMD. In general, a noise process whose auto correlation function does not die out rapidly as a function of lag time is undesirable. It is possible to show that the auto correlation function of the Phase Locked Loop Thermal Noise Jitter is of the exponential cosine type which decays very rapidly as a function of lag time. Thus it is possible to approximate doppler loop thermal jitter by a white noise process with a δ function auto correlation function [5].

b. *Reference Oscillator Phase Instability.* If in equation (V-1) we allow the phase error ϕ to vary over the signal time of flight, τ, we have
Doppler Phase Error = (Const.)

$$\{[\phi(t_{i+1}) - \phi(t_{i+1} - \tau_{i+1})] - [\phi(t_i) - \phi(t_i - \tau_i)_3\} \tag{V-4}$$

Develet has shown that the RMS error in range rate due to a drifting oscillator is

$$\sigma_{\dot{\varrho}}(\text{osc. drift}) = \begin{cases} \dfrac{c}{\sqrt{2}} \dfrac{1}{2\pi K_1 K_2 f_r} \dfrac{1}{T_s} \left(\dfrac{\tau}{T_c}\right)^{1/2} & 0 \leqslant \tau \leqslant T_s \\[3mm] \dfrac{c}{\sqrt{2}} \dfrac{1}{2\pi K_1 K_2 f_r} \dfrac{1}{T_s} \left(\dfrac{T_s}{T_c}\right)^{1/2} & T_s \leqslant \tau \leqslant \infty \end{cases} \tag{V-5}$$

where T_c, the coherence time, is the time for the rms phase drift of the oscillator to reach 1 rad.
Coherence times in the tracking system are on the order of 6 sec.
Thus for $T_s < \tau$

$$\sigma_{\dot{\varrho}}(\text{oscillator drift}) \doteq \frac{4.2}{\sqrt{T_s}} \text{ mm/sec.} \tag{V-6}$$

For the near encounter phase of the mission when $\tau \approx 300$ sec.

$$\sigma_{\dot{\varrho}}(\text{oscillator drift}) \doteq \frac{73.5}{T_s} \text{ mm/sec., } T_s > 300 \text{ sec.} \tag{V-7}$$

This situation applies only when the flight path analysis team is attempting to fit long arcs, i.e., >300 sec. The actual effect on the PMD of long term drift is a bit worse than indicated here and will be touched upon later [6].

C. RANGING

1. *System Operation*

The DSN Ranging System measures the round trip propagation time of a signal from the ground transmitter to the spacecraft and back to the ground receiver. The accuracy and resolution of the system is independent of the slant range or velocity of the spacecraft. The measurement is made continuously and may be sampled on demand. The unit of measurement is called the RU (range unit) and has the dimensions of time ($1 \text{ RU} \doteq 3.5$ nsec).

Mark 1A versus Planetary Ranging System. There are two basic types of ranging systems in the DSN. The operational Mark 1A Ranging System is standard at DSN stations equipped with 85-ft antennas. It has an unambiguous range of 800 000 km.

A developmental planetary ranging system exists only at the 210-ft antenna at Goldstone, California. The unambiguous range of this system is 398×10^6 km.

The fundamental technique of ranging is the same for both systems. The mechanization of the Planetary Ranging System is much more elaborate than that of the Mark 1A and incorporates many more capabilities which are required to measure very long propagation times.

Only the Mark 1A will be discussed in any detail herein.

Referring to the block diagram of Figure V-7, the transmitter coder generates a code with particular properties which is biphase modulated onto the transmitted signal. The spacecraft transponder receives and retransmits the signal and code coherently.

The code consists of a 992 kHz clock combined with a pseudo random shift register sequence (PN Code). The class of PN Codes used, called acquirable codes, has the characteristic property of two level auto correlation; that is, when the code is auto-correlated, the correlation is at a maximum; when it is out of correlation, the auto correlation is a uniform small value regardless of the lag position. The sequence is sufficiently long to guarantee unambiguous code matching over the range measured.

At initial acquisition the clock component of the received code is acquired and tracked by a phase locked loop in the receiver. The PN code component of the received code is compared with a locally generated replica of the code.

The received and locally generated codes are matched by digitally shifting the receiver code and measuring the correlation between them at each relative shift position until a maximum is obtained. The correlation properties of PN sequences guarantee an unequivocable maximum.

The total shift of the receiver code is a measure of the ranging time delay at the start of acquisition. This shift is monitored and the appropriate number of RU are added into the coarse ranger accumulator whenever a shift is made. The code shift tallies are added concurrently with the algebraic adding of clock doppler and RF doppler tallies which record phase shift due to target motion as well as phase shift due to the fact that the total phase difference between the receiver code and the received code may not have been an integral number of clock bits.

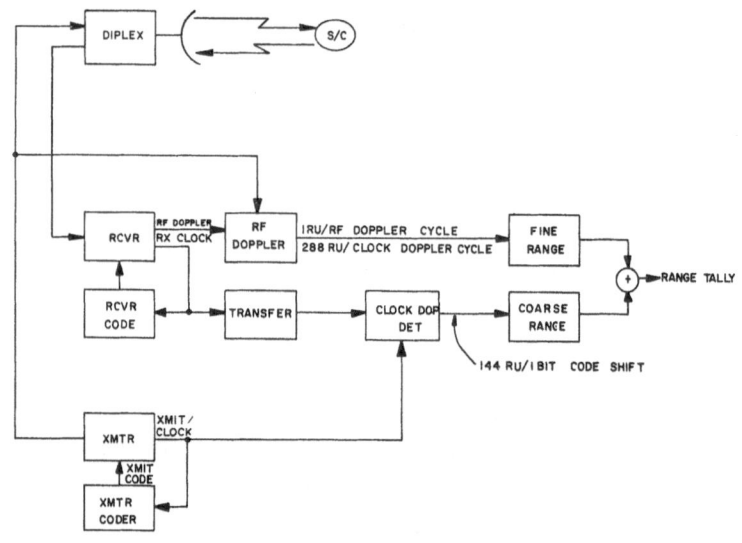

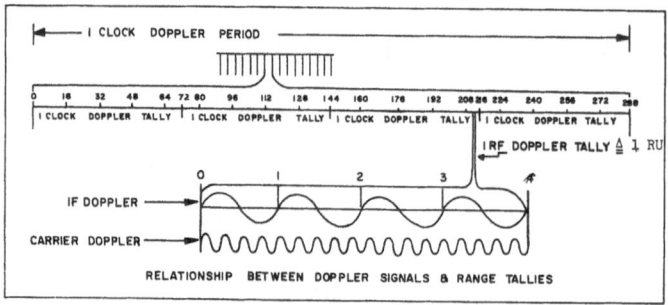

Fig. V-7.

The ranging time delay continues to be updated by means of the clock and RF doppler tally for as long as the system remains locked. The resolution of the system may be seen as follows: (Figure V-7).

$$1 \text{ RU} = 1 \text{ RF doppler cycle} = 288 \text{ RF doppler tallies} = 1 \text{ Clock Doppler}$$

$$= \frac{1}{496 \times 10^3 \text{ Hz}}$$

$$1 \text{ RU} \doteq \frac{1}{288} \frac{1}{496 \times 10^3} \doteq 7.02 \text{ nanoseconds of round trip light time}$$

$$1 \text{ RU} \doteq 3.5 \text{ nanoseconds of 1-way light time}.$$

2. *System Errors*

The only significant errors in the ranging measurement are phase errors due to unknown phase delays in the ground system or spacecraft and phase errors due to unknown phase drifts in the ground system. The rms sum of such errors for a typical system are on the order of 18 to 35 nsec (5 to 10 m of range).

The preceding is true because the ranging measurement is basically a digital measurement. One either has range lock or one does not. If one does have it, the Range Tally is as good as the system phase delays.

However, the probability of achieving range lock is a function of more conventional receiver parameters, such as system temperature and received power.

Code correlation as performed in the DSN Ranging System is a form of correlation detection. The maximum likelihood detector for these codes includes an integrator. The probability of code acquisition is directly related to integration time. Hence, we have such curves as those of Figures V-8 and V-9. Figure V-8 shows ranging signal to noise ratio as a function of received carrier power for an appropriate range of system temperatures. Probe radial distances and days from launch are shown for a typical MM69 trajectory.

Figure V-9 gives a family of plots of code acquisition probabilities as a function of ranging signal to noise ratio, for a number of integration settings. Integration setting is related to integration time as shown on the figure.

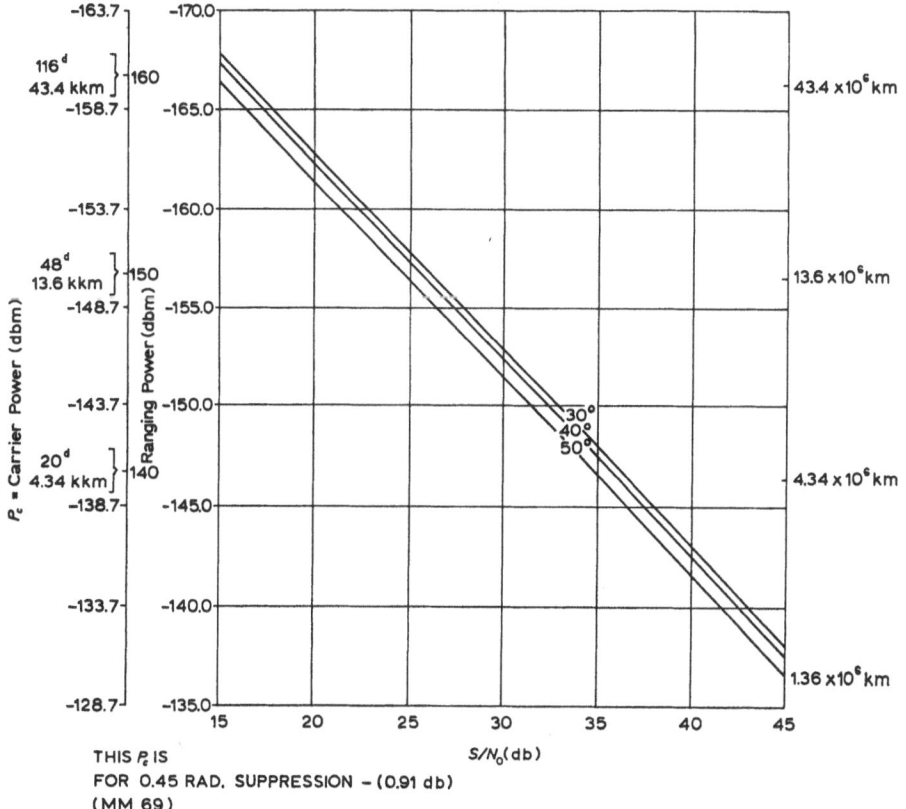

Fig. V-8.

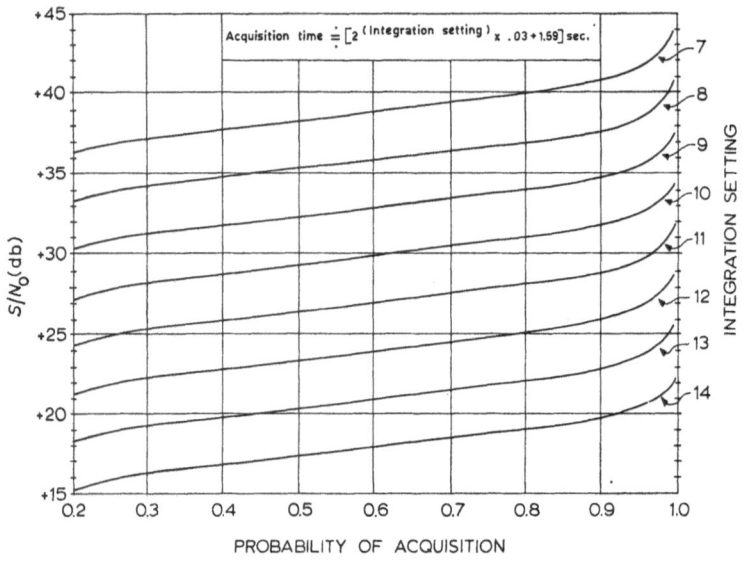

Fig. V-9.

As an example of the use of the curves, the acquisition time for the ranging code at 20 days is given. Assume the system temperature is 36 K. The ranging SNR is 36.7 db. An acquisition probability better than 0.99 can be achieved with an integration setting of 9. The corresponding acquisition time is 17 sec. It is interesting to compare this number with a typical integration time for the planetary ranging system, to achieve the same measurement accuracy. It is on the order of 1800 to 18 000 sec for a target at a range of 3×10^8 km.

It should be pointed out that in the example the earth-probe distance is well beyond the unambiguous range of the Mark 1A code at that point. However, time correlation, which also requires code acquisition, could be performed.

3. *Time Correlation*

If the range and range rate are assumed to be accurately known, it is possible to use the ranging system to determine the phase difference between the frequency standards of two distant stations.

Figure V-10 illustrates the technique. Station 1 transmits to the spacecraft and receives the ranging code in the normal fashion. Station 2 acquires the downlink signal with station 1 range code superimposed. At each station a rubidium derived 10 MHz counter is started by the same *local* 1-sec pulse. The counting process is stopped at each station by the detection of a certain received code component which causes the Mark 1A Ranging System to generate a marker pulse. The difference in counter elapsed time should be equal to the round trip signal time of flight if the two rubidiums are in synch. If they are not, the time difference is related to the phase difference between the two standards.

This system is vulnerable to uncalibrated ranging system time delays and errors in

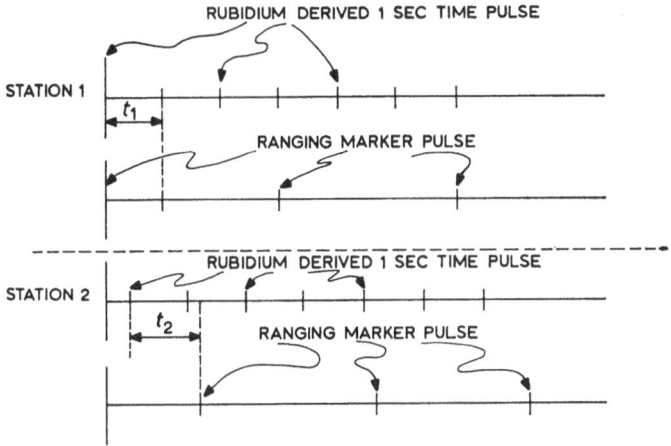

Fig. V-10.   (see Reference 7).

$$(t_2 - t_1) = \varepsilon_t + \tau + \varepsilon_{sys} \quad \text{where} \quad \begin{cases} \varepsilon & = \text{Time synch. error} \\ \tau & = \text{Total propagation time} \\ \varepsilon_{sys} & = \text{System errors} \end{cases}$$

estimates of probe position and velocity. The $1\sigma_{error}$ of this measurement is approximately 5 μsec.

D. CHARGED PARTICLES

Due to the fact that the medium through which the signal travels is ionized, it has a non-unity index of refraction. This has the effect of *decreasing* the signal phase propagation time and thus decreasing the number of doppler cycles counted for a given range difference, while it *increases* the signal group propagation time. This causes the ranging time delay to be increased for a given range.

The effect is separated into two parts; that due to the ionosphere of the earth, and that due to the interplanetary plasma. These effects produce a range difference error which is on the order of 2.0 m over an averaging time of 10^4 sec.

Modeling the ionosphere and correcting the raw metric data for ionospheric errors put the error in the region 0.1 to 1.0 m/10^4 sec. The fact that range data is retarded while doppler is advanced gives rise to the interesting possibility of calibrating the ray bending due to charged particles. It is estimated that the combination of ionosphere modeling and range/range rate calibration would reduce the ray bending error to the region 0.05 to 0.5 m/10^4 sec. Thus far, hardware stability limitations have limited the usefulness of the technique, but it will be attempted on MM69 [8].

E. SUMMARY OF RAW METRIC DATA ACCURACIES

Time   5 to 50 μsec -1 σ
Frequency 5×10^{-11}
Equivalent range rate 0.1 mm/sec at $L+5$ days to 0.5 mm/sec at encounter
(60 sec Av time)   (1 Hz doppler = 0.065 m/sec)
Equivalent range 15 m (1 RU = 1.05 m).

VI. Processed Metric Data

A. ORBIT DETERMINATION – STATEMENT OF PROBLEM

Due to errors introduced during the powered portion of flight and other sources (e.g., incompleteness in our model of the real world as characterized by the equations of motion we use), the actual flight path of the spacecraft differs from our preflight trajectory; i.e., the initial conditions, at least, of our D.E's of motion are different from those of the nominal preflight trajectory. The basic problem of orbit determination is to determine the actual initial conditions or initial state vector of the trajectory based on observations of spacecraft position, velocity, etc.; i.e., based on the raw metric data.

Three major factors must be considered. First, the relationship between the RMD and the initial state vector is highly non-linear. Second, the system is highly over-determined; that is, there are many more observations than required to determine uniquely the initial state vector. Third, the RMD is corrupted with noise.

B. THE JET PROPULSION LABORATORY/ORBIT DETERMINATION PROGRAM BRIEFLY CHARACTERIZED

Thus we must perform a linearization process, then solve an over-determined system in such a way that the effect of RMD noise is minimized in some sense.

The linearization is fundamentally a generalization of the Newton-Raphson method.

The solution of the data noise corrupted over-determined system is by weighted least squares.

The linearization may be stated as follows:

$$\Delta(\text{observable at } t_i) = \frac{\partial}{\partial q_1}(\text{observable at } t_i)\, \Delta q_1$$

$$+ \cdots + \frac{\partial}{\partial q_n}(\text{observable at } t_i)\, \Delta q_n \qquad \text{(VI-1)}$$

where $q_1, \ldots, q_n$ are the components of the initial state vector; for example, 3 position and 3 velocity coordinates at t_0, the time of injection into planetary transfer orbit.

The weighted least squares solution is described as follows: Let A be an $n \times N$ matrix of analytic partials whose elements are

$$A_{ij} = \frac{\partial}{\partial q_i}(\text{observable at } t_i') \quad \left\{ \begin{array}{l} n \text{ is the number of RMD samples} \\ N \text{ is the number of components} \\ \text{of the initial state vector} \end{array} \right. \qquad \text{(VI-2)}$$

Δy is an $n \times 1$ column matrix such that

$$y_i = \Delta(\text{observable at } t_i) \qquad \text{(VI-3)}$$

If ΔQ is an $N \times 1$ initial state difference vector, then VI-1 becomes

$$\Delta y = A \Delta Q \qquad \text{(VI-4)}$$

The weighted least square solution for VI-4 is given by

$$\varDelta Q = J^{-1} A^T W \varDelta y$$

where W is the diagonal weighting matrix, and $J = A^T W A$, the normal matrix.
Figure VI-1 gives a flow chart of the basic orbit determination procedure. The basic

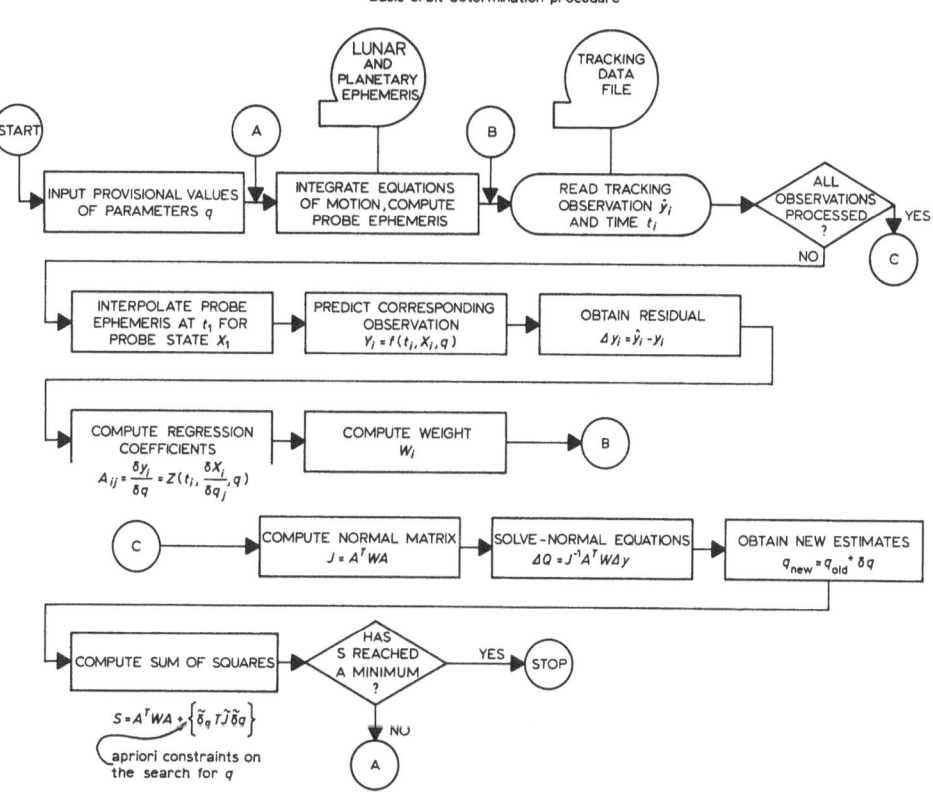

Fig. VI-1 (see Reference 10).

Newton-Raphson iterative process is illustrated by the path from (A) thru (C) thru
the decision diamond back to (A) [9].

The parameter vector Q in general includes many more parameters than the
trajectory initial state vector. The Orbit Determination Program (ODP) recognizes
three categories of parameters:

(1) 'Solve For' parameters: Those parameters whose estimates are obtained from
the least squares fit.

(2) 'Consider' parameters: Those parameters whose a priori estimates are not
corrected but whose errors are considered when computing the statistics for Q.

(3) 'Exactly constrained' parameters: Parameters that are functionally related to
the 'Solve For' and 'Consider' parameters.

The ODP user groups the parameters into two subsets:

(1) Those which directly affect the computation of the path of the probe but only indirectly influence the tracking data computation; e.g., trajectory initial state vector, masses and gravitational potential field coefficients of celestial bodies, time transformation parameters, solar pressure constant, etc.

(2) Those which appear in tracking data equations thus directly affecting the computation of observables; e.g., station location parameters.

Estimates of physical constants obtained in this way are only as good as our model; i.e., our 'fitters universe', since errors in our model will appear in our estimates of the physical constants. Such estimates of physical constants must be carefully interpreted. Nevertheless, use of such techniques over the past several years has yielded significant improvements in the AU, earth/moon/mass ratio, and the masses of Venus and Mars, to mention only a few.

C. ORBIT DETERMINATION PROGRAM CAPABILITY

There are presently two orbit determination programs in use for MM69.

1. *Double Precision Orbit Determination Program (DPODP)*

Some general capabilities of the DPODP are

(a) Processes 17 different RMD types, any number of observations.

(b) Solves for or 'considers' 50 parameters (i.e., J matrix may be 50×50).

(c) All computations are double precision (i.e., 8 bit characteristic and 64 bit mantissa).

The standard ODP support sequence for spacecraft guidance during a mission is as follows:

(a) Compute and transmit station pointing predictions.

(b) Accept, edit, and format for computation RMD.

(c) Estimate guidance parameters and physical constants.

(d) Obtain parameter statistics at maneuver and target encounter times.

(e) Using parameter estimates and statistics, compute spacecraft guidance commands.

(f) Using parameter estimates, compute spacecraft trajectory history.

2. *Single Precision Orbit Determination Program (SPODP)*

As the name suggests, the SPODP differs from the DPODP in computational precision. Although the inversion of the J matrix is done double precision, all other computations are single precision (i.e., 8 bit characteristics, 28 bit mantissa). Also, the 'fitters universe' is somewhat less sophisticated. The maximum size of the J matrix is 20×20. The principal advantage of the SPODP at present over the DPODP is operational experience. For this reason only the SPODP is committed to MM69, even though as we shall see, the mission requirements cannot be met using the SPODP alone.

D. INTERACTION BETWEEN ORBIT DETERMINATION PROGRAM AND RAW METRIC DATA

The ODP is a highly complex, though indispensable, tool for spacecraft navigation.

It interacts with the RMD in many ways, some of which are quite obvious, others of which are rather subtle and not well understood. It has been shown that the ODP is an optimum low-pass linear filter. The optimality is taken in the Weiner sense and depends on how well the W matrix characterizes the noise on the RMD.

Since the ODP acts like a low-pass filter, it is most vulnerable to low frequency, long correlation time noise, whose signature on the signal is similar to the signal itself. It rejects quite successfully high frequency noise. (Here high frequency noise may be considered to be noise whose auto correlation function, $R(\tau)$, is negligible for τ greater than the round-trip-signal-time-of-flight.)

Such low frequency noise may arise from the RMD, but this is rather unlikely. It is very likely to arise as a result of differences between the 'fitters universe' and the 'real universe'. Thus errors in the astrodynamic constants, station location, inadequacies in the D.E's of motion, etc., very often give rise to errors whose effects may be looked upon as low frequency noise process with very long time constants.

Such problems are handled by many techniques based on long experience with the OD process. One such technique is of interest here since it provides us with a relationship between RMD accuracy and PMD accuracy. The effective RMD variance is given by

$$\sigma^2(\text{EFF}) = \sum_{i=1}^{n} \sigma_i^2 \max\left(1, \frac{T_{l_i}}{T_s}\right) \quad \text{where} \quad \max\left(1, \frac{T_{l_i}}{T_s}\right) = \begin{cases} 1, & \frac{T_{l_i}}{T_s} \leqslant 1 \\ \frac{T_{l_i}}{T_s}, & \frac{T_{l_i}}{T_s} > 1 \end{cases}$$

and $\begin{cases} \sigma_i^2 \text{ is the variance of the } i\text{th error source} \\ T_{l_i} \text{ is the correlation width of the } i\text{th error source} \\ T_s \text{ is the sample spacing of the data points}. \end{cases}$ (VI-7)

The RMD effective variance is a measure of the noise on the raw data plus the degradation introduced by the software that processes the data.

It is the variance the ODP sees the data as having. Thus if one generates noise-free RMD (i.e., the W matrix equals the identity matrix) from certain trajectories, and computes orbits from these data, there will in general be a distribution of errors at the target due to software errors. The statistics of this target error distribution will match target statistics of orbits determined using a W matrix with main diagonal not equal to 1. The effective variance of the data type is the value of the main diagonal of the W matrix required to yield the same target statistics, all other variables being held constant.

TABLE VI-1

For example, using the SPODP, a miss statistic at Mars on a certain trajectory was 52 Km using 'perfect data'. The value of σ^2 (doppler) required to match this was 3×10^{-3} m/sec. RMD σ (doppler) $= 6.5 \times 10^{-3}$ m/sec/Hz $\times 5 \times 10^{-2}$ Hz error $= 32.5 \times 10^{-5}$ m/sec. At a T_s of 60 sec, $T_l = (3/32.5) \times 10^{-3} \times 10^5 \times 60 = 554$ sec effective correlation width, even tho the actual correlation width is $\doteq .01$ sec. For the DPODP σ (eff. doppler) $= 1 \times 10^{-3}$ m/sec and T_l becomes 177 sec.

VII. Mission Accuracy Analysis

A. TARGET COORDINATE SYSTEMS

The aiming point near Mars is determined by the magnitude and direction of the **B** vector (Figure VII-1). Errors at the aiming point are described in several related coordinate systems.

1. *Miss Ellipse*

Assuming that probe position errors due to orbit determination errors are normally distributed, the probe position error frequency function at the time, t_E, of passage through the **R**, **T** plane is given by

$$p(x, y, z) = \frac{1}{(2\pi)^{3/2}} \frac{1}{\text{Det } \Lambda^{3/2} (t_E)} \exp - x^t \Lambda^{-1} (t_E) x$$

where

$$x = \begin{Bmatrix} x(t_E) \\ y(t_E) \\ z(t_E) \end{Bmatrix} \quad \text{and} \quad \Lambda(t_E)$$

is the covariance matrix of probe position errors at t_E.

The relationship $X^T \Lambda^{-1} x =$ (constant) defines ellipsoids whose surfaces are surfaces of constant probability. The 1σ miss ellipse is the section on the **R**, **T** plane of such an ellipsoid of constant probability. SMAA and SMIA are the magnitudes of the semi-major and semi-minor axes respectively of the miss ellipse. θ_{MEO} is the orientation of the miss ellipse with respect to the **R**, **T** system.

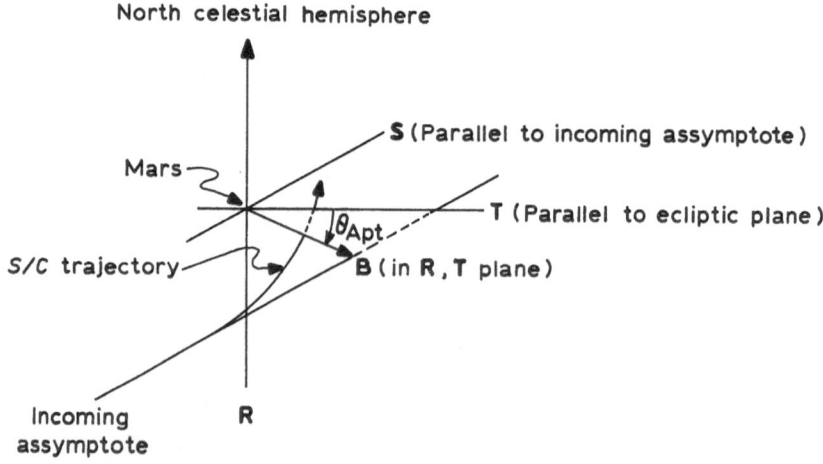

Fig. VII-1.

2. $\sigma_{B \cdot T}$ and $\sigma_{B \cdot R}$

$\sigma_{B \cdot T}^2$ and $\sigma_{B \cdot R}^2$ are the diagonal elements of the covariance matrix of errors in $B \cdot T$ and $B \cdot R$. These quantities are used to specify the allowable miss in the mission requirements. They are related to SMAA and SMIA as follows:

$$\Lambda_{R, T} = \begin{Bmatrix} \sigma_{B \cdot T}^2 & R(B \cdot T, B \cdot R) \\ R(B \cdot T, B \cdot R) & \sigma_{B \cdot R}^2 \end{Bmatrix}$$

$$= \begin{Bmatrix} (SMIA)^2 \cos^2 \theta_{MEO} + (SMAA)^2 \sin^2 \theta_{MEO} & |-[(SMAA)^2 - (SMIA)^2] \sin \theta_{MEO} \cos \theta_{MEO} \\ (\text{Sym. term}) & (SMIA)^2 \sin^2 \theta_{MEO} + (SMAA)^2 \cos^2 \theta_{MEO} \end{Bmatrix}$$

$$(\text{VII-2})$$

3. σ_B and σ_{PB}

σ_B^2 and σ_{PB}^2 are the diagonal elements of the covariance matrix of errors in the B direction itself, and perpendicular to B. σ_{PB} is used in the mission requirements to represent the accuracy requirements of UV spectrometer slit alignment. The B, PB covariance matrix is related to SMAA and SMIA as follows:

$$\Lambda_{B, PB} = \begin{Bmatrix} \sigma_B^2 R(B, PB) \\ R(B, PB) \sigma_{PB}^2 \end{Bmatrix}$$

$$= \begin{Bmatrix} (SMIA)^2 \cos^2 \eta + (SMAA)^2 \sin^2 \eta & [(SMAA)^2 - (SMIA)^2] \sin \eta \cos \eta \\ (\text{Sym. term}) & (SMIA)^2 \sin^2 \eta + (SMAA)^2 \cos^2 \eta \end{Bmatrix}$$

where

$$\eta = \frac{\pi}{2} - (\theta_{MEO} - \theta_{APT}).$$

B. MISSION ACCURACY REQUIREMENTS

The mission navigational accuracy requirements are given in JPL Project Document 91, 'Mariner Mars 1969 Project Mission Plan and Requirements', dated February 1969. These requirements are specified for two representative launch dates and assume a midcourse maneuver at Launch plus 5 days and a second maneuver at Encounter minus 30 days. 3σ bounds are quoted and the document states that they are to be interpreted non-parametrically, i.e., "The actual error must be less than the bounds prescribed 99.7% of the time. The probabilities associated with other sigma levels will be 68.2% 1σ and 95.4% 2σ only if these error functions are Gaussian." Since the navigational accuracy studies used in this paper do assume that the error functions are Gaussian, Table VII-1 tabulates the 3σ errors which would be associated with the mission requirements if the error functions were Gaussian. This is done to allow comparison between requirements and estimated capabilities.

1. UV Spectrometer Slit Alignment

The pacing requirement on accuracy of navigation at encounter is that of alignment of the Ultraviolet Spectrometer slit. The experimenter has stated that the slit alignment error should be less than 3σ with a 90% probability.

TABLE VII-1

Mission requirements – 3σ (Design goal)

	σ_B (km)	σ_{PB} (km)
$I+5$ days	800	600
$I+30$ days (1)	600	600
E-6 hours (2)	150	100

(1) Midcourse execution errors at $I+5$: $\sigma_x = \sigma_y = \sigma_z = 0.1$ m/sec
(2) Midcourse execution errors at E-30: $\sigma_x = \sigma_y = \sigma_z = 0.1$ m/sec

The UVS looks at the atmosphere at the lighted limb of the planet in its primary experiment. The general objective of the experiment is to find out what gasses are present at various altitudes, and to construct a model of the atmosphere. The UVS slit must be aligned parallel to the surface of the planet so that a constant altitude is being viewed to achieve the objective.

If ε_A is the alignment angle, error due to OD errors, then

$$\sigma_A(\deg) = \frac{\sigma_{PB}^2 + \sigma_E^2}{r} (57.3) \qquad (\text{VII-4})$$

σ_{PB} is used here because, as will be discussed later, the error perpendicular to the probe-target vector after the spacecraft is in the Mars sphere of influence, will be the error in the PB direction at t_E plus ephemeris error, σ_E. σ_E is less than 20 km in the PB direction. r is the magnitude of the probe-target vector at the desired data-taking point. (About 100 km above the planet surface, so that r 3485 km.)

If the slit alignment errors are normally distributed, and the UVS pointing error $\sigma I = 1°$ then the probability that the total alignment error will be less than $3°$ is given by

$$\Phi\left[x = \frac{3°}{(\sigma_A^2 + 1°)^{1/2}}\right] \quad \text{where} \quad \phi[\ ] = \int_{-x}^{x} \text{normal } fr \cdot f(y)\, dy.$$

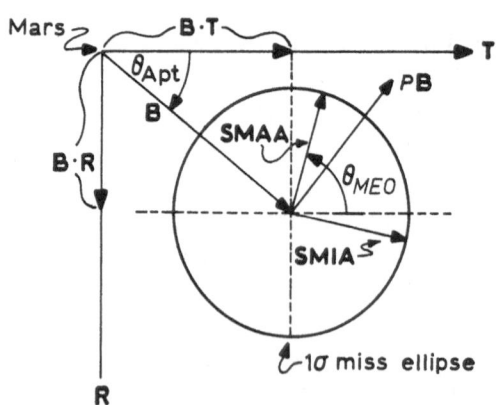

Fig. VII-2.

Conversely,

$$X_{0.90} = \frac{3°}{(\sigma_A^2 + 1)^{1/2}} = 1.645$$

yields the value of σ_A which OD is allowed. Thus σ_A (allowed) = 0.0265 rad, and the total allowed error perpendicular to the probe-target vector is given by

$$(\sigma_{PB}^2 + \sigma_E^2)^{1/2} = (0.0265)(3485 \text{ km}) = 92.4 \text{ km}. \qquad \text{(VII-5)}$$

Splitting the allowable error equally between OD and ephemeris gives approximately 65 km to each. This presents a severe requirement even on the capabilities of the DPODP.

2. Typical Mission Plan

a. *Tracking Coverage.* As discussed in the section on trajectory characteristics, the geocentric declination of the probe has a large negative value for the first portion of the mission. Thus most Orbit Determination accuracy studies assume only two stations tracking. The actual mission plan provides for some coverage of each spacecraft at each DSN longitude for each day of the mission.

All stations will range on the spacecraft as long as ranging threshold permits. DSS 14 will provide planetary ranging every two weeks for each spacecraft. Special ranging coverage will be provided for midcourse maneuvers.

b. *Maneuver Strategy.* The penultimate user of navigation parameters extracted from the RMD by the ODP is the maneuver strategist. He must formulate and execute a maneuver policy to achieve one or more of the following goals:

(1) Maximize the immediate probability of guidance success (i.e., the probability of being in the target zone after guidance correction).

(2) Minimize the immediate expected value of the square of the miss in B space.

(3) Minimize the expected square of the magnitude of the next midcourse correction.

(4) Minimize the impact probability.

Each maneuver aiming point selection strategy is subject to an impact probability or a planetary quarantine constraint, as well as limitations on the magnitude and direction of turns that can be made, the magnitude of velocity increment, the time of the maneuver, and many others.

Naturally, the ultimate user of the data, the spacecraft, must perform in a nominal fashion.

The maneuver policy chosen for MM69 is to minimize the magnitude of the velocity required at first maneuver and maximize the probability of guidance success after one maneuver.

Not only does the ODP provide the input for the maneuver computation but also it evaluates the success of the maneuver in the sense that it evaluates the post maneuver trajectory. From this information it can be determined to some accuracy whether or not the maneuver was successful. In fact, it is true that the value quoted for miss at the target due to midcourse executions errors is much less than the miss due to OD errors. This is some comfort since it allows us to assume we *can* get there from here even through we may not know for sure whether or not we *did*.

Figure VII-3 gives a feeling for the size of miss attributable to maneuver execution.

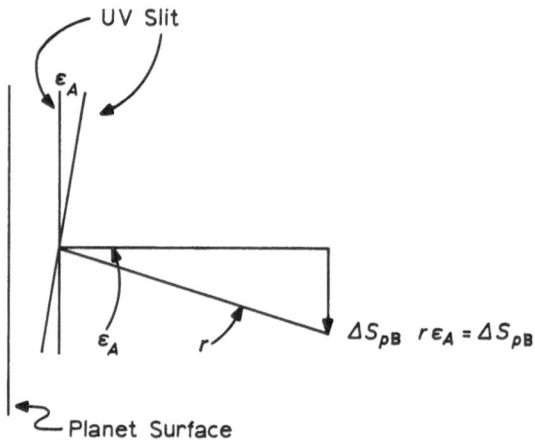

Fig. VII-3.

C. ACCURACIES OBTAINABLE WITH SPODP

Table VII-2 summarizes the most recent orbit determination study using the SPODP.

Comparison with Table VII-1 shows that the mission accuracy requirements cannot be met at encounter using the SPODP.

D. ENCOUNTER ORBIT DETERMINATION ACCURACY ANALYSIS

1. *Freeing the ODP*

It is the practice in Orbit Determination to pick some epoch near encounter and

TABLE VII-2

	Pre midcourse		Post midcourse		Pre encounter	
	Traj. 1	Traj. 2.	Traj. 1	Traj. 2	Traj. 1	Traj. 2
SMAA (km)	435.1	428.4	423.5	424.9	419.1	420.5
SMIA (km)	418.3	377.7	365.2	363.6	205.8	205.3
θ MEO (deg)	49.8	91.3	89.2	89.8	96.1	96.1
σ_B	424.3	388.5	365.4	378.2	417.3	418.7
$\sigma_{B \cdot R}$	428.1	428.4	423.4	424.9	209.4	208.9
$\sigma_{B \cdot T}$	425.3	377.7	365.2	363.6	215.0	246.9
σ_{pB}	429.2	418.6	423.3	412.0	414.5	397.5
σ_{t_E}	90.0	90.0	60.0	60.0	8.0	8.0

Trajectory 1 launched 2-25-63, arrive Mars 7-31-69
Trajectory 2 launched 3-27-69, arrive Mars 8-5-69

discard all apriori information about probe position and velocity. This has the effect of 'freeing' the orbit determination process from accumulated modeling errors and allows the probe trajectory to be moved in a discontinuous fashion so as to achieve better agreement between predicted and observed data as the probe nears the target planet.

The epoch chosen for MM69 (and for previous planetary flights) is $E-5$ days. It is possible to approximate the errors in orbit determinations from this time on to near encounter by a very simple model.

2. The Dominant Error Sources and Their Effects

It can be shown that the dominant contributers to SMIA and SMAA can be combined by simple RMS averaging to provide a reasonably accurate representation of SMIA and SMAA for an orbit determined over the period from $E-5$ days to $E-1$ days. Thus

$$(\text{SMAA})^2 \doteq \left(\frac{\partial \text{SMAA}}{\partial r_s}\right)^2 \sigma_{r_s}^2 + \left(\frac{\partial \text{SMAA}}{\partial \dot{\varrho}}\right)^2 \sigma_{\dot{\varrho}}^2 \qquad \text{(VII-6)}$$

$$(\text{SMIA})^2 \doteq \left(\frac{\partial \text{SMIA}}{\partial \lambda}\right)^2 \sigma_{k_\lambda}^2 + \left(\frac{\partial \text{SMIA}}{\partial \dot{\varrho}}\right)^2 \sigma_{\dot{\varrho}}^2 \qquad \text{(VII-7)}$$

where r_s is the radial distance from the earth's spin axis to the tracking station. $\dot{\varrho}$ is probe topocentric radial velocity measured at the tracking station. $\lambda 15$ station longitude, and $k_\lambda = r_s \Delta \lambda$. Using Equations VII-3, VII-6, VII-7, we may write the approximate expressions for $\sigma_\mathbf{B}^2$ and $\sigma_\mathbf{PB}^2$ in terms of $\sigma_{r_s}^2$ $\sigma_{k_\lambda}^2$ and $\sigma_{\dot{\varrho}}^2$.

$\doteq$	$\sigma_{r_s}^2$	$\sigma_{k_\lambda}^2$	$\sigma_{\dot{\varrho}}^2$
$\sigma_\mathbf{B}^2$	$\left(\dfrac{\partial \text{SMIA}}{\partial r_s}\right)^2 \cos^2\eta$	$\left(\dfrac{\partial \text{SMAA}}{\partial \lambda}\right)^2 \sin^2\eta$	$\left[\left(\dfrac{\partial \text{SMIA}}{\partial \dot{\varrho}}\right)^2 \cos^2\eta + \left(\dfrac{\partial \text{SMAA}}{\partial \dot{\varrho}}\right)^2 \sin^2\eta\right]$
$\sigma_\mathbf{PB}^2$	$\left(\dfrac{\partial \text{SMIA}}{\partial r_s}\right)^2 \sin^2\eta$	$\left(\dfrac{\partial \text{SMAA}}{\partial \lambda}\right)^2 \cos^2\eta$	$\left[\left(\dfrac{\partial \text{SMIA}}{\partial \dot{\varrho}}\right)^2 \sin^2\eta + \left(\dfrac{\partial \text{SMAA}}{\partial \dot{\varrho}}\right)^2 \cos\eta\right]$

$$\text{(VII-8)}$$

Table VII-3 gives the squares of the partials. For trajectory 1 they yield results accurate to about 15%; for trajectory 2, 6%.

3. Navigational Accuracy at Encounter −1 Day

Table VII-4 gives the numerical values of the elements in the array of Equation VII-8.

TABLE VII-3

÷	$\left(\dfrac{\partial}{\partial r_s}, \text{km/m}\right)^2$ $\left(\dfrac{\partial}{\partial \lambda}, \text{km/m}\right)^2$	$\left(\dfrac{\partial}{\partial \dot{\varrho}}, \text{km/mm/sec}\right)^2$
SMAA	1685.5	309.7
SMAA	415.5	29.2

TABLE VII-4

	Traj 1, $\eta = -4.7°$			Traj 2, $\eta = 22°$		
÷	$\sigma_{r_s}^2$	σ_λ^2	$\sigma_{\dot{\varrho}}^2$	$\sigma_{r_s}^2$	σ_λ^2	$\sigma_{\dot{\varrho}}^2$
σ_B^2	11.3	412.7	31.1	236.5	357.2	68.6
σ_{PB}^2	1674.5	2.8	307.8	1448.9	58.3	270.3

Table VII-5 gives the effective standard deviations of the major error sources as they are seen by the SPODP and the DPODP. Thus for the DPODP, because of increased numerical precision and a more realistic model, station location parameters and RMD range rate are utilized several times more effectively. This results in dramatic improvement in the estimation of the critical parameters B and PB. Table VII-6 gives the relevant comparisons. Table VII-6 shows that the navigational accuracy requirements can be met using the DPODP.

TABLE VII-5

	$\sigma_{\mu s}/\sigma_\lambda$	$\sigma_{\dot{\varrho}}$
SPODP	5 to 10 m	3 to 6 mm/sec
DPODP	1.5 to 3 m	1 to 3 mm/sec

TABLE VII-6a

Trajectory 1

	SPODP (Note 1)	DPODP (worst case)	DPODP (best case)
σ_B (km)	215	64	35
σ_{PB} (km)	414	134	81
P_x (Note 2)	0.34	0.80	0.98

Note 1:   Numbers in this column are taken from Table VII-2, not computed from Equations VII-8.

Table VII-6 shows that the navigational accuracy requirements can be met using the DPODP.

TABLE VII-6b

Trajectory 2

	SPODP (Note 1)	DPODP (worst case)	DPODP (best case)
B (km)	247	77	44
PB (km)	395	126	71
P_x (Note 2)	0.35	0.84	0.98

Note 2: $P_x = \int\limits_{-x}^{x} \text{normal } f_r \cdot f\left(\dfrac{3°}{(\sigma_{PB}^2 + 1)^{1/2}}\right) dx$

See Section VII-A

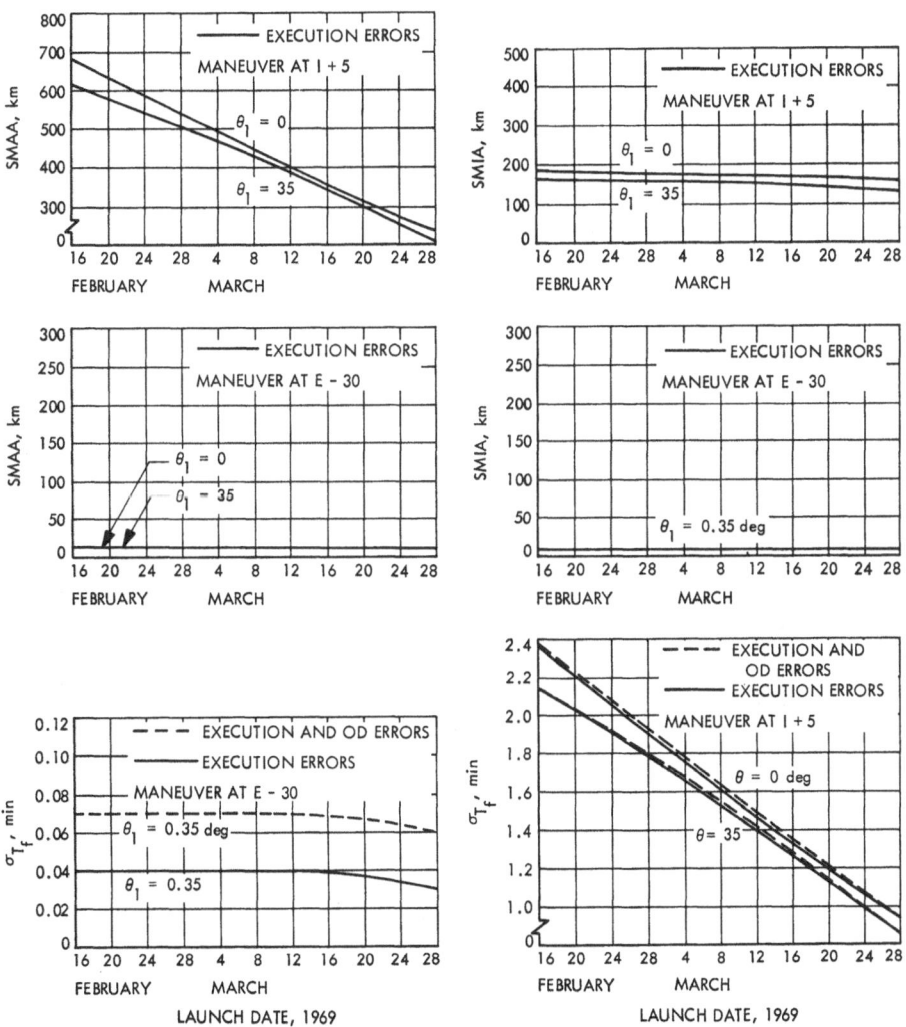

Fig. VII-4.

4. Near Encounter Orbit Determination

In the near vicinity of the planet, when the probe is being strongly influenced by the planetary gravitational field, the orbit determination process is capable of measuring very accurately the radial distance from the probe to the planet center. The ODP gives a very weak solution for the direction perpendicular to the probe-target vector.

We can approximate the effect of near encounter tracking on the uncertainty in the **B, PB** coordinates as follows:

Let $\Lambda_{\mathbf{B},\,p\mathbf{B}}(E-1)$ be the covariance matrix in **B, PB** at $E-1$, and let $\Lambda_{\mathbf{B}p\mathbf{B}}(E)$ be the covariance matrix very near to closest approach due to tracking from $E-1$ to a point

$$\Lambda_{\mathbf{B}p\mathbf{B}}(E) = \begin{bmatrix} \varepsilon & 0 \\ 0 & S_\infty \end{bmatrix}$$

where $\varepsilon \ll 1$ and S_∞ is very large.

The covariance matrix of the combined estimate of the tracking from $E-5$ days to E, and the tracking from $E-1$ day to closest approach is given by

$$\Lambda_{\mathbf{B},\,P\mathbf{B}}(\text{combined}) = \{\Lambda_{\mathbf{B},\,P\mathbf{B}}^{-1}(E-1) + \Lambda_{\mathbf{B},\,P\mathbf{B}}^{-1}(E)\}^{-1} \qquad \text{(VII-9)}$$

$$\Lambda_{\mathbf{B},\,P\mathbf{B}}(\text{combined}) \doteq \left\{ \begin{array}{ll} \varepsilon\left(1 - \dfrac{\varepsilon}{\sigma_{\mathbf{B}}^2}\right) & \text{correlation terms} \\[2ex] \text{correlation terms} & \sigma_{P\mathbf{B}}^2\left(1 - \dfrac{\sigma_p^2}{S_\infty}\right) \end{array} \right\} \qquad \text{(VII-10)}$$

Equation VII-10 paraphrases the previous statement. We get no improvement in $\sigma_{p\mathbf{B}}$, while $\sigma_{\mathbf{B}}$ is improved a great deal.

5. Ephemeris

It should be noted that Mars ephemeris errors are omitted from the computation of P_x in Table VII-6. They are also omitted from Equation VII-9 which would read

$$\Lambda(\text{Combined}) = [\{\Lambda(E-1) + \Lambda(\text{Ephemeris})\}^{-1} + \Lambda(E)]^{-1}$$

if ephemeris errors were included.

This is because an extremely good ephemeris is being used for MM69 whose 1σ solid of concentration at Mars is approximately a sphere of radius 18 km, which introduces negligible errors into the OD process. The ephemeris was obtained by combining all of the available planetary observational data for the last 60 years and performing what is essentially an orbit determination process for all of the major bodies of the solar system simultaneously deriving in all 64 constants of near motion of the planets. The result is a unique dynamically consistent ephemerides for the solar system.

Acknowledgements

We acknowledge the assistance of the members of the technical staff of the Tracking and Data Acquisition organization as well as the Precision Navigation staff of the Jet Propulsion Laboratory in preparing this paper. We are also indebted to K. W. Linnes and V. C. Clarke, Jr. for their critical reviews.

References

[1] Schurmeier, H. M., 'The Mission of Mariner Mars 1969', AIAA Paper M68-1050, 21 October 1968.

[2] Norris, H. W., 'The Mariner Spacecraft-1969 Model', AIAA Paper M68-1140, 21 October 1968.

[3] Trask, D. W. and Muller, P. M., 'Timing DSIF 2-Way Doppler Inherent Accuracy Limitations', JPL Space Programs Summary 37-39, Vol. III, p. 7.

[4] Develet, J. A., 'Fundamental Accuracy Limitations in a 2-Way Coherent Doppler Measurement System', *IRE Trans. Space Electron. Telemetry* (September 1961).

[5] Develet, J. H., 'Fundamental Sensitivity Limitations for Second Order Phase Lock Loops', IRSV Spring Meeting, May 1961.

[6] Curkendall, D. W., 'The Influence of Oscillator Instability on Orbit Accuracy', JPL Space Programs Summary 37-41, Vol. III, p. 42.

[7] Martin, W. L., 'Resolving the DSN Clock Synchronization Error', JPL Space Programs Summary 37-39, Vol. III, p. 49.

[8] Trask, D. W. and Efron, L., 'DSIF 2-Way Doppler Inherent Accuracy Limitations: Changed Particles', JPL Space Programs Summary 37-41, Vol. III, p. 49.

[9] Warner, M., 'The Double Precision Orbit Determination Program', JPL Space Programs Summary 37-47, Vol. II, p. 35.

[10] Warner, M., JPL Space Programs Summary 37-47, Vol. II, p. 37.

[11] Proceedings of the Seventh International Symposium on Space Technology and Science, Tokyo 1967.

[12] Kohlhase, C.W., Astronautics/Aeronautics, July 1969.

OPTIMAL VALUE FOR THE ANTENNA SPACING
OF A THREE ANTENNA INTERFEROMETER

V. V. LEBEDEV

U.S.S.R.

Abstract. The results of an article by W. Kendall [1] are considered. It is shown that the antenna spacing of a three-antenna interferometer is not independent of the required accuracy of angle measurements. An expression is obtained for the optimum value of the antenna spacing.

A somewhat peculiar conclusion, namely, that the accuracy of unambiguous measurements does not depend on interferometer antenna spacing, has been drawn in an interesting article by W. B. Kendall 'Unambiguous Accuracy of an Interferometer Angle-Measuring System'. At first glance, this conclusion is justified by contradictory demands for the minimization of sighting errors and for resolving ambiguities. As the spacing increases, the sighting accuracy increases too but conditions for resolving ambiguities become worse.

However, the essentially non-linear dependence of the error probabilities on antenna spacing does not permit one to hope for such a fortunate mutual cancellation of their tendencies of variation. More careful consideration of the results obtained by W. B. Kendall shows that the optimum antenna spacing of the interferometer is dependent of the required sighting accuracy.

Let us make use of the expression for the sighting error probability P_e

$$P_e = 1 - \theta \left(\frac{\delta}{\sqrt{2}\,\sigma_{\sin \theta}} \right) \theta \left(\frac{R}{2\sqrt{2}} \right)$$

where, in the previous author's notations,

$$\theta(x) = \frac{2}{\sqrt{\pi}} \int\limits_0^x \exp(-t^2)\, dt$$

δ is the permitted value of the sighting error,

$$\sigma_{\sin \theta} = \frac{1}{2N\pi \sqrt{\dfrac{E}{N_0}}} \sqrt{\frac{1 - \varrho^2}{1 - 2\varrho M/N + (M/N)^2}},$$

$$R = 2\pi \sqrt{\frac{E}{N_0}} \frac{D_{\min}}{\sqrt{1 - 2\varrho M/N + (M/N)^2}},$$

$\theta(R/2\sqrt{2})$ is the probability of resolving the ambiguity correctly, and ϱ is the correlation coefficient.

Substituting the expressions for $\sigma_{\sin \theta}$ and R we obtain a formula for P_e which

G. A. Partel (ed.), Proceedings of the Second International Conference on Space Engineering. All rights reserved.

is more convenient for analysis than that used by the previous author:

$$P_e = 1 - \theta \left(\frac{\delta}{\sqrt{2}} \frac{2\pi N}{\sqrt{1 - \varrho^2}} \sqrt{\frac{E}{N_0}} \sqrt{1 - 2\varrho \frac{M}{N} + \left(\frac{M}{N} \right)^2} \right)$$

$$\times \theta \left(\frac{\pi}{2} \sqrt{\frac{E}{N_0}} \frac{D_{min}}{\sqrt{1 - 2\varrho \frac{M}{N} + \left(\frac{M}{N} \right)^2}} \right)$$

where E/N_0 is the signal to noise ratio,

N is the distance between the extreme antennas measured in wavelengths,

M is the distance between the intermediate antennas, and

D_m is the ambiguity resolution efficiency function.

If we note that $D_m N \simeq \frac{1}{2}$ is independent of N for optimal arrangement of antennas, the dependence of the sighting error value δ on $\sqrt{(N^2 - MN + M^2)}$ and therefore also on $N(M = \frac{1}{2}; \varrho = \frac{1}{2})$ is clearly revealed.

Making use of the author's assertion that the only maximum of $\theta(x) \theta(a/x)$ is at $x = \sqrt{a}$ we obtain

$$\sqrt{N^2 - MN + M^2} = \sqrt{\frac{\sqrt{0.75}}{2} \frac{ND_{min}}{\delta}}$$

whence

$$N_{opt} = \frac{1}{2\sqrt{\delta}} \quad \text{for} \quad \delta \ll 1.$$

That N_{opt} is indeed a critical value is illustrated by Figures 1 and 2.

Thus, W. B. Kendall's results enable us to state the following:

(a) the spacing of the interferometer antennas depends on the sighting error. The optimal value of N is given above;

(b) optimal arrangement of antennas is the same for all signal-to-noise ratios;

(c) for greater E/N_0 values and greater accuracies $1/\delta$ smaller deviations of the antenna spacing its optimal value is allowed.

In conclusion, we would like to consider whether W. B. Kendall's δ-criterion is correct. It is natural to use it for monopulse sighting-angle measurements. However,

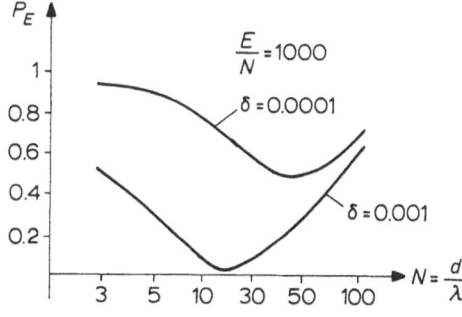

Fig. 1.

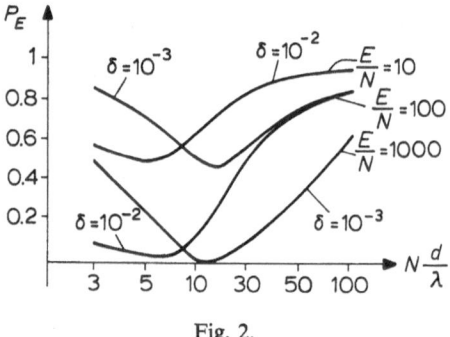

Fig. 2.

in real systems, even for monopulse structure of the radar, practically always a successive processing of many sighting measurements is effected. In such situations the ambiguity error may be perceived as the appearance of another target and, anyway, simply rejected. Then it is more reasonable to specify the value of the allowed ambiguity error separately and minimize P_e with $\theta(R/2\sqrt{2})$ fixed.

Reference

[1] W. B. Kendall, *JEEE Trans.* Nr. 2 (1965) SET-11, 62–70.

PART IX

TESTING, SUPPORT AND ACCESSORIES, II

Chairman: B. N. Petrov, U.S.S.R.

SPACE PROGRAMME IN INDIA

H. G. S. MURTHY

Thumba Equatorial Rocket Launching Station, India

1. Introduction

The scientific study of meteorology and of the variations in the earth's magnetism were started more than a century ago in India. It is now over 100 years ago the British scientist Allan Broun, with Indian assistants, made his important contribution to geomagnetism. (J. Allan Broun, F.R.S., was the Director of the Trivandrum Observatory from 1852 to 1865; and made some fundamental contributions to geomagnetism by his work on the diurnal variations of the magnetic declination at Trivandrum near the magnetic equator.) With a long established tradition of research in meteorology, ionospheric physics, geo-magnetism, cosmic rays, astrophysics and solar physics, India has a good scientific base and a deep interest in the exploration of outer space. Science and technology provide the very basis upon which the future of a country rests. It is not possible for a developing country like India to progress in this field, unless one can provide the opportunities of making discoveries in the fields of scientific research, which would assist in increasing the economy and the standard of living of the people. Space research provides many areas in which it will yield results of great practical interest and importance, and it is highly desirable that any country should undertake such research to advance knowledge in science and technology.

It is well-known that experiments are being conducted on communication via satellites. Satellite communication is expected to offer a dependable wide-band system of telecommunication, free from the usual atmospheric interferences, over large distances which may well supersede our present means of intercontinental communication and make intercontinental or worldwide television possible. This certainly is an area in which we should prepare ourselves from now to take advantage of the new developments in technology. Another area in which space research will make important and practical contribution is in the study of weather. This is a subject of importance not only for communication but also for agriculture.

Realising the above facts, India decided to undertake activities in space science and technology, consistent with our resources. In April 1962, Indian National Committee for Space Research (INCOSPAR) was constituted by the Department of Atomic Energy (DAE), Government of India, under the chairmanship of Dr. Vikram A. Sarabhai, for the promotion of space research and international cooperation in exploration of space for peaceful purposes.

2. Thumba Equatorial Rocket Launching Station (TERLS)

In 1963, an Equatorial Rocket Launching Station was established at Thumba, near Trivandrum, which lies practically on the magnetic equator and on the West Coast

G. A. Partel (ed.), Proceedings of the Second International Conference on Space Engineering. All rights reserved.

of India. The main objective was to carry out scientific studies of the upper atmosphere in equatorial regions, from 30 to 200 km altitude, using sounding rockets. With the active assistance and cooperation from France, USA and USSR, the TERLS became operational on November 21, 1963, when the first sounding rocket carrying a scientific payload was successfully launched from Thumba. TERLS has received UN sponsorship as an international range and was dedicated to the UN by the Prime Minister of India on 2nd February, 1968. Bilateral agreements between India and many countries were drawn up providing for international cooperation for carrying out scientific experiments from Thumba.

TERLS is now fully equipped as an international facility, to carry out rocket experiments. The facilities provided include vehicle and payload assembly, testing and check-out, ground support including, radar tracking, telemetry, a DOVAP and an AN/GMD-I equipment, range safety and met. support, range timing, telephone and radio communications, and range clearance equipments like, helicopter and sea vessel. At TERLS, certain ground-based experiments are also provided as back-up for the rocket experiments. These include measurement of ionospheric drifts, recording of ionograms using an automatic ionospheric recorder, measurement of cosmic radio noise absorption by riometer technique and measurement of total electron content in the ionosphere by recording the Faraday fading of Beacon Satellite signals and a magnetometer. Shortly, a back-scatter experiment is being added.

Since its inception Thumba has carried out 74 rocket experiments using single and two-stage rockets like Judi-Dart, Nike-Apache, Centaure (French) and boosted ARCAS. These experiments have been designed and carried out from Thumba for the study of the dynamics and composition of the upper atmosphere in the equatorial latitudes. The payloads include vapour cloud to measure winds in the neutral atmosphere using sodium, trimethyl aluminium and barium techniques; electron-ion probes and plasma noise probes for study of the nature of the ionosphere; UV detectors and ion-mass spectrometers to measure the composition of the ionosphere; proton precision magnetometers to investigate the electro-jet; X-Ray astronomy payload for measurement of X-Ray flux emanated from distant stars like Scorpio, Centaurus, Crab etc. in the celestial sky, and chaff to measure winds in the mesosphere. TERLS participated in the world synoptic launchings carried out in 1965 for measurement of upper atmospheric winds using sodium techniques. In addition to carrying out rocket experiments, TERLS has a Development Wing to give support to the experimenters. This division has undertaken development of special devices such as ejectable and split nose cone systems, de-spin mechanism, a chaff dart for met. programme etc. A payload recovery system suitable for sea recovery and an antenna deployment system for 'D' region experiment with radio propagation unit are now in hand. A number of electronic flight hardware such as telemetry transmitters, VCOs, DC-DC converters, electronic timers etc., have been developed and successfully used. Additional accessories for the ground support system such as radar, have been developed. These include open-sight for tracking radar, a computer and X–Y plotter.

An International Seminar on 'Sounding Rocket Techniques and Experiments' was organised by INCOSPAR at Kodaikanal and Thumba, with the assistance from UN, in January 1965. Several scientists from India and abroad participated.

Being an international facility, it is possible for all member countries who wish to conduct space research, to do so using the facilities at Thumba, under bilateral agreements acceptable to the participating countries. Naturally, it helps member states, which are unable because of economic or technological factors, or the unsuitability of their territory, to support rocket experiments through their cooperative efforts. In fact, this theme of international collaboration, which would be beneficial for various countries, particularly those developing countries, was greatly appreciated at the last UN Conference in Vienna held in August, 1968.

3. Experimental Satellite Communications Earth Station (ESCES)

To enable India to gain competence in global satellite communications, the Experimental Satellite Communications Earth Station has been established at Ahmedabad by INCOSPAR with assistance from the UN Special Fund. An important objective of ESCES is to train scientists and engineers of India and of other developing countries to enable them to make use of satellite communications. A modest beginning in space communication was made in late August 1967 when ESCES successfully transmitted and received messages and television pictures from NASA's ATS-2 satellite and established contact with Japan.

4. Space Science & Technology Centre (SSTC)

In 1965, the Atomic Energy Commission approved a proposal of INCOSPAR, to establish a Space Science and Technology Centre with major responsibility for developing sounding rockets of superior performance and for acquiring expertise in aerospace engineering and scientific payload construction for rockets and satellites. The Centre was also to provide back-up support to space research through ground-based experiments. The Space Science and Technology Centre (SSTC) has been set up at Veli Hill, by the side of TERLS. A large group of engineers has been engaged at the SSTC to work in different disciplines of rocket technology such as propellant engineering, propulsion, structural engineering, aerodynamics, materials engineering and quality control, control and guidance, technical physics, electronics, mechanical and systems engineering. The first India-made rocket of the Rohini RH-75 series was successfully flight-tested in November 1967.

5. Rocket Manufacture in India

To establish rocket manufacture in India, the Department of Atomic Energy has set up a Rocket Propellant Plant (RPP), at Thumba for manufacture of the propellant required for the French Centaure two-stage sounding rockets under licence from M/s Sud Aviation of France. This plant will be able to take up manufacture of large size propellant charges for indigenous rockets being developed in SSTC.

The hardware for the Centaure rocket has been manufactured in the Central Workshops of Bhabha Atomic Research Centre (BARC) in Bombay. A Rocket Fabrication Facility, which can undertake manufacture of hardware for various types of rockets has been approved by the Atomic Energy Commission, and will be set up at Thumba, near TERLS. This will be commissioned by April 1970.

6. Radio Astronomy

An extensive programme to pursue studies in radio astronomy has been undertaken both at the Tata Institute of Fundamental Research (TIFR), Bombay and at Physical Research Laboratory (PRL), Ahmedabad.

Scientists of PRL and other scientific institutions in India are engaged on development of scientific payloads to carry out rocket experiments from Thumba. In addition, scientists from France, USA, USSR, West Germany and Japan have already participated in collaborative programmes with India under bilateral agreements and experiments have been successfully carried out from Thumba.

7. Indian Rocket Society

With the increased tempo of space research activities in India, an Indian Rocket Society was formed to promote the work on space engineering and technology. This society was affiliated to the International Astronautical Federation (IAF), as a Voting Member. The society was inaugurated on 3rd February, 1968 in Trivandrum (India), by Dr. Luigi G. Napolitano of Italy, the then President of the IAF. A number of scientists and engineers from various countries have given talks to members of the society.

8. Future Plan

India proposes to set up another rocket launching station on the East Coast. This facility will be used in the initial stages to prove the indigenous rockets and ultimately have a capability for launching satellites.

A preliminary feasibility study of developing a modest scientific satellite is being worked out.

Satellite television can be a very powerful tool of mass communication and for promoting national integration. Its use for popularising agricultural productivity, in promoting general education and disseminating information about population control is of great significance. To demonstrate convincingly the relative advantages of this system, a pilot project was jointly undertaken in January 1967 by the Department of Atomic Energy, the All India Radio, the Indian Agricultural Research Institute and the Delhi Administration and the experiment was successful.

A Satellite Communication Project on a national basis is being actively considered and a study group of engineers has been formed to go into the details.

THE CHURCHILL RESEARCH RANGE
AND CANADA'S SPACE RESEARCH PROGRAM

R. S. RETTIE

National Research Council of Canada, Ottawa, Canada

Abstract. The Churchill Research Range, located on the shores of Hudson Bay in the geographic centre of Canada, is unique in many respects. It is located in the middle of the zone of maximum occurrence of the aurora borealis and is ideally suited for scientific investigations of this and related geophysical phenomena. Sounding rockets have been launched from the Churchill Research Range for over a decade and currently nearly 80 major scientific rockets and nearly 200 meteorological rockets are launched annually on behalf of Canadian, American, and other scientists.

In view of its sub-arctic location, most of these rockets are fired from within buildings, and all other equipment must be similarly protected. The paper will deal with the operation of the Churchill Research Range, operations at other northern launch sites in Canada, and the relationship of these to the Canadian scientific sounding rocket program.

1. Introduction

The Churchill Research Range, a joint Canadian/American project in Northern Canada is probably the most active scientific sounding rocket range in the world, launching 60 to 80 scientific payloads into the upper atmosphere each year. Additionally, many meteorological rockets and large balloons are handled each year but no satellites and essentially no test vehicles.

2. Location and History

The Churchill Research Range was located in 1956 at Fort Churchill, Manitoba, on the shores of Hudson Bay in Northern Canada because this location lies near the centre line of the auroral zone, the circumpolar belt in which auroral activity is of maximum intensity and occurs most frequently. Figure 1 shows the centre line of the zone, and the locations of the north geographic, geomagnetic and magnetic dip poles, and of Churchill. Also shown are the paths of totality for the solar eclipses of 1970, 1972 and 1979 and the fortuitously favourable position of Churchill.

Due to the tilt of the geomagnetic axis, the auroral zone lies further south in Canada than in Europe, and, of all possible locations for a rocket range, only Churchill had rail, sea and air transportation available. Large sea and uninhabited land areas would provide safe zones for the impact of spent rockets and payloads. A Canadian Army Base (Figure 2) could provide support functions far in excess of those available at the small settlement and port of Churchill about 5 miles away and with a population of about 3000 including nearby Eskimo and Indian villages (Figure 3). The Defense Research Board Northern Laboratory at Fort Churchill had been studying auroral and related phenomena for a decade.

Thus, in the mid-1950's when the possibilities of using sounding rockets to probe

G. A. Partel (ed.), Proceedings of the Second International Conference on Space Engineering. All rights reserved.

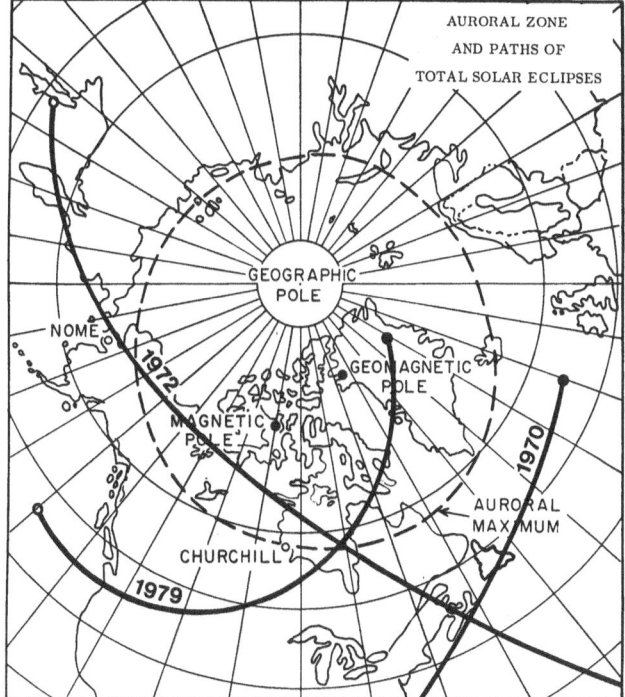

Fig. 1.   Auroral zones and paths of total solar eclipses.

the aurora borealis (Figure 4) became apparent, a group of American and Canadian scientists proposed the establishment of a rocket launching site at Fort Churchill. By agreement between the two governments and with the active assistance of the armies of both countries, a capability to launch Aerobee and Nike-booster rockets was constructed in time for operation during the IGY in 1957 and 1958. Nearly 200 sounding rockets carried scientific payloads into the upper atmosphere during this period. The range was operated as an offshoot of the US Army's White Sands Range until a serious fire destroyed much of the facility in February 1961.

The scientific results of previous launches had been sufficiently valuable that arrangements were soon made for rebuilding and reopening the range in late 1962 in time for the solar eclipse program of July 1963. The range was now operated by the United States Air Force. In the next two years it became very clear that civilian organizations were the major users of the range and that Canadian use had increased to a point where Canadian management was warranted. For various reasons, the Canadian Army had left Fort Churchill in 1964, and in the following year an inter-governmental agreement designated two civilian cooperating agencies to control the range, – the National Aeronautics and Space Administration for the USA and the National Research Council for Canada. Costs were to be shared equitably and NRC was given the responsibility for operation and management of the range. The actual transfer took place on 1 January 1966.

Fig. 2.   Fort Churchill.

3. Activities and Practices

The National Research Council of Canada, through its Space Research Facilities
Branch, has a small staff of about 15 people at the Churchill Research Range who
schedule rocket launches, carry out some auroral research which is interdependent
with the needs of rocket experimenters for ground support measurements, and
monitor the activities of the operating contractor for the range. Since 1962 the
operating contractor for the USAF and later for NRC has been the Aerospace
Services Division of Pan American World Airways who employ about 200 persons
at the CRR.

   Studies of auroral and related geophysical activities are the most usual objectives
of major sounding rocket launches at the CRR. Other upper atmospheric studies of
chemical and physical constitution, airglow, noctilucent clouds, etc. are also carried
out, and the CRR launches, three or more rocket-sondes per week as part of the
Meteorological Rocket Network. Because of the concentration on auroral studies,
which require darkness and clear weather, the range is more active in the late winter
months than during the rest of the year. Each scientist has his own objective of studying a particular type of phenomena, the occurrence of which may be relatively rare.
The scientist has the final say in regard to launching his rocket and may wait for days

Fig. 3.   Town of Churchill.

for a suitable event (some rockets have been ready on the launcher as often as 20 nights). Range Safety requirements may prevent a launch at an otherwise suitable time but no one except the project scientist determines the suitability of a geophysical event and orders the launching of the rocket.

In each of the last three years, the CRR has launched 60 to 80 major scientific rockets of which about half have been related to auroral phenomena. Considering that the average number of attempts before geophysical and weather conditions are acceptable is three or four and that on many evenings no attempt can even be started because of bad weather, it is seen that scientific groups tend to all want the same night and have to queue up for their turns. It is not uncommon to have four or five groups present at any one time. The CRR thus needs considerable flexibility in types of launchers and in its telemetry equipment.

A further consequence of concentration on particular auroral events is the need for rapid response on the part of the launch crews when the project scientist decides it is time to launch. When it is realized that the desired auroral event may only last a few minutes and that the rocket takes perhaps three of these to reach the desired altitude, the time which can be allowed between the decision to launch and the actual launch is very short. Extended holds at $T-5$ min, $T-180$ sec, $T-90$, and even $T-30$ sec are the rule, the time chosen being determined by the characteristics of the launch, the experiments and other support equipment, and on the level of auroral activity at the moment. Because of strain on crews, holds at short times to launch

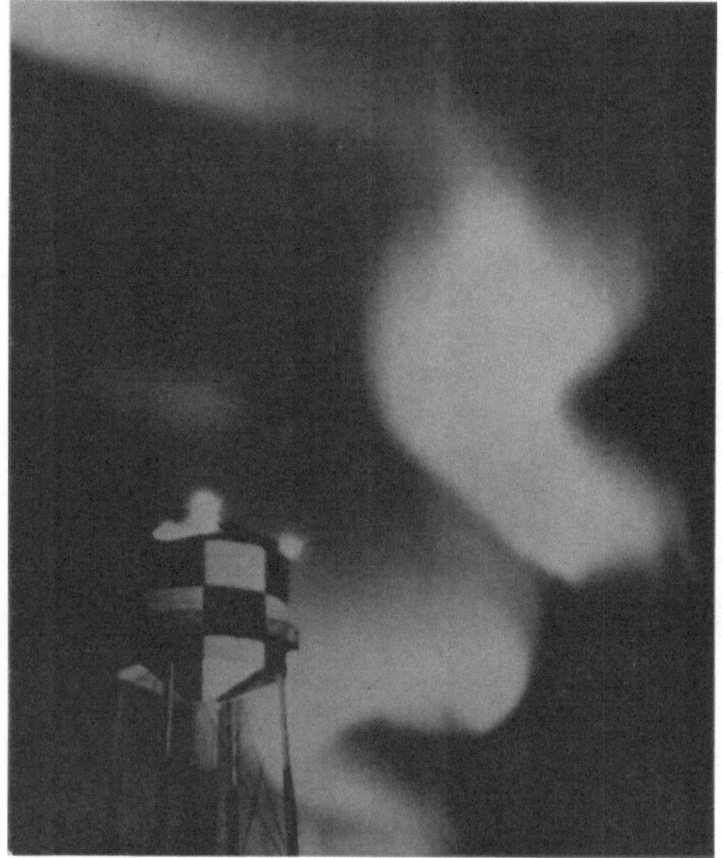

Fig. 4. Aurora borealis.

cannot be very long and are frequently converted to a standby condition as, for example, being prepared to pick up the count at T − 30 sec after 5 min warning. Again, because of the rapid changes characteristic of auroral activity, an ability must exist to stop the count at any time before T − 0. Holds within the last few seconds of a count usually necessitate recycling the count to an earlier point and occasionally far back in the count if supplies of magnetic tape or recording charts become depleted and reloading is needed. A change in auroral conditions after T − 0 may be fortunate or unfortunate, and, in the former case, unexpected information is obtained while, in the latter case, nothing can usually be done to salvage the experiment.

Often coupled with these requirements for flexibility in time of launch and rapid reaction to changing geophysical conditions are very low temperatures and winds up to limits determined by launcher structural problems, ballistic performance of rockets and safety of personnel. Temperatures are frequently too low to allow propellant grain and payloads to cold soak for appreciable periods while wind chill factors and cold may prevent efficient crew operations. Consequently, enclosed launchers and passageways, heat shield chimneys, and quick opening rocket exit and blast relief doors are used at the CRR, and the complex interactions between the capabilities of these devices and the requirements of the scientists must be thoroughly understood by the launch crews.

4. CRR Headquarters

It is against a background of operations such as these that the CRR has been designed and has evolved with great flexibility and a capability to overcome bad weather conditions. Its headquarters, transport, supply and mechanical support services are located in the ex-army camp which also contains housing, schools, a hospital, a technical training school for Eskimos, and detachments of various Government agencies having activities in the central Arctic. Also located in the camp are laboratories and work spaces for permanent staff, for rocket user groups before they proceed to the range head, for balloon user groups, and for visiting scientists of varied disciplines who come on a space available basis. The commercial airport adjacent to the camp is used by the Air Section of the CRR whose duties include recovery of parachuted scientific packages and logistic transport to outstations. In the summertime, high altitude research balloons are flown from the airport under the auspices of the CRR (Figure 5). Some 20 to 30 US Navy Skyhook balloons of up to 10 000 000 ft^3 capacity is the usual workload with the range supplying tracking and some telemetry on an 'if available' basis. Recovery of balloon experiments is carried out several hundreds of miles to the West by contractors to the US Navy. Other smaller balloons are flown at various times, usually by the experimenters themselves, and usually as part of an experiment involving rockets as well, perhaps to determine geophysical parameters and thus allow a decision to proceed with a rocket launch.

5. Remote Sites

The launch head of the CRR is located about 11 miles east of Fort Churchill and 10 miles further on is Twin Lakes, accessible by the only road in the area and used as a remote camera and optical site, for sound ranging on grenade explosions in the upper atmosphere, and for radio interferometric measurements of rocket-borne transmitters in order to determine the effects of ionospheric irregularities, particularly those associated with columns of auroral ionization (Figure 6).

Other remote camera sites include Belcher (57 miles south), O'Day (73 miles south) and Sipiwisk (290 miles southwest), accessible by rail and helicopter, and North Knife River (51 miles southwest) and Caribou River (52 miles northwest), accessible by helicopter and special fixed wing aircraft and with great difficulty over land, rivers, and the Hudson Bay. All these sites are equipped with shelters for personnel and equipment and with electric generators. Other locations for triangulation of upper atmospheric activities are available at Fort Churchill itself, at the Auroral Observatory of the CRR on the road to the launch site, and at Poste-de-la-Baleine, Quebec, at a distance of about 650 miles on the opposite side of Hudson Bay where the Space

Fig. 5.   Scientific high altitude balloon.

Research Facilities Branch of the NRC operates the Great Whale River Geophysical Station (Figure 7) in a location magnetically conjugate to Byrd Station in the Antarctic. These last three locations are manned continuously but the others are available only on demand and choices are made on the basis of the scientific requirements and on the capabilities of the CRR and its visitors to man and equip such stations.

6. Auroral Observatory

Starting east from Fort Churchill, one reaches, after 3 miles, the Auroral Observatory of the CRR (Figure 8). This facility provides a great number of observing domes and huts for spectrometers, photometers, etc., and resident scientists and visitors find reasonably comfortable conditions for their equipment and themselves. An Auroral Radar is also operated on 48 MHz, riometers at 30 MHz on a polar axis and on an axis which covers frequently used rocket trajectories and various magnetometers

Fig. 6.   Twin Lakes Outstation.

Fig. 7.   Great Whale Geophysical Station

Fig. 8.   Auroral Observatory.

together with remoted indications as required and available from certain outstations. Another riometer is located at the launch head and other magnetometers at the Magnetic Observatory of the Dominion Observatory, northwest of the camp.

The Auroral Observatory is thus a kind of geophysical nerve centre of the CRR and some rocket project scientists prefer to control their launches from here rather than at the launch site. The visual observing domes are also somewhat more suitable for an overall picture than the one tower at the range head. Also located at the Auroral Observatory is a Real Time Telemetry Station specially built as a cooperative project between the Federal Republic of Germany and Canada for reception of data from the FRG satellite AZUR expected to be launched later in 1969 (Figure 9).

Between the Auroral Observatory and the CRR Radar Site, some 9 miles from Fort Churchill, is located a Partial Reflection Ionosphere Sounder operated by the Communications Research Centre of the Canadian Department of Communications, and transmitting and receiving systems for various Northern Communications circuits. Some of these are instrumented to provide data of value in analyzing Polar Cap Absorption events. To the south of Fort Churchill itself is a Vertical Incidence Ionosonde also operated by the Department of Communications. Thus, radio frequency measurements of ionospheric quantities are readily available for comparison with rocket-borne measurements.

7. Radar

At the CRR Radar Site (Figure 10) are located four radars, one AN/MPS 19 and two FPQ/11 for tracking and one MPS504 for surveillance. All operate at S band

Fig. 9.   Real time telemetry antenna for FRG satellite AZUR.

(2700–2900 MHz). The tracking radars have the following technical characteristics.

AN/MPS 19 and FPQ 11 radars
Power   500 kW
Antenna   8 ft parabolic (MPS 19)
                                           14 ft parabolic (FPQ 11)
Tracking Accuracy-Range   ± 20 yds
                    Angles   ± 10 mils (MPS 19)
                                     ± 0.15 mils (FPQ 11).

All tracking radars may be used for either skin or beacon tracking. All radars are capable of skin tracking a 0.15 m² target to 224 nautical miles and with the addition of a beacon in the payload, the maximum range can be extended to about 1000 miles. The FPQ 11 radars provide both analog and digital data, the MPS 19 only analog data. Each radar provides remote positioning data which may be used for pointing other radars. Acquisition aids include optical tracking, infrared, closed circuit television and telemetry. Signals from the telemetry transmitter on the rocket are received on four antennas connected as interferometers and mounted on a tracking pedestal to which the radars may be slaved. The radar output data, in digital form, may be reduced at the CRR using the IBM 1800 computer system in the Meteorological Section into forms more suitable for scientific use by referring the origin of the coordinates to the launch site or other suitable location, smoothing the radar data, extrapolating beyond radar coverage, correcting for earth curvature, etc.

Fig. 10.   Radar site.

8. Operations, Telemetry

The launch head of the Churchill Research Range comprises an operations building, a meteorological building, a blockhouse and power station, and seven launchers (Figure 11). The operations building contains the main and backup telemetry ground stations, a users payload preparation area 60 ft by 100 ft capable of providing space for half a dozen payloads simultaneously, the communications centre of the range, a cafeteria, and workshops for plant maintenance.

The main and backup telemetry stations in the operations building provide three independent telemetry systems (Figure 12). Stations 2A and 2B are two systems which may be used to support simultaneous launchings. For missions of extreme complexity, the facilities of the two systems may be combined. The backup station 2C is used to provide redundant recording of the composite video signals from receivers and antennas separate from those of the primary station. The telemetry systems are based on IRIG standards. FM/FM telemetry at P Band frequencies with proportional bandwidth subcarrier modulation is used on most flights. PAM and PDM time division multiplex signals of up to 90 channels may be decommutated. PCM support at present is limited to reception and recording of the RF link only. Telemetry data is not reduced at CRR. Flight results are given to the project scientist in the form of magnetic tapes and quick-look data in the form of permanent oscillograph and direct write recordings.

Fig. 11.   Launch site.

Fig. 12.   Telemetry station operations building.

9. Meteorology

Meteorological support consists of the sampling and analysis of surface pilot balloon and radiosonde data (Figure 13). Surface wind and low level wind data up to 200 ft are obtained by the use of standard aerovanes mounted on a microwave communications tower. Single theodolite pilot balloon tracking provides winds up to 5000 ft. The rawinsonde system, a balloon-borne radiosonde tracked by an AN/GMD-1 1680 MHz radio direction finder, extends the wind data to about 100 000 ft. Meteorological rockets launched on the World Meteorological weather rocket program provide further wind speed and direction to about 250 000 ft. An IBM 1800 computer located in the Meteorological Building is used in the reduction of radar data, impact prediction and all rawinsonde data reduction.

10. Blockhouse and Launchers

The blockhouse from which rockets are controlled has consoles for appropriate combinations of the seven launchers and has three independent positions where payload checkout equipment can be connected for pre-launch tests. The adjacent powerhouse houses four 250 kVA diesel generators, a steam plant and associated equipment. A personnel passageway leads to rocket preparation areas with rocket passageways continuing to the enclosed launchers. Launcher facilities are described as follows:

Fig. 13.   Meteorological station.

Pad 1 Universal Launcher (Figure 14)

This boom-type launcher is enclosed in a heated building when in the horizontal position. The roof opens to permit the elevation of the boom and rocket but only slowly at some minutes before launching. The launcher is therefore equipped with a foamed plastic split heat shield chimney which surrounds the rocket and through which heated air can be blown. This heat shield can be opened as late as one minute before launching. Azimuth and elevation can be controlled locally or remotely from the blockhouse. The launcher is designed to handle vehicles up to 60 ft in length and up to 10 000 pounds in weight. Interchangeable rail systems allow this launcher to be used for a great variety of hanging rockets (i.e. mounted under the boom before elevation) and it is the most flexible of the CRR launchers.

Pad 2 Nike Launcher (Inside) (Figure 15)

A modified Nike Launcher with remote and local controls is enclosed in a special building with a quick-acting roof opening. Top-riding rockets of the Nike Booster class and the Black Brant III can be elevated to the vertical within the building.

Fig. 14.   Pad 1 Universal Launcher.

Fig. 15.   Pad 2 Nike Launcher.

Pad 3 Aerobee Launcher (Figure 16)

An Aerobee Tower Launcher, 120 ft high, has been enclosed somewhat more thoroughly than similar launchers in more temperate climates. The top 60 ft of the Tower is not enclosed. Once more control of azimuth and elevation can be local or remote. The rail system is suitable for the three-fin Aerobee 150 which is the only liquid propellant rocket launched at the CRR, and filling liquid and high pressure gas facilities are located in adjacent buildings.

Pad 4 Arcas Launcher (Figure 17)

A standard breech loading tube-type launcher, manually controlled for azimuth and elevation, is located in an annex to the Aerobee preparation building. A manually

Fig. 16.   Pad 3 Aerobee Launcher.

operated sliding roof allows the launcher to be elevated. A similar launcher for Judi-Darts can be located in this same room or, if required, in other launcher areas.

Pad 5 Outside Black Brant Launcher (Figure 18)

A boom-type launcher for Black Brant vehicles on a fixed azimuth of 108.3° is located outside the blockhouse. No heat shield is available but a portable shelter can be used to protect personnel and the vehicle during launch preparations. A plastic bag over the rocket itself can be heated by forced hot air to provide protection after the shelter is removed and the launcher elevated.

Pad 6 Nike Launcher (Outside) (Figure 19)

This launcher is similar to the Pad 2 launcher but has only local elevation and azimuth controls. It is equipped with a foamed plastic heat shield similar to that on the Universal Launcher and this can be opened and closed locally or remotely. This launcher is located outside the blockhouse. Other pads in the same area have been surveyed and are capable of accepting Nike Launchers which can be brought in for special occasions. Four such launchers were installed for the 1963 eclipse program and it is expected that similar action will be needed in 1972.

Pad 7 Auroral Launcher (Figure 20)

This specially designed launcher handles all hanging rockets except the Javelin and

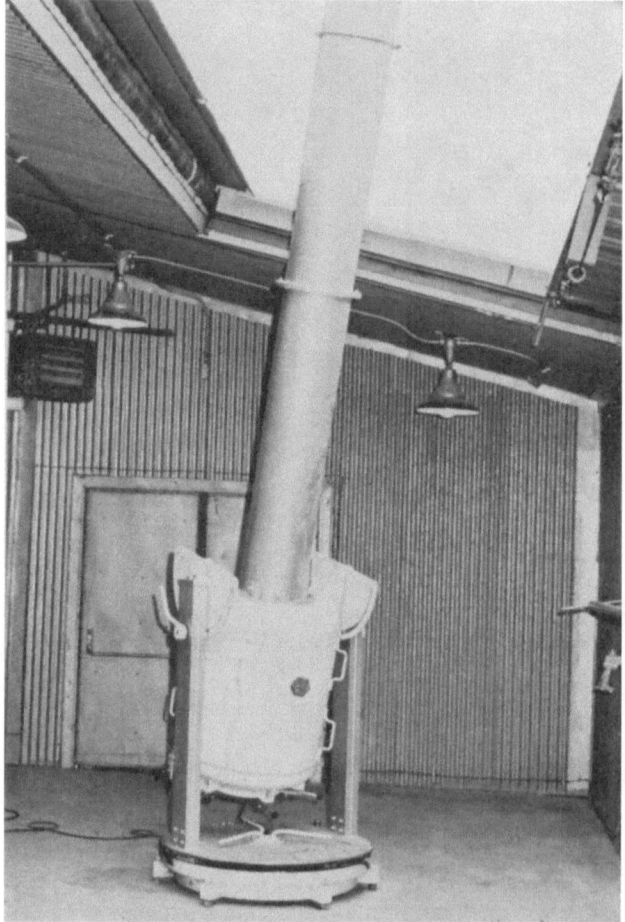

Fig. 17.   Pad 4 Arcas Launcher.

is limited in length to 38 ft and in weight to 4000 pounds. Rockets can be elevated and positioned completely within the building by local or remote control and quick-acting roof doors and blast release doors are opened during the final seconds before launching. This launcher is of tripod construction, mounted on a circular track. After the boom and rail system are elevated to near vertical, the top of the boom does not move. Instead the bottom is moved radially or circumferentially with the tripod around the circular track. This somewhat complex construction allows a much smaller roof area, much quicker acting doors both opening and closing for reheat whether or not the rocket has been launched, and provides an apparently improved performance as regards wind turbulence effects created by building structures.

The list of rockets that are launched from the Churchill Research Range include

Fig. 18.   Pad 5 Outside Black Brant Launcher.

the following:

Javelin (Argo D-4)	Black Brant VB
Aerobee 150	Black Brant VA
Nike Tomahawk	Black Brant II
Nike Iroquois	Black Brant III
Nike Javelin	Black Brant IV
Nike Apache	Skua
Nike Cajun	Boosted Arcas
Judi Dart	Arcas

An appropriate selection from the available launchers is made as required and many of these vehicles can be launched from any one of several launchers. The Boosted Arcas is launched either from a rail or from the Arcas tube launcher, depending on the particular type of rocket. The Skua is also launched from a launcher tube which is merely strapped onto any of the boom-type launchers as required.

Beyond the launcher complex at the beginning of the road to Twin Lakes are ocated the heated storage bunkers for rocket motors, igniters and pyrotechnics.

Fig. 19.   Pad 6 Outside Nike Launcher.

11. Impact Areas (Figure 21)

The Churchill Research Range has a land impact zone south of the range head approximately 40 miles wide by 100 miles long. This is used for smaller rockets and for any recovery packages. Recovery is practical only in the wintertime because of very extensive water and swamp areas. Recovery from Hudson Bay is not favourably regarded. Other impact areas include a nearby section of Hudson Bay reaching out 100 miles plus other areas totalling almost the whole of Hudson Bay, the availability of these areas being determined on a seasonal basis dependent upon shipping activity from Fort Churchill to Hudson Straits.

12. Other Locations (Figure 21)

Although the Canadian scientific sounding rocket program is primarily oriented towards the launchings from the Churchill Research Range, a few rockets are launched elsewhere each year. Expeditions have been carried out in each of the last three years

Fig. 20.   Pad 7 Auroral Launcher.

to Resolute Bay on Cornwallis Island in the centre of the Arctic Archipelago at 75°N, 95°W, near the magnetic dip pole. In addition to studies of the polar cap ionosphere, this location allows certain measurements, such as Galactic X-ray Astronomy, to be made from rockets without interference from trapped particles in the Van Allen radiation belt. Rudimentary launching facilities have been placed near the airport of this supply centre for the Canadian Arctic. A fixed azimuth boom launcher capable of handling under-hung vehicles such as the Black Brant III and a standard closed breech Arcas tube launcher have been installed. Small buildings provide space for rocket assembly, telemetry, and checkout equipment brought in with each expedition. Meteorological facilities are provided by the Meteorological Service of Canada. The only trajectory information currently available is derived from Doppler type measurements on the telemetry frequencies. In addition to the Canadian Black Brant III rockets, American colleagues have launched Arcasondes and are proposing to launch some Nike boosted vehicles.

A further expedition is planned for March 7, 1970 when four Black Brant III rockets will be launched from East Quoddy, some 60 miles northeast of Halifax,

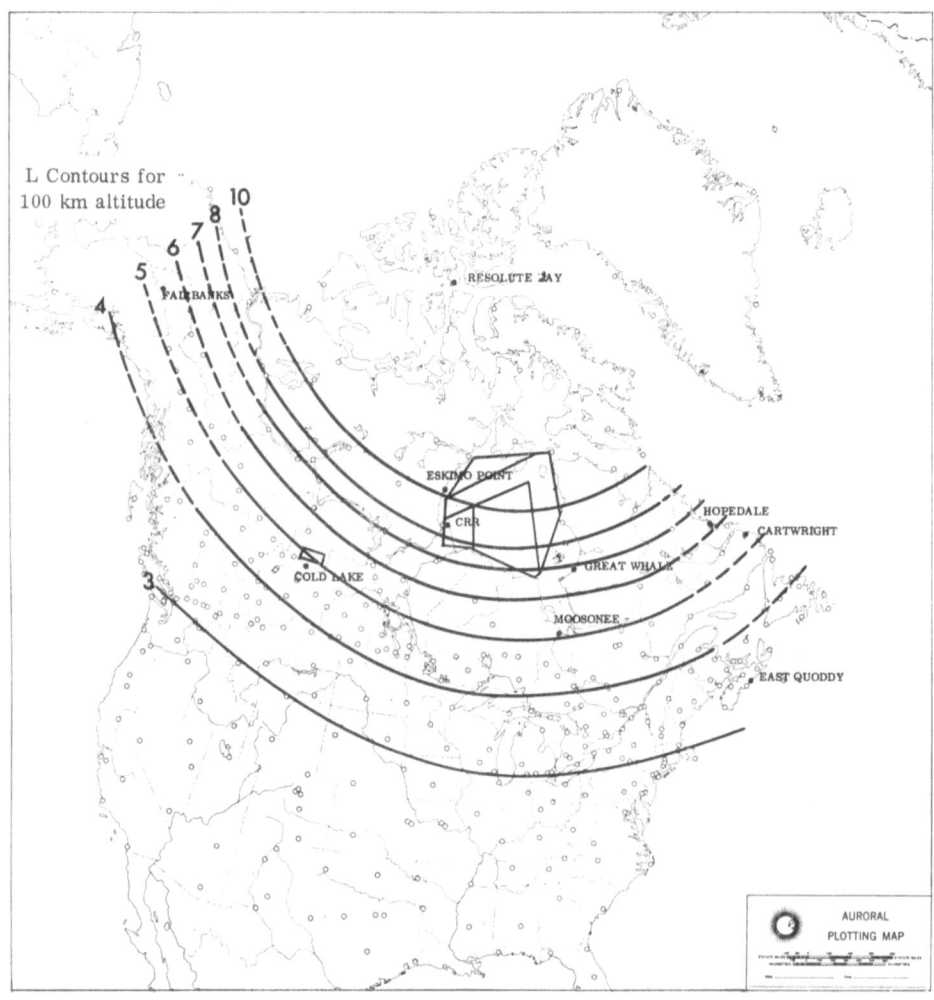

Fig. 21.   Impact areas CRR and magnetic contours.

Nova Scotia during the total eclipse of the Sun on that date. This location has the interesting property that it may well be used again on the 10th of July, 1972 during another total eclipse. The path of this latter eclipse not only intersects that of the 1970 eclipse path in Nova Scotia but also passes across the middle of Hudson Bay. Great Whale River Geophysical Station will be on the edge of totality at ground level and rockets launched at the Churchill Research Range can readily pass through the shadow region at altitude. One particular experiment, a joint Canadian/American/ British experiment will probably require the establishment of a temporary launch site at Eskimo Point, 150 miles north of Churchill on the west coast of Hudson Bay in order that the rocket payload may fly along the shadow path rather than at right angles to it.

13. International Cooperation

The Churchill Research Range is a joint American/Canadian undertaking but its facilities are available to scientists from other countries. Already several rockets have been launched for scientists from the Federal Republic of Germany. The Canadian sounding rocket program in itself has also some international activities. Guest scientists have made arrangements with Canadian colleagues to have experiments incorporated in Canadian payloads. The rules are simply that these experiments shall provide either complementary information or critical comparisons between methods of measurements. The procedures, which are rigidly enforced, are that all contacts between guest experimenters and the National Research Council are through the Canadian host scientist except when merely technical details of payload integration have to be worked out. Guests from France, Germany, Sweden, the United Kingdom and the United States have taken part in Canadian rocket- or balloon-borne experiments at the Churchill Research Range.

Acknowledgements

The author wishes to thank members of his staff, both at headquarters in Ottawa and at the Churchill Research Range, not only for providing material for this paper but for an excellent job of developing and operating the Churchill Research Range, and in particular, Mr. G. O. Berringer of the Range Section at SRFB, Headquarters, who has provided much of the written material on technical matters.

DESIGN AND DYNAMIC CHARACTERISTICS INVESTIGATION
OF A HIGH PERFORMANCE SPHERICAL BEARING

E. GIMELLI and D. DINI

Nuova San Giorgio S.p.A., Divisione Servosistemi ed Elettronica

and

R. C. MICHELINI and R. GHIGLIAZZA

Istituto di Meccanica Applicata alle Macchine, Università di Genova

Abstract. The note presents some design considerations for a multipad externally pressurized spherical bearing fed with incompressible fluid.

The bearing is mounted on a ground equipment designed for the tests on the attitude-control system of satellites; it allows considerable pitch and roll movements and a complete freedom in yaw.

The test equipment can be utilized both for the tests of three axes stabilized and spin stabilized satellites.

A great amount of experimental work has been done on this particular bearing; and its utmost dynamic characteristics are herein presented.

The complete stability in operating conditions is pointed out, even with the quite low damping properties of the system.

1. Brief Description of the Test Equipment in which the Spherical Bearing is Utilized

In the framework of the program of the SIRIO satellite (an experimental communication satellite developed under the responsibility of the Italian industrial consortium CIA), a special equipment for testing the attitude control system of satellites was designed and is now in production.

The test equipment consists essentially of two parts:

a spherical hydrostatic bearing capable of carrying, with minimum torque interactions, the complete satellite or a full scale inertial model of it.

a three-axes servoed suspension supporting the above said bearing and capable of following or reproducing the attitude movements of the satellite about its center of mass.

The general mechanization of the test equipment is sketched in Figure 1; this drawing shows the spin-stabilized SIRIO satellite carrying an Apogee motor in operative conditions; for the purpose of the test, the motor is replaced by the inertially equivalent satellite adaptor.

The satellite is mounted, through the adaptor, on the spherical bearing, in such a way that its center of mass coincides with the bearing center of rotation (this operation is carried out as a normal static balancing, on the spherical bearing itself).

The bearing allows the satellite to move freely about three axes, for a limited amplitude; but the entire suspension with its servocommanded motion allows the satellite a considerable freedom of the motion about $\pm 20°$ in pitch and roll, and continuous rotation in yaw.

G. A. Partel (ed.), Proceedings of the Second International Conference on Space Engineering. All rights reserved.

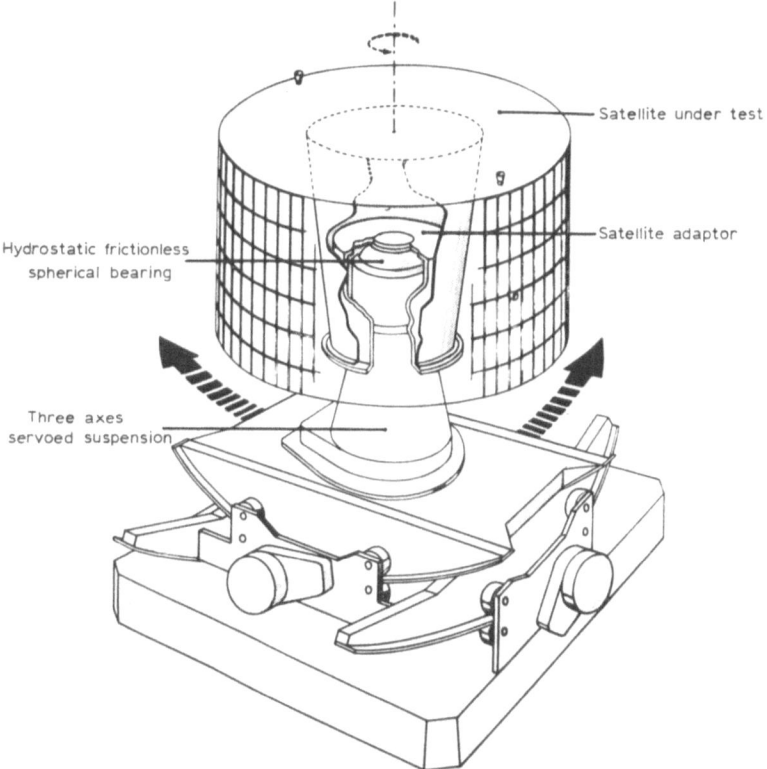

Fig. 1.   Three degrees of freedom simulator for the attitude control system tests.

In effect, a set of proper pick-ups measures the relative motion of the two parts of the spherical bearing; these measures go to control, after trigonometrical transformations, the three electromechanical servos of the suspension, in order to make the entire equipment to follow the satellite motion.*

With this mechanization, the test equipment becomes a very flexible tool for a variety of tests (from inertial measurements to the complete dynamic tests under real jet operation) and presents some marked advantages over the two classical types of attitude control system test equipments (gas bearing and flight simulators) as it has been discussed in [1].

Between the various problems arising from the design of the test equipment, some one relating to the design of the three axes servoed suspension, are dealt with in another paper of this Congress; in this paper we discuss the design of the spherical bearing and summarize the results of a long experimentation.

* In the use with spin stabilized satellites, as the SIRIO satellite, the suspension can be slightly modified in order to maintain the spin-speed stated value, against the air drag effects, if the tests are done for simplicity, in ambient pressure: in this case between the inner element of the servoed suspension (the spinning shaft) and the spherical bearing 'moving' element, a light homokinetic joint is added to apply about the main inertia axis of the satellite the small torques necessary to counteract the drag torque.

2. Requirements for the Spherical Bearing

2.1. Load carrying capacity

The first requirement is dictated by the need to sustain a satellite weighing up to 450 kg; the bearing must be able to carry such load both when in vertical position and when in non-vertical position, with a maximum angle of about 25° with respect to the vertical; in the latter case the transversal load of the bearing is about 200 kg.

2.2. Spurious torques

The bearing must minimize the spurious torques transmitted between the two parts of the bearing.

With the adoption of a servoed basis the relative motion of the two parts of the bearing is minimized due to the fact that the cup of the bearing follows the motion of the ball; therefore the viscous torques associated with the relative motion are only those due to the dynamic errors of the servoed suspension.

On the other hand, the fact that the bearing works in non-vertical positions can give rise to other spurious torques, also of viscous nature, due to the non-symmetry of the oil efflux in the bearing, and careful design to avoid these effects is necessary.

2.3. Axial and lateral stiffness and stability of the bearing

The adoption of a servoed equipment, permits the use of an oil bearing instead of an air bearing because the relative rotation is minimized.

As a counter part, the bearing becomes an important element of the chain of the servomechanisms that are, in the present case, relatively high band-width servos (of the order of 5–6 Hz).

Therefore important features of the bearing are its lateral and vertical stiffness, as well as the stability of the bearing clearance.

3. General Description of the Spherical Bearing

3.1. Type of bearing

The bearing is of the so-called 'hydrostatic' type, that is with external pressurization, and is supplied with hydraulic oil of proper viscosity.

The active part of the bearing is an assembly of eight pads, each consisting of a recess and a sill; the geometry of the pads is such as to cover an annular portion of the spherical surface of the cup forming the 'fixed' part of the bearing.

The chosen disposition of the pads gives lateral rigidity to the bearing and in the same time leaves free the central portion of the cup in which the shaft of the spherical homokinetic joint recalled in the note on the previous page, has to pass.

This central hollow part of the cup also collects and sends to the reservoir the oil flowing through the bearing for recirculation.

The various pads are independently fed with pressurized oil; the supply independence of the various pads enables the bearing to withstand lateral loads and is ensured by interposing along the feeding pipe of each recess a compensating element, consisting, in our case, of a calibrated restrictor.

A simplified diagram of the hydraulic system is given in Figure 2.

In operation the oil pressure lifts the ball by an amount that is a function of the parameters of the bearing; the oil film thickness over the bearing is in general non-uniform and depends on the displacement of the two parts of the bearing as well as on the difference in diameter of cup and ball; as a matter of fact, two types of coupling are possible: the so-called 'fitted' bearings in which the ball and cup diameters are identical; and the 'clearance' bearings in which the ball diameter is slightly smaller than the cup diameter.

In the first case, during the operation the ball assumes a certain eccentricity e with respect to the cup; and the film thickness decreases from the center to the edge of the bearing.

In the second case the film thickness from the center to the edge of the bearing can diverge or converge depending on the eccentricity that the ball assumes during operation; in particular it is constant and equal to the clearance C, when the eccentricity is zero.

The 'fitted' bearings are more used than the 'clearance' ones, since it is easier to lap the ball and cup together than to make a ball only few hundredths of a millimeter

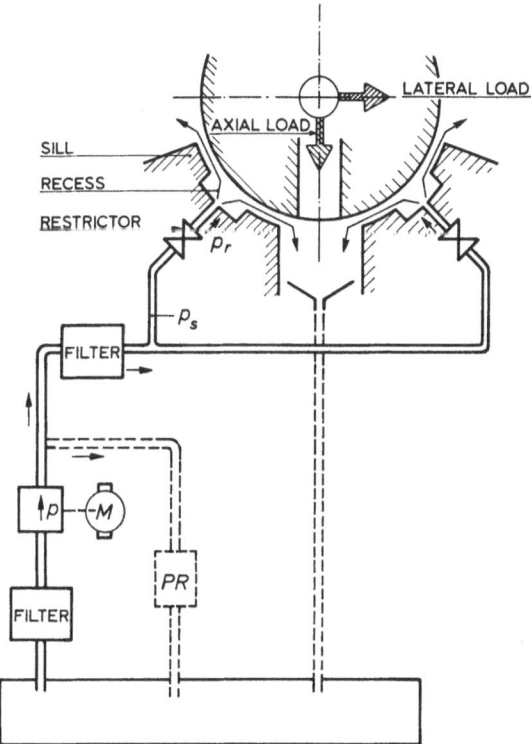

Fig. 2.   Simplified schematic of the hydraulic system.

smaller than the cup; however when the bearing has to carry heavy lateral loads, and therefore the included angle of the cup is necessarily large (± 60 deg or more) it has the disadvantage of too thin film thickness at the edge of the cup.

In our case a 'clearance' type bearing was chosen and the value of the clearance was about 5–6 hundredths of millimeters.

Lastly, as regards the dimension of the sphere, two factors have been considered:

necessity of having, inside the sphere, sufficient space for the homokinetic joint and for a torque transducer,

compromise between the dimensions of the bearing, and the values of the oil pressure necessary to lift the load.

On the basis of the above requirements a nominal diameter of 160 mm was chosen.

3.2. DESIGN PARAMETERS

For describing the pad geometry a coordinate system is adopted as shown in Figure 3.

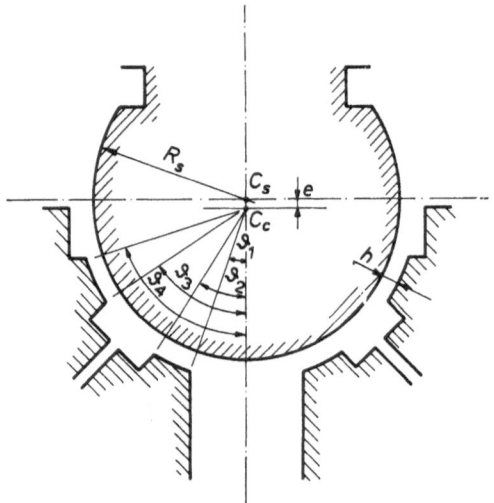

Fig. 3.   Definition of the geometrical parameters.

The following other primary parameters are defined:

h	film thickness	(m)
$p_0 (=0)$	discharge pressure	(N/m²)
p_r	bearing recess pressure	(N/m²)
p_s	bearing supply pressure	(N/m²)
A_p	total projected bearing pad area	(m²)
Q	volume flow of lubricant	(m³/sec)
μ	absolute viscosity of the fluid	(N·sec/m²)
W	load acting on the bearing	(N)
R_s	radius of the sphere	(m)

3.3. FUNCTIONAL CHARACTERISTICS

The main functional characteristics of the bearing may be expressed in the following way:

3.3.1. *Load Carrying Capacity*

This may be expressed by the relation:

$$W = a_f \cdot A_p \cdot p_r \tag{1}$$

where:

A_p is the total area of the pad projected on a plane perpendicular to the load direction; for axial or nearly axial load A_p can be expressed analytically by the relation:

$$A_p = \pi R_s^2 (\sin^2 \theta_4 - \sin^2 \theta_1). \tag{2}$$

a_f is a proper dimensionless 'load coefficient', that is function of the geometrical characteristics of the pad; on the hypothesis that the pressure drops linearly along the sill (the hypothesis has been proved to be quite adequate for our purposes), a_f can be expressed by the relation:

$$a_f = \frac{1}{(\sin^2 \theta_4 - \sin^2 \theta_1)} \left[\frac{\cos \theta_3 - \cos \theta_4}{\ln \left(\tan \dfrac{\theta_4}{2} \Big/ \tan \dfrac{\theta_3}{2} \right)} - \frac{\cos \theta_1 - \cos \theta_2}{\ln \left(\tan \dfrac{\theta_2}{2} \Big/ \tan \dfrac{\theta_1}{2} \right)} \right]. \tag{3}$$

Relation (1) is valid for axial loads and consequently for uniform distribution of pressure in the symmetrically disposed pads.

When the load has lateral components, the relation is still valid, considering for each pad a proper component of the load, as seen in the following sections.

3.3.2. *Flow through the Bearing*

As for the load capacity, the total flow of the bearing may be expressed as a function of the primary parameters of the bearing and of a proper dimensionless flow coefficient:

$$Q = p_r \cdot q_f \cdot \frac{h^3}{\mu}. \tag{4}$$

Equation (4) represents the assumption of linearity between pressure and flow: this assumption may be considered sufficiently approximate when the pads have relatively large sills.

To get closer to the physical behaviour of the bearing, the equation should be written in inverse form:

$$h = \left(\frac{Q \, \mu}{p_r \, q_f} \right)^{1/3} \tag{4'}$$

the film thickness being the dependent variable.

The equation may be applied to the total flow when h is constant all over the bearing pad, i.e., when the clearance bearing works in a centered position; for working conditions near the centered one, it can be used with a certain approximation, assuming for h a proper average value; when we are far from these conditions analogous equations may be applied for proper portions of the bearing.

In the centered position, the value of q_f may be expressed analytically.

$$q_f = \frac{\pi}{6} \frac{1}{\ln\left(\tan\dfrac{\theta_4}{2}\tan\dfrac{\theta_3}{2}\right)} + \frac{1}{\ln\left(\tan\dfrac{\theta_2}{2}\tan\dfrac{\theta_1}{2}\right)}. \tag{5}$$

3.3.3. *Viscous Friction of the Bearing*

When two corresponding elements A_i of the surfaces of the cup and of the ball have a relative velocity $\bar{V}_i$ the consequent contribution to the interacting torque may be expressed by the relation:

$$\bar{M}_i = \mu A_i \frac{\bar{V}_i \wedge \bar{r}_i}{h_i} \tag{6}$$

when:

A_i element of surface of the ball
$\bar{V}_i$ its velocity relative to the cup
$\bar{r}_i$ radius vector of the surface element A_i with respect to the center of the sphere
h_i film thickness of the element A_i.

To get total torque interaction, given a certain relative speed $\bar{\omega}$ and given the position of the ball (knowledge of the distribution of h_i over the surface) it is necessary to integrate the above function on the entire surface.

The analytical integration is quite difficult, and we found some simplified expressions for particular cases of motion, i.e., rotation about the longitudinal axis and about a 'transversal axis'; these expressions are compared with the experimental results in Section 5.

3.3.4. *Characteristics of the Compensating Elements*

Various types of compensating elements can be inserted along the feeding pipe of each recess; these characteristics greatly influence the behaviour of the bearing in so far as concerns flow, vertical and lateral stiffness.

Let us now examine the main types of compensating elements and their characteristics.

3.3.4.1. *'Linear Restrictors'*. They are characterized by a linear relation between pressure drop and flow (laminar flow):

$$Q_l = \frac{K_l}{\mu} \Delta P. \tag{7}$$

Where:

Q_l is the flow through the linear compensating element

$\Delta P = $ equal to $P_s - P_r$

$\dfrac{K_l}{\mu}$ is the 'conductance' of the restrictor; the term K_l represents the geometric characteristic.

The linear restrictors may be built as capillary tubes of suitable diameter and length; when it is desirable to vary easily the conductance of the restrictors, they can be made in various ways, for instance employing adjustable conical pins.

3.3.4.2. *'Sharp Edged Orifice' Restrictors.* They are characterized by a relation between pressure and flow of the type:

$$Q_0 = K_0 \, \Delta P^{1/2} \tag{7'}$$

where K_0, in first approximation is given by:

$$K_0 = 0.6 \, \frac{\pi d_0^2}{4} \left(\frac{2}{\varrho}\right)^{1/2},$$

where d_0 is the diameter of the orifice, ϱ the mass density of the fluid.

3.3.4.3. *'Constant Flow' Restrictors.* They are characterized by the relation

$$Q_v = \text{const} \tag{7''}$$

and are made by pressure-compensated flow control valves.

4. Elements of the Design

In Section 3 we saw the expression of the main parameters of the bearing as functions of the operating conditions.

Now we have to see how the bearing goes to work in certain conditions and the design criteria that have to be followed to obtain the desired performance.

For having a general picture of how the various factors interact in the performance of the bearing we draw a block diagram of the bearing functions.

We look separately at two types of feeding circuits: constant pressure and constant flow supply.

4.1. BLOCK DIAGRAM FOR 'CONSTANT PRESSURE' SUPPLY

Figure 4a is a functional block diagram of the behaviour of the bearing when supplied with a constant pressure fluid: for the sake of simplicity we now consider the case of axial load; it is therefore considered that the pressure is equal in the various pads.

In the diagram both 'linear' and 'sharp edged' restrictors have been considered.

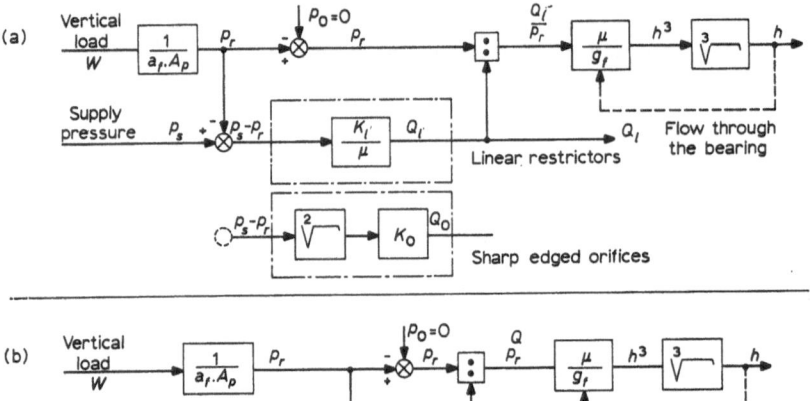

Fig. 4.   Block diagrams. (a) – constant pressure. (b) – constant flow.

As seen in the diagram, the two main inputs of the system are the load and the feeding pressure.

Load entity W determines univocally the value of the recess pressure p_r, according to Equation (1); the parameters a_f and A_p are geometrical characteristics of the pad configuration, and are therefore constant quantities.

On the other hand, the values of p_s and p_r define the pressure drop through each restrictor and therefore determine univocally the flow Q, as function of the flow coefficient of the restrictors according to Equation (7) (or 7' if the restrictors are of the orifice type).

Finally, since the flow through the bearing pads is the same as the flow through the restrictors and since the pressure variation across the bearing is defined, in consequence, the conductance of the bearing, that is the value h of the film thickness, according to an equation of the type of the (4'), is defined.

4.1.1. Vertical Stiffness with Linear Restrictors

We can express the film thickness as a function of the inputs of the system (load, pressure), of the geometrical parameters of the bearing (a_f, A_p, q_f) and of the conductance of the restrictors K_l.

From the equation (4') we have:

$$h_l = \left(\frac{q \cdot \mu}{q_f \cdot p_r}\right)^{1/3}.$$ (4')

Substituting Equation (7) we have:

$$h_l = \frac{K_l^{1/3}}{q_f^{1/3}} \left(\frac{P_s - P_r}{P_r}\right)^{1/3}.$$ (8)

We have finally:

$$h_l = \frac{K_l^{1/3}}{q_f^{1/3}} \left(\frac{P_s \cdot a_f \cdot A_p - W}{W} \right)^{1/3}.$$ (9)

We can now define the *vertical stiffness of the bearing* as:

$$R_l = - \frac{dW}{dh_l}.$$ (10)

From the equation (9) we have:

$$R_l = - \frac{dW}{dh_l} = 3W^{4/3} \frac{q_f^{1/3}}{K_l^{1/3}} \frac{(a_f \cdot A_p \cdot p_s - W)^{2/3}}{a_f \cdot A_p \cdot p_s}.$$ (11)

The first note is that the stiffness varies with the load, if pressure p_s and the other parameters are constant.

For a given value of p_s we find that the load for which the stiffness is maximum, is given by:

$$W = \tfrac{2}{3} a_f \cdot A_p \cdot p_s$$ (12)

and the value of the stiffness at this point is given by:

$$R_{1m} = 0.836 \left(\frac{q_f}{K_l} \right)^{1/3} a_f \cdot A_p \cdot p_s$$

(see Figure 5, curve 1).

On the other hand, for a given load, the stiffness varies with the supply pressure p_s: if the values of W and K_l are constant we find from (11) that the maximum of stiffness is given for:

$$p_s = \frac{3W}{a_f \cdot A_p}$$ (13)

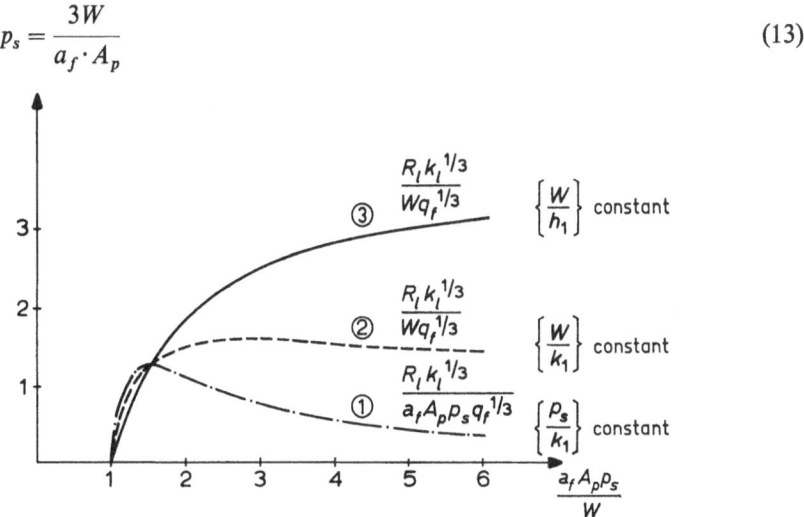

Fig. 5. Vertical rigidity with constant pressure supply linear restrictors.

and its value is given by:

$$R'_{lm} = 1.59 \cdot \left(\frac{q_f}{K_l}\right)^{1/3} \cdot W \tag{14}$$

(see Figure 5, curve 2).

We have to notice however that, as p_s increases, the value of h_l will increase according to (9), unless we compensate h with a corresponding diminution of K_l; making this compensation we further improve the stiffness of the bearing (see curve 3 of Figure 5).

In general the value of h_l has a lower limit due to the tolerances of construction of the pads, or to the viscous friction associated with the bearing motion as in Equation (6); (an upper limit can also be given by the value of the oil flow through the bearing).

The value of the stiffness can be expressed as function of W and h_l by the equation:

$$R_l = \frac{3W}{h_l}\left(1 - \frac{W}{a_f \cdot A_p \cdot p_s}\right) \tag{11'}$$

which shows, for W and h_l constant, an asymptotic limit having the value:

$$\frac{3W}{h_l}.$$

In conclusion, we can improve the stiffness with higher values of p_s and lower values of K_l, taking into account the limits for h_l and the fact that, for values of p_s higher than 3 or 4 times the value of $W/af \cdot Ap$, the improvement of R_l tends to become insignificant.

4.1.2. Vertical Stiffness with 'Orifice Restrictors'

In the same way as before we can express h as function of the inputs of the system and the geometrical parameters of the bearing; we obtain for the film thickness:

$$h_0 = \left(\frac{K_0\mu}{q_f}\right)^{1/3} \cdot (a_f A_p)^{1/6} \cdot \left(\frac{a_f \cdot A_p \cdot p_s - W}{W^2}\right)^{1/6} \tag{15}$$

and for the vertical stiffness:

$$R_0 = -\frac{dW}{dh_0} = 6W^{4/3} \cdot \left(\frac{q_f}{K_0\mu}\right)^{1/3} \cdot \frac{1}{(a_f A_p)^{1/6}} \cdot \frac{(a_f \cdot A_p \cdot p_s - W)^{5/6}}{2a_f \cdot A_p \cdot p_s - W}. \tag{16}$$

As before, for a given pressure the stiffness varies with the load, and has a maximum when the load value is:

$$W = 0.691 a_f \cdot A_p \cdot p_s. \tag{17}$$

On the other hand, for a given load, the stiffness varies with the supply pressure p_s, up to a maximum given for:

$$P_s = \frac{3.5W}{a_f \cdot A_p}.$$ (18)

Also in this case we can consider the possibility of lowering the value of K_0 as p_s increases, in order to have constant film thickness.

We can express R_0 as function of W and h_0 by the equation:

$$R_0 = \frac{3W}{h_0} \left(2 \, \frac{a_f \cdot A_p \cdot p_s - W}{2a_f \cdot A_p \cdot p_s - W} \right)$$ (16′)

which shows an asymptotic value of $3 W/h_0$ as Equation (11′).

For values of pressure of the order of 3–4 times the value of $W/a_f \cdot A_p$, we have in this case some advantage over linear restrictors.

4.2. Block diagram for 'constant flow' supply

We do not consider here the case in which every recess is fed independently by a flow control valve, because, in spite of some advantages with respect to the other systems, it was not considered in our development owing to its major complication and dimensions.

We look at the case in which the main supply is flow-controlled, and the various recesses are hydraulically separated by linear restrictors.

The functional block diagram is drawn in Figure 4b.

It is evident that the vertical stiffness in this case can be higher than for constant pressure: in fact for constant flow, a variation of load produces a variation of h only through a variation of p_r; in the case of constant pressure, we have in the same time a variation of Q and p_r that produce a variation of h.

We can express analytically, as in the preceding Section 4.1, the value of the film thickness and the vertical stiffness, starting from the inputs and the geometrical characteristics of the bearing.

We have for the film thickness:

$$h_v = \left(\frac{Q_v}{W}\right)^{1/3} \cdot \left(\frac{\mu a_f \cdot A_p}{q_f}\right)^{1/3}$$ (19)

and for the vertical stiffness:

$$R_v = -\frac{dW}{dh_v} = 3W^{4/3} \left(\frac{q_f}{Q_v \mu a_f \cdot a_p}\right)^{1/3}.$$ (20)

R_v can be also expressed as function of h:

$$R_v = \frac{3W}{h_v}.$$ (20′)

Equations (20′), compared with (11′) and (16′) shows that this type of compensation gives some advantage as far as stiffness is concerned when we work in the lower region of the values of the supply pressure.

4.3. BEHAVIOUR WITH LATERAL LOADS

The same type of considerations as for the case of axial load can be made in this case; however now the equations become much more complicated due to the non-uniform distribution of the pressures arising in the various pads to counteract the lateral component of the load, and to the non-uniform film thickness in the various pads.

We can make some simplifying hypotheses, in order to describe the approximate behaviour of the bearing:

(a) the various pads do not interact with each other in the clearance between sphere and cup (in other words the oil flow of each pad does not influence the flow of the adjacent ones)

(b) the film thickness is considered constant over each pad, taking suitable average values h_{mi}.

In this way the flow through each pad may be expressed by an equation of the type:

$$Q_i = \frac{p_{ri} \cdot q_f' h_{mi}^3}{\mu}. \tag{21}$$

For each pad we consider a suitable load coefficient a_f' and a suitable projected area A_p' such that by multiplying for the recess pressure of the pad p_{ri} we have the value of the force exerted on the sphere, F_i, radially aligned and applied in the center of the surface of the pad:

$$F_i = a_f' \cdot A_p' \cdot p_{ri}. \tag{22}$$

With reference to the notations of Figure 6, the behaviour of our 8-pads bearing is expressed by the following system of equations, on the hypothesis of constant pressure supply, linear restrictors.

(1) $\quad \sum_1^8 (F_i \cos \vartheta_m) = W_a$

(2) $\quad \sum_1^8 (F_i \cos \alpha_i \sin \vartheta_m) = W_l$

(3) $\quad \sum_1^8 (F_i \sin \alpha_i \sin \vartheta_m) = 0 \tag{23}$

(4–11) $\quad p_{ri} = \dfrac{p_s K_1}{1 + q_f' h_{mi}^3}$

(12–19) $\quad h_{mi} \cong C + e_v \cos \vartheta_m + e_0 \cos \alpha_i.$

From these equations, and from the Equations (21) and (22) we obtain:

the 3 components of the sphere displacement $e_v, e_0, e_{0'} = 0$ (this last condition is implicit in the choice of the origin for α, and in the form of equation 23/12–19)

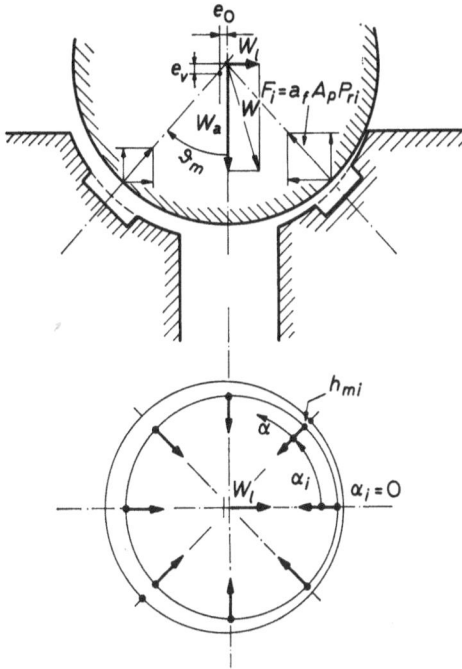

Fig. 6. Pressure forces and film thickness for lateral loads.

the corresponding values of the film thickness h_{mi}
the values of p_{ri}
the values of Q_i,
as functions of the load components, the supply pressure, the restrictor conductance and the geometrical characteristics of the bearing.

4.4. VISCOUS FRICTION OF THE BEARING

4.4.1. *Rotation about x Axis*

The integral of Equation (6) over the active part of the bearing leads to an expression of the type:

$$M_x = 2\pi\mu R_s^3 \omega_x \left(\int_{\vartheta_1}^{\vartheta_4} \frac{\sin^3 \vartheta}{C + e_v \cos \vartheta}\, d\vartheta - 0.22 \int_{\vartheta_2}^{\vartheta_3} \frac{\sin^3 \vartheta}{C + e_v \cos \vartheta}\, d\vartheta \right). \qquad (24)$$

4.4.2. *Rotation about y Axis*

An approximate expression for the viscous torque is given by:

$$M_y = 4\eta\mu R_s^3 \omega_y \int_0^{\vartheta_4} \frac{\cos^3 \vartheta \cdot \arccos \dfrac{\cos \vartheta_4}{\cos \vartheta}}{C + e_v \cos \vartheta}\, d\vartheta \qquad (25)$$

Fig. 7.   Prototype of the spherical bearing.

where the coefficient η represents a proper average ratio between the sills surface and the total pad surface of the bearing.

5. Experimental Results

For the development of the spherical bearing a prototype was produced and extensive experimentation was carried out (see photos of Figure 8).

Some aspects are still under investigation, but the main results obtained up to now have well confirmed the basic assumptions and the theory sketched in the preceding sections; in some case the experimental work completed the theoretical investigation, as for instance in cases in which exact analytical solutions were very difficult to derive.

The completed part of the experiments concerns the constant pressure type pressurization and compensation with adjustable linear restrictors.

Some experiments with constant flow supply will be carried out, always with linear restrictors. The choice of linear restrictors has been made because, in spite of some less rigidity, they are simpler to be made in an adjustable version; further with comparative values of flow and pressure, they have less tendency to clog than the orifice type restrictors.

On the other hand the values of stiffness that we obtained were quite adequate for our needs.

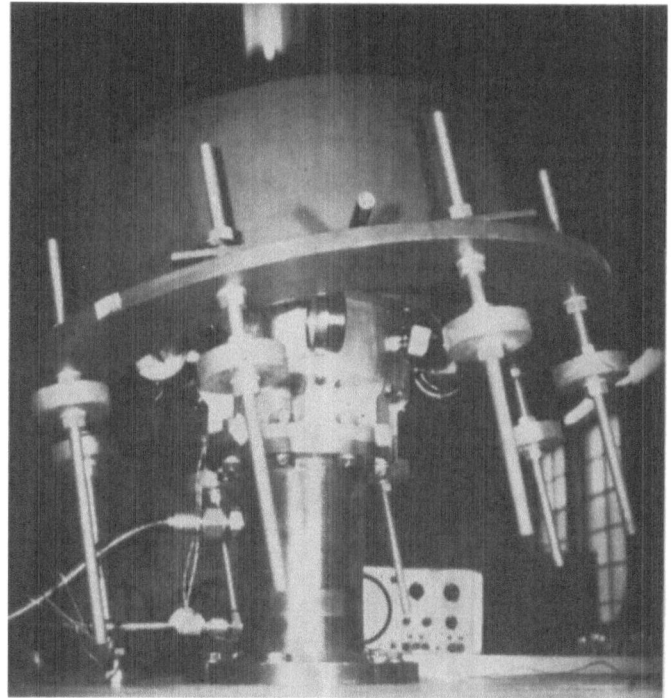

Fig. 8. Tests with a load simulating the satellite.

The main results of the experiment, regarding load carrying capacity, restrictor performance, load displacement characteristics, as well as internal and external damping of the bearing are summarized in the following sub-sections.

5.1. LOAD CARRYING CAPACITY (VERTICAL LOAD)

For the geometry of our bearing, the theoretical value of the product $a_f \cdot A_p$ given by Equations (2) and (3) is:

$$a_f \cdot A_p = 99.5 \times 10^{-4} \, \text{m}^2.$$

The experimental value, checked in a variety of conditions (h varied between 5 and 13×10^{-5} m; p_s/p_r from 1.35 to 6.5; load from 1500 to 4000 N), had an average value of 100×10^{-4} and maximum dispersions of $\pm 1\%$; this shows that the assumption of linear pressure drop along the sills is quite adequate.

5.2. LINEARITY OF THE RESTRICTORS

Variable restrictors made of conical pins were used; the linear low between flow and $p_s - p_r$ is given by Figure 9.

The measurements are made for various conditions and show the close compliance with Equation (7).

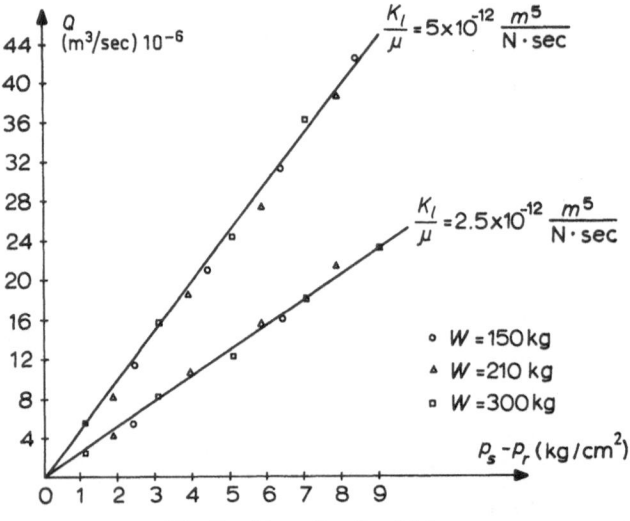

Fig. 9.   Linearity of restrictors.

5.3. LOAD-DISPLACEMENT CHARACTERISTICS

5.3.1. *Axial Load*

Relation (9) between the film thickness, the restrictor conductance, the supply pressure and the load has been checked for various operating conditions.

For the film thickness we adopted average values derived from the measured values of the sphere vertical eccentricity e_v, the value of the bearing clearance C and the geometrical shape of the pad.

$$h_m = \frac{\left[C + e_v \cos \dfrac{\vartheta_1 + \vartheta_2}{2}\right] \sin \dfrac{\vartheta_1 + \vartheta_2}{2} + \left[Ce_v \cos \dfrac{\vartheta_3 + \vartheta_4}{2}\right] \sin \dfrac{\vartheta_3 + \vartheta_4}{2}}{\sin \dfrac{\vartheta_1 + \vartheta_2}{2} + \sin \dfrac{\vartheta_3 + \vartheta_4}{2}}.$$

In Figure 10 are summarized the results in a plot of h_m towards $(p_s - p_r)/p_r$ for various values of K_1.

It should be noted that the various points of the logarithmic diagram lie approximately on two straight lines representing two values of the coefficient K_1; the slope of the straight lines is $\frac{1}{3}$ and their distance corresponds to a ratio of $\sqrt[3]{4}$, in accord with equation (9).

5.3.2. *Lateral Load*

Various tests with lateral load were performed: the lateral loads were those representing an inclination of the bearing with respect to the vertical of up to 20°.

The functioning conditions of the bearing were those giving a good performance with respect to vertical stiffness.

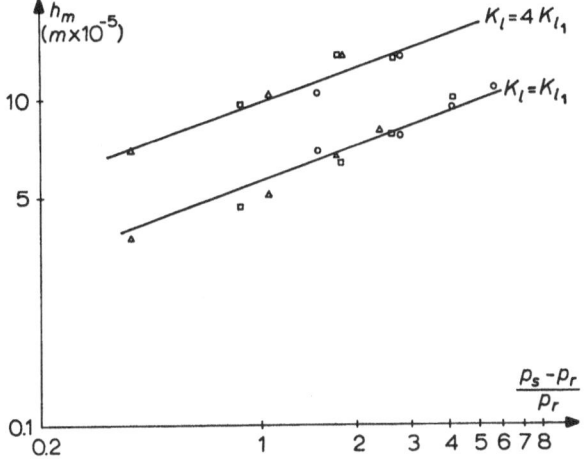

Fig. 10.   Vertical load characteristics.

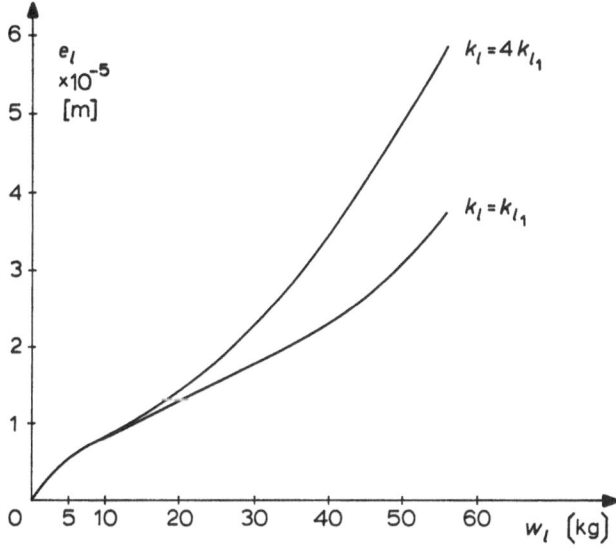

Fig. 11.   Lateral displacement versus load.

Figure 11 represents lateral displacement versus load. The p_s value was about 4 times the value of p_r when lateral load was zero.

We note that the increase in the hydraulic resistance of the restrictors (diminution of K_l) gives an improvement in the lateral rigidity especially for higher loads (the effect is of the same types as for vertical load).

We note also that near to zero and near to the maximum load, the stiffness slightly decreases: the average values corresponding to the conditions of load of Figure 11 are of the order of 10^7 N/m.

Figure 12 shows the pressure distribution in the various pads corresponding to the load conditions of Figure 11.

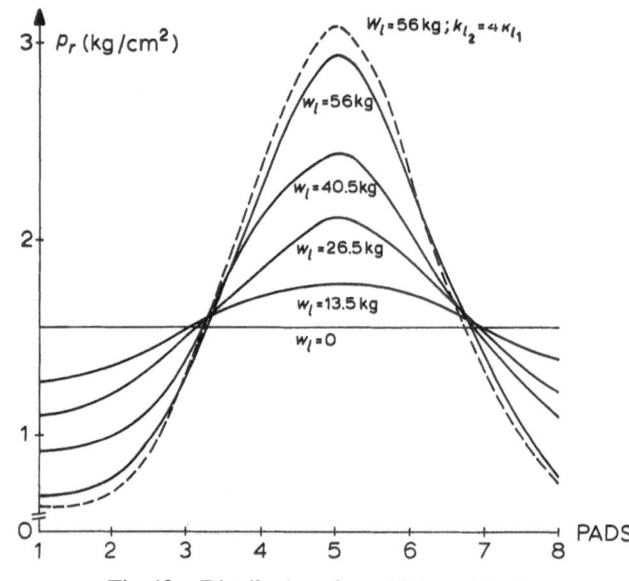

Fig. 12.   Distribution of p_r with lateral load.

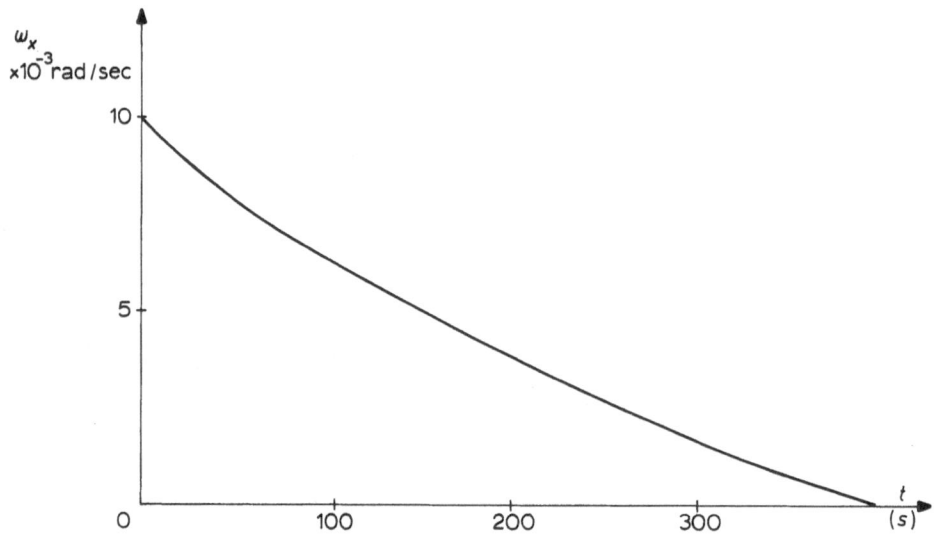

Fig. 13.   Velocity decay about x axis (small velocities).

5.4. VISCOUS FRICTION OF THE BEARING

In Figures 13–15 are summarized the results of various tests made to check the theoretical data about the rotational damping of the bearing.

The main comment is that for relatively high and medium values of angular velocity the behaviour is in good accord with the theoretical values; for very low values of velocity a different law of friction appears; in other words, instead of having a pure viscous torque (proportional to the velocity) a small component independent of velocity appears.

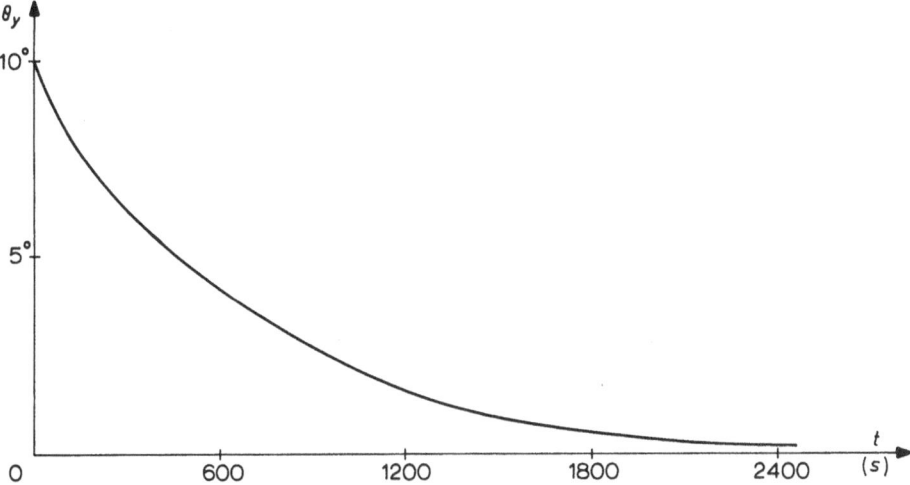

Fig. 14.   Amplitude decay about y axis (sinus motion – large angles).

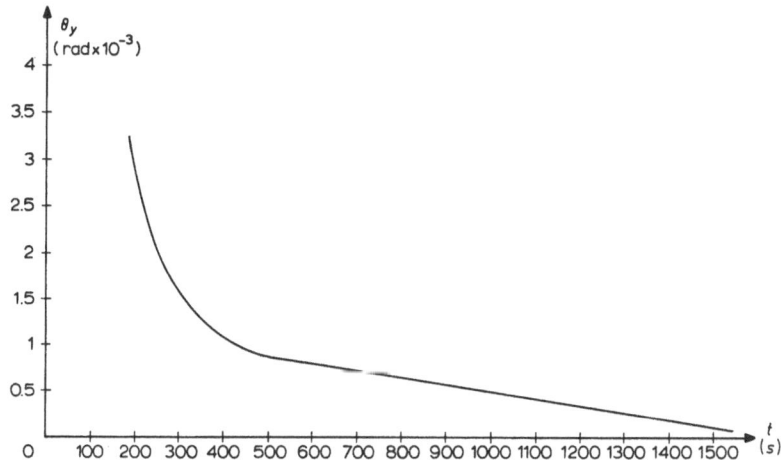

Fig. 15.   Amplitude decay about y axis (sinus motion – small angles).

This residual torque appeared with different values about the x and y axes: for x axis the order of magnitude in average operating conditions was about 1.4 gr × cm.; for y axis the order of magnitude was about 0.05 gr × cm.

We are still investigating this phenomenon, which might limit the use of oil bearings.

The difference between the two axes seems to be due to the particular configuration of the set of pads, which is substantially annular about the x axis.

Anyway it is quite likely that, with a type of pads having lower ratio between sill area and recess area we should reach, for both axes, residual values of the order of the lowest one found, which is comparable to the minimum of controllable torques for air bearings also.

5.5. TRANSLATIONAL DAMPING OF THE BEARING

We have seen that the bearing behaves substantially as an elastically restrained system with variable parameters.

Apart from the problem of the stiffness (or in other words of the natural frequency of the bearing), the internal damping characteristics of the bearing was of course controlled too.

We found, by means of responses to steps of applied load, quite satisfactory behaviour in all the operating conditions, as expected.

References

[1] Rossi, L. C. and Gimelli, E., 'Metodi per le prove a terra del sistema di controllo di assetto di satelliti stabilizzati per rotazione', Congresso AIDA-AIR – Milano 5–8 Ottobre 1967.
[2] Rippel, H. C., 'Conical and Spherical Pads' – Hydrostatic Bearings – Part 7, in *Machine Design*, October 24 (1963) 185–192.
[3] Michelini, R. C. and Ghigliazza, R., 'Optimum Geometrical Design of Multipad Externally Pressurized Journal Bearings', *Meccanica* III, no. 4 (1968) 231–241.
[4] Ghigliazza, R. and Michelini, R. C., 'Comparative Investigation of Friction in Externally Pressurized Journal Bearings' in *Wear*, no. 12 October (1968) 241–251.
[5] O'Donoghue, J. P. and Rowe, W. B., 'Hydrostatic Bearing Design', *Tribology* 2, no. 1 (1969) 25–71.
[6] Wilcock, D. F. and Bevier, W. E., 'Externally Pressurized Bearings as Vibration Attenuators' *Trans. ASME*, Series F, July (1968) 614–617.
[7] Bogdanov, O. I. and Danil'tsev, V. G., 'Design of Hydrostatic Spherical Bearings with Centrally Located Lubricant Feed Chamber', *Izvestiya Vysikh Uchebnykh Zavedenii, Mashinostroenie* no. 10 (1965) 52–55.

A POLYVALENT DATA ACQUISITION SYSTEM FOR GROUND SPACE TESTS AT FIAT AEROSPACE LABORATORY

E. PADOVANO

Sezione Velivoli, FIAT, Italy

Abstract. For ground test data acquisition on spacecraft, a highly sophisticated electronic instrumentation system has been developed.

Every kind of test has access to an unique instrumentation system, physically grouped but ready to perform simultaneously measurements of several tests.

The accomplishment of such an instrumentation involved a detailed work of standardisation of transducers, interfaces, recorders and plotters.

The capability of the system covers nearly all fields of measurement on ground-simulated space tests; acquisition possibilities range from several thousand channels at quasi-static condition, to several hundred at high frequency.

1. Introduction

This paper describes a D.A.S. (Data acquisition system) developed at FIAT Aerospace Laboratory for structural, thermal, vacuum and dynamic tests.

It is almost impossible to cover all the measurement requirements that arise in testing space components by means of a defined and specific D.A.S.

However a flexible D.A.S., which can be easily adopted for various experimental work, is a powerful research tool.

The main criterion in selecting the instrumentation to build up a D.A.S. has been that of the compatibility and standardisation between the individual groups in order to obtain a very flexible system.

This flexibility allows an ever changing configuration of the system that easily follows the requirements of the various tests.

The numerical capability of the system is of the order of 3500 channels shared among different types of transducers and mastered by six data loggers.

The global maximum scanning speed ranges from 53 channels per second, for a total of 316 channels, to 17 channels per second for 3560 channels.

2. System Description

This paragraph describes in detail the system component starting with the transducers up to the test data.

2.1. TRANSDUCERS AND TRANSMITTERS

Almost any kind of transducer can be connected to the system. The greatest number of them are strain gages or strain gage based transducers. Others are thermocouples, thermistors, inductive and piezoelectric transducers, event markers. They all have easy access to the group of signal conditioners.

G. A. Partel (ed.), Proceedings of the Second International Conference on Space Engineering. All rights reserved.

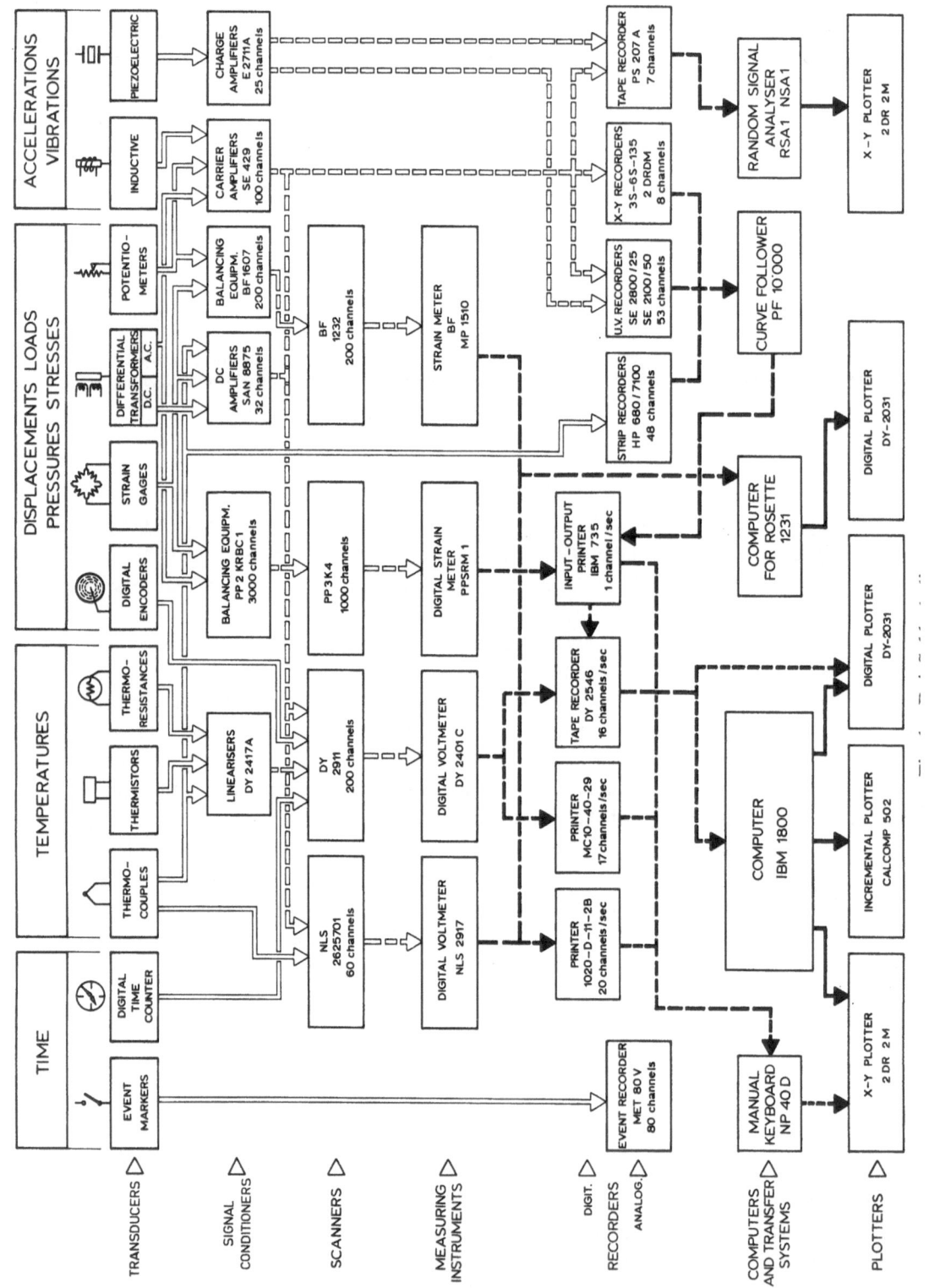

2.2. Signal conditioning equipment

The strain gage bridges or strain gage based transducers are fitted to:

(a) three thousand channels balancing equipment at 1000 cps carrier frequency (see Figure 2).

Fig. 2.   D.L. for 3000 strain gages.

Fig. 3.   Carrier amplifiers.

(b) one hundred amplifiers at 3000 cps carrier frequency, complete with demodulators (see Figure 3).

(c) two hundred channels balancing equipment in a D.C. constant current system (see Figure 4).

All these conditioners allow for the complete balancing in resistance and capacity.

Standard 120 ohm strain gages are usually employed but also 300 and 500 ohm strain gages can be used where necessary by means of the amplifiers of point (b).

The displacement transducers are:

D.C. differential transformers which are supplied by constant voltage regulators.

A.C. differential transformers that can be connected to the amplifiers of point (b) or to a series of transducer converters which provide the supply and demodulation for the signal.

Potentiometer transmitters that are handled like strain gages by means of bridge conditioners (Figure 5) and the amplifier of point (b) above.

Fig. 4.   D.L. for 200 strain gages.

Fatigue gages are connected to D.C. supplied balancing bridges which are provided with an automatic periodic zero control system (Figure 6).

Piezoelectric force or acceleration transducers have a set of 34 charge amplifiers with provision for calibration and sensitivity control.

DEFLECTION POTENTIOMETER CONDITIONER

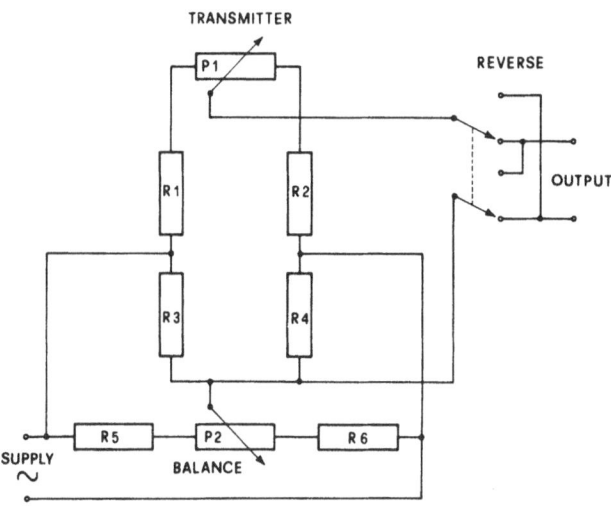

Fig. 5.

FATIGUE GAGE CONDITIONER

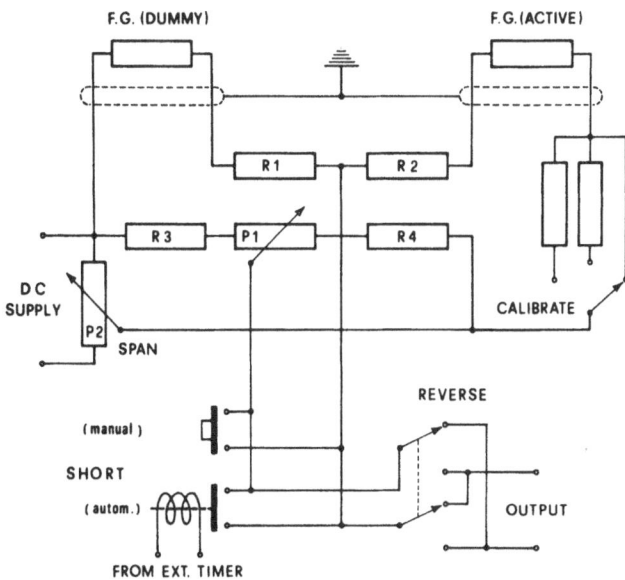

Fig. 6.

2.3. SCANNING EQUIPMENT

The total automatic scanning capability of the system is 3460 channels. They are shared among the different Data Loggers according to the nature of the transducers (Table I).

TABLE I

Data acquisition system characteristics

Data logger type	No. of D.L.	Max number of channels	Max scanning speed	Time to complete scan	Channels scanned in least time			
–			Ch/sec	sec	sec			
–	–	–	–	–	6	12.5	125	200
Peekel	3	3 × 1000	3 × 8 = 24	125	154	300	3000	3000
H & P	1	200	16	12.5	96	200	200	200
B & F	1	200	1	200	6	12	125	200
N.L.S.	1	60	10	6	60	60	60	60
Totals	6	3460	–	–	316	572	3385	3460
Max scanning speed – Ch/sec					53	46	27	17

Strain gages or strain gage based transducers are connected to:

a) three 1000 channel scanners each having a speed ranging from 0.1 to 8 per second.

b) one 200 channel scanner with a fixed speed of 1 channel per second; one 200 channel scanner and one 60 channel scanner with speeds of 1 to 16 and 1 to 10 channels per second respectively which can receive any kind of signal from the various conditioners described.

The scanning capability in terms of speed and associated number of channels is shown in Figures 7 and 8.

2.4. MEASURING INSTRUMENTS

Each of the six Data Loggers has its own A.D. converter. Each datum consists of: channel number, sign and measurement.

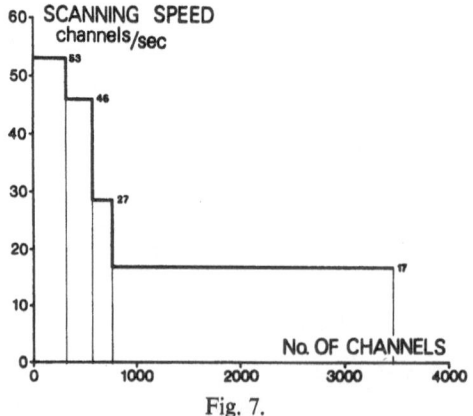

Fig. 7.

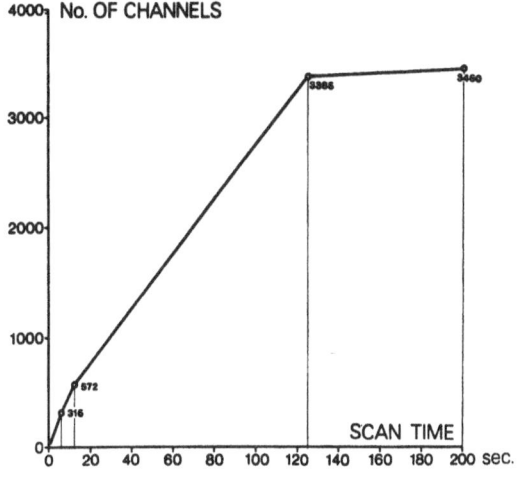

Fig. 8.

The latter has 4 to 6 digits according to the Data Logger. All converters have auto-ranging capability and a standardised output in BCD 8421 state 1 negative.

The signals which come from signal conditioners and that must remain in analogic form, are directly sent to recorders where they are accepted in parallel up to about 150 channels.

2.5. RECORDING EQUIPMENT

The digital recording equipment consists of:
for the three data loggers of point (a)
3 typewriting machines with maximum speed of 1 data/sec each
3 magnetic tape incremental recorders with max speed of 8 data/sec each (see Figure 9).

These can be run in parallel at the same time in order to obtain the printed data required for instantaneous vision and on a magnetic tape for the subsequent reduction. The other three data loggers (200 channels for strain gages, 200 and 60 channels for various signals) have strip chart high speed printers with parallel magnetic tape incremental recorders. The maximum recording speed is 1, 16 and 10 channels per second respectively.

All the tape recordings have NRZI coding and of course, BCD 8421 therefore they can be reduced without changing the reading programme.

Analog data are directly sent from the conditioning equipment to the recorders via the scanners.
The recorders available are:
Four Galvanometric U.V. recorders with a max capacity of 153 channels. Their frequency range is from D.C. to 5000 Hz flat response.

Ten strip chart recorders with a max capacity of 48 channels. Their max frequency response is 2 Hz.

Fig. 9.   Printer and magnetic tape recorder.

One event recorder with 80 individual time sequence determinations (see Figure 10) on which time of the order of 0.002 sec can be estimated.

Ten X-Y plotters are used for visualization of the most significant data.

2.6. DATA EVALUATION EQUIPMENT

The basic rule of the system is to have the digitised raw data corresponding in engineering units to the measured quantities (kg, mm, °C, etc.).

This means that the data are quickly understandable because all conversions needed are performed in the conditioning part of the system.

The linearization of non-linear transducers (thermocouples, thermistors, etc.) and the correction of the zero offset are performed before digitizing the data by controlling the sample period of a Digital Volt Meter and repositioning the decimal point; introducing a preset count.

The conversion of the deflection potentiometer signals is accomplished by the amplifier gain control in order to make the deflection numerically equal to the output in mv.

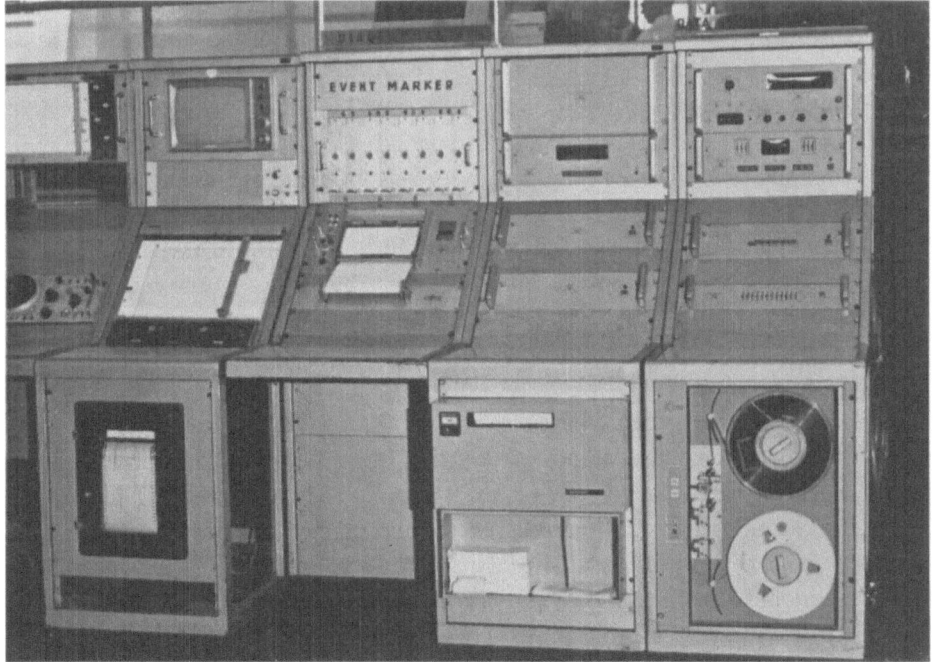

Fig. 10. Event marker recorder.

A similar conversion is performed for the dynamometers and other transducers.

The fatigue gages signal coming from the conditioners (see figure 6) gives calibrated linear increments in the useful range of resistance change. The conditioners are designed to obtain the engineering units conversion by means of a small regulation of the D.C. supply.

Once the raw data are recorded in engineering units on magnetic tape they are ready for the subsequent reduction and this is performed by an IBM 1800 Data Acquisition and Control System which is located near the Laboratory.

The latter is employed to read the tapes, make the necessary calculations and print the reduced data in suitable tabulated form, while a high speed plotter traces the curves for the most significant channels.

A direct connection between the D.A.S. and the IBM computer is now in course. This means that the 'on line' reduction of the test data will soon be available.

3. Functional Description

The block diagram of Figure 1 shows the main functional configuration of the complete system.

The input transducer signals may arrive from several different tests at the same time. The flexibility of the system is such that fatigue, static or thermal tests can be carried out simultaneously by selecting the appropriate inputs.

This selection depends primarily on the kind of measurements to be performed and on the availability of data paths along the different data loggers which compose the system.

Although the theoretical number of channels is 3460 the actual number available is determined by the compatibility of the transducers with the conditioning equipment and the feasibility of the connections between the different components of the system.

It should be borne in mind that the D.A.S. is not intended as a physically assembled unit but rather as a number of sub systems which can perform their task in a very high number of configurations in order to adapt themselves to the requirements of various specific tests.

4. Conclusions

The D.A.S. is distinguished by the following characteristics.

(1) The equipment that compose the system is easily interfaced in order to satisfy any special test requirement.

(2) The scanning speed and number of channels are chosen according to the kind of test to be carried out (i.e. static or dynamic).

(3) Nearly all the channels are capable of accepting all types of strain gages (one to four active arms), thermocouples, load cells, deflection potentiometers, etc.

(4) Raw data are in engineering units, no conversion or range switching is required; they may be printed and recorded on magnetic tape or taken out for 'quick look' plotting.

(5) The D.A.S. will be connected to a digital computer for on line data reduction, tabulation and plotting. The computer will also perform limit and alarm functions.

(6) Future instrumentation development is foreseen in order to increase the scanning speed vs. number of channels envelope.

MÉTHODES ET MOYENS D'ESSAIS RELATIFS AU DÉVELOPPEMENT DES PROPULSEURS

M. DOUAT

S.E.R.E.B., France

Résumé. Le développement de l'expérimentation au sol d'un ensemble propulsif doit aboutir à satisfaire l'objectif suivant:
(1) connaître les performances d'ensemble,
(2) s'assurer que les conditions de fonctionnement et d'ambiance ne donnent pas lieu à des phénomènes dangereux qui compromettent le fonctionnement.
L'exposé traitera uniquement du développement des propulseurs à poudre et, dans ce cadre, on exposera le principe de certains travaux que la S.E.R.E.B. a été amenée à effectuer:
(1) mesure des performances dans les phases stationnaires et transitoires,
(2) simulation d'altitude,
(3) grands taux de détente,
(4) écoulement interne,
(5) excitation vibratoire.
On se limitera à exposer les besoins, les principes et les méthodes, leur validité et leur limitation.

1. Introduction

Cette communication n'est pas un exposé exhaustif des différents travaux expérimentaux qui ont pu être faits dans le développement des propulseurs, mais il nous a paru préférable de donner un aperçu de travaux particuliers dans le contexte d'ensemble qui nous a toujours guidé:

assurer au sol les conditions de vol les plus réalistes pour être certain de pouvoir garantir les performances et le fonctionnement,

effectuer cette expérimentation dans un souci de simplicité et d'économie sans compromettre pour autant la qualité.

Nous nous limiterons volontairement à la propulsion à poudre, et nous exposerons dans ce cadre les besoins et les exigences qui nous ont paru à priori indispensables de satisfaire, en ajoutant ensuite celles qui ont été nécessaires à la suite des enseignements des essais en vol.

Nous exposerons ensuite les solutions expérimentales que nous avons retenues.

Si nous suivons l'évolution d'un engin à plusieurs étages, un certain nombre de phénomènes essentiels sont à considérer:

il faut que les propulseurs fonctionnent correctement, c'est-à-dire que outre leur chargement et les problèmes associés, les problèmes de performance soient parfaitement connus et, par là même, optimisés. Ceci veut dire que l'organisation de l'écoulement interne, l'estimation des pertes, l'impulsion spécifique soient déterminées avec le plus de rigueur,

il faut que les différents phénomènes de culot soient contrôlés et en particulier le couplage éventuel entre l'écoulement externe et interne,

G. A. Partel (ed.), Proceedings of the Second International Conference on Space Engineering. All rights reserved.

il faut que les séparations soient correctes, sous l'aspect séquentiel, pilotage, et électrique. Ce dernier point que nous n'avions pas pris en compte dans l'inventaire des besoins, nous a été imposé par les essais en vol,

il faut que le pilotage soit correct dans les différentes conditions de vol,

il faut enfin, que si l'on veut atteindre un but donné, les conditions initiales de la phase ballistique soient connues et contrôlées.

Du recensement de ces besoins, nous dégagerons les points essentiels suivants : mesure des performances propulsives, et efficacité du pilotage, simulation d'altitude, mesure des performances propulsives, et efficacité du pilotage,

simulation d'altitude,

fonctionnement des arrêts de poussée, et problèmes d'interaction,

étude expérimentale de la séparation,

étude de l'écoulement interne.

2. Mesure des performances propulsives

La connaissance des performances propulsives consiste à déterminer le système résultant des forces élémentaires de pression interne sur le système propulsif, dans un système d'axes lié à l'engin. Si le pilotage est assuré par des tuyères mobiles, 2 à 2 symétriques, on est en droit de penser qu'il y a une corrélation directe entre la position de l'organe mobile et la résultante des forces de contact, de sorte que dans ce cas, les efforts transverses sont contrôlés par la connaissance des caractéristiques du dispositif de commande. Il n'en est plus de même, par contre, si le pilotage est réalisé par un contrôle non mécanique de la poussée, injection par exemple, ou s'il n'y a pas de pilotage du tout, comme par exemple pour le 3ème étage du lanceur 'DIAMANT'.

Dans ces cas, il y a lieu de connaître les forces transverses, et le problème est donc de mesurer des forces faibles en présence de forces proportionnellement très grandes, dont le rapport est de l'ordre de 1 à 100. Il est évident par ailleurs que pour ne pas multiplier les essais, nous avons profité de ces tirs pour connaître avec précision la poussée.

Ajoutons de plus que ces essais ne peuvent toujours se faire à l'air libre, car il faut évidemment que l'écoulement dans les tuyères soit toujours significatif, et que pour des tuyères à grand taux de détente, ce qui est le cas pour les étages supérieurs, il y a lieu de travailler en simulation d'altitude.

Dans ces dispositifs, nous avons retenu des principes élémentaires simples suivants :

nous avons considéré que le système propulsif indéformable avait ses 6 degrés de liberté bridés par des liaisons se ramenant au mieux à 6 appuis simples, où ne passeraient que des forces pures,

nous avons éliminé toute cause d'irréversibilité mécanique, jeux, frottements secs, sources incontrôlables d'infidélités. Nous avons retenu des liaisons purement élastiques,

les efforts doivent être mesurés le plus près possible de l'endroit où ils prennent naissance, pour éviter des moments de transfert importants, qui surchargent les dynamomètres en détériorant la précision effective, nous avons considéré que si le dispositif est fidèle et de conception saine, la précision sera définie par la qualité de l'étalonnage. Ce point a donc été particulièrement étudié. Il s'agit en fait d'appliquer des forces connues en direction et module. Nous avons été amenés à réaliser des cardans élastiques définissant un point géométrique de passage de l'effort, parfaitement accessible et contrôlable par voie optique. Chaque élément d'application d'effort était donc constitué d'une liaison entre l'élément balance et une référence absolue, terminés par des cardans élastiques, dans laquelle on insère un élément actif (vérin) associé à un dynamomètre étalon. La connaissance absolue du vecteur effort appliqué dans le système d'axe sol permet de déterminer les coefficients d'influence.

Si le système est infiniment raide, il est manifeste que la matrice de correspondance efforts-signaux est d'order 6. Par contre, du fait des déformations élastiques, le système n'est plus linéaire, des termes quadratiques apparaissent, et il y a lieu de surveiller le poids de ces termes qui comprennent en général la poussée sur les signaux des termes transverses. Dans ce cas, le problème n'étant plus linéaire, il faut convenir d'un état de référence qui est celui pour lequel le tarage a été accompli. Nous donnons ci-après les valeurs des termes linéaires qui caractérisent le découplage élastique entre les liaisons, et le tableau des termes quadratiques qui découlent des rigidités principales des liaisons.

Par exemple, voici les résultats acquis sur un banc sur lequel ont été obtenues les performances du 3ème étage du lanceur 'DIAMANT'.

Matrice linéaire,

termes quadratiques,

caractéristique impulsion spécifique.

Nous ajouterons que la base des mesures est à jauges extensométriques alimentées en courant continu, amplificateur continu symétrique, enregistrement magnétique. Dépouillement différé auromatique et traitement numérique. Etalonnage électrique automatique avant et après essai.

3. Problèmes de simulation d'altitude

Pour que la mesure des performances soit correcte au sol, il faut que l'écoulement dans la tuyère soit correctement établi. De plus, le couplage de l'écoulement sortant du propulseur avec l'écoulement général, ou des problèmes spécifiques d'éclatement de jet et de répartition de pression de culot, impose de travailler en altitude, avec ou sans écoulement extérieur.

Le problème du couplage de l'écoulement externe et de l'écoulement interne n'a paru intéressant que dans les phases où il y avait des risques de perturbation importants, phase transsonique et séparation.

Pour le transsonique, nous avons simulé en soufflerie à échelle importante ∅ 550, des maquettes simulant l'engin, à des altitudes et pression équivalentes de ceux du vol. Nous avons réalisé des maquettes chargées de poudre donnant, avec les mêmes gaz, les mêmes pressions de combustions et locales. Les maquettes équipées de capteurs de pression et de fluxmètres ont permis de mesurer les paramètres locaux essentiels.

Pour les séparations, nous avons effectué des essais en soufflerie en similitude.

Par contre, la mesure des performances et des problèmes de pilotage pour des tuyères à grand taux de détente exige des essais en simulation d'altitude avec propulseur échelle grandeur.

La pression génératrice est la pression de combustion qui est de l'ordre de 50 bars pour les propulseurs à poudre.

Les conditions de fonctionnement de la tuyère exigent des pressions d'éjection inférieures à la pression atmosphérique, mais nous avons poussé au maximum la dépression ambiante pour qualifier par la même occasion des aménagements arrières et nous avons visé des pressions de 10 mb, soit des altitudes de l'ordre de 25 km.

Le dispositif que nous avons adopté est une trompe à débit nul, les flux primaires étant les gaz propulsifs. Le tube éjecteur a été dimensionné pour chaque propulseur. Le propulseur crée ou entretient lui-même le vide dans l'enceinte qui l'enveloppe. Il est monté à l'intérieur sur un banc dynamométrique. Un vide initial est créé, le tube éjecteur étant fermé par une membrane de mylar qui est détruite lors de la mise à feu. Le problème du désamorçage en fin de propulsion est de loin le plus gênant.

La remontée de l'onde de choc de recompression de l'éjecteur supersonique induit sur le spécimen une surpression au culot extrêmement brutale. Jusqu'à ce jour les effets n'ont pas été dommageables ni à l'installation ni au spécimen. C'est dans de telles installations que sont mis au point les systèmes de pilotage par injection sur étages supérieurs et que sont mesurées les performances d'ensemble.

Il est toutefois certain que ce dispositif a des limitations intrinsèques qui ne peuvent atteindre de très grand taux de détente. Cette limitation tient au fait qu'on amorce une tuyère avec une mauvaise récupération entre deux niveaux de pression totale, l'une à l'aval qui est la pression atmosphérique, l'autre à l'amont qui est la pression du propulseur. Comme cette pression n'est jamais très haute de l'ordre de 50 bars, on est limité à des rapports de détente qui ne peuvent dépasser 1000. Or, dans le cas du fonctionnement des propulseurs, phase d'arrêt de poussée en particulier, il y a des configurations où l'on a besoin d'examiner les écoulements dans une configuration où les rapports de détente sont beaucoup plus élevés. A l'ouverture des arrêts de poussée, les gaz de combustion se détendent pratiquement dans le vide, les rapports de détente sont supérieurs à 20 000. Avec de tels rapport de détente, qui s'approchent de la détente limite, la partie avant est plongée dans un écoulement qui l'enveloppe et il y a interaction aérodynamique entre les deux corps.

C'est cette simulation que nous avons essayé de réaliser et, corrélativement, nous avons voulu chiffrer la valeur de cette interaction.

Réaliser une détente aussi élevée exige une température minimale génératrice telle qu'il n'y ait pas liquéfaction, d'autre part le coefficient γ intervenant, nous avons décidé de conserver les mêmes gaz. Nous avons donc ainsi été amenés à faire un essai en similitude, dans laquelle les corps étaient en similitude géométrique, les gaz identiques. Nous avons augmenté la pression de combustion, qui a été portée à 250 bars, et mis l'ensemble dans un caisson couplé à des sphères à vide de 5000 m^3, où régnait une pression initiale de $\frac{1}{100}$ d'atmosphère. L'écoulement était déclenché par rupture de diaphragmes claquant à la pression nominale du propulseur de façon à éviter la phase transitoire d'établissement.

Dans ces conditions nous avons obtenu des rapports de détente allant jusqu'à 26 000, qui ont permis d'explorer les effets de l'interaction en fonction du rapport de détente. La mesure des efforts stationnaires a été délicate nous l'échelon d'efforts, pendant l'expérience. Nous avons réalisé un système mécanique à jauges à bande passante la plus élevée possible (600 Hz). Un dépouillement mécanographique a permis ensuite de faire un calcul fin. Les résultats montrent qu'au début du mouvement relatif, les jets de gaz ont un effet décélérateur puis accélérateur, jusqu'à une distance de plusieurs diamètres.

4. Phase d'arrêt de poussée

Jusqu'ici nous avons abordé des problèmes quasi stationnaires. Par contre, il est des phases propulsives qui correspondent à des phases transitoires dont la connaissance est importante. Nous pensons en particulier à la phase d'arrêt de poussée qui contrôle la séparation de la partie amont de la partie aval, et qui conditionne en fait l'état initial (position et vitesse) de la trajectoire de cette partie amont. Nous avons vu que, du fait des grands taux de détente, il y avait une interaction des deux corps l'un sur l'autre, mais cet effet vient en correction sur la phase fondamentale de séparation. Il est évident que cette séparation doit être franche, connue, datée. Il s'agit de phénomènes de l'ordre de quelques millisecondes et nous nous sommes proposés de connaître ce transitoire, assurant la transition entre deux états stationnaires avant et après fonctionnement des arrêts de poussée. Ce transitoire tient compte du propulseur, de son aménagement, de sorte que nous avons éliminé toute étude définitive sur maquette, et nous avons procédé à l'expérimentation à l'échelle grandeur. On se trouve donc en présence de masses très importantes, soumises à des forces et des variations de force dépassant 1000 KN.

Pour que la réponse soit significative, il faut que cet échelon d'effort n'excite pas de fréquences du système dans la bande spectrale du phénomène à étudier, bande passante qui va du continu à des fréquences élevées (plusieurs centaines de hertz au moins).

Ces considérations simples nous ont conduits à éliminer tout dispositif dynamo-métrique classique qui imposerait obligatoirement des fréquences propres basses, peu amorties qui auraient masqué ce phénomène. Faute de pouvoir découpler par le haut le dispositif expérimental du phénomène à étudier, nous l'avons découplé par le bas, et réalisé un système à fréquence nulle, c'est-à-dire libre.

Le dispositif expérimental que nous avons réalisé donc dans les phases suivantes:
allumage du propulseur et combustion jusqu'à t_0 fixé à l'avance,

à t_0, largage du propulseur sur un rail par découpes d'attaches pyrotechniques. Le propulseur se trouve alors en mouvement uniformément accéléré,

à $t_0 + \Delta t$ (correspondant à une course de 1 mètre environ), mise à feu du dispositif d'arrêt de poussée, décélération de l'ensemble et inversion de vitesse.

Ayant ainsi créé des conditions dynamiques significatives, le problème ensuite etait essentiellement centré sur le problème mesure. Nous avons adopté les solutions suivantes:

mesure de l'accélération longitudinale,

mesure du déplacement longitudinal.

L'accélération longitudinale a été mesurée par accéléromètres passant le continu et des fréquences assez hautes (1000 Hz). Nous avons utilisé plusieurs capteurs à jauges et piézo-électriques. Malgré des précautions importantes prises dans la mise en oeuvre des capteurs piézo-électriques passant le continu, nous avons obtenu les résultats les plus fidèles et les plus significatifs avec les capteurs à jauge. Il apparaît que les capteurs piézo-électriques en présence des gaz ionisés et des charges électrostatiques qui se développent sont d'un emploi particulièrement délicat.

La mesure du déplacement longitudinal a été obtenu soit par verni et entre partie mobile et référence fixe, soit par photodiode. La précision de la mesure a été correcte, la mesure du temps associé également. Par contre, le traitement de cette mesure exige une double dérivation. Ce point a été particulièrement travaillé, par voie numérique avec support d'informations accélérométriques au même point. L'analyse spectrale de ces dernières a permis de. dégager le spectre du signal. L'étude de ce spectre a défini le filtre numérique passe-bas de l'information utile à conserver, compatible évidemment avec la cadence d'échantillonnage des mesures de déplacement. C'est à travers l'image de ce filtre qu'ont été exploitées les mesures de déplacement pour obtenir la loi de vitesse et ensuite la loi d'accélération.

Les résultats au stade de la double dérivation ne sont pas toujours exploitables mais au stade de la première dérivation avec recoupement de l'intégration première des informations accélérométriques nous avons obtenu de bons recoupements. Les essais en vol en ont confirmé la valeur.

5. Étude expérimentale de la séparation

Nous parlerons simplement des dispositions que nous avons adoptées pour l'étude de la séparation d'étages, des problèmes particuliers que les essais en vol nous ont amenés à considérer. Le but que nous nous sommes fixés à priori est de qualifier la séquence, de s'analyser les différents phénomènes locaux, de juger des perturbations éventuelles que brisait l'engin. En faisant le bilan des forces relatives du 1er étage par rapport au deuxième on a constaté que le bilan forces propulsives-poids 1er étage avait une résultante sensiblement égale au poids de ce 1er étage mais de sens opposé. Ceci nous a amenés à concevoir l'expérimentation ci-après.

Le deuxième étage actif est fixe, les tuyères orientées vers le haut, il est surmonté de l'interétage, lui-même surmonté d'un mannequin inerte simulant en poids les forces résultantes du 1er étage.

Les équipements fonctionnels nominaux sont mis en oeuvre. La séquence nominale est testée, à l'échelle 1. Les perturbations transverses sur le 2ème étage sont mesurées par des liaisons dynamométriques particulièrement raides.

L'étude du séquentiel, des perturbations, du mouvement du mannequin nous a permis de qualifier correctement cette phase.

Par contre, en vol, nous avons rencontré des difficultés électriques au moment de la séparation des étages. Des parasites importants ont perturbé le fonctionnement du calculateur. Nous avons pensé que ces perturbations pouvaient venir de l'état d'ionisation des gaz de propulseur. A cette fin, nous avons réalisé une expérimentation sur maquette tendant à mettre en évidence cette cause probable.

L'expérimentation a porté sur une expérimentation maquette (échelle $\frac{1}{4}$) où la dynamique de séparation était le plus fidèlement simulée. Le chargement était de même nature que le chargement réel, de façon à avoir les mêmes gaz, à la même température, et avoir dans les conditions de séparation la même loi de pression et d'écartement.

L'objectif était de mettre en évidence dans la dynamique de séparation l'apparition de potentiels entre étages qui pourraient créer, en fonction de la distance de séparation et de l'état d'ionisation, des gaz des charges disruptives. La présence de tels phénomènes par suite des potentiels variables pouvait par induction créer des parasites importants qui perturbaient le fonctionnement électrique normal.

La difficulté dans cette expérimentation a consisté essentiellement à isoler correctement les divers éléments par rapport au sol, à mesurer des potentiels élevés sans perturber le phénomène, à avoir des temps de réponse faibles pour mettre en évidence la présence éventuelle d'arcs.

Nous avons utilisé une méthode par pont capacitif, à grande impédance d'entrée, symétrique et l'enregistrement étant fait par enregistreur magnétique à entrée assymétrique, nous avons incorporé un amplificateur qui a permis l'adaptation.

Un certain nombre d'expériences ont été faites, et nous avons ainsi mis en évidence des apparitions de potentiel électrostatique importantes avec charges disruptives, qui reproduisaient en séquence les perturbations de vol.

Ce diagnostic confirmé a permis de prendre les dispositions nécessaires pour se prémunir contre de tels incidents.

6. Écoulement interne

Les expériences que nous avons décrites jusqu'ici ont trait en majeure partie à du matériel grandeur, et sauf dans certains cas, n'ont pas présenté un caractère d'étude. Par contre, nous avons été amenés à faire certaines expérimentations plus fines, moins spécifiques et c'est de cet aspect que nous voudrions maintenant parler.

Nous aborderons les problèmes d'écoulement interne dans les propulseurs.

Un propulseur est un générateur de gaz qui doit en ayant le moins de pertes, assurer correctement l'alimentation des tuyères. Les écoulements ne sont pas figés, ils sont en général biphasés, ils sont à haute température et haute pression. Les vitesses locales sont importantes et en plus des problèmes de rayonnement, les problèmes d'aérothermique sont au premier plan des préoccupations. L'analyse de l'écoulement, compte tenu de coefficients de remplissages élevés, est particulièrement intéressante en début de combustion. Nous nous sommes, dans ce cas, proposés d'analyser les pertes, de caractériser l'alimentation et les survitesses, de chiffrer les fluctuations, de mesurer également les flux de façon à pouvoir vérifier la validité des formules de convection. Pour ce programme, nous avons lancé une série d'expérimentations allant du coup de sonde qualitatif à une étude quantitative à échelle 1. Faute de pouvoir travailler avec des gaz réels, nous avons effectué ces essais à l'air, froid puis chaud, à des niveaux de pression compatibles avec les installations disponibles, mais suffisants pour amorcer très largement les tuyères, et avoir à chaud en particulier un nombre de Reynolds rapporté au diamètre du col identique au Reynolds réel.

En tant qu'essais qualitatifs, nous avons d'abord fait des expériences de visualisation au tunnel hydrodynamique de l'O.N.E.R.A. suivant les méthodes développées par cet organisme. L'écoulement dans le propulseur est très largement subsonique de sorte que l'hypothèse incompressible est fondée dans cette zone. Pour simplifier les maquettes nous avons travaillé en écoulement plan.

Puis nous avons travaillé à échelle 1, à froid d'abord.

Les visualisations pariétales, s'appuyant sur les visualisations précédentes, ont permis de connaître l'organisation des écoulements. Nous avons mis en évidence des zones tourbillonnaires, corrélées effectivement avec les érosions importantes détectées en tir.

Les mesures de pression nous ont permis de connaître les pertes.

La mesure des pressions instationnaires, leur spectre, ont permis d'optimiser les profils, les formes et les réglages. Elle a été également un critère intéressant de recoupement avec les zones de flux important.

Pour obtenir directement les flux, nous avons été amenés à mettre en place un bruleur propane kérosène qui réchauffait l'air jusqu'à des températures de 500 °C, et à des pressions allant jusqu'à 5 bars. Les flux ont été mesurés par des fluxmètres-calorimètres. Dans ces conditions nous avons également mesuré les pressions locales en statique et dynamique et nous avons même fait de l'anémométrie à chaud pour connaître les fluctuations de vitesse en certain points particuliers.

Nous avons trouvé des coefficients de convection plus élevés que ne laissaient espérer les formules classiques de Bartz ou de Collburn (de l'ordre de 50%). Avec ces coefficients d'échange, la connaissance des grandeurs locales, l'estimation du rayonnement, nous avons pu par le calcul retrouver des épaisseurs ablatées et leur répartition comparables à celles observées dans des tirs réels. Nous avons conscience que ces expériences analytiques fines sont un outil précieux pour la connaissance de l'écoulement et l'estimation des caractéristiques aérothermiques internes.

Cet exposé général portant sur les problèmes essentiellement liés aux performances et à la connaissance des phénomènes aérothermiques n'est qu'un volet des problèmes posés par les propulseurs à poudre. Nous avons passé sous silence tous ceux qui sont plus particulièrement attachés à l'aspect technologique.

Pour conclure, nous préciserons que toute installation d'essais et tout essai sur propulseur en combustion doivent être conditionnés par les aspects suivants:

choix très précis des objectifs et connaissance correcte des limitations de l'expérimentation caractère irréversible d'un tir et fiabilité des installations

qualité des informations et acquisition intégrale et fidèle

traitement différé, particulièrement fin pour tout ce qui concerne l'aspect des performances, de préférence d'ailleurs par voie numérique.

Annexe: Comparaison de deux tuyères sur un même propulseur

On a comparé une tuyère à divergent conique et une tuyère coquetier montées successivement sur le même propulseur.

Les résultats de mesure d'impulsion spécifique ont été:

Divergent évolutif	Divergent conique
268.69	266.07
268.65	266.98
269.56	266.41
270.37	267.29
269.27	267.28
268.71	266.02
Moyenne 269.38	266.67
$= 0.65$	$= 0.55$

Le test de Snedecor n'interdit pas de considérer que les deux échantillons ont même variance.

Le test de Student Fisher montre qu'au seuil $\frac{1}{1000}$ les moyennes sont significativement différentes.

NEW CAPACITIVE PRESSURE TRANSDUCER
FOR AEROSPATIAL USE

E. TURCI

Filotecnica Salmoiraghi, Italy

Introduction

The transducer we have developed was designed to satisfy specific requirements for aerospatial use.

Although the operational ranges are conceptionally attainable either at high or low pressures, we have considered the solution of the problem at high pressure in particular.

Performance levels achieved are as follows:

(a) Accuracy: 0.5% full scale at ambient temperature
(b) Overall dimensions for pressures up to 200 kg/cm^2: height 75 mm, max diameter at cylindrical mounting base 60 mm.
(c) Typical full scale output voltage: 5 VDC
(d) Temperature of hot gases under pressure for few minutes period: approximately 100–150 °C.

In Figure 1 is shown a view of the instrument in order to give an idea of the overall dimensions.

The most original feature of the transducer, by which we were able to arrive at the solution of the problem of linearity and hot gas temperature under pressure, is the adoption of a conceptionally linear coupling of the two armatures which form the variable capacitance of the transducer, achieving at the same time a sufficiently long travel of the membrane to reduce the thermal effects of hot gases on the membrane itself.

Principle of Operation

The principle of operation of the transducer may be summarized as follows:

The component transducing the pressure information to electrically proportional data is a corrugated diaphragm which has a travel of approximately 0.5 mm max.

At the center of the diaphragm a plate is located which bears on its top surface a series of metal laminations, vertical and parallel to each other, which insert themselves into a similar series of vertical and parallel laminations without touching them based on another plate electrically isolated and fixed by means of a dielectric support to the case.

In Figure 2 is shown an exploded view of the instrument in order to give an idea of the metal lamination fitting.

It is evident that this 'comb' coupling of the two plates gives way to a capacitance variable with the membrane displacement and thus proportional to the pressure.

G. A. Partel (ed.), Proceedings of the Second International Conference on Space Engineering. All rights reserved.

It is now interesting to note that the use, already adopted by others, of two flat surfaces facing each other and utilized as capacitance variable with the displacament involves considerable complications inasmuch as:

Fig. 1.   View of the instrument.

Fig. 2.   Exploded view of the instrument in order to give an idea of the metal laminating fitting.

(a) A very short air space is required between the armatures and an extremely short travel is necessary to attain a satisfactory response linearity, unless the armature surface is much larger than the one adapted by us.

This solution would be in all cases unacceptable due to the excessive increase in dimensions.

(b) The transducer would be much more sensitive to thermal variations and less easily compensated.

The variable capacitance produced through the plate is coupled through a fixed capacitor to a multivibrator operating at high frequency.

The electrical signal obtained at the capacitor terminals is fed to a high input impedance circuit, then full wave rectified and finally fed to one of the two differentials input of an integrated linear amplifier.

3. Corrugated Diaphragm Considerations

The component transducing the pressure information to electrically proportional data is a corrugated diaphragm, so the necessity is clear of particular care in designing this component.

In Figure 3 is shown a view of the diaphragm with the plate for a 200 kg/cm² range.

A corrugated diaphragm is a shell of very complex shape, and therefore its calculation involves considerable difficulties even for the range of small deflections.

These difficulties increase significantly when large deflections are taken into account.

The use of an approximate method is satisfactory with experimental results.

The mathematical model of the corrugated diaphragm is arrived at from this consideration: consider a corrugated diaphragm with periodical corrugations loaded by an uniformly distributed pressure.

Fig. 3.   View of the diaphragm with the plate for 200 kg/cm² range.

Cut an element with finite dimensions from the corrugated diaphragm and compare it with an element of a flat diaphragm.

The flat diaphragm element bends and extends in the radial and peripheral directions when loaded.

While the flat element presents equal rigidity both in the radial and peripheral directions, the geometrical shape of the corrugated element is such that its rigidities in these two directions are different: the resistance of the element to bending and stretching will be much smaller in the radial direction than in the peripheral direction.

Thus a corrugated diaphragm is anisotropic, due to its particular geometry.

A flat anisotropic diaphragm can accordingly be used as a mathematical model for the corrugated diaphragm.

The elastic coefficient of an anisotropic material is so determined that the tensile and flexural rigidities of the anisotropic plate are equal to the corresponding parameters of a corrugated diaphragm.

The thickness of the anisotropic diaphragm will be taken as equal to the thickness of the corrugated diaphragm.

The final consideration necessary for the mathematical model realisation is: the tensile rigidity in the radial direction of the corrugated diaphragm is considerably smaller than the rigidity of a flat isotropic diaphragm, while the flexural rigidities of the two diaphragms in the same direction are very similar.

The equation describing the deflection as a function of pressure is cubic just as was that of a flat diaphragm.

The linear term can be found by solving deflections in the range of very small displacements.

The theoretical and practical characteristic of the deflections as a function of pressure is shown in Table I.

We see from Table I that the experienced and calculated unlinearity values for small displacements and for calculated unlinearity values less than 0.5% are practically the same.

TABLE I

Typical characteristic of the displacement as a function of pressure

| thickness = 1.80 mm | | diameter = 34 mm | |
| C72-50 Rockwell | | | |

kg/cm^2	experienced mm	experienced unlinearity %	calculated unlinearity
0	0.000	0	0
40	0.170	0	0
80	0.340	0	0
120	0.484	−3	−0.75
160	0.603	−9	−1.4
200	0.712	−16	−3

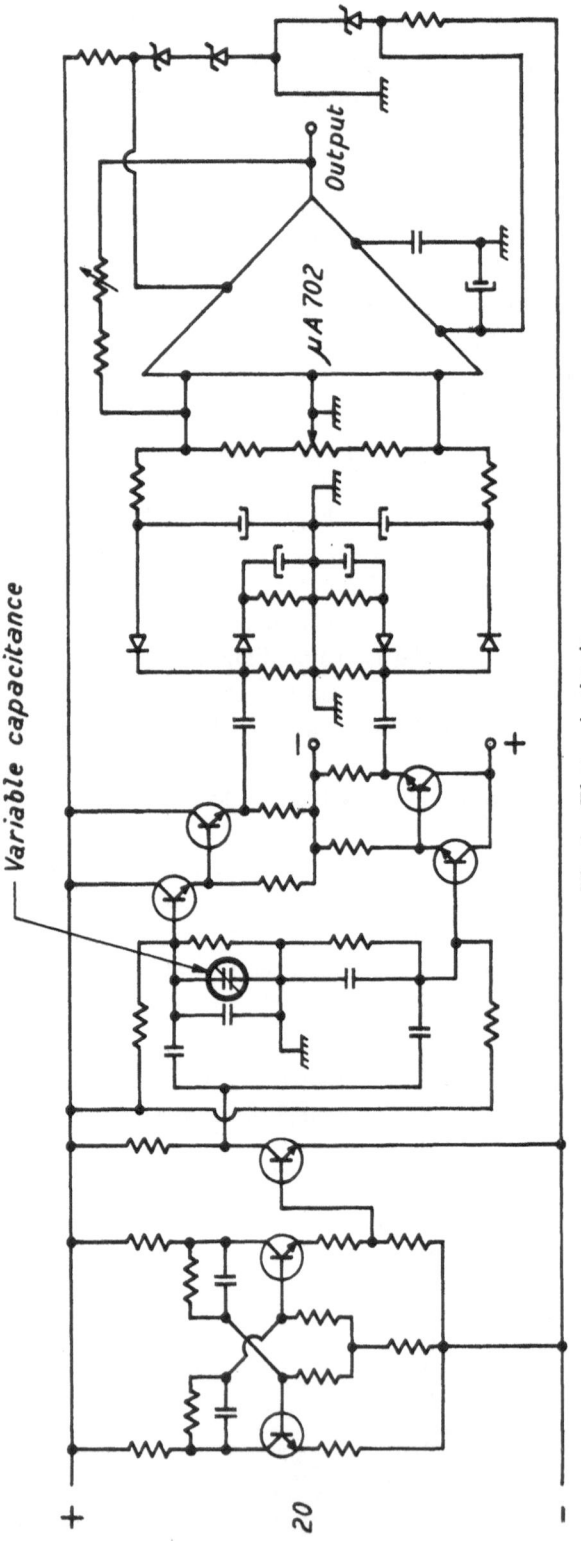

Variable capacitance

Fig. 4. Electronic circuit.

For wider displacements the experienced unlinearity values are more and more larger than the calculated values.

4. Circuit Considerations (see Figure 4)

The following elucidates the design considerations discussed in Section 2.

The variable capacitance produced through the plate is coupled through a fixed capacitor to a multivibrator operating at high frequency. The electric signal obtained at the capacitor terminals is fed to an high input impedance circuit, then full wave rectified, and finally fed to one of the two differential inputs of an integrated linear amplifier.

TABLE II

Typical characteristic of the output voltage as a function of pressure

kg/cm²	theoretical V	experienced V	unlinearity %
0	0	0	0
20	0.534	0.530	−0.1
40	1.062	1.065	+0.1
60	1.600	1.605	+0.1
80	2.130	2.150	+0.5
100	2.670	2.670	0
120	3.200	3.190	−0.25
140	3.730	3.700	−0.7
150	4.000	3.925	−2

Fig. 5.   View of the electronic package.

In order to attain a satisfactory thermal compensation together with good repeat-ability, we have designed a second circuit network similar to that described above except that the input signal, instead of being applied to the variable capacitor, is applied to a thermally stable fixed capacitor, while the output is applied to the other differential input terminal of the integrated amplifier.

In Table II is shown a typical characteristic of the output voltage as a function of pressure and in Figure 5 is shown a view of the electronic package.

For space applications such as rocket motor evaluations etc. a rapid response time is necessary.

Tests have been made on the transducer using a step input generated by a shock tube to test the rise time.

Repeated tests of this nature have shown the response time to be less than 2–3 m sec.

5. Importance of the High Pressure Transducer in Space Applications – Conclusions

The resonant combustion in solid or liquid fuelled rockets is a very difficult, not yet resolved, problem.

The importance of a high pressure transducer in space applications consists, above all, in the testing of rocket motors to examine the oscillating combustions in the motors.

An incorrected combustion may produce pressure peaks of many times the nominal value with very rapid response time.

The design we have presented is tested under these conditions and is able to with-stand all the conditions to be found in space applications.

PART X

SPEECHES DELIVERED AT THE BANQUET

MANNED SPACE STATIONS

Gateway to Our Future in Space

Space stations orbiting the Earth have long been a dream of space engineers and scientists. Many designs and concepts have been proposed, and many justifications have been advanced for their use. In recent years, a series of technical developments have taken place which have led scientists in America's National Aeronautics and Space Administration to begin detailed planning for a large manned space station.

Among the technical advances which make NASA confident it can build a space station are, first, NASA has had several successful launches of the Saturn V booster rocket. Second, the cumulative experience NASA has gained from manned space flight has helped them to define the environment that will be necessary for extended work and survival in space. And, third, NASA has now after considerable study determined the kinds of activities in science and technology that could be performed in a space station.

I will discuss briefly why NASA believes that a space station is a logical and desirable program for the middle of the next decade, and what some of the characteristics of the space station might be.

A great deal of fictional and speculative writing has been done about space stations. These works have become confused with the more scholarly writings on the subject. And as a result, there is a great deal of misunderstanding about space stations. Let me, therefore, offer a definition of a space station before I address the question, "Why a space station?" The definition I am offering to you was first expressed by Dr. Robert Gilruth, NASA's able director of the Manned Spacecraft Center in Houston, Texas. He described a space station as "... a location in space which is developed to support men and equipment on a permanent basis in order to take full advantage of the economies of size, centralization and permanency. In other words, a space station is a *base* in space." It would serve as a central location in which there would be: power, working and living space, logistical supplies, experimental equipment, communications, and data processing systems.

NASA scientists and, I'm sure, space scientists throughout the world, are interested in space stations for a variety of reasons. To begin with, there would be economic benefits which would accrue from the long life-span of a space station. The basic equipment required to support human life in the station and the specialized scientific equipment with which to conduct experiments could operate in space for long periods of time. And if they did so in a long lifetime space station, they would not represent exorbitant investments. With proper logistics, many pieces of equipment could be

G. A. Partel (ed.), Proceedings of the Second International Conference on Space Engineering. All rights reserved.

repaired or refurbished in space. Some could even be modified on the basis of gains in technology or to allow for changes in the objectives of certain experiments.

With regard to logistics, the placing of a permanent station in space to which man could return again and again would allow us to store expendables and supplies and, thereby, minimize the number of purely logistical flights. These flights would be the largest single cost item in a space station program, and they must be kept to the minimum.

Another special value of a large, permanent space station would be in regard to power systems. The size of the station would allow the use of large panels for solar cells or it could be designed to provide separation, shielding, and radiator areas for a nuclear power system. It has been generally estimated that a space station should be able to provide somewhere between fifty and one hundred kilowatts of power to insure the operation of the experimental equipment and to provide an acceptable living environment in space.

There are also important advantages from the standpoint of personnel duties. When the size of the crew reaches ten to twelve, we can begin to realize an economy from the fact that a smaller percentage of the total crew worktime will be spent on the functions of operating and maintaining the station. There are also economies to be gained from specialization. Experiments could be carried out by genuine experts, and there would be little need to spend time cross-training members of the crew to be able to do a variety of jobs. In a large space station, designed for a large crew, there would be more opportunity for providing space for personal comfort and convenience and for insuring that conditions would be more earth-like than we could afford in smaller space vehicles.

I would like now to describe for you in some detail what the NASA Space Station might look like. I wish to stress, the space station I am describing is simply typical of what might be built. NASA has by no means approved a specific design at this time.

To begin with, the space station would feature an orbital workshop which would contain about 4000 cubic feet per man. That space would permit the use of supporting equipment that would make it possible for the astronauts to live and work in an environment far more like the earth's than the environment in the closely packed spacecraft of this decade. The station would be designed for breathing air at 10 to 14 pounds-per-square-inch pressure, and one-half to one G of artificial gravity in the major living and operational areas. I will discuss the question of artificial G at some length in just a moment.

The space station should have the capability to accommodate at least a 50-man crew, with the prospect of expansion to at least 100 by the addition of new sections set up separately.

The station must operate at a higher inclination than those angles used in our manned space flights so far. Studies have shown that about 50 deg of inclination, or somewhat higher, is a good compromise for operational and observational purposes.

That inclination would provide a near minimum of radiation background, which is important for minimizing the effects of radiation on the crew and on the sensitive film used in astronomical observations. It is also sufficiently high to cover the majority of the occupied land masses.

The space station will also have both attached and station-keeping modules for special experiments. For example, we can assume that telescopes would be housed either in a module tethered to the parent vehicle or flying nearby in a station-keeping mode. This will mean that it will be necessary to develop a technique for visiting these modules for data gathering and maintenance. It would be desirable to be able to bring the smaller modules back into the space station, as if it were a hangar, for repair and special servicing.

The space station would consist of three main parts, an APOLLO spacecraft attached to the Workshop which is connected in turn to a spent stage of a Saturn booster by a tubular docking adapter with an airlock. The workshop would be outfitted with equipment, much of which would be originally stored in the docking adapter. The Workshop would have a floor, approximately in the middle, which would be installed before launch. The equipment would be set up on that floor so as to form four rooms. The power source would be solar cells extending from the side of the spent Saturn stage.

Three of the four rooms in the Workshop would serve as sleeping compartments, waste management, an food preparation. At NASA's Marshall Space Flight Center, there is a full-scale mockup of the Workshop. Tests will be performed in the three rooms to determine the habitability characteristics required to support the astronauts in the space station.

In the mockup will be equipment for a variety of medical test programs. There will be, for example, an Erogometer, a bicycle-like device, and there will be a Barany chair for carrying out medical experiments in vestibular activities.

The Workshop in the space station would also hold the equipment for the major scientific experiments. One experiment that has been discussed is a solar observation experiment. The observing device will be part of an unmanned launch, and it will rendezvous with the space station later on. It will be docked under the control of an astronaut using remote control devices.

This experiment will provide vital information relative to future space station programs. To begin with, it will be the first chance for man to conduct a really complex scientific experiment in space; and, of course, we will be able to observe man's ability to perform such functions. Second, we will be developing techniques for sending equipment and supplies to a space station in an unmanned mode. This mode will be an important method for logistic support in the future.

The NASA scientists have given a great deal of thought concerning the need for artificial gravity and for a test program to define its characteristics. There is little doubt that astronauts would be able to adapt to a zero gravity condition during a three to six months tour in a space station. But one of the goals of the space station should be to provide the crew with living and working conditions as near to normal as possible, in the interest of basic efficiency and task effectiveness.

If some artificial gravity could be provided, it would improve three categories of functions or activities: first, processes involving fluids; second, those connected with locomotion and orientation; and, third, those related to man/machine interfaces. If there were artificial gravity, fluid processes such as those associated with personal hygiene, cleaning, food preparation, chemistry, and others, could be performed in a manner identical to that which we are accustomed to here on Earth.

With the establishment of artificial gravity, the astronauts could also walk with their hands free, and there would be little need for training in adaptation to the standard Zero G conditions in space.

Artificial gravity would also provide for normal man/machine relations with a wide variety of equipment, again minimizing the need for adaptive training. As a corollary, artificial G would permit the use of equipment developed for use in earth laboratories that would otherwise not be useful in space.

For these reasons, the NASA scientists have concluded that the characteristics of artificial gravity are important objectives for a near-term earth orbital activity. I will discuss briefly some of the parameters which should be studied by special experiments before we can be firm about a conceptual design which would meet the space station objectives.

Artificial Gravity has two important parameters for any given level: the rate at which the vehicle is rotated and the radius at which the man is located. We have identified in various tests a so-called 'Comfort Zone' in which man can tolerate rotation. Parabolic profiles flown by aircraft have demonstrated that most of the problems of locomotion and fluid transfer can be overcome by a gravity as low as three-tenths G. Since rotational simulators on the ground have indicated that the average person can accept four revolutions per minute, we are led to a minimum rotational arm of about 50 feet. However, the ground tests are limited because they always have a one G field affecting the results, and we cannot be certain that when that field is non-existent, the rotational forces will not affect the subconscious adaptive mechanisms more strongly. Thus we are interested in even lower rotational rates and higher gravity forces, even as high as one G.

These considerations have led NASA to consider a system which involves the rotation of an APOLLO spacecraft with an experimental module, both anchored to a spent booster stage, in such a way that various radii can be achieved.

With the information from such an experiment, we could be ready to construct the type of space accommodations that will be necessary to meet the comfort and work requirements for astronauts in space stations.

I will now discuss briefly the activities to be conducted on board the space station. In general, these activities fall under two headings: observational experiments which include astronomy and earth sensing, and onboard experiments such as bioscience and biomedical experiments and high energy physics.

The sciences which make use of the advantages of observation from space will benefit most from space station operations. Astronomy especially can be expected to produce dramatic results from being able to record observations outside of our earth's dirty and shimmering atmosphere.

NASA space station officials talk chiefly about a forty-inch objective type telescope. This is a big and awkward device which will require a large space station as a base. But this forty-inch telescope is just the beginning. The National Academy of Sciences has seriously proposed orbiting a 120-in. telescope. The concept is that such a telescope would be flown adjacent to a space station and that it would be visited periodically for service and data retrieval.

The development of remote sensing equipment for the observation of the earth's surface and, probably, subsurface will be one of the major tasks in readying our space station.

To date, our principal experiments with earth observation have been by photography. Despite the fact that the photographs have been taken with simple equipment, we have obtained dramatically new and astonishingly useful views of our home planet. The unique capabilities of space photography result from the broad synoptic panoramas of large land features which can be captured in a single photograph. Large mountain systems, for example, can be studied in their entirety, and other important geophysical relationships can be established over continental areas. Photographs from space could also be important in regard to oceanography. The views we have obtained so far of such ocean phenomena as the Atlantic Gulf Stream tell us clearly that regular photographs from space could be extremely valuable in the expansion of the science of oceanography.

But photography reveals only a fraction of the information about the Earth that can be obtained from space. What is needed is the capability to measure and monitor the earth's surface constantly over the entire electromagnetic spectrum. There is no difficulty in identifying the kinds of sensors that would be needed for electromagnetic earth sensing. The difficulty will lie in building a space station capable of housing these sensors.

There is little doubt that earth sensing from space holds great promise for future applications of vast benefit to a wide variety of professions and occupations, as well as to the scientific community. The great challenge for the nations of the world is to design the international management organizations to exploit this potential. I can foresee the day when international space stations will house crews from many nations, working with cooperating ground facilities throughout the world.

The second category of activities and experiments that could be conducted from a manned space station would be in the areas of bioscience, biomedicine, and high energy physics.

The unique environment of space with its Zero Gravity offers many possibilities for new avenues of approach in research in the growth processes of animals and plants. NASA scientists have conceived of a space laboratory complete with animal colonies and plant farms.

In regard to high energy physics, it is interesting to recall that the study of nature's fundamental particles began with observations of cosmic radiation, made from balloons and mountain tops. Space station capabilities re-open the natural environment as a primary source of high-energy particles. By placing a large cosmic-ray facility in a space station, we could extend the basic particle energies available for investigation by three orders of magnitude – from 70 to 70 000 GeV (Billion Electron Volts). To make such investigations, we would need cryogenically-cooled, superconducting magnets. If we had a space station facility with such equipment, we might be able to find new clues to such mysteries as antimatter in the cosmos. Operating such a laboratory would be a very complex affair, involving the most advanced techniques for recording and analyzing data. Its character would dictate the presence of scientists, not just astronauts.

Looking beyond the first Post-APOLLO Orbital Workshop that I have been discussing, it is not difficult to imagine still another generation of space stations, larger and more sophisticated than the Orbital Workshop. One concept is of a space station with 240-ft radius to the extremes of the living quarters, and as much as a 375-ft radius to the balance and power sections on the extreme ends. Such a space station would be launched in three parts by three Saturn V rockets, with docking and assembly in space. The station would have a gross weight of about one million pounds and would carry as much as 100 000 pounds of experimental equipment.

At the hub of this station would be a zero G laboratory with about 45 000 cubic feet of working space. The large radius would permit near one G at 3.5 rpm. Having a large inertia, this station would require 7000 pounds of propellant just for the spin-up to 3.5 rpm.

This central laboratory would accommodate the basic biomedical and bioscience test equipment. It would also include complete machine shops and repair rooms, as well as the living quarters and command and data reduction areas.

Such a space station would have growth possibilities. An additional launch of a Saturn V could double the accommodations by delivering additional living space, with the spent booster stage serving as a balance. It would also be possible to retain the empty S-11 stage to add additional working and living space.

In summary, I have indicated to you the place of the space station in our future space plans; the reasons why such stations will be important links with the future; the prototype steps we are taking in our present programs; and, finally a conceptual design of a relatively large space station of the future.

Such a station would merge the space technology of the period with elements of both pure science and practical science into a single facility which can be logistically operated at a single place in space. It would thrust basic research into a new and stimulating environment.

Permanency would be its predominant asset, and it would provide a large source of energy and a generous amount of utilizable volume. Its equipment could be continually re-used, its expendables re-supplied, and its resources enhanced.

The tens of thousands of hours of operational experience with men and equipment in the space environment that it would give us would assure that such a space station would be a true gateway into the exciting space era of the future.

USAF   GENERAL B. A. SCHRIEVER

Mr President, Ladies and Gentlemen and Esteemed Colleagues!

Allow me, on behalf of the Soviet scientists and experts, to greet you – the participants of the 'Second International Conference of Space Engineering'.

I wish to pay tribute to the fruitful initiative of the Conference Organization Committee, which has taken so much trouble drawing up the scientific programme, and carrying out many difficult arrangements inherent in organizing such international gatherings as this.

I should like, particularly, to extend our gratitude to the distinguished President of the Committee, Com.te Glauco Partel. I should equally like to convey our appreciation to the 'Centro Studi Missilistici' (STM), and the French AERA, which sponsored the conference.

We are very grateful to our Italian colleagues for giving us the opportunity to be present at a conference in this beautiful country, and in a city famous for great scientific traditions.

We thank the Mayor of Venice for the cordial hospitality and attention, which has been accorded to the Soviet delegation.

The object of the conference is to develop new branches of science and technology employed in the exploration of outer space both for scientific studies and practical use. The conference-programme is tightly-scheduled and hard-working; but it takes in all the basic trends of modern science and technology.

The progress of the first few days of the Conference has shown the great urgency of the problems contained in the reports; and these have stirred up keen interest amongst the participants.

This Conference has made it possible for scientists and experts from various countries to meet and discuss burning scientific questions.

We have noticed, with satisfaction, that the newspapers of Venice are paying considerable attention to our Conference. But it is a pity that 'Il Gazzettino' today called me "The Soviet Von Braun". If the head of each Soviet delegation of scientists were called a 'Von Braun', there would be hundreds of 'Von Brauns' in the USSR!

Over the last 12 years – the years of the cosmic era – mankind has achieved new victories over the elements: he has conquered terrestrial space, and has begun going out into outer space. The great achievements of the USSR and the USA in the exploration and conquest of outer space are well known.

More and more countries are now beginning the exploration of outer space. France and Italy have developed their national programmes and are successfully

G. A. Partel (ed.), Proceedings of the Second International Conference on Space Engineering. All rights reserved.

carrying them through. The number of cosmic centres and space research laboratories is being multiplied daily.

Space research and technology are the new fields for present-day science and modern engineering. And success in these fields will give great profit in all fields of science and engineering.

The cooperation between the Soviet Union and other Socialist countries, as well as with France, in the peaceful use of outer space, is progressing fruitfully.

We hope that Italian international collaboration will be wider and that its scientific connections, in particular with our country, will grow.

There are some other space research programmes, which are being carried out by joint efforts of many countries. Joint efforts, by various nations, to solve complex scientific-technical problems in the exploration and use for peaceful purposes of outer space have already yielded fruits: and one can hope for new and substantial successes in further development of joint actions.

Allow me to wish the sponsors and all the participants of this Conference all success, a fruitful exchange of knowledge and the development of personal contacts.

I propose a toast to the President of the Organization Committee, to the ladies present, to the progress of the Conference and to the joint efforts of all the scientists engaged in exploring outer space for peaceful purposes!

Academy of Sciences, U.S.S.R.   B. N. PETROV

ASTROPHYSICS AND SPACE SCIENCE LIBRARY

Edited by

J. E. Blamont, R. L. F. Boyd, L. Goldberg, C. de Jager, Z. Kopal, G. H. Ludwig, R. Lüst,
B. M. McCormac, H. E. Newell, L. I. Sedov, Z. Švestka, and W. de Graaff

1. C. de Jager (ed.), *The Solar Spectrum. Proceedings of the Symposium held at the University of Utrecht, 26–31 August, 1963.* 1965, XIV + 417 pp.   Dfl. 50.—

2. J. Ortner and H. Maseland (eds.), *Introduction to Solar Terrestrial Relations. Proceedings of the Summer School in Space Physics held in Alpbach, Austria, July 15–August 10, 1963 and Organized by the European Preparatory Commission for Space Research.* 1965, IX + 506 pp.   Dfl. 65.—

3. C. C. Chang, and S. S. Huang (eds.), *Proceedings of the Plasma Space Science Symposium. Held at the Catholic University of America, Washington, D.C., June 11-14, 1963.* 1965, IX + 377 pp.
Dfl. 68.—

4. Zdeněk Kopal, *An Introduction to the Study of the Moon.* 1966, XII + 464 pp.   Dfl. 72.—

5. Billy M. McCormac (ed.), *Radiation Trapped in the Earth's Magnetic Field. Proceedings of the Advanced Study Institute, Held at the Chr. Michelsen Institute, Bergen, Norway, August 16– September 3, 1965.* 1966. XII + 901 pp.   Dfl. 130.—

6. A. B. Underhill, *The Early Type Stars.* 1966, XIII + 282 pp.   Dfl. 56.—

7. Jean Kovalevsky, *Introduction to Celestial Mechanics.* 1967, VIII + 127 pp.   Dfl. 30.—

8. Zdeněk Kopal and Constantine L. Goudas (eds.), *Measure of the Moon. Proceedings of the Second International Conference on Selenodesy and Lunar Topography held in the University of Manchester, England, May 30–June 4, 1966.* 1967, XVIII + 479 pp.   Dfl. 90.—

9. J. G. EMMING (ed.), *Electromagnetic Radiation in Space. Proceedings of the Third ESRO Summer School in Space Physics, held in Alpbach, Austria, from 19 July to 13 August, 1965.* 1968, VIII + 307 pp.   Dfl. 58.—

10. R. L. Carovillano, John, F. McClay, and Henry R. Radoski (eds.), *Physics of the Magnetosphere. Based upon the Proceedings of the Conference held at Boston College, June 19–28, 1967.* 1968, X + 686 pp.   Dfl. 130.—

11. Syun-Ichi Akasofu, *Polar and Magnetospheric Substorms.* 1968, XVIII + 280 pp.   Dfl. 55.—

12. Peter M. Millman (ed.), *Meteorite Research. Proceedings of a Symposium on Meteorite Research held in Vienna, Austria, 7–13 August, 1968.* 1969, XV + 941 pp.   Dfl. 160.—

13. Margherita Hack (ed.), *Mass Loss from Stars. Proceedings of the Second Trieste Colloquium on Astrophysics, 12–17 September, 1968.* 1969, XII + 345 pp.   Dfl. 65.—

14. N. D'Angelo (ed.), *Low-Frequency Waves and Irregularities in the Ionosphere. Proceedings of the 2nd ESRIN-ESLAB Symposium, held in Frascati, Italy, 23–27 September, 1968.* 1969, VII + 218 pp.   Dfl. 43.—

16. S. Fred Singer, *Manned Laboratories in Space. Second International Orbital Laboratory Symposium.* 1969, XIII + 133 pp.   Dfl. 30.—

SOLE DISTRIBUTORS FOR U.S.A. AND CANADA:

Vols. 2–6, and 8: Gordon & Breach Inc., 150 Fifth Ave., New York, N.Y. 10011
Vols. 7 and 9ff.: Springer Verlag New York, Inc., 175 Fifth Ave., New York, N.Y. 10011